Current Topics in Microbiology and Immunology

Volume 398

More information about this series at http://www.springer.com/series/82

Marc Stadler · Petra Dersch
Editors

How to Overcome the Antibiotic Crisis

Facts, Challenges, Technologies and Future Perspectives

Responsible series editor: Klaus Aktories

Springer

Editors
Marc Stadler
Department Microbial Drugs
Helmholtz Centre for Infection Research
Braunschweig, Niedersachsen
Germany

Petra Dersch
Department Molecular Infection Biology
Helmholtz Centre for Infection Research
Braunschweig, Niedersachsen
Germany

ISSN 0070-217X ISSN 2196-9965 (electronic)
Current Topics in Microbiology and Immunology
ISBN 978-3-319-49282-7 ISBN 978-3-319-49284-1 (eBook)
DOI 10.1007/978-3-319-49284-1

Library of Congress Control Number: 2016959514

Printed on acid-free paper

This Springer imprint is published by Springer Nature
The registered company is Springer International Publishing AG
The registered company address is: Gewerbestrasse 11, 6330 Cham, Switzerland

Preface

The worrying increase of nosocomial infections caused by multi-resistant pathogenic bacterial strains, such as methicillin-resistant *Staphylococcus aureus*, vancomycin-resistant enterococci (VRE), fluoroquinolone-resistant *Pseudomonas aeruginosa* (FQRPA) is generally regarded as one of the major global challenges in human medicine. Since the new threat has been caused by multiple factors, a multi-disciplinary collaborative approach will be imperative to keep these new super-bugs at bay. The current volume was compiled by the principal investigators of the Helmholtz Centre for Infection Research (Braunschweig, Germany) and the associated universities and research institutes, where expertise in all important fields, ranging from infection biology and microbiology to epidemiology, immunology and drug discovery is readily available.

The authors of this book have contributed articles that cover all core disciplines and a range of current topics and emerging approaches in infection and drug research at the centre and world-wide. They address several aspects of basic research, such as the evolution and spread of resistance in bacterial pathogens and the influence of currently used antibiotics on the gut microbiota. Contributions on the search for novel biochemical targets that block pathogenicity and the development of alternatives to the state of the art treatments, such as antivirulence therapies, combine basic and applied aspects, leading to our strategy in the search for new antibiotics and their preclinical and clinical development. Several papers are dedicated to this important topic, ranging from strategies to discover new drugs based on biodiversity and genome mining and the optimisation of novel lead structures. Others address the accessibility of anti-infectives by means of medicinal chemistry, total synthesis, and biotechnological technologies, and include aspects of drug delivery in an alternative approach to overcome the cell wall of gram-negative bacteria as an important biological barrier that prevents most antibiotics from reaching their target site in those pathogens. The development of vaccines as valid alternative therapeutic approach to antibiotics therapy is also outlined. Moreover, aspects of clinical research, such as epidemiological studies of important bacterial pathogens and the development of innovative technologies for detection and diagnostics of the "bad bugs" are also illustrated by articles in this ebook.

The following 17 contributions are from microbiologists, immunologists and natural product chemists of the Helmholtz Centre for Infection Research (HZI) in Braunschweig, Germany, its two branches, the Helmholtz Institute for Pharmaceutical Research Saarland (HIPS) in Saarbrücken and the TWINCORE in Hannover, and neighboring partner universities (Leibniz Institute DSMZ, Veterinary School Hannover, Leibniz University Hannover). The HZI is one of the largest research centres world-wide that is dedicated to infection biology, immunology, and drug research. The institute, which was formerly known as the Gesellschaft für Biotechnologische Forschung (GBF), has recently celebrated its 50th anniversary. The GBF has become renowned for its biotechnological production facilities and above all, the accomplishments of its natural product chemistry departments. Those included the discovery and development of the anticancer drug, Ixabepilone, the antifungal soraphens and many other compounds derived from gliding bacteria. Ten years ago, the focus of the GBF has changed to infection research, followed by the renaming of the institute to HZI in 2006. Over the past years, the HZI has refined its research portfolio and strengthened its expertise amongst others by founding joint institutes with strong partner institutions. The HIPS in Saarbrücken was founded in 2009 with the Saarland University as a daughter institute of the HZI, and has significantly intensified HZI drug research, e.g. through recruitments with substantial industrial expertise. At the TWINCORE in Hannover, a joint institute with the Hannover Medical School, basic research scientist and clinicians work side by side since 2008 to transfer latest research findings to new therapies or diagnostic procedures for patients. With the translational scope in mind, the HZI's strategy combines basic research at all resolution levels with clinically oriented and pharmaceutical research, focusing on clinically relevant bacterial and viral pathogens, the immune system and immune interventions as well as anti-infectives. Antibiotics research is a field of topical significance due to multi-resistant bacteria increasingly challenging modern medicine. Thus, microbiologists, cell biologists, epidemiologists and immunologists of the HZI collaborate closely with the drug research departments, as well as with renowned scientists from the neighboring and international partner universities and research institutes to address this task. The current volume addresses the challenges afflicted with the threat of an antibiotic crisis and thus covers the important scientific activities of the HZI directed to combat bacterial pathogens, but not its excellent research on viral diseases and important host defense mechanisms.

Part I Antibiotic Resistance: Problems and New Opportunities

The first part of this two-part volume includes some of the most recent aspects and developments in infection biology, antibiotic resistance and host defense. E. Medina and D. Pieper [1] highlight current problems emerging multidrug resistant pathogens with a special focus on methicillin-resistant *Staphylococcus aureus*, MDR- and XDR-resistant *Mycobacterium tuberculosis* and ESBL-producing gram-negative pathogens. They discuss new possibilities to prevent and control infections of these dangerous pathogens, reduce the emergence of antibiotic resistance and develop new drugs. The following article by Nübel et al.

[2] emphasizes the importance to address emergence and spread of antimicrobial resistance using bacterial population genomics. Recent progress in DNA sequencing technologies and genomic studies now allows us to follow the evolution of antibiotic resistances. This approach not only illustrated that global populations of certain drug-resistant bacterial pathogens are dominated by a few clones, it also revealed features that were crucial for their spreading success. Most studies on the epidemiology of multidrug-resistant variants of clinically important pathogens such as methicillin-resistant *S. aureus* (MRSA) were focused on the emergence and spread in hospitals and other health care settings. The contribution by Mehraj et al. [3] presents more recent epidemiological studies, which analyzed carriage patterns in community settings, providing new insights on risk factors for colonization and new ideas for strategies to prevent infections. The high and increasing rate of antibiotic resistance hinders conventional use of antibiotics and increases morbidity and mortality due to ineffective treatments of infections. This situation demands fast and precise diagnostics of pathogens and their antibiotic resistance profile. Hornischer and Häußler [4] highlight new approaches and possibilities implementing modern omics technologies for the development of biomarker-driven molecular test systems for early diagnostics and resistance profiling for targeted therapy and a more effective stewardship of antibiotic agents. Use of antimicrobial agents in veterinary medicine is essential to control infectious diseases in domestic animals, but it also increases emergence of antimicrobial resistance. Seitz et al. [5] outline current aspects and problems related to the use of antimicrobial agents in animal farms, in particular swine husbandries with emphasis on resistance in *Streptococcus suis*, a major pathogen in swine. Current research on the intestinal microbiome revealed that a consequence of antibiotic treatment is a drastic change of the composition of the microbiota. The article by Thiemann et al. [6] provides current information about how this can be associated with an enhanced susceptibility towards gastrointestinal infections and metabolic disorders, and how it can also increase abundance of antibiotic resistance genes with bacterial commensals. To successfully combat against antibiotic resistance, novel treatment options and alternative antimicrobial therapies are urgently required. One novel strategy is to target and interfere with crucial bacterial pathogenicity factors or virulence-associated traits to bypass the evolutional pressure on the bacterium to develop resistance. Numerous potential drug targets for antivirulence therapies which have been identified over the last years, as well as structure-based tailoring of intervention strategies and established screening assays for small molecule inhibitors of such pathways are presented in the article by Mühlen and Dersch [7]. More specific approaches to block pathogenesis by interference with the flagella apparatus and the associated chemosensory system are highlighted in the contribution by Erhardt [8]. Finally, Schulze and colleagues [9] outline the importance of vaccination to control infectious diseases and emphasize a new immunization strategy using a non-invasive mucosal and transdermal application system to increase vaccination efficacy.

Part II Natural Compound Research and Anti-infective Development
The second part of this volume mainly focuses on various aspects that relate to the discovery and development of antibiotics. As outlined by Mohr [10], this field of research has been historically dominated by natural products, since the vast majority of the molecules that were turned into therapeutically useful antibiotics have been derived from fermentation of bacteria and fungi. The chapter by Landwehr et al. [11] accordingly treats the most important groups of bacterial secondary metabolite producers, i.e., Actinobacteria and Myxobacteria, and describes the current scenario that is being employed to screen these organisms using sophisticated methodologies that are dominated by state-of-the art techniques, including the development of special isolation protocols, e.g. for organisms from extreme habitats. The following chapter by Karwehl and Stadler [12] is dedicated to fungi, emphasizing the importance of combining biodiversity-driven approaches to identify producer organisms of novel lead compounds. An immense diversity has recently been recognized in fungi since modern methods of molecular phylogenetics have become available, and these organisms are highly likely to yield further innovative lead compounds. The chapter by Hermann et al. [13] focuses on three case studies involving highly promising current exploratory research projects on new antibiotics that are derived from bacteria. The utility of synthetic biotechnology, structural biology and genomics research for optimising natural lead structures and discovery of their mode-of-action is amply illustrated, based on the model compounds, griselimycin, aminochelocardin and cytobactamid. The outcome of these projects gives some hope that it will finally be possible to overcome the void in antibiotics discovery, using a combination of rational methods of lead structure generation that could not be imagined even ten years previously. However, these compounds will still need to be adopted by industrial partners since the resources of a public research institute are insufficient to cover the high costs involved with clinical drug development. Klahn and Brönstrup [14] have added to this approach, emphasizing on the importance of molecular target evaluation and give various examples for developmental candidates and investigational drugs mostly based on natural scaffolds but also give some examples for synthetic compounds that are now under development. Kalesse et al. [15] illustrate the utility of total synthesis of antibiotics, using selected "historical" examples. This approach will continue to provide a valid alternative to the biotechnological production of natural antibiotics, especially for less complex molecules that can easily be built up and modified synthetically but are produced by the microbes at rather low titers. An overview on the current status of the global pipeline of antibiotics in clinical trials is given by Hesterkamp [16]. Finally, a strategy to tackle one of the greatest challenge of the 21th century in novel antibiotics discovery, i.e. the improvement of pharmaceutical properties of such antibiotic drugs to overcome the barrier of the cell wall of the gram-negative pathogens by employing sophisticated drug delivery technologies, is presented by Graef et al. [17].

We dedicate this volume to two esteemed colleagues who have been instrumental in the implementation of the aforementioned strategic re-orientation of the GBF: Jürgen Wehland, the former Scientific Director, and Gursharan Singh Chhatwal, the former head of department of Microbial Pathogenicity who passed away recently.

That the HZI has now developed to one of the global strongholds of infection biology and anti-infective research is without doubt due to their great efforts. We are very thankful to all authors for their contribution in this special volume and hope that it will increase interest in antibacterial research and stimulate work on antibacterial drug discovery that is urgently needed.

1. Medina EI, Pieper DH (2016) Tackling threats and future problems of multidrug-resistant bacteria.
2. Nübel U (2016) Emergence and spread of antimicrobial resistance: recent insights from bacterial population genomics.
3. Mehraj J, Witte W, Akmatov MK, Layer F, Werner G, Krause G (2016) Epidemiology of *Staphylococcus aureus* nasal carriage patterns in the community.
4. Hornischer K, Häußler S (2016) Diagnostics and resistance profiling of bacterial pathogens.
5. Seitz M, Valentin-Weigand P, Willenborg J (2016) Use of antibiotics and antimicrobial resistance in veterinary medicine as exemplified by the swine pathogen *Streptococcus suis*.
6. Thiemann S, Smit N, Strowig T (2016) Antibiotics and the intestinal microbiome: individual responses, resilience of the ecosystem and the susceptibility to infections.
7. Mühlen S, Dersch P (2016) Anti-virulence strategies to target bacterial infections.
8. Erhardt M (2016) Strategies to block bacterial pathogenesis by interference with motility and chemotaxis.
9. Schulze K, Ebensen T, Riese P, Prochnow B, Lehr CM, Guzman CA (2016) New horizons in the development of novel needle-free immunization strategies to increase vaccination efficacy
10. Mohr KI (2016) History of antibiotics research.
11. Landwehr W, Wolf C, Wink C (2016) Actinobacteria and Myxobacteria – Two of the most important bacterial resources for novel antibiotics.
12. Karwehl S, Stadler M (2016) Exploitation of fungal biodiversity for discovery of novel antibiotics.
13. Herrmann J, Lukežič T, Kling A, Baumann S, Hüttel S, Petković H, Müller R (2016) Strategies for the discovery and development of new antibiotics from natural products: Three Case Studies.
14. Klahn P, Brönstrup M (2016) New structural templates for clinically validated and novel targets in antimicrobial drug research and development.
15. Kalesse M, Böhm A, Kipper A, Wandelt V (2016) Synthesis of antibiotics.
16. Hesterkamp T (2016) Antibiotics clinical development and pipeline.
17. Graef F Gordon S, Lehr CM (2016) Anti-infectives in drug delivery—overcoming the Gram-negative bacterial cell envelope.

Braunschweig, Germany

Marc Stadler
Petra Dersch
Dirk Heinz

Contents

Part I
Antibiotic Resistance: Problems and New Opportunities

Tackling Threats and Future Problems of Multidrug-Resistant Bacteria

Eva Medina and Dietmar Helmut Pieper

Abstract With the advent of the antibiotic era, the overuse and inappropriate consumption and application of antibiotics have driven the rapid emergence of multidrug-resistant pathogens. Antimicrobial resistance increases the morbidity, mortality, length of hospitalization and healthcare costs. Among Gram-positive bacteria, *Staphylococcus aureus* (MRSA) and multidrug-resistant (MDR) *Mycobacterium tuberculosis*, and among the Gram-negative bacteria, extended-spectrum beta-lactamase (ESBLs)-producing bacteria have become a major global healthcare problem in the 21st century. The pressure to use antibiotics guarantees that the spread and prevalence of these as well as of future emerging multidrug-resistant pathogens will be a persistent phenomenon. The unfeasibility of reversing antimicrobial resistance back towards susceptibility and the critical need to treat bacterial infection in modern medicine have burdened researchers and pharmaceutical companies to develop new antimicrobials effective against these difficult-to-treat multidrug-resistant pathogens. However, it can be anticipated that antibiotic resistance will continue to develop more rapidly than new agents to treat these infections become available and a better understanding of the molecular, evolutionary and ecological mechanisms governing the spread of antibiotic resistance is needed. The only way to curb the current crisis of antimicrobial resistance will be to develop entirely novel strategies to fight these pathogens such as combining antimicrobial drugs with other agents that counteract and obstruct the antibiotic resistant mechanisms expressed by the pathogen. Furthermore, as many antibiotics are often inappropriately prescribed, a more personalized approach based on precise diagnosis tools will ensure that proper treatments can be promptly

E. Medina (✉)
Infection Immunology Research Group, Helmholtz Centre for Infection Research, Inhoffenstrasse 7, 38124 Braunschweig, Germany
e-mail: eva.medina@helmholtz-hzi.de

D.H. Pieper
Microbial Interactions and Processes Research Group, Helmholtz Centre for Infection Research, Inhoffenstrasse 7, 38124 Braunschweig, Germany
e-mail: dietmar.pieper@helmholtz-hzi.de

Current Topics in Microbiology and Immunology (2016) 398: 3–34
DOI 10.1007/82_2016_492

Published Online: 24 February 2016

applied leading to more targeted and effective therapies. However, in more general terms, also the overall use and release of antibiotics in the environment needs to be better controlled.

List of Abbreviations

CA-MRSA	Community-acquired methicillin-resistant *Staphylococcus aureus*
CDC	Center for Disease Control and Prevention
CRE	Carbapenem-resistant *Enterobacteriacea*
DOTS	Direct Observed Treatment Short-Course
ESBL	Extended spectrum β-lactamase
ESKAPE	*Enterococcus faecium, Staphylococcus aureus, Klebsiella pneumoniae, Acinetobacter baumannii, Pseudomonas aeruginosa, Enterobacter* spp.
FDA	Food and Drug Administration
H-MRSA	Hospital-acquired methicillin-resistant *Staphylococcus aureus*
hVISA	Heterogeneous vancomycin-intermediate *Staphylococcus aureus*
IDSA	Infectious Diseases Society of America
IMP	Metallo-β-lactamase active on imipenem
KPC	*Klebsiella pneumoniae* carbapenemases
LPS	Lipopolysaccharide
MDR	Multidrug-resistant
MDR-TB	Multidrug-resistant tuberculosis
MIC	Minimum inhibitory concentration
MRAB	Multidrug-resistant *Acinetobacter baumannii*
MRSA	Methicillin-resistant *Staphylococcus aureus*
MSSA	Methicillin-susceptible *Staphylococcus aureus*
MYSTIC	Meropenem Yearly Susceptibility Test Information Collection
NDM	New Delhi β-lactamase
PBP2a	Penicillin-binding protein 2a
PDR	Pandrug-resistant
SIM	Seoul imipenemase
SME	*Serratia marcescens* enzyme
TDR-TB	Totally drug-resistant tuberculosis
VIM	Verona integron-encoded metallo-β-lactamase
VISA	Vancomycin intermediate-resistant *Staphylococcus aureus*
VRSA	Vancomycin-resistant *Staphylococcus aureus*
WHO	World Health Organization
XDR	Extensive drug resistant
XDR-TB	Extensive drug resistant tuberculosis

Contents

1 Introduction

The discovery of antibiotics has been one of the most significant advances in the modern medicine. These drugs have saved millions of lives not only by treating infections but also by preventing bacterial infections in individuals with weakened immune system such as those undergoing chemotherapeutic treatments against cancer or after organ transplantation. As we live in the antibiotic era, it is easy to assume that these drugs are permanently available in the arsenal of medicines. Lamentably, the use or rather the misuse of antibiotics has been accompanied by the rapid emergence of antibiotic resistance (Carlet et al. 2011). Furthermore, the use of antibiotics in the food animals industry has significantly fueled the rise of antibiotic resistant pathogens (Van Boeckel et al. 2015). We should not forget that antibiotics are not a human invention as they are produced by microorganisms such as *Streptomyces* spp. or fungi as weapons to eliminate bacteria that are competing for the same resources (Waksman et al. 1942; Waksman and Woodruff 1941, 1942). In turn, bacteria have developed mechanisms to counterattack the noxious effect of these antimicrobial agents as an adaptive trait to survive by out-competing their microbial neighbors in the environment. The biological pressure imposed by the continuous exposure to different antibiotics during clinical application has led to the cumulative acquisition of resistant traits in major human pathogens resulting in multidrug-resistant (MDR) bacteria, which are practically impossible to treat. Methicillin-resistant *Staphylococcus aureus* (MRSA) is probably the best-known example, but the list keeps growing. Among the emerging MDR bacteria are the extended spectrum β-lactamase-producing (ESBL) *Klebsiella pneumoniae* and *Escherichia coli*, carbapenem-resistant *Enterobacteriaceae* (CRE) and multidrug-resistant *Acinetobacter baumannii* (MRAB). Furthermore, some antibiotic resistance genes such as those encoding TEM β-lactamases (see below) exhibit an enormous plasticity and are capable to evolve and produce variants that are capable to confer resistance to a wider range of antibiotics (Salverda et al. 2010). The term "ESKAPE" has been introduced by the CDC to highlight the six pathogens that increasingly escape the effects of antibiotics (*Enterococcus faecium*, *S. aureus*, *K. pneumoniae*, *A. baumannii*, *Pseudomonas aeruginosa* and *Enterobacter* spp.) (Pendleton et al. 2013).

Even more alarming is the fact that these antibiotic-resistant bacteria are able to propagate very quickly not only in the health care facilities but also in the community across the globe. It is unquestionable that antibiotic resistance is life threatening and patients with infections caused by resistant pathogens have higher mortality rates and increased human suffering. Due to the prolonged hospitalization and decreased productivity, the economic burden associated with antibiotic resistance is enormous and take a staggering toll in the health care system. For example, the total costs attributable to extended hospitalization due to MRSA in the 31 countries that participated in the European Antimicrobial Resistance Surveillance System were 44 million Euros in 2007 (de Kraker et al. 2011). However, reliable estimates of the true burden are difficult to obtain. In order to sustain the effective lifetime of antibiotics, many hospitals have opted to reserve certain drugs such as vancomycin and imipenem and use them only as a last resource when other antimicrobials have proved ineffective. The downside of this policy is the risk associated with the delayed administration of effective antibiotics in infected patients. Moreover, the limited use of reserve antibiotics provides low incentives due to poor returns on investments and they are, therefore, nonprofitable for the pharmaceutical industry. Indeed, the lack of new classes of antibiotics coming to the market and the scarcity of new drugs in the pharmaceutical pipeline further aggravates the problem of antibiotic resistance. Only three new classes of antibiotics for treatment of human infections have been introduced to the market in this millennium (Spellberg et al. 2008; Bassetti et al. 2013). New antibiotics capable to defeat MDR pathogens are desperately needed, as otherwise we will soon face a future where an ordinary infection can be deadly. However, the development of new antibiotics will not provide an ultimate solution to the antibiotic crisis since the countdown to the emergence of resistance begins at the moment that the new antibiotic is introduced for clinical use. It is clear that antibiotic resistance is not a single-solution problem and, in addition to the steady development of new antibiotics, other strategies including effective vaccines and faster and more sensitive diagnostics can contribute to reduce the emergence of resistance. Thus, solutions to the problem of antibiotic resistance, in the long-run, will require a concerted mass effort of the medical and scientific community, government bodies and pharmaceutical industry.

2 The Serious Menace of "Superbugs"

At present, a major concern in bacterial infections are the so-called "Superbugs", which are bacterial pathogens that have developed resistance against different antibiotic classes commonly used for their treatment (Nordmann et al. 2007).

The resistant mechanisms of Superbugs can be mediated by the acquisition of genes that enable them to survive the exposure to antibiotics or by mutations that have evolved as part of a natural selection process. Superbugs are currently a global epidemic and are widespread around the planet. MRSA, MDR *Mycobacterium tuberculosis, K. pneumoniae, Streptococcus pneumoniae, Clostridium difficile, Neisseria gonorrhoeae, A. baumannii, Salmonella* spp., *E. coli, P. aeruginosa* and vancomycin-resistant *Enterococcus* spp., are typical examples of Superbugs. The historical background and current epidemiology of some of these multiresistant pathogens will be described in more detail in the next sections.

2.1 *Methicillin-Resistant* Staphylococcus aureus *(MRSA)*

The Gram-positive bacterium *S. aureus*, a member of the bacterial class "Bacilli", is historically the most notorious superbug (Peacock and Paterson 2015; Stryjewski and Corey 2014; Uhlemann et al. 2014; Otto 2012; Gould et al. 2012; Moellering 2012). *S. aureus* is an extraordinarily successful human pathogen, which is largely due to its versatility and enormous capacity to acquire antibiotic resistances. In the pre-antibiotic era, mortality associated with *S. aureus* exceeded 80 % (Skinner and Keefer 1941). Although *S. aureus* is a common commensal of humans and can be frequently isolated from the anterior nares (von Eiff et al. 2001) as well as other nasal regions (Kaspar et al. 2015), it can also cause a variety of infections ranging from mild skin infections to life-threatening diseases such as pneumonia, osteomyelitis, sepsis, endocarditis and bacteremia (David and Daum 2010). Nasal colonization is a risk factor for infections in both hospital and community settings (Wertheim et al. 2005; Kluytmans and Wertheim 2005; Keene et al. 2005; von Eiff et al. 2001).

2.1.1 Historical Background

The introduction of penicillin in 1941 resulted in a dramatic reduction of staphylococcal infections mortality rate and in improved prognosis of infected patients. Penicillin is a so-called β-lactam antibiotic (agents that contain a β-lactam ring in their molecular structure) that inhibits the cross-linking of peptidoglycan in the bacterial cell wall. The β-lactam antibiotics bind to and inhibit the cross-linking transpeptidase (also termed penicillin-binding protein) involved in bacterial cell wall biosynthesis and induce cell lysis (Yocum et al. 1979). Unfortunately, *S. aureus* strains resistant to penicillin were reported already one year after this new antimicrobial was introduced (Rammelkamp and Maxon 1942). Resistance to penicillin is due to the acquisition of β-lactamase activity (Murray and Moellering 1978; Bondi and Dietz 1945). The enzyme β-lactamase is encoded by the gene *blaZ*, which is part of a transposable element within a large plasmid. This enzyme hydrolyzes the β-lactam ring thereby rendering the β-lactam antibiotic inactive

(Kernodle 2000). Penicillin was commonly used in the 1950s and 60s and soon after that, more than 80 % of staphylococcal isolates were resistant to penicillin (Finland 1955; Barber and Rozwadowska-Dowzenko 1948). Penicillin-resistant strains were first emerging in hospitals, spreading later to the community, where they became prevalent (Chambers 2001). This prompted a search for β-lactamase-resistant drugs that led to the production of semisynthetic penicillins including methicillin and other derivatives such as oxacillin, cloxacillin, dicloxacillin, flucloxacillin and nafcillin. Methicillin was introduced for clinical use in 1959 and the first MRSA strain was reported in 1961 (Jevons 1961). The mechanism responsible for resistance to methicillin involves the acquisition of a transpeptidase termed PBP2a or PBP2' that has a reduced affinity for penicillin and other β-lactam antibiotics and is encoded by the *mecA* gene (Chambers 1997; Hartman and Tomasz 1981). This gene is located on the mobile genomic element staphylococcal cassette chromosome SCCmec, which carries a copy of *mecA* along with the genes encoding site-specific recombinases (*ccr*) required for the integrations and excision of the mobile element (Katayama et al. 2000). Soon after the initial description of SCCmec, several structurally different SCCmec elements were described based on the combination of the *ccr* gene allotype and the *mec* class that they carry (see http://www.sccmec.org/). Currently, eleven different SCCmec types are known (http://www.sccmec.org/Pages/SCC_TypesEN.html).

After the first case of MRSA was reported in the United Kingdom in 1961 (Jevons 1961), MRSA has spread widely in Europe by the 1970s and in the USA by the late 1980s (Enright et al. 2002; Maple et al. 1989). By the end of the 1990s, highly epidemic clones of MRSA had become the most frequent causative agents of *S. aureus* disease in both hospitals and communities in the USA (DeLeo and Chambers 2009; Voss et al. 1994; Panlilio et al. 1992). Although MRSA is not more virulent than methicillin-sensitive *S. aureus* (MSSA), MRSA infections are more difficult to treat than those caused by MSSA. The increasing prevalence of MRSA forced a significant escalation in the utilization of vancomycin (Levine 2006). Vancomycin is a glycopeptide that also exerts its antimicrobial effect against *S. aureus* by inhibiting the cell wall synthesis (Gardete and Tomasz 2014; Hiramatsu 2001). In contrast to β-lactams, which inhibit transpeptidase, the susceptibility of Gram-positive organisms to vancomycin is due to its high affinity for d-alanyl-d-alanine (D-Ala-D-Ala), which constitutes the terminal residue of the disaccharide-pentapeptide building blocks of the cell wall attached to a polyisoprene lipid carrier called lipid II. The first MRSA with decreased susceptibility to vancomycin was isolated in Japan in 1997 (Hiramatsu et al. 1997b). This strain exhibited only a modest increased minimum inhibitory concentration (MIC) value for vancomycin and became therefore known as vancomycin intermediate-resistant *S. aureus* (VISA; MIC = 4–8 μg/ml). The increased resistance to vancomycin of VISA strains is considered to have emerged from vancomycin-susceptible *S. aureus* (VSSA) through multiple genetic alterations that appear during vancomycin therapy, including mutations in regulator genes (Meehl et al. 2007). Typically, mutations result in thickening of the cell wall due to either accumulation of excess amounts of peptidoglycan, aberrant separation of daughter cells after cell division or altered rates of autolysis (Howden et al. 2010; Hiramatsu 2001; Walsh and Howe 2002; Sieradzki and Tomasz 2003; Gardete and Tomasz 2014). In those variants with thickened cell wall, more vancomycin molecules are assumed to be trapped in the increased number of peptidoglycan layers before reaching its target in the cell membrane (Hiramatsu 2001). An additional type of vancomycin resistance in *S. aureus,* the so-called hetero-VISA (hVISA) was reported from Japan in 1997 (Hiramatsu et al. 1997a). These strains are classified as susceptible strains (MIC < 4 μg/ml) but can generate VISA (MIC ≥ 8 μg/ml) at high frequency through mutations in the *rpoB* gene, which encodes the RNA polymerase β subunit (Matsuo et al. 2015). The first vancomycin-resistant *S. aureus* strain (VRSA; MIC ≥ 16 mg/ml) was reported in the USA in 2002 (Sievert et al. 2008). The VRSA strains have acquired the *vanA* vancomycin resistance gene located within transposon Tn*1546* on a plasmid which was acquired from vancomycin-resistant *Enterococcus faecalis* (Gardete and Tomasz 2014). Tn1546 is composed of nine genes, out of which five genes including the VanRS regulatory system, the VanH α-ketoacid dehydrogenase, the VanA ligase and the VanX dipeptidase are involved in vancomycin resistance (Courvalin 2006). VanH reduces pyruvate to D-lactate and VanA catalyzes ester bond formation between D-alanine and D-lactate with the resulting D-Ala-D-Lac depsipeptide replacing the D-Ala-D-Ala dipeptide in peptidoglycan synthesis. This substitution decreases the affinity of the molecule for glycopeptides considerably and prevents vancomycin binding (Arthur et al. 1996). By the end of 2013, VRSA strains have been isolated from several countries in the world (Kos et al. 2012; Limbago et al. 2014; Saha et al. 2008; Azimian et al. 2012; Melo-Cristino et al. 2013). Although dissemination of VRSA has so far been limited and there has been no secondary transmission of any of the VRSA strains to the patients family members or other contacts, the multiple mechanisms of *S. aureus* to develop resistance to vancomycin represent a warning for the public health system, and the potential for spread of such clinical isolates should not be underestimated.

More recently, daptomycin, a cyclic lipopeptide derived from *Streptomyces roseosporus* with bactericidal activity against a broad range of Gram-positive bacteria (Tedesco and Rybak 2004; Steenbergen et al. 2005), has been used to treat MRSA infections (Fowler et al. 2006; Moellering 2008). Daptomycin was approved by the US Food and Drug Administration in 2003 for treatment of skin and soft tissue infections and in 2006 for therapy of *S. aureus* bacteremia and endocarditis. Daptomycin disrupts the function of the cytoplasmic membrane in Gram-positive organisms, causing leakage of potassium and other ions, ultimately leading to loss of membrane potential and cell death (Jung et al. 2004; Cotroneo et al. 2008; Silverman et al. 2003). Because of its unique mechanism of action, it was assumed that MRSA would not easily develop resistance to daptomycin. However, the first case of daptomycin treatment failure was reported in 2005 (Vikram et al. 2005; Hayden et al. 2005). To the current knowledge, the mechanisms of daptomycin resistance seem to be very diverse, involving perturbations in the cell membrane as well as in the cell wall (Bayer et al. 2013).

2.1.2 Epidemiology

The first epidemic MRSA clone was identified in the UK in 1980 and was largely limited to Europe. Novel lineages of MRSA have since then become a worldwide pandemic in all continents, causing infections in both the health care setting and in the community (Enright et al. 2002). Overall, the vast majority of MRSA isolated worldwide belong to a limited number of clones, some of which are associated with global epidemics. Globally disseminated MRSA strains causing outbreaks and epidemics in health care facilities are of particular concern. For example, Epidemic MRSA-15 (EMRSA-15) is currently the most successful healthcare-associated MRSA clone spreading rapidly within and between hospitals, as well as to different countries (Holden et al. 2013). EMRSA-15 which belongs to multilocus sequence type ST22 and carries a type IV SCCmec element, was initially isolated in the UK in 1991 (Richardson and Reith 1993) and by 2000, it accounted for over 60 % of MRSA nosocomial bacteremias in this country (Johnson et al. 2001). The development of resistance to fluoroquinolone during the mid-1980s that was caused by point mutations generating amino acid substitutions in topoisomerase IV and gyrase A, has been suggested to be a key in the success of EMRSA-15 (Holden et al. 2013).

Although MRSA is nowadays the most common multidrug-resistant pathogen causing nosocomial infections, the prevalence of hospital-acquired MRSA shows considerable geographical variation (Sievert et al. 2013; Moellering 2012). For example, the percentages of isolates reported as MRSA in Europe in 2013 ranged from 0 % in Iceland to 64.5 % in Romania (European Centre for Disease Prevention and Control 2013). Although MRSA infections have declined in recent years (Meyer et al. 2014; Khatib et al. 2013; Wyllie et al. 2011a, b; Kallen et al. 2010), seven out of 30 European countries from where data were available, reported MRSA percentages above 25 % in 2013 (European Centre for Disease Prevention and Control 2013). Hence, MRSA remains a public health priority in Europe. The success of countries such as the Netherlands or the Scandinavian countries in the prevention of MRSA transmission in healthcare facilities has been largely due to the implementation of a stringent "Search and Destroy (S&D) policy" (Bootsma et al. 2006). This policy involves the identification and isolation of MRSA carriers and of patients at increased risk for MRSA carriage as well as minimization of the risk of MRSA transmission to other individuals within the healthcare facility by wearing of personal protective equipment and disinfection of the room after discharge of these patients.

By the late 1990s, MRSA infections started to spread within the community and affected patients without previous healthcare exposure (Herold et al. 1998; DeLeo et al. 2010; Vandenesch et al. 2003; Stegger et al. 2014). In contrast to hospital-acquired MRSA (H-MRSA), which typically carry the SCCmec types I, II and III, the initial community-acquired MRSA (CA-MRSA) strains in the USA carried the SCCmec type IV and produced Panton-Valentine leukocidin (PVL) (DeLeo et al. 2010; Ma et al. 2002), a potent pore-forming cytotoxin. These strains of the ST1 lineage (also known as USA400) were promptly replaced by the strain USA300 (ST8 lineage), which is currently the predominant CA-MRSA in the

USA (DeLeo et al. 2010) and has spread throughout the world (Nimmo 2012). Prevention and control of MRSA infections in non-clinical arenas may create a bigger challenge compared to the reduction of the spread of MRSA in healthcare settings. Strategies to prevent spreading of MRSA infections in the community involve hand hygiene, avoiding sharing of personal items and adherence to contact precautions during care of individuals with known MRSA infections.

*2.2 MDR- and XDR-*Mycobacterium tuberculosis

Mycobacterium tuberculosis is the causal agent of tuberculosis, a devastating infection of the lungs with an outsized burden of morbidity and mortality (Murray et al. 2014). More than 50 years after the introduction of the first anti-tuberculosis drugs, this infection remains still a leading cause of death. In 2013, about 9 million people were infected with *M. tuberculosis* and 1.5 million succumbed to the infection (World Health Organization 2014b). Furthermore, an estimated one third of the world population is infected with *M. tuberculosis* that exists in a latent stage and at risk of disease reactivation (World Health Organization 2014b; Wayne and Sohaskey 2001). Chemotherapy of tuberculosis involves a combination of several drugs including isoniazid, rifampicin, pyrazinamide, and either ethambutol or streptomycin for several months (Keshavjee and Farmer 2012). In recent years, the increased prevalence of multidrug-resistant tuberculosis (MDR-TB) caused by strains which are resistant to at least two of the first-line anti-tuberculosis drugs i.e. isoniazid and rifampicin (Iseman and Madsen 1989), has worsened the situation and prompted the use of second line drugs with the subsequent emergence of extensively drug-resistant *M. tuberculosis* (XDR-TB) (World Health Organization 2014b; Bastos et al. 2014; Abubakar et al. 2013). The XDR-TB strains, defined as being resistant against isoniazid, rifampicin, fluoroquinolone and to at least one second-line drug including kanamycin, capreomycin or amikacin (World Health Organization 2014b; Bastos et al. 2014; Abubakar et al. 2013), have become a major public health concern. It is estimated that about 5 % of tuberculosis patients are infected with MDR strains and that 9 % of MDR-TB patients acquire or develop XDR-TB (World Health Organization 2014b). In spite of the fact that global rates of new tuberculosis infections are decreasing since 2005, MDR- and XDR-TB cases have increased, with an estimated 450,000 new cases in 2012 (World Health Organization 2014b).

2.2.1 Historical Background

Streptomycin became the standard treatment for tuberculosis in the 1940s (Hinshaw et al. 1946; Daniel 2006). Although many patients receiving streptomycin had a rapid clinical improvement, treatment failures were also common and streptomycin-resistant *M. tuberculosis* strains could soon be isolated from non-responding patients (Crofton and Mitchison 1948). In the 1950s, isoniazid was tested against tuberculosis with great success (Selikoff and Robitzek 1952) and was followed by pyrazinamide (1952), cycloserine (1952), ethionamide (1956), rifampicin (1957), and ethambutol (1962) (Keshavjee and Farmer 2012). With the introduction of new antibiotics to treat *M. tuberculosis* infections, new resistance mechanism emerged. For example, resistance to rifampicin was detected soon after the administration of the antibiotic to tuberculosis patients (Manten and Van Wijngaarden 1969). The rapid development of resistance during monotherapy of these antibiotics led to combination therapies including several antibiotics, which remains the cornerstone of tuberculosis treatment. The first empirical treatment consisted of a four-drugs regimen administered for one year (Mitchison and Davies 2012; Crofton 1959; Fox et al. 1999). This long therapy resulted in poor compliance to treatment and the emergence of MDR-TB strains (Iseman and Madsen 1989). Even more disturbing was the emergence of XDR-TB. But it can always be worse and the first case of totally drug-resistant tuberculosis (designated TDR-TB, although this terminology need to be officially endorsed by the WHO) was observed in Italy in 2003 (Migliori et al. 2007) and later in Iran (Velayati et al. 2009). TDR-TB strains exhibit *in vitro* resistance to all first- and second-line drugs tested including isoniazid, rifampicin, streptomycin, ethambutol, pyrazinamide, ethionamide, cycloserine, ofloxacin, amikacin, ciprofloxacin, capreomycin and kanamycin (Parida et al. 2015). In 2012, new TDR-TB cases were reported in India (Udwadia 2012). Recently, the three new drugs bedaquiline, delamanid and linezolid have been approved by the US Food and Drug Administration (FDA) and the European Medicines Agency that may offer therapeutic solutions, at least temporarily, for treatment of TDR-TB (Parida et al. 2015).

2.2.2 Mechanisms of Drug Resistance of MRD-TB Strains

Drug resistance in *M. tuberculosis* is generally caused by bacterial chromosomal mutations arising during sub-optimal drug therapy (Almeida Da Silva and Palomino 2011). Isoniazid is a pro-drug that needs to be activated by the *M. tuberculosis* KatG enzyme, a bifunctional enzyme with both catalase and peroxidase activity (Metcalfe et al. 2008; Zhang et al. 1992). Isoniazid activation leads to inhibition of the synthesis of mycolic acids, 2-alkyl-3-hydroxy long-chain fatty acids which are major and specific components of the mycobacterial cell wall (Winder and Collins 1970). The target of the activated isoniazid, assumed to be an isonicotinic-acyl radical (Rozwarski et al. 1998, 1999), is the long-chain enoyl-acyl carrier protein reductase InhA, an enzyme essential for mycolic acid biosynthesis (Banerjee et al. 1994). Activated isoniazid binds covalently to the nicotinamide ring of the NADH found within the active site of this enzyme, creating a highly effective inhibitor (Rozwarski et al. 1998). Mutations in the genes encoding KatG and InhA or the respective promoter regions (Ramaswamy et al. 2003; Hazbon et al. 2006) have been shown to be typically responsible for the resistant phenotype. The S315T gene mutation in *katG* is the most frequent in MDR-TB strains and is associated with high-levels of resistance (MIC > 1 μg/ml) (Hazbon et al. 2006). This mutant retains peroxidase/catalase activity, however, it is highly deficient in forming an isoniazid-NAD-adduct (Yu et al. 2003).

Rifampicin is one of the most effective antibiotics to treat tuberculosis and acts by inhibiting transcription through the targeting of the DNA-dependent RNA polymerase. RNA polymerase consists of four different subunits encoded by the *rpoA*, *rpoB*, *rpoC* and *rpoD* genes, respectively. Rifampicin binds to the β-subunit thereby inhibiting the elongation of mRNA (Wehrli 1983). Resistance to rifampicin is conferred by mutations in the *rpoB* gene that alter residues of the rifampicin-binding site of the RNA polymerase β-subunit, leading to low affinity binding (Telenti et al. 1993a, b). Interestingly, more than 90 % of *M. tuberculosis* strains resistant to rifampicin are also resistant to other antibiotics including isoniazid (Garcia de Viedma et al. 2002). Therefore, rifampicin resistance is used as a surrogate marker for MDR-TB.

2.2.3 Epidemiology

The global incidence of *M. tuberculosis* infections slowly declined from 1997 to 2001, but an increase in the incidence rate was observed in 2001, most probably fanned by the expansion of HIV pandemic (Sulis et al. 2014). Nevertheless, a reduction in the number of tuberculosis infections has been observed since 2002, reaching 2.2 % reduction between 2010 and 2011 (Sulis et al. 2014). The introduction of the DOTS (Direct Observed Treatment Short-Course) by the WHO was most probably the driven instrument responsible for this decline (Kochi 1997). This program was introduced to overcome the poor adherence and premature interruption of tuberculosis treatment that was responsible for prolonged illness and development of drug resistance. Key elements of this strategy are bacteriological diagnosis and direct supervision of drug intake by patients (Kochi 1997). A total of 46 million of tuberculosis patients worldwide were treated in DOTS programs between 1995 and 2010 with an overall rate of treatment success of 87 % in 2009 (Glaziou et al. 2011). With outbreaks of MDR-TB reported in the early 1990s, mainly among HIV-infected individuals (Cohn et al. 1997), a new program called DOTS-Plus was initiated as a more rigorous treatment strategy (World Health Organization 2001). The prevalence of MDR-TB in several countries including China, India, Russia and South Africa was 460,000–870,000 in 2010 and 58 countries reported at least one case of XDR-TB in 2011 (Lienhardt et al. 2012).

2.3 *Emerging Extended-Spectrum β-Lactamase Producing Pathogens (ESBLs)*

Gram-negative bacteria, specifically *Enterobacteriaceae* and other γ-Proteobacteria such as *Pseudomonas* spp. are currently an important cause of hospital-acquired infections including pneumonia, surgical site infections, urinary tract infections and bloodstream infections. Under antibiotic selection pressure, these organisms are highly efficient at developing or acquiring genes that confer resistance to multiple antibiotics and the emergence of multidrug-resistant strains, which are practically resistant to all approved antibiotics, is currently an enormous public health concern (Hawkey and Jones 2009). The excessive use of broad-spectrum, orally administered β-lactams antibiotics including ampicillin, amoxicillin or cephalexin has provided a favorable context for the selection of extended-spectrum β-lactamases-producing (ESBLs) γ-Proteobacteria in both hospital and community settings. Whereas the resistance against β-lactams antibiotics in *S. aureus* as described above is often mediated by modification of proteins that are the target of β-lactams antibiotics (such as modified transpeptidases), β-lactam resistance in Gram-negative organisms is largely mediated by the production of a diverse set of β-lactamases. These enzymes hydrolyze the lactam ring and are often capable to attack a broad spectrum of β-lactam antibiotics (Bush and Fisher 2011). The most frequent ESBLs harboring bacteria are the *Enterobacteriaceae E. coli* and *K. pneumoniae* as well as Pseudomonadales such as *A. baumannii* and *P. aeruginosa* (Livermore 2012). In addition to β-lactams, ESBLs-producing pathogens are frequently resistant also to non-β-lactam antibiotics since the β-lactamases are encoded on rapidly spreading mobile elements carrying resistance determinants for multiple antimicrobial families including the aminoglycosides, tetracyclines, and fluoroquinolones (Paterson and Bonomo 2005). As a result, a large proportion of the ESBLs-producing bacteria are indeed multidrug-resistant.

2.3.1 Historical Background

Different classification schemes have been used to discriminate between the vast variety of β-lactamases. One of these schemes, based on phylogenetic analyses proposed by Ambler (1980), uses a molecular classification and divides β-lactamases into four classes, where classes A, C and D include enzymes that hydrolyze their substrates by forming an acyl enzyme through an active site serine, whereas class B enzymes are zinc-dependent metalloenzymes. The very first ESBLs seem to have evolved from the narrow-spectrum β-lactamases of the so called TEM and SHV lineages of *E. coli and Klebsiella* spp. (Turner 2005) which, like penicillinase from *S. aureus,* belong to the class A β-lactamases. TEM-1 is the most commonly encountered β-lactamase in Pseudomonadales. Mutations in the active site of these ancestor enzymes enlarged the number of substrates that the β-lactamases could hydrolyze. Bacteria carrying these mutations were selected by antibiotic pressure and the encoding genes rapidly spread via mobile genetic elements (Kliebe et al. 1985; Sirot et al. 1987; Sougakoff et al. 1988). TEM and SHV ESBLs comprise several members conferring resistance not only to penicillins and first generation cephalosporins (Turner 2005), but also to second generation cephalosporins and monobactams. The CTX-M β-lactamase type seems to have originated from chromosomally encoded enzymes of *Kluyvera* spp. from where the genes were then recruited onto plasmids and transferred especially to *E. coli* and *K. pneumoniae* (Poirel et al. 2002). CTX-M-type ESBLs, characterized by their potent hydrolytic activity against cefotaxime, were initially reported in the second half of the 1980s, and their rate of spreading and prevalence in most parts of the world has increased dramatically since 1995 (Bonnet 2004).

In addition to ESBL class A enzymes, class C enzymes, also termed extended spectrum AmpC-type β-lactamases, can mediate resistance to a number of β-lactams including third-generation cephalosporins (Jacoby 2009). AmpC β-lactamases were known to be chromosomal encoded cephalosporinases in *P. aeruginosa* (Lodge et al. 1990) and many *Enterobacteriaceae* (Jaurin and Grundstrom 1981), but have been then recruited on plasmids (Philippon et al. 2002) and spread among the *Enterobacteriaceae* (Harris 2015; Munier et al. 2010). This development was reported to coincide with a change from a chromosomal gene controlled by a weak promoter to a plasmid with strong constitutive expression. Strains carrying this plasmid have been isolated from nosocomial, community onset, and healthcare-associated infections and are often associated with a high mortality rate (Rodriguez-Bano et al. 2012). Plasmids carrying AmpC β-lactamases may also frequently carry genes encoding for additional resistances including resistance to aminoglycosides, chloramphenicol, quinolones, sulfonamide, tetracycline, and trimethoprim as well as genes encoding for other β-lactamases (Jacoby 2009).

An additional element favoring the development of multiresistance is the permeability of the bacterial outer membrane to β-lactams, which is determined by the presence of porin channels through which β-lactams penetrate as well as by the presence of efflux pumps, which transport them out of the cell. Decreasing the number of porin entry channels or increasing the expression of efflux pumps can reduce the permeability of the bacterial outer membrane to certain antibiotics and favor the acquisition of mutiresistance. For example, resistance to carbapenem in *P. aeruginosa* involves the combination of overproduction of AmpC β-lactamase, decreased production of the OprD porin channel for imipenem entry, and activation of MexAB-OprM and other efflux systems (Lister et al. 2009).

The class D β-lactamases are typically also termed OXA-type β-lactamases based on their high activity against cloxacillin and oxacillin (Bush et al. 1995). They were originally observed predominantly in *P. aeruginosa* (Philippon et al. 2002) but have been detected in many γ-Proteobacteria and are widespread in *E. coli* strains (Livermore 1995). Even though most of these lactamases do not hydrolyze second-generation cephalosporins at a significant rate, OXA-10 hydrolyzes some of them and its acquisition results in strains with a reduced susceptibility to these antibiotics (Paterson and Bonomo 2005).

Enzymes termed carbapenemases are the most versatile β-lactamases, as many of them recognize almost all hydrolyzable β-lactams (Livermore and Woodford 2006; Queenan and Bush 2007). Carbapenemases have been identified inside the above described class A, B, and D lactamases. The first described carbapenemases were chromosomally encoded class A enzymes and described from *Serratia marcescens* isolates (Yang et al. 1990) and thus termed SME (*S. marcescens* enzyme). Later on, carbapenemases have often been identified on transferable plasmids and since the end of the 1990s, the *K. pneumoniae* carbapenemases (KPC), also identified as class A lactamases, began to emerge (Yigit et al. 2001). In fact, KPC carbapenemases hydrolyze β-lactams of all classes, and belong to the most wide-spread carbapenemases. Also widely disseminated became the carbapenemases of the class B metalloenzyme type including the VIM (Verona integron-encoded metallo β-lactamase) (Lauretti et al. 1999), IMP (metallo β-lactamase active on imipenem) (Watanabe et al. 1991) and the NDM (New Delhi metallo-β-lactamase) (Dortet et al. 2014) variants.

2.3.2 Epidemiology

The incidence of infections caused by ESBLs-producing bacteria has dramatically increased in recent years (Peleg and Hooper 2010). ESBLs-producing *K. pneumoniae* can cause clonal outbreaks in healthcare facilities such as in intensive care units of hospitals (Harris et al. 2007). These outbreaks are primarily caused by TEM- or SHV-producing *K. pneumoniae*. On the other hand, *E. coli* strains carrying predominantly CTX-M β-lactamases generally cause community-acquired infections (Rodriguez-Bano et al. 2006).

The epidemiology of ESBLs-producing pathogens is complex and the prevalence of bacteria producing ESBLs varies across the world (World Health Organization 2014a). Data from the global surveillance program MYSTIC (Meropenem Yearly Susceptibility Test Information Collection), which has collected data for ESBL-producing pathogens in Europe (1997–2004) and the USA (1999–2004), indicates a high prevalence of *Klebsiella* spp. among the ESBL-producing isolates in both Europe and the USA (Goossens and Grabein 2005). However, the prevalence and types of ESBLs-producing pathogens found in the USA are in strong contrast to the epidemiology observed in the rest of the world. Thus, whereas strains producing CTX-M β-lactamase are becoming widespread in Europe and Asia, they are still relatively uncommon in the USA (Livermore and Hawkey Livermore and Hawkey 2005). Furthermore, whereas an increased prevalence of *E. coli* and *Klebsiella* spp. was observed in Europe during this time period, increased prevalence of *Enterobacter* spp. was seen in the USA (Goossens and Grabein 2005). One representative example linked to ESBL-producing *Enterobacteriaceae* is the recent and fast global dissemination of the highly virulent ciprofloxacin-resistant clonal group of *E. coli* of sequence type ST131 and surface O-antigen type O25:H4. These pathogens cause urinary tract infections and carry the CTX-M variant CTX-M-15 β-lactamase (Clermont et al. 2008; Nicolas-Chanoine et al. 2008), which is now spread in all human and animal compartments as well as the environment all over the world (Canton et al. 2012).

Carbapenemase-producing *Enterobacteriaceae* were only occasionally observed until the 1990s, but since the late 1990s, KPC-producing *K. pneumoniae* spread globally and are endemic in the USA, Israel, Greece, and Italy, with a carbapenem resistance prevalence of 60.5 % being recorded in 2012 in Greece (Doi and Paterson 2015). In particular, *K. pneumoniae* ST258 and related strains are currently globally spread (Doi and Paterson 2015). A further group of carpapenemases of epidemiological interest are the class D OXA-type carpapenemases. OXA-48 was first found in *K. pneumoniae* strain 11978 isolated in Turkey in 2001 (Poirel et al. 2004). Up to now, OXA-48 has been mainly observed in *K. pneumoniae* but could be potentially transferred to other *Enterobacteriaceae* (Poirel et al. 2012).

NDM (New Delhi metallo-β-lactamase) was first identified in 2008 in a *K. pneumoniae* isolate recovered from a patient previously hospitalized in New Delhi, India (Yong et al. 2009). Since then, NDM carbapenemases came into focus worldwide due to the rapid dissemination among *Enterobacteriaceae* as well as to *Acinetobacter* spp. (Dortet et al. 2014). Recently, prevalence rates of NDM-producing *Enterobacteriaceae* of up to 18 % were reported in hospitals in India and Pakistan (Perry et al. 2011; Lascols et al. 2011). Importantly, the gene encoding this enzyme was not only detected in *Enterobacteriaceae* isolated from hospitals, but also in environmental samples (Walsh et al. 2011). In addition to pathogenic *Enteobacteriaceae*, NDM was also observed in *A. baumannii* (El-Sayed-Ahmed et al. 2015) and NDM producing *A. baumannii* are reported to be on the top list of carbapenemase producers in Europe (Dortet et al. 2014).

2.4 *"Pandrug-Resistant" (PDR) Pathogens: The Example of* Acinetobacter baumannii

Pandrug-resistant (PDR) pathogens are those that have become resistant against all commercially available antimicrobial agents and remain susceptible only to older and potentially more toxic agents such as colistin and tigecycline (Magiorakos et al. 2012; Poulikakos et al. 2014). Consequently, PDR pathogens cause infections for which limited and suboptimal treatment options exist (Falagas et al. 2005). Clinically relevant PDR bacteria include *A. baumannii*, *P. aeruginosa* and *K. pneumoniae*.

During the last decade, multidrug-resistant *A. baumannii* has emerged as an important cause of healthcare-associated infections worldwide (Peleg et al. 2008). The remarkable success of *A. baumannii* as nosocomial pathogen is bestowed by its extraordinary capacity to develop or acquire antibiotic resistance determinants and by its capacity to adapt and survive for long periods in the hospital environment (Bergogne-Berezin and Towner 1996; Fournier and Richet 2006). Critically-ill patients with impaired host defenses are at high risk to acquire *A. baumannii* and can develop a wide range of infections including pneumonia, bacteremia, meningitis, urinary tract infections, and wound infections (Falagas et al. 2008). *A. baumannii* is in the list of the most difficult-to-treat antimicrobial-resistant γ-Proteobacteria (Talbot et al. 2006). Prior to the 1970s, common antibiotics such as aminoglycosides, β-lactams, and tetracyclines were still effective as therapy for *A. baumannii* infections (Bergogne-Berezin and Towner 1996). *A. baumannii* is reported to possess an intrinsic class D oxacillinase belonging to the OXA-51-like enzymes (Turton et al. 2006), which has only a weak activity against carbapenems and is not active against second generation cephalosporins (Heritier et al. 2005). However, strains of *A. baumannii* highly resistant to carbapenems and virtually to all β-lactams started to emerge in the 1990s (Go et al. 1994; Evans et al. 2013). The first β-lactamases with carbapenemase activity termed ARI-1 (*Acinetobacter* resistant to imipenem), a class D OXA-23 type enzyme, was described in 1993 from a strain previously isolated in 1985. Since then, it has been asserted that the most widespread carbapenemase activities in *A. baumannii* are due to carbapenem-hydrolysing class D β-lactamase of the OXA-23, OXA-24 and OXA-58 lineages (Poirel and Nordmann 2006). However, the carbapenem resistance level provided by these enzymes in *A. baumannii* is assumed to be much lower compared to that of class B metallo-β-lactamases (Poirel and Nordmann 2006). Until recently, three groups of metallo-β-lactamases dominated in *Acinetobacter*, which are the VIM, IMP, and SIM (Seoul imipenemase) groups (Queenan and Bush 2007) and a recent study reported an increasing prevalence and identified 26.5 % of imipenem-resistant *Acinetobacter* isolates carrying these lactamases in Korea (Yong et al. 2006). Importantly, NDM-producing *Acinetobacte*r strains are also known since a few years (Pfeifer et al. 2011; Dortet et al. 2014), and analysis of the genetic environments of the genes encoding NMD-1 indicated that *Acinetobacter* spp. has been a reservoir of the respective lactamase encoding genes before they have been spread to *Enterobacteriaceae* (Dortet et al. 2012). Interestingly, almost all NDM producers including *Enterobacteriaceae*, *Acinetobacter*, and *Pseudomonas*, showed a simultaneous resistance to various other antibiotics such as aminoglycosides, quinolones, and macrolides (Dolejska et al. 2013).

In addition to the acquisition of antimicrobial-inactivating enzymes, other mechanisms of antimicrobial resistance in *A. baumannii* includes: (1) decreased outer membrane permeability caused by the loss or reduced expression of porins; (2) overexpression of multidrug efflux pumps; and (3) mutations that change targets or cellular functions such as alterations in penicillin-binding proteins (Gordon and Wareham 2010; Fernandez-Cuenca et al. 2003). The rapid global spread of *A. baumannii* strains resistant to all first-line antimicrobials has posed an enormous therapeutic challenge to clinicians (Peleg et al. 2008; Falagas and Bliziotis 2007; Visca et al. 2011). The only antibiotics that remain effective against PDR *A. baumannii* strains are colistin and tigecycline (Gordon and Wareham 2010). As resistance to tigecycline has been already reported by many countries (Kulah et al. 2009; Navon-Venezia et al. 2007), the old drug colistin is currently the last resource for the treatment of PDR *A. baumannii* infections (Falagas and Kasiakou 2005). Colistin was introduced for clinical use in the 1950s and acts against Gram-negative bacteria by interacting with the lipid A moiety of lipopolysaccharide (LPS) molecules in the bacterial outer membrane resulting in increased membrane permeability, leakage of cell contents and cell death (Davis et al. 1971). The use of colistin was abandoned in the early 1970s due to its reported toxicity (Lim et al. 2010). However, the emergence of PDA *A. baumannii* strains has led to the resurgence of colistin as last resource to treat these infections (Falagas and

Kasiakou 2005). Unfortunately, the first colistin-resistant PDA *A. baumannii* was reported in 1999 (Hejnar et al. 1999) and since then colistin-resistant strains have been reported in several countries around the world, with the highest resistance rate reported in Europe (Gales et al. 2006). The mechanism of resistance of *Acinetobacter* strains to colistin remains unclear (Cai et al. 2012). Some studies have suggested the loss of LPS after inactivation of a lipid A biosynthesis gene resulting in an extensive alteration of the bacterial surface, including the expression of transport systems (Moffatt et al. 2010; Henry et al. 2012). Other studies indicated that mutations in the PmrAB two-component regulatory system that results in modification of the LPS molecule are responsible for resistance against colistin (Adams et al. 2009; Beceiro et al. 2011). The current worldwide emergence of PDR *A. baumannii* strains resistant to colistin is distressing and highlights the need for new approaches to control this dangerous pathogen.

3 Reversing the Global Trend of Antibiotic Resistance in the Future

In order to win the battle against MDR pathogens, a concerted mass effort led by the medical community, governments, pharmaceutical industry and social education is necessary. The trend of physicians prescribing broad-spectrum antibiotics for any inflammatory process needs to be reversed and fast actions needs to be taken to reduce the use of antibiotics by the modern animal food industry. Control programs against resistant pathogens will involve rational antibiotic prescription, reduced use of antimicrobials as prophylaxis, compliance with antibiotic therapy and appropriate hospital hygiene. Furthermore, development of faster diagnostic tools and rapid and more accurate antimicrobial profiling can lead to a more precise and targeted antibiotic therapy. Therefore, to curb the global trend of antibiotic resistance in the future will require a multifaceted approach including (1) prevention and control of infections caused by antibiotic-resistant pathogens, (2) development of new drugs, and (3) prolongation of the effectiveness of current and new antibiotics.

3.1 Prevention and Control of Infections Caused by Antibiotic-Resistant Pathogens

Stringent monitoring of antimicrobial susceptibility trends by long-term surveillance systems at a local, regional, or global level is required to assess and monitor the scope and magnitude of antibiotic resistance. Surveillance data are critical for implementing prevention efforts, selection of treatment and for guiding new drug development. A deeper understanding of the molecular basis of antimicrobial resistance is required to facilitate the development of new diagnostic molecular methods for the fast detection of specific resistance mechanisms. The fast detection of pathogen resistance patterns can guarantee the implementation of the correct therapeutic treatment, minimizing the need for broad-spectrum antibiotics and, in turn, reduce the emergence of antimicrobial resistance. Prevention of spreading and cross-transmission of antibiotic resistant pathogens must be based on a many-sided strategy that should include appropriate screening policies to detect colonized patients, improved hospital hygiene and specific contact precautions.

3.2 Development of New Drugs

Despite the urgent need for new antimicrobial drugs for clinical use, antibiotic development has slowed dramatically over the past 30 years and many pharmaceutical companies have abandoned antibiotic discovery programs. Currently, the are 36 new antibiotics in the development phase, eight in phase 1 clinical trails, 20 in phase 2, and 8 in phase 3 (The Pew Charitable Trusts 2015; Hesterkamp 2015). As not all of them will be approved for clinical use, it is clear that the number of new antibiotics in the development phase will not be enough to meet the current needs. Furthermore, among the new drugs currently in the pipeline, only a few seem to target specifically resistant pathogens (Cormican and Vellinga 2012; Spellberg et al. 2013). For these reasons, various agencies and governmental organizations have reiterated their calls for the development of novel antibiotics. For example, the Infectious Diseases Society of America (IDSA) has recently launched a program called "*the '10 × '20' initiative*" with the aim to develop ten new, safe and effective antibiotics by 2020 (Infectious Diseases Society of America 2010). In the same line, the European Innovative Medicine Initiative (http://www.imi.europa.eu), which supports collaborative research projects between the pharmaceutical industry and academic institutions, combines public and private funding to boost antimicrobial drug research in Europe. The primary call-to-action of these initiatives is to increase the number of promising antibiotics in development and to achieve one of the century most important challenges, that is the effective treatment of all bacterial infections in the coming years.

3.3 Prolongation of the Effectiveness of Current and New Antibiotics by Preventing the Emergence of Antibiotic Resistance

A bigger challenge than to generate new antimicrobial agents is the challenge of preserving the beneficial lifetime of the current and newly generated antibiotics. Several strategies can be pursued to prevent the emergence of antibiotic resistance. In this regard, implementation of antibiotic stewardship programs is a critical issue for the preservation of antibiotics efficacy (Carlet et al. 2012). This involves optimal selection of the antibiotic or antibiotic combination for each specific pathogen as well as right dosage and duration of the antimicrobial treatment. Restricted use of antibiotics with a broad spectrum of action would also help to curtail the development of antibiotic resistance. As prolonged administration of antibiotics appears to be a factor promoting the emergence of antibiotic resistance, recent efforts have been made to implement shorter courses of antibiotic treatment (Schrag et al. 2001; Pugh et al. 2010, 2015). In fact, the optimal duration of antibiotic therapy for different infections may be uncertain and an optimized treatment time may result in reduction of the selection pressure exerted on the pathogen being treated, with concomitant decreased propensity for development of resistance.

One of the biggest problems in antibiotic resistance arises from the massive amount of antibiotics released into the environment through their excessive use, not only in human, but also in veterinary settings. Thus, the environment can be regarded as an important breeding and selection reservoir for antibiotic resistance genes (Canton 2009). Specifically, wastewater systems and effluents from the pharmaceutical industry or animal husbandry may contain high bacterial loads and antibiotics and thus may select for antibiotic resistances and favor the spread of respective genes in the environment (Gullberg et al. 2011; Mira et al. 2015). Even more concerning is the observation that the presence of multiple concentrations of antibiotics in an environment may result in the selection of a high diversity of new antibiotic-resistant genotypes within the microbial populations (Mira et al. 2015). Obviously, the presence of antibiotic resistant bacteria in the environment represents a serious risk to humans and more efforts should be made to reduce the levels of antibiotics contamination in this setting.

4 Conclusions

Antibiotic resistance is one of the most serious threats to human health today. Believing that the increasing problem of antibiotic resistance can be solely solved by the development of new antimicrobial drugs is illusory. The threat of resistance will always accompany any new drug introduced for clinical use. The only possible, sustainable solution is to keep pace with it. This will involve introducing profound changes in the use of these drugs, including stewardship programs for rational use and improve targeted therapy. Furthermore, implementation of adequate preventive measures such as vaccines and faster diagnostic tools as well as improving hygiene and reducing the use of antibiotics in animals, will be the only way for preserving the usefulness of antibiotics for future generations and ensure a healthy future for the world's population. Finally, it is of critical importance to acquire a more comprehensive understanding of the molecular, evolutionary and ecological mechanisms governing the spread of antibiotic resistance.

References

Abubakar I, Zignol M, Falzon D, Raviglione M, Ditiu L, Masham S, Adetifa I, Ford N, Cox H, Lawn SD, Marais BJ, McHugh TD, Mwaba P, Bates M, Lipman M, Zijenah L, Logan S, McNerney R, Zumla A, Sarda K, Nahid P, Hoelscher M, Pletschette M, Memish ZA, Kim P, Hafner R, Cole S, Migliori GB, Maeurer M, Schito M, Zumla A (2013) Drug-resistant tuberculosis: time for visionary political leadership. Lancet Infect Dis 13(6):529–539

Adams MD, Nickel GC, Bajaksouzian S, Lavender H, Murthy AR, Jacobs MR, Bonomo RA (2009) Resistance to colistin in *Acinetobacter baumannii* associated with mutations in the PmrAB two-component system. Antimicrob Agents Chemother 53(9):3628–3634

Almeida Da Silva PE, Palomino JC (2011) Molecular basis and mechanisms of drug resistance in *Mycobacterium tuberculosis*: classical and new drugs. J Antimicrob Chemother 66(7):1417–1430

Ambler RP (1980) The structure of beta-lactamases. Philos Trans R Soc Lond B Biol Sci 289 (1036):321–331

Arthur M, Reynolds P, Courvalin P (1996) Glycopeptide resistance in enterococci. Trends Microbiol 4(10):401–407

Azimian A, Havaei SA, Fazeli H, Naderi M, Ghazvini K, Samiee SM, Soleimani M, Peerayeh SN (2012) Genetic characterization of a vancomycin-resistant *Staphylococcus aureus* isolate from the respiratory tract of a patient in a university hospital in northeastern Iran. J Clin Microbiol 50 (11):3581–3585

Banerjee A, Dubnau E, Quemard A, Balasubramanian V, Um KS, Wilson T, Collins D, de Lisle G, Jacobs WR Jr (1994) *inhA*, a gene encoding a target for isoniazid and ethionamide in *Mycobacterium tuberculosis*. Science 263(5144):227–230

Barber M, Rozwadowska-Dowzenko M (1948) Infection by penicillin-resistant staphylococci. Lancet 2(6530):641–644

Bassetti M, Merelli M, Temperoni C, Astilean A (2013) New antibiotics for bad bugs: where are we? Ann Clin Microbiol Antimicrob 12:22

Bastos ML, Hussain H, Weyer K, Garcia-Garcia L, Leimane V, Leung CC, Narita M, Pena JM, Ponce-de-Leon A, Seung KJ, Shean K, Sifuentes-Osornio J, Van der Walt M, Van der Werf TS, Yew WW, Menzies D (2014), Collaborative Group for Meta-analysis of Individual Patient Data in M-T Treatment outcomes of patients with multidrug-resistant and extensively drug-resistant tuberculosis according to drug susceptibility testing to first- and second-line drugs: an individual patient data meta-analysis. Clin Infect Dis 59(10):1364–1374

Bayer AS, Schneider T, Sahl HG (2013) Mechanisms of daptomycin resistance in *Staphylococcus aureus*: role of the cell membrane and cell wall. Ann N Y Acad Sci 1277:139–158

Beceiro A, Llobet E, Aranda J, Bengoechea JA, Doumith M, Hornsey M, Dhanji H, Chart H, Bou G, Livermore DM, Woodford N (2011) Phosphoethanolamine modification of lipid A in colistin-resistant variants of *Acinetobacter baumannii* mediated by the pmrAB two-component regulatory system. Antimicrob Agents Chemother 55(7):3370–3379

Bergogne-Berezin E, Towner KJ (1996) *Acinetobacter* spp. as nosocomial pathogens: microbiological, clinical, and epidemiological features. Clin Microbiol Rev 9(2):148–165

Bondi A Jr, Dietz CC (1945) Penicillin resistant staphylococci. Proc Soc Exp Biol Med 60:55–58

Bonnet R (2004) Growing group of extended-spectrum beta-lactamases: the CTX-M enzymes. Antimicrob Agents Chemother 48(1):1–14

Bootsma MC, Diekmann O, Bonten MJ (2006) Controlling methicillin-resistant *Staphylococcus aureus*: quantifying the effects of interventions and rapid diagnostic testing. Proc Nati Acad Sci USA 103(14):5620–5625

Bush K, Fisher JF (2011) Epidemiological expansion, structural studies, and clinical challenges of new beta-lactamases from gram-negative bacteria. Ann Rev Microbiol 65:455–478

Bush K, Jacoby GA, Medeiros AA (1995) A functional classification scheme for beta-lactamases and its correlation with molecular structure. Antimicrob Agents Chemother 39(6):1211–1233

Cai Y, Chai D, Wang R, Liang B, Bai N (2012) Colistin resistance of *Acinetobacter baumannii*: clinical reports, mechanisms and antimicrobial strategies. J Antimicrob Chemother 67 (7):1607–1615

Canton R (2009) Antibiotic resistance genes from the environment: a perspective through newly identified antibiotic resistance mechanisms in the clinical setting. Clin Microbiol Infect 15 (Suppl 1):20–25

Canton R, Gonzalez-Alba JM, Galan JC (2012) CTX-M enzymes: origin and diffusion. Front Microbiol 3:110

Carlet J, Collignon P, Goldmann D, Goossens H, Gyssens IC, Harbarth S, Jarlier V, Levy SB, N'Doye B, Pittet D, Richtmann R, Seto WH, van der Meer JW, Voss A (2011) Society's failure to protect a precious resource: antibiotics. Lancet 378(9788):369–371

Carlet J, Jarlier V, Harbarth S, Voss A, Goossens H, Pittet D, Participants of the 3rd World Healthcare-Associated Infections Forum (2012) Ready for a world without antibiotics? The pensieres antibiotic resistance call to action. Antimicrob Resist Infect Control 1(1):11

Chambers HF (1997) Methicillin resistance in staphylococci: molecular and biochemical basis and clinical implications. Clin Microbiol Rev 10(4):781–791

Chambers HF (2001) The changing epidemiology of *Staphylococcus aureus*? Emerg Infect Dis 7 (2):178–182

Clermont O, Lavollay M, Vimont S, Deschamps C, Forestier C, Branger C, Denamur E, Arlet G (2008) The CTX-M-15-producing *Escherichia coli* diffusing clone belongs to a highly virulent B2 phylogenetic subgroup. J Antimicrob Chemother 61(5):1024–1028

Cohn DL, Bustreo F, Raviglione MC (1997) Drug-resistant tuberculosis: review of the worldwide situation and the WHO/IUATLD global surveillance project. International union against tuberculosis and lung disease. Clin Infect Dis 24(1):S121–130

Cormican M, Vellinga A (2012) Existing classes of antibiotics are probably the best we will ever have. BMJ 344:e3369

Cotroneo N, Harris R, Perlmutter N, Beveridge T, Silverman JA (2008) Daptomycin exerts bactericidal activity without lysis of *Staphylococcus aureus*. Antimicrob Agents Chemother 52 (6):2223–2225

Courvalin P (2006) Vancomycin resistance in gram-positive cocci. Clin Infect Dis 42(Suppl 1): S25–34

Crofton J (1959) Chemotherapy of pulmonary tuberculosis. BMJ 1(5138):1610–1614

Crofton J, Mitchison DA (1948) Streptomycin resistance in pulmonary tuberculosis. BMJ 2 (4588):1009–1015

Daniel TM (2006) The history of tuberculosis. Respir Med 100(11):1862–1870

David MZ, Daum RS (2010) Community-associated methicillin-resistant *Staphylococcus aureus*: epidemiology and clinical consequences of an emerging epidemic. Clin Microbiol Rev 23 (3):616–687

Davis SD, Iannetta A, Wedgwood RJ (1971) Activity of colistin against *Pseudomonas aeruginosa*: inhibition by calcium. J Infect Dis 124(6):610–612

de Kraker ME, Davey PG, Grundmann H, BURDEN study group (2011) Mortality and hospital stay associated with resistant *Staphylococcus aureus* and *Escherichia coli* bacteremia: estimating the burden of antibiotic resistance in Europe. PLoS Med 8(10):e1001104

DeLeo FR, Chambers HF (2009) Reemergence of antibiotic-resistant *Staphylococcus aureus* in the genomics era. J Clin Invest 119(9):2464–2474

DeLeo FR, Otto M, Kreiswirth BN, Chambers HF (2010) Community-associated meticillin-resistant *Staphylococcus aureus*. Lancet 375(9725):1557–1568

Doi Y, Paterson DL (2015) Carbapenemase-producing Enterobacteriaceae. Semin Respir Crit Care Med 36(1):74–84

Dolejska M, Villa L, Poirel L, Nordmann P, Carattoli A (2013) Complete sequencing of an IncHI1 plasmid encoding the carbapenemase NDM-1, the ArmA 16S RNA methylase and a resistance-nodulation-cell division/multidrug efflux pump. J Antimicrob Chemother 68(1):34–39

Dortet L, Nordmann P, Poirel L (2012) Association of the emerging carbapenemase NDM-1 with a bleomycin resistance protein in Enterobacteriaceae and *Acinetobacter baumannii*. Antimicrob Agents Chemother 56(4):1693–1697

Dortet L, Poirel L, Nordmann P (2014) Worldwide dissemination of the NDM-type carbapenemases in gram-negative bacteria. Biomed Res Int 2014:249856

El-Sayed-Ahmed MA, Amin MA, Tawakol WM, Loucif L, Bakour S, Rolain JM (2015) High prevalence of bla(NDM-1) carbapenemase-encoding gene and 16S rRNA armA methyltransferase gene among *Acinetobacter baumannii* clinical Isolates in Egypt. Antimicrob Agents Chemother 59(6):3602–3605

Enright MC, Robinson DA, Randle G, Feil EJ, Grundmann H, Spratt BG (2002) The evolutionary history of methicillin-resistant *Staphylococcus aureus* (MRSA). Proc Natl Acad Sci USA 99 (11):7687–7692

European Centre for Disease Prevention and Control (2013) EARSS Annual Reports. In http://ecdc.europa.eu/en/publications/_layouts/forms/Publication_DispForm.aspx?List=4f55ad51-4aed-4d32-b960-af70113dbb90&ID=1205. Accessed 22 Oct 2015

Evans BA, Hamouda A, Amyes SG (2013) The rise of carbapenem-resistant *Acinetobacter baumannii*. Curr Pharm Des 19(2):223–238

Falagas ME, Bliziotis IA (2007) Pandrug-resistant gram-negative bacteria: the dawn of the post-antibiotic era? Int J Antimicrob Agents 29(6):630–636

Falagas ME, Kasiakou SK (2005) Colistin: the revival of polymyxins for the management of multidrug-resistant gram-negative bacterial infections. Clin Infect Dis 40(9):1333–1341

Falagas ME, Bliziotis IA, Kasiakou SK, Samonis G, Athanassopoulou P, Michalopoulos A (2005) Outcome of infections due to pandrug-resistant (PDR) gram-negative bacteria. BMC Infect Dis 5:24

Falagas ME, Karveli EA, Siempos II, Vardakas KZ (2008) Acinetobacter infections: a growing threat for critically ill patients. Epidemiol Infect 136(8):1009–1019

Fernandez-Cuenca F, Martinez-Martinez L, Conejo MC, Ayala JA, Perea EJ, Pascual A (2003) Relationship between beta-lactamase production, outer membrane protein and penicillin-binding protein profiles on the activity of carbapenems against clinical isolates of *Acinetobacter baumannii*. J Antimicrob Chemother 51(3):565–574

Finland M (1955) Emergence of antibiotic-resistant bacteria. N Engl J Med 253(21):909–922

Fournier PE, Richet H (2006) The epidemiology and control of *Acinetobacter baumannii* in health care facilities. Clin Infect Dis 42(5):692–699

Fowler VG Jr, Boucher HW, Corey GR, Abrutyn E, Karchmer AW, Rupp ME, Levine DP, Chambers HF, Tally FP, Vigliani GA, Cabell CH, Link AS, DeMeyer I, Filler SG, Zervos M, Cook P, Parsonnet J, Bernstein JM, Price CS, Forrest GN, Fatkenheuer G, Gareca M, Rehm SJ, Brodt HR, Tice A, Cosgrove SE, *S. aureus* Endocarditis and Bacteremia Study Group (2006) Daptomycin versus standard therapy for bacteremia and endocarditis caused by *Staphylococcus aureus*. N Engl J Med 355(7):653–665

Fox W, Ellard GA, Mitchison DA (1999) Studies on the treatment of tuberculosis undertaken by the British Medical Research Council tuberculosis units, 1946–1986, with relevant subsequent publications. Int J Tuberc Lung Dis 3(10 Suppl 2):S231–279

Gales AC, Jones RN, Sader HS (2006) Global assessment of the antimicrobial activity of polymyxin B against 54 731 clinical isolates of gram-negative bacilli: report from the SENTRY antimicrobial surveillance programme (2001-2004). Clin Microbiol Infect 12 (4):315–321

Garcia de Viedma D, del Sol Diaz Infantes M, Lasala F, Chaves F, Alcala L, Bouza E (2002) New real-time PCR able to detect in a single tube multiple rifampin resistance mutations and high-level isoniazid resistance mutations in *Mycobacterium tuberculosis*. J Clin Microbiol 40 (3):988–995

Gardete S, Tomasz A (2014) Mechanisms of vancomycin resistance in *Staphylococcus aureus*. J Clin Invest 124(7):2836–2840

Glaziou P, Floyd K, Korenromp EL, Sismanidis C, Bierrenbach AL, Williams BG, Atun R, Raviglione M (2011) Lives saved by tuberculosis control and prospects for achieving the 2015 global target for reducing tuberculosis mortality. Bull World Health Organ 89(8):573–582

Go ES, Urban C, Burns J, Kreiswirth B, Eisner W, Mariano N, Mosinka-Snipas K, Rahal JJ (1994) Clinical and molecular epidemiology of acinetobacter infections sensitive only to polymyxin B and sulbactam. Lancet 344(8933):1329–1332

Goossens H, Grabein B (2005) Prevalence and antimicrobial susceptibility data for extended-spectrum beta-lactamase-and AmpC-producing Enterobacteriaceae from the MYSTIC Program in Europe and the United States (1997–2004). Diagn Microbiol Infect Dis 53(4):257–264

Gordon NC, Wareham DW (2010) Multidrug-resistant *Acinetobacter baumannii*: mechanisms of virulence and resistance. J Antimicrob Chemother 35(3):219–226

Gould IM, David MZ, Esposito S, Garau J, Lina G, Mazzei T, Peters G (2012) New insights into meticillin-resistant *Staphylococcus aureus* (MRSA) pathogenesis, treatment and resistance. Int J Antimicrol Agents 39(2):96–104

Gullberg E, Cao S, Berg OG, Ilback C, Sandegren L, Hughes D, Andersson DI (2011) Selection of resistant bacteria at very low antibiotic concentrations. PLoS Pathog 7(7):e1002158

Harris PN (2015) Clinical management of infections caused by Enterobacteriaceae that express extended-spectrum beta-lactamase and AmpC enzymes. Semin Respir Crit Care Med 36 (1):56–73

Harris AD, Perencevich EN, Johnson JK, Paterson DL, Morris JG, Strauss SM, Johnson JA (2007) Patient-to-patient transmission is important in extended-spectrum beta-lactamase-producing *Klebsiella pneumoniae* acquisition. Clin Infect Dis 45(10):1347–1350

Hartman B, Tomasz A (1981) Altered penicillin-binding proteins in methicillin-resistant strains of *Staphylococcus aureus*. Antimicrob Agents Chemother 19(5):726–735

Hawkey PM, Jones AM (2009) The changing epidemiology of resistance. J Antimicrob Chemother 64(Suppl 1):i3–10

Hayden MK, Rezai K, Hayes RA, Lolans K, Quinn JP, Weinstein RA (2005) Development of Daptomycin resistance in vivo in methicillin-resistant *Staphylococcus aureus*. J Clin Microbiol 43(10):5285–5287

Hazbon MH, Brimacombe M, Bobadilla del Valle M, Cavatore M, Guerrero MI, Varma-Basil M, Billman-Jacobe H, Lavender C, Fyfe J, Garcia-Garcia L, Leon CI, Bose M, Chaves F, Murray M, Eisenach KD, Sifuentes-Osornio J, Cave MD, Ponce de Leon A, Alland D (2006) Population genetics study of isoniazid resistance mutations and evolution of multidrug-resistant *Mycobacterium tuberculosis*. Antimicrob Agents Chemother 50(8):2640–2649

Hejnar P, Kolar M, Hajek V (1999) Characteristics of *Acinetobacter* strains (phenotype classification, antibiotic susceptibility and production of beta-lactamases) isolated from haemocultures from patients at the teaching Hospital in Olomouc. Acta Univ Palackianae Olomucensis Facultatis Med 142:73–77

Henry R, Vithanage N, Harrison P, Seemann T, Coutts S, Moffatt JH, Nation RL, Li J, Harper M, Adler B, Boyce JD (2012) Colistin-resistant, lipopolysaccharide-deficient *Acinetobacter baumannii* responds to lipopolysaccharide loss through increased expression of genes involved in the synthesis and transport of lipoproteins, phospholipids, and poly-beta-1,6-N-acetylglucosamine. Antimicrob Agents Chemother 56(1):59–69

Heritier C, Poirel L, Fournier PE, Claverie JM, Raoult D, Nordmann P (2005) Characterization of the naturally occurring oxacillinase of *Acinetobacter baumannii*. Antimicrob Agents Chemother 49(10):4174–4179

Herold BC, Immergluck LC, Maranan MC, Lauderdale DS, Gaskin RE, Boyle-Vavra S, Leitch CD, Daum RS (1998) Community-acquired methicillin-resistant *Staphylococcus aureus* in children with no identified predisposing risk. JAMA 279(8):593–598

Hesterkamp T (2015) Antibiotics clinical development and pipeline. Curr Top Microbiol Immunol. doi:10.1007/82_2015_451

Hinshaw HC, Feldman WH, Pfuetze KH (1946) Treatment of tuberculosis with streptomycin; a summary of observations on one hundred cases. JAMA 132(13):778–782

Hiramatsu K (2001) Vancomycin-resistant *Staphylococcus aureus*: a new model of antibiotic resistance. Lancet Infect Dis 1(3):147–155

Hiramatsu K, Aritaka N, Hanaki H, Kawasaki S, Hosoda Y, Hori S, Fukuchi Y, Kobayashi I (1997a) Dissemination in Japanese hospitals of strains of *Staphylococcus aureus* heterogeneously resistant to vancomycin. Lancet 350(9092):1670–1673

Hiramatsu K, Hanaki H, Ino T, Yabuta K, Oguri T, Tenover FC (1997b) Methicillin-resistant *Staphylococcus aureus* clinical strain with reduced vancomycin susceptibility. J Antimicrob Chemother 40(1):135–136

Holden MT, Hsu LY, Kurt K, Weinert LA, Mather AE, Harris SR, Strommenger B, Layer F, Witte W, de Lencastre H, Skov R, Westh H, Zemlickova H, Coombs G, Kearns AM, Hill RL, Edgeworth J, Gould I, Gant V, Cooke J, Edwards GF, McAdam PR, Templeton KE, McCann A, Zhou Z, Castillo-Ramirez S, Feil EJ, Hudson LO, Enright MC, Balloux F, Aanensen DM, Spratt BG, Fitzgerald JR, Parkhill J, Achtman M, Bentley SD, Nubel U (2013) A genomic portrait of the emergence, evolution, and global spread of a methicillin-resistant *Staphylococcus aureus* pandemic. Genome Res 23(4):653–664

Howden BP, Davies JK, Johnson PD, Stinear TP, Grayson ML (2010) Reduced vancomycin susceptibility in *Staphylococcus aureus*, including vancomycin-intermediate and heterogeneous vancomycin-intermediate strains: resistance mechanisms, laboratory detection, and clinical implications. Clin Microbiol Rev 23(1):99–139

Infectious Diseases Society of America I (2010) The 10 '20 Initiative: pursuing a global commitment to develop 10 new antibacterial drugs by 2020. Clin Infect Dis 50(8):1081

Iseman MD, Madsen LA (1989) Drug-resistant tuberculosis. Clin Chest Med 10(3):341–353

Jacoby GA (2009) AmpC beta-lactamases. Clin Microbiol Rev 22(1):161–182

Jaurin B, Grundstrom T (1981) *ampC* cephalosporinase of *Escherichia coli* K-12 has a different evolutionary origin from that of beta-lactamases of the penicillinase type. Proc Natl Acad Sci USA 78(8):4897–4901

Jevons MP (1961) "Celbenin"-resistant staphylococci [letter]. Br Med J 1:124–125

Johnson AP, Aucken HM, Cavendish S, Ganner M, Wale MC, Warner M, Livermore DM, Cookson BD, participants UE (2001) Dominance of EMRSA-15 and -16 among MRSA causing nosocomial bacteraemia in the UK: analysis of isolates from the European Antimicrobial Resistance Surveillance System (EARSS). J Antimicrob Chemother 48 (1):143–144

Jung D, Rozek A, Okon M, Hancock RE (2004) Structural transitions as determinants of the action of the calcium-dependent antibiotic daptomycin. Chem Biol 11(7):949–957

Kallen AJ, Mu Y, Bulens S, Reingold A, Petit S, Gershman K, Ray SM, Harrison LH, Lynfield R, Dumyati G, Townes JM, Schaffner W, Patel PR, Fridkin SK, Active Bacterial Core surveillance MIotEIP (2010) Health care-associated invasive MRSA infections, 2005–2008. JAMA 304(6):641–648

Kaspar U, Kriegeskorte A, Schubert T, Peters G, Rudack C, Pieper DH, Wos-Oxley M, Becker K (2015) The culturome of the human nose habitats reveals individual bacterial fingerprint patterns. Environ Microbiol. doi:10.1111/1462-2920.12891

Katayama Y, Ito T, Hiramatsu K (2000) A new class of genetic element, staphylococcus cassette chromosome mec, encodes methicillin resistance in *Staphylococcus aureus*. Antimicrob Agents Chemother 44(6):1549–1555

Keene A, Vavagiakis P, Lee MH, Finnerty K, Nicolls D, Cespedes C, Quagliarello B, Chiasson MA, Chong D, Lowy FD (2005) *Staphylococcus aureus* colonization and the risk of infection in critically ill patients. Infect Control Hosp Epidemiol 26(7):622–628

Kernodle DS (2000) Gram-positive pathogens, vol mechanisms of resistance to β-lactam antibiotics. American Society for Microbiology, Washington, DC, USA, pp 609–620

Keshavjee S, Farmer PE (2012) Tuberculosis, drug resistance, and the history of modern medicine. N Engl J Med 367(10):931–936

Khatib R, Sharma M, Iyer S, Fakih MG, Obeid KM, Venugopal A, Fishbain J, Johnson LB, Segireddy M, Jose J, Riederer K (2013) Decreasing incidence of *Staphylococcus aureus* bacteremia over 9 years: greatest decline in community-associated methicillin-susceptible and hospital-acquired methicillin-resistant isolates. Am J Infect Control 41(3):210–213

Kliebe C, Nies BA, Meyer JF, Tolxdorff-Neutzling RM, Wiedemann B (1985) Evolution of plasmid-coded resistance to broad-spectrum cephalosporins. Antimicrob Agents Chemother 28 (2):302–307

Kluytmans JA, Wertheim HF (2005) Nasal carriage of *Staphylococcus aureus* and prevention of nosocomial infections. Infection 33(1):3–8

Kochi A (1997) Tuberculosis control–is DOTS the health breakthrough of the 1990s? World Health Forum 18(3–4):225–232 (discussion 233–247)

Kos VN, Desjardins CA, Griggs A, Cerqueira G, Van Tonder A, Holden MT, Godfrey P, Palmer KL, Bodi K, Mongodin EF, Wortman J, Feldgarden M, Lawley T, Gill SR, Haas BJ, Birren B, Gilmore MS (2012) Comparative genomics of vancomycin-resistant *Staphylococcus aureus* strains and their positions within the clade most commonly associated with Methicillin-resistant *S. aureus* hospital-acquired infection in the United States. mBio 3(pii): e00112-12

Kulah C, Celebi G, Aktas E, Mengeloglu Z, Comert F, Ankarali H (2009) Unexpected tigecycline resistance among *Acinetobacter baumannii* Isolates: high minor error rate by Etest. J Chemother 21(4):390–395

Lascols C, Hackel M, Marshall SH, Hujer AM, Bouchillon S, Badal R, Hoban D, Bonomo RA (2011) Increasing prevalence and dissemination of NDM-1 metallo-beta-lactamase in India: data from the SMART study (2009). J Antimicrob Chemother 66(9):1992–1997

Lauretti L, Riccio ML, Mazzariol A, Cornaglia G, Amicosante G, Fontana R, Rossolini GM (1999) Cloning and characterization of blaVIM, a new integron-borne metallo-beta-lactamase gene from a *Pseudomonas aeruginosa* clinical isolate. Antimicrob Agents Chemother 43 (7):1584–1590

Levine DP (2006) Vancomycin: a history. Clin Infect Dis 42(Suppl 1):S5–12

Lienhardt C, Glaziou P, Uplekar M, Lonnroth K, Getahun H, Raviglione M (2012) Global tuberculosis control: lessons learnt and future prospects. Nat Rev Microbiol 10(6):407–416

Lim LM, Ly N, Anderson D, Yang JC, Macander L, Jarkowski A 3rd, Forrest A, Bulitta JB, Tsuji BT (2010) Resurgence of colistin: a review of resistance, toxicity, pharmacodynamics, and dosing. Pharmacotherapy 30(12):1279–1291

Limbago BM, Kallen AJ, Zhu W, Eggers P, McDougal LK, Albrecht VS (2014) Report of the 13th vancomycin-resistant *Staphylococcus aureus* isolate from the United States. J Clin Microbiol 52(3):998–1002

Lister PD, Wolter DJ, Hanson ND (2009) Antibacterial-resistant *Pseudomonas aeruginosa*: clinical impact and complex regulation of chromosomally encoded resistance mechanisms. Clin Microbiol Rev 22(4):582–610

Livermore DM (1995) Beta-Lactamases in laboratory and clinical resistance. Clin Microbiol Rev 8 (4):557–584

Livermore DM (2012) Current epidemiology and growing resistance of gram-negative pathogens. Korean J Intern Med 27(2):128–142

Livermore DM, Hawkey PM (2005) CTX-M: changing the face of ESBLs in the UK. J Antimicrob Chemother 56(3):451–454

Livermore DM, Woodford N (2006) The beta-lactamase threat in *Enterobacteriaceae*, *Pseudomonas* and *Acinetobacter*. Trends Microbiol 14(9):413–420

Lodge JM, Minchin SD, Piddock LJ, Busby SJ (1990) Cloning, sequencing and analysis of the structural gene and regulatory region of the *Pseudomonas aeruginosa* chromosomal ampC beta-lactamase. Biochem J 272(3):627–631

Ma XX, Ito T, Tiensasitorn C, Jamklang M, Chongtrakool P, Boyle-Vavra S, Daum RS, Hiramatsu K (2002) Novel type of staphylococcal cassette chromosome mec identified in community-acquired methicillin-resistant *Staphylococcus aureus* strains. Antimicrob Agents Chemother 46(4):1147–1152

Magiorakos AP, Srinivasan A, Carey RB, Carmeli Y, Falagas ME, Giske CG, Harbarth S, Hindler JF, Kahlmeter G, Olsson-Liljequist B, Paterson DL, Rice LB, Stelling J, Struelens MJ, Vatopoulos A, Weber JT, Monnet DL (2012) Multidrug-resistant, extensively drug-resistant and pandrug-resistant bacteria: an international expert proposal for interim standard definitions for acquired resistance. Clin Microbiol Infect 18(3):268–281

Manten A, Van Wijngaarden LJ (1969) Development of drug resistance to rifampicin. Chemotherapy 14(2):93–100

Maple PA, Hamilton-Miller JM, Brumfitt W (1989) World-wide antibiotic resistance in methicillin-resistant *Staphylococcus aureus*. Lancet 1(8637):537–540

Matsuo M, Hishinuma T, Katayama Y, Hiramatsu K (2015) A mutation of RNA polymerase beta' subunit (RpoC) converts heterogeneously vancomycin-intermediate *Staphylococcus aureus* (hVISA) into "slow VISA". Antimicrob Agents Chemother 59(7):4215–4225

Meehl M, Herbert S, Gotz F, Cheung A (2007) Interaction of the GraRS two-component system with the VraFG ABC transporter to support vancomycin-intermediate resistance in *Staphylococcus aureus*. Antimicrob Agents Chemother 51(8):2679–2689

Melo-Cristino J, Resina C, Manuel V, Lito L, Ramirez M (2013) First case of infection with vancomycin-resistant *Staphylococcus aureus* in Europe. Lancet 382(9888):205

Metcalfe C, Macdonald IK, Murphy EJ, Brown KA, Raven EL, Moody PC (2008) The tuberculosis prodrug isoniazid bound to activating peroxidases. J Biol Chem 283(10):6193–6200

Meyer E, Schroder C, Gastmeier P, Geffers C (2014) The reduction of nosocomial MRSA infection in Germany: an analysis of data from the hospital infection surveillance system (KISS) between 2007 and 2012. Deutsches Ärzteblatt Int 111(19):331–336

Migliori GB, De Iaco G, Besozzi G, Centis R, Cirillo DM (2007) First tuberculosis cases in Italy resistant to all tested drugs. Euro Surveill 12(5):E070517 070511

Mira PM, Meza JC, Nandipati A, Barlow M (2015) Adaptive landscapes of resistance genes change as antibiotic concentrations change. Mol Biol Evol 32(10):2707–2715

Mitchison D, Davies G (2012) The chemotherapy of tuberculosis: past, present and future. Int J Tuberc Lung Dis 16(6):724–732

Moellering RC Jr (2008) Current treatment options for community-acquired methicillin-resistant *Staphylococcus aureus* infection. Clin Infect Dis 46(7):1032–1037

Moellering RC Jr (2012) MRSA: the first half century. J Antimicrob Chemother 67(1):4–11

Moffatt JH, Harper M, Harrison P, Hale JD, Vinogradov E, Seemann T, Henry R, Crane B, St Michael F, Cox AD, Adler B, Nation RL, Li J, Boyce JD (2010) Colistin resistance in *Acinetobacter baumannii* is mediated by complete loss of lipopolysaccharide production. Antimicrob Agents Chemother 54(12):4971–4977

Munier GK, Johnson CL, Snyder JW, Moland ES, Hanson ND, Thomson KS (2010) Positive extended-spectrum-beta-lactamase (ESBL) screening results may be due to AmpC beta-lactamases more often than to ESBLs. J Clin Microbiol 48(2):673–674

Murray BE, Moellering RC Jr (1978) Patterns and mechanisms of antibiotic resistance. Med Clin North Am 62(5):899–923

Murray CJ, Ortblad KF, Guinovart C, Lim SS, Wolock TM et al (2014) Global, regional, and national incidence and mortality for HIV, tuberculosis, and malaria during 1990–2013: a systematic analysis for the global burden of disease study 2013. Lancet 384(9947):1005–1070

Navon-Venezia S, Leavitt A, Carmeli Y (2007) High tigecycline resistance in multidrug-resistant *Acinetobacter baumannii*. J Antimicrob Chemother 59(4):772–774

Nicolas-Chanoine MH, Blanco J, Leflon-Guibout V, Demarty R, Alonso MP, Canica MM, Park YJ, Lavigne JP, Pitout J, Johnson JR (2008) Intercontinental emergence of *Escherichia coli* clone O25: H4-ST131 producing CTX-M-15. J Antimicrob Chemother 61(2):273–281

Nimmo GR (2012) USA300 abroad: global spread of a virulent strain of community-associated methicillin-resistant *Staphylococcus aureus*. Clin Microbiol Infect 18(8):725–734

Nordmann P, Naas T, Fortineau N, Poirel L (2007) Superbugs in the coming new decade; multidrug resistance and prospects for treatment of *Staphylococcus aureus*, Enterococcus spp. and *Pseudomonas aeruginosa* in 2010. Curr Opin Microbiol 10(5):436–440

Otto M (2012) MRSA virulence and spread. Cell Microbiol 14(10):1513–1521

Panlilio AL, Culver DH, Gaynes RP, Banerjee S, Henderson TS, Tolson JS, Martone WJ (1992) Methicillin-resistant *Staphylococcus aureus* in U.S. hospitals, 1975–1991. Infect Control Hosp Epidemiol 13(10):582–586

Parida SK, Axelsson-Robertson R, Rao MV, Singh N, Master I, Lutckii A, Keshavjee S, Andersson J, Zumla A, Maeurer M (2015) Totally drug-resistant tuberculosis and adjunct therapies. J Intern Med 277(4):388–405

Paterson DL, Bonomo RA (2005) Extended-spectrum beta-lactamases: a clinical update. Clin Microbiol Rev 18(4):657–686

Peacock SJ, Paterson GK (2015) Mechanisms of methicillin resistance in *Staphylococcus aureus*. Ann Rev Biochem 84:577–601

Peleg AY, Hooper DC (2010) Hospital-acquired infections due to gram-negative bacteria. N Engl J Med 362(19):1804–1813

Peleg AY, Seifert H, Paterson DL (2008) *Acinetobacter baumannii*: emergence of a successful pathogen. Clin Microbiol Rev 21(3):538–582

Pendleton JN, Gorman SP, Gilmore BF (2013) Clinical relevance of the ESKAPE pathogens. Expert Rev Anti Infect Ther 11(3):297–308

Perry JD, Naqvi SH, Mirza IA, Alizai SA, Hussain A, Ghirardi S, Orenga S, Wilkinson K, Woodford N, Zhang J, Livermore DM, Abbasi SA, Raza MW (2011) Prevalence of faecal carriage of Enterobacteriaceae with NDM-1 carbapenemase at military hospitals in Pakistan, and evaluation of two chromogenic media. J Antimicrob Chemother 66(10):2288–2294

Pfeifer Y, Witte W, Holfelder M, Busch J, Nordmann P, Poirel L (2011) NDM-1-producing *Escherichia coli* in Germany. Antimicrob Agents Chemother 55(3):1318–1319

Philippon A, Arlet G, Jacoby GA (2002) Plasmid-determined AmpC-type beta-lactamases. Antimicrob Agents Chemother 46(1):1–11

Poirel L, Nordmann P (2006) Carbapenem resistance in *Acinetobacter baumannii*: mechanisms and epidemiology. Clin Microbiol Infect 12(9):826–836

Poirel L, Kampfer P, Nordmann P (2002) Chromosome-encoded Ambler class A beta-lactamase of *Kluyvera georgiana*, a probable progenitor of a subgroup of CTX-M extended-spectrum beta-lactamases. Antimicrob Agents Chemother 46(12):4038–4040

Poirel L, Heritier C, Tolun V, Nordmann P (2004) Emergence of oxacillinase-mediated resistance to imipenem in *Klebsiella pneumoniae*. Antimicrob Agents Chemother 48(1):15–22

Poirel L, Potron A, Nordmann P (2012) OXA-48-like carbapenemases: the phantom menace. J Antimicrob Chemother 67(7):1597–1606

Poulikakos P, Tansarli GS, Falagas ME (2014) Combination antibiotic treatment versus monotherapy for multidrug-resistant, extensively drug-resistant, and pandrug-resistant *Acinetobacter* infections: a systematic review. Eur J Clin Microbiol Infect Dis 33(10):1675–1685

Pugh RJ, Cooke RP, Dempsey G (2010) Short course antibiotic therapy for Gram-negative hospital-acquired pneumonia in the critically ill. J Hosp Infect 74(4):337–343

Pugh R, Grant C, Cooke RP, Dempsey G (2015) Short-course versus prolonged course antibiotic therapy for hospital-acquired pneumonia in critically ill adults. Cochrane Database Syst Rev 8: CD007577

Queenan AM, Bush K (2007) Carbapenemases: the versatile beta-lactamases. Clin Microbiol Rev 20(3):440–458

Ramaswamy SV, Reich R, Dou SJ, Jasperse L, Pan X, Wanger A, Quitugua T, Graviss EA (2003) Single nucleotide polymorphisms in genes associated with isoniazid resistance in *Mycobacterium tuberculosis*. Antimicrob Agents Chemother 47(4):1241–1250

Rammelkamp CH, Maxon T (1942) Resistance of *Staphylococcus aureus* to the action of penicillin. Proc Royal Soc Exp Biol Med 51:386–389

Richardson JF, Reith S (1993) Characterization of a strain of methicillin-resistant *Staphylococcus aureus* (EMRSA-15) by conventional and molecular methods. J Hosp Infect 25(1):45–52

Rodriguez-Bano J, Navarro MD, Romero L, Muniain MA, de Cueto M, Rios MJ, Hernandez JR, Pascual A (2006) Bacteremia due to extended-spectrum beta -lactamase-producing *Escherichia coli* in the CTX-M era: a new clinical challenge. Clin Infect Dis 43(11):1407–1414

Rodriguez-Bano J, Miro E, Villar M, Coelho A, Gozalo M, Borrell N, Bou G, Conejo MC, Pomar V, Aracil B, Larrosa N, Aguero J, Oliver A, Fernandez A, Oteo J, Pascual A, Navarro F (2012) Colonisation and infection due to Enterobacteriaceae producing plasmid-mediated AmpC beta-lactamases. J Infect 64(2):176–183

Rozwarski DA, Grant GA, Barton DH, Jacobs WR Jr, Sacchettini JC (1998) Modification of the NADH of the isoniazid target (InhA) from *Mycobacterium tuberculosis*. Science 279 (5347):98–102

Rozwarski DA, Vilcheze C, Sugantino M, Bittman R, Sacchettini JC (1999) Crystal structure of the *Mycobacterium tuberculosis* enoyl-ACP reductase, InhA, in complex with NAD+ and a C16 fatty acyl substrate. J Biol Chem 274(22):15582–15589

Saha B, Singh AK, Ghosh A, Bal M (2008) Identification and characterization of a vancomycin-resistant *Staphylococcus aureus* isolated from Kolkata (South Asia). J Med Microbiol 57(Pt 1):72–79

Salverda ML, De Visser JA, Barlow M (2010) Natural evolution of TEM-1 beta-lactamase: experimental reconstruction and clinical relevance. FEMS Microbiol Rev 34(6):1015–1036

Schrag SJ, Pena C, Fernandez J, Sanchez J, Gomez V, Perez E, Feris JM, Besser RE (2001) Effect of short-course, high-dose amoxicillin therapy on resistant pneumococcal carriage: a randomized trial. JAMA 286(1):49–56

Selikoff IJ, Robitzek EH (1952) Tuberculosis chemotherapy with hydrazine derivatives of isonicotinic acid. Dis Chest 21(4):385–438

Sieradzki K, Tomasz A (2003) Alterations of cell wall structure and metabolism accompany reduced susceptibility to vancomycin in an isogenic series of clinical isolates of *Staphylococcus aureus*. J Bacteriol 185(24):7103–7110

Sievert DM, Rudrik JT, Patel JB, McDonald LC, Wilkins MJ, Hageman JC (2008) Vancomycin-resistant *Staphylococcus aureus* in the United States, 2002–2006. Clin Infect Dis 46(5):668–674

Sievert DM, Ricks P, Edwards JR, Schneider A, Patel J, Srinivasan A, Kallen A, Limbago B, Fridkin S, National Healthcare Safety Network (NHSN) Team and Participating NHSN Facilities (2013) Antimicrobial-resistant pathogens associated with healthcare-associated infections: summary of data reported to the National Healthcare Safety Network at the Centers for Disease Control and Prevention, 2009–2010. Infect Control Hosp Epidemiol 34 (1):1–14

Silverman JA, Perlmutter NG, Shapiro HM (2003) Correlation of daptomycin bactericidal activity and membrane depolarization in *Staphylococcus aureus*. Antimicrob Agents Chemother 47 (8):2538–2544

Sirot D, Sirot J, Labia R, Morand A, Courvalin P, Darfeuille-Michaud A, Perroux R, Cluzel R (1987) Transferable resistance to third-generation cephalosporins in clinical isolates of *Klebsiella pneumoniae*: identification of CTX-1, a novel beta-lactamase. J Antimicrob Chemother 20(3):323–334

Skinner D, Keefer CS (1941) Significance of bacteremia caused by *Staphylococcus aureus*. Arch Intern Med 68:851–875

Sougakoff W, Goussard S, Gerbaud G, Courvalin P (1988) Plasmid-mediated resistance to third-generation cephalosporins caused by point mutations in TEM-type penicillinase genes. Rev Infect Dis 10(4):879–884

Spellberg B, Guidos R, Gilbert D, Bradley J, Boucher HW, Scheld WM, Bartlett JG, Edwards J Jr, Infectious Diseases Society of A (2008) The epidemic of antibiotic-resistant infections: a call to action for the medical community from the Infectious Diseases Society of America. Clin Infect Dis 46(2):155–164

Spellberg B, Bartlett JG, Gilbert DN (2013) The future of antibiotics and resistance. N Engl J Med 368(4):299–302

Steenbergen JN, Alder J, Thorne GM, Tally FP (2005) Daptomycin: a lipopeptide antibiotic for the treatment of serious gram-positive infections. J Antimicrob Chemother 55(3):283–288. doi:10.1093/jac/dkh546

Stegger M, Wirth T, Andersen PS, Skov RL, De Grassi A, Simoes PM, Tristan A, Petersen A, Aziz M, Kiil K, Cirkovic I, Udo EE, del Campo R, Vuopio-Varkila J, Ahmad N, Tokajian S, Peters G, Schaumburg F, Olsson-Liljequist B, Givskov M, Driebe EE, Vigh HE, Shittu A, Ramdani-Bougessa N, Rasigade JP, Price LB, Vandenesch F, Larsen AR, Laurent F (2014) Origin and evolution of European community-acquired methicillin-resistant *Staphylococcus aureus*. mBio 5(5):e01044–01014

Stryjewski ME, Corey GR (2014) Methicillin-resistant *Staphylococcus aureus*: an evolving pathogen. Clin Infect Dis 58(Suppl 1):S10–19

Sulis G, Roggi A, Matteelli A, Raviglione MC (2014) Tuberculosis: epidemiology and control. Mediterr J Hematol Infect Dis 6(1):e2014070

Talbot GH, Bradley J, Edwards JE, Jr., Gilbert D, Scheld M, Bartlett JG, Antimicrobial Availability Task Force of the Infectious Diseases Society of America (2006) Bad bugs need drugs: an update on the development pipeline from the antimicrobial availability task force of the infectious diseases society of America. Clin Infect Dis 42 (5):657–668

Tedesco KL, Rybak MJ (2004) Daptomycin. Pharmacotherapy 24(1):41–57

Telenti A, Imboden P, Marchesi F, Lowrie D, Cole S, Colston MJ, Matter L, Schopfer K, Bodmer T (1993a) Detection of rifampicin-resistance mutations in *Mycobacterium tuberculosis*. Lancet 341(8846):647–650

Telenti A, Imboden P, Marchesi F, Schmidheini T, Bodmer T (1993b) Direct, automated detection of rifampin-resistant *Mycobacterium tuberculosis* by polymerase chain reaction and single-strand conformation polymorphism analysis. Antimicrob Agents Chemother 37 (10):2054–2058

The Pew Charitable Trusts P (2015) Tracking the pipeline of antibiotics in development. http://www.pewtrusts.org/en/research-and-analysis/issue-briefs/2014/03/12/tracking-the-pipeline-of-antibiotics in development

Turner PJ (2005) Extended-spectrum beta-lactamases. Clin Infect Dis 41(Suppl 4):S273–275

Turton JF, Woodford N, Glover J, Yarde S, Kaufmann ME, Pitt TL (2006) Identification of *Acinetobacter baumannii* by detection of the blaOXA-51-like carbapenemase gene intrinsic to this species. J Clin Microbiol 44(8):2974–2976

Udwadia ZF (2012) Totally drug-resistant tuberculosis in India: who let the djinn out? Respirology 17(5):741–742

Uhlemann AC, Otto M, Lowy FD, DeLeo FR (2014) Evolution of community- and healthcare-associated methicillin-resistant *Staphylococcus aureus*. Infect Genet Evol 21:563–574

Van Boeckel TP, Brower C, Gilbert M, Grenfell BT, Levin SA, Robinson TP, Teillant A, Laxminarayan R (2015) Global trends in antimicrobial use in food animals. Proc Natl Acad Sci USA 112(18):5649–5654

Vandenesch F, Naimi T, Enright MC, Lina G, Nimmo GR, Heffernan H, Liassine N, Bes M, Greenland T, Reverdy ME, Etienne J (2003) Community-acquired methicillin-resistant *Staphylococcus aureus* carrying panton-valentine leukocidin genes: worldwide emergence. Emerg Infect Dis 9(8):978–984

Velayati AA, Masjedi MR, Farnia P, Tabarsi P, Ghanavi J, Ziazarifi AH, Hoffner SE (2009) Emergence of new forms of totally drug-resistant tuberculosis bacilli: super extensively drug-resistant tuberculosis or totally drug-resistant strains in Iran. Chest 136(2):420–425

Vikram HR, Havill NL, Koeth LM, Boyce JM (2005) Clinical progression of methicillin-resistant *Staphylococcus aureus* vertebral osteomyelitis associated with reduced susceptibility to daptomycin. J Clin Microbiol 43(10):5384–5387

Visca P, Seifert H, Towner KJ (2011) Acinetobacter infection—an emerging threat to human health. IUBMB Life 63(12):1048–1054

von Eiff C, Becker K, Machka K, Stammer H, Peters G (2001) Nasal carriage as a source of *Staphylococcus aureus* bacteremia. Study group. N Engl J Med 344(1):11–16

Voss A, Milatovic D, Wallrauch-Schwarz C, Rosdahl VT, Braveny I (1994) Methicillin-resistant *Staphylococcus aureus* in Europe. Eur J Clin Microbiol Infect Dis 13(1):50–55

Waksman SA, Woodruff HB (1941) *Actinomyces antibioticus*, a new soil organism antagonistic to pathogenic and non-pathogenic bacteria. J Bacteriol 42(2):231–249

Waksman SA, Woodruff HB (1942) Selective antibiotic action of various substances of microbial origin. J Bacteriol 44(3):373–384

Waksman SA, Horning ES, Spencer EL (1942) The production of two antibacterial substances. Fumigacin and Clavacin. Science 96(2487):202–203

Walsh TR, Howe RA (2002) The prevalence and mechanisms of vancomycin resistance in *Staphylococcus aureus*. Ann Rev Microbiol 56:657–675

Walsh TR, Weeks J, Livermore DM, Toleman MA (2011) Dissemination of NDM-1 positive bacteria in the New Delhi environment and its implications for human health: an environmental point prevalence study. Lancet Infect Dis 11(5):355–362

Watanabe M, Iyobe S, Inoue M, Mitsuhashi S (1991) Transferable imipenem resistance in *Pseudomonas aeruginosa*. Antimicrob Agents Chemother 35(1):147–151

Wayne LG, Sohaskey CD (2001) Nonreplicating persistence of *Mycobacterium tuberculosis*. Ann Rev Microbiol 55:139–163

Wehrli W (1983) Rifampin: mechanisms of action and resistance. Rev Infect Dis 5(Suppl 3):S407–411

Wertheim HF, Melles DC, Vos MC, van Leeuwen W, van Belkum A, Verbrugh HA, Nouwen JL (2005) The role of nasal carriage in *Staphylococcus aureus* infections. Lancet Infect Dis 5 (12):751–762

Winder FG, Collins PB (1970) Inhibition by isoniazid of synthesis of mycolic acids in *Mycobacterium tuberculosis*. J Gen Microbiol 63(1):41–48

World Health Organization W (2001) Guidelines for establishing DOTS-plus pilot projects for the management of multidrug-resistant tuberculosis (MDR-TB)

World Health Organization W (2014a) Antimicrobial resistance global report on surveillance. http://www.who.int/drugresistance/documents/surveillancereport/en/

World Health Organization W (2014b) Global tuberculosis report 2014. http://www.who.int/tb/publications/global_report/en/ Accessed 22 Oct 2015

Wyllie D, Paul J, Crook D (2011a) Waves of trouble: MRSA strain dynamics and assessment of the impact of infection control. J Antimicrob Chemother 66(12):2685–2688

Wyllie DH, Walker AS, Miller R, Moore C, Williamson SR, Schlackow I, Finney JM, O'Connor L, Peto TE, Crook DW (2011b) Decline of meticillin-resistant *Staphylococcus aureus* in Oxfordshire hospitals is strain-specific and preceded infection-control intensification. BMJ Open 1(1):e000160

Yang YJ, Wu PJ, Livermore DM (1990) Biochemical characterization of a beta-lactamase that hydrolyzes penems and carbapenems from two *Serratia marcescens* isolates. Antimicrob Agents Chemother 34(5):755–758

Yigit H, Queenan AM, Anderson GJ, Domenech-Sanchez A, Biddle JW, Steward CD, Alberti S, Bush K, Tenover FC (2001) Novel carbapenem-hydrolyzing beta-lactamase, KPC-1, from a carbapenem-resistant strain of *Klebsiella pneumoniae*. Antimicrob Agents Chemother 45 (4):1151–1161

Yocum RR, Waxman DJ, Rasmussen JR, Strominger JL (1979) Mechanism of penicillin action: penicillin and substrate bind covalently to the same active site serine in two bacterial D-alanine carboxypeptidases. Proc Natl Acad Sci USA 76(6):2730–2734

Yong D, Choi YS, Roh KH, Kim CK, Park YH, Yum JH, Lee K, Chong Y (2006) Increasing prevalence and diversity of metallo-beta-lactamases in *Pseudomonas* spp., *Acinetobacter* spp., and Enterobacteriaceae from Korea. Antimicrob Agents Chemother 50(5):1884–1886

Yong D, Toleman MA, Giske CG, Cho HS, Sundman K, Lee K, Walsh TR (2009) Characterization of a new metallo-beta-lactamase gene, bla(NDM-1), and a novel erythromycin esterase gene carried on a unique genetic structure in *Klebsiella pneumoniae* sequence type 14 from India. Antimicrob Agents Chemother 53(12):5046–5054

Yu S, Girotto S, Lee C, Magliozzo RS (2003) Reduced affinity for isoniazid in the S315T mutant of *Mycobacterium tuberculosis* KatG is a key factor in antibiotic resistance. J Biol Chem 278 (17):14769–14775

Zhang Y, Heym B, Allen B, Young D, Cole S (1992) The catalase-peroxidase gene and isoniazid resistance of *Mycobacterium tuberculosis*. Nature 358(6387):591–593

Emergence and Spread of Antimicrobial Resistance: Recent Insights from Bacterial Population Genomics

Ulrich Nübel

Abstract Driven by progress of DNA sequencing technologies, recent population genomics studies have revealed that several bacterial pathogens constitute 'measurably evolving populations'. As a consequence, it was possible to reconstruct the emergence and spatial spread of drug-resistant bacteria on the basis of temporally structured samples of bacterial genome sequences. Based on currently available data, some general inferences can be drawn across different bacterial species as follows:

(1) Resistance to various antibiotics evolved years to decades earlier than had been anticipated on the basis of epidemiological surveillance data alone.
(2) Resistance traits are more rapidly acquired than lost and commonly persist in bacterial populations for decades.
(3) Global populations of drug-resistant pathogens are dominated by very few clones, yet the features enabling such spreading success have not been revealed, aside from antibiotic resistance.
(4) Whole-genome sequencing proved very effective at identifying bacterial isolates as parts of the same transmission networks.

Contents

U. Nübel (✉)
DZIF Group on Microbial Genome Research, Leibniz Institute DSMZ, Braunschweig, Germany
e-mail: ulrich.nuebel@dsmz.de

U. Nübel
Technical University Braunschweig, Braunschweig, Germany

U. Nübel
German Center for Infection Research (DZIF), Partner Site Hannover-Braunschweig, Braunschweig, Germany

Current Topics in Microbiology and Immunology (2016) 398: 35–54
DOI 10.1007/82_2016_505

Published Online: 21 September 2016

1 Introduction

Widespread antibiotic resistance among pathogenic bacteria threatens the achievements of modern medicine. Even routine surgery bears the risk of healthcare-associated infections that are difficult to treat and outbreaks of drug-resistant tuberculosis exhibit a mortality last seen in the pre-antibiotic era. Antibiotic resistance has become a global crisis, because resistant pathogens move across continents alongside their hosts. Recently, unprecedented insights into the mechanisms and dynamics of the emergence, evolution, and spatial spread of antibiotic-resistant bacterial pathogens were provided by large-scale bacterial whole-genome sequence analyses. This progress was driven by advancements of high-throughput sequencing technologies. In this review, I will highlight some of the latest insights into the genomic epidemiology of several major antibiotic-resistant bacterial pathogens and try to identify a number of general inferences. I will focus on a limited number of examples where extensive genomic datasets exist.

2 Measurably Evolving Bacterial Populations

During the past ten years, progress in DNA sequencing technologies has improved sequencing speed and throughput by several orders of magnitude, and, at the same time, has caused a sharp decrease of sequencing costs (van Dijk et al. 2014). Thanks to this development, datasets encompassing near-complete genome sequences from multiple closely related bacterial isolates became available for several species of bacterial pathogens, enabling studies into their 'population genomics', i.e., investigations of population genetics at genomic scale. One of the most remarkable findings of early bacterial population genomics studies was that pathogenic bacteria proved to have accumulated evolutionary change in their genomes over unexpectedly short time periods (Harris et al. 2010; Nübel et al. 2010). Novel sequence variation was discovered in bacterial genomic sequences that had been sampled only months to years apart, which meant that bacteria constituted 'measurably evolving populations', similar to rapidly evolving RNA viruses (Drummond et al. 2003). Short-term evolutionary rates range from 6 base substitutions per genome per year in *Staphylococcus aureus* to 0.4 base substitutions per genome per year in *Mycobacterium tuberculosis*. Most other bacterial pathogens exhibit evolutionary rates within this range, and average rates commonly have been found to be largely uniform among diverse strains within a given species (Didelot et al. 2012). These findings have enabled time-calibrated phylogenetic analyses, which have provided useful insights into the temporal dynamics of pathogen evolution and spread. On the basis of appropriate sets of genomic sequences serially sampled through time, it is now possible to date evolutionary events in the bacterial population, such as the acquisition of mutations that confer antimicrobial drug resistance, to reconstruct the temporal succession of a pathogen's spatial dispersal, and to model past population size changes (Holden et al. 2013; Bentley and Parkhill 2015; Biek et al. 2015).

3 Methicillin-Resistant *Staphylococcus aureus*

S. aureus is an important human pathogen causing opportunistic infections that range from self-limiting skin infections to life-threatening pneumonia, bloodstream infection, or endocarditis. *S. aureus* has demonstrated its ability to acquire resistance to all classes of antibiotics. In particular, methicillin-resistant *S. aureus* (MRSA) is a notorious cause of large outbreaks of infections in both health care and community settings. The first MRSA was reported already in 1961 (Jevons 1961), which was within a year after the introduction into clinical use of methicillin—a penicillinase-stable beta-lactam antibiotic. Since then, MRSA have emerged multiple times through acquisition of variants of the transferable genetic element SCC*mec*, which renders staphylococci resistant to all beta-lactams by encoding a peptidoglycan transpeptidase enzyme with low affinity for beta-lactam antibiotics (Robinson and Enright 2004; Nübel et al. 2008). A small number of healthcare-associated MRSA clones are very widely disseminated internationally, such that associated infections occur at pandemic scale (Crisóstomo et al. 2001; Chambers and Deleo 2009). These healthcare-associated MRSA clones very rarely cause infections outside of hospitals or nursing homes. In contrast, genetically unrelated, community-associated MRSA strains exist that cause infections in otherwise healthy individuals without contact to hospital patients or staff. Globally, healthcare-associated MRSA infections by far outnumber those in the community, even though outbreaks with one specific community-associated MRSA named 'USA300', have reached an epidemic level in the USA (David and Daum 2010; DeLeo et al. 2010), and regions with low prevalence of healthcare-associated MRSA exist (Johnson 2011). In addition, livestock-associated MRSA colonize fattening pigs and broiler chickens at high frequency, and these MRSA strains cause infections in humans occasionally (Cuny et al. 2013).

Population genomic analyses of major MRSA clonal lineages have provided detailed records of their emergence, spatial spread, and continued evolution, including the acquisition and loss of antibiotic resistance traits. One healthcare-associated MRSA strain, dubbed 'EMRSA-15', has dominated MRSA populations in Europe, Australia, and parts of Asia for the last twenty years. This strain was reported for the first time in 1991 (Richardson and Reith 1993), yet datings based on genomic sequence variation later revealed that its emergence in the UK had begun already during the mid-1970s, almost two decades before its first reported discovery (Holden et al. 2013). Genome-based coalescent analyses further documented that EMRSA-15 had acquired resistance to fluoroquinolone drugs in the early 1980s and traced the emergence of this trait to a region around the city of Birmingham in the UK. This result suggested that resistance to fluoroquinolones had emerged during clinical trials that were performed in several hospitals in the area, prior to licensing of the drug (Holden et al. 2013). It is remarkable that fluoroquinolone resistance seems to have been instrumental for the subsequent spread of this MRSA over large geographic areas, within the UK and beyond. Another, unrelated MRSA strain (EMRSA-16) emerged in the UK during the same time period after acquisition of fluoroquinolone resistance on the basis of two identical point mutations (McAdam et al. 2012). Indeed, all major healthcare-associated MRSA are resistant to fluoroquinolones, and the rapid increase of fluoroquinolone use during the 1990s apparently has promoted the dispersal of these strains, even though this class of antibiotic is not applied for treating staphylococcal infections. Rather, widespread application of fluoroquinolone drugs appears to promote colonization with resistant MRSA due to skin secretion of the antibiotic and suppression of the normal bacterial flora (Hawkey 1997; Weber et al. 2003). Genomic analyses further suggested that this development was paralleled in the USA by community-associated MRSA USA300, which acquired fluoroquinolone resistance in the early 1990s, just prior to a strong population expansion that occurred during a time of heavy outpatient usage of fluoroquinolone drugs (Uhlemann et al. 2014; Alam et al. 2015).

The evolution of antibiotic resistance of MRSA is driven by selection pressure from antibiotic usage. As outlined, resistance of *S. aureus* to various antibiotics emerged very shortly after the introduction of novel drugs in several cases. Accordingly, regional and country-specific differences among drug use regimens are reflected by the geographic dissemination of resistance (Holden et al. 2013; Reuter et al. 2016). Genome-based phylogenetic analyses further showed that resistance traits generally are more rapidly acquired than lost in *S. aureus*. Once acquired, genomic features causing resistance, including point mutations and intra or extrachromosomal mobile genetic elements, commonly persist in the MRSA population for decades (Holden et al. 2013). As a result, the average number of resistance traits per isolate increases with isolation date (Holden et al. 2013; Strommenger et al. 2014). In addition, the accumulation of point mutations in specific core genes over time may result in reduced susceptibility to increasing antibiotic concentrations. This was demonstrated for healthcare-associated MRSA ST239, which is multidrug-resistant (MDR), globally distributed, and highly prevalent in Asia and Australia (Baines et al. 2015).

In addition to a few pandemic MRSA, many other MRSA strains have evolved that display similar resistance but have as yet not spread beyond local scales. There has been a long-standing debate, as to which additional characteristics may determine the extraordinary spreading success of some MRSA strains, and if it would be possible to predict it. Traits considered in this regard have been virulence (DeLeo and Chambers 2009), tolerance to desiccation and other stresses (Knight et al. 2012; Baldan et al. 2015), resistance to antiseptics (Cooper et al. 2012), or growth rate (Knight et al. 2012), among other factors. However, comparative Genomics has as yet revealed little convergence of genomic traits among major MRSA strains aside from antibiotic resistance. Therefore, it remains unclear to what extent those widely disseminated strains have been favored by any selective edge. In addition to selection, undoubtedly, stochasticity and genetic drift are dominant forces shaping the population structure of *S. aureus* (Fraser et al. 2005). Prospective pathogen surveillance applying bacterial whole-genome sequencing will enable early detection of newly emerging and spreading strains ('high-risk clones') that pose particular threats to public health (Aanensen et al. 2016; Reuter et al. 2016).

Genomic analyses also have provided detailed insights into MRSA spread over local scales. Compared to more conventional typing techniques, including multilocus sequence typing, *spa* typing, or DNA macrorestriction (pulsed-field gel electrophoresis), genome sequencing provides much better power to discriminate related isolates (Nübel et al. 2011). Accordingly, the ability to delineate outbreaks of infections proved superior to conventional typing, which is particularly useful in hospital settings, where endemic strains of healthcare-associated MRSA commonly predominate. Moreover, genome-based phylogenetic analyses enabled the inference of the temporal progression of outbreaks (Köser et al. 2012; Nübel et al. 2013; Tong et al. 2015). While hospital patients and clinical staff provide the primary reservoir for healthcare-associated MRSA, members of community households may be asymptomatically colonized and represent a long-term reservoir for maintenance and transmission of MRSA in the community (Uhlemann et al. 2014; Alam et al. 2015). Eradication of community-associated MRSA will therefore require decolonization of household members in addition to those individuals suffering infections, similar to the measures recommended for curbing MRSA spread in the healthcare setting (Uhlemann et al. 2014; Alam et al. 2015).

A better understanding of within-host diversity will be required for the inference of transmission pathways (Worby et al. 2014; Paterson et al. 2015; Tong et al. 2015). Importantly, the common practice of analyzing single isolates from each case may provide misleading information on genetic distances between pathogens from different hosts (Worby et al. 2014). This is because *S. aureus* isolates sampled from individual hosts may exhibit multiple genomic differences despite their recent clonal origin (Young et al. 2012; Golubchik et al. 2013; Paterson et al. 2015). Within-host diversity may fluctuate temporally and the accumulation of mutations may accelerate during the transition from nasal colonization to disease (Young et al. 2012).

4 *Mycobacterium tuberculosis*

M. tuberculosis (sensu stricto) is an obligate human pathogen causing pulmonary disease, with no reservoir in other host species or the environment. As many as two billion individuals worldwide are estimated to be latently infected with *M.*

tuberculosis, and 1.5 million people die from tuberculosis every year (WHO 2015). The current epidemic of tuberculosis is driven by coinfection with HIV and drug resistance (Dippenaar and Warren 2015). Antimicrobial resistance in *M. tuberculosis* has been increasing for decades due to inefficient healthcare systems, poor patient compliance with several-month chemotherapy, and between-patient pharmacokinetic variability (Cegielski 2010; Srivastava et al. 2011). Outbreaks with MDR *M. tuberculosis*, which is resistant to at least isoniazid and rifampicin, have been reported in the USA since 1985 (Cegielski 2010). In 2014, approximately 8 % of new tuberculosis cases worldwide were caused by MDR *M. tuberculosis*, and 10 % of MDR cases were extensively drug resistant (XDR), i.e., additionally resistant to fluoroquinolones and at least one second-line antibiotic (kanamycin, capreomycin, or amikacin) (WHO 2015).

The so-called *Beijing* strain of *M. tuberculosis* currently causes >25 % of tuberculosis cases worldwide (Luo et al. 2015). Population genomic analyses have pinpointed the geographic origin of *Beijing* to East Asia, where the strain is endemic and shows greatest genomic diversity (Luo et al. 2015; Merker et al. 2015). While the *Beijing* strain has evolved long before the antibiotic era, two MDR clones of *Beijing* have expanded massively throughout Central Asia and Russia (Merker et al. 2015). In a recent study based on genome sequences from 1000 *M. tuberculosis* isolates from the Samara region in Russia, 64 % of isolates were affiliated to the *Beijing* lineage, and 87 % of these were MDR (Casali et al. 2014). Alarmingly, 16 % of MDR isolates were XDR and, furthermore, 82 % of isolates from the two epidemic MDR *Beijing* clones were already resistant to kanamycin, hence, lacked resistance to fluoroquinolones only to qualify for the XDR phenotype. Genome-based phylogenetic analyses indicated that kanamycin resistance-causing mutations had been acquired multiple times independently, reflecting the strong selective pressure on *M. tuberculosis* that is caused by widespread kanamycin usage in the therapy of MDR tuberculosis in Russia (Casali et al. 2014). This development was paralleled by an unrelated MDR strain (called the '*M*' strain) in Argentina, which had been discovered in the early 1990s as the causative agent of a large tuberculosis outbreak among HIV patients in a hospital in Buenos Aires. Time-calibrated phylogenetic analyses revealed that the *M* strain had acquired resistance to ethambutol, pyrazinamide, and kanamycin already around 1979 (95 % confidence intervals, 1975–1983), and that it had expanded and spread widely for the thirty years to follow (Eldholm et al. 2015). From this pre-XDR strain, the XDR phenotype evolved at least six times independently (Eldholm et al. 2015). Since phylogenetic analyses indicated the explosive radiation and rapid spread of these MDR and kanamycin-resistant strains, resistance-causing mutations do not seem to impair their propensity to cause disease and transmit (Casali et al. 2014; Eldholm et al. 2015; Merker et al. 2015). Rather, widespread antibiotic use provides a selective advantage to drug-resistant strains and expedites their expansion. In addition, resistance mutations in *M. tuberculosis* frequently are associated with additional mutations in the same or interacting genes, which have been suggested to compensate any deleterious effects of resistance (Comas et al. 2012; Casali et al. 2014).

Whole-genome sequencing proved superior to conventional genotyping for discriminating closely related *M. tuberculosis* strains. In conjunction with epidemiological data, genomic analyses enabled the reconstruction of tuberculosis outbreaks. A retrospective investigation of a tuberculosis outbreak in Northern Germany, which had affected 86 people over 14 years, demonstrated that previous genotyping had provided misleading results by clustering unrelated isolates (Roetzer et al. 2013). In contrast, genome sequence data was in good accordance with epidemiological contact tracing and identified a total of eight transmission chains, each consisting of two to eight patients (Roetzer et al. 2013). A maximum of three SNP differences were detected between the genomes from isolates that had been transmitted from one patient to another. Genome analyses further revealed that this outbreak had led to the expansion of a specific *M. tuberculosis* strain (dubbed the '*Hamburg*' clone) (Roetzer et al. 2013). During the expansion of the *Hamburg* clone, base substitutions had been accumulated in its genome at a rate of 0.4 mutations per year. Similar substitution rates have since been reported for other *M. tuberculosis* strains (Eldholm et al. 2015; Guerra-Assunção et al. 2015a, b). Considering an experimentally determined per-generation mutation rate for *M. tuberculosis* (Ford et al. 2011), the base substitution rate suggests that the *Hamburg* clone during the outbreak had replicated at approximately 400 generations per year, which is similar to the maximum laboratory growth rate of *M. tuberculosis* (Gutierrez-Vazquez 1956). While increased transmissibility apparently was independent from peculiarities in the patients' social environment, the specific genomic features that may have caused the particular fitness of this strain could not be pinpointed (Roetzer et al. 2013). Similarly, the integration of *M. tuberculosis* genome sequences from another outbreak in Malawi with disease onset dates enabled the construction of transmission trees, which revealed striking differences in transmissibility among *M. tuberculosis* clonal lineages that were independent from the host's population structure or HIV infection (Guerra-Assunção et al. 2015a, b). While the specific mechanistic reasons for the differential spreading success of *M. tuberculosis* strains have remained futile, it may be caused by differences in virulence (as transmission of tuberculosis requires disease), aside from antimicrobial resistance (Brites and Gagneux 2015). In the Malawi setting, in addition, *M. tuberculosis* genome sequencing enabled to distinguish relapse from reinfection in 54 % of cases of recurrent tuberculosis. Reinfection turned out to be less frequent than relapse and associated with HIV infection (Guerra-Assunção et al. 2015a, b).

5 Multidrug-Resistant *Escherichia coli* and *Klebsiella pneumoniae*

Several species of gram-negative bacteria are common causes of healthcare-associated infections. The current trend of increasing drug resistance among these pathogens severely limits therapeutic options for the treatment of infections. *Escherichia coli* and *Klebsiella pneumoniae* are members of the *Enterobacteriaceae* family. They are natural constituents of the gut microbiome and may cause a variety of opportunistic infections, including infections of the urinary tract, wounds, soft tissues and the bloodstream, and pneumonia. Drug resistance spreads quickly among these bacteria, as resistance-conferring genes are encoded on plasmids and other mobile genetic elements, which can be transferred horizontally between bacterial cells and shared even among distantly related bacterial species (Carattoli 2009).

Extended-spectrum beta-lactamases (ESBL) are enzymes that confer resistance to most beta-lactam antibiotics, including third-generation cephalosporins. They were first detected in the 1980s, shortly after the introduction of broad-spectrum cephalosporins into clinical practice, and hundreds of variants have since been reported. At a global scale, currently, CTX-M enzymes are the most prevalent type of ESBL. For unknown reasons, these have largely supplanted TEM and SHV enzymes since the turn of the millennium (Woerther et al. 2013). Of note, the fecal carriage of ESBL-producing *Enterobacteriaceae* outside of healthcare institutions has increased since and provides a reservoir for these pathogens (Woerther et al. 2013). The drugs of choice against ESBL-producing gram-negative pathogens are carbapenems, which are beta-lactams that are inefficiently hydrolyzed by ESBL enzymes. Soon after their introduction in the 1980s, the first carbapenem-resistant *Enterobacteriaceae* were discovered, yet these did not spread widely. After the turn of the millennium, however, *Enterobacteriaceae* producing carbapenemase enzymes that efficiently hydrolyzed carbapenems caused multiple major outbreaks of hospital-associated infections with high mortality in the USA and in Southern Europe (Temkin et al. 2014). While several carbapenemase families have evolved convergently, variants of four enzyme families (KPC, VIM, NDM, and OXA) have spread widely and bacteria producing these are endemic in many countries today. The therapy of infections with carbapenemase-producing enterobacteria usually is based on colistin, even though this drug has poor effectivity and nephrotoxic side effects (Temkin et al. 2014). Colistin resistance occasionally has been reported in carbapenemase-producing bacteria (see below).

Some MDR clones of *Enterobacteriaceae* gained global distribution recently (Woodford et al. 2011). Since 2008, *E. coli* ST131 producing the ESBL CTX-M-15 has increasingly been reported as a cause of urinary tract infections and bacteremia. Population genomic analyses revealed that a single subclone of ST131—dubbed *H30*-Rx—had acquired CTX-M-15 after it had become resistant to fluoroquinolones (Price et al. 2013; Petty et al. 2014). *H30*-Rx isolates collected in Europe, North America, Australasia, and India were all very closely related, and *H30*-Rx likely has spread globally within less than 25 years (Stoesser et al. 2016). The gene encoding CTX-M-15 was part of a transposable element, which in the different *H30*-Rx isolates was found to be inserted at several distinct positions on the bacterial chromosome or on an extrachromosomal plasmid, respectively. Some isolates even had two copies of the transposon on both the chromosome and the plasmid, and several others had lost it together with the CTX-M-15 gene (Price et al. 2013; Petty et al. 2014). While this finding documented the transposon's repeated intragenomic mobilization, it apparently got acquired by the predecessor of *H30*-Rx only once, presumably through transfer from a plasmid. Hence, wide geographic dissemination of *E. coli* ST131 producing CTX-M-15 was driven by clonal bacterial expansion from a single focus, rather than frequent lateral transfer, which is quite remarkable, especially since bacterial genotyping by DNA macrorestriction (pulsed-field gel electrophoresis) previously had suggested the opposite (Price et al. 2013). It seems obvious that drug resistance has enabled this expansion, even though it currently remains difficult to understand why other *E. coli* strains with equal resistance do not proliferate at similar rates (Price et al. 2013). Recently, carbapenem-resistant *E. coli* ST131 were reported that produced either one carbapenemase of all four prevalent classes (Peirano et al. 2011; Morris et al. 2012; Accogli et al. 2014; Kutumbaka et al. 2014; Johnson et al. 2015). At least two of these reported strains simultaneously produced CTX-M-15 (Accogli et al. 2014), but their phylogenetic affiliations to the *H30*-Rx clone were not investigated. Even more alarmingly, plasmid-encoded colistin resistance has also been detected in *E. coli* ST131 (Hasman et al. 2015). While that particular ST131 strain, isolated from chicken meat in Denmark, was still susceptible to carbapenems, the combination of colistin resistance and carbapenem resistance was found in another *E. coli* isolate (unrelated to ST131) from a human infection, suggesting the potential for 'pandrug-resistant' *E. coli* that may cause untreatable infections (Falgenhauer et al. 2016).

K. pneumoniae ST258 producing the KPC-3 carbapenemase emerged in the USA in the late 1990s and spread globally since (Adler and Carmeli 2011; Munoz-Price et al. 2013). Commonly, this strain is susceptible only to colistin, tigecycline, and gentamicin, but colistin-resistant variants have been reported (Bogdanovich et al. 2011). Bloodstream infections with this MDR strain have high attributable mortality (Munoz-Price et al. 2013). Comparative genomic analyses indicated that ST258 is the result of a recombination event that caused the transfer of a 1.1-Mbp fragment between two distantly related clonal complexes (Chen et al. 2014). A second recombination involving a 52-kilobase fragment replaced the gene encoding the capsular antigen in one clade of the ST258 radiation (Chen et al. 2014). Two studies investigating genomic diversity across the entire species confirmed horizontal genetic transfer as a common mechanism for generating both large-scale genomic variation and capsular diversity of *K. pneumoniae* (Holt et al. 2015; Wyres et al. 2015). Diversification of the antigenic profile through homologous recombination may be the pathogen's strategy to evade host immunity, similar to *Streptococcus pneumoniae* (Croucher and Klugman 2014). Extensive capsule diversity and its rapid evolution may hinder novel pathogen control strategies involving vaccination or phage-derived substances (Wyres et al. 2015).

The KPC-3 carbapenemase gene is located on a transposon that may be carried by various plasmids, together with additional resistance traits (DeLeo et al. 2014; Temkin et al. 2014). However, despite the proven motility of the transposon and the plasmids through horizontal genetic transfer, the major driver of carbapenemase spread over local and regional (i.e., countrywide) scales appears to be the clonal expansion of a limited number of epidemic strains (Temkin et al. 2014; Adler et al. 2016).

The distribution of other resistance genes in *K. pneumoniae* varies between different regions (countries), however, presumably reflecting differences in antibiotic usage, which suggests that the adaptation to local conditions through acquisition of resistance is a faster process than long-distance spread (Holt et al. 2015).

Genomic sequencing in combination with detailed epidemiological data revealed that an outbreak of infections with carbapenemase-producing *K. pneumoniae* ST258 had spread more widely within a hospital than initially expected, explaining a resurgence several weeks after its apparent eradication (Snitkin et al. 2012). Another strain of carbapenem-resistant *K. pneumoniae* was revealed to have persisted for months on several hospital units associated with neonatal care, causing occasional clusters of infections (Stoesser et al. 2014). Similarly, a prolonged outbreak with ESBL-producing *K. pneumoniae* on a neonatal ward could be demonstrated by genome-based phylogenetic analyses to have lasted several years before it was discovered and to have been more complex than suspected (Haller et al. 2015). In all these settings, the unperceived spread likely had been driven by repeated person-to-person transmission among silently colonized patients and/or staff members which did not themselves develop any disease (Snitkin et al. 2012; Stoesser et al. 2014; Haller et al. 2015). In addition, genomic data revealed that the carbapenemase-producing strain had acquired colistin resistance during the course of the outbreak on the basis of three independent evolutionary events in different phylogenetic clades, which partially may explain therapeutic failure (Snitkin et al. 2012).

6 Emergence of Resistance

It has long been established that, in many cases, antimicrobial resistance emerged very soon after novel drugs had been introduced to clinical use (Richardson and Reith 1993; Chambers and Deleo 2009; Cegielski 2010; Woerther et al. 2013). However, where analysis was feasible based on the availability of historic isolates, population genomics invariably indicated that resistance had evolved even significantly earlier than anticipated from epidemiological surveillance data. Newly emerging, resistant strains of *M. tuberculosis*, MRSA, and *E. coli* had spread for more than 10 years and reached wide geographic distribution prior to their discovery through molecular surveillance (Holden et al. 2013; Price et al. 2013; Eldholm et al. 2015). One fluoroquinolone-resistant MRSA strain had emerged already during clinical drug trials (Holden et al. 2013).

Across different bacterial species, populations of drug-resistant pathogens are dominated by descendants from very few clones that have the ability to rapidly spread over large areas. In genomic data, this is reflected by an extremely shallow diversity among globally collected isolates, due to their common descent from a recent, explosive evolutionary radiation (Holden et al. 2013; Price et al. 2013;

Merker et al. 2015). Global spread of these 'pandemic strains' and the concomitant displacement of locally preexisting competitors appears to succeed regularly every few decades (Holden et al. 2013; Stoesser et al. 2016). This is a remarkable commonality among antibiotic-resistant bacterial pathogens, in spite of striking differences in terms of their biology and epidemiology. The reasons for the spreading success of particular strains have remained elusive, however, despite a great interest in this matter. Obviously, antimicrobial resistance may provide selective advantages, and resistance to fluoroquinolones seems to have been particularly instrumental, since it often directly preceded expansions. Across different species, however, only a small minority of equally resistant strains has spread widely, and hence, additional selective features have been suspected to contribute to the success of these strains (DeLeo and Chambers 2009; Brites and Gagneux 2015; Mathers et al. 2015). Despite the availability of considerably large genomic datasets, however, very little convergence has as yet been discovered among different pandemic strains, aside from drug resistance (Holden et al. 2013; Brites and Gagneux 2015).

Despite their rapid spatial spread, widely dispersed strains of MRSA, *K. pneumoniae*, and *M. tuberculosis* frequently have adapted to local conditions by acquiring additional resistance traits even faster than spreading over long distance (Casali et al. 2012; Holden et al. 2013; Eldholm et al. 2015; Holt et al. 2015). Remarkably, a reversal of resistance has been observed much more rarely and, hence, seems to be a comparatively slow process (Casali et al. 2012; Holden et al. 2013). This long-term trend among diverse species of pathogenic bacteria will need to be considered when devising strategies to mitigate resistance, for example, through combination therapy (Baym et al. 2016).

7 Detection of Outbreaks and Epidemics

Genome sequencing provides ultimate discriminatory power for genotyping of bacterial strains and it is very effective at identifying isolates as part of the same transmission network (Eyre et al. 2012; Snitkin et al. 2012; Nübel et al. 2013). Moreover, and in contrast to conventional molecular typing, bacterial genome sequences carry high-resolution phylogenetic signal that can reveal detailed insights into the temporal progression of outbreaks and epidemics. By now, it has been demonstrated for several species of bacterial pathogens—including those discussed above—that genomic data can inform analyses of outbreaks at local scales, e.g., within single hospitals or local communities. Maximum insight was usually gained when sequence data was integrated with epidemiological data, such as temporal contact patterns among patients and medical staff (Snitkin et al. 2012; Nübel et al. 2013; Roetzer et al. 2013; Walker et al. 2013; Didelot et al. 2014; Haller et al. 2015). Genome sequencing may even enable the reconstruction of patient-to-patient transmission pathways at great detail (Morelli et al. 2012; Jombart et al. 2014; Croucher and Didelot 2015). As, technically, results from such an analysis could be provided over clinically relevant time scales, it would be very useful as a diagnostic tool for informing infection control measures (Eyre et al. 2012; Sintchenko and Holmes 2015). Clearly, however, inference of transmission will require an improved understanding of bacterial diversity within individual hosts, which available data suggests may vary between individuals, fluctuate with time, and be costly to measure (Golubchik et al. 2013; Worby et al. 2014; Black et al. 2015; Paterson et al. 2015; Stoesser et al. 2015). Critically, the current common practice of sampling and characterizing single bacterial isolates per individual may obscure transmission links (Worby et al. 2014; Stoesser et al. 2015).

8 Outlook

Genomic data enables the prediction of clinically relevant aspects of a strain's phenotype, including its antimicrobial resistance (Gordon et al. 2014; Bradley et al. 2015). Therefore, the genome-based surveillance of antibiotic-resistant bacteria holds great potential for the early discovery of relevant outbreaks and of causative bacterial strains, which would be of great value for their duly containment. Fortunately, due to advancements of sequencing technology and the associated decrease of sequencing costs, the integration of genomic approaches into diagnostic microbiology is making progress, which foreseeably will result in the generation of abundant microbial genome sequence data (Didelot et al. 2012; Reuter et al. 2013). To benefit public health and to enable rapid responses to emerging threats, however, effective early warning systems will require the rapid sharing of data beyond individual laboratories, institutions, and nations. At the least, this data should include the sequences and some minimum associated epidemiological information (Aarestrup and Koopmans 2016). Web-based data repositories with open access to data and analysis tools may facilitate data sharing and joint interpretation (Aanensen et al. 2016; Aarestrup and Koopmans 2016).

Beyond applications to molecular epidemiology, however, rich sequence datasets allow for studies of the evolution of bacterial genomes at unprecedented detail (Feil 2015). Integrating large genomic datasets with experimental and epidemiological data recently enabled the identification of statistical associations between genomic mutations and specific microbial phenotypes, including antibiotic resistance (Farhat et al. 2013; Earle et al. 2016), toxicity (Laabei et al. 2014), and host specificity (Sheppard et al. 2013). Future genome-wide association studies can be expected to reveal abundant fundamental insights into bacterial genome function, including the complexity of multifactorial traits and epistatic interactions (Cui et al. 2015; Laabei et al. 2015).

References

Aanensen DM, Feil EJ, Holden MT, Dordel J, Yeats CA, Fedosejev A, Goater R, Castillo-Ramirez S, Corander J, Colijn C, Chlebowicz MA, Schouls L, Heck M, Pluister G, Ruimy R, Kahlmeter G, Ahman J, Matuschek E, Friedrich AW, Parkhill J, Bentley SD, Spratt BG, Grundmann H, European SRLWG (2016) Whole-genome sequencing for routine pathogen surveillance in public health: a population snapshot of invasive *Staphylococcus aureus* in Europe. MBio 7(3):e00444-16

Aarestrup FM, Koopmans MG (2016) Sharing data for global infectious disease surveillance and outbreak detection. Trends Microbiol 24(4):241–245

Accogli M, Giani T, Monaco M, Giufre M, Garcia-Fernandez A, Conte V, D'Ancona F, Pantosti A, Rossolini GM, Cerquetti M (2014) Emergence of *Escherichia coli* ST131 sub-clone H30 producing VIM-1 and KPC-3 carbapenemases, Italy. J Antimicrob Chemother 69(8):2293–2296

Adler A, Carmeli Y (2011) Dissemination of the *Klebsiella pneumoniae* carbapenemase in the health care settings: tracking the trails of an elusive offender. MBio 2(6):e00280-11

Adler A, Khabra E, Paikin S, Carmeli Y (2016) Dissemination of the bla_{KPC} gene by clonal spread and horizontal gene transfer: comparative study of incidence and molecular mechanisms. J Antimicrob Chemother

Alam MT, Read TD, Petit 3rd RA, Boyle-Vavra S, Miller LG, Eells SJ, Daum RS, David MZ (2015) Transmission and microevolution of USA300 MRSA in U.S. households: evidence from whole-genome sequencing. MBio 6(2):e00054

Baines SL, Holt KE, Schultz MB, Seemann T, Howden BO, Jensen SO, van Hal SJ, Coombs GW, Firth N, Powell DR, Stinear TP, Howden BP (2015) Convergent adaptation in the dominant global hospital clone ST239 of methicillin-resistant *Staphylococcus aureus*. MBio 6(2):e00080

Baldan R, Rancoita PM, Di Serio C, Mazzotti M, Cichero P, Ossi C, Biancardi A, Nizzero P, Saracco A, Scarpellini P, Cirillo DM (2015) Epidemic MRSA clone ST22-IV is more resistant to multiple host- and environment-related stresses compared with ST228-I. J Antimicrob Chemother 70(3):757–765

Bentley SD, Parkhill J (2015) Genomic perspectives on the evolution and spread of bacterial pathogens. Proc Biol Sci 282(1821):20150488

Biek R, Pybus OG, Lloyd-Smith JO, Didelot X (2015) Measurably evolving pathogens in the genomic era. Trends Ecol Evol 30(6):306–313

Black PA, de Vos M, Louw GE, van der Merwe RG, Dippenaar A, Streicher EM, Abdallah AM, Sampson SL, Victor TC, Dolby T, Simpson JA, van Helden PD, Warren RM, Pain A (2015) Whole genome sequencing reveals genomic heterogeneity and antibiotic purification in *Mycobacterium tuberculosis* isolates. BMC Genom 16(1):857

Bogdanovich T, Adams-Haduch JM, Tian GB, Nguyen MH, Kwak EJ, Muto CA, Doi Y (2011) Colistin-resistant, *Klebsiella pneumoniae* carbapenemase (KPC)-producing *Klebsiella pneumoniae* belonging to the international epidemic clone ST258. Clin Infect Dis 53(4):373–376

Bradley P, Gordon NC, Walker TM, Dunn L, Heys S, Huang B, Earle S, Pankhurst LJ, Anson L, de Cesare M, Piazza P, Votintseva AA, Golubchik T, Wilson DJ, Wyllie DH, Diel R, Niemann S, Feuerriegel S, Kohl TA, Ismail N, Omar SV, Smith EG, Buck D, McVean G, Walker AS, Peto TE, Crook DW, Iqbal Z (2015) Rapid antibiotic-resistance predictions from genome sequence data for Staphylococcus aureus and *Mycobacterium tuberculosis*. Nat Commun 6:10063

Brites D, Gagneux S (2015) Co-evolution of *Mycobacterium tuberculosis* and *Homo sapiens*. Immunol Rev 264(1):6–24

Carattoli A (2009) Resistance plasmid families in *Enterobacteriaceae*. Antimicrob Agents Chemother 53(6):2227–2238

Casali N, Nikolayevskyy V, Balabanova Y, Ignatyeva O, Kontsevaya I, Harris SR, Bentley SD, Parkhill J, Nejentsev S, Hoffner SE, Horstmann RD, Brown T, Drobniewski F (2012) Microevolution of extensively drug-resistant tuberculosis in Russia. Genome Res 22(4):735–745

Casali N, Nikolayevskyy V, Balabanova Y, Harris SR, Ignatyeva O, Kontsevaya I, Corander J, Bryant J, Parkhill J, Nejentsev S, Horstmann RD, Brown T, Drobniewski F (2014) Evolution and transmission of drug-resistant tuberculosis in a Russian population. Nat Genet 46(3):279–286
Cegielski JP (2010) Extensively drug-resistant tuberculosis: "there must be some kind of way out of here". Clin Infect Dis 50(Suppl 3):S195–S200
Chambers HF, Deleo FR (2009) Waves of resistance: *Staphylococcus aureus* in the antibiotic era. Nat Rev Microbiol 7(9):629–641
Chen L, Mathema B, Pitout JD, DeLeo FR, Kreiswirth BN (2014) Epidemic *Klebsiella pneumoniae* ST258 is a hybrid strain. MBio 5(3):e01355-01314
Comas I, Borrell S, Roetzer A, Rose G, Malla B, Kato-Maeda M, Galagan J, Niemann S, Gagneux S (2012) Whole-genome sequencing of rifampicin-resistant *Mycobacterium tuberculosis* strains identifies compensatory mutations in RNA polymerase genes. Nat Genet 44 (1):106–110
Falgenhauer L, Waezsada SE, Yao Y, Imirzalioglu C, Kasbohrer A, Rösler U, Michael GB, Schwarz S, Werner G, Kreienbrock L, Chakraborty T, Consortium, R (2016) Colistin resistance gene mcr-1 in extended-spectrum beta-lactamase-producing and carbapenemase-producing Gram-negative bacteria in Germany. Lancet Infect Dis S1473-3099(16): 00009-00008
Cooper BS, Kypraios T, Batra R, Wyncoll D, Tosas O, Edgeworth JD (2012) Quantifying type-specific reproduction numbers for nosocomial pathogens: evidence for heightened transmission of an Asian sequence type 239 MRSA clone. PLoS Comput Biol 8(4):e1002454
Crisóstomo MI, Westh H, Tomasz A, Chung M, Oliveira DC, de Lencastre H (2001) The evolution of methicillin resistance in *Staphylococcus aureus*: similarity of genetic backgrounds in historically early methicillin-susceptible and -resistant isolates and contemporary epidemic clones. Proc Natl Acad Sci USA 98(17):9865–9870
Croucher NJ, Didelot X (2015) The application of genomics to tracing bacterial pathogen transmission. Curr Opin Microbiol 23:62–67
Croucher NJ, Klugman KP (2014) The emergence of bacterial "hopeful monsters". MBio 5(4): e01550-01514
Cui Y, Yang X, Didelot X, Guo C, Li D, Yan Y, Zhang Y, Yuan Y, Yang H, Wang J, Wang J, Song Y, Zhou D, Falush D, Yang R (2015) Epidemic clones, oceanic gene pools, and eco-LD in the free living marine pathogen *Vibrio parahaemolyticus*. Mol Biol Evol 32(6):1396–1410
Cuny C, Köck R, Witte W (2013) Livestock associated MRSA (LA-MRSA) and its relevance for humans in Germany. Int J Med Microbiol 303(6–7):331–337
David MZ, Daum RS (2010) Community-associated methicillin-resistant *Staphylococcus aureus*: epidemiology and clinical consequences of an emerging epidemic. Clin Microbiol Rev 23 (3):616–687
DeLeo FR, Chambers HF (2009) Reemergence of antibiotic-resistant Staphylococcus aureus in the genomics era. J Clin Invest 119(9):2464–2474
DeLeo FR, Otto M, Kreiswirth BN, Chambers HF (2010) Community-associated meticillin-resistant *Staphylococcus aureus*. Lancet 375(9725):1557–1568
DeLeo FR, Chen L, Porcella SF, Martens CA, Kobayashi SD, Porter AR, Chavda KD, Jacobs MR, Mathema B, Olsen RJ, Bonomo RA, Musser JM, Kreiswirth BN (2014) Molecular dissection of the evolution of carbapenem-resistant multilocus sequence type 258 *Klebsiella pneumoniae*. Proc Natl Acad Sci USA 111(13):4988–4993
Didelot X, Bowden R, Wilson DJ, Peto TE, Crook DW (2012) Transforming clinical microbiology with bacterial genome sequencing. Nat Rev Genet 13(9):601–612
Didelot X, Gardy J, Colijn C (2014) Bayesian inference of infectious disease transmission from whole-genome sequence data. Mol Biol Evol 31(7):1869–1879
Dippenaar A, Warren RM (2015) Fighting an old disease with next-generation sequencing. Elife 4: e06782
Drummond AJ, Pybus OG, Rambaut A, Forsberg R, Rodrigo AG (2003) Measurably evolving populations. Trends Ecol Evol 18:481
Earle SG, Wu C-H, Charlesworth J, Stoesser N, Gordon NC, Walker AS, Spencer CCA, Iqbal Z, Clifton DA, Hopkins KL, Woodford N, Smith EG, Imail N, Llewelyn MJ, Peto TE,

Crook DW, McVean G, Walker AS, Wilson D (2016) Identifying lineage effects when controlling for population structure improves power in bacterial association studies. Nature Microbiology 1:e16041

Eldholm V, Monteserin J, Rieux A, Lopez B, Sobkowiak B, Ritacco V, Balloux F (2015) Four decades of transmission of a multidrug-resistant *Mycobacterium tuberculosis* outbreak strain. Nat Commun 6:7119

Eyre DW, Golubchik T, Gordon NC, Bowden R, Piazza P, Batty EM, Ip CL, Wilson DJ, Didelot X, O'Connor L, Lay R, Buck D, Kearns AM, Shaw A, Paul J, Wilcox MH, Donnelly PJ, Peto TE, Walker AS, Crook DW (2012) A pilot study of rapid benchtop sequencing of *Staphylococcus aureus* and *Clostridium difficile* for outbreak detection and surveillance. BMJ Open 2(3):e001124

Farhat MR, Shapiro BJ, Kieser KJ, Sultana R, Jacobson KR, Victor TC, Warren RM, Streicher EM, Calver A, Sloutsky A, Kaur D, Posey JE, Plikaytis B, Oggioni MR, Gardy JL, Johnston JC, Rodrigues M, Tang PK, Kato-Maeda M, Borowsky ML, Muddukrishna B, Kreiswirth BN, Kurepina N, Galagan J, Gagneux S, Birren B, Rubin EJ, Lander ES, Sabeti PC, Murray M (2013) Genomic analysis identifies targets of convergent positive selection in drug-resistant *Mycobacterium tuberculosis*. Nat Genet 45(10):1183–1189

Feil EJ (2015) Toward a synthesis of genotypic typing and phenotypic inference in the genomics era. Future Microbiol 10(12):1897–1899

Ford CB, Lin PL, Chase MR, Shah RR, Iartchouk O, Galagan J, Mohaideen N, Ioerger TR, Sacchettini JC, Lipsitch M, Flynn JL, Fortune SM (2011) Use of whole genome sequencing to estimate the mutation rate of *Mycobacterium tuberculosis* during latent infection. Nat Genet 43 (5):482–486

Fraser C, Hanage WP, Spratt BG (2005) Neutral microepidemic evolution of bacterial pathogens. Proc Natl Acad Sci USA 102(6):1968–1973

Golubchik T, Batty EM, Miller RR, Farr H, Young BC, Larner-Svensson H, Fung R, Godwin H, Knox K, Votintseva A, Everitt RG, Street T, Cule M, Ip CL, Didelot X, Peto TE, Harding RM, Wilson DJ, Crook DW, Bowden R (2013) Within-host evolution of *Staphylococcus aureus* during asymptomatic carriage. PLoS ONE 8(5):e61319

Gordon NC, Price JR, Cole K, Everitt R, Morgan M, Finney J, Kearns AM, Pichon B, Young B, Wilson DJ, Llewelyn MJ, Paul J, Peto TE, Crook DW, Walker AS, Golubchik T (2014) Prediction of *Staphylococcus aureus* antimicrobial resistance by whole-genome sequencing. J Clin Microbiol 52(4):1182–1191

Guerra-Assunção JA, Crampin AC, Houben RM, Mzembe T, Mallard K, Coll F, Khan P, Banda L, Chiwaya A, Pereira RP, McNerney R, Fine PE, Parkhill J, Clark TG, Glynn JR (2015a) Large-scale whole genome sequencing of *M. tuberculosis* provides insights into transmission in a high prevalence area. Elife 4:e05166

Guerra-Assunção JA, Houben RM, Crampin AC, Mzembe T, Mallard K, Coll F, Khan P, Banda L, Chiwaya A, Pereira RP, McNerney R, Harris D, Parkhill J, Clark TG, Glynn JR (2015b) Recurrence due to relapse or reinfection with *Mycobacterium tuberculosis*: a whole-genome sequencing approach in a large, population-based cohort with a high HIV infection prevalence and active follow-up. J Infect Dis 211(7):1154–1163

Gutierrez-Vazquez JM (1956) Studies on the rate of growth of mycobacteria. I. Generation time of *Mycobacterium tuberculosis* on several solid and liquid media and effects exerted by glycerol and malachite green. Am Rev Tuberc 74(1):50–58

Haller S, Eller C, Hermes J, Kaase M, Steglich M, Radonic A, Dabrowski PW, Nitsche A, Pfeifer Y, Werner G, Wunderle W, Velasco E, Abu-Sin M, Eckmanns T, Nübel U (2015) What caused the outbreak of ESBL producing *Klebsiella pneumoniae* in a neonatal intensive care unit, Germany 2009 to 2012? Reconstructing transmission with epidemiological analysis and whole-genome sequencing. BMJ Open 5:e007397

Harris SR, Feil EJ, Holden MT, Quail MA, Nickerson EK, Chantratita N, Gardete S, Tavares A, Day N, Lindsay JA, Edgeworth JD, de Lencastre H, Parkhill J, Peacock SJ, Bentley SD (2010) Evolution of MRSA during hospital transmission and intercontinental spread. Science 327 (5964):469–474

Hasman H, Hammerum AM, Hansen F, Hendriksen RS, Olesen B, Agerso Y, Zankari E, Leekitcharoenphon P, Stegger M, Kaas RS, Cavaco LM, Hansen DS, Aarestrup FM, Skov RL (2015) Detection of mcr-1 encoding plasmid-mediated colistin-resistant Escherichia coli isolates from human bloodstream infection and imported chicken meat, Denmark 2015. Euro Surveill 20(49):1–5

Hawkey PM (1997) Quinolones in sweat and quinolone resistance. Lancet 349(9046):148–149

Holden MT, Hsu LY, Kurt K, Weinert LA, Mather AE, Harris SR, Strommenger B, Layer F, Witte W, de Lencastre H, Skov R, Westh H, Zemlickova H, Coombs G, Kearns AM, Hill RL, Edgeworth J, Gould I, Gant V, Cooke J, Edwards GF, McAdam PR, Templeton KE, McCann A, Zhou Z, Castillo-Ramirez S, Feil EJ, Hudson LO, Enright MC, Balloux F, Aanensen DM, Spratt BG, Fitzgerald JR, Parkhill J, Achtman M, Bentley SD, Nübel U (2013) A genomic portrait of the emergence, evolution and global spread of a methicillin resistant *Staphylococcus aureus* pandemic. Genome Res 23:653–664

Holt KE, Wertheim H, Zadoks RN, Baker S, Whitehouse CA, Dance D, Jenney A, Connor TR, Hsu LY, Severin J, Brisse S, Cao H, Wilksch J, Gorrie C, Schultz MB, Edwards DJ, Nguyen KV, Nguyen TV, Dao TT, Mensink M, Minh VL, Nhu NT, Schultsz C, Kuntaman K, Newton PN, Moore CE, Strugnell RA, Thomson NR (2015) Genomic analysis of diversity, population structure, virulence, and antimicrobial resistance in *Klebsiella pneumoniae*, an urgent threat to public health. Proc Natl Acad Sci USA 112(27):E3574-3581

Jevons MP (1961) Celbenin-resistant staphylococci. Brit Med J 1:124–125

Johnson, AP (2011) Methicillin-resistant *Staphylococcus aureus*: the European landscape. J Antimicrob Chemother 66(Suppl 4):iv43-iv48

Johnson TJ, Hargreaves M, Shaw K, Snippes P, Lynfield R, Aziz M, Price LB (2015) Complete genome sequence of a carbapenem-resistant extraintestinal pathogenic *Escherichia coli* strain belonging to the sequence type 131 H30R subclade. Genome Announc 3(2):e00272-15

Jombart T, Cori A, Didelot X, Cauchemez S, Fraser C, Ferguson N (2014) Bayesian reconstruction of disease outbreaks by combining epidemiologic and genomic data. PLoS Comput Biol 10(1): e1003457

Baym M, Stone LK, Kishony, R (2016) Multidrug evolutionary strategies to reverse antibiotic resistance. Science 351(6268):aad3292

Knight GM, Budd EL, Whitney L, Thornley A, Al-Ghusein H, Planche T, Lindsay JA (2012) Shift in dominant hospital-associated methicillin-resistant *Staphylococcus aureus* (HA-MRSA) clones over time. J Antimicrob Chemother 67(10):2514–2522

Köser CU, Holden MTG, Ellington MJ, Cartwright EJP, Brown NM, Ogilvy-Stuart AL, Hsu LY, Chewapreecha C, Croucher NJ, Harris SR, Sanders M, Enright MC, Dougan G, Bentley SD, Schulz-Trieglaff OB, Smith GP, Peacock SJ (2012) Rapid whole-genome sequencing for investigation of a neonatal MRSA outbreak. N Engl J Med 366:2267–2275

Kutumbaka KK, Han S, Mategko J, Nadala C, Buser GL, Cassidy MP, Beldavs ZG, Weissman SJ, Morey KE, Vega R, Samadpour M (2014) Draft genome sequence of blaNDM-1-positive *Escherichia coli* O25b-ST131 clone isolated from an environmental sample. Genome Announc 2(3):e00462-14

Laabei M, Recker M, Rudkin JK, Aldeljawi M, Gulay Z, Sloan TJ, Williams P, Endres JL, Bayles KW, Fey PD, Yajjala VK, Widhelm T, Hawkins E, Lewis K, Parfett S, Scowen L, Peacock SJ, Holden M, Wilson D, Read TD, van den Elsen J, Priest NK, Feil EJ, Hurst LD, Josefsson E, Massey RC (2014) Predicting the virulence of MRSA from its genome sequence. Genome Res 24(5):839–849

Laabei M, Uhlemann AC, Lowy FD, Austin ED, Yokoyama M, Ouadi K, Feil E, Thorpe HA, Williams B, Perkins M, Peacock SJ, Clarke SR, Dordel J, Holden M, Votintseva AA, Bowden R, Crook DW, Young BC, Wilson DJ, Recker M, Massey RC (2015) Evolutionary trade-offs underlie the multi-faceted virulence of *Staphylococcus aureus*. PLoS Biol 13(9): e1002229

Luo T, Comas I, Luo D, Lu B, Wu J, Wei L, Yang C, Liu Q, Gan M, Sun G, Shen X, Liu F, Gagneux S, Mei J, Lan R, Wan K, Gao Q (2015) Southern East Asian origin and coexpansion of *Mycobacterium tuberculosis* Beijing family with Han Chinese. Proc Natl Acad Sci USA 112 (26):8136–8141

Mathers AJ, Peirano G, Pitout JD (2015) The role of epidemic resistance plasmids and international high-risk clones in the spread of multidrug-resistant Enterobacteriaceae. Clin Microbiol Rev 28(3):565–591

McAdam PR, Templeton KE, Edwards GF, Holden MT, Feil EJ, Aanensen DM, Bargawi HJ, Spratt BG, Bentley SD, Parkhill J, Enright MC, Holmes A, Girvan EK, Godfrey PA, Feldgarden M, Kearns AM, Rambaut A, Robinson DA, Fitzgerald JR (2012) Molecular tracing of the emergence, adaptation, and transmission of hospital-associated methicillin-resistant *Staphylococcus aureus*. Proc Natl Acad Sci USA 109(23):9107–9112

Merker M, Blin C, Mona S, Duforet-Frebourg N, Lecher S, Willery E, Blum MG, Rusch-Gerdes S, Mokrousov I, Aleksic E, Allix-Beguec C, Antierens A, Augustynowicz-Kopec E, Ballif M, Barletta F, Beck HP, Barry CE 3rd, Bonnet M, Borroni E, Campos-Herrero I, Cirillo D, Cox H, Crowe S, Crudu V, Diel R, Drobniewski F, Fauville-Dufaux M, Gagneux S, Ghebremichael S, Hanekom M, Hoffner S, Jiao WW, Kalon S, Kohl TA, Kontsevaya I, Lillebaek T, Maeda S, Nikolayevskyy V, Rasmussen M, Rastogi N, Samper S, Sanchez-Padilla E, Savic B, Shamputa IC, Shen A, Sng LH, Stakenas P, Toit K, Varaine F, Vukovic D, Wahl C, Warren R, Supply P, Niemann S, Wirth T (2015) Evolutionary history and global spread of the *Mycobacterium tuberculosis* Beijing lineage. Nat Genet 47(3):242–249

Morelli MJ, Thebaud G, Chadoeuf J, King DP, Haydon DT, Soubeyrand S (2012) A Bayesian inference framework to reconstruct transmission trees using epidemiological and genetic data. PLoS Comput Biol 8(11):e1002768

Morris D, McGarry E, Cotter M, Passet V, Lynch M, Ludden C, Hannan MM, Brisse S, Cormican M (2012) Detection of OXA-48 carbapenemase in the pandemic clone *Escherichia coli* O25b:H4-ST131 in the course of investigation of an outbreak of OXA-48-producing *Klebsiella pneumoniae*. Antimicrob Agents Chemother 56(7):4030–4031

Munoz-Price LS, Poirel L, Bonomo RA, Schwaber MJ, Daikos GL, Cormican M, Cornaglia G, Garau J, Gniadkowski M, Hayden MK, Kumarasamy K, Livermore DM, Maya JJ, Nordmann P, Patel JB, Paterson DL, Pitout J, Villegas MV, Wang H, Woodford N, Quinn JP (2013) Clinical epidemiology of the global expansion of *Klebsiella pneumoniae* carbapenemases. Lancet Infect Dis 13(9):785–796

Nübel U, Roumagnac P, Feldkamp M, Song JH, Ko KS, Huang YC, Coombs G, Ip M, Westh H, Skov R, Struelens MJ, Goering RV, Strommenger B, Weller A, Witte W, Achtman M (2008) Frequent emergence and limited geographic dispersal of methicillin-resistant *Staphylococcus aureus*. Proc Natl Acad Sci USA 105:14130–14135

Nübel U, Dordel J, Kurt K, Strommenger B, Westh H, Shukla SK, Zemlickova H, Leblois R, Wirth T, Jombart T, Balloux F, Witte W (2010) A timescale for evolution, population expansion, and spatial spread of an emerging clone of methicillin-resistant *Staphylococcus aureus*. PLoS Pathog 6(4):e1000855

Nübel U, Strommenger B, Layer F, Witte W (2011) From types to trees: reconstructing the spatial spread of *Staphylococcus aureus* based on DNA variation. Int J Med Microbiol 301:614–618

Nübel U, Nachtnebel M, Falkenhorst G, Benzler J, Hecht J, Kube M, Brocker F, Moelling K, Buhrer C, Gastmeier P, Piening B, Behnke M, Dehnert M, Layer F, Witte W, Eckmanns T (2013) MRSA transmission on a neonatal intensive care unit: epidemiological and genome-based phylogenetic analyses. PLoS ONE 8(1):e54898

Paterson GK, Harrison EM, Murray GG, Welch JJ, Warland JH, Holden MT, Morgan FJ, Ba X, Koop G, Harris SR, Maskell DJ, Peacock SJ, Herrtage ME, Parkhill J, Holmes MA (2015) Capturing the cloud of diversity reveals complexity and heterogeneity of MRSA carriage, infection and transmission. Nat Commun 6:6560

Peirano G, Schreckenberger PC, Pitout JD (2011) Characteristics of NDM-1-producing *Escherichia coli* isolates that belong to the successful and virulent clone ST131. Antimicrob Agents Chemother 55(6):2986–2988

Petty NK, Ben Zakour NL, Stanton-Cook M, Skippington E, Totsika M, Forde BM, Phan MD, Gomes Moriel D, Peters KM, Davies M, Rogers BA, Dougan G, Rodriguez-Bano J, Pascual A,

Pitout JD, Upton M, Paterson DL, Walsh TR, Schembri MA, Beatson SA (2014) Global dissemination of a multidrug resistant *Escherichia coli* clone. Proc Natl Acad Sci USA 111 (15):5694–5699

Price LB, Johnson JR, Aziz M, Clabots C, Johnston B, Tchesnokova V, Nordstrom L, Billig M, Chattopadhyay S, Stegger M, Andersen PS, Pearson T, Riddell K, Rogers P, Scholes D, Kahl B, Keim P, Sokurenko EV (2013) The epidemic of extended-spectrum-beta-lactamase-producing *Escherichia coli* ST131 is driven by a single highly pathogenic subclone, H30-Rx. MBio 4(6):e00377-00313

Reuter S, Ellington MJ, Cartwright EJ, Koser CU, Torok ME, Gouliouris T, Harris SR, Brown NM, Holden MT, Quail M, Parkhill J, Smith GP, Bentley SD, Peacock SJ (2013) Rapid bacterial whole-genome sequencing to enhance diagnostic and public health microbiology. JAMA Intern Med 173(15):1397–1404

Reuter S, Torok ME, Holden MT, Reynolds R, Raven KE, Blane B, Donker T, Bentley SD, Aanensen DM, Grundmann H, Feil EJ, Spratt BG, Parkhill J, Peacock SJ (2016) Building a genomic framework for prospective MRSA surveillance in the United Kingdom and the Republic of Ireland. Genome Res 26(2):263–270

Richardson JF, Reith S (1993) Characterization of a strain of methicillin-resistant *Staphylococcus aureus* (EMRSA-15) by conventional and molecular methods. J Hosp Infect 25(1):45–52

Robinson DA, Enright MC (2004) Multilocus sequence typing and the evolution of methicillin-resistant *Staphylococcus aureus*. Clin Microbiol Infect 10(2):92–97

Roetzer A, Diel R, Kohl TA, Ruckert C, Nübel U, Blom J, Wirth T, Jaenicke S, Schuback S, Rusch-Gerdes S, Supply P, Kalinowski J, Niemann S (2013) Whole genome sequencing versus traditional genotyping for investigation of a *Mycobacterium tuberculosis* outbreak: a longitudinal molecular epidemiological study. PLoS Med 10(2):e1001387

Sheppard SK, Didelot X, Meric G, Torralbo A, Jolley KA, Kelly DJ, Bentley SD, Maiden MC, Parkhill J, Falush D (2013) Genome-wide association study identifies vitamin B5 biosynthesis as a host specificity factor in *Campylobacter*. Proc Natl Acad Sci USA 110(29):11923–11927

Sintchenko V, Holmes EC (2015) The role of pathogen genomics in assessing disease transmission. BMJ 350:h1314

Snitkin ES, Zelazny AM, Thomas PJ, Stock F, Group NCSP, Henderson DK, Palmore TN, Segre JA (2012) Tracking a hospital outbreak of carbapenem-resistant *Klebsiella pneumoniae* with whole-genome sequencing. Sci Transl Med 4(148):148ra116

Srivastava S, Pasipanodya JG, Meek C, Leff R, Gumbo T (2011) Multidrug-resistant tuberculosis not due to noncompliance but to between-patient pharmacokinetic variability. J Infect Dis 204 (12):1951–1959

Stoesser N, Giess A, Batty EM, Sheppard AE, Walker AS, Wilson DJ, Didelot X, Bashir A, Sebra R, Kasarskis A, Sthapit B, Shakya M, Kelly D, Pollard AJ, Peto TE, Crook DW, Donnelly P, Thorson S, Amatya P, Joshi S (2014) Genome sequencing of an extended series of NDM-producing *Klebsiella pneumoniae* isolates from neonatal infections in a Nepali hospital characterizes the extent of community- versus hospital-associated transmission in an endemic setting. Antimicrob Agents Chemother 58(12):7347–7357

Stoesser N, Sheppard AE, Moore CE, Golubchik T, Parry CM, Nget P, Saroeun M, Day NP, Giess A, Johnson JR, Peto TE, Crook DW, Walker AS, Modernizing Medical Microbiology Informatics G (2015) Extensive within-host diversity in fecally carried extended-spectrum-beta-lactamase-producing *Escherichia coli i*solates: implications for transmission analyses. J Clin Microbiol 53(7):2122–2131

Stoesser N, Sheppard AE, Pankhurst L, De Maio N, Moore CE, Sebra R, Turner P, Anson LW, Kasarskis A, Batty EM, Kos V, Wilson DJ, Phetsouvanh R, Wyllie D, Sokurenko E, Manges AR, Johnson TJ, Price LB, Peto TE, Johnson JR, Didelot X, Walker AS, Crook DW, Modernizing Medical Microbiology Informatics G (2016) Evolutionary history of the global emergence of the *Escherichia coli* epidemic clone ST131. MBio 7(2):e02162

Strommenger B, Bartels MD, Kurt K, Layer F, Rohde SM, Boye K, Westh H, Witte W, De Lencastre H, Nübel U (2014) Evolution of methicillin-resistant *Staphylococcus aureus* towards increasing resistance. J Antimicrob Chemother 69(3):616–622

Temkin E, Adler A, Lerner A, Carmeli Y (2014) Carbapenem-resistant enterobacteriaceae: biology, epidemiology, and management. Ann NY Acad Sci 1323:22–42

Tong SY, Holden MT, Nickerson EK, Cooper BS, Koser CU, Cori A, Jombart T, Cauchemez S, Fraser C, Wuthiekanun V, Thaipadungpanit J, Hongsuwan M, Day NP, Limmathurotsakul D, Parkhill J, Peacock SJ (2015) Genome sequencing defines phylogeny and spread of methicillin-resistant Staphylococcus aureus in a high transmission setting. Genome Res 25 (1):111–118

Uhlemann AC, Dordel J, Knox JR, Raven KE, Parkhill J, Holden MT, Peacock SJ, Lowy FD (2014) Molecular tracing of the emergence, diversification, and transmission of *S. aureus* sequence type 8 in a New York community. Proc Natl Acad Sci USA 111(18):6738–6743

van Dijk EL, Auger H, Jaszczyszyn Y, Thermes C (2014) Ten years of next-generation sequencing technology. Trends Genet 30(9):418–426

Walker TM, Ip CL, Harrell RH, Evans JT, Kapatai G, Dedicoat MJ, Eyre DW, Wilson DJ, Hawkey PM, Crook DW, Parkhill J, Harris D, Walker AS, Bowden R, Monk P, Smith EG, Peto TE (2013) Whole-genome sequencing to delineate *Mycobacterium tuberculosis* outbreaks: a retrospective observational study. Lancet Infect Dis 13(2):137–146

Weber SG, Gold HS, Hooper DC, Karchmer AW, Carmeli Y (2003) Fluoroquinolones and the risk for methicillin-resistant *Staphylococcus aureus* in hospitalized patients. Emerg Infect Dis 9 (11):1415–1422

WHO (2015) Global tuberculosis report. World Health Organization, Geneva

Woerther PL, Burdet C, Chachaty E, Andremont A (2013) Trends in human fecal carriage of extended-spectrum beta-lactamases in the community: toward the globalization of CTX-M. Clin Microbiol Rev 26(4):744–758

Woodford N, Turton JF, Livermore DM (2011) Multiresistant Gram-negative bacteria: the role of high-risk clones in the dissemination of antibiotic resistance. FEMS Microbiol Rev 35(5): 736–755

Worby CJ, Lipsitch M, Hanage WP (2014) Within-host bacterial diversity hinders accurate reconstruction of transmission networks from genomic distance data. PLoS Comput Biol 10(3): e1003549

Wyres KL, Gorrie C, Edwards DJ, Wertheim HF, Hsu LY, Van Kinh N, Zadoks R, Baker S, Holt KE (2015) Extensive capsule locus variation and large-scale genomic recombination within the *Klebsiella pneumoniae* clonal group 258. Genome Biol Evol 7(5):1267–1279

Young BC, Golubchik T, Batty EM, Fung R, Larner-Svensson H, Votintseva AA, Miller RR, Godwin H, Knox K, Everitt RG, Iqbal Z, Rimmer AJ, Cule M, Ip CL, Didelot X, Harding RM, Donnelly P, Peto TE, Crook DW, Bowden R, Wilson DJ (2012) Evolutionary dynamics of *Staphylococcus aureus* during progression from carriage to disease. Proc Natl Acad Sci USA 109(12):4550–4555

Epidemiology of *Staphylococcus aureus* Nasal Carriage Patterns in the Community

Jaishri Mehraj, Wolfgang Witte, Manas K. Akmatov, Franziska Layer, Guido Werner and Gérard Krause

Abstract *Staphylococcus aureus* (*S. aureus*) is a Gram-positive opportunistic pathogen that colonizes frequently and asymptomatically the anterior nares of humans and animals. It can cause different kinds of infections and is considered to be an important nosocomial pathogen. Nasal carriage of *S. aureus* can be permanent or intermittent and may build the reservoir for autogenous infections and cross-transmission to other individuals. Most of the studies on the epidemiology of *S. aureus* performed in the past were focused on the emergence and dissemination of methicillin-resistant *Staphylococcus aureus* (MRSA) in healthcare settings. There are, however, a number of more recent epidemiological studies have aimed at analysing carriage patterns over time in the community settings providing new insights on risk factors for colonization and important data for the development of strategies to prevent infections. This chapter aims to give a review of current epidemiological studies on *S. aureus* carriage patterns in the general community and put them into perspective with recent, yet unpublished, investigations on the *S. aureus* epidemiology in the general population in northern Germany.

J. Mehraj · M.K. Akmatov · G. Krause (✉)
Department of Epidemiology, Helmholtz Centre for Infection Research, Inhoffenstraße 7, 38124 Brunswick, Germany
e-mail: Gerard.Krause@helmholtz-hzi.de

J. Mehraj · G. Krause
Hannover Medical School, Hannover, Germany

W. Witte · F. Layer · G. Werner
The Robert Koch Institute, Wernigerode Branch, Wernigerode, Germany

M.K. Akmatov
TWINCORE Centre for Experimental and Clinical Infection Research, Hannover, Germany

Current Topics in Microbiology and Immunology (2016) 398: 55–88
DOI 10.1007/82_2016_497

Published Online: 28 June 2016

Contents

1 Introduction

Staphylococcus aureus (*S. aureus*) is a widely disseminated colonizer of the skin and mucosa of mammals and birds which is also a frequently observed pathogen in the community and hospital settings. Depending on predispositions of the host affected and on particular virulence-associated traits of the bacterium, *S. aureus* is able to cause a wide range of infections such as impetigo and exfoliative dermatitis and deep seated skin and soft tissue infections (SSTI) such as furunculosis and abscesses, osteomyelitis, pneumonia, sepsis, endocarditis, staphylococcal toxic shock syndrome, or necrotizing fasciitis (Kluytmans et al. 1997; Yu et al. 1986; Dufour et al. 2002; Wertheim et al. 2005a). Severe and invasive infections such as pneumonia and sepsis can have a fatal outcome. A substantial proportion of these infections are of endogenous origin (Wertheim et al. 2005b; von Eiff et al. 2001). In many countries, *S. aureus* is still among the most frequent nosocomial pathogens causing healthcare-associated infections and putting an extra burden on healthcare-associated expenditures (Bereket et al. 2012). By the end of the 1870s, *S. aureus* was identified as an important infectious agent yet its role as a colonizer of healthy humans was only taken into account 50 years later (Williams 1963). In humans and several animal species, it colonizes the skin and mucosae. The anterior nares are the most frequent carriage site, and other less frequently colonized sites are mucosa in the oropharynx, the skin in the frontotemporal region, the axillae, the perineum, and the vagina (Williams 1963; Armstrong-Esther and Smith 1976; Guinan et al. 1982). Whether real gastrointestinal carriage exists or not is still under debate.

Different factors mediate adhesion of *S. aureus* to the epithelium of the anterior nares like cell wall-associated factors and receptors of the epithelial cells. Cell wall-associated factors include glycosylated teichoic acid and clumping factor B, and receptors are cytokeratin-10 and loricin (Winstel et al. 2015; Walsh et al. 2004; Mulcahy et al. 2012). Adhesion molecules on staphylococcus and receptor proteins at the site of infection in the human host are ubiquitous. Although it is biologically plausible that certain bacterial genotypes are more likely associated with persistent carriage than others, robust data on this association are scarce (Eriksen et al. 1995; VandenBergh et al. 1999). There are limited data on genetic dispositions to nasal carriage of *S. aureus* with respect to host compatibility antigens (HLA) and vitamin D receptor polymorphisms (Kinsman et al. 1983; Messaritakis et al. 2014).

Nasal carriage is important for endogenous infections and for transmission to other individuals, as colonization of extra nasal sites often originates from the nasal reservoir (Wertheim et al. 2005b). In this regard, special attention was paid to methicillin-resistant *S. aureus* (MRSA) which has become a public health issue during the past three decades (Köck et al. 2014). This applies to hospital-associated MRSA (HA-MRSA) as well as to community-associated MRSA (CA-MRSA), and livestock-associated MRSA (LA-MRSA).

Already observations in the 1950s and 1960s indicated the importance of the nasal reservoir of *S. aureus* for the development of nosocomial and community-acquired infections (Williams 1963), which is now established by studies based on molecular typing (Chow and Yu 1989; Corbella et al. 1997; Kluytmans et al. 1995; Luzar et al. 1990). Numerous cross-sectional studies revealed that approximately 25–30 % of the healthy individuals are colonized with *S. aureus* (Williams 1963; Casewell and Hill 1986; Kluytmans et al. 1997; Werthheim et al. 2005a; Kolata et al. 2011). *S. aureus* carriage has been extensively studied in healthy adults and in hospitalized patients by numerous studies (Wertheim et al. 2004). The early studies described individuals carrying *S. aureus* in their nares as carriers or non-carriers, dependent upon demonstration of this bacterium at the time of screening (Gould and McKillop 1954). Later longitudinal studies showed that the carrier state changes over time, which led to discrimination of three different *S. aureus* carriage patterns: persistent carriage, transient carriage, and non-carriage (Williams 1963). More recent studies confirm these observations (Eriksen et al. 1995; VandenBergh et al. 1999; van Belkum et al. 2009). However, criteria used to identify these carriage patterns differ among various studies conducted in community settings.

Persistent carriage poses the risk of severe infections in case of predisposing conditions with high case fatality rates (Gupta et al. 2013). Higher colonization density of persistent carriage can enhance transmission of *S. aureus* among individuals in the community. As nasal MRSA carriage is a major risk factor for nosocomial MRSA infections, most of the guidelines and recommendations for prevention of MRSA infections recommend nasal screening at the time of hospital admission (Tacconelli and Johnson 2011; Humphreys 2008). Furthermore, the number of surgical site infections with *S. aureus* acquired in hospitals can obviously be reduced by rapid screening and subsequent decolonization of nasal *S. aureus* carriage on admission (Bode et al. 2010). Thus, active surveillance is an important measure of infection control and prevention (Gupta et al. 2013).

Although risk factors for *S. aureus* nasal carriage patterns have been identified by many of the earlier investigations, more comprehensive knowledge about risk factors associated with *S. aureus* carriage, as well as about drug resistance patterns and typing characteristics of the isolates, would be helpful for targeted screening and decolonization of patients (Köck et al. 2014; McKinnell et al. 2013). Therefore, epidemiological studies focusing on the dynamics of *S. aureus* carriage patterns will be helpful for future designs of infection prevention, intervention measures, and control strategies both in community and in hospital care settings.

2 Molecular Typing of *S. aureus*

Typing of isolates is essential for deeper insights into the dynamics of carriage patterns; in the following, we give a short overview of typing methods.

The clonal relatedness among *S. aureus* strains is determined through many different typing methods (Strommenger et al. 2006; Palavecino 2014). Among the "fragment-based" procedures, *Sma*I macrorestriction pattern analysis (pulsed-field gel electrophoresis: PFGE) has the highest discriminatory power and is considered most reliable and reproducible when performed in a standardized manner (Tenover et al. 1995). Furthermore, amplified fragment length polymorphism (AFLP) is revealed as discriminative and also as reproducible when performed with appropriate fragment separation (Melles et al. 2009). Older genotyping techniques include restriction fragment length polymorphism (RFLP) analysis of the genes which contain repetitive stretches and encode *S. aureus* proteins such as protein A and coagulase (Frenay et al. 1996; Goh et al. 1992).

Macrorestriction pattern analysis suffers from several disadvantages with respect to high workload, sometimes lack of interlaboratory comparability of the results and lack of electronic portability. This was overcome by introduction of sequence-based typing. Multilocus sequence typing is based on polymorphisms in seven housekeeping genes, and these results allowed to discriminate clonal lineages and clonal complexes and give evolutionary insights (Feil et al. 2003; Palavecino 2014). During the past decade, *spa* typing based on polymorphisms in the (variable number tandem repeat region) X-region of the *spa* gene became the most frequently used tool for first-line typing (Frenay et al. 1994). In our study, we used Ridom StaphType software for *spa* sequence analysis. Based upon repeat pattern (BURP) algorithms can be used to attribute *spa* types into *spa* clonal complexes (*spa*-CCs) (Harmsen et al. 2003; Mellmann et al. 2007). With a few exceptions due to homoplasy, *spa* types can be mapped to the corresponding clonal lineages defined by multilocus sequence typing (MLST) to a large extent. Therefore, MLST-based clonal complexes (MLST CC) can be deduced from *spa* types (Strommenger et al. 2006 (see also www.spa.ridom.de/spatypes.shtml)). MRSA acquire resistance to nearly all ß-lactam antibiotics by acquisition of the *mec*A gene that is contained by SCC*mec* elements. Typing of SCC*mec* elements can give first indications on relatedness of MRSA attributed to the same clonal lineage (Palavecino 2014). The analysis of whole genome single nucleotide polymorphisms revealed as extremely fruitful for the studies on the evolution of MRSA clonal lineages and their phylogeography (Holden et al. 2013). Its usefulness for epidemiological typing was also shown (Leopold et al. 2014). Further developments with respect to applicability to larger numbers of isolates and standardized analysis of the results in the sense of definition of "types" will lead to broad use in molecular epidemiology and surveillance.

3 Epidemiological Studies on *S. aureus* Carriage Patterns

Epidemiological studies reporting carriage patterns of *S. aureus* have defined persistent nasal carriage in different ways. Variety of definitions probably has contributed to differences in carriage proportions across the Europe. Generally, carriers were categorized into two main categories as transient or intermittent carriers and persistent or long-term carriers. These categories were explained by the duration of the time period at which a carrier is colonized and by the number and percentage of swabs that are positive for *S. aureus* at different sampling times during the study. Furthermore, persistent carriers were characterized as a special group of individuals participating in colonization studies as indicated from certain characteristics such as clearance of *S. aureus* from nose and having certain host genetic factors (van Belkum et al. 2009; van den Akker et al. 2006).

In previous studies, transient carriage has been reported in ranges from 9 to 69 % and persistent carriage from 9 to 37 % in various non-hospitalized populations (Miller et al. 1962; Hoeffler et al. 1978; Eriksen et al. 1995; VandenBergh et al. 1999; van den Akker et al. 2006; Nouwen et al. 2004; van Belkum et al. 2009; Muthukrishnan et al. 2013). Most of these studies used various methods for analysing *S. aureus* carriage patterns and have reported different estimates of persistent and transient carriage in their study population (details are provided in Table 1). In the following, we will focus on keynote publications in this field.

The study performed in Denmark on healthy individuals in 1992 and 1993 over a 19-month period discriminated between different categories of carriage by means of a carrier index which was defined as the number of positive swabs divided by the number of total swabs for each person. Persons with less than six swabs (seven persons) were excluded from the analysis. For persistent carriers, this index was between 0.9 and 1.0, for transient carriers between 0.5 and 0.8, for occasional carriers between 0.1 and 0.4, and 0 for non-carriers. In 12 swabbing rounds, a median of 14 swabs (range 6–21) was obtained per participant. There were differences in the interval of swabbing and correspondingly the number of swabs (ranging from 6 to 21) taken from each individual. Among the 104 persons enrolled in this study, 14.4 % were categorized as persistent carriers, 16.3 % as intermittent carriers, 52.9 % as occasional carriers, and 16.3 % as non-carriers (Eriksen et al. 1995).

Of particular interest are the results from two consecutive studies performed in the Netherlands in 1988 and 1995. In the first study on carriage among 91 healthcare workers, 10–12 nasal swabs were taken weekly over a three-month period and individuals positive for *S. aureus* in ≥ 80 % swabs were classified as persistent carriers. The study reported 36 % persistent carriers, 17 % transient carriers, and 47 % non-carriers by using carrier index of 0.8 for defining persistent

Table 1 Persistent carriage proportions reported among non-hospitalized individuals in previous longitudinal studies

Study settings	Sample size and study population	Number of swabs per participant	Swab collection interval and follow-up duration	Persistent carrier proportion	Transient carrier proportion	Non-carrier proportion	Study period, other main observation	References
Oxford, United Kingdom	1123 adults from general practices and students	14 swabs	Bimonthly for up to 36 months (median two years)	33 %	46 %	21 %	Dec 2008–Dec 2009, *S. aureus* acquisition and loss observed	Miller et al. (2014)
Florida, USA	109 healthy individuals at University of Central Florida	Median 4 swabs (range 2–18)	varying periods of swab collection in three years	24 %	32 %	44 %	61 carriers and 48 non-carriers—were followed longitudinally	Muthukrishnan et al. (2013)
Saint-Etienne, France	90 healthy healthcare workers	7 swabs	On days 0, 2, 4 and weekly on day 7, 15, 23, 31 in five weeks	24 %	15 %	61 %	Mar–Apr 2010, different algorithm developed for defining persistent carriage	Verhoeven et al. (2012)
Rotterdam, The Netherlands	72 healthy individuals	2 swabs	weekly for 2 weeks	33 %	14 %	53 %	–	Manenschijn et al. (2012)
Rotterdam, The Netherlands	51 volunteers	6 swabs	swabs at 1, 2, 4, 8, 16, and 22 weeks after inoculation	24 %	47 %	29 %	Participants were artificially innoculated and followed up for loss of carriage	van Belkum et al. (2009)

(continued)

Table 1 (continued)

Study settings	Sample size and study population	Number of swabs per participant	Swab collection interval and follow-up duration	Persistent carrier proportion	Transient carrier proportion	Non-carrier proportion	Study period, other main observation	References
Rotterdam, The Netherlands	3851 elderly persons (2929 with complete data)	2 swabs	weekly for 2 weeks	18 % (19 %)	9 % –	73 % (81 %)	Apr 1997–Dec 1999, (no intermittent carriers in complete data)	van den Akker et al. (2006)
Rotterdam, The Netherlands	51 healthy persons (derivation cohort)	12 swabs	weekly for 12 weeks	29 %	31 %	39 %	(a) Sep 1995–Mar 1996, Swabs taken by physicians	Nouwen et al. (2004)
	106 healthy persons (validation cohort)	12 swabs	weekly for 12 weeks	30 %	16 %	54 %	(b) Oct 1997–Apr 1998	
Rotterdam, The Netherlands	91 staff members of University hospital	12 swabs in 1988 and 1 swab in 1995	weekly for 12 weeks and 1 after 8 years	36 %	17 %	47 %	1988 and 1995, Persistent carriers followed after 8 years	Van denBergh et al. (1999)
Copenhagen, Denmark	104 staff members, laboratory technicians, academics, secretory	median 14 swabs (range 6–21)	- over 19 month period	14 %	69 % (16 % intermittent and 53 % occasional)	17 %	Jan 1992 and Aug 1993	Eriksen et al. (1995)

(continued)

Table 1 (continued)

Study settings	Sample size and study population	Number of swabs per participant	Swab collection interval and follow-up duration	Persistent carrier proportion	Transient carrier proportion	Non-carrier proportion	Study period, other main observation	References
Fukuoka, Japan	120 outpatients and 51 students and 12 laboratory staff	12–24 swabs from volunteers and 4–6 times from outpatients	Volunteers 1–2/month for 1 year, outpatients once every 4–6 months for 2 years	12 % students, 21 % outpatients	19 % students, 26 % outpatients	69 % students, 53 % outpatients	Volunteers (Apr 1990–May 1991) and outpatients (Apr 1990–May 1992)	Hu et al. (1995)
Ife, Nigeria	50 pharmacy students	maximum 20 swabs per person	Biweekly for 15 months	26 %	28 % (16 % sporadic non-carrier and 12 % intermittent)	46 %		Lamikanra and Olusanya (1988)
Cologne, Germany and Krakow, Poland	261 nursing students and laboratory personnel	6 swabs	Weekly for 6 weeks	9 %	20 %	71 %	Combined results of Germany and Poland	Hoeffler et al. (1978)
	103 students, 71 laboratory personnel	–	–	6 %	25 %	69 %	(a) Cologne, Germany	
	36 students, 51 laboratory personnel	–	–	14 %	12 %	74 %	(b) Krakow, Poland	
Guildford, Surrey, UK	50 university personnel	average 27 (range 9–30)	Weekly for 30 weeks	34 %	64 %	2 %	1970–1972	Armstrong-Esther and Smith (1976)

(continued)

Table 1 (continued)

Study settings	Sample size and study population	Number of swabs per participant	Swab collection interval and follow-up duration	Persistent carrier proportion	Transient carrier proportion	Non-carrier proportion	Study period, other main observation	References
Salt lake city, Utah, US	127 hospital staff and medical students	5 swabs or more	Biweekly for 10 weeks to 6 years	18 %	39 % (15 % intermittent, 24 % sporadic)	43 %	Apr 1962–Jun 1968	Maxwell et al. (1969)
Different places, UK	515 army recruits	2 swabs	7–8weeks in 7- to 8-week follow-up	37 %	24 %	39 %	Sep 1959–Jul 1960 At 61 different places	Miller et al. (1962)
Leyden, The Netherlands	243 family members of patients	6–7 swabs	Biweekly, 8-week follow-up	25 %	25 %	50 %	Jan–Dec 1955	Goslings and Buchli (1958)
Edinburgh, Scotland	520 medical students	12–30 swabs or more	Weekly for three months to one year	24 %	57 %	19 %	1951–1953 Medical student of three successive classes	Gould and Mckillop (1954)

carriage (VandenBergh et al. 1999). In the second follow-up after eight years, the persistent carriage was observed in 71 % of the same participants (12/17) who were identified as persistent carriers in the first phase of the study.

Using the results of this study in the Netherlands as "derivation cohort", a second two-phased study was performed in the same country which included the quantification of colony-forming units per swab. This study was aimed to discriminate between persistent, intermittent, or non-carriers by means of the quantity of colonization and thus to minimize the number of swabs ("culture rule" Nouwen et al. 2004). The first study phase was based on 12 quantitative cultures of swabs from 51 healthy volunteers. Individuals with 11 or 12 cultures positive for *S. aureus* were classified as persistent carriers, and those with negative results in all cultures were classified as non-carriers. All others were classified as intermittent carriers. By means of logistic regression and receiver operating characteristic (ROC) curves, a rule based on the quantities of colonization ("culture rule") was derived. In a second phase, this culture rule was subsequently validated in 106 participants of an ongoing study in 3882 elderly persons, again with the use of 12 quantitative nasal cultures taken during a three-month period. The positive predictive value for persistent carriage by the use of two consecutive positive culture results was 79 % in both cohorts. The number of positive culture results combined with the quantity of *S. aureus* in these cultures provided the best differentiation between persistent and intermittent or non-carriers. Area under the ROC curve was 0.98 (95 % confidence interval [CI], 0.95–1.0) for the derivation cohort and 0.93 (95 % CI, 0.88–0.99) for the validation cohort, by using two culture results (Nouwen et al. 2004). The authors concluded that combining qualitative and quantitative results of the two nasal swab cultures can accurately predict the persistent *S. aureus* carriage state.

This rule was used later on in a study in Rotterdam, Netherlands, on 51 volunteers between the ages of 18–65. The study aimed for a deeper understanding of carriage types based on eradication of existing colonization by topical application of muprirocin. This antibiotic is preferentially used for nasal decolonization (Casewell and Hill 1986). This study used subsequent inoculation with a mixture of strains attributed to different clonal lineages and follow-up of recolonization (van Belkum et al. 2009). Nasal swabs were taken at 1, 2, 4, 8, 16, and 22 weeks post-inoculation. Among the participants, 24 % were persistent carriers, 47 % transient carriers, and 29 % non-carriers. The median survival of *S. aureus* was more than 154 days among persistent carriers compared to 14 days among intermittent carriers and four days among non-carriers. Interestingly, this study found that persistent carriers reselected their autologous strain from the inoculum. In addition, among persistent carriers, immunoglobulin (Ig) A and G levels were different from those in transient and non-carriers (van Belkum et al. 2009). These findings underlined that *S. aureus* carriers can be divided into at least two types of nasal carriers which include persistent carriers and others.

An algorithm based on one or two nasal swab samples and quantitative culture was also reported from France (Verhoeven et al. 2012). A total of seven swabs were obtained from 90 healthcare workers in Saint-Etienne. First three swabs were collected every two days (days 0, 2, 4) and thereafter weekly (days 7, 15, 23, 31) for a five-week duration. Participants with ≥ 80 % positive swabs were considered persistent carriers and all with negative culture non-carriers. Based on quantitative results, an algorithm using one or two sampling times was built. Carriage was classified as persistent if the first sample yielded 10^3 colony-forming units per swab or $>10^2$ CFU/swab and $<10^{3/}$swab when a second sample was taken. Sensitivity was 83.1 % and specificity 95.6 % in case of using one sampling. In cases of two samplings, sensitivity and specificity were increased to 95.5 and 94.9 %, respectively (Verhoeven et al. 2012). Verhoeven and colleagues considered this algorithm accurate to identify persistent nasal carriers of *S. aureus*. They identified 24 % persistent carrier, 15 % transient carriers, and 61 % non-carriers.

Another interesting study on carriage patterns conducted in Florida also used carrier index and included *spa* typing in their analysis. A total of 109 healthy individuals including 61 carriers and 48 non-carriers were followed up for a period of three years. Swabs were collected at varying time intervals in the period of three years with median 4 swabs (range 2–18). All persons with carrier index of exactly 0 and 1 were defined as persistent and non-carriers, respectively, while participants with carrier index between 0 and 1 were considered intermittent carriers in this study. They observed 24 % persistent carriers, 32 % transient carriers, and 44 % non-carriers. However, there was no consistency observed in their sample collection procedures as samples were collected at different time points and in varying quantity (Muthukrishnan et al. 2013).

A more recent study conducted at Oxfordshire, UK, included typing of the isolates into the definitions of persistent carriage (Miller et al. 2014). General practitioners recruited 1123 adult participants, and *S. aureus* nasal carriage was observed in 571 individuals from primary care and was investigated bimonthly for up to a 36-month time period. In this study, 57 % of the participants carried *S. aureus* consistently, and among this group, 49 % carried isolates exhibiting the same *spa* type over the two-year study period. Highly transient carriage with new *S. aureus* acquisitions was observed for a median of two years. As with an increasing length of follow-up, the proportion of long-term carriers was continuously declined throughout the study. Truly, persistent carriage seemed to be unlikely in this study population.

Complementary to this approach, the Helmholtz Center for Infections Research (HZI) conducted a prospective cohort study to determine carriage patterns in the general population of Braunschweig, a city with 246,742 inhabitants in northern Germany (Akmatov et al. 2014; Mehraj et al. 2014). The HZI randomly selected 2026 male and female inhabitants between the ages of 20–69, through the resident's registration office of the city, and asked them to send a self-collected nasal swab to the HZI each month. The HZI applied user instructions for taking the swab and respective validation of the self-collection procedure as reported previously (Akmatov et al. 2014). The HZI observed 33 % transient carriers, 8 % persistent carriers, and 59 % non-carriers among 360 participants in our study who provided all seven self-collected nasal swabs in the period from July 2012 to January 2013 (unpublished results). Of 2520 swabs from 360 participants processed, *S. aureus* was isolated in 564 and MRSA in 16 samples. Monthly point prevalence estimates ranged from 19 to 26 % for methicillin-sensitive *S. aureus* (MSSA) and 0.3 to 1.4 % for MRSA without indication of a trend. Of the 360 participants, 215 (60 %) were negative in all swabs and 145 (40 %) positive for *S. aureus* at one or more time points (unpublished results).

In this study, carrier index 0 described non-carriers (all of the 7 samples taken monthly were negative for *S. aureus*), and carrier index 1 means that all of the 7 samples were positive. Transient carriers had scores between 0 and 1. The HZI also used one category as "carrier" which includes all persistent and transient carriers regardless of the number of positive swabs. The HZI checked all seven swabs for persistence. It is recognized previously that more than or equal to 7 swabs are required for differentiating non-carriers from transient carriers (Nouwen et al. 2004; Verhoeven et al. 2012).

The large proportion of *S. aureus* carriers with a monthly average of 23 % observed in the HZI study has also to be expected for individuals admitted to healthcare settings even if they do not belong to a specific high-risk group. Forty percent of the participants of this study carried *S. aureus* at different time points over the six months. Screening of patients at hospital admission is important for early detection of introduction of MRSA into hospitals (McKinnell et al. 2013; Köck et al. 2014). These observations raise concern about the negative predictive value of screening results from sampling at only one time. Previous studies were mostly conducted among populations associated with healthcare settings, which may explain why persistent carriage in our study population (8 %) was lower as compared to previous estimates. The MRSA carriage among the study participants was comparably low ranging from 0.3 to 1.3 % at each time of sampling and also lower in comparison with more recent data from Germany for MRSA screening of patients admitted to hospitals, e.g. 2.8 and 3.1 %, which is again likely due to the fact that the HZI study population was less likely to have a selection bias for carriage (Reich-Schupke et al. 2010; Herrmann et al. 2013).

4 Risk Factors Associated with *S. aureus* Carriage

Basically, we have to discriminate between (i) general preconditions such as male sex, increasing age, alcoholism, and diabetes mellitus, (ii) healthcare-related exposures such as hospital stay, surgery, and antibiotic intake, and (iii) occupational exposition (healthcare workers, farmers, veterinarians). As already mentioned above, there are very few studies which analysed risk factors associated with specific carriage patterns and especially in the community. Those few studies reported age, sex, and job categories but did not report details of other potential clinical risk factors such as allergies, diabetes, animal contact, previous hospitalization or surgery, and antibiotic intake.

In one study conducted in Guildford, Surrey, UK, transient carriage was observed frequently in the children of age group 10–19 years (Armstrong-Esther and Smith 1976). Some studies observed no difference in age, profession, or job category (VandenBergh et al. 1999; Eriksen et al. 1995). Miller et al. (2014) also could not identify particular risk factors of persistent carriage in a large longitudinal study on *S. aureus* carriage. With respect to sex, one study performed in Europe reported a higher proportion of persistent carriage among male participants (Eriksen et al. 1995), while some studies performed in Nigeria, Africa, and UK did not observe differences in carriage proportions among male and females (Lamikanra and Olusanya 1988; Armstrong-Esther and Smith 1976). In other general preconditions, smoking status was associated with reduced carriage in some cross-sectional studies (Qu et al. 2010; Herwaldt et al. 2004; Olsen et al. 2012).

In order to investigate risk factors for different carriage patterns, we asked the participants to complete a questionnaire on demographic characteristics as well as selected exposures in the last 12 months prior to the study as described previously (Mehraj et al. 2014). Upon completion of the study, we also requested participants to complete a questionnaire on exposures during the last six months. We observed that male person was significantly more frequently colonized as compared to female participants and that colonization proportions were high in younger participants. Persistent carriage was associated with male gender, presence of allergies, and occupational animal contact, whereas transient carriage was associated with male sex, antibiotic intake, and young age.

Previous studies indicated that antibiotic intake may increase the proportion of transient carriage, as only temporarily suppresses *S. aureus* colonization (van Belkum et al. 2009; Miller et al. 2014). Suppression of colonization can occur through local antiseptics such as octenidine and lavasept, but only few systematic antibiotics such as rifampicin eradicate *S. aureus* carriage. In our study, the association of occupational animal contact with persistent carriage is an important observation. MRSA transmission from livestock to farmers as well as veterinarians and from horses to veterinary personnel is well known (van Loo et al. 2007). Studies performed at pig farms in the Netherlands and Germany reported nasal colonization with LA-MRSA CC398 for more than 80 % of the farmers. Co-colonization with methicillin-susceptible *S. aureus* (MSSA) was rare (Cuny et al. 2009; van Cleef et al. 2011a). This observation suggests frequent replacement of the original colonizer by LA-MRSA CC398, probably due to heavy exposure to MRSA-contaminated dust particles in the staples. It was also observed that when exposure to livestock was interrupted (e.g. by holidays), carriage persisted in about 50 % of the participants. Cases of colonization and infections in humans in North Rhine Westphalia in Germany without occupational exposure such as farms and slaughter houses as well as veterinary care centres suggest establishing of LA-MRSA CC398 as nasal colonizer in the community (Deiters et al. 2015). There are similar observations from other European countries (Cuny et al. 2015). As LA-MRSA CC398 is not less virulent than human-adapted *S. aureus* in general (isolates from SSTI and from septicaemia, (Becker et al. 2015), its role as nasal colonizer independent from livestock exposure needs further attention.

As far as MSSA transmission between livestock and humans is concerned, there is limited data available (Osadebe et al. 2013). The isolates from participants of our study with occupational animal contact exhibited *spa* type t091 which attribute them to clonal lineage ST7. MRSA ST7 was recently reported from chicken and chicken meat in Poland (Krupa et al. 2014). Further studies of MSSA isolates from livestock in northern Germany would be desirable for a better interpretation of this finding. Continuous exposure to high-risk environments such as working in hospitals, nursing care, and occupational animal contact (as observed in our study) may sustain persistent carriage by the host (Humphreys 2008; Garcia-Graells et al. 2013). Transmission of *S. aureus* between pet animals and their owners has also been demonstrated. However, the prevalence of *S. aureus* colonization of pet animals seems to be low as shown by studies in Germany and in the USA (Walther et al. 2012; Kottler et al. 2010).

5 *Spa* Types and Clonal Complexes of *S. aureus* Among Transient and Persistent Carriers

Generally, a diversity of strain types has been observed in community-based studies previously although the population of *S. aureus* colonizing healthy humans is rather clonal (Day et al. 2001; Feil et al. 2003; Sakwinska et al. 2009). Previous studies observed clonal complexes (CC) 5, CC8, CC15, CC22, CC25, CC30, and CC45 in the isolates of nasal colonization from different continents (Ko et al. 2008; Melles et al. 2008; Ruimy et al. 2008; Day et al. 2001; Feil et al. 2003; Sakwinska et al. 2009; Rolo et al. 2012). Other clonal complexes such as CC7, CC9, CC12, CC59, and CC121 were rarely found in previous studies. A population analysis of *S. aureus* nasal colonizers by means of molecular typing is especially meaningful for recognizing the emergence of strains, with particular virulence and antibiotic resistance traits as permanent colonizer in the community. Furthermore, changes of colonization associated with antibiotic therapy of the carrier and with exposure of humans to particular environments such as the nosocomial setting and livestock farming need to be explored.

The dynamics of nasal carriage can be influenced by the population structure of the colonizer with respect to clonal complexes as shown by a recently reported study in Oxfordshire, UK (Miller et al. 2014). Isolates from this study were attributed to CC 30, CC15, CC22, and CC8. *S. aureus* acquisition rates were higher for CC 15 or CC8 compared to other CCs. CC22 (MRSA) was more prevalent in long-term carriers than in intermittent carriers. It was reported that carriage of different *spa* types/CCs was more frequent in long-term carrier individuals. Persistent carriers were more likely to carry CC22 and less likely to carry particularly CC15 than intermittent carriers. Slow loss of isolates exhibiting particular *spa* types was noted among long-term *S. aureus* carriers, over a 4- to 6-month time period. Extending the follow-up period led to observation of decreasing persistent carriage as carriers lost the strain or did not carry the same strain over time.

The authors observed that pre-existing colonization did not affect *S. aureus* loss in general; co-carriage of isolates attributed to different clonal lineages was significantly associated with subsequent loss of the original strain. This observation underlines the highly dynamic nature of carried populations and that nasal competition of particular strains is an important factor in co-carriage in vivo. In this study, 15 % of long-term carriers at species level did not carry the same *spa* type consistently; in a study from Switzerland, this was observed for 8 % of the persistent carriers.

In view of the results from our study mentioned above, data from an investigation on temporal stability of *S. aureus* colonization of 267 healthy individuals performed in Switzerland which was based on two times of sampling (Sakwinska et al. 2010) should be interpreted with caution. At the time of the first sampling, 87 of the 89 carriers were colonized with the one strain, two carried two strains. After on the average 6–15 months later, 53 individuals (60 %) were still colonized with the same strain, 5 (6 %) carried a different strain, 1 (1.1 %) was colonized with a different strain, and 2 (2.2 %) carried an additional strain; swabs from 27 (30 %) were negative for *S. aureus*. The authors observed changes of the *spa* types in 6 from the above-mentioned 53 participants (one due to a point mutation, 5 due to repeat deletion). These isolates exhibited the same AFLP pattern and MLST; attribution to MLST-based clonal complexes did not change. We have to consider these results when *spa* typing is used in longitudinal studies on *S. aureus* colonization.

Co-colonization and replacement of the original colonizer strain is a significant factor if the newcomer has particular virulence and antibiotic resistance traits such as CA-MRSA. This might apply to the success of CA-MRSA USA 300 in the United States of America for which nasal colonization was reported for a substantial number of patients who were affected by deep SSTI (Hidron et al. 2005).

The situation seems to be different for *S. aureus* isolates from SSTI attributed to CC121 and containing *luk*-PV (coding for the Panton–Valentine toxin) which represent a particular "pathotype" (Kurt et al. 2013). This subpopulation is rare among nasal isolates in general, but in cases of nasal colonization, this reservoir is associated with recurrent furunculosis (Demos et al. 2012), and decolonization is important for the treatment (Davido et al. 2013).

Clonal complexes can share different proportion among carriage isolates in different countries. There is a relatively little number of CCs which are worldwide disseminated. Occasionally, clones may expand locally. A comparison of isolates from nasal colonization from four countries France, Algeria, Moldova, and Cambodia revealed global dissemination of certain clones as CC30, CC45, CC121, CC15, CC5, and CC8 were observed in the samples of all four countries. Previous reports have noted that CC30 frequently occurs in 20–25 % of the isolates throughout the Europe (Ruimy et al. 2009, 2010).

Differences between European countries were shown by a large cross-sectional study on *S. aureus* nasal carriage by den Heijer et al. (2013). That study was carried out in the nationwide family doctor network APRES in Austria, Belgium, Croatia, France, Hungary, Spain, Sweden, the Netherlands, and the UK (den Heijer et al. 2013). A total of 32,206 patients were recruited by family doctors in those countries. Identical or related *spa* types were exhibited by isolates from several patients attending several practices in each country. They identified 53 different *spa* types from 91 MRSA isolates, and t002 and t008 were the most prevalent among them. In few countries, one cluster predominated, while in majority of the countries, heterogeneity remained prominent even after clustering strains into *spa*-CCs. More than 60 % of the MRSA strains were clustered in *spa* clonal complex 022/032 (MRSA clonal complex 22 including EMRSA-15) in UK, and more than half of the MRSA belonged to *spa* clonal complex 008 (corresponds to MRSA clonal complex 8) in France. These findings correspond to observations from a previous study on geographic distribution of *spa* types among *S. aureus* isolates from invasive infections in Europe (Grundmann et al. 2010).

LA-MRSA CC398 is more frequent as nasal colonizer and infectious agent in countries with conventional livestock farming. Within these countries, their proportion among all MRSA from humans is also significantly higher in areas with high livestock density (van Rijen et al. 2008; van Cleef et al. 2011a, b; Köck et al. 2013).

We also observed differences in *spa* type distribution among isolates from transient and persistent carriers from our own study samples. From 145 carrier participants, 580 *S. aureus* strains were isolated at different time points during the study. This includes 197 strains from 28 persistent carriers and 383 strains from 117 transient carriers. A total of 262 isolates were selected for *spa* typing which included 29 % (57/197) isolates from persistent carriers and 54 % (205/383) from transient carriers. We randomly selected up to three isolates from persistent carriers and did not find changes of the *spa* types. In the selection of strains from transient carriers, we assumed that those participants who lost the carriage and become positive again for *S. aureus* carriage may have acquired a new strain. Therefore, we selected strains for *spa* typing before and after every negative swab culture. We selected all *S. aureus* isolates from MRSA carriers in order to check which strain types replaced the carriage of MRSA strains. Among 12.8 % (11/86) participants who were checked at least two times in the whole study period, we observed different *S. aureus spa* types at different time points during the study as shown in Table 2.

We applied BURP analysis in 157 isolates after removing the duplicate isolates (in case of demonstration of the same type in the same individual). Among them, 82 different MSSA *spa* types were found that clustered into 12 distinct groups. Four clusters had no founder. Among 157 isolates, 19 *spa* types (23.1 %) could not be clustered into a *spa*-CC and classified as singletons. We observed *spa* types belonging to CC45, CC15, CC30, CC8, and CC22, which are in line with similar observations on *S. aureus* carriage patterns in Europe (Miller et al. 2014; Votintseva et al. 2014; Sangvik et al. 2011). Most of MSSA *spa* types 14 (17.1 %) belonged to MLST CC45 and 11 (13.4 %) to MLST CC15.

Among the MSSA, we also observed t091, t021, and t008 *spa* types among participants who reported occupational animal contact.

Of particular interest is the finding of MSSA exhibiting *spa* type t571 (CC398) from one of our study participants who was positive for *S. aureus* carriage in six out of seven swabs. This *spa* type is indicative for the ancestral MSSA subpopulation from which LA-MRSA CC398 has been evolved (Uhlemann et al. 2012a). MRSA t571 and CC398 are reported from clusters of severe invasive infections in the USA, France, and Belgium. Our observation is in accordance with a previous report

Table 2 Distribution of *Staphylococcus aureus spa* types, *spa* clonal complexes, and MLST clonal complexes during time period of six months among study participants of Braunschweig, northern Germany

No. of persons	Jul 12	Aug 12	Sep 12	Oct 12	Nov 12	Dec 12	Jan 13	*S. aureus* isolated	*spa* types obtained	*spa*-CC	MLST CC
P1	t772							7	1	*spa*-CC 015	CC 45
P2	t275	t275		t275				7	3	*spa*-CC 012	CC 30
P3	t242	t242						7	2	*spa*-CC 002	CC 5
P4	t706	t706						7	2	*spa*-CC 330/065	CC 45
P5	t2275	t2275		t2275				7	3	*spa*-CC 330/065	CC 45
P6	t091	t091						7	2	*spa*-CC 499	CC 7
P7	t012							7	1	*spa*-CC 012	CC 30
P8	t159	t159	t159					7	3	Singletons	CC 121
P9	t160							7	1	No founder	ST 12
P10	t091			t091				7	2	*spa*-CC 499	CC 7
P11	t2313	t2313						7	2	Singletons	
P12	t018							7	1	*spa*-CC 012	CC 30
P13	t084			t084				7	2	*spa*-CC 499	CC 15
P14	t073	t073		t073				7	3	*spa*-CC 015	CC 45
P15	t056							7	1	No founder	ST 101
P16	t254			t254				7	2	*spa*-CC 499	CC 15
P17	t10983							7	1	Singletons	
P18	t1862	t1862						7	2	*spa*-CC 005	CC 22
P19	t160	t160		t160				7	3	No founder	ST 12
P20	t002	t002						7	2	*spa*-CC 002	CC 5
P21	t2032							7	1	*spa*-CC 304	CC 8
P22	t012	t012						7	2	*spa*-CC 012	CC 30
P23	t084							7	1	*spa*-CC 499	CC 15
P24	t021			t021				7	2	*spa*-CC 012	CC 30
P25	t091							7	1	*spa*-CC 499	CC 7
P26	t6311			t6311				7	2	*spa*-CC 330/065	CC 45
P27	t2032							7	1	*spa*-CC 304	CC 8
P28*	t032, t330	t008	t330	t330	t330	t330	t330	7	7	*spa*-CC 005, 304, 330/065	CC 22, CC 8, CC 45
P29	t550							6	1	*spa*-CC 015	CC 45
P30	t005	t005						6	2	*spa*-CC 005	CC 22
P31	t13588							6	1	No founder	ST 101
P32	t026			t026				6	2	Excluded	
P33	t1977						t1977	6	2	Excluded	
P34	t1433	t1433					t1433	6	3	*spa*-CC 005	CC 22
P35	t021						t021	6	2	*spa*-CC 012	CC 30
P36	t13383						t13383	6	2	*spa*-CC 015	CC 45
P37*	t3741				t13450			6	2	No founder, Singletons	
P38	t5337						t5337	6	2	Singletons	
P39	t1057			t1057				6	2	Singletons	
P40	t021			t021				6	2	*spa*-CC 012	CC 30
P41	t6115			t6115				6	2	Excluded	
P42	t1925		t1925					6	2	No founder	CC 23
P43	t13587		t13587					6	2	Singletons	
P44	t012						t012	6	2	*spa*-CC 012	CC 30
P45		t571						6	1	Singletons	CC 398
P46	t032	t032	t032	t032	t032			5	5	*spa*-CC 005	CC 22
P47	t360			t360			t360	5	3	*spa*-CC 499	CC 15
P48	t1541					t1541		5	2	Singletons	
P49*	t002				t2167			5	2	*spa*-CC 002	CC 5
P50	t2277			t2277		t2277		5	3	*spa*-CC 015	CC 45
P51	t012			t012			t012	5	3	*spa*-CC 012	CC 30
P52	t091			t091			t091	5	3	*spa*-CC 499	CC 7
P53	t032			t032				5	2	*spa*-CC 005	CC 22
P54		t1313		t1313				5	2	No founder	
P55		t346		t346				5	2	*spa*-CC 499	CC 15
P56*		t008		t159				5	2	*spa*-CC 304, Singletons	CC 8, CC 121
P57		t11462						5	1	Singletons	

(continued)

Table 2 (continued)

P58		t056					t056	5	2	No founder	ST 101
P59		t13384					t13384	5	2	*spa*-CC 499	CC 15
P60				t091				5	1	*spa*-CC 499	CC 7
P61	t012							4	1	*spa*-CC 012	CC 30
P62	t7088	t7088	t7088					4	3	Singletons	
P63	t11363	t11363		t11363				4	3	*spa*-CC 304	CC 8
P64	t032	t032			t032	t032		4	4	*spa*-CC 005	CC 22
P65	t091				t091			4	2	*spa*-CC 499	CC 7
P66*	t015				t550			4	2	*spa*-CC 015	CC 45
P67	t019				t019		t019	4	3	*spa*-CC 012	CC 30
P68	t304	t304		t304		t304		4	4	*spa*-CC 304	CC 8
P69	t005		t005			t005		4	3	*spa*-CC 005	CC 22
P70	t018		t018			t018		4	3	*spa*-CC 012	CC 30
P71	t025			t025		t025	t025	4	4	*spa*-CC 005	CC 22
P72	t026			t026				4	2	Excluded	
P73		t091			t091			4	2	*spa*-CC 499	CC 7
P74*		t002		t1094			t1094	4	3	*spa*-CC 002	CC 5
P75*			t008		t3283			4	2	*spa*-CC 304, Singletons	CC 8
P76			t084			t084		4	2	*spa*-CC 499	CC 15
P77				t008				3	1	*spa*-CC 304	CC 8
P78	t4508							3	1	*spa*-CC 499	CC 15
P79	t021			t021				3	2	*spa*-CC 012	CC 30
P80	t015			t015				3	2	*spa*-CC 015	CC 45
P81	t132				t132			3	2	Excluded	
P82	t012				t012			3	2	*spa*-CC 012	CC 30
P83	t491			t491				3	2	*spa*-CC 499	CC 15
P84	t368			t368				3	2	*spa*-CC 499	CC 15
P85	t537		t537					3	2	Singletons	
P86	t1265		t1265		t1265			3	3	*spa*-CC 002	CC 5
P87			t4308			t4308		3	2	Singletons	
P88				t5804			t5804	3	2	*spa*-CC 499	CC 15
P89				t065		t065		3	2	*spa*-CC 330/065	CC 45
P90*					t1094		t002	2	2	*spa*-CC 002	CC 5
P91				t159				2	1	Singletons	CC 121
P92			t065		t065			2	2	*spa*-CC 330/065	CC 45
P93*			t1094			t002		2	2	*spa*-CC 002	CC 5
P94			t012				t012	2	2	*spa*-CC 012	CC 30
P95		t771					t771	2	2	No founder	ST 12
P96		t008				t008		2	2	*spa*-CC 304	CC 8
P97		t098				t098		2	2	Singletons	
P98*		t1974		t116				2	2	*spa*-CC 015	CC 45
P99	t160							2	1	No founder	ST 12
P100	t026							2	1	Excluded	
P101	t499						t499	2	2	*spa*-CC 499	CC 15
P102	t078						t078	2	2	No founder	CC 25
P103	t021					t021		2	2	*spa*-CC 012	CC 30
P104*		t032				t021		2	2	*spa*-CC 005, 012	CC 22, CC 30
P105				t015		t015		2	2	*spa*-CC 015	CC 45
P106			t084			t084		2	2	*spa*-CC 499	CC 15
P107			t190		t190			2	2	*spa*-CC 304	CC 8
P108	t032							1	1	*spa*-CC 005	CC 22
P109	t13449							1	1	Singletons	
P110	t012							1	1	*spa*-CC 012	CC 30
P111		t032						1	1	*spa*-CC 005	CC 22
P112			t012					5	1	*spa*-CC 012	CC 30
P113				t026				4	1	Excluded	
P114				t084				4	1	*spa*-CC 499	CC 15
P115					t065			3	1	*spa*-CC 330/065	CC 45
P116					t14401			3	1	Singletons	
P117					t11026			3	1	*spa*-CC 304	CC 8
P118					t1171			2	1	*spa*-CC 304	CC 8
P119				t091				3	1	*spa*-CC 499	CC 7

(continued)

Table 2 (continued)

P120			t056					2	1	No founder	ST 101
P121		t018						2	1	*spa*-CC 012	CC 30
P122		t14399						2	1	Singletons	
P123						t14400		2	1	*spa*-CC 015	CC 45
P124						t7690		2	1	*spa*-CC 005	CC 22
P125							t8909	1	1	*spa*-CC 499	CC 15
P126							t091	1	1	*spa*-CC 499	CC 7
P127							t005	1	1	*spa*-CC 005	CC 22
P128							t008	1	1	*spa*-CC 304	CC 8
P129						t3002		1	1	*spa*-CC 005	CC 22
P130						t026		1	1	Excluded	
P131						t002		1	1	*spa*-CC 002	CC 22
P132						t275		1	1	*spa*-CC 012	CC 30
P133						t1925		1	1	No founder	CC 25
P134						t008		1	1	*spa*-CC 304	CC 8
P135						t021		1	1	*spa*-CC 012	CC 30
P136						t056		1	1	No founder	ST 101
P137						t008		1	1	*spa*-CC 304	CC 8
P138						t13588		1	1	No founder	ST 101
P139						t2427		1	1	*spa*-CC 304	CC 8
P140					t056			1	1	No founder	ST 101
P141				t14398				1	1	Singletons	
P142			t012					1	1	*spa*-CC 012	CC 30
P143		t211						1	1	*spa*-CC 304	CC 8
P144		t211						1	1	*spa*-CC 304	CC 8
P145	NT							1	0		

**spa* type change observed, ‡t330 is a MSSA. Total of 145 participants were positive for *S. aureus* carriage, and three persons were positive for both MRSA and MSSA strains
Abbreviations: *Spa-CC spa* clonal complex, *MLST CC* Multilocus sequence typing-based clonal complexes, *ST* Sequence type, *NT* Nontypeable, *MSSA* Methicillin-sensitive *Staphylococcus aureus*, *MRSA* Methicillin-resistant *Staphylococcus aureu*

Color code	MRSA	MSSA	Other bacteria

on a very low proportion among isolates from MSSA infections in Germany (Cuny et al. 2013a, b).

Three participants were positive for *S. aureus* isolates with *spa* type t159 in swab samples from different months. Isolates exhibiting *spa* type t159 (CC121) are mainly associated with two entities of skin and soft tissue infections (superficial: impetigo, exfoliative dermatitis, and more invasive: furunculosis, more rarely necrotizing pneumonia).

MRSA obtained in our study exhibited *spa* types t032 and t025. These are attributed to CC22 which represents the second most frequent one among HA-MRSA from Germany. This finding would be consistent with possible spread of MRSA from healthcare settings into the community as observed in other studies (Miller et al. 2014; Votintseva et al. 2014; Espadinha et al. 2013). There are also other CCs of HA-MRSA found in Germany such as CC5, ST5 (*spa type* t002) and CC5, ST225 (*spa* type t003) also known as Rhine-Hesse prototype/EMRSA-3/New York clone. In addition, CC8, ST8 (*spa* type t008: northern German MRSA/USA300 cMRSA), CC22, ST22 (*spa* types t022, t025, t032: Barnim MRSA/EMRSA-15), and CC45, ST45 (*spa* types t004: Berlin MRSA prototype/USA600), ST228 (*spa type* t001: Southern German

MRSA) were also identified in previous studies. In contrast to these, the occurrence of CC8, ST239 (*spa* types t030 and t037) is rare in German hospitals. On the other hand, CA-MRSA from Germany containing *luk*-PV were mainly attributed to ST8 (*spa* type t008 corresponds to CA-MRSA "USA300") andST80 (*spa* type t044), whereas ST1 (*spa* type t127) and ST152 (*spa* type t355) were rarely found in the community (Witte et al. 2008a; Chaberny et al. 2008; Pfingsten-Würzburg et al. 2011; Layer et al. 2012; Herrmann et al. 2013). Diversity of MRSA *spa* types is also common in community settings of other European countries. For example, t002 (CC5) was noted in France, Hungary, Croatia, Spain, and UK; t008 in Austria, Belgium, and France; and t032 (ST22) found in Germany, Hungary, and UK (Mehraj et al. 2014; den Heijer et al. 2013; Andersen et al. 2012, 2013; Zanelli et al. 2002; Skråmm et al. 2011).

Of particular interest in our findings are changes in *spa* types over time which indicates replacements of the colonizing strains among carriers. The HZI study identified a change in *spa* types of 12.8 % carriers at different sampling points. This change could be due to multiple strain carriages in an individual or within-host evolution during asymptomatic carriage (Votintseva et al. 2014; Golubchik et al. 2013; Uhlemann et al. 2012b). Most of the persistent carriers kept the same *spa* type during the study period as also observed in previous studies (Miller et al. 2014; Sangvik et al. 2011). We observed changes of clonally unrelated *spa* types in five participants during the study period. Among these five, one participant had *spa* type t032 (CC22) and t008 (CC8), which were replaced by t330 (CC45). Four participants had reported antibiotic use and their *spa* types changed as t3741 (No founder) was replaced by t13450 (singleton), t008 (CC8) by t159 (singleton), t008 (CC8) by t3283 (singleton), and t032 (CC22) by t021 (CC30). Interestingly in six other participants, changes in *spa* type appeared only because of a single genetic event (like repeat deletion/insertion). In a result, t015 (CC45) replaced by t550 (CC45) and t1974 (CC45) by t116 (CC45) and both reported antibiotic intake. In same way, t002 (CC5) was replaced by t2167 (CC5), t002 (CC5) by t1094 (CC5), or vice versa. In this situation, the possibility of within-host evolution like micromutation (point mutation and small insertions/deletions) cannot be over looked. Among these 11 participants, ten had an upper respiratory tract infection, eight visited outpatient clinics, and six reported antibiotic intake in the last year.

In the HZI study, most of the isolates from transient carriers exhibited the same *spa* type. Interestingly, in a study conducted in Norway, *spa* type t084 was associated with transient carriage (Sangvik et al. 2011). Higher variations among colonizing strains in transient carriers were also reported previously (VandenBergh et al. 1999). Exchange of *S. aureus* clones was observed significantly higher among intermittent carriers than in persistent carriers (van Belkum et al. 2009). There could be multiple factors behind this scenario. It is possible that some individuals more likely carry identical or multiple *S. aureus* strain types, due to their genetic susceptibility to *S. aureus* carriage or due to a compromised immune system (Ruimy et al. 2010; Quinn and Cole 2007). It is also possible that *S. aureus* went below the detection level due to certain conditions such as antibiotic intake or co-colonization with other competing microflora and became active again after some time when these conditions resolved. Antibiotic pressure can induce a viable but non-culturable state in *S. aureus* (Pasquaroli et al. 2013; Zandri et al. 2012). Such kind of situation may also apply to nasal decolonization by application of mupirocin ointment and the observation of regrowth of *S. aureus* after some time (Bode et al. 2010). A study performed in New York City estimated that multiple genotypes of *S. aureus* are carried by 7 % of the participants on the basis of mathematical modelling (Cespedes et al. 2005). Participants of this cohort were 121 drug users as part of the hospital-associated methadone maintenance program in New York and 28 hospital patients. Multiple strain carriage was also observed in a multicentre cohort study conducted among intensive care unit (ICU) patients in six different European countries (Bloemendaal et al. 2009). For strain typing, they used *spa* typing and multilocus variable number of tandem repeat analysis (MLVA) sequence types. In their analysis, they found that 1 % of participants carried at least two genetically closely related isolates and 6 % carried genetically distinct strains (Bloemendaal et al. 2009).

6 Antibiotic Resistance Patterns of *S. aureus* from Nasal Colonization in the Community

The extent of antimicrobial resistance has become a global public health concern during the last decade (ECDC 2014; WHO 2014; Kaplan and Laing 2004). Usually, surveillance focuses on isolates from infections in hospital settings, and in part on emergence and spread of particular pathogens (e.g. CA-MRSA, LA-MRSA) in the community. Many bacterial infections, especially those associated with *S. aureus*, are caused by the patients' own commensal microbiota. Therefore, knowledge about antibiotic resistance in these colonizers is highly important for the formulation of strategies against resistance development.

Antibiotic resistance development is based on two major components: bacterial strains containing appropriate antibiotic resistance genes (acquisition of genes coding for resistance mechanisms and mutations in housekeeping genes conferring resistance) and selective pressure exerted by antibiotic usage. Temporary colonization with resistant bacterial strains and rare genetic events in a primarily susceptible bacterial population is a natural phenomenon. These genetic events can be resistance mutations and transfer of mobile genetic elements. Resistance mutations mostly affect the drug target sites, whereas mobile genetic elements contain genes which are responsible for several kinds of resistance mechanisms. It largely depends on the selective pressure of antibiotics that will lead to resistance development in bacterial strains. Selective pressure of antibiotics affects not only the pathogen at the site of infection but also commensal microbiome of the host, particularly in the case, when the concentration of the particular antibiotic at the colonization site is sufficient for suppressing growth of susceptible bacteria. Besides direct selection, there is also co-selection for antibiotic resistance traits which do not correspond to the antibiotic used. In this case, multiresistant bacterial strains replace the original susceptible colonizing strains. Multiresistance can be based on the acquisition of mobile genetic elements containing several resistance genes or on acquisition of a resistance gene coding for a mechanisms that confers resistance against several antibiotic classes.

Acquisition of MRSA by *S. aureus* nasal carriers in the community was reported from a longitudinal study over three years in the USA (Muthukrishnan et al. 2013). Transfer of resistance genes can occur in the nasopharyngeal microbiome as shown for transferable mupirocin resistance (Hurdle et al. 2005) for erythromycin resistance (Nys et al. 2005). A more detailed study, based on inoculation of a mixture of strains containing different transferable genetic elements in a gnotobiotic pig model, was conducted by McCarthy et al. (2014). This study has shown extensive exchange of transferable genetic elements in the nasal microbiome of pigs.

During the past decade, transferable resistance to oxazolidinones, lincosamides, and streptogramin B compounds conferred by *cfr* emerged first in coagulase-negative staphylococci (CoNS) from livestock (Witte and Cuny 2011). Later on, it was detected in CoNS and also in MRSA from nosocomial infections in humans (Mendes et al. 2014). In this situation, dissemination among staphylococci as colonizers in the community isolates cannot be excluded (Cuny et al. 2016). In this context, topical use of the recently introduced oxazolidinone retapamulin in dermatology (McNeil et al. 2014) will require attention.

One of the important features of the studies on antibiotic resistance in *S. aureus* strains (from infections and colonization) is their focus on isolates from nosocomial infections. Studies on antibiotic resistance patterns of *S. aureus* as nasal colonizer of humans in the community are infrequent so far. Most studies used the time interval of less than 48 h after admission for *S. aureus*/MRSA isolation from infection and colonization for the categorization as community origin. While, we should keep in mind that hospital acquired MRSA carriage can persist over several months (Mattner et al. 2010) independent from several risk factors (Sanford et al. 1994). Furthermore, MRSA are in the focus of most studies and surveillance activities, and less attention is paid to MSSA and their resistance traits. Of particular interest are observations of the dissemination of a Panton–Valentine leukocidin (PVL) toxin-positive *S. aureus* subpopulation of CC121 clone that had acquired resistance to fusidic acid (transferable, *fus*-B), a few isolates were also resistant to mupirocin and to retapamulin (Rijnders et al. 2012).

Only a few epidemic virulent clonal lineages MRSA are widely disseminated such as ST5, ST8, ST22, ST239 worldwide, and ST225 in Europe (Witte et al. 2008b; Lindsay 2013). So far, the reasons for this endemicity in the nosocomial setting are unknown. The same applies to their comparatively weak capacity for establishing as a competitive colonizer in the community.

During the past 15 years, the proportion of epidemic HA-MRSA exhibiting broad antibiotic resistance phenotypes such as ST247 ST228 among all MRSA from colonization and infections declined in most of the European countries. They were replaced by less broad resistant epidemic clonal lineages (e.g. ST22, ST45, and ST225). Whether this dynamics is associated with changes in the nasal microbiome that are associated with hospital stay needs to be shown in future. The same applies to the success of particular CA-MRSA clonal lineages that are examples for the evolution of an efficient combination of virulence and resistance traits (Otto 2013).

Antibiotic susceptibility analysis of colonization isolates from the APRES network mentioned above revealed 1.3 % MRSA isolates among a total of 6093 carriers. Highest frequency (2.1 %) of MRSA occurrence was observed in Belgium. Among MSSA, 73 % of the isolates were resistant to benzylpenicillin as test substance for all β-lactamase susceptible penicillins including aminopenicillins and acylureidopenicillins when testing staphylococci. These penicillins were the most frequently administered antibiotics in the countries enrolled in the study. Here, we have to consider co-selection in case of treatment of infections caused by other pathogens. The second most common resistance was 1.6 % on average for macrolide with the highest frequency (16.9 %) in France. Frequencies of resistance to other antibiotics remained below 10 % (van Bijnen et al. 2015). In Taiwan, MRSA carriage was 3.8 % among screened patients at the time of hospital admission after exclusion of previous hospital stay (Lu et al. 2011). A cross-sectional study carried out in school children in Jordan revealed 7 % carriage of MRSA (Alzoubi et al. 2014). Typing of isolates in this study was attributed to the CC22 subpopulation (*spa* type t223) which is prevalent in Jordan and Palestine.

In our prospective study reported here, we performed antibiotic susceptibility testing of selected isolates. Antibiotic susceptibility was tested for 17 substances from 15 antibiotic classes including antistaphylococcal β-lactams (penicillin, oxacillin), macrolides (erythromycin), lincosamides (clindamycin), fluoroquinolones (ciprofloxacin, moxifloxacin), phosphonic acids (fosfomycin), glycopeptides (vancomycin), lipopeptides (daptomycin), and glycylcyclines (tigecycline). Results from 157 *S. aureus* isolates revealed that 71 (45.2 %) isolates were sensitive to all tested antibiotics, 75 (47.7 %) resistant to benzylpenicillin (test substance for all ß-lactamase susceptible penicillins), 16 (10.2 %) to fluoroquinolones, 10 (6.3 %) to erythromycin, and 2 (1.3 %) to clindamycin (Table 3). High proportions of penicillin resistance were already observed among MSSA colonization and infections in Germany since decades (German reference centre for staphylococci and enterococci, unpublished). It indicates the high volume of usage of ß-lactam antibiotics outside hospitals in Germany (GERMAP 2012) and corresponds to the results from the APRES study (van Bijnen et al. 2015).

Resistance to fluoroquinolones also needs attention and could be explained by co-selection of resistant *S. aureus* as nasal colonizer when these antibiotics are used to treat infections caused by other bacterial pathogens. Fluoroquinolones reach concentration levels in nasal secretions which are appropriate for selection of resistant mutants (Darouiche et al. 1990; Blondeau et al. 2004). In Denmark, resistance to erythromycin and clindamycin was less frequent among *S. aureus* from nasal colonization 30 years ago than now (Eriksen et al. 1995). This is very likely due to the increase of erythromycin consumption by the end of 1989 (Westh et al. 1989; van Bijnen et al. 2014). In our study in Germany, we observed that all seven MRSA were attributed to CC22 and resistant to penicillin and

Table 3 Antimicrobial susceptibility testing of selected 157 *Staphylococcus aureus* strains isolated from the nasal swabs of study participants of Braunschweig, northern Germany

Antibiotic resistance	Frequency	Percentage
PEN	75	47.77
CIP	16	10.19
MFL	15	9.55
ERY	10	6.37
OXA	7	4.46
TET	3	1.91
CLI	2	1.27
FUS	2	1.27
DAP	1	0.64
Sensitive	71	45.23

Abbreviations: *CIP* Ciprofloxacin, *CLI* Clindamycin, *DAP* Daptomycin, *ERY* Erythromycin, *FUS* Fusidic acid, *MFL* Moxifloxacin, *OXA* Oxacilin, *PEN* Penicillin, *TET* Tetracycline
All isolates tested were susceptible to vancomycin, tigecycline, gentamicin, linezolid, fosfomycin, rifampicin, and trimethoprim/sulphamethoxazole

fluoroquinolones. In addition to these antibiotics, one MRSA was also resistant to erythromycin and clindamycin. Usually, HA-MRSA ST22 contains the plasmid-located *erm* (C) gene which confers resistance to erythromycin and clindamycin when constitutively expressed. We have to assume that it was lost by MRSA CCC22 from nasal colonization. In short, antibiotic resistance patterns highlight that there are still sufficient treatment alternatives for *S. aureus* infections originating from nasal colonization. MRSA dissemination and emergence of resistance to newly introduced antimicrobial agents is a growing health concern. In this scenario, screening for colonization in the community and surveillance programs is of important value at local, country, or international levels.

7 Limitations of the Studies on Carriage Patterns

A main limitation of the carriage pattern studies is that samples are only taken from the nose neglecting other body sites for which colonization has been documented, such as throat, intertriginous skin areas, and the upper vagina. In some studies, sampling interval and duration of follow-up was either too short or too long or the number of swab culture was too small. Furthermore, study populations were often convenient samples (e.g. students or hospital personnel) or too small and thus not representative for the general population (Eriksen et al. 1995; Hoeffler et al. 1978; Hu et al. 1995; Manenschijn et al. 2012; Maxwell et al. 1969). There are very few studies conducted for more than six-month duration which took samples at regular intervals and systematically investigated *S. aureus* genotypes among persistent and transient carriers (Muthukrishnan et al. 2013; Miller et al. 2014; Votintseva et al. 2014).

The differences observed in carriage patterns can also be due to sampling and culturing techniques, sample handling, etc. One study showed that the number of carriers in a given population is also dependent on the swab material, transport medium, the medium for cultivation, and incubation period (Eriksen et al. 1994).

Carrier index method can also increase or decrease persistent carriage with changes in the cut-off of the index. For example, decrease in carrier index or decrease in required proportion of positive swabs will ultimately lead to an increase in the number of persistent carriers. Decreased proportion of persistent and non-carriers is also possible if follow-up time is increased or interval of swab samples is decreased. Intermittent carriers can be misclassified as persistent carriers or non-carriers if follow-up time is short or swab culture interval is too long. In our study, we tried to address these limitations through recruiting participants from randomly selected population and defining persistence carriage with seven swabs positive for *S. aureus* during six-month period.

8 Conclusion and Future Perspective

Studies on *S. aureus* carriage patterns have provided an important insight into the dynamics of *S. aureus* carriage in the community. Most of the studies have tried to define the persistence in an accurate way and its associated features in detail. At same time, some studies were descriptive and the underlying biological conditions are still poorly understood so far. Nevertheless, experiences and data from these studies are important prerequisites to design target surveillance systems in order to detect emerging strains with particular resistance or virulence such as HA-MRSA, CA-MRSA, and LA-MRSA. Risk factors identified in carriage pattern studies will be helpful in the formulation of public health policies and screening criteria.

Acknowledgments We are thankful to all staff of the epidemiology department for organizational support in the study. We are indebted to all study participants for their participation over six months. We also want to thank Ms. Angie Massicotte, from Research Institute of the McGill University Health Centre, Montreal, Quebec, Canada for her review and valuable suggestions which have led to significant improvements in the presentation of this manuscript.

References

Akmatov MK, Mehraj J, Gatzemeier A, Strömpl J, Witte W, Krause G, Pessler F (2014) Serial home-based self-collection of anterior nasal swabs to detect *Staphylococcus aureus* carriage in a randomized population-based study in Germany. Int J Infect Dis 25:4–10

Alzoubi HM, Aqel AA, Al-Sarayreh SA, Al-Zayadneh E (2014) Methicillin-resistant *Staphylococcus aureus* nasal carriage among primary school-aged children from Jordan: prevalence, antibiotic resistance and molecular characteristics. J Egypt Public Health Assoc 89:114–118

Andersen PS, Pedersen JK, Fode P, Skov RL, Fowler VG Jr, Stegger M, Christensen K (2012) Influence of host genetics and environment on nasal carriage of *Staphylococcus aureus* in danish middle-aged and elderly twins. J Infect Dis 206:1178–1184

Andersen PS, Larsen LA, Fowler VG Jr, Stegger M, Skov RL, Christensen K (2013) Risk factors for *Staphylococcus aureus* nasal colonization in Danish middle-aged and elderly twins. Eur J Clin Microbiol Infect Dis 32:1321–1326

Armstrong-Esther CA, Smith JE (1976) Carriage patterns of *Staphylococcus aureus* in a healthy non-hospital population of adults and children. Ann Hum Biol 3:221–227

Becker K, Ballhausen B, Kahl BC, Köck R (2015) The clinical impact of livestock-associated methicillin-resistant *Staphylococcus aureus* of the clonal complex 398 for humans. Vet Microbiol (S0378-1135:30082-1)

Bereket W, Hemalatha K, Getenet B, Wondwossen T, Solomon A, Zeynudin Kannan S (2012) Update on bacterial nosocomial infections. Eur Rev Med Pharmacol Sci 16:1039–1044

Bloemendaal AL, Fluit AC, Jansen WT, Vriens MR, Ferry T, Amorim JM, Pascual A, Stefani S, Papaparaskevas J, Borel Rinkes IH, Verhoef J (2009) Colonization with multiple *Staphylococcus aureus* strains among patients in European intensive care units. Infect Control Hosp Epidemiol 30:918–920

Blondeau JM, Hansen G, Metzler K, Hedlin P (2004) The role of PK/PD parameters to avoid selection and increase of resistance: mutant prevention concentration. J Chemother 3:1–19

Bode LG, Kluytmans JA, Wertheim HF, Bogaers D, Vandenbroucke-Grauls CM, Roosendaal R, Troelstra A, Box AT, Voss A, van der Tweel I, van Belkum A, Verbrugh HA, Vos MC (2010) Preventing surgical-site infections in nasal carriers of *Staphylococcus aureus*. N Engl J Med 362:9–17

Casewell MW, Hill RL (1986) Elimination of nasal carriage of *Staphylococcus aureus* with mupirocin ('pseudomonic acid')–a controlled trial. J Antimicrob Chemother 17:365–372

Cespedes C, Said-Salim B, Miller M, Lo SH, Kreiswirth BN, Gordon RJ, Vavagiakis P, Klein RS, Lowy FD (2005) The clonality of *Staphylococcus aureus* nasal carriage. J Infect Dis 191:444–452

Chaberny IF, Bindseil A, Sohr D, Gastmeier P (2008) A point-prevalence study for MRSA in a German university hospital to identify patients at risk and to evaluate an established admission screening procedure. Infection 36:526–532

Chow JW, Yu VL (1989) *Staphylococcus aureus* nasal carriage in hemodialysis patients. Its role in infection and approaches to prophylaxis. Arch Intern Med 149:1258–1262

Corbella X, Domínguez MA, Pujol M, Ayats J, Sendra M, Pallares R, Ariza J, Gudiol F (1997) *Staphylococcus aureus* nasal carriage as a marker for subsequent staphylococcal infections in intensive care unit patients. Eur J Clin Microbiol Infect Dis 16:351–357

Cuny C, Nathaus R, Layer F, Strommenger B, Altmann D, Witte W (2009) Nasal colonization of humans with methicillin-resistant *Staphylococcus aureus* (MRSA) CC398 with and without exposure to pigs. PLoS ONE 4(8):e6800

Cuny C, Layer F, Köck R, Werner G, Witte W (2013a) Methicillin susceptible *Staphylococcus aureus* (MSSA) of clonal complex CC398, t571 from infections in humans are still rare in Germany. PLoS ONE 8:e83165

Cuny C, Köck R, Witte W (2013b) Livestock associated MRSA (LA-MRSA) and its relevance for humans in Germany. Int J Med Microbiol 303:331–337

Cuny C, Wieler L, Witte W (2015) Livestock-associated MRSA: The impact on humans. Antibiotics 4:521–543

Cuny C, Arnold P, Hermes J, Eckmanns T, Mehraj J, Schoenfelder S, Ziebuhr W, Zhao Q, Wang Y, Feßler AT, Krause G, Schwarz S, Witte W (2016) Occurrence of cfr-mediated multiresistance in staphylococci from veal calves and pigs, from humans at the corresponding farms, and from veterinarians and their family members. Vet Microbiol (30076-1)

Darouiche R, Perkins B, Musher D, Hamill R, Tsai S (1990) Levels of rifampin and ciprofloxacin in nasal secretions: correlation with MIC90 and eradication of nasopharyngeal carriage of bacteria. J Infect Dis 162:1124–1127

Davido B, Dinh A, Salomon J, Roux AL, Gosset-Woimant M, Pierre I, Perronne C, Bernard L (2013) Recurrent furunculosis: Efficacy of the CMC regimen–skin disinfection (chlorhexidine), local nasal antibiotic (mupirocin), and systemic antibiotic (clindamycin). Scand J Infect Dis 45:837–841

Day NP, Moore CE, Enright MC, Berendt AR, Smith JM, Murphy MF, Peacock SJ, Spratt BG, Feil EJ (2001) A link between virulence and ecological abundance in natural populations of *Staphylococcus aureus*. Science 292:114–116

Deiters C, Günnewig V, Friedrich AW, Mellmann A, Köck R (2015) Are cases of methicillin-resistant *Staphylococcus aureus* clonal complex (CC) 398 among humans still livestock-associated? Int J Med Microbiol 305:110–113

Demos M, McLeod MP, Nouri K (2012) Recurrent furunculosis: a review of the literature. Br J Dermatol 167:725–732

den Heijer CD, van Bijnen EM, Paget WJ, Pringle M, Goossens H, Bruggeman CA, Schellevis FG, Stobberingh EE; APRES Study Team (2013) Prevalence and resistance of commensal *Staphylococcus aureus*, including meticillin-resistant *Staphylococcus aureus*, in nine European countries: A cross-sectional study. Lancet Infect Dis 13:409–415

Dufour P, Gillet Y, Bes M, Lina G, Vandenesch F, Floret D, Etienne J, Richet H (2002) Community-acquired methicillin-resistant Staphylococcus aureus infections in France: emergence of a single clone that produces Panton-Valentine leukocidin. Clin Infect Dis 35:819–824

Eriksen NHR, Espersen F, Rosdahl VT, Jensen K (1994) Evaluation of methods for the detection of nasal carriage of *Staphylococcus aureus*. APMIS 102:407–412

Eriksen NHR, Espersen F, Rosdahl VT, Jensen K (1995) Carriage of *Staphylococcus aureus* among 104 healthy persons during a 19-month period. Epidemiol Infect 115:51–60

Espadinha D, Faria NA, Miragaia M, Lito LM, Melo-Cristino J, de Lencastre H, Network Médicos Sentinela (2013) Extensive dissemination of methicillin-resistant *Staphylococcus aureus* (MRSA) between the hospital and the community in a country with a high prevalence of nosocomial MRSA. PLoS ONE 8:e59960

European Centre for Disease Prevention and Control (2014) Antimicrobial resistance surveillance in Europe 2013. Annual Report of the European Antimicrobial Resistance Surveillance Network (EARS-Net). ECDC, Stockholm. Available at: http://ecdc.europa.eu/en/publications/Publications/antimicrobial-resistance-europe-2014.pdf

Feil EJ, Cooper JE, Grundmann H, Robinson DA, Enright MC, Berendt T, Peacock SJ, Smith JM, Murphy M, Spratt BG, Moore CE, Day NP (2003) How clonal is *Staphylococcus aureus*? J Bacteriol 185:3307–3316

Frenay HM, Theelen JP, Schouls LM, Vandenbroucke-Grauls CM, Verhoef J, van Leeuwen WJ, Mooi FR (1994) Discrimination of epidemic and nonepidemic methicillin-resistant *Staphylococcus aureus* strains on the basis of protein A gene polymorphism. J Clin Microbiol 32:846–847

Frenay HM, Bunschoten AE, Schouls LM, van Leeuwen WJ, Vandenbroucke-Grauls CM, Verhoef J, Mooi FR (1996) Molecular typing of methicillin-resistant *Staphylococcus aureus* on the basis of protein A gene polymorphism. Eur J Clin Microbiol Infect Dis 15:60–64

Garcia-Graells C, van Cleef BA, Larsen J, Denis O, Skov R, Voss A (2013) Dynamic of livestock-associated methicillin-resistant *Staphylococcus aureus* CC398 in pig farm households: a pilot study. PLoS ONE 8:e65512

GERMAP (2012) Federal Office of Consumer Protection and Food Safety, Paul-Ehrlich-Gesellschaft fur Chemotherapie e.V., Infectiology Freiburg. GERMAP 2012—Report on the consumption of antimicrobials and the spread of antimicrobial resistance in human and veterinary medicine in Germany. Antiinfectives Intelligence, Rheinbach. Available at: http://www.bvl.bund.de/SharedDocs/Downloads/08_PresseInfothek/Germap_2012_en.pdf?__blob=publicationFile&v=3

Goh SW, Byrne SK, Zhang JL, Chow AW (1992) Molecular typing of *Staphylococcus aureus* on the basis of coagulase gene polymorphisms. J Clin Microbiol 30:1642–1645

Golubchik T, Batty EM, Miller RR, Farr H, Young BC, Larner-Svensson H, Fung R, Godwin H, Knox K, Votintseva A, Everitt RG, Street T, Cule M, Ip CL, Didelot X, Peto TE, Harding RM,

Wilson DJ, Crook DW, Bowden R (2013) Within-host evolution of *Staphylococcus aureus* during asymptomatic carriage. PLoS ONE 8:e61319

Goslings WR, Buchli K (1958) Nasal carrier rate of antibiotic-resistant staphylococci; influence of hospitalization on carrier rate in patients, and their household contacts. AMA Arch Intern Med 102:691–715

Gould JC, Mckillop EJ (1954) The carriage of *Staphylococcus pyogenes* var. aureus in the human nose. J Hyg (Lond) 52:304–310

Grundmann H, Aanensen DM, van den Wijngaard CC, Spratt BG, Harmsen D, Friedrich AW, European Staphylococcal Reference Laboratory Working Group (2010) Geographic distribution of *Staphylococcus aureus* causing invasive infections in Europe: a molecular-epidemiological analysis. PLoS Med 7:e1000215

Guinan ME, Dan BB, Guidotti RJ, Reingold AL, Schmid GP, Bettoli EJ, Lossick JG, Shands KN, Kramer MA, Hargrett NT, Anderson RL, Broome CV (1982) Vaginal colonization with *Staphylococcus aureus* in healthy women: a review of four studies. Ann Intern Med 96:944–947

Gupta K, Martinello RA, Young M, Strymish J, Cho K, Lawler E (2013) MRSA nasal carriage patterns and the subsequent risk of conversion between patterns, infection, and death. PLoS ONE 8:e53674

Harmsen D, Claus H, Witte W, Rothgänger J, Claus H, Turnwald D, Vogel U (2003) Typing of methicillin-resistant *Staphylococcus aureus* in a university hospital setting by using novel software for spa repeat determination and database management. J Clin Microbiol 41:5442–5448

Herrmann M, Petit C, Dawson A, Biechele J, Halfmann A, von Müller L, Gräber S, Wagenpfeil S, Klein R, Gärtner B (2013) Methicillin-resistant *Staphylococcus aureus* in Saarland, Germany: A statewide admission prevalence screening study. PLoS ONE 8:e73876

Herwaldt LA, Cullen JJ, French P, Hu J, Pfaller MA, Wenzel RP, Perl TM (2004) Preoperative risk factors for nasal carriage of *Staphylococcus aureus*. Infect Control Hosp Epidemiol 25:481–484

Hidron AI, Kourbatova EV, Halvosa JS, Terrell BJ, McDougal LK, Tenover FC, Blumberg HM, King MD (2005) Risk factors for colonization with methicillin-resistant *Staphylococcus aureus* (MRSA) in patients admitted to an urban hospital: emergence of community-associated MRSA nasal carriage. Clin Infect Dis 41:159–166

Hoeffler U, Bulanda M, Heczko PB, Pulverer G (1978) A comparison of staphylococcal nasal carrier rates in Germany and Poland. Med Microbiol Immunol 164:285–290

Holden MT, Hsu LY, Kurt K, Weinert LA, Mather AE, Harris SR, Strommenger B, Layer F, Witte W, de Lencastre H, Skov R, Westh H, Zemlicková H, Coombs G, Kearns AM, Hill RL, Edgeworth J, Gould I, Gant V, Cooke J, Edwards GF, McAdam PR, Templeton KE, McCann A, Zhou Z, Castillo-Ramírez S, Feil EJ, Hudson LO, Enright MC, Balloux F, Aanensen DM, Spratt BG, Fitzgerald JR, Parkhill J, Achtman M, Bentley SD, Nübel U (2013) A genomic portrait of the emergence, evolution, and global spread of a methicillin-resistant *Staphylococcus aureus* pandemic. Genome Res 23(4):653–664

Hu L, Umeda A, Kondo S, Amako K (1995) Typing of *Staphylococcus aureus* colonizing human nasal carriers by pulsed-field gel electrophoresis. J Med Microbiol 42:127–132

Humphreys H (2008) Can we do better in controlling and preventing methicillin-resistant *Staphylococcus aureus* (MRSA) in the intensive care unit (ICU)? Eur J Clin Microbiol Infect Dis 27:409–413

Hurdle JG, O'Neill AJ, Mody L, Chopra I, Bradley SF (2005) In vivo transfer of high-level mupirocin resistance from *Staphylococcus epidermidis* to methicillin-resistant *Staphylococcus aureus* associated with failure of mupirocin prophylaxis. J Antimicrob Chemother 56:1166–1168

Kaplan W, Laing R (2004) Priority medicines for Europe and the world. Geneva: World Health Organization. Available at: http://apps.who.int/iris/bitstream/10665/68769/1/WHO_EDM_PAR_2004.7.pdf

Kinsman OS, McKenna R, Noble WC (1983) Association between histocompatability antigens (HLA) and nasal carriage of *Staphylococcus aureus*. J Med Microbiol 16:215–220

Kluytmans JA, Mouton JW, Izerman EP, Vandenbroucke-Grauls CM, Maat AW, Wagenvoort JH, Verbrugh HA (1995) Nasal carriage of *Staphylococcus aureus* as a major risk factor for wound infections after cardiac surgery. J Infect Dis 171:216–219

Kluytmans JA, van Belkum A, Verbrugh H (1997) Nasal carriage of *Staphylococcus aureus*: epidemiology, underlying mechanisms, and associated risks. Clin Microbiol Rev 10:505–520

Ko KS, Lee JY, Baek JY, Peck KR, Rhee JY, Kwon KT, Heo ST, Ahn KM, Song JH (2008) Characterization of *Staphylococcus aureus* nasal carriage from children attending an outpatient clinic in Seoul, Korea. Microb Drug Resist 14:37–44

Köck R, Schaumburg F, Mellmann A, Köksal M, Jurke A, Becker K, Friedrich AW (2013) Livestock-associated methicillin-resistant *Staphylococcus aureus* (MRSA) as causes of human infection and colonization in Germany. PLoS ONE 8(2):e55040

Köck R, Becker K, Cookson B, van Gemert-Pijnen JE, Harbarth S, Kluytmans J, Mielke M, Peters G, Skov RL, Struelens MJ, Tacconelli E, Witte W, Friedrich AW (2014) Systematic literature analysis and review of targeted preventive measures to limit healthcare-associated infections by meticillin-resistant *Staphylococcus aureus*. Euro Surveil 19:23–49

Kolata J, Bode LG, Holtfreter S, Steil L, Kusch H, Holtfreter B, Albrecht D, Hecker M, Engelmann S, van Belkum A, Voelker U, Broeker BM (2011) Distinctive patterns in the human antibody response to *Staphylococcus aureus* bacteremia in carriers and non-carriers. Proteomics 11:3914–3927

Kottler S, Middleton J, Perry J, Weese J, Cohn L (2010) Prevalence of Staphylococcus aureus and methicillin-resistant Staphylococcus aureus carriage in three populations. J Vet Intern Med 24:132–139

Krupa P, Bystroń J, Bania J, Podkowik M, Empel J, Mroczkowska A (2014) Genotypes and oxacillin resistance of *Staphylococcus aureus* from chicken and chicken meat in Poland. Poult Sci 93:3179–3186

Kurt K, Rasigade JP, Laurent F, Goering RV, Žemličková H, Machova I, Struelens MJ, Zautner AE, Holtfreter S, Bröker B, Ritchie S, Reaksmey S, Limmathurotsakul D, Peacock S, Cuny C, Layer F, Witte W, Nübel U (2013) Subpopulations of *Staphylococcus aureus* clonal complex 121 are associated with distinct clinical entities. PLoS ONE 8:e58155

Lamikanra A, Olusanya OI (1988) A long-term study of the nasal carriage of *Staphylococcus aureus* in healthy Nigerian students. Trans R Soc Trop Med Hyg 82:500–502

Layer F, Cuny C, Strommenger B, Werner G, Witte W (2012) Current data and trends on methicillin-resistant *Staphylococcus aureus* (MRSA). Bundesgesundheitsblatt Gesundheitsforschung Gesundheitsschutz 55:1377–1386

Leopold SR, Goering RV, Witten A, Harmsen D, Mellmann A (2014) Bacterial whole-genome sequencing revisited: portable, scalable, and standardized analysis for typing and detection of virulence and antibiotic resistance genes. J Clin Microbiol 52:2365–2370

Lindsay JA (2013) Hospital-associated MRSA and antibiotic resistance-what have we learned from genomics? Int J Med Microbiol 303:318–323

Lu SY, Chang FY, Cheng CC, Lee KD, Huang YC (2011) Methicillin-resistant *Staphylococcus aureus* nasal colonization among adult patients visiting emergency department in a medical center in Taiwan. PLoS ONE 6:e18620

Luzar MA, Coles GA, Faller B, Slingeneyer A, Dah GD, Briat C, Wone C, Knefati Y, Kessler M, Peluso F (1990) *Staphylococcus aureus* nasal carriage and infection in patients on continuous ambulatory peritoneal dialysis. N Engl J Med 322:505–509

Manenschijn L, Jetten AM, van Wamel WJ, Tavakol M, Koper JW, van den Akker EL, van Belkum A, van Rossum EF (2012) Long-term cortisol levels are not associated with nasal carriage of *Staphylococcus aureus*. Eur J Clin Microbiol Infect Dis 31:97–100

Mattner F, Biertz F, Ziesing S, Gastmeier P, Chaberny IF (2010) Long-term persistence of MRSA in re-admitted patients. Infection 38:363–371

Maxwell JG, Ford CR, Peterson DE, Mitchell CR (1969) Long-term study of nasal staphylococci among hospital personnel. Am J Surg 118:849–854

McCarthy AJ, Loeffler A, Witney AA, Gould KA, Lloyd DH, Lindsay JA (2014) Extensive horizontal gene transfer during *Staphylococcus aureus* co-colonization in vivo. Genome Biol Evol 6:2697–2708

McKinnell JA, Miller LG, Eells SJ, Cui E, Huang SS (2013) A systematic literature review and meta-analysis of factors associated with methicillin-resistant *Staphylococcus aureus* colonization at time of hospital or intensive care unit admission. Infect Control Hosp Epidemiol 34:1077–1086

McNeil JC, Hulten KG, Kaplan SL, Mason EO (2014) Decreased susceptibilities to Retapamulin, Mupirocin, and Chlorhexidine among *Staphylococcus aureus* isolates causing skin and soft tissue infections in otherwise healthy children. Antimicrob Agents Chemother 58:2878–2883

Mehraj J, Akmatov MK, Strömpl J, Gatzemeier A, Layer F, Werner G, Pieper DH, Medina E, Witte W, Pessler F, Krause G (2014) Methicillin-sensitive and methicillin-resistant *Staphylococcus aureus* nasal carriage in a random sample of non-hospitalized adult population in northern Germany. PLoS ONE 9:e107937

Melles DC, Tenover FC, Kuehnert MJ, Witsenboer H, Peeters JK, Verbrugh HA, van Belkum A (2008) Overlapping population structures of nasal isolates of *Staphylococcus aureus* from healthy Dutch and American individuals. J Clin Microbiol 46:235–241

Melles DC, Schouls L, François P, Herzig S, Verbrugh HA, van Belkum A, Schrenzel J (2009) High-throughput typing of *Staphylococcus aureus* by amplified fragment length polymorphism (AFLP) or multi-locus variable number of tandem repeat analysis (MLVA) reveals consistent strain relatedness. Eur J Clin Microbiol Infect Dis 28:39–45

Mellmann A, Weniger T, Berssenbrügge C, Rothgänger J, Sammeth M, Stoye J, Harmsen D (2007) Based Upon Repeat Pattern (BURP): an algorithm to characterize the long-term evolution of *Staphylococcus aureus* populations based on spa polymorphisms. BMC Microbiol 7:98

Mendes RE, Flamm RK, Hogan PA, Ross JE, Jones RN (2014) Summary of linezolid activity and resistance mechanisms detected during the 2012 LEADER surveillance program for the United States. Antimicrob Agents Chemother 58:1243–1247

Messaritakis I, Samonis G, Dimopoulou D, Maraki S, Papadakis JA, Daraki V, Fragaki M, Choulaki C, Andrianaki AM, Kofteridis DP (2014) *Staphylococcus aureus* nasal carriage might be associated with vitamin D receptor polymorphisms in type 2 diabetes. Clin Microbiol Infect 20:920–925

Miller DL, McDonald JC, Jevons MP, Williams RE (1962) Staphylococcal disease and nasal carriage in the Royal Air Force. J Hyg (Lond) 60:451–465

Miller RR, Walker AS, Godwin H, Fung R, Votintseva A, Bowden R, Mant D, Peto TE, Crook DW, Knox K (2014) Dynamics of acquisition and loss of carriage of *Staphylococcus aureus* strains in the community: The effect of clonal complex. J Infect 68:426–439

Mulcahy ME, Geoghegan JA, Monk IR, O'Keeffe KM, Walsh EJ, Foster TJ, McLoughlin RM (2012) Nasal colonisation by *Staphylococcus aureus* depends upon clumping factor B binding to the squamous epithelial cell envelope protein loricrin. PLoS Pathog 8:e1003092

Muthukrishnan G, Lamers RP, Ellis A, Paramanandam V, Persaud AB, Tafur S, Parkinson CL, Cole AM (2013) Longitudinal genetic analyses of *Staphylococcus aureus* nasal carriage dynamics in a diverse population. BMC Infect Dis 13:221

Nouwen JL, Ott A, Kluytmans-Vandenbergh MF, Boelens HA, Hofman A, van Belkum A, Verbrugh HA (2004) Predicting the *Staphylococcus aureus* nasal carrier state: derivation and validation of a "culture rule". Clin Infect Dis 39:806–811

Nys S, Tjhie JH, Bartelds AI, Heijnen ML, Peeters MF, Stobberingh EE (2005) Erythromycin resistance in the commensal throat flora of patients visiting the general practitioner: a reservoir for resistance genes for potential pathogenic bacteria. Int J Antimicrob Agents 26:133–137

Olsen K, Falch BM, Danielsen K, Johannessen M, Ericson Sollid JU, Thune I, Grimnes G, Jorde R, Simonsen GS, Furberg AS (2012) *Staphylococcus aureus* nasal carriage is associated with serum 25-hydroxyvitamin D levels, gender and smoking status: The Tromso Staph and Skin Study. Eur J Clin Microbiol Infect Dis 31:465–473

Osadebe LU, Hanson B, Smith TC, Heimer R (2013) Prevalence and characteristics of *Staphylococcus aureus* in Connecticut swine and swine farmers. Zoonoses Public Health 60:234–243

Otto M (2013) Community-associated MRSA: what makes them special? Int J Med Microbiol 303:324–330

Palavecino EL (2014) Clinical, epidemiologic, and laboratory aspects of methicillin-resistant *Staphylococcus aureus* infections. Methods Mol Biol 1085:1–24

Pasquaroli S, Zandri G, Vignaroli C, Vuotto C, Donelli G, Biavasco F (2013) Antibiotic pressure can induce the viable but non-culturable state in *Staphylococcus aureus* growing in biofilms. J Antimicrob Chemother 68:1812–1817

Pfingsten-Würzburg S, Pieper DH, Bautsch W, Probst-Kepper M (2011) Prevalence and molecular epidemiology of meticillin-resistant *Staphylococcus aureus* in nursing home residents in northern Germany. J Hosp Infect 78:108–112

Qu F, Cui E, Guo T, Li H, Chen S, Liu L, Han W, Bao C, Mao Y, Tang YW (2010) Nasal colonization of and clonal transmission of methicillin-susceptible *Staphylococcus aureus* among Chinese military volunteers. J Clin Microbiol 48:64–69

Quinn GA, Cole AM (2007) Suppression of innate immunity by a nasal carriage strain of *Staphylococcus aureus* increases its colonization on nasal epithelium. Immunology 122:80–89

Reich-Schupke S, Geis G, Reising M, Altmeyer P, Stücker M (2010) MRSA in dermatology–Prospective epidemiological study in employees and patients of a dermatological department of a university hospital. J Dtsch Dermatol Ges 8:607–613

Rijnders MI, Wolffs PF, Hopstaken RM, den Heyer M, Bruggeman CA, Stobberingh EE (2012) Spread of the epidemic European fusidic acid-resistant impetigo clone (EEFIC) in general practice patients in the south of The Netherlands. J Antimicrob Chemother 67:1176–1180

Rolo J, Miragaia M, Turlej-Rogacka A, Empel J, Bouchami O, Faria NA, Tavares A, Hryniewicz W, Fluit AC, de Lencastre H; CONCORD Working Group (2012) High genetic diversity among community-associated *Staphylococcus aureus* in Europe: results from a multicenter study. PLoS ONE 7:e34768

Ruimy R, Maiga A, Armand-Lefevre L, Maiga I, Diallo A, Koumare AK, Ouattara K, Soumare S, Gaillard K, Lucet JC, Andremont A, Feil EJ (2008) The carriage population of *Staphylococcus aureus* from Mali is composed of a combination of pandemic clones and the divergent Panton-Valentine leukocidin-positive genotype ST152. J Bacteriol 190:3962–3968

Ruimy R, Armand-Lefevre L, Barbier F, Ruppé E, Cocojaru R, Mesli Y, Maiga A, Benkalfat M, Benchouk S, Hassaine H, Dufourcq JB, Nareth C, Sarthou JL, Andremont A, Feil EJ (2009) Comparisons between geographically diverse samples of carried *Staphylococcus aureus*. J Bacteriol 191:5577–5583

Ruimy R, Angebault C, Djossou F, Dupont C, Epelboin L, Jarraud S, Lefevre LA, Bes M, Lixandru BE, Bertine M, El Miniai A, Renard M, Bettinger RM, Lescat M, Clermont O, Peroz G, Lina G, Tavakol M, Vandenesch F, van Belkum A, Rousset F, Andremont A (2010) Are host genetics the predominant determinant of persistent nasal *Staphylococcus aureus* carriage in humans? J Infect Dis 202:924–934

Sakwinska O, Kuhn G, Balmelli C, Francioli P, Giddey M, Perreten V, Riesen A, Zysset F, Blanc DS, Moreillon P (2009) Genetic diversity and ecological success of *Staphylococcus aureus* strains colonizing humans. Appl Environ Microbiol 75:175–183

Sakwinska O, Blanc DS, Lazor-Blanchet C, Moreillon M, Giddey M, Moreillon P (2010) Ecological temporal stability of *Staphylococcus aureus* nasal carriage. J Clin Microbiol 48:2724–2728

Sanford MD, Widmer AF, Bale MJ, Jones RN, Wenzel RP (1994) Efficient detection and long-term persistence of the carriage of methicillin-resistant *Staphylococcus aureus*. Clin Infect Dis 19:1123–1128

Sangvik M, Olsen RS, Olsen K, Simonsen GS, Furberg AS, Sollid JU (2011) Age-and gender-associated *Staphylococcus aureus* spa types found among nasal carriers in a general population: the Tromsø Staph and Skin Study. J Clin Microbiol 49:4213–4218

Skråmm I, Moen AE, Bukholm G (2011) Nasal carriage of *Staphylococcus aureus*: frequency and molecular diversity in a randomly sampled Norwegian community population. APMIS 119:522–528

Strommenger B, Kettlitz C, Weniger T, Harmsen D, Friedrich AW, Witte W (2006) Assignment of Staphylococcus isolates to groups by spa typing, SmaI macrorestriction analysis, and multilocus sequence typing. J Clin Microbiol 44:2533–2540

Tacconelli E, Johnson AP (2011) National guidelines for decolonization of methicillin-resistant *Staphylococcus aureus* carriers: the implications of recent experience in the Netherlands. J Antimicrob Chemother 66:2195–2198

Tenover F, Arbeit RD, Goering RV, Mickelsen PA, Murray BE, Persing DH, Swaminathan B (1995) Interpreting chromosomal DNA restriction patterns produced by pulsed-field gel electrophoresis: criteria for bacterial strain typing. J Clin Microbiol 33:2233–2239

Uhlemann AC, Porcella SF, Trivedi S, Sullivan SB, Hafer C, Kennedy AD, Barbian KD, McCarthy AJ, Street C, Hirschberg DL, Lipkin WI, Lindsay JA, DeLeo FR, Lowy FD (2012a) Identification of a highly transmissible animal-independent *Staphylococcus aureus* ST398 clone with distinct genomic and cell adhesion properties. MBio 3 (e00027-12)

Uhlemann AC, Kennedy AD, Martens C, Porcella SF, DeLeo FR, Lowy FD (2012b) Towards an understanding of the evolution of *Staphylococcus aureus* strain USA300 during colonization in community households. Genome Biol Evol 4:1275–1285

van Belkum A, Verkaik NJ, de Vogel CP, Boelens HA, Vereer J, Nouwen JL, Verbrugh HA, Wertheim HF (2009) Reclassification of *Staphylococcus aureus* nasal carriage types. J Infect Dis 199:1820–1826

van Bijnen EM, Paget J, de Lange-de Klerk ES, den Heijer CD, Versporten A, Stobberingh EE, Goossens H, Schellevis FG; collaboration with the APRES Study Team (2015) Antibiotic exposure and other risk factors for antimicrobial resistance in nasal commensal *Staphylococcus aureus*: An ecological study in 8 European countries. PLoS ONE 10:e0135094

van Bijnen EM, Paget WJ, den Heijer CD, Stobberingh EE, Bruggeman CA, Schellevis FG; APRES Study Team (2014) Primary care treatment guidelines for skin infections in Europe: congruence with antimicrobial resistance found in commensal *Staphylococcus aureus* in the community. BMC Fam Pract 15:175

van Cleef BA, Graveland H, Haenen AP, van de Giessen AW, Heederik D, Wagenaar JA, Kluytmans JA (2011a) Persistence of livestock-associated methicillin-resistant *Staphylococcus aureus* in field workers after short-term occupational exposure to pigs and veal calves. J Clin Microbiol 49:1030–1033

van Cleef BA, Monnet DL, Voss A, Krziwanek K, Allerberger F, Struelens M, Zemlickova H, Skov RL, Vuopio-Varkila J, Cuny C, Friedrich AW, Spiliopoulou I, Pászti J, Hardardottir H, Rossney A, Pan A, Pantosti A, Borg M, Grundmann H, Mueller-Premru M, Olsson-Liljequist B, Widmer A, Harbarth S, Schweiger A, Unal S, Kluytmans JA (2011b) Livestock-associated methicillin-resistant *Staphylococcus aureus* in humans Europe. Emerg Infect Dis 17:502–505

van den Akker EL, Nouwen JL, Melles DC, van Rossum EF, Koper JW, Uitterlinden AG, Hofman A, Verbrugh HA, Pols HA, Lamberts SW, van Belkum A (2006) *Staphylococcus aureus* nasal carriage is associated with glucocorticoid receptor gene polymorphisms. J Infect Dis 194:814–818

van Loo I, Huijsdens X, Tiemersma E, de Neeling A, van de Sande-Bruinsma N, Beaujean D, Voss A, Kluytmans J (2007) Emergence of methicillin-resistant *Staphylococcus aureus* of animal origin in humans. Emerg Infect Dis 13:1834–1839

van Rijen MM, Van Keulen PH, Kluytmans JA (2008) Increase in a Dutch hospital of methicillin-resistant *Staphylococcus aureus* related to animal farming. Clin Infect Dis 46:261–263

VandenBergh MF, Yzerman EP, van Belkum A, Boelens HA, Sijmons M, Verbrugh HA (1999) Follow-up of *Staphylococcus aureus* nasal carriage after 8 years: redefining the persistent carrier state. J Clin Microbiol 37:3133–3140

Verhoeven PO, Grattard F, Carricajo A, Lucht F, Cazorla C, Garraud O, Pozzetto B, Berthelot P (2012) An algorithm based on one or two nasal samples is accurate to identify persistent nasal carriers of *Staphylococcus aureus*. Clin Microbiol Infect 18:551–557

von Eiff C, Becker K, Machka K, Stammer H, Peters G (2001) Nasal carriage as a source of *Staphylococcus aureus* bacteremia. New Engl J Med 344:11–16

Votintseva AA, Miller RR, Fung R, Knox K, Godwin H, Peto TE, Crook DW, Bowden R, Walker AS (2014) Multiple-strain colonization in nasal carriers of *Staphylococcus aureus*. J Clin Microbiol 52:1192–1200

Walsh EJ, O'Brien LM, Liang X, Hook M, Foster TJ (2004) Clumping factor B, a fibrinogen-binding MSCRAMM (microbial surface components recognizing adhesive matrix molecules) adhesin of *Staphylococcus aureus*, also binds to the tail region of type I cytokeratin 10. J Biol Chem 279:50691–50699

Walther B, Hermes J, Cuny C, Wieler LH, Vincze S, Abou Elnaga Y, Stamm I, Kopp PA, Kohn B, Witte W, Jansen A, Conraths FJ, Semmler T, Eckmanns T, Lübke-Becker A (2012) Sharing more than friendship—nasal colonization with coagulase-positive staphylococci (CPS) and co-habitation aspects of dogs and their owners. PLoS ONE 7:e35197

Wertheim HF, Vos MC, Ott A, van Belkum A, Voss A, Kluytmans JA, van Keulen PH, Vandenbroucke-Grauls CM, Meester MH, Verbrugh HA (2004) Risk and outcome of nosocomial *Staphylococcus aureus* bacteraemia in nasal carriers versus non-carriers. Lancet 364:703–705

Wertheim HF, Verveer J, Boelens HA, van Belkum A, Verbrugh HA, Vos MC (2005a) Effect of mupirocin treatment on nasal, pharyngeal, and perineal carriage of *Staphylococcus aureus* in healthy adults. Antimicrob Agents Chemother 49:1465–1467

Wertheim HF, Melles DC, Vos MC, van Leeuwen W, van Belkum A, Verbrugh HA, Nouwen JL (2005b) The role of nasal carriage in *Staphylococcus aureus* infections. Lancet Infect Dis 5:751–762

Westh H, Rosdahl VT, Friis H (1989) Erythromycin resistance in Danish *Staphylococcus aureus* hospital strains with emphasis on erythromycin consumption. APMIS 97:1121–1124

Williams RE (1963) Healthy carriage of *Staphylococcus aureus*: its prevalence and importance. Bacteriol Rev 26:56–71

Winstel V, Kühner P, Salomon F, Larsen J, Skov R, Hoffmann W, Peschel A, Weidenmaier C (2015) Wall teichoic acid glycosylation governs *Staphylococcus aureus* nasal colonization. MBio 6:e00632

Witte W, Cuny C (2011) Emergence and spread of cfr-mediated multiresistance in staphylococci: an interdisciplinary challenge. Future Microbiol 6:925–931

Witte W, Braulke C, Strommenger B (2008a) Community-associated methicillin-resistant *Staphylococcus aureus* ST8 ("USA300") in an HIV-positive patient in Cologne, Germany, February 2008. Euro Surveill 13:8080

Witte W, Cuny C, Klare I, Nübel U, Strommenger B, Werner G (2008b) Emergence and spread of antibiotic-resistant Gram-positive bacterial pathogens. Int J Med Microbiol 298:365–377

World Health Organization (2014) Antimicrobial resistance: global report on surveillance 2014. WHO Geneva. Available at: http://apps.who.int/iris/bitstream/10665/112642/1/9789241564748_eng.pdf

Yu VL, Goetz A, Wagener M, Smith PB, Rihs JD, Hanchett J, Zuravleff JJ (1986) *Staphylococcus aureus* nasal carriage and infection in patients on hemodialysis: Efficacy of antibiotic prophylaxis. New Engl J Med 315:91–96

Zandri G, Pasquaroli S, Vignaroli C, Talevi S, Manso E, Donelli G, Biavasco F (2012) Detection of viable but non-culturable staphylococci in biofilms from central venous catheters negative on standard microbiological assays. Clin Microbiol Infect 18:259–261

Zanelli G, Sansoni A, Zanchi A, Cresti S, Pollini S, Rossolini GM, Cellesi C (2002) *Staphylococcus aureus* nasal carriage in the community: a survey from central Italy. Epidemiol Infect 129:417–420

Diagnostics and Resistance Profiling of Bacterial Pathogens

Klaus Hornischer and Susanne Häußler

Abstract Worldwide infectious disease is one of the leading causes of death. Despite improvements in technology and healthcare services, morbidity and mortality due to infections have remained unchanged over the past few decades. The high and increasing rate of antibiotic resistance is further aggravating the situation. Growing resistance hampers the use of conventional antibiotics, and substantial higher mortality rates are reported in patients given ineffective empiric therapy mainly due to resistance to the agents used. These infections cause suffering, incapacity, and death and impose an enormous financial burden on both healthcare systems and on society in general. The accelerating development of multidrug resistance is one of the greatest diagnostic and therapeutic challenges to modern medicine. The lack of new antibiotic options underscores the need for optimization of current diagnostics, therapies, and prevention of the spread of multidrug-resistant organisms. The so-called -omics technologies (genomics, transcriptomics, proteomics, and metabolomics) have yielded large-scale datasets that advanced the search for biomarkers of infectious diseases in the last decade. One can imagine that in the future the implementation of biomarker-driven molecular test systems will transform diagnostics of infectious diseases and will significantly accelerate the identification of the bacterial pathogens at the infected host site. Furthermore, molecular tests based on the identification of markers of antibiotic resistance will dramatically change resistance profiling. The replacement of culturing methods by molecular test systems for early diagnosis will provide the basis not only for a prompt and targeted therapy, but also for a much more effective stewardship of antibiotic agents and a reduction of the spread of multidrug resistance as well as the appearance of new antibiotic resistances.

K. Hornischer · S. Häußler (✉)
Dept. Molecular Bacteriology, Helmholtz Centre for Infection Research, Braunschweig, Germany
e-mail: susanne.haeussler@helmholtz-hzi.de

K. Hornischer
e-mail: klaus.hornischer@helmholtz-hzi.de

Current Topics in Microbiology and Immunology (2016) 398: 89–102
DOI 10.1007/82_2016_494

Published Online: 28 June 2016

Contents

1 The Urgent Need for Early Diagnostics

Hospital-acquired infections caused by antibiotic-resistant opportunistic pathogens are on the increase worldwide and pose one of the greatest challenges to modern medicine. Multidrug resistance dramatically impacts on human health, public health, and the principles and practice of clinical medicine.

Management of nosocomial infections, especially those caused by multidrug-resistant Gram-negative bacteria, requires a multipronged strategy that includes not only the development of new antimicrobial compounds and a rational use of antimicrobial agents, but also early diagnosis for targeted treatment and the implementation of effective infection control principles.

It is common knowledge that one of the best ways to fight cancer is to catch it in the early stages, when it is more successfully treatable. Similarly, the prognosis of infectious diseases can be dramatically improved by the administration of an early and targeted antimicrobial therapy. Thus, early diagnosis is one of the most powerful weapons in the battlefield of multidrug-resistant infections.

However, despite of the clinical need, diagnostics of infectious diseases have experienced little fundamental change over the years. Many infectious diseases are still diagnosed by their clinical presentation. Especially in the early phase of the disease clinical symptoms and signs of clinical manifestations are often unspecific. They can vary and retrieve a list of various differential diagnoses making definitive diagnosis a challenge. As a consequence proper decisions for adequate treatment are difficult. For example the detection of tuberculosis (TB), which is caused by the Gram-positive *Mycobacterium tuberculosis*, remains complex, partly because of inaccurate diagnostic methods. An estimated third of all new TB cases are still being missed in spite of substantial efforts to increase case detection closer to the point at which clinical disease symptoms become evident. This remains a major obstacle to global control of TB (Wallis et al. 2010; Zignol et al. 2012; Lawn et al. 2013).

The diagnosis of an infectious disease is confirmed by clinical evaluation and microbiological testing for infectious agents. Bacterial species identification and resistance testing still predominantly rely on culture-dependent methods. As a consequence clinical microbiology is still labor-intensive and slow. Gains in terms of turnaround times of culture for identification and susceptibility testing remain limited.

The need for fast and reliable early diagnosis is most urgent in severe life-threatening infections. Sepsis is one of the major complications in intensive care units (ICUs). Mortality as a consequence of sepsis and septic shock is high (Garnacho-Montero et al. 2003; MacArthur et al. 2004). The Surviving Sepsis Campaign Guidelines Committee (2012) strongly recommends administering adequate antimicrobial therapy within the first hour of hospital admission in patients diagnosed with sepsis. It has been reported that 83 % of septic shock patients survived if treated with adequate antibiotic therapy within the first 30 min from the onset of shock symptoms. The survival rate was 6 % poorer when treatment started within the next 30 min. Mortality then increased by 7 % for each additional hour of delay until treatment started (Kumar et al. 2006).

For acute infections any delay of antibiotic treatment can be fatal. As a consequence it is common practice to implement early treatment with antimicrobial agents in serious infections, for which rapid killing is essential. Since the therapy cannot be based on results from microbiological diagnostics, an "empiric" therapy with broad-spectrum antibiotics is initiated. This therapeutic strategy usually follows general recommendations and the healthcare unit's experience with infections in the unit during the recent past (Fitousis et al. 2010).

However, the initial therapy, although frequently life-saving, may prove to be overtreatment, because the patient's symptoms are derived from a non-infectious cause, or a more specific, narrow-spectrum antibiotic would have been sufficient to treat the infection. Such an overtreatment is critical for various reasons. The negative consequences of a long-term treatment with broad-spectrum antibiotic agents are severe side effects of the medication such as renal and hepatic dysfunction as well as an impact on future resistance development. Therefore, the clinical guidelines in use suggest "de-escalation" approaches. This means that the initially administered broad-spectrum antimicrobials should be—as soon as the results of pathogen identification and characterization (especially susceptibility profiling) are available—changed to a more narrowly focused regime (Soo Hoo et al. 2005).

The slowness of diagnosis not only increases antibiotic use but also hampers antibiotic stewardship. "To combat antibiotic resistant infections, we must not only develop new drugs, but also learn to use existing drugs more effectively" (Farhat et al. 2013). A prerequisite for an effective antibiotic stewardship is detailed knowledge of the infecting pathogen. Thus, fast and accurate diagnostic methods have to be developed to also ensure implementation of rational strategies for antibiotic use (Livermore and Wain 2013).

In a publication of 2011 the Infectious Diseases Society of America (IDSA) calls for a shift from the more conservative established repertoire of diagnostic methods to molecular diagnostics. They argue that the century-old methods of culture on solidified nutrients used for the identification and characterization of pathogens are antiquated, too slow, and that they do not cover all possible pathogens compared to the recently developed molecular methods.

However, although many predicted a dramatic impact of newly developed molecular methods on routine medical microbiological diagnostics, these methods have not made it into clinical application so far. Nevertheless, there is clearly a need for clinical microbiology to change. Clinical microbiology needs to provide more impact on the management of infectious diseases. This also includes an improved antimicrobial stewardship which helps to reduce the spread and impact of antibiotic resistance. Furthermore, future microbiological diagnostics should accommodate the general drive toward a more cost-efficient medicine.

2 Biomarker Identification

The implementation of robust and cost-effective molecular diagnostic test systems that reliably detect infectious diseases early and deliver detailed information on pathogen determinants will be a groundbreaking advancement. Early and reliable biomarker-driven diagnosis of infectious diseases that includes information on the distribution of resistance markers as well as on the clonal relatedness of multidrug-resistant bacteria has the potential to transform diagnostics and reduce morbidity and mortality of the patients as it will enable new strategies for individualized therapy and control of resistance.

A biomarker is defined as a biological characteristic which can be measured and evaluated objectively. It can be molecular, anatomic, physiological, or biochemical. Biomarkers act as indicators of normal biological processes, pathogenic processes, or pharmacological responses to a therapeutic intervention (Biomarkers Definitions Working Group 2001; European Commission 2010).

Biomarkers can be subdivided into diagnostic biomarkers, which detect a specific disease, preferably in an early phase; susceptibility or risk biomarkers, which are indicators of disease prognosis; prognostic biomarkers, which monitor the development and the extent of a disease; and predictive biomarkers, which monitor the clinical response and the toxicity to a given treatment (Biomarkers Definitions Working Group 2001; European Commission 2010).

Biomarkers of infectious diseases can be characteristic of both host and pathogen. Biomarker molecules can be detected in the bloodstream and in other body fluids such as urine and pus. They form a record of the physiological state of the host organism (Liotta et al. 2003; Rai 2007).

In clinical applications biomarkers are used for the initial diagnosis of a disease, they guide clinicians in the choice for a targeted treatment and deliver information on the prognosis of disease development and outcome. Diagnostic markers are most effective if they detect disease in an early stage, preferentially in the asymptomatic phase (Classen et al. 2008; Zolg and Langen 2004).

Biomarkers have to be highly specific for a certain disease condition. Only the combination of a number of markers will produce the required reliability of an "application-specific fingerprint" (Classen et al. 2008; Liotta et al. 2003; He and Chiu 2003).

The development of biomarkers for clinical application is subdivided into a phase for discovery and characterization (discovery phase), the development of an appropriate assay (prototype developmental phase), and the implementation of clinical laboratories (product development phase). The ideal biomarker assay is sensitive, specific, and cost-effective, it is fast, and its outcome is not sensitive to the operator or the laboratory in which it is conducted (Classen et al. 2008; Zolg and Langen 2004).

Biomarkers are usually detected in the so-called case–control studies. The basic principle underlying these studies is the comparison of a group of individuals carrying a specific phenotype (the "case" group) against another group of individuals, which lack this phenotype (the "control" group). The phenotype, or trait, is a disease, which the members of the case group are carrying, whereas the control group members are healthy with respect to the disease in question. Another example would be a number of bacterial strains in the case group, which carry a resistance trait, whereas the members of the control group do not. The task of the case–control study is to identify markers that are significantly different in these two groups, for instance SNPs which are statistically overrepresented in one of the groups, genes which are over- or under-expressed, and proteins, peptides, or metabolites which are present in elevated concentration.

In the past many biomarkers have been identified as a by-product of research trying to understand biological pathways or through the so-called candidate gene approaches. In such an approach an a priori hypothesis is the basis for the attempt to identify an association between the genetic variation within pre-specified genes and a disease, or a phenotype. The high-throughput technologies developed recently (the so-called -omics technologies)—genomics, transcriptomics, proteomics, and metabolomics—generated a boost in biomarker development and identification. The -omics technologies include large-scale methods for the purification, identification, and characterization of DNA, RNA, proteins, and metabolites. The huge amount of data produced in -omics experiments provide the foundation for systematic, comprehensive, and unbiased studies.

2.1 *Genomic Markers*

Most projects aiming at the identification of biomarkers through genomics approaches use the so-called next-generation sequencing (NGS) or even single-molecule DNA sequencing technology, which is sometimes referred to as "third-generation sequencing" (Mardis 2008, 2013; van Dijk et al. 2014).

Information on genome-wide data allows the search for the presence and/or the absence of genes as well as for nucleotide polymorphisms, especially single nucleotide polymorphisms (SNPs), insertions, deletions, and microsatellites which are correlated with a specific phenotype.

The large-scale genome projects for biomarker identification are frequently called genome-wide association studies (GWAS), or WGAS, for whole-genome association studies. During such a study markers are tested for a non-random distribution between cases and controls, i.e. genomic variations, which are over- or under-represented in a case group compared to a control group. Although the association of genomic variations with a disease or phenotype seems to be obvious, case–control studies face several statistical problems which have to be properly addressed. One of the major problems in the statistical evaluation of GWAS data is the fact that frequently the population of interest includes subgroups of individuals that are on average more related to each other than to other members of the wider population (Balding 2006).

Here an association between a phenotype and a marker may be detected which is not causative but rather based on relationship, a so-called spurious association. Thus, a careful selection of cases and controls for a homogenous population structure is important for a proper statistical analysis (Pritchard and Donnelly 2001).

Similarly in bacterial species which rarely recombine and thus are highly clonal, the population structure has to be taken into account in GWAS (Read and Massey 2014; Chen and Shapiro 2015). A frequently applied approach to detect bacterial subpopulations is the construction of a phylogenetic tree which is usually based on sequence variations of housekeeping genes or the core genome of the given bacterial species (Leekitcharoenphon et al. 2014).

2.2 Transcriptomics Experiments

With transcriptomics a molecular pattern or transcriptional fingerprint is produced, which can serve as an RNA-based biomarker. Clustering of gene expression profiles with significantly up- or down-regulated genes provides an efficient method for the description and investigation of even complex phenotypes. Since today transcriptomics experiments are based on sequencing RNA molecules, transcriptome experiments also provide information about sequence variations of the transcribed genes at the single nucleotide level.

2.3 Proteomic Biomarker Research

The term proteome refers to the entire set of proteins which are expressed by a cell, tissue, or organism at a certain time point under defined conditions. Some peptides and proteins are released from cells and tissues—in case of infection both from host and pathogen—into the surrounding matrix, either as mature proteins, or as products of cleavage and degradation. They can therefore be detected in body fluids, such as blood, urine, cerebrospinal fluid, etc. Disease alters the composition of the proteome in these body fluids, so that proteins can be used as biomarkers for the diseased state. The predominant clinical sample for protein biomarker detection is blood serum (Liotta et al. 2003).

Main technologies used for proteomics are especially high-resolution mass spectrometry (MS) methods, but other techniques are also applied, such as microarray technology (Rai 2007; Classen et al. 2008).

Quantification of protein levels is critical for the identification and measurement of protein biomarkers, because a given protein or peptide usually is present in samples of both healthy and diseased tissues. A single protein marker, however, will usually not be sensitive and specific enough for a safe assessment. In most cases a protein profile, a signature, will be used (He and Chiu 2003; Dittmar and Selbach 2015). There are examples, though, in which single markers are successfully applied, such as procalcitonin, which is in clinical use as a marker for bacterial infection and sepsis (Reinhart and Hartog 2010).

2.4 *Metabolomics*

The metabolome is defined as the complete set of small molecules which is found in a biological sample. It contains endogenous metabolites and signaling molecules such as amino acids, organic acids, fatty acids, nucleic acids, sugars, vitamins, and others, but also exogenous chemicals introduced by diet, infecting organisms, drugs, toxins, and antibiotics (Ackermann et al. 2006; Collino et al. 2013).

Metabolomics involves the identification of a metabolic fingerprint of a biological sample, such as in urine, blood, and exhaled breath (Banoei et al. 2014). A special field is lipidomics, which is a lipid-targeted metabolomic approach, elucidating molecular mechanisms which are involved in the disruption of lipid metabolic pathways (Collino et al. 2013; Zhao et al. 2015).

Techniques used in metabolomics for the identification of biomarkers are usually based on MS coupled to either gas (GC-MS) or liquid chromatography (LC-MS) and proton nuclear magnetic resonance (^{1}H NMR) spectroscopy (Collino et al. 2013; Banoei et al. 2014).

3 Pathogen Identification and Characterization

The responsibility of clinical microbiology laboratories is the detection and characterization of the causal agent, to identify the bacterial species, to perform antibiotic susceptibility testing, and to provide a surveillance infrastructure. The diagnostic laboratories report to clinicians in the hospitals or to practitioners in order to guide them in their treatment choice or the implementation of hygiene measures. Today antibiotic susceptibility testing involves phenotypic test methods which mostly rely on semiautomated systems. There are only few molecular resistance profiling assays on the market which provide rapid results. However, those assays do not cover the entire "resistome" of the pathogens and are far too expensive to be introduced into routine diagnostic microbiology. Thus, technical advances in today's routine diagnostic microbiology are still focussed on culture-dependent methods. Nevertheless, there are continuous efforts to optimize molecular genotyping assays.

3.1 Detection and Identification of Pathogens

The detection of most bacteria is commonly performed by the use of culture-based methods. Organism culture can be performed at low cost. Although still the gold standard for the identification of bacterial pathogens, with a turnaround time of at least 12 hours the method is too time-consuming for early diagnosis of bacterial infections (Köser et al. 2012). Furthermore, many microbes are difficult to cultivate. These include important pathogens such as mycobacteria or chlamydia. Problems also occur when the species composition of complex bacterial communities has to be resolved, for instance in stool microbiota. Therefore many scientists propose to shift pathogen detection from culture-based to molecular-based diagnostic methods (Infectious Diseases Society of America 2011).

For the identification of microorganisms semiautomated systems are used in clinical microbiology since the 1970s. They combine a number of phenotypic assays and identify bacterial species by morphological and biochemical tests, supplemented as needed by specialized tests such as serotyping and antibiotic inhibition patterns (Endimiani et al. 2011). More recently a fast and reliable method for the identification and differentiation of bacterial species has been introduced into clinical microbiology. The development of matrix-assisted laser desorption ionization time-of-flight MS (MALDI-TOF MS) devices has revolutionized the routine identification of microorganisms in clinical microbiology laboratories. The technique generates characteristic mass spectral fingerprints as unique signatures for each microbe. These biomarker peaks result from signals corresponding to highly abundant conserved proteins (or typical fragments thereof), mostly ribosomal proteins or proteins with housekeeping function (Croxatto et al. 2012). A small number of unique peaks (5–10 markers) in a spectrum is usually enough to identify a microbe at the genus and species level; larger numbers of characteristic peaks frequently even serve subspecies identification and thus may be used for strain typing purposes. Advantages of MS are the high-resolution speed of about one sample per minute and a high degree of automation (Rodrigues Ribeiro Teles et al. 2010). Furthermore, MS-based species identification is robust, cost-effective, fast, and does not require a predefined target. This allows the identification of organisms from very diverse groups, including fastidious bacteria, anaerobes, mycobacteria, fungi, viral pathogens, and even some kinds of pathogenic algae. The application of MS methods to date still requires a single colony from culture, although work is in progress to achieve bacterial identification directly from clinical samples (Croxatto et al. 2012; Randell 2014; Kostrzewa et al. 2013).

For more than a decade PCR-based methods have been used for species identification through amplification of species-specific or strain-specific DNA sequences. The sequences targeted for such a PCR-based sequence typing are housekeeping genes or the 16S rRNA gene and the 16S-23S rRNA gene spacer region. Primer pairs specific for pathogenic microbes are used in multiplex PCR assays. For the detection of uncharacterized or unknown species a pair of universal primers can be used, most commonly designed to amplify the 16S rRNA gene sequence (Cai et al. 2014).

3.2 *Epidemiology*

Next to hygienic measures the most powerful weapons against multidrug-resistant infections are early diagnosis of antimicrobial resistance and the implementation of effective surveillance and outbreak detection infrastructures. Investigations on the epidemiology of the bacterial pathogens may uncover the spread and outbreak reservoirs of multiresistant bacterial pathogens (Schürch and Siezen 2010).

Molecular genotyping methods have provided new approaches for enhanced surveillance and outbreak detection. These include multilocus sequence typing (MLST, Maiden et al. 1998) that detects sequence variations within housekeeping genes as universal discriminatory markers to infer genetic relationships between various strains. The established sequence typing techniques such as MLST cannot, however, produce the same discriminative power as whole-genome sequencing (WGS) could (Köser et al. 2012).

3.3 *Resistance Profiling*

One of the key elements of pathogen characterization is antimicrobial resistance profiling. The knowledge about resistances against and susceptibility to antimicrobial agents determines the course of patient treatment. Antibiotic susceptibility test methods include agar and broth dilution as well as disk diffusion methods. There are also commercial semiautomated antimicrobial susceptibility testing systems. These tests mostly detect the direct effect of antimicrobial drugs on growing microorganisms. The disadvantage of conservative testing strategies is that they are time-consuming.

The successful use of MS, especially MALDI-TOF MS, for the identification of pathogens has encouraged further developments for pathogen characterization and resistance profiling via MS. Rather than a direct detection of resistance conferring proteins, such as antibiotic-degrading enzymes, antimicrobial resistance can be detected through characteristic peaks of antibiotic degradation products, that is exemplified by the carbapenem antibiotics (Hrabák et al. 2011) and β-lactamase antibiotics (Sparbier et al. 2012). The advantage of such an approach is that the direct result of the antibiotic-degrading activity is traced.

Detection systems that are based on the identification of resistance conferring genes by the use of DNA microarrays have been developed for more than a decade (Call et al. 2003; Strauss et al. 2014). There are also PCR-based resistance tests available. They test either for the presence of known resistance genes or for the presence of point mutations that confer resistance. PCR-based testing methods provide a fast time-to-result. This is particularly important for resistance profiling in *M. tuberculosis* where culture-based resistance profiling may require several weeks (Arnold et al. 2005). Other examples for molecular resistance profiling is the identification of methicillin-resistant *Staphylococcus aureus* (MRSA) by the detection of the *mecA* gene (van Belkum 2003) and the detection of carbapenemase genes directly from blood culture bottles using real-time PCR (Francis et al. 2012).

Although already applied for susceptibility testing in a number of scientific studies, whole genome sequencing (WGS) has not yet been introduced into clinical microbiology. Reasons for this are high costs, long processing times, and the requirement for trained and experienced personnel. A general problem for resistance profiling based on sequence matching is that the prediction of antimicrobial susceptibility is dependent upon the availability of a current and curated database of reference sequences. Even if no known resistance gene or pathway is detected, resistance against a particular antibiotic drug cannot be excluded. Thus, for at least some time there will be a need for phenotypic verification of the results obtained by genotypic resistance profiling (Dunne et al. 2012; Köser et al. 2012; Didelot et al. 2012). Transcriptome analysis provides information on the expression levels of genes, and thus may aid the detection of resistance mechanisms that rely on the up- or down-regulation of genes, as for instance the up-regulation of antibiotic efflux pumps.

3.4 Virulence Factor Detection

A wide variety of virulence factors mediate bacterial pathogenicity and can cause severe invasive infections. Knowledge of virulence factors and their mode of action is important and can serve as prognostic markers at an early stage of the disease and may provide information which is crucial to the decision for the treatment approach. Virulence determinants might also be important targets for antimicrobial therapy: Targeting bacterial virulence inhibits pathogenesis without placing a high selection pressure on the target pathogen (Cegelski et al. 2008).

4 Point-of-Care (POC) Testing

The majority of the infectious diseases are treatable, but simple, accurate, and stable tests for their diagnosis are lacking. The challenge is to design tests for early diagnosis, which are selective, specific, and fast on the one hand, but on the other hand have to be easy-to-handle and technically robust.

The turnaround time of such diagnostic devices and tests has to be short. It is important that the time which is required to perform a medical test is short enough so that on the basis of the test results a safe decision of the required therapy can be made and the patient can be send home with the proper medication.

A timely completion of testing with subsequent clinical decision and prompt initiation of appropriate therapy is not only reducing morbidity and mortality in patients, but also helps to slow down the emergence of virulent and resistant strains (Myers et al. 2011; Pai et al. 2012) and reduces costs for healthcare system and society.

For this reason the World Health Organization (WHO) has called for the development of accurate and affordable diagnostic tests and describes the optimal diagnostic test with the acronym ASSURED. The ASSURED criteria are **A**ffordable, **S**ensitive (yielding few false-negative results), **S**pecific (yielding few false-positive results), **U**ser-friendly (simply to perform and requiring minimal training), **R**apid (enabling treatment at first visit) and robust (not requiring storage in a cold chain), **E**quipment-free, **D**eliverable (to those who need it) (WHO 2004).

5 Conclusions

Although the development of antibiotic drugs has to be urgently accelerated, the development of fast and reliable diagnostic methods is of equal importance. A shift from the traditional, established, but time-consuming methodologies of pathogen identification to molecular biological test systems may be achieved in the next few years. It seems obvious that molecular diagnostic methods are faster and offer much more clinically relevant information about the pathogen than the traditional culture-based methods, although comparative studies are required to prove this.

Several promising molecular methods for pathogen identification and characterization are already in use, albeit yet rather in research laboratories, including PCR- and microchip-based technologies and next generation sequencing (NGS). Mass spectrometry (MS) has already made its way into diagnostic labs.

Whole-genome sequencing (WGS) has been introduced into clinical diagnostics just recently. The potential of WGS for clinical microbiology is enormous, offering possibilities not only for the identification of the species and clonal relatedness of an organism, but also for the detection of resistance markers and virulence factors, all in a single processing step. Genome-wide SNP detection, generation of expression profiles via RNA-seq, and extraction of epidemiological data—features which cannot, or only to a limited extent, be produced by other analysis and diagnostic techniques—are possible. As WGS does not require homogeneous samples, metagenomics approaches will become applicable. Disadvantages are still the high costs for the sequencing device, relatively high running costs and run times, and the requirement of well-trained personnel to functionalize and interpret the genome-wide data (Didelot et al. 2012; Dunne et al. 2012; Köser et al. 2012).

Nevertheless, the speed in the development of WGS techniques and the progress made recently makes the vision of small easy-to-use tabletop WGS machines for clinical use plausible. The development of hand-held sequencers such as the MinION designed by Oxford Nanopore Technologies (Kilianski et al. 2015) or the Freedom4 developed at Otago University demonstrates that there is a good chance that WGS might develop into a gold standard for clinical microbiology in the next few years.

References

Ackermann BL, Hale JE, Duffin KL (2006) The role of mass spectrometry in biomarker discovery and measurement. Curr Drug Metab 5:525–539

Arnold C, Westland L, Mowat G et al (2005) Single-nucleotide polymorphism-based differentiation and drug resistance detection in Mycobacterium tuberculosis from isolates or directly from sputum. Clin Microbiol Infect 11:122–130

Balding DJ (2006) A tutorial on statistical methods for population association studies. Nat Rev Genet 7:781–791

Banoei MM, Donnelly SJ, Mickiewicz B et al (2014) Metabolomics in critical care medicine: a new approach to biomarker discovery. Clin Invest Med 37:E363–E376

Biomarkers Definitions Working Group (2001) Biomarkers and surrogate endpoints: preferred definitions and conceptual framework. Clin Pharmacol Ther 69:89–95

Cai HY, Caswell JL, Prescott JF (2014) Nonculture molecular techniques for diagnosis of bacterial disease in animals: a diagnostic laboratory perspective. Vet Pathol 51:341–350

Call DR, Bakko MK, Krug MJ et al (2003) Identifying antimicrobial resistance genes with DNA microarrays. Antimicrob Agents Chemother 47:3290–3295

Cegelski L, Marshall GR, Eldridge GR et al (2008) The biology and future prospects of antivirulence therapies. Nat Rev Microbiol 6:17–27

Chen PE, Shapiro BJ (2015) The advent of genome-wide association studies for bacteria. Curr Opin Microbiol 25:17–24

Classen S, Staratschek-Jox A, Schultze JL (2008) Use of genome-wide high-throughput technologies in biomarker development. Biomark Med 2:509–524

Collino S, Martin FP, Rezzi S (2013) Clinical metabolomics paves the way towards future healthcare strategies. Br J Clin Pharmacol 75:619–629

Croxatto A, Prod'hom G, Greub G (2012) Applications of MALDI-TOF mass spectrometry in clinical diagnostic microbiology. FEMS Microbiol Rev 36:380–407

Dellinger RP, Levy MM, Rhodes A et al, Surviving Sepsis Campaign Guidelines Committee including the Pediatric Subgroup (2012) Surviving sepsis campaign: international guidelines for management of severe sepsis and septic shock: 2012. Crit Care Med 41:580–637

Didelot X, Bowden R, Wilson DJ et al (2012) Transforming clinical microbiology with bacterial genome sequencing. Nat Rev Genet 13:601–612

Dittmar G, Selbach M (2015) SILAC for biomarker discovery. Proteomics Clin Appl 9:301–306

Dunne WM Jr, Westblade LF, Ford B (2012) Next-generation and whole-genome sequencing in the diagnostic clinical microbiology laboratory. Eur J Clin Microbiol Infect Dis 31:1719–1726

Endimiani A, Hujer KM, Hujer AM et al (2011) Are we ready for novel detection methods to treat respiratory pathogens in hospital-acquired pneumonia? Clin Infect Dis 52(Suppl 4):S373–S383

European Commission (2010) Biomarkers for patient stratification. http://ec.europa.eu/research/health/pdf/biomarkers-for-patient-stratification_en.pdf

Farhat MR, Shapiro BJ, Kieser KJ et al (2013) Genomic analysis identifies targets of convergent positive selection in drug-resistant Mycobacterium tuberculosis. Nat Genet 45:1183–1189

Fitousis K, Moore LJ, Hall J et al (2010) Evaluation of empiric antibiotic use in surgical sepsis. Am J Surg 200:776–782

Francis RO, Wu F, Della-Latta P et al (2012) Rapid detection of Klebsiella pneumoniae carbapenemase genes in enterobacteriaceae directly from blood culture bottles by real-time PCR. Am J Clin Pathol 137:627–632

Garnacho-Montero J, Garcia-Garmendia JL, Barrero-Almodovar A et al (2003) Impact of adequate empirical antibiotic therapy on the outcome of patients admitted to the intensive care unit with sepsis. Crit Care Med 31:2742–2751

He QY, Chiu JF (2003) Proteomics in biomarker discovery and drug development. J Cell Biochem 89:868–886

Hrabák J, Walková R, Studentová V et al (2011) Carbapenemase activity detection by matrix-assisted laser desorption ionization-time of flight mass spectrometry. J Clin Microbiol 49:3222–3227

Infectious Diseases Society of America (2011) An unmet medical need: rapid molecular diagnostics tests for respiratory tract infections. Clin Infect Dis 52(Suppl 4):S384–S395

Kilianski A, Haas JL, Corriveau EJ et al (2015) Bacterial and viral identification and differentiation by amplicon sequencing on the MinION nanopore sequencer. Gigascience 4:12

Köser CU, Ellington MJ, Cartwright EJ et al (2012) Routine use of microbial whole genome sequencing in diagnostic and public health microbiology. PLoS Pathog 8:e1002824

Kostrzewa M, Sparbier K, Maier T et al (2013) MALDI-TOF MS: an upcoming tool for rapid detection of antibiotic resistance in microorganisms. Proteomics Clin Appl 7:767–778

Kumar A, Roberts D, Wood KE et al (2006) Duration of hypotension before initiation of effective antimicrobial therapy is the critical determinant of survival in human septic shock. Crit Care Med 34:1589–1596

Lawn SD, Mwaba P, Bates M et al (2013) Advances in tuberculosis diagnostics: the Xpert MTB/RIF assay and future prospects for a point-of-care test. Lancet Infect Dis 13:349–361

Leekitcharoenphon P, Nielsen EM, Kaas RS et al (2014) Evaluation of whole genome sequencing for outbreak detection of Salmonella enterica. PLoS ONE 9:e87991

Liotta LA, Ferrari M, Petricoin E (2003) Clinical proteomics: written in blood. Nature 425:905

Livermore DM, Wain J (2013) Revolutionising bacteriology to improve treatment outcomes and antibiotic stewardship. Infect Chemother 45:1–10

MacArthur RD, Miller M, Albertson T et al (2004) Adequacy of early empiric antibiotic treatment and survival in severe sepsis: experience from the MONARCS trial. Clin Infect Dis 38:284–288

Maiden MC, Bygraves JA, Feil E et al (1998) Multilocus sequence typing: a portable approach to the identification of clones within populations of pathogenic microorganisms. Proc Natl Acad Sci USA 95:3140–3145

Mardis ER (2008) Next-generation DNA sequencing methods. Annu Rev Genomics Hum Genet 9:387–402

Mardis ER (2013) Next-generation sequencing platforms. Annu Rev Anal Chem 6:287–303

Myers FB, Henrikson RH, Xu L et al (2011) A point-of-care instrument for rapid multiplexed pathogen genotyping. Conf Proc IEEE Eng Med Biol Soc 2011:3668–3671

Pai NP, Vadnais C, Denkinger C et al (2012) Point-of-care testing for infectious diseases: diversity, complexity, and barriers in low- and middle-income countries. PLoS Med 9: e1001306

Pritchard JK, Donnelly P (2001) Case-control studies of association in structured or admixed populations. Theor Popul Biol 60:227–237

Rai AJ (2007) Biomarkers in translational research: focus on discovery, development and translation of protein biomarkers to clinical immunoassays. Expert Rev Mol Diagn 7:545–553

Randell P (2014) It's a MALDI but it's a goodie: MALDI-TOF mass spectrometry for microbial identification. Thorax 69:776–778

Read TD, Massey RC (2014) Characterizing the genetic basis of bacterial phenotypes using genome-wide association studies: a new direction for bacteriology. Genome Med 6:109

Reinhart K, Hartog CS (2010) Biomarkers as a guide for antimicrobial therapy. Int J Antimicrob Agents 36:S17–S21

Rodrigues Ribeiro Teles FS, de Távora Pires, Tavira LA, Pina da Fonseca LJ (2010) Biosensors as rapid diagnostic tests for tropical diseases. Crit Rev Clin Lab Sci 47:139–169

Schürch AC, Siezen RJ (2010) Genomic tracing of epidemics and disease outbreaks. Microb Biotechnol 3:628–633

Soo Hoo GW, Wen YE, Nguyen TV et al (2005) Impact of clinical guidelines in the management of severe hospital-acquired pneumonia. Chest 128:2778–2787

Sparbier K, Schubert S, Weller U et al (2012) Matrix-assisted laser desorption ionization-time of flight mass spectrometry-based functional assay for rapid detection of resistance against β-lactam antibiotics. J Clin Microbiol 50:927–937

Strauss C, Endimiani A, Perreten V (2014) A novel universal DNA labeling and amplification system for rapid microarray-based detection of 117 antibiotic resistance genes in Gram-positive bacteria. J Microbiol Methods 108:25–30

van Belkum A (2003) Molecular diagnostics in medical microbiology: yesterday, today and tomorrow. Curr Opin Pharmacol 3:497–501

van Dijk EL, Auger H, Jaszczyszyn Y et al (2014) Ten years of next-generation sequencing technology. Trends Genet 30:418–426

Wallis RS, Pai M, Menzies D et al (2010) Biomarkers and diagnostics for tuberculosis: progress, needs, and translation into practice. Lancet 375:1920–1937

World Health Organization (2004) Mapping the landscape of diagnostics for sexually transmitted infections. http://www.who.int/tdr/publications/documents/mapping-landscape-sti.pdf

Zhao YY, Cheng XL, Lin RC et al (2015) Lipidomics applications for disease biomarker discovery in mammal models. Biomark Med 9:153–168

Zignol M, van Gemert W, Falzon D et al (2012) Surveillance of anti-tuberculosis drug resistance in the world: an updated analysis, 2007–2010. Bull World Health Organ 90:111–119D

Zolg JW, Langen H (2004) How industry is approaching the search for new diagnostic markers and biomarkers. Mol Cell Proteomics 3:345–354

Use of Antibiotics and Antimicrobial Resistance in Veterinary Medicine as Exemplified by the Swine Pathogen *Streptococcus suis*

Maren Seitz, Peter Valentin-Weigand and Jörg Willenborg

Abstract Use of antimicrobial agents in veterinary medicine is essential to control infectious diseases, thereby keeping animals healthy and animal products safe for the consumer. On the other hand, development and spread of antimicrobial resistance is of major concern for public health. *Streptococcus* (*S.*) *suis* reflects a typical bacterial pathogen in modern swine production due to its facultative pathogenic nature and wide spread in the pig population. Thus, in the present review we focus on certain current aspects and problems related to antimicrobial use and resistance in *S. suis* as a paradigm for a bacterial pathogen affecting swine husbandry worldwide. The review includes (i) general aspects of antimicrobial use and resistance in veterinary medicine with emphasis on swine, (ii) genetic resistance mechanisms of *S. suis* known to contribute to bacterial survival under antibiotic selection pressure, and (iii) possible other factors which may contribute to problems in antimicrobial therapy of *S. suis* infections, such as bacterial persister cell formation, biofilm production, and co-infections. The latter shows that we hardly understand the complexity of factors affecting the success of antimicrobial treatment of (porcine) infectious diseases and underlines the need for further research in this field.

Contents

M. Seitz · P. Valentin-Weigand (✉) · J. Willenborg
Institute for Microbiology, University of Veterinary Medicine,
Bischofsholer Damm 15, 30173 Hannover, Germany
e-mail: Peter.Valentin@tiho-hannover.de

M. Seitz
e-mail: Maren.Seitz@tiho-hannover.de

J. Willenborg
e-mail: Joerg.Willenborg@tiho-hannover.de

Current Topics in Microbiology and Immunology (2016) 398: 103–122
DOI 10.1007/82_2016_506

Published Online: 21 September 2016

1 Introduction to the Use of Antimicrobial Agents in Swine Husbandry

Antimicrobial agents are widely used in the treatment of food-producing animals (Wassenaar 2005), since they are important to control infectious diseases, thereby allowing economic production of healthy animals and animal products which are safe for the consumer. However, use of antimicrobials bear a risk for the development of antimicrobial resistance in bacteria (Hao et al. 2014), which is of major concern in public health. The "One-Health" concept should promote the integration of public health, veterinary disease, food, and environmental surveillance to strengthen the detection and control of bacterial resistance.

The enactment of several political restrictions on the use of antimicrobials is an attempt to restrict the amount of antibiotics in veterinary medicine in order to reduce the increase in antimicrobial resistance. In the EU the use of antimicrobials as growth promotors was banned in 2006. Taiwan and Mexico have similar restrictions but with some exceptions for several antibiotics. Nevertheless, in many countries, including the USA, Australia, and many Asian countries, antimicrobials are still frequently used for growth-enhancing purposes (Maron et al. 2013). Sweden was the pioneer in antimicrobial control policies and restricted use as early as 1986, followed by Denmark in 2000 (Aarestrup 2012). Following, the use of antimicrobials in food-producing animals without veterinary prescriptions was prohibited in the EU, Taiwan, Mexico, and also in Japan, Hong Kong, and Brazil. More recently, several nationwide antibiotic resistance strategies and systems have been established to monitor, control, and reduce antimicrobial use in food-producing animals more efficiently, such as in Germany the *Deutsche Antibiotika-Resistenzstrategie* (DART) (2010), the Swedish Veterinary Antimicrobial Resistance Monitoring (SVARM) (2012), the Austrian report on antimicrobial resistance (AURES) (2012), the Monitoring of Antimicrobial Resistance and Antibiotic Usage in Animals in the Netherlands (MARAN) (2013), and the Danish Integrated Antimicrobial Resistance Monitoring and Research Program (DANMAP) (2013), respectively (van Rennings et al. 2014, 2015). Of note, in Germany veterinarians are allowed to dispense medical products such as the delivery of antimicrobials to the animal owner for a distinct purpose via self-administration, but consumption of antimicrobials has to be documented by means of the German pharmaceutical law. On the other hand, in contrast to some other (European) countries, prophylactic administration of antimicrobials is prohibited in Germany.

Nevertheless, a wide range of antibiotics are used in pig husbandry for targeted prophylactic (individual pigs), metaphylactic (herd treatment), and therapeutic use after veterinary prescription (Barton 2014). In a recent study on the antimicrobial use on 495 pig farms in Germany by van Rennings et al. (2015), tetracyclines were most commonly used, followed by beta-lactams and trimethoprim-sulfonamides (van Rennings et al. 2014, 2015). According to the WHO (2011) highest priority critically important antimicrobials used for specific infections in humans, like fluochinolones, cephalosporins, glycopeptides, and aminoglycoside antimicrobials, were infrequently used. In contrast, macrolides, which are also especially important for human medicine, were still moderately used in 2011 in Germany and Austria (Trauffler et al. 2014; van Rennings et al. 2014, 2015). In good agreement, sales quantities of antimicrobial substances showed the same ranking in 2011 in Germany (van Rennings et al. 2014, 2015). A similar distribution of antimicrobial consumption was observed in 10 investigated European countries in 2007, in which tetracyclines were most often sold in relation to the biomass of food-producing animals (including pigs, poultry, and cattle). In all investigated countries, except for Switzerland and the UK, beta-lactams were ranked second and trimethoprim-sulfonamides third. Most of the antimicrobial substances were sold in the Netherlands, followed by France. In the UK, Czech Republic, Switzerland, and Germany sales volumes were comparable and at the third position, followed by Denmark. Antimicrobials were less frequently used in Finland, Sweden, and Norway, which reflects a less intensive pig industry in these countries (Grave et al. 2010). In the Australian and Canadian pig industry sulfonamides, tetracyclines, and penicillines were the most widely used antimicrobial compounds in the recent years (Jordan et al. 2009; Rajic et al. 2006). According to a statement by the German "Federal Office of Consumer Protection and Food Safety" in 2014 the delivery quantities of all antimicrobial substances in Germany were 15 % lower than in 2011 with penicillins being more frequently used than tetracyclines. However, the overall decrease should be interpreted with care since quantities of third generation cephalosporines and fluorchinolones increased. Especially fluorchinolones were licensed for much lower dosages (2.5–10 mg per kg body weight) when compared to, e.g., tetracyclines (up to 80 mg per kg body weight) in cases of antibiotic therapy (Wallmann et al. 2015). This in an example how such data may be underestimated in terms of effective antibiotic treatments that actually occurred in the field. In order to more precisely interpret annual delivery quantities of antimicrobial substances in the future (also retrospectively), surveillance programs need to define comparable standards. Good examples are the definitions 'treatment unit' (number of treatments per active ingredient per animal and per day) and 'treatment frequency' (the number of treatment days per active ingredient per animal), respectively, which allow to record the actual use of antimicrobial substances in animal husbandry in a more comparable and reliable manner (van Rennings et al. 2015).

Based on projections more than 1300 tons of antimicrobial substances were used in food-producing animals (pigs, broiler chicken, cattle) in Germany in 2011, of which the majority (974.5 tons) was administrated to pigs (van Rennings et al. 2014). On pig farms, respiratory diseases were by far the most reported indication for the consumption of antimicrobials by pigs of all age groups. For piglets, weaners, and fattening pigs, intestinal diseases are the second most indication for antimicrobial use, whereas reproductive disorders are more important in sows (van Rennings et al. 2015). Based on different age groups within the pig population, piglets were treated on average most frequently with 14.7 of 100 days (standardized period), followed by weaners (6.6/100), fatting pigs (3.7/100), and at least sows (0.9/100) on German pig farms in 2011 (van Rennings et al. 2015). Most of the antimicrobials were administrated via the oral route as feed components (98 %), whereas only 2 % were applied by intramuscular injection (Callens et al. 2012; Timmerman et al. 2006; van Rennings et al. 2015). For in-feed applications higher doses are necessary to reach efficient treatment concentrations, leading to a considerable higher antimicrobial consumption. Of importance, a study on consumption of antimicrobial substances in Belgium revealed that 50–75 % of all orally administrated antimicrobials were underdosed, whereas 90 % of injectable antimicrobials were overdosed (Timmerman et al. 2006). Taken together, tetracyclines, beta-lactams, and trimethoprim-sulfonamides are the most commonly used antimicrobial substances in pig husbandry, and piglets were most frequently treated, at least in Germany.

2 General Aspects of Antimicrobial Resistance in Veterinary Medicine

The use of antimicrobial substances and possibly associated increasing resistance of bacteria is an emerging problem in animal husbandry worldwide. Resistant bacteria are not only a major concern for animal health but have also been identified as a risk factor for the development of antimicrobial resistance in human infections. Consumption of animal products or exposure to resistant bacteria by direct contact might function as a source of antimicrobial resistance. For each approved antimicrobial agent, an individual dosage (mg/kg body weight), application route, application duration, and, if applicable, reapplications are defined, based on the specific minimal inhibitory concentration (MIC, in μg/ml) breakpoints and pharmacokinetic data of the respective antimicrobial substance. The Clinical and Laboratory Standards Institute (CLSI) publishes MIC breakpoints of antimicrobials commonly used in veterinary medicine for several, but not all pathogens. According to the CLSI criteria, MIC breakpoints are interpreted by the laboratory as "susceptible," "resistant," or "intermediate." A classification as "susceptible" indicates that the respective antimicrobial substance used in the recommended dosage for the particular animal and bacterial species is sufficient to inhibit bacterial growth at the infected tissue site (Leekha et al. 2011). As an example, for tetracycline (which is frequently used in pig husbandry) a "susceptible" MIC breakpoint of ≤ 0.5 μg/ml, an "intermediate" MIC breakpoint of 1 μg/ml, and a "resistant" breakpoint of ≥ 2 μg/ml were defined for *Streptococcus suis*. Moreover, the MIC is commonly used in diagnostic laboratories to determine resistance of bacteria for further treatment advices (Andrews 2001). Nevertheless, the MIC has some limitations. For instance, the required concentrations determined in vitro may be not sufficient to inactivate the respective bacteria in vivo, or the achieved concentration of the antimicrobial substance in the target tissue is too low to inhibit bacterial growth efficiently. Furthermore, the MIC is always species-specific or even specific for an individual bacterial strain, serotype, or isolate (Leekha et al. 2011). Gradually increasing MIC values of bacterial isolates are of interest and may be indicative of an emerging antimicrobial resistance phenotype of a bacterial species (Callens et al. 2013).

3 *Streptococcus suis* as a Major Pathogen in Swine

S. suis is the most important bacterial cause of meningitis in pigs worldwide and a major porcine pathogen leading to high economical losses in pig husbandry. Efficient treatment and prevention of *S. suis* infections is of crucial interest in pig husbandry and, thus, contributes substantially to antimicrobial use in swine (Goyette-Desjardins et al. 2014). The pathogen causes diseases in pigs of different ages, including suckling and weaning piglets as well as growers. Importantly, *S. suis* is a colonizer of mucosal surfaces, in particular the upper respiratory tract and, thus, healthy carrier piglets are the major reservoir (Clifton-Hadley and Alexander 1980). Notably, *S. suis* might also cause zoonotic infections, such as meningitis, septicemia, endocarditis and other diseases in humans. Close contact with pigs, wild boars or pork is considered to be an important risk factor for this zoonosis (Arends and Zanen 1988; Baums et al. 2007). In Asia, *S. suis* is classified as an emerging zoonotic pathogen as it was involved in two large outbreaks of severe human infections in 1998 and 2005 in China, and it is currently considered as one of the most important causes of bacterial meningitis in adults (Mai et al. 2008; Wertheim et al. 2009). For further details on the zoonotic potential of *S. suis*, the reader is referred to recent reviews (Fulde and Valentin-Weigand 2013; Gottschalk et al. 2007; Lun et al. 2007). Pigs of any age can be infected, but susceptibility generally decreases with age following weaning. Outbreaks occur usually due to the introduction of a carrier into the herd. Within a carrier herd, outbreaks occur especially in young animals disposed to stress conditions. As *S. suis* is a facultative pathogen, different biotic and abiotic factors such as virus infections, corrosive gases, and crowding are thought to promote *S. suis* diseases in modern swine production.

Several modes of transmission between animals in a herd have been suggested. Piglets born to sows with genital infection may acquire the infection vertically

(Amass et al. 1997). Transmission can also occur by contact (Berthelot-Herault et al. 2001), or by infection through the navel, genital, or alimentary tract (Staats et al. 1997). Colonized pigs will usually harbor the bacteria in their tonsils. Some animals will remain healthy carriers, whilst other will sooner or later develop clinical signs (Gottschalk and Segura 2000). Morbidity rarely exceeds 5 %, although it can reach more than 50 % in cases of poor hygiene and concurrent disease (Staats et al. 1997). With appropriate treatment mortality is usually low (ca. 5 %), but can be up to 20 % in untreated herds. The preferred antimicrobials used for treatment of *S. suis* infections in pigs are beta-lactams (penicillin, ceftriaxone, or ceftiofur) and fluoroquinolones (enrofloxacin) (Day et al. 2015; Yao et al. 2014). In humans similar antimicrobial agents are used often in combination with dexamethasone administration (Mai et al. 2008).

4 Antimicrobial Resistance in *S. suis*

A comprehensive review on the resistance patterns for *S. suis* isolates from various countries of the most commonly use antimicrobial substances in the pig industry was published in 2013 (Varela et al. 2013). This retrospective study includes various data of different years and time periods (1992–2008) as well as different isolates, serotypes, and sources, including porcine and human isolates from clinically healthy as well as diseased individuals. For more detailed data the reader is referred to this review (Varela et al. 2013).

Resistance of *S. suis* against tetracycline is common in all countries worldwide, but geographical differences occur (Soares et al. 2014; Varela et al. 2013). High levels of tetracycline resistance were found in North America, Asia, and some European countries like Germany, Spain, and Portugal ranging from 86.9 up to 100 % resistance. In all other European countries a moderate tetracycline resistance was reported (40.3–73.3 %). In contrast, due to early use prohibition, tetracycline resistance is very low in Sweden (7.7 %) (Aarestrup et al. 1998). Nevertheless, up to 90 % resistance for tetracycline was reported for human *S. suis* isolates, especially in Asia where a significant increase in resistance to tetracycline was found in human patients suffering from meningitis (Hoa et al. 2011; Ma et al. 2008; Strangmann et al. 2002). One hundred percent tetracycline resistance was also reported in a recent study in *S. suis* strains isolated from a Chinese pig farm, and a significant association between the prevalence of the *tetO* gene and sequence type 1 isolates was observed (Huang et al. 2015). Interestingly, resistance of *S. suis* to beta-lactams, like penicillin and ampicillin, the second most frequently used antimicrobial subclass, was relatively low in all investigated countries (0–27 and 0–23 %, respectively). Likewise, all tested human isolates from clinically healthy and diseased patients were susceptible to penicillin (Huang et al. 2015; Marie et al. 2002; Varela et al. 2013).

Notably, increasing macrolide and lincosamide resistance rates were found worldwide. Varela et al. 2013 observed a high prevalence of erythromycin (macrolide) and clindamycin (lincosamide) resistance in North America (67.2–82 and 40–91 %, respectively). In European countries erythromycin resistance showed a regional variation ranging from 8 up to 75 % resistance and clindamycin resistance was only investigated in the Netherlands (8 %) and Spain (87.4 %). Nevertheless, erythromycin as well as clindamycin resistance was low in human patients and healthy carrier pigs in Germany (Varela et al. 2013). *S. suis* isolates from Asian pigs and humans were moderately resistant ($\sim$20 %), except for Hong Kong where more than 90 % of the tested isolates were resistant (Varela et al. 2013). Moreover, a high level of resistance to tylosin was found in Spain (89.4 %) and the USA ($\sim$80 %). Rapidly increasing resistant rates to macrolides (erythromycin, azithromycin, tyosin, and tilmicosin) of *S. suis* strains isolated between 2005 and 2010 were also found in China and erythromycin resistance in *S. suis* sequence type 1 strains was significantly associated with the prevalence of the *ermB* gene (Huang et al. 2015).

Moderate antimicrobial susceptibility of *S. suis* to trimethoprim-sulfonamides, the third commonly used antimicrobial in pig industry, was found worldwide (0–59.1 %) (Varela et al. 2013). In contrast all tested *S. suis* strains isolated from clinically healthy pigs in Brazil between 2009 and 2010 were resistant to trimethoprim-sulfonamides (Soares et al. 2014).

None of the studies revealed significant differences between the proportion of resistant strains isolated from clinically diseased and clinically healthy pigs, thus antimicrobial resistance seems not to be related to virulence (Huang et al. 2015; Marie et al. 2002). Nevertheless, in a recent study on antimicrobial susceptibility of *S. suis* isolates to commonly used antimicrobials it was found that among all tested strains nearly all isolates (99.61 %) were classified as multidrug resistant (resistant to $\geq$3 antimicrobial agents), of which 9 strains were resistant to all 16 tested antimicrobial substances (Soares et al. 2014). Recently, the plasmid-born multi-resistant genes *cfr*, which encodes a methyltransferase that modifies 23S rRNA at A2503, and *fexA* a chloramphenicol/florfenicol efflux protein were identified in a *S. suis* isolate. These genes confer resistance to five chemically unrelated antimicrobial classes, including phenicols, lincosamides, oxazolidinones, pleuromutilins, and streptogramin. Likewise, this isolate exhibited high MIC values for florfenicol, clindamycin, tiamulin, tetracycline, and erythromycin (Long et al. 2006; Wang et al. 2013b). Such genetic elements may cause and transfer multi-resistant phenotypes to *S. suis*.

Geographical variations of antimicrobial resistant patterns of *S. suis* may be due to the use of different antimicrobial substances and/or regional different use prohibition of several antibiotics. Furthermore, a various actual or investigated serotype distribution has to be taken into consideration. Nevertheless, classification of different resistotypes of *S. suis* did not correlate with specific *S. suis* serotypes (Soares et al. 2014).

5 Mechanisms and Spread of Antimicrobial Resistance in *S. suis*

The different molecular mechanisms determining bacterial resistance can be roughly divided into three groups according to their mode of action: (i) minimization of the intracellular concentration of the antimicrobial due to intrinsic means (for example antibiotics (e.g., vancomycin) that cannot cross the outer membranes of Gram-negatives) or by active efflux of the compound, (ii) genetic or posttranslational modification of the bacterial target attacked by the antibiotic, and (iii) inactivation of the antibiotic itself by hydrolysis or modification (Blair et al. 2015). Despite the emergence of (multi)-drug-resistant *S. suis* strains in the recent past (Ge et al. 2012; Hu et al. 2011; Palmieri et al. 2011b), molecular mechanisms of resistance have not been well documented for *S. suis* so far. Nevertheless, examples for resistance determinants belonging to one of these groups have been described for *S. suis*. Especially the high percentage of tetracycline and macrolide resistance found in *S. suis* strains worldwide is attributed to genes encoding for ribosomal modification reactions or efflux pumps. Examples are *erm* (macrolide), *tet*(M), *tet* (O), *tet*(Q), *tet*(T), *tet*(W) (all for tetracycline resistance; capital letter code distinguishes different classes of respective tetracycline resistance genes) and *mef* (macrolide), *tet*(K), *tet*(L) (all for tetracycline resistance) genes, respectively (Hoa et al. 2011; Palmieri et al. 2011a). Furthermore, *S. suis* may inactivate lincosamides since a reconstituted *lnu*(E) gene product is able to catalyze the nucleotidylation of lincomycin (Zhao et al. 2015). For further details on the molecular mechanisms involved in antimicrobial resistance of *S. suis* to antimicrobial substances mentioned in the previous chapter the reader is referred to comprehensive review articles (Palmieri et al. 2011b; Varela et al. 2013). Here, we focus on elements that may be crucial for effective antibiotic treatment of *S. suis* infections or even spread of antibiotic resistance genes in the future. Hence, Wang et al. (2013) identified a multiresistance (*cfr*) gene in *S. suis* for which homologous genes of other pathogenic bacteria have been reported to confer resistance to five chemically unrelated antimicrobial classes, including phenicols, lincosamides, oxazolidinones, pleuromutilins, and streptogramin A (Shen et al. 2013). The exogenous localisation of the *cfr* gene on a plasmid, named pStrcfr, augments the probability of spreading among other *S. suis* strains, although this has not yet been documented. Additional means of rapid spread of antimicrobial resistance traits in *S. suis* are mobile genetic elements carrying resistance genes, such as transposons (Hoa et al. 2011; Takamatsu et al. 2003) and integrative and conjugative elements (ICE) (Holden et al. 2009; Palmieri et al. 2013, 2012). ICEs are self-transmissible, mobile genetic elements that retain the ability to undergo excision of the bacterial nucleoid and transfer by conjugation to other bacteria. For example, the S. *suis* ICE*Ssu*32457 element carries tetracycline, erythromycin, and aminoglycoside resistance genes and is transferable at high frequency within species and, even more alarming, between species (*Streptococcus agalactiae*, *Streptococcus pyogenes*, and *Streptococcus pneumoniae*) level (Marini et al. 2015; Palmieri et al. 2012).

Taken together the existence of *S. suis* genes coding for resistance against intensively used antimicrobials in veterinary medicine such as teracyclines, sulfonamides, and macrolides (van Rennings et al. 2015; Varela et al. 2013), as well as the occurrence of a multiresistance gene and the possibility of rapid spread of resistances by mobile genetic elements suggest that *S. suis* may serve as a 'resistance reservoir' for other veterinary pathogens or even human pathogenic streptococci.

6 Phenotypic Antimicrobial Tolerance of *S. suis*

As described above genetic resistance is accomplished by specific genes conferring antimicrobial resistance. In contrast, phenotypic tolerance describes the phenomenon that a certain bacterial subpopulation within a clonal but antibiotic sensitive bulk of a bacterial population can survive antimicrobial concentrations vastly exceeding the MIC. This subpopulation, predominately consisting of so-called persister cells, is genetically identical but phenotypically heterogeneous. Physiologically, persisters are thought to constitute a dormant or quiescent cell state that makes them temporally tolerating comparatively high concentrations of antimicrobials. Nevertheless, persisters can switch back from the dormant and antibiotic tolerating into a replicating and antibiotic sensitive stage. Persister cell formation can occur in two ways. First, bacterial cells can switch stochastically to a persistent state prior to any environmental change (Balaban et al. 2013; Kussell and Leibler 2005). Second, environmental sensing of a stochastically built subpopulation ensures the acquisition of an antibiotic tolerating phenotype in response to an environmental trigger (phenotype diversification), which resembles the mechanism of type I persisters (Balaban et al. 2004; Kotte et al. 2014). Since the first report by Bigger in 1944, bacterial persister cells have been described for numerous pathogenic species, including *Escherichia coli*, *Staphylococcus aureus*, *Pseudomonas aeruginosa*, *Mycobacterium tuberculosis*, and *S. suis* (Lewis 2007, 2010; Willenborg et al. 2014). Although the clinical relevance of *S. suis* persisters is yet speculative, persister cell formation and their multidrug tolerance has been demonstrated in vitro (Willenborg et al. 2014). In line with the observations of other persister studies, *S. suis* persisters were found to be predominately formed in the stationary growth phase by a mechanism resembling that of type I persisters. These persisters were able to tolerate, e.g., concentrations of 100-fold MIC amoxicillin, which is one of the mostly used antibiotics in pig husbandry in Germany (van Rennings et al. 2015). Additional studies are needed to characterize persister cells of *S. suis* in more detail and to evaluate their relevance for antibiotic tolerance of *S. suis* in the pig.

In general, the mechanism of phenotypic heterogeneity is considered a bet-hedging strategy and ensures survival of a bacterial population in adverse conditions, such as antibiotic treatments (Veening et al. 2008). In line with this, phenotypic heterogeneity might also help to explain the different subsets of bacterial cells, including persisters, found in antibiotic tolerating biofilms (Lewis 2008; Spoering and Lewis 2001; Stewart and Franklin 2008). Biofilms are matrix-enclosed microbial populations that colonize biological or nonbiological surfaces (Hall-Stoodley et al. 2004). Biofilm formation allows bacterial survival under inhospitable conditions and can protect pathogenic bacteria from clearance by the host immune system and/or by antibiotics. For many bacterial species persister cells have been found in biofilms (Cohen et al. 2013; Keren et al. 2004; Lewis 2007, 2008; Shapiro et al. 2011; Singh et al. 2009; Spoering and Lewis 2001). This indicates a clinical relevance of persister cells and their impact on persistent and/or recurrent infections after antibiotic therapy (Lewis 2007; Shapiro et al. 2011; Spoering and Lewis 2001). Although clinical data on recurrent *S. suis* infections or in vivo biofilm formation in pigs are missing, biofilm formation has also been shown for certain *S. suis* strains in vitro. In comparison to planktonic grown bacteria, biofilms exhibited strongly elevated MBC (minimal bactericidal concentration) values for penicillin G and ampicillin (Bonifait et al. 2008; Grenier et al. 2009a; Olson et al. 2002). In this regard, one may speculate that *S. suis* persister cells within biofilm structures may contribute to the high drug tolerance of colonizing biofilms, as has been described for *Pseudomonas aeruginosa* persister cells (Spoering and Lewis 2001). To date, there is no experimental evidence for biofilm formation of *S. suis* in vivo, but the observation that the *S. suis* colonization rate of porcine tonsils is nearly 100 % suggests a contribution of streptococcal biofilm formation to this phenotype (Brisebois et al. 1990; MacInnes et al. 2008). Thus, a study reporting unsuccessful elimination of the tonsillar carrier state of *S. suis* in swine by antibiotic treatment with penicillin, ampicillin, or ceftiofur may be an indirect hint for the formation of antibiotic-tolerating biofilms of *S. suis* (Amass et al. 1996). Though more studies are needed to elucidate the role of *S. suis* biofilms and persisters in vivo, we think that, in addition to genetic resistance, phenotypic variations of *S. suis* conferring antibiotic tolerance should be considered as possible explanations for noneffective antibiotic therapies.

7 Possible Impact of Co-infections on Antimicrobial Therapy and Resistance of *S. suis*

Current studies on the failure of antibacterial therapy and resistance mechanisms of bacteria are mainly focusing on mono-infections, i.e., single pathogens, though patients are often co-infected with different pathogens. Furthermore, interactions of pathogens with commensal microbiota have been shown to be involved in bacterial colonization and pathogenicity. However, possible effects of interactions between different pathogens and between pathogens and commensal microorganisms on the efficiency of antimicrobial therapy have just begun to be studied in more detail (Birger et al. 2015).

Regarding *S. suis* infections in pigs, the upper respiratory tract is a reservoir for a heterogeneous community of (potentially) pathogenic microorganisms and commensals. Pneumonia often represents a multifactorial disease complex caused by polymicrobial infections, thus termed porcine respiratory disease complex (PRDC) (Opriessnig et al. 2011). Often primary viral agents, such as porcine reproductive and respiratory syndrome virus (PRRSV), porcine circovirus type 2 and swine influenza virus (SIV), predispose pigs to secondary infections and pneumonia by opportunistic bacterial pathogens, such as *Pasteurella multocida*, *Mycoplasma hyopneumoniae,* and *S. suis* (Fablet et al. 2011). Until now, co-infection studies of *S. suis* mainly focused on the interaction of *S. suis* with two common viruses, namely PRRSV and SIV. However, very little is known about co-infections of *S. suis* with other pathogens in general and the role of such co-infections in antimicrobial treatment success of associated diseases.

Recently, a comprehensive review on the potential impact of co-infections on the antimicrobial therapy and resistance in humans was published, providing an overview of potential mechanisms of interactions during co-infections (Birger et al. 2015). Interactions of co-infecting pathogens can be either synergistic, thereby promoting evolution of antibiotic resistance, or they are antagonistic, impeding resistance development (Birger et al. 2015). These effects may be explained by different mechanisms. First, modulation of the immune response by the primary infecting agent may lead to an impaired antibacterial innate immune response (Metzger and Sun 2013). For instance, impaired immune-control in HIV-positive patients facilitates infection and persistence of antibiotic-resistant *Mycobacterium tuberculosis* strains in humans (Bifani et al. 2008). Second, modulation of resources like nutrients, the availability of basement membrane proteins/receptors, or the presence of different host niches and cells can affect resistance development. This has been reported for human influenza virus *Streptococcus pneumoniae* co-infection, in which the "denudation" of the respiratory epithelial layer exposes fibronectin as a bacterial adhesion receptor and supplies (cleaved) sialic acids as a carbon source for rapid bacterial growth (Bosch et al. 2013; McCullers and Bartmess 2003). Both types of mechanisms may facilitate the proliferation and persistence of co-infecting pathogens including subpopulations of antibiotic-resistant bacteria. Furthermore, secondary bacterial infections of virus-infected patients lead to an increased use of antimicrobials, which may further contribute to a positive collateral selection and better fitness of resistant bacteria.

In a study on antimicrobial treatment of nursery pigs co-infected by PRRSV and *S. suis*, penicillin and ampicillin failed to minimize disease. Only administration of ceftiofur appeared to be adequate to treat co-infected nursery pigs (Day et al. 2015; Tantituvanont et al. 2009). A main characteristic of PRRSV infection is its damaging effect on the host immune system, which facilitates the spread of *S. suis* (Thanawongnuwech et al. 2000). Enhanced bacterial proliferation may lead to an increase in antibiotic-tolerant bacteria thus promoting resistance development. We and others recently showed that pre-infection of respiratory epithelial cells by SIV is also beneficial for *S. suis*, since bacterial adherence, colonization, and invasion of deeper tissues was efficiently promoted (Meng et al. 2015; Wang et al. 2013a; Wu et al. 2015). *S. suis* is to some extent able to invade different types of host cells like epithelial and endothelial cells as well as phagocytic cells. An intracellular stage might serve as a niche for antibiotic escape as discussed for intracellular persisting *Staphylococcus aureus* (Thwaites and Gant 2011). As a consequence of viral propagation along the respiratory tract, ciliostatic effects occur, resulting in a reduced muco-ciliary clearance and a concerted damage of the respiratory epithelium (Meng et al. 2015). Such cell degrading processes may allow access to cellular basement receptors like integrins for *S. suis* adherence and may provide nutrients for better growth.

As mentioned in the previous paragraph, a further mechanism for antibiotic resistance propagation is the formation of biofilms, in which bacteria might, at least partially, be protected from antimicrobial substances. Limited penetration of substances in biofilms and/or a reduced growth rate of biofilm-forming bacteria may lead to a lowered antimicrobial sensitivity of the bacteria. These biofilms are formed by one bacterial species or can consist of a multispecies community in the case of co-infections (Elias and Banin 2012). *S. suis* has the ability to form biofilms and was found to be more resistant to beta-lactams in mono-species biofilms than in planktonic cultures (Bonifait et al. 2008; Grenier et al. 2009b). However, whether or not *S. suis* can be a constituent of a multispecies biofilm is not known yet.

A key mechanism of the acquisition and spread of antimicrobial resistance is the exchange of resistance genes by horizontal gene transfer (Palmieri et al. 2011a, c). Resistance can be transferred within one species, between co-colonizing pathogens or between the pathogen and members of the microbiota. Additionally, the production and release of substances inhibiting antimicrobial agents, e.g., beta-lactamases, by co-infecting or commensal bacteria might protect otherwise susceptible bacteria against certain antimicrobials, e.g., beta-lactams (Weimer et al. 2011).

In addition to direct interactions of co-infecting pathogens, polymicrobial infections can increase the inflammatory response, resulting in a pH change in the inflamed tissue and a decreased tissue perfusion. This might affect the antimicrobial absorption leading to an inefficient antimicrobial concentration at the infection site, thus facilitating development of resistant bacteria. Such significant changes in the pharmacokinetic properties were found for ceftiofur in *S. suis*-PRRSV co-infected pigs (McCullers and Bartmess 2003).

It is yet unknown how the described mechanisms are involved in positive selection and the spread of antibiotic-resistant *S. suis* strains and/or the development of de novo resistance in co-infected pigs. Nevertheless, co-infections and the role of the microbiota should be taken into account in treatment strategies and the development of new antimicrobial substances.

8 Conclusions and Outlook

In the present review, we focused on certain known and putative factors, which may contribute to problems in the control of porcine *S. suis* infections by antimicrobial therapy (depicted in Fig. 1). The pathogen is well-suited as an example for livestock husbandry and antimicrobial resistance in veterinary medicine for several reasons; first, it is a paradigm for a bacterial pathogen of global importance in swine husbandry, second, *S. suis* infections are widely controlled by antimicrobial treatments, and third, it can also be transmitted to and cause diseases in humans. From the pathogen point of view, genetic resistance mechanisms, such as efflux pumps,

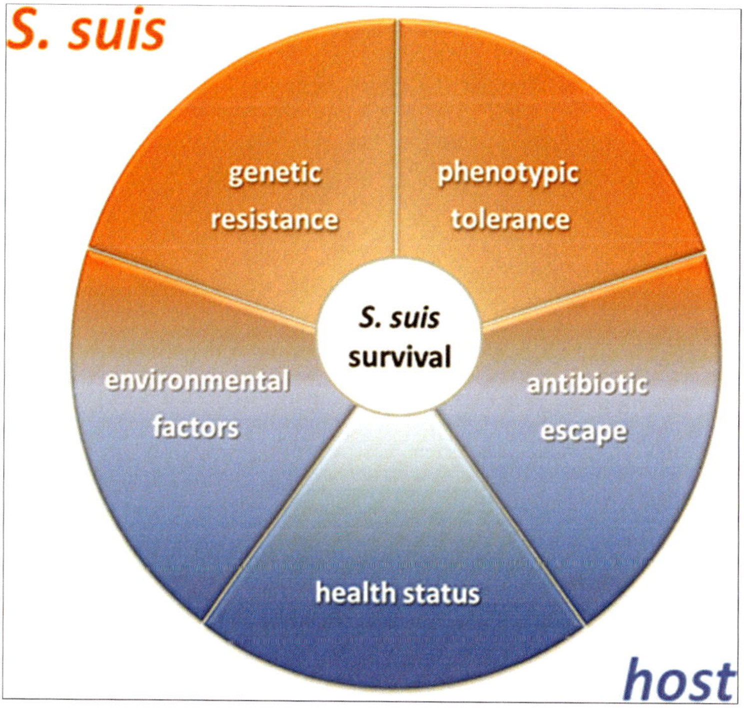

Fig. 1 Factors possibly involved in survival of *S. suis* exposed to antimicrobial therapy. From the pathogen point of view, **genetic resistance** mechanisms such as efflux pumps, inactivation of the antibiotic or modification of an antimicrobial target are the main proven factors that contribute to *S. suis* survival under certain antibiotic selection pressure. In addition, **tolerance** to antibiotic treatments by phenotypic alterations, such as biofilm production or formation of persister cell subpopulations have been shown in vitro. A possible intracellular transition stage, even of a very low number of bacteria, may protect *S. suis* from sufficient antibacterial concentrations, thus reflecting a putative **escape** mechanism. Overall, the **health status** of the host may affect efficient antibiotic therapy, e.g., an impaired immune response due to co-infections can lower antimicrobial effectivity. Certain **environmental** factors may not only influence the host health and immune status but may also trigger the development of phenotypic tolerance and/or genetic resistance of *S. suis*

inactivation of the antibiotic agent or modification of an antimicrobial target, surely contribute to bacterial survival under antibiotic selection pressure. However, problems in antibiotic therapy cannot be explained solely by the presence of resistance genes and/or the detection of in vitro resistance to antimicrobials. Most likely, additional factors play a role in vivo. For example, though yet only shown in vitro, tolerance to antibiotic treatment by phenotypic alterations such as formation of persister cells and biofilm production may well be of importance and, thus, should be further investigated in the future. Another unresolved question is the relevance of a possible intracellular stage of *S. suis* in host cells, as a mean for the pathogens to escape from antimicrobial agents which usually target extracellular bacteria. Finally, host factors, such as competing commensal microorganisms or other co-infecting pathogens may affect not only morbidity and mortality of *S. suis* infections, but also the efficacy of antibiotic treatment.

At present we just begin to understand the complex mechanisms that limit the success of antimicrobial treatment, in general and also of *S. suis* infections in veterinary medicine. In the future, a combined multidisciplinary effort is needed to develop improved antimicrobial strategies and to better control the spread of antimicrobial resistance.

Acknowledgments The work presented here was supported by grants from the Deutsche Forschungsgemeinschaft to PVW (grant GO983-3/1 within the Priority Programme SPP1316, grant Va23917-1).

References

Aarestrup F (2012) Sustainable farming: get pigs off antibiotics. Nature 486(7404):465–466. doi:10.1038/486465a

Aarestrup FM, Rasmussen SR, Artursson K, Jensen NE (1998) Trends in the resistance to antimicrobial agents of *Streptococcus suis* isolates from Denmark and Sweden. Vet Microbiol 63(1):71–80. doi:10.1016/S0378-1135(98)00228-4

Amass SF, Wu CC, Clark LK (1996) Evaluation of antibiotics for the elimination of the tonsillar carrier state of *Streptococcus suis* in pigs. J Vet Diagn Inv 8(1):64–67 (Official Publication of the American Association of Veterinary Laboratory Diagnosticians, Inc)

Amass SF, SanMiguel P, Clark LK (1997) Demonstration of vertical transmission of *Streptococcus suis* in swine by genomic fingerprinting. J Clin Microbiol 35(6):1595–1596

Andrews JM (2001) Determination of minimum inhibitory concentrations. J Antimicrob Chemoth 48:5–16

Arends JP, Zanen HC (1988) Meningitis caused by *Streptococcus suis* in humans. Rev Infect Dis 10(1):131–137

Balaban NQ, Merrin J, Chait R, Kowalik L, Leibler S (2004) Bacterial persistence as a phenotypic switch. Science 305(5690):1622–1625. doi:10.1126/science.1099390

Balaban NQ, Gerdes K, Lewis K, McKinney JD (2013) A problem of persistence: still more questions than answers? Nat Rev Microbiol 11(8):587–591

Barton MD (2014) Impact of antibiotic use in the swine industry. Curr Opin Microbiol 19:9–15. doi:10.1016/j.mib.2014.05.017

Baums CG, Verkuhlen GJ, Rehm T, Silva LM, Beyerbach M, Pohlmeyer K, Valentin-Weigand P (2007) Prevalence of *Streptococcus suis* genotypes in wild boars of Northwestern Germany. Appl Environ Microbiol 73(3):711–717. doi:10.1128/AEM.01800-06

Berthelot-Herault F, Gottschalk M, Labbe A, Cariolet R, Kobisch M (2001) Experimental airborne transmission of *Streptococcus suis* capsular type 2 in pigs. Vet Microbiol 82(1):69–80

Bifani P, Mathema B, Kurepina N, Shashkina E, Bertout J, Blanchis AS, Moghazeh S, Driscoll J, Gicquel B, Frothingham R, Kreiswirth BN (2008) The evolution of drug resistance in *Mycobacterium tuberculosis*: from a mono-rifampin-resistant cluster into increasingly multidrug-resistant variants in an HIV-seropositive population. J Infect Dis 198(1):90–94. doi:10.1086/588822

Birger RB, Kouyos RD, Cohen T, Griffiths EC, Huijben S, Mina M, Volkova V, Grenfell B, Metcalf CJ (2015) The potential impact of coinfection on antimicrobial chemotherapy and drug resistance. Trends Microbiol 23(9):537–544. doi:10.1016/j.tim.2015.05.002

Blair JM, Webber MA, Baylay AJ, Ogbolu DO, Piddock LJ (2015) Molecular mechanisms of antibiotic resistance. Nat Rev Microbiol 13(1):42–51. doi:10.1038/nrmicro3380

Bonifait L, Grignon L, Grenier D (2008) Fibrinogen induces biofilm formation by *Streptococcus suis* and enhances its antibiotic resistance. Appl Environ Microbiol 74(15):4969–4972. doi:10.1128/AEM.00558-08

Bosch AATM, Biesbroek G, Trzcinski K, Sanders EAM, Bogaert D (2013) Viral and bacterial interactions in the upper respiratory tract. Plos Pathog 9(1):e1003057. doi:10.1371/journal.ppat.1003057 (ARTN)

Brisebois LM, Charlebois R, Higgins R, Nadeau M (1990) Prevalence of *Streptococcus suis* in four to eight week old clinically healthy piglets. Can J Vet Res (Revue canadienne de recherche veterinaire) 54(1):174–177

Callens B, Persoons D, Maes D, Laanen M, Postma M, Boyen F, Haesebrouck F, Butaye P, Catry B, Dewulf J (2012) Prophylactic and metaphylactic antimicrobial use in Belgian fattening pig herds. Prev Vet Med 106(1):53–62. doi:10.1016/j.prevetmed.2012.03.001

Callens BF, Haesebrouck F, Maes D, Butaye P, Dewulf J, Boyen F (2013) Clinical resistance and decreased susceptibility in *Streptococcus suis* Isolates from clinically healthy fattening pigs. Microb Drug Resist 19(2):146–151. doi:10.1089/mdr.2012.0131

Clifton-Hadley FA, Alexander TJ (1980) The carrier site and carrier rate of *Streptococcus suis* type II in pigs. Vet Rec 107(2):40–41

Cohen NR, Lobritz MA, Collins JJ (2013) Microbial persistence and the road to drug resistance. Cell Host Microbe 13(6):632–642. doi:10.1016/j.chom.2013.05.009

Day DN, Sparks JW, Karriker LA, Stalder KJ, Wulf LW, Zhang J, Kinyon JM, Stock ML, Gehring R, Wang C, Ellingson J, Coetzee JF (2015) Impact of an experimental PRRSV and *Streptococcus suis* coinfection on the pharmacokinetics of ceftiofur hydrochloride after intramuscular injection in pigs. J Vet Pharmacol Ther 38(5):475–481. doi:10.1111/jvp.12209

Elias S, Banin E (2012) Multi-species biofilms: living with friendly neighbors. FEMS Microbiol Rev 36(5):990–1004. doi:10.1111/j.1574-6976.2012.00325.x

Fablet C, Marois C, Kuntz-Simon G, Rose N, Dorenlor V, Eono F, Eveno E, Jolly JP, Le Devendec L, Tocqueville V, Queguiner S, Gorin S, Kobisch M, Madec F (2011) Longitudinal study of respiratory infection patterns of breeding sows in five farrow-to-finish herds. Vet Microbiol 147(3–4):329–339. doi:10.1016/j.vetmic.2010.07.005

Fulde M, Valentin-Weigand P (2013) Epidemiology and pathogenicity of zoonotic streptococci. Curr Top Microbiol 368:49–81. doi:10.1007/82_2012_277

Ge Y, Wu J, Xia Y, Yang M, Xiao J, Yu J (2012) Molecular dynamics simulation of the complex PBP-2x with drug cefuroxime to explore the drug resistance mechanism of *Streptococcus suis* R61. PLoS ONE 7(4):e35941. doi:10.1371/journal.pone.0035941

Gottschalk M, Segura M (2000) The pathogenesis of the meningitis caused by *Streptococcus suis*: the unresolved questions. Vet Microbiol 76(3):259–272

Gottschalk M, Segura M, Xu J (2007) *Streptococcus suis* infections in humans: the Chinese experience and the situation in North America. Anim Health Res Rev 8(1):29–45. doi:10.1017/S1466252307001247

Goyette-Desjardins G, Auger JP, Xu J, Segura M, Gottschalk M (2014) *Streptococcus suis*, an important pig pathogen and emerging zoonotic agent-an update on the worldwide distribution based on serotyping and sequence typing. Emerg Microbes Infect 3(6):e45. doi:10.1038/emi.2014.45

Grave K, Torren-Edo J, Mackay D (2010) Comparison of the sales of veterinary antibacterial agents between 10 European countries. J Antimicrob Chemoth 65(9):2037–2040. doi:10.1093/jac/dkq247

Grenier D, Grignon L, Gottschalk M (2009a) Characterisation of biofilm formation by a *Streptococcus suis* meningitis isolate. Vet J 179(2):292–295. doi:10.1016/j.tvjl.2007.09.005

Grenier D, Grignon L, Gottschalk M (2009b) Characterisation of biofilm formation by a *Streptococcus suis* meningitis isolate. Vet J 179(2):292–295. doi:10.1016/j.tvjl.2007.09.005

Hall-Stoodley L, Costerton JW, Stoodley P (2004) Bacterial biofilms: from the natural environment to infectious diseases. Nat Rev Microbiol 2(2):95–108. doi:10.1038/nrmicro821

Hao H, Cheng G, Iqbal Z, Ai X, Hussain HI, Huang L, Dai M, Wang Y, Liu Z, Yuan Z (2014) Benefits and risks of antimicrobial use in food-producing animals. Front Microbiol 5:288. doi:10.3389/fmicb.2014.00288

Hoa NT, Chieu TTB, Nghia HDT, Mai NTH, Anh PH, Wolbers M, Baker S, Campbell JI, Chau NW, Hien TT, Farrar J, Schultsz C (2011) The antimicrobial resistance patterns and associated determinants in *Streptococcus suis* isolated from humans in southern Vietnam, 1997–2008. Bmc Infect Dis 11:6. doi:10.1186/1471-2334-11-6 (Artn)

Holden MT, Hauser H, Sanders M, Ngo TH, Cherevach I, Cronin A, Goodhead I, Mungall K, Quail MA, Price C, Rabbinowitsch E, Sharp S, Croucher NJ, Chieu TB, Mai NT, Diep TS, Chinh NT, Kehoe M, Leigh JA, Ward PN, Dowson CG, Whatmore AM, Chanter N, Iversen P, Gottschalk M, Slater JD, Smith HE, Spratt BG, Xu J, Ye C, Bentley S, Barrell BG, Schultsz C, Maskell DJ, Parkhill J (2009) Rapid evolution of virulence and drug resistance in the emerging zoonotic pathogen *Streptococcus suis*. PLoS ONE 4(7):e6072. doi:10.1371/journal.pone.0006072

Hu P, Yang M, Zhang A, Wu J, Chen B, Hua Y, Yu J, Chen H, Xiao J, Jin M (2011) Comparative genomics study of multi-drug-resistance mechanisms in the antibiotic-resistant *Streptococcus suis* R61 strain. PLoS ONE 6(9):e24988. doi:10.1371/journal.pone.0024988

Huang JH, Shang KX, Kashif J, Wang LP (2015) Genetic diversity of *Streptococcus suis* isolated from three pig farms of China obtained by acquiring antibiotic resistance genes. J Sci Food Agr 95(7):1454–1460. doi:10.1002/jsfa.6841

Jordan D, Chin JJC, Fahy VA, Barton MD, Smith MG, Trott DJ (2009) Antimicrobial use in the Australian pig industry: results of a national survey. Aust Vet J 87(6):222–229. doi:10.1111/j.1751-0813.2009.00430.x

Keren I, Kaldalu N, Spoering A, Wang Y, Lewis K (2004) Persister cells and tolerance to antimicrobials. FEMS Microbiol Lett 230(1):13–18

Kotte O, Volkmer B, Radzikowski JL, Heinemann M (2014) Phenotypic bistability in *Escherichia coli*'s central carbon metabolism. Mol Syst Biol 10:736. doi:10.15252/msb.20135022

Kussell E, Leibler S (2005) Phenotypic diversity, population growth, and information in fluctuating environments. Science 309(5743):2075–2078. doi:10.1126/science.1114383

Leekha S, Terrell CL, Edson RS (2011) General principles of antimicrobial therapy. Mayo Clin Proc 86(2):156–167. doi:10.4065/mcp.2010.0639

Lewis K (2007) Persister cells, dormancy and infectious disease. Nat Rev Microbiol 5(1):48–56. doi:10.1038/nrmicro1557

Lewis K (2008) Multidrug tolerance of biofilms and persister cells. Curr Top Microbiol Immunol 322:107–131

Lewis K (2010) Persister cells. Annu Rev Microbiol 64:357–372. doi:10.1146/annurev.micro.112408.134306

Long KS, Poehlsgaard J, Kehrenberg C, Schwarz S, Vester B (2006) The Cfr rRNA methyltransferase confers resistance to phenicols, lincosamides, oxazolidinones, pleuromutilins, and streptograrnin A antibiotics. Antimicrob Agents Chemother 50(7):2500–2505. doi:10.1128/Aav.00131.06

Lun ZR, Wang QP, Chen XG, Li AX, Zhu XQ (2007) *Streptococcus suis*: an emerging zoonotic pathogen. Lancet Infect Dis 7(3):201–209. doi:10.1016/S1473-3099(07)70001-4

Ma E, Chung PH, So T, Wong L, Choi KM, Cheung DT, Kam KM, Chuang SK, Tsang T, Streptococ CSG (2008) *Streptococcus suis* infection in Hong Kong: an emerging infectious disease? Epidemiol Infect 136(12):1691–1697. doi:10.1017/S0950268808000332

MacInnes JI, Gottschalk M, Lone AG, Metcalf DS, Ojha S, Rosendal T, Watson SB, Friendship RM (2008) Prevalence of *Actinobacillus pleuropneumoniae*, *Actinobacillus suis*, *Haemophilus parasuis*, *Pasteurella multocida*, and *Streptococcus suis* in representative Ontario swine herds. Can J Vet Res 72(3):242–248

Mai NT, Hoa NT, Nga TV, le Linh D, Chau TT, Sinh DX, Phu NH, Chuong LV, Diep TS, Campbell J, Nghia HD, Minh TN, Chau NV, de Jong MD, Chinh NT, Hien TT, Farrar J, Schultsz C (2008) *Streptococcus suis* meningitis in adults in Vietnam. Clin Infect Dis 46 (5):659–667. doi:10.1086/527385

Marie J, Morvan H, Berthelot-Herault F, Sanders P, Kempf I, Gautier-Bouchardon AV, Jouy E, Kobisch M (2002) Antimicrobial susceptibility of *Streptococcus suis* isolated from swine in France and from humans in different countries between 1996 and 2000. J Antimicrob Chemoth 50(2):201–209. doi:10.1093/jac/dkf009

Marini E, Palmieri C, Magi G, Facinelli B (2015) Recombination between *Streptococcus suis* ICESsu32457 and *Streptococcus agalactiae* ICESa2603 yields a hybrid ICE transferable to *Streptococcus pyogenes*. Vet Microbiol 178(1–2):99–104. doi:10.1016/j.vetmic.2015.04.013

Maron DF, Smith TJ, Nachman KE (2013) Restrictions on antimicrobial use in food animal production: an international regulatory and economic survey. Globalization Health 9:48. doi:10.1186/1744-8603-9-48

McCullers JA, Bartmess KC (2003) Role of neuraminidase in lethal synergism between influenza virus and *Streptococcus pneumoniae*. J Infect Dis 187(6):1000–1009. doi:10.1086/368163

Meng F, Wu NH, Nerlich A, Herrler G, Valentin-Weigand P, Seitz M (2015) Dynamic virus-bacterium interactions in a porcine precision-cut lung slice coinfection model: swine influenza virus paves the way for *Streptococcus suis* infection in a two-step process. Infect Immun 83(7):2806–2815. doi:10.1128/IAI.00171-15

Metzger DW, Sun K (2013) Immune dysfunction and bacterial coinfections following influenza. J Immunol 191(5):2047–2052. doi:10.4049/jimmunol.1301152

Olson ME, Ceri H, Morck DW, Buret AG, Read RR (2002) Biofilm bacteria: formation and comparative susceptibility to antibiotics. Can J Vet Res (Revue canadienne de recherche veterinaire) 66(2):86–92

Opriessnig T, Gimenez-Lirola LG, Halbur PG (2011) Polymicrobial respiratory disease in pigs. Anim Health Res Rev 12(2):133–148. doi:10.1017/S1466252311000120

Palmieri C, Princivalli MS, Brenciani A, Varaldo PE, Facinelli B (2011a) Different genetic elements carrying the tet(W) gene in two human clinical isolates of *Streptococcus suis*. Antimicrob Agents Ch 55(2):631–636. doi:10.1128/Aac.00965 10

Palmieri C, Varaldo PE, Facinelli B (2011b) *Streptococcus suis*, an emerging drug-resistant animal and human pathogen. Front Microbiol 2:235. doi:10.3389/fmicb.2011.00235

Palmieri C, Varaldo PE, Facinelli B (2011c) *Streptococcus suis*, an emerging drug-resistant animal and human pathogen. Front Microbiol 2:235. doi:10.3389/Fmicb.2011.00235 (Artn)

Palmieri C, Magi G, Mingoia M, Bagnarelli P, Ripa S, Varaldo PE, Facinelli B (2012) Characterization of a *Streptococcus suis* tet(O/W/32/O)-carrying element transferable to major streptococcal pathogens. Antimicrob Agents Chemother 56(9):4697–4702. doi:10.1128/AAC.00629-12

Palmieri C, Magi G, Creti R, Baldassarri L, Imperi M, Gherardi G, Facinelli B (2013) Interspecies mobilization of an erm(T)-carrying plasmid of *Streptococcus dysgalactiae* subsp. equisimilis by a coresident ICE of the ICESa2603 family. J Antimicrob Chemoth 68(1):23–26. doi:10.1093/jac/dks352

Rajic A, Reid-Smith R, Deckert AE, Dewey CE, McEwen SA (2006) Reported antibiotic use in 90 swine farms in Alberta. Can Vet J 47(5):446–452

Shapiro JA, Nguyen VL, Chamberlain NR (2011) Evidence for persisters in *Staphylococcus epidermidis* RP62a planktonic cultures and biofilms. J Med Microbiol 60(Pt 7):950–960. doi:10.1099/jmm.0.026013-0

Shen J, Wang Y, Schwarz S (2013) Presence and dissemination of the multiresistance gene cfr in gram-positive and gram-negative bacteria. J Antimicrob Chemother 68(8):1697–1706. doi:10.1093/jac/dkt092

Singh R, Ray P, Das A, Sharma M (2009) Role of persisters and small-colony variants in antibiotic resistance of planktonic and biofilm-associated *Staphylococcus aureus*: an in vitro study. J Med Microbiol 58(Pt 8):1067–1073. doi:10.1099/jmm.0.009720-0

Soares TCS, Paes AC, Megid J, Ribolla PEM, Paduan KD, Gottschalk M (2014) Antimicrobial susceptibility of *Streptococcus suis* isolated from clinically healthy swine in Brazil. Can J Vet Res 78(2):145–149

Spoering AL, Lewis K (2001) Biofilms and planktonic cells of *Pseudomonas aeruginosa* have similar resistance to killing by antimicrobials. J Bacteriol 183(23):6746–6751. doi:10.1128/JB.183.23.6746-6751.2001

Staats JJ, Feder I, Okwumabua O, Chengappa MM (1997) *Streptococcus suis*: past and present. Vet Res Commun 21(6):381–407

Stewart PS, Franklin MJ (2008) Physiological heterogeneity in biofilms. Nat Rev Microbiol 6 (3):199–210. doi:10.1038/nrmicro1838

Strangmann E, Froleke H, Kohse KP (2002) Septic shock caused by *Streptococcus suis*: case report and investigation of a risk group. Int J Hyg Environ Health 205(5):385–392. doi:10.1078/1438-4639-00165

Takamatsu D, Osaki M, Sekizaki T (2003) Chloramphenicol resistance transposable element TnSs1 of *Streptococcus suis*, a transposon flanked by IS6-family elements. Plasmid 49(2):143–151

Tantituvanont A, Yimprasert W, Werawatganone P, Nilubol D (2009) Pharmacokinetics of ceftiofur hydrochloride in pigs infected with porcine reproductive and respiratory syndrome virus. J Antimicrob Chemoth 63(2):369–373. doi:10.1093/jac/dkn496

Thanawongnuwech R, Brown GB, Halbur PG, Roth JA, Royer RL, Thacker BJ (2000) Pathogenesis of porcine reproductive and respiratory syndrome virus-induced increase in susceptibility to *Streptococcus suis* infection. Vet Pathol 37(2):143–152. doi:10.1354/Vp.37-2-143

Thwaites GE, Gant V (2011) Are bloodstream leukocytes Trojan Horses for the metastasis of *Staphylococcus aureus*? Nat Rev Microbiol 9(3):215–222. doi:10.1038/nrmicro2508

Timmerman T, Dewulf J, Catry B, Feyen B, Opsomer G, de Kruif A, Maes D (2006) Quantification and evaluation of antimicrobial drug use in group treatments for fattening pigs in Belgium. Prev Vet Med 74(4):251–263. doi:10.1016/j.prevetmed.2005.10.003

Trauffler M, Obritzhauser W, Raith J, Fuchs K, Kofer J (2014) The use of the "highest priority critically important antimicrobials" in 75 Austrian pig farms—evaluation of on-farm drug application data. Berl Munch Tierarztl 127(9–10):375–383. doi:10.2376/0005-9366-127-375

van Rennings L, von Munchhausen C, Hartmann M, Ottilie H, Honscha W, Kasbohrer A, Kreienbrock L (2014) Antibiotic usage and antibiotic sales in Germany in 2011-the situation of drug usage in veterinary medicine. Berl Munch Tierarztl 127(9–10):366–374. doi:10.2376/0005-9366-127-366

van Rennings L, von Munchhausen C, Ottilie H, Hartmann M, Merle R, Honscha W, Kasbohrer A, Kreienbrock L (2015) Cross-sectional study on antibiotic usage in pigs in Germany. Plos One 10(3). doi:10.1371/journal.pone.0119114 (UNSP e0119114)

Varela NP, Gadbois P, Thibault C, Gottschalk M, Dick P, Wilson J (2013) Antimicrobial resistance and prudent drug use for *Streptococcus suis*. Anim Health Res Rev 14(1):68–77. doi:10.1017/S1466252313000029

Veening JW, Smits WK, Kuipers OP (2008) Bistability, epigenetics, and bet-hedging in bacteria. Annu Rev Microbiol 62:193–210. doi:10.1146/annurev.micro.62.081307.163002

Wallmann J, Hauck R, Schwarz C, Klee S, Kaspar H, Heberer T (2015) Antibiotics: legislation and data acquisition in Germany. J Verbrauch Lebensm 10(3):251–257. doi:10.1007/s00003-015-0954-8

Wang Y, Gagnon CA, Savard C, Music N, Srednik M, Segura M, Lachance C, Bellehumeur C, Gottschalk M (2013a) Capsular sialic acid of *Streptococcus suis* serotype 2 binds to swine influenza virus and enhances bacterial interactions with virus-infected tracheal epithelial cells. Infect Immun 81(12):4498–4508. doi:10.1128/IAI.00818-13

Wang Y, Li DX, Song L, Liu Y, He T, Liu HB, Wu CM, Schwarz S, Shen JZ (2013b) First report of the multiresistance gene cfr in *Streptococcus suis*. Antimicrob Agents Ch 57(8):4061–4063. doi:10.1128/Aac.00713-13

Wassenaar TM (2005) Use of antimicrobial agents in veterinary medicine and implications for human health. Crit Rev Microbiol 31(3):155–169. doi:10.1080/10408410591005110

Weimer KED, Juneau RA, Murrah KA, Pang B, Armbruster CE, Richardson SH, Swords WE (2011) Divergent mechanisms for passive pneumococcal resistance to beta-lactam antibiotics in the presence of *Haemophilus influenzae*. J Infect Dis 203(4):549–555. doi:10.1093/infdis/jiq087

Wertheim HF, Nguyen HN, Taylor W, Lien TT, Ngo HT, Nguyen TQ, Nguyen BN, Nguyen HH, Nguyen HM, Nguyen CT, Dao TT, Nguyen TV, Fox A, Farrar J, Schultsz C, Nguyen HD, Nguyen KV, Horby P (2009) *Streptococcus suis*, an important cause of adult bacterial meningitis in northern Vietnam. PLoS ONE 4(6):e5973. doi:10.1371/journal.pone.0005973

Willenborg J, Willms D, Bertram R, Goethe R, Valentin-Weigand P (2014) Characterization of multi-drug tolerant persister cells in *Streptococcus suis*. BMC Microbiol 14:120. doi:10.1186/1471-2180-14-120

Wu NH, Meng F, Seitz M, Valentin-Weigand P, Herrler G (2015) Sialic acid-dependent interactions between influenza viruses and *Streptococcus suis* affect the infection of porcine tracheal cells. J Gen Virol. doi:10.1099/jgv.0.000223

Yao J, Shang K, Huang J, Ran W, Kashif J, Wang L (2014) Overexpression of an ABC transporter and mutations of GyrA, GyrB, and ParC in contributing to high-level ciprofloxacin resistance in *Streptococcus suis* type 2. Biosci Trends 8(2):84–92

Zhao Q, Wang M, Xu D, Zhang Q, Liu W (2015) Metabolic coupling of two small-molecule thiols programs the biosynthesis of lincomycin A. Nature 518(7537):115–119. doi:10.1038/nature14137

Antibiotics and the Intestinal Microbiome: Individual Responses, Resilience of the Ecosystem, and the Susceptibility to Infections

Sophie Thiemann, Nathiana Smit and Till Strowig

Abstract The intestinal microbiota is a diverse ecosystem containing thousands of microbial species, whose metabolic activity affects many aspects of human physiology. Large-scale surveys have demonstrated that an individual's microbiota composition is shaped by factors such as diet and the use of medications, including antibiotics. Loss of overall diversity and in some cases loss of single groups of bacteria as a consequence of antibiotic treatment in humans has been associated with enhanced susceptibility toward gastrointestinal infections and with enhanced weight gain and obesity in young children. Moreover, the extensive use of antibiotics has led to an increased abundance of antibiotic resistance genes (ARGs) within commensal bacteria that can be transferred to invading pathogens, which complicates the treatment of bacterial infections. In this review, we provide insight into the complex interplay between the microbiota and antibiotics focussing on (i) the effect of antibiotics on the composition of the microbiota, (ii) the impact of antibiotics on gastrointestinal infections, and (iii) finally the role of the microbiota as reservoir for ARGs. We also discuss how targeted manipulation of the microbiota may be used as an innovative therapeutic approach to reduce the incidence of bacterial infections as well as resulting complications.

Contents

S. Thiemann · N. Smit · T. Strowig (✉)
Helmholtz Centre for Infection Research, Brunswick, Germany
e-mail: till.strowig@helmholtz-hzi.de

Current Topics in Microbiology and Immunology (2016) 398: 123–146
DOI 10.1007/82_2016_504

Published Online: 16 September 2016

1 Introduction

All plants and animals including humans are colonized by complex microbial ecosystems, also termed microbiota or microbiome, comprising thousands of different species and strains of bacteria, archaea and eukaryotes (Leu et al. 2008; Bulgarelli et al. 2013). At this point, it has been widely recognized that these communities contribute significantly to the physiology of their host and that disturbances in the composition of the microbiota may lead or contribute to disease of the host. High-throughput surveys of these communities using next-generation sequencing (NGS) approaches have revealed tremendous biodiversity within the microbiota that reflect the multidimensional interactions between the host, microbial communities, and environmental factors (Lozupone et al. 2012). In parallel, numerous factors that shape the microbiome have been identified including the exposure to microbes, diet composition, and medications including antibiotics. In this review, we will focus on the influence of antibiotic treatment on the microbiome of humans and model organisms. Specifically, we will describe how antibiotics affect the composition of the microbiome and how they affect the function of the microbiome in preventing infection with pathogens. Finally, we will discuss the role of the microbiome in the spread of antibiotic resistance genes (ARGs).

2 The Impact of Antibiotics on the Composition of the Microbiota

Humans and their microbiota are exposed to antibiotics directly due to their use as medications, but also indirectly through their use as drugs and growth promotion factor in food-producing animals. If how a specific antibiotic will act on the microbiota is determined by a multitude of factors including the administration route, pharmacokinetics and excretion route of the specific antibiotic, but also the site of the microbial community, e.g., oral, skin, or intestinal microbiome. Initial studies employing NGS-based methods for community assessment demonstrated that the abundance of up to 30 % of bacterial species was affected in the intestine of healthy adults after being treated with the broad-spectrum antibiotic ciprofloxacin (Dethlefsen et al. 2008). This was associated with decreases in taxonomic richness, i.e., the number of species, and diversity, i.e., the structure of the community, that largely recovered within a month after treatment. Notably, the short-term loss of microbial diversity was not associated with any reported immediate gross impairment of intestinal function in healthy subjects suggesting that the functional repertoire after limited antibiotic exposure was not equally affected. Comparable short-term effects on the composition and diversity of the microbiome were observed in two other early studies focusing on amoxicillin and clindamycin (De La Cochetière et al. 2005; Jernberg et al. 2007). Beyond the short-term effects, measurable long-term shifts (up to 1–2 years post antibiotic exposure) in the composition of the intestinal microbiome were reported for cephalosporin and clindamycin, although these studies included relatively few patients and alterations in microbiome composition were variable between individuals (Dethlefsen and Relman 2011; Jernberg et al. 2007). A recent randomized placebo-controlled clinical trial following 66 healthy individuals for one year after receiving a standard course of either ciprofloxacin, clindamycin, amoxicillin, minocycline, or a placebo was published (Zaura et al. 2015). They provided strong support for short-term effects (up to one month), yet only clindamycin had statistically significant effects 4 months after treatment. After 1 year, no significant changes in the intestinal community were detected for any of the antibiotics compared to the placebo control group. Notably, in the same study communities in the saliva were only affected briefly (after 1 week), but recovered rapidly to their original state after 1 month. This highlights the fact that communities at different sites have potentially distinct resilience to disturbances, i.e., their ability to return to their original state (Fig. 1). In another study, specific compositions of the microbiome and particular diversity of the ecosystem were found to be predictive for the expansion of the opportunistic pathogen *Enterobacter cloacae* in the intestine after treatment with cephalosporin (Raymond et al. 2016). Overall, differences in the dynamic behavior and resilience of the microbiota between the mentioned studies likely reflect the tremendous variability within the intestinal ecosystem between individuals leading to differences in the impact of antibiotics on the microbiota (Lozupone et al. 2012). While these studies have been performed in healthy adults, which had not received antibiotics for at least 3 months, the behavior of the microbiota is likely to be different and distinct in populations that have intrinsically a more instable microbiome such as individuals that received multiple courses of antibiotics within short periods of time as well as infants, young children, and elderly individuals.

Indeed, epidemiological studies have suggested that early-life use of antibiotics, in particular the class of macrolides, is associated with increased risk of asthma and weight gain leading to obesity and metabolic diseases (Cox and Blaser 2015; Saari et al. 2015). A recent study involving 142 children in Finland provided first support that early childhood antibiotic use is linked to clinical phenotypes via alterations of the intestinal microbiota composition (Korpela et al. 2016). This study demonstrated that use of macrolides such as azithromycin and clarithromycin are associated with larger and long-lasting impact on the composition of the microbiome than use of penicillin correlating with enhanced risks of asthma and obesity. These antibiotic-induced perturbations are likely stronger in (young) children than in adults, since this is a phase in which the microbiome assembles and displays lower resilience (Lozupone et al. 2012). Several studies suggest that already the first colonization in newborns may be impacted by antibiotics as intrapartum antibiotic prophylaxis in expecting mothers to prevent Group B Streptococcal disease has

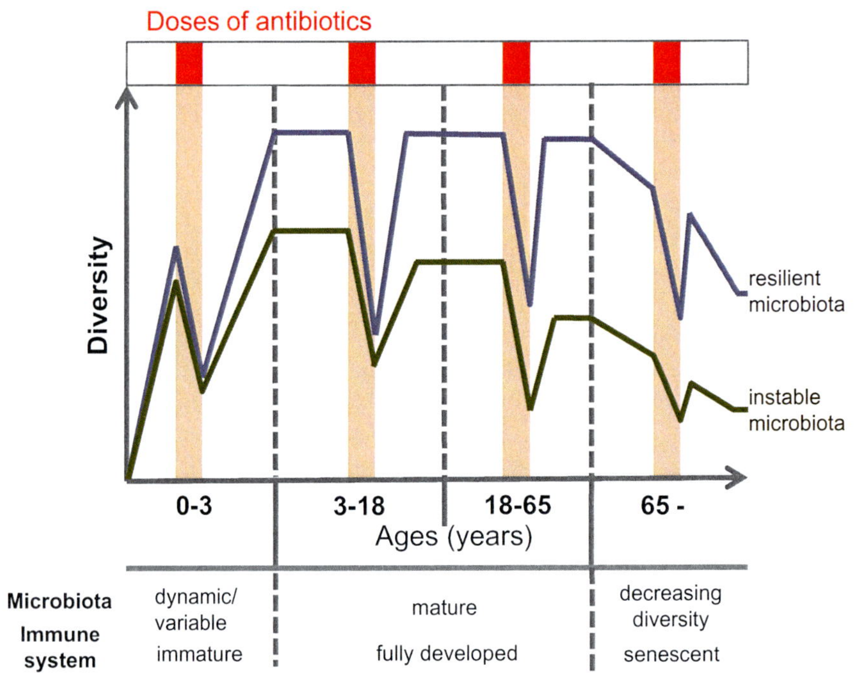

Fig. 1 Impact of antibiotics on microbial diversity. The gut microbiota develops during the first years of life and affects the maturation of the immune system. Antibiotic treatment in childhood and adulthood may lead in some individuals to long-lasting disruption of the ecosystem reducing bacterial diversity resulting in an instable community and negative impact on host physiology. In other individuals with a resilient microbiome, microbial species recover to form a stable microbial community

been demonstrated to result in reduced abundance of *Bifidobacteria*, typically found in high frequency in newborns, and to decrease overall diversity (Aloisio et al. 2014). Similarly, treatment with antibiotics within the first week of life has been reported to cause reduced abundances of *Bifidobacteria* and *Lactobacillus* and in turn, increased frequencies of the opportunistic pathogens *Enterococcus* and Enterobacteriaceae for at least 8 weeks (latest time point investigated) (Tanaka et al. 2009; Fouhy et al. 2012).These studies typically included small numbers of patients and limited follow-up; hence, it is impossible to draw long-term conclusions of how use of antibiotics directly after birth impacts the microbiome. Since most children have received at least one course of antibiotics by age 5, larger longitudinal studies that include frequent microbiota monitoring, documentation of antibiotic use and multiple other clinical parameters are required to firmly demonstrate the effect different classes of antibiotics have on their microbiome (Rutten et al. 2015; Zoch et al. 2015). Albeit these limitations, it has already been recommended by some researchers that macrolides should be avoided in pediatric patients to prevent harming the microbiome if feasible and safe according to the clinical indication (Korpela et al. 2016).

Similar to young children, elderly individuals (>65 years old) represent another population group that have potentially higher susceptibility to antibiotic-induced disturbances of the microbiome. Overall, it has been found that individuals living in long-term care facilities are characterized by reduced diversity of the microbiome compared to individuals, which do not require assistance (Claesson et al. 2012; Jeffery et al. 2016). While antibiotic use in the latter group induced stronger changes, their microbiome also recovered better than the one of individuals in long-term care facilities (Jeffery et al. 2016). Since individuals in these facilities face an increased risk to nosocomial infections, the impaired or delayed recovery of the microbiome may affect its function preventing infections with opportunistic pathogens (further discussed below).

3 The Influence of the Gut Microbiota on Antibiotic-Resistant Gastrointestinal Pathogens

Besides its role in modulating the metabolic state of the host, one of the most important functions of the mammalian microbiota is to protect the host of exogenous pathogens in a process called colonization resistance (Bohnhoff and Miller 1962). The phenomenon was already described five decades ago when Bohnhoff and Miller (1962) showed that mice treated with antibiotics have a 10,000-fold higher susceptibility to *Salmonella enterica* serovar Typhimurium. It is thought that colonization resistance in the classical sense is largely mediated by direct inhibition of invading pathogens by commensal bacteria and their metabolites, e.g., the competition for nutrients or release of inhibitory substances such as bacteriocins by the commensal microbiota (Fig. 2a). For example, it has been demonstrated that the *Lactococcus lactis*-produced nisin (class I bacteriocin) targets a wide range of Gram-negative bacteria including *Clostridium difficile* (Le Lay et al. 2016) (Fig. 2a). Other bacteria such as *Bifidobacterium bifidum* secrete soluble inhibitory factors as galactooligosaccharides, which inhibit the growth of *Salmonella* (Searle et al. 2009). However, through culture-independent techniques we are now beginning to understand that protection of the host is not only mediated by direct inhibition of the pathogens but also through the priming of protective components of the immune system by commensals, i.e., immune-mediated colonization resistance (Buffie and Pamer 2013). A well-studied example is a group of bacteria, called segmented filamentous bacteria (SFB), which are able to induce T helper (T_H17) cell differentiation (Fig. 2a). During infection these cells provide the cytokine IL-17, which in turn leads to the production of antimicrobial peptides and the recruitment of neutrophils (Ivanov et al. 2009; Cua and Tato 2010). This mechanism contributes to the protection of mice against the enteropathogen *Citrobacter rodentium*, a mouse model of enterohaemorrhagic and enteropathogenic *Escherichia coli* infection, EHEC and EPEC, respectively (Ivanov et al. 2009). Moreover, SFBs induce specific and nonspecific IgA production by B cells enhancing opsonization and bactericidal elimination to prevent pathogen translocation (Talham et al. 1999). Another important mechanism of how the gut microbiota prevents the infection of pathogens is contributing to the maintenance of the intestinal barrier, specifically the mucus layer, separating luminal bacteria and intestinal epithelial cells (IECs) (Fig. 2a). Mucus is mainly composed of different glycoproteins, collectively termed mucins, which are produced and secreted by goblet cells. While the mucus layer consists of a single layer in the small intestine, it is composed in the stomach and the colon of two layers. The inner layer is a compact dense layer, firmly attached to the epithelial cells and considered largely sterile (Johansson et al. 2008). In contrast to the inner layer, the outer layer is looser, which allows endogenous bacteria to enter, forming a distinct microbial niche that serves as a carbon source for specific intestinal bacteria (Johansson et al. 2011; Li et al. 2015). Specifically, the enzyme 2-α-L-fucosyltransferase 2 (FUT2), expressed by goblet cells, adds fucose residues to intestinal mucins (Hurd et al. 2005). Mucin-bound fucose serves then as a substrate for protective bacteria of the gut microbiota, therefore enriching bacterial diversity that in turn, prevent outgrowth of invading pathogens (Pickard et al. 2014). The microbiota is not only influenced by the mucus layer, but is also important for proper development of the mucus layer. Germ-free mice have a thinner mucus layer in the small intestine and in the colon it is more penetrable by bacteria than in conventional raised mice. Transfer of a diverse microbiota to germ-free mice restores normal mucus properties to levels observed in conventional raised mice, but this process takes several weeks (Johansson et al. 2015). Moreover, bacterial products such as lipopolysaccharides (LPS), a major component of the outer membrane of Gram-negative bacteria, are by itself sufficient to increase the thickness of the mucus layer in germ-free mice (Petersson et al. 2011). Disruption of the intestinal ecosystem as a result of antibiotic treatment leads to impairment of the mucosal barrier, which increases the chance of pathogens to invade (Wlodarska et al. 2011) (Fig. 2b). Similarly, it was demonstrated in mice that the fucosylation status of mucin is furthermore depended on the microbiota, as SFB or LPS induce secretion of IL-22 by type 3 innate lymphoid cells (ILCs), which is required for FUT2 expression (Goto et al. 2014). Accordingly, mice lacking *Fut2* show a higher susceptibility toward *Salmonella* Typhimurium and *C. rodentium* (Goto et al. 2014; Pickard et al. 2014). In addition, administration of fucosylated molecules to *C. rodentium* infected *Il22ra1*$^{-/-}$ mice restored richness and diversity of the intestinal community, which attenuated enteric infection (Pham et al. 2014).

In summary, microbiota-mediated protection of the host is achieved by preventing outgrowth of gastrointestinal pathogens (direct/classical colonization resistance) and by inducing the mucus layer as well as by stimulating multiple protective immune pathways (indirect/immune-mediated colonization resistance). However, the use of antibiotics changes the situation dramatically: The main issue is that antibiotic treatment targets not only pathogenic bacteria, but also a broad range of commensals. The microbial composition is shifted in a way that direct and

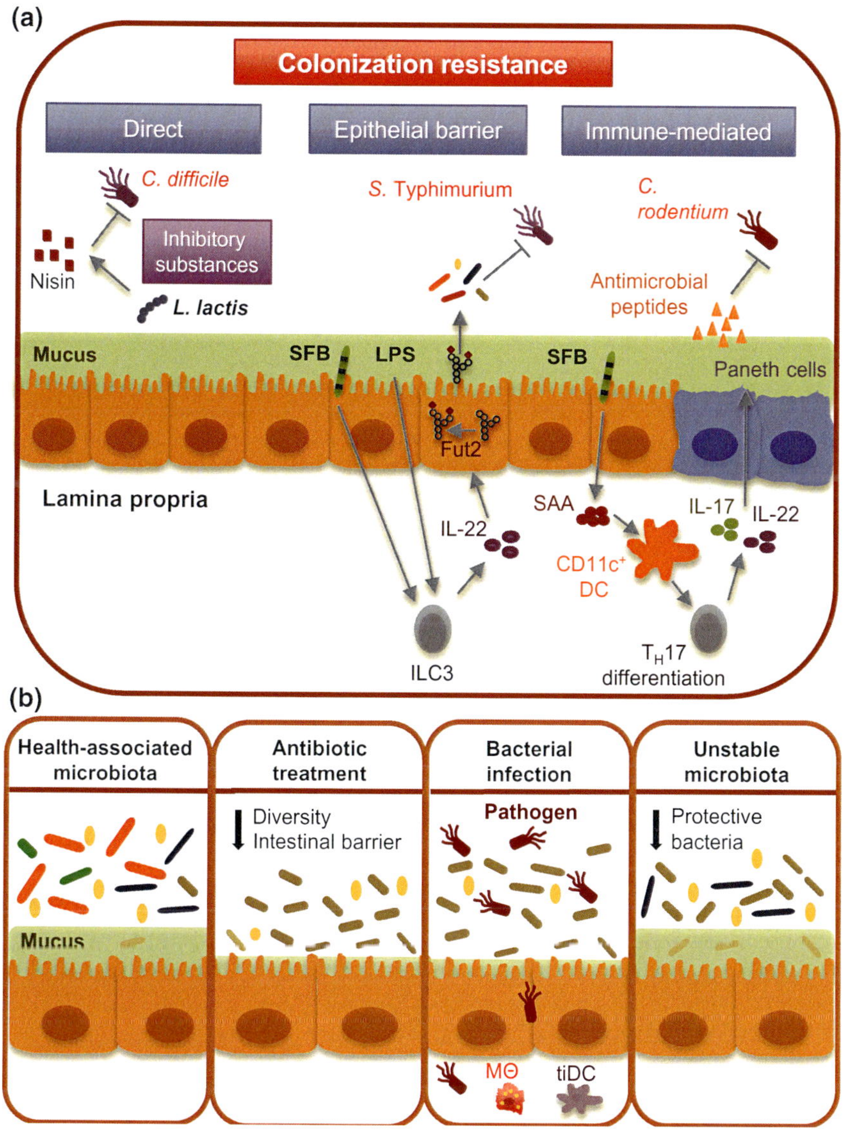

(a)
Colonization resistance
Direct
Epithelial barrier
Immune-mediated
C. difficile
Inhibitory substances
Nisin
L. lactis
S. Typhimurium
C. rodentium
Antimicrobial peptides
Mucus
SFB
LPS
SFB
Paneth cells
Fut2
Lamina propria
IL-22
SAA
IL-17
IL-22
CD11c+ DC
ILC3
TH17 differentiation
(b)
Health-associated microbiota
Mucus
Antibiotic treatment
Diversity
Intestinal barrier
Bacterial infection
Pathogen
MΘ
tiDC
Unstable microbiota
Protective bacteria

◀ **Fig. 2** The effects of antibiotic treatment on colonization resistance toward enteric infections. **a** Commensal intestinal bacteria protect the host against invading pathogens via different ways of defense mechanisms. Direct mechanisms are for example when commensals produce inhibitory substances such as bacteriocins (e.g., *L. lactis*-produced nisin), which inhibit the growth of specific pathogenic bacteria. Another way is the maintenance of the epithelial barrier: Commensal bacteria can induce the expression of Fut2 by the activation of ILC3 and therefore IL-22, which in turn affects the bacterial diversity conferring resistance against *Salmonella* spp. infection. In addition, activating of the immune system by the gut microbiota may lead to resistance. For example, SFB induce IL-17 and IL-22 via SAA that activates the expression of antimicrobial peptides by paneth cells, which in turn targets *Citrobacter rodentium*. *Fut2* Fucosyltransferase 2, *IL* Interleukin, *ILC* Innate lymphoid cells, *LPS* Lipopolysaccharide, *SAA* Serum amyloid A, *SFB* Segmented filamentous bacteria. **b** A healthy gut microbiota is characterized by the presence of diverse bacteria supporting a tight and protective epithelial barrier. Antibiotic treatment affects diversity by inducing the loss of antibiotic-sensitive commensal bacteria and the outgrowth of resistant bacteria with altered functionality resulting in disruption of the epithelial barrier (e.g., decreased mucus production, enhanced epithelial permeability). Altered colonization resistance enables invasion of pathogens, induction of inflammatory responses, and intestinal and/or systemic pathology. Repeated use of antibiotics may lead to an unstable microbiota with a different microbial structure and function, predisposing the host to chronic infections. *MΦ* Macrophage, *tiDC* Tissue residing dendritic cells

immune-mediated colonization resistance is decreased. Furthermore, antibiotic treatment not only has an impact on the composition of the community, but also alters the functions of specific commensal bacteria (Maurice et al. 2013). As a consequence, antibiotic-resistant pathogens, which are commonly encountered in hospital settings, take advantage of the reduced colonization resistance to expand in the lumen and then penetrate the intestinal barrier, which may result in bacteremia and sepsis (Ubeda et al. 2010). Antibiotic-resistant bacterial strains of *C. difficile*, vancomycin-resistant *Enterococcus* (VRE), and Gram-negative Enterobacteriaceae nowadays cause the most common nosocomial infections. In the past years, studies shed light on the impact of the intestinal microbiota on different bacterial infections, which will be discussed in the next sections.

3.1 Clostridium difficile

Clostridium difficile is a Gram-positive, anaerobic, rod-shaped bacterium that is able to form endospores. It was first described in 1935 as a member of the microbiota in healthy neonates (Hall and O'Toole 1935). Today, *Clostridium difficile* infection (CDI) is estimated to be the most common infectious cause of antibiotic-related diarrhea as well as the most common healthcare-associated infection, which occurs after hospitalization or antibiotic treatment (Larentis et al. 2015; Lessa et al. 2015; Khanna et al. 2015).

Specifically, most patients acquire CDI in hospitals and other health care facilities after receiving antibiotics for other indications such as urinary tract infections. Administration of antibiotics influences the abundance of distinct commensal members (see below) creating an environment in which *C. difficile* can colonize and expand (Buffie et al. 2012; Lewis et al. 2015). After germination of transmitted spores, *C. difficile* produces two exotoxins, TcdA and TcdB, in the vegetative state. These toxins break the intestinal barrier by targeting host GTPases Rho, Rac, and Cdc42 that influence polymerization of actin required for many cellular processes including maintenance of tight junctions between IECs (Hecht et al. 1988; Just et al. 1994). That leads to a mild or severe diarrhea and may result in a life-threatening toxic megacolon. Standard treatment against CDI is the administration of antibiotics like metronidazole or vancomycin (Cohen et al. 2010), but 10–35 % of patients develop recurrent CDI causing significant clinical complications and mortality. Notably, increasing numbers of severe CDI cases are caused by a strain (NAP1/027/B1) that is associated with increased virulence and has acquired different antimicrobial resistances against clindamycin, cephalosporins, and fluoroquinolones representing a novel threat to public health (Loo et al. 2005; Redelings et al. 2007; O'Connor et al. 2009; Archbald-Pannone et al. 2014). As an alternative to antibiotic treatment, fecal transplantation (FT) was shown to be efficient in treating even recurring antibiotic-resistant CDI highlighting the potential of FT to treat infections (see Sect. 5).

To explore the mechanisms of pathogenicity, host defense, and the role of the microbiota, CDI is widely studied using mouse models (Best et al. 2012). As in patients mice seem to require a pre-exposure to antibiotics for colonization of *C. difficile* (Chen et al. 2008). Interestingly, it was shown that disease severity correlates with the loss of Lachnospiraceae and an increase in the abundance of Enterobacteriaceae (Reeves et al. 2011). Monocolonization of germ-free mice with isolates from the genus Lachnospiraceae shows a decrease of *C. difficile* colonization demonstrating that specific species of commensal bacteria are sufficient to exert strong colonization resistance towards *C. difficile* (Reeves et al. 2012). Recent studies have shown that this resistance is mediated by the conversion of host-derived primary to secondary bile acids by commensals inhibiting vegetative growth of *C. difficile* (Wilson 1983; Sorg and Sonenshein 2008). *Clostridium scindens* was identified as a key microbe that confers resistance to CDI by producing secondary bile acids (Buffie et al. 2015).

Besides direct colonization resistance, *C. difficile* colonization is also inhibited by immune-mediated mechanisms: Undefined commensal bacteria activate NOD1, MYD88 and interleukin-1β (IL-1β) signaling which increase the recruitment of neutrophils enhancing bactericidal activity (Jarchum et al. 2012; Hasegawa et al. 2012). Stimulation of Toll-like receptor 5 (TLR5) by commensal-derived flagellin also contributes to defense against *C. difficile* by inducing production of antimicrobial peptides (Jarchum et al. 2011). Along these lines, the potential of microbiome-targeted approaches to combat *C. difficile* is high and sound evidence exists that transfer of the microbiota from a healthy individual to a diseased individual is able to cure recurrent CDI (Kelly et al. 2012; Brandt et al. 2012). However, transplantation of bulk communities that potentially include opportunistic pathogens poses significant risk factors. Additional studies are required to understand the dynamic changes in microbial communities as a consequence of antibiotic treatment and identify specific members of the intestinal microbiota, which are highly effective against CDI eventually replacing experimental fecal transplantation.

3.2 *Vancomycin-Resistant* Enterococcus *Spp.*

Enterococcus spp. are Gram-positive cocci and some species are members of the commensal microbiota. Among them *E. faecalis* and *E. faecium* are the two most prevalent enterococci in the human intestine (Gilmore et al. 2014). Under healthy conditions, they are nonpathogenic bacteria, which compromise less than 1 % of the intestinal microbiota (Sghir et al. 2000). However, administration of antibiotics to treat unrelated infections in humans can facilitate the expansion of antibiotic-resistant *Enterococcus* spp., which can then translocate into the tissue and cause bloodstream infections. *Enterococcus* spp. are considered very robust bacteria, which can survive high temperatures (60 °C, up to 30 min), live in salty conditions (6.5 % NaCl) as well as in a broad pH range (4.5–10.0) (Fisher and Phillips 2009). Their robustness as well as intrinsic and acquired antibiotic resistances to broad-spectrum antibiotics such as penicillins, cephalosporins, or vancomycin renders them difficult to treat. Consequently, *Enterococcus* spp. have become a significant nosocomial pathogen worldwide, particularly in older or immune-compromised patients (Sievert et al. 2013). The acquisition of vancomycin resistance by enterococci (VRE), in particular by *E. faecalis and E. faecium*, was first described in 1986 and has steadily rising infection rates (Uttley et al. 1988). Today, in the United States more than 80 % of *E. faecium* isolates from hospitals carry a vancomycin resistance (Hidron et al. 2008). Subsequent disruption of the microbiota by antibiotics in patients carrying antibiotic-resistant *Enterococcus* spp. predisposes them to their expansion (Ubeda et al. 2010). In order to understand how antibiotics enhance the risk of complications caused by VRE, mouse models are currently employed. In mice, antibiotic treatment decreases the expression of regenerating islet-derived protein 3 gamma (RegIIIγ), a potent antimicrobial peptide with anti-VRE activity whose production in Paneth cells is regulated by the commensal microbiota (Brandl et al. 2008). RegIIIγ expression can be restored by TLR5 agonist flagellin in antibiotic treated mice eliminating VRE from mice (Kinnebrew et al. 2010). In light of these findings, attention has turned to manipulating the microbiota to interfere with VRE colonization. Studies in mice have demonstrated that commensal bacteria of the *Barnesiella* genus are associated suppressing VRE (Ubeda et al. 2013), but it is not known which *Barnesiella* species protect against VRE colonization and if the protection is linked to inducing RegIIIγ production. The high hurdles to develop probiotic-based therapies against VRE were demonstrated by a recent clinical trial showing that *Lactobacillus rhamnosous*, a probiotic intestinal bacterium with anti-VRE activity in vitro, failed to interfere with VRE colonization in patients (Doron et al. 2015). In summary, further studies in animal models will be required to develop microbiota-based strategies to interfere with VRE colonization to eventually reduce *Enterococcus* spp.-induced complications in hospitalized patients.

3.3 *Enterobacteriaceae*

In the past decades Gram-negative bacteria, especially of the family Enterobacteriaceae, have acquired multiple antibiotic resistances including resistances to third generation cephalosporins resulting in so-called "multidrug resistant Gram-negative rods." Advanced resistances are in part mediated by extended spectrum beta-lactamases (ESBL), which enable bacteria to cleave distinct beta-lactam antibiotics, such as penicillins and cephalosporins. Importantly, ESBLs are found worldwide in nosocomial infection-causing *Klebsiella pneumoniae* and *E. coli.* Additionally to ESBL, in the past years Enterobacteriaceae could achieve also resistances to carbapeneme or/and fluoroquinolones. Especially, further complications are observed due the increasing prevalence of carbapenem-resistant Enterobacteriaceae. Carbapenemases not only cleave as ESBLs beta-lactam antibiotics, but also carbapenem antibiotics, which currently are used to treat ESBL-expressing bacteria. In particular, carbapenemase-positive *K. pneumoniae* are causing "difficult to treat" infections and its infections are particularly problematic to handle in hospitalized patients with comorbidities (Nordmann et al. 2011). Notably, the intestinal microbiota is considered the main reservoir of Carbapenemases-positive Enterobacteriaceae in humans (Nordmann et al. 2011). Consequently, beyond its classical role of entry site for enteropathogenic Enterobacteriaceae such as *Salmonella* spp. and *Campylobacter* spp., the intestine and the associated microbiota are gaining intensified attention to devise strategies that interfere with colonization of multiple antibiotic-resistant bacteria (see next section).

Using different mouse models, recent studies shed light on the influence of the gut microbiota on infections with Gram-negative Enterobacteriaceae, such as *E. coli*, *K. pneumoniae,* and *S. enterica* serovar Typhimurium. Disruption of the intestinal microbiota caused by different antibiotic treatments, as mentioned above, alters host susceptibility to the aforementioned pathogens just as in the case of *C. difficile* or VRE (Bohnhoff and Miller 1962; Sekirov et al. 2008). In case of *S.* Typhimurium, members of the phylum Bacteroidetes, family Porphyromonadaceae, are associated with colonization resistance toward infection (Ferreira et al. 2011). Also immune-mediated mechanism of colonization resistance against *S.* Typhimurium are known: Undefined commensal bacteria are able to stimulate Paneth cell-specific Myd88 dependent activation of antimicrobial peptides such as RegIIIγ (Vaishnava et al. 2008). Furthermore, commensal bacteria can activate transepithelial dendrites, which capture Enterobacteriaceae as *S.* Typhimurium in the lumen (Niess et al. 2005). In both cases it is still unclear, which commensal bacteria are responsible for the immune-mediated protection. The colonization of the mouse-restricted pathogen *C. rodentium*, an in vivo model for EPEC and EHEC, was also reported to be dependent on the intestinal microbiota. Activation of IL-22 by the mouse commensal SFB was associated with enhanced resistance to *C. rodentium* through the production of the antimicrobial peptides RegIIIβ and RegIIIγ (Ivanov et al. 2009; Willing et al. 2011). Furthermore, increases in the abundances of Bacteroidaceae, Lachnospiraceae, and an unclassified family of Clostridiales show a higher resistance to *C. rodentium* (Willing et al. 2011). In case of *K. pneumoniae*, germ-free mice are significantly higher colonized in their gastrointestinal tract than specific pathogen-free (SPF) mice, suggesting that colonization is dependent of the intestinal microbiota (Lau et al. 2008). In addition, antibiotic treatment leads to higher colonization of *K. pneumoniae* in mice compared to untreated controls (Caballero et al. 2015). It remains elusive which intestinal members regulate the colonization of *K. pneumoniae* and how they interact with each other. Further complications may arise if two antibiotic-resistant bacteria such as VRE and carbapenem-resistant *K. pneumoniae* encounter each other after antibiotic treatment. In mouse models, the high colonization levels of both pathogens is unaffected by the presence of the other species since they occupy different niches (Caballero et al. 2015). Hence, co-infection with multiple antibiotic-resistant bacteria highlights even more the need to identify specific commensal members, which could confer resistance to bacterial infections, in particular with antibiotic-resistant pathogens.

In summary, administration of antibiotics leads to a dramatic change in the bacterial composition in the intestine, shifting the microbiota to an imbalance in which pathogens have an increased chance to colonize and expand within the gastrointestinal tract causing in severe cases systemic bacteremia. Especially, antibiotic-resistant bacteria demonstrate a major problem due to reduced treatment options for infected patients. The complex interplay between host and microbiota as well as pathogens is far from understood and it is still under investigation, which clinically relevant bacterial infections, besides the mentioned ones, are regulated by commensal bacteria. To translate these findings into new therapies the next step will be to identify the specific bacteria and responsible metabolites important for colonization resistance.

4 The Gut Microbiota as a Reservoir of Antibiotic Resistance Genes

The relationship between bacteria and antibiotics can be interpreted as a molecular arms race in which the introduction of novel antibiotics is inevitably followed by the evolution, selection, and spread of ARGs. Already shortly after the discovery of antibiotics and their introduction into clinical practice, resistant bacteria were described (Fleming 2001; Aminov 2010). Indeed, the current widespread and often little regulated use of antibiotics in many countries has caused a rampant spread of ARG against many classes of antibiotics. That may even cause problems in treating common bacterial infections, e.g., urinary tract infection caused by Enterobacteriaceae expressing extended spectrum β-lactamase that confers resistance against penicillins and cephalosporins. Moreover, bacteria that are resistant against multiple classes of antibiotics, like some strains of *C. difficile* and *S. enterica,* are more frequently encountered in the hospital setting (Sebaihia et al. 2006; Pokharel et al. 2006; Wani et al. 2009).

Numerous studies over the past years have revealed that the intestinal microbiota serves as reservoir for ARGs, which can be transferred to pathogens that enter the intestines via contaminated food or water (Salyers et al. 2004; Sommer et al. 2009; Hu et al. 2013) (Table 1). While ARGs can be initially acquired by mutations in specific genes, e.g., in those that encode for an increased efflux or a decreased intake of a specific antibiotic (Oethinger et al. 2000; Webber and Piddock 2003), they can be subsequently transferred from bacteria to bacteria using horizontal gene transfer (HGT). A high cell density as observed in the intestinal ecosystem is favorable for HGT (Aminov 2011). Notably, the bacteria within the microbiota of an individual contain a pool of related ARGs, although the encoding bacteria are not closely related to each other. This supports that there is an high incidence of HGT between members of the microbiota; moreover, even bacteria that are taxonomically distinct can still share a group of related ARGs (Dobrindt et al. 2010; Frye et al. 2011; Broaders et al. 2013). There are multiple mechanisms to obtain DNA via HGT: conjugation (by plasmids, conjugative transposons, and conjugative integrons), transduction (bacteriophages), and transformation (naked DNA) (Choi and Kim 2009; Harrison and Brockhurst 2012; Arutyunov and Frost 2013; Huddleston 2014). Bacteria can express special genes specifically for absorbing DNA for instance during stress situations which bacteria may face on a regular basis in the intestine. Identified stress situation include antibiotic treatment itself, nutrient-limited conditions or DNA damage in the bacterial genome (Michod et al. 1988; Chandler 1992; Finkel and Kolter 2001; Saleeby et al. 2003; Bushman et al. 2011; Stecher et al. 2012; Huddleston et al. 2013).

In a large study in 2013, a comparison of the intestinal metagenomes of 275 individuals revealed the presence of ARGs against 53 different antibiotics (Ghosh et al. 2013). Notably, there were differences in the detected abundances of ARGs in different countries corresponding partly to the spectrum of antibiotics dominantly prescribed in a country. A study comparing the presence of ARGs of Chinese, Danish, and Spanish participants showed that Chinese participants harbor more resistance genes than seen in other participants correlating with frequency of antibiotic use within this population (Ghosh et al. 2013). But even in Amerindians, a group of people who have presumably never been in contact with the developed world and synthetic antibiotics before, a significant presence of ARGs was found in their microbiota, i.e., 28 different functional ARGs conferring resistance against, among others, cephaolosporins (Clemente et al. 2015). In developed countries, the relative abundances of ARGs are the highest for antibiotics that have been used the longest and also for ones used in veterinary medicine as well as for growth promotion providing a rationale in which higher exposure to antibiotics results in increased selection pressure to acquire ARG by the microbiota (Hu et al. 2013;

Table 1 Examples of bacteria with acquired ARGs, which can interact with commensal bacteria

Bacteria	Mechanism	Resistance
Escherichia coli (Machado and Sommer 2014; Singh et al. 2005)	Plasmids and conjugative integrons	MD
Enterococcus faecalis, Enterococcus faecium (Garnier et al. 2000)	Conjugative transposons	van
Lactobacilli, Streptococci (Gevers et al. 2003; Mathur and Singh 2005; Milazzo et al. 2006)	Conjugation	tet, ery, pen, chl, van, kan
Campylobacter jejuni (De Boer et al. 2002; Jeon et al. 2008)	Transformation	tet, met, MD
Salmonella enterica serovar Typhimurium (Schmieger and Schicklmaier 1999; Threlfall 2000)	Bacteriophages	amp, tet, chl, str, spe, sul
Helicobacter pylori (Taylor and Chau 1997)	Transformation	cla, ery
Klebsiella pneumonia (Ramirez et al. 2014; Tolmasky et al. 1988; Woloj et al. 1986)	Plasmids, conjugative transposons	cep, car, ami, flu, amik
Citrobacter freundii, Enterobacter cloacae (Conlan et al. 2014)	Plasmids, conjugative transposons	car
Clostridium difficile (Mullany et al. 2015, 1996; Sebaihia et al. 2006)	Conjugative transposons	tet, chl, mac, lin, strg
Staphyloccocus aureus (Hachler et al. 1987)	Conjugative transposons	ery, mac, lin, strg
Bacteroides fragilis (Husain et al. 2014)	Conjugative transposons	MD
Acinetobacter baumannii (Nigro et al. 2015)	Plasmids	car, amik

Note that not all strains of a bacterial species contain these resistance genes
Abbreviations: *ami* aminoglycosides, *amik* amikacin, *amp* ampicillin, *car* carbapenems, *cep* cephalosporins, *chl* chloramphenicol, *cla* clarithromycin, *ery* erythromycin, *flu* fluoroquinolone, *kan* kanamycin, *lin* lincomycin, *mac* macrolide, *MD* multidrug, *met* methicillin, *pen* penicillin, *spe* spectinomycin, *str* streptomycin, *strg* streptogramin B, *sul* sulphonamides, *tet* tetracyclin, *van* vancomycin

Forslund et al. 2013). Yet, the occurrence of ARGs in humans with no previous contact with synthetic antibiotics such as the Amerindians can be explained either by consumption of foods that contain natural antibiotics as defense against bacteria or by exchanging genes with bacteria that are in contact with antibiotic-producing soil microbes (D'Costa et al. 2011; Forsberg et al. 2012). Similarly, also children with no prior exposure to antibiotics already harbor ARGs in their microbiota. While these may be partly passed on from the mother, also ARGs, which were not present in the mother, were found potentially having been acquired from the environment shortly after birth (Gosalbes et al. 2016; Gueimonde et al. 2006; Zhang et al. 2011). Finally, elderly, especially those living in a nursing home, harbor large pools of ARGs, likely caused by a large number of antibiotic treatments, which the individuals had been subjected during their lifetime (Neuhauser et al. 2003; Lautenbach et al. 2009).

Moreover, globalization and increased international travel are also impacting the spread of ARG via the microbiome. In multiple studies, it has been found that individuals that travel to countries in which endemic antibiotic-resistant bacteria circulate, can return with these ARGs in their microbiota. For instance, a study found that three individuals who traveled from France to India acquired carbapenem-resistant Enterobacteriaceae, but cleared them within 1–3 months after return (Ruppe et al. 2014). A larger scaled study found that 113 of 370 Dutch travelers (of which 92 % traveled to Asia or Africa) acquired multidrug-resistant Enterobacteriaceae (Paltansing et al. 2013). Importantly, 19 individuals still carried the bacteria after 6 months providing evidence for the hypothesis that the microbiota is a reservoir, but also highlighting that comparative studies are needed to understand why some individuals clear these bacteria. Another study investigated the acquisition of quinolone resistance in the microbiota and found in Swedish travelers returning from developing countries an increased abundance of bacteria carrying resistance genes (Johnning et al. 2015).

Besides the human microbiome, other tremendous reservoirs for ARG are animals, particularly livestock. Over several decades antibiotics were given to large groups of animals in agriculture and aquaculture not only to treat and prevent diseases, but also to enhance their growth (Cabello 2006; Frye et al. 2011). Indeed, the microbiota of animals from farms who are using antibiotics for growth promotion has an higher amount of ARG. Already in 1976 a study found proof of the transfer of a resistance gene in *E. coli* from poultry to humans (Levy et al. 1976). Similarly, increased abundances of resistant bacteria are found in the microbiota of farmers who are using antibiotics for growth promotion (Howells and Joynson 1975; Gorbach 2001; Wenzel 2002). Other studies also show the transfer of fluoroquinolone and ciprofloxacin resistance genes from animals to humans through food ingestion (Endtz et al. 1991; Bertrand et al. 2006). Consequently, since 2006 there has been a total ban on growth promoting antibiotics for food animals in the European Union, whereas antibiotics to treat diseases are still allowed. A significant decline in tetracycline and vancomycin resistance in the animals as well as in humans has been observed since (Pantosti et al. 1999; Aarestrup et al. 2001; Accountability NSUG 2011). As part of the concept of "One Health" recognizing environmental factors as being important modulators of human health incorporating knowledge from veterinary sciences will be essential to curtail the threat of ARG globally.

5 Further Directions: Fecal Transplantation to Cure Enteric Infections and Correct Antibiotic-Induced Disturbances of the Microbiota?

The discovery of the numerous effects that the microbiota and specific commensal bacteria have on the human physiology has generated an increasing interest in exploring manipulations of the microbiota for therapeutic purposes. While this is not a new idea, already more than 50 years ago, Hentges and Freter (1962) suggested the need for the intestinal manipulation to fight against enteric infections, culture-independent surveys and advanced model systems provide now solid evidence to drive preclinical as well as clinical development and testing. Different directions are currently explored that are either based on transfer of complex communities from one individual to another (→ fecal transplantation/transfer: FT) or providing novel types of prebiotics/probiotics. Much attention focuses now on developing options to prevent recurrent CDI, which is strongly linked to disturbed states of the microbiota induced by antibiotics. Transplant of fresh or frozen complex communities to individuals with recurrent CDI has response rates of roughly 90 %, at least as high if not higher than what is seen with antibiotics (Gough et al. 2011; Landy et al. 2011). Moreover, the development of frozen "stool banks" (in analogy to blood banks) has been pioneered, which promises to largely simplify the screening of feces samples for potential bacterial and viral pathogens (Paramsothy et al. 2015). Yet, much is still unknown about why in some patients with recurrent CDI FT fails. One general concern for FT is that the community within every fecal sample is distinct and difficult to fully identify even using NGS-based approaches. Besides, it is even possible that a bacterium completely changes its behavior after transplantation into a new environment because of the host genotype, its diet, or the residual microbiota, turning form a beneficial bacterium to a pro-inflammatory pathobiont. Therefore, many efforts are underway to identify a mixture of beneficial bacteria, which can help to fight against enteric infections, by administering small consortia of cultivable bacteria rather to transfer the whole unidentified microbiota. In small-scale studies, it was shown that a mixture of 10 isolated bacteria including anaerobes and aerobes was effective to cure CDI in 5 patients (Tvede and Rask-Madsen 1989). In addition, a more recent study showed that a group of 33 bacteria can cure CDI (Petrof et al. 2013). In the meantime several companies already have started clinical trials with "semi-standardized" donor material to treat recurrent CDI including a successful phase II trial by Rebiotix (Orenstein et al. 2016). Notably, Rebiotix is now planning a clinical trial for microbiota restoration therapy to prevent CDI. Moreover, they have provided evidence that administration of their product promotes clearance of VRE, which is beneficial for CDI patients. Therefore, manipulation of the microbiota might be an approach to "decolonize" an individual from antibiotic-resistant opportunistic pathogens. This is supported by studies, which showed that species of a mixture of donor bacteria are lost over time or do not fully colonize (Lawley et al. 2012; Petrof et al. 2013), alteration through addition of microbes increase the microbial diversity in the recipients and may occupy niches, which are normally taken by the enteric pathogen.

In summary, we are just starting to identify which intestinal members can cure or prevent enteric infections, but fecal transplantation is an effective intermediate treatment option for some indications until the development of probiotics with defined consortia of beneficial bacteria.

6 Conclusions

Antibiotics have a tremendous impact on the composition of the microbiome imprinting in some, but not all, individuals long-term taxonomic and functional changes. Firm evidence has linked antibiotic-induced disturbances in specific functions of microbiome to enhanced susceptibility to gastrointestinal infections such as CDI. In the meantime numerous clinical studies are being conducted to understand how the microbiota affects and contributes to other human diseases such as metabolic syndrome or IBD and whether they are modulated or induced in these cases by antibiotics. Generally, well-adjusted global antibiotic stewardship in human and veterinary medicine is an essential mean for the future to prevent the spread of antibiotic resistances, but also to prevent damage to the microbiome.

References

Aarestrup FM, Seyfarth AM, Emborg HD et al (2001) Effect of abolishment of the use of antimicrobial agents for growth promotion on occurrence of antimicrobial resistance in fecal enterococci from food animals in Denmark. Antimicrob Agents Chemother 45:2054–2059

Accountability NUSG (2011) Antibiotic resistanste: agencies have made limited progress addressing antibiotic use in animals

Aloisio I, Mazzola G, Corvaglia LT et al (2014) Influence of intrapartum antibiotic prophylaxis against group B *Streptococcus* on the early newborn gut composition and evaluation of the anti-*Streptococcus* activity of *Bifidobacterium* strains. Appl Microbiol Biotechnol 98:6051–6060

Aminov RI (2010) A brief history of the antibiotic era: lessons learned and challenges for the future. Front Microbiol 1:134

Aminov RI (2011) Horizontal gene exchange in environmental microbiota. Front Microbiol 2:158

Archbald-Pannone LR, Boone JH, Carman RJ et al (2014) *Clostridium difficile* ribotype 027 is most prevalent among inpatients admitted from long-term care facilities. J Hosp Infect 88:218–221

Arutyunov D, Frost LS (2013) F conjugation: back to the beginning. Plasmid 70:18–32

Bertrand S, Weill F-X, Cloeckaert A et al (2006) Clonal emergence of extended-spectrum beta-lactamase (CTX-M-2)-producing *Salmonella enterica* serovar Virchow isolates with reduced susceptibilities to ciprofloxacin among poultry and humans in Belgium and France (2000 to 2003). J Clin Microbiol 44:2897–2903

Best EL, Freeman J, Wilcox MH (2012) Models for the study of *Clostridium difficile* infection. Gut Microbes 3:145–167

Bohnhoff M, Miller CP (1962) Enhanced susceptibility to Salmonella infection in streptomycin-treated mice. J Infect Dis 111:117–127

Brandl K, Plitas G, Mihu CN et al (2008) Vancomycin-resistant enterococci exploit antibiotic-induced innate immune deficits. Nature 455:804–807

Brandt LJ, Aroniadis OC, Mellow M et al (2012) Long-term follow-up of colonoscopic fecal microbiota transplant for recurrent *Clostridium difficile* infection. Am J Gastroenterol 107:1079–1087

Broaders E, Gahan CGM, Marchesi JR (2013) Mobile genetic elements of the human gastrointestinal tract: potential for spread of antibiotic resistance genes. Gut Microbes 4:271–280

Buffie CG, Pamer EG (2013) Microbiota-mediated colonization resistance against intestinal pathogens. Nat Rev Immunol 13:790–801

Buffie CG, Jarchum I, Equinda M et al (2012) Profound alterations of intestinal microbiota following a single dose of clindamycin results in sustained susceptibility to *Clostridium difficile*-induced colitis. Infect Immun 80:62–73

Buffie CG, Bucci V, Stein RR et al (2015) Precision microbiome reconstitution restores bile acid mediated resistance to *Clostridium difficile*. Nature 517:205–208

Bulgarelli D, Schlaeppi K, Spaepen S et al (2013) Structure and functions of the bacterial microbiota of plants. Annu Rev Plant Biol 64:807–838

Bushman FD, Minot S, Sinha R et al (2011) The human gut virome: inter-individual variation and dynamic response to diet. Genome Res 21:1616–1625

Caballero S, Carter R, Ke X et al (2015) Distinct but spatially overlapping intestinal niches for vancomycin-resistant *Enterococcus faecium* and carbapenem-resistant *Klebsiella pneumoniae*. PLoS Pathog 11:e1005132

Cabello FC (2006) Heavy use of prophylactic antibiotics in aquaculture: a growing problem for human and animal health and for the environment. Environ Microbiol 8:1137–1144

Chandler MS (1992) The gene encoding cAMP receptor protein is required for competence development in Haemophilus influenzae Rd. Proc Natl Acad Sci USA 89:1626–1630

Chen X, Katchar K, Goldsmith JD et al (2008) A mouse model of *Clostridium difficile*-associated disease. Gastroenterology 135:1984–1992

Choi K-H, Kim K-J (2009) Applications of transposon-based gene delivery system in bacteria. J Microbiol Biotechnol 19:217–228

Claesson MJ, Jeffery IB, Conde S et al (2012) Gut microbiota composition correlates with diet and health in the elderly. Nature 488:178–184

Clemente JC, Pehrsson EC, Blaser MJ et al (2015) The microbiome of uncontacted Amerindians. Sci Adv 1:e1500183–e1500183

Cohen SH, Gerding DN, Johnson S et al (2010) Clinical practice guidelines for *Clostridium difficile* infection in adults: 2010 update by the society for healthcare epidemiology of America (SHEA) and the infectious diseases society of America (IDSA). Infect Control Hosp Epidemiol 31:431–455

Conlan S, Thomas PJ, Deming C, et al (2014) Single-molecule sequencing to track plasmid diversity of hospital-associated carbapenemase-producing Enterobacteriaceae. Sci Transl Med 6:254ra126–254ra126. doi:10.1126/scitranslmed.3009845

Cox LM, Blaser MJ (2015) Antibiotics in early life and obesity. Nat Rev Endocrinol 11:182–190

Cua DJ, Tato CM (2010) Innate IL-17-producing cells: the sentinels of the immune system. Nat Rev Immunol 10:479–489

D'Costa VM, King CE, Kalan L et al (2011) Antibiotic resistance is ancient. Nature 477:457–461

de Boer P, Wagenaar JA, Achterberg RP et al (2002) Generation of *Campylobacter jejuni* genetic diversity in vivo. Mol Microbiol 44:351–359

De La Cochetière MF, Durand T, Lepage P et al (2005) Resilience of the dominant human fecal microbiota upon short-course antibiotic challenge. J Clin Microbiol 43:5588–5592

Dethlefsen L, Relman DA (2011) Incomplete recovery and individualized responses of the human distal gut microbiota to repeated antibiotic perturbation. Proc Natl Acad Sci USA 108 Suppl 1:4554–4561

Dethlefsen L, Huse S, Sogin ML, Relman DA (2008) The pervasive effects of an antibiotic on the human gut microbiota, as revealed by deep 16S rRNA sequencing. PLoS Biol 6:e280

Dobrindt U, Chowdary MG, Krumbholz G, Hacker J (2010) Genome dynamics and its impact on evolution of *Escherichia coli*. Med Microbiol Immunol 199:145–154

Doron S, Hibberd PL, Goldin B et al (2015) Effect of *Lactobacillus rhamnosus* GG administration on vancomycin-resistant *Enterococcus* colonization in adults with comorbidities. Antimicrob Agents Chemother 59:4593–4599

Endtz HP, Ruijs GJ, van Klingeren B et al (1991) Quinolone resistance in *campylobacter* isolated from man and poultry following the introduction of fluoroquinolones in veterinary medicine. J Antimicrob Chemother 27:199–208

Ferreira RBR, Gill N, Willing BP et al (2011) The intestinal microbiota plays a role in *Salmonella*-induced colitis independent of pathogen colonization. PLoS ONE 6:e20338

Finkel SE, Kolter R (2001) DNA as a nutrient: novel role for bacterial competence gene homologs. J Bacteriol 183:6288–6293

Fisher K, Phillips C (2009) The ecology, epidemiology and virulence of *Enterococcus*. Microbiology (Reading, Engl) 155:1749–1757

Fleming A (2001) On the antibacterial action of cultures of a penicillium, with special reference to their use in the isolation of B. influenzae. 1929. World Health Organization

Forsberg KJ, Reyes A, Wang B et al (2012) The shared antibiotic resistome of soil bacteria and human pathogens. Science 337:1107–1111

Forslund K, Sunagawa S, Kultima JR et al (2013) Country-specific antibiotic use practices impact the human gut resistome. Genome Res 23:1163–1169

Fouhy F, Guinane CM, Hussey S et al (2012) High-throughput sequencing reveals the incomplete, short-term recovery of infant gut microbiota following parenteral antibiotic treatment with ampicillin and gentamicin. Antimicrob Agents Chemother 56:5811–5820

Frye JG, Lindsey RL, Meinersmann RJ et al (2011) Related antimicrobial resistance genes detected in different bacterial species co-isolated from swine fecal samples. Foodborne Pathog Dis 8:663–679

Garnier F, Taourit S, Glaser P et al (2000) Characterization of transposon Tn1549, conferring VanB-type resistance in *Enterococcus* spp. Microbiology (Reading, Engl) 146 (Pt 6):1481–1489

Gevers D, Huys G, Swings J (2003) In vitro conjugal transfer of tetracycline resistance from *Lactobacillus* isolates to other Gram-positive bacteria. FEMS Microbiol Lett 225:125–130

Ghosh TS, Gupta SS, Nair GB, Mande SS (2013) In silico analysis of antibiotic resistance genes in the gut microflora of individuals from diverse geographies and age-groups. PLoS ONE 8: e83823

Gilmore MS, Clewell DB, Ike Y et al (2014) *Enterococcus* diversity, origins in nature, and gut colonization. Massachusetts Eye and Ear Infirmary, Boston

Gorbach SL (2001) Antimicrobial use in animal feed—time to stop. N Engl J Med 345:1202–1203

Gosalbes MJ, Valles Y, Jimenez-Hernandez N et al (2016) High frequencies of antibiotic resistance genes in infants' meconium and early fecal samples. J Dev Orig Health Dis 7:35–44

Goto Y, Obata T, Kunisawa J et al (2014) Innate lymphoid cells regulate intestinal epithelial cell glycosylation. Science 345:1254009–1254009

Gough E, Shaikh H, Manges AR (2011) Systematic review of intestinal microbiota transplantation (fecal bacteriotherapy) for recurrent *Clostridium difficile* infection. Clin Infect Dis 53:994–1002

Gueimonde M, Salminen S, Isolauri E (2006) Presence of specific antibiotic (tet) resistance genes in infant faecal microbiota. FEMS Immunol Med Microbiol 48:21–25

Hachler H, Berger-bachi B, Kayser FH (1987) Genetic characterization of a *Clostridium difficile* erythromycin- clindamycin resistance determinant that is transferable to *Staphylococcus aureus*. Microbiology 31:1039–1045

Hall IC, O'Toole E (1935) Intestinal flora in new-born infants: with a description of a new pathogenic anaerobe, *Bacillus difficilis*. Am J Dis Child 49:390–402

Harrison E, Brockhurst MA (2012) Plasmid-mediated horizontal gene transfer is a coevolutionary process. Trends Microbiol 20:262–267

Hasegawa M, Kamada N, Jiao Y et al (2012) Protective role of commensals against *Clostridium difficile* infection via an IL-1β-mediated positive-feedback loop. J Immunol 189:3085–3091
Hecht G, Pothoulakis C, LaMont JT, Madara JL (1988) *Clostridium difficile* toxin A perturbs cytoskeletal structure and tight junction permeability of cultured human intestinal epithelial monolayers. J Clin Invest 82:1516–1524
Hentges DJ, Freter R (1962) In vivo and in vitro antagonism of intestinal bacteria against *Shigella flexneri*. I. Correlation between various tests. J Infect Dis 110:30–37
Hidron AI, Edwards JR, Patel J et al (2008) NHSN annual update: antimicrobial-resistant pathogens associated with healthcare-associated infections: annual summary of data reported to the National Healthcare Safety Network at the Centers for Disease Control and Prevention, 2006–2007. Infect Control Hosp Epidemiol 29:996–1011
Howells CH, Joynson DH (1975) Possible role of animal feeding-stuffs in spread of antibiotic-resistant intestinal coliforms. 1:156–157
Hu Y, Yang X, Qin J et al (2013) Metagenome-wide analysis of antibiotic resistance genes in a large cohort of human gut microbiota. Nat Commun 4:2151
Huddleston JR (2014) Horizontal gene transfer in the human gastrointestinal tract: potential spread of antibiotic resistance genes. Infect Drug Resist 7:167–176
Huddleston JR, Brokaw JM, Zak JC, Jeter RM (2013) Natural transformation as a mechanism of horizontal gene transfer among environmental *Aeromonas* species. Syst Appl Microbiol 36:224–234
Hurd EA, Holmén JM, Hansson GC, Domino SE (2005) Gastrointestinal mucins of Fut2-null mice lack terminal fucosylation without affecting colonization by *Candida albicans*. Glycobiology 15:1002–1007
Husain F, Veeranagouda Y, Boente R et al (2014) The Ellis Island Effect: A novel mobile element in a multi-drug resistant *Bacteroides fragilis* clinical isolate includes a mosaic of resistance genes from Gram-positive bacteria. Mobile Genetic Elements 4:e29801
Ivanov II, Atarashi K, Manel N et al (2009) Induction of intestinal Th17 cells by segmented filamentous bacteria. Cell 139:485–498
Jarchum I, Liu M, Lipuma L, Pamer EG (2011) Toll-like receptor 5 stimulation protects mice from acute *Clostridium difficile* colitis. Infect Immun 79:1498–1503
Jarchum I, Liu M, Shi C et al (2012) Critical role for MyD88-mediated neutrophil recruitment during *Clostridium difficile* colitis. Infect Immun 80:2989–2996
Jeffery IB, Lynch DB, O'Toole PW (2016) Composition and temporal stability of the gut microbiota in older persons. ISME J 10:170–182
Jeon B, Muraoka W, Sahin O, Zhang Q (2008) Role of Cj1211 in natural transformation and transfer of antibiotic resistance determinants in *Campylobacter jejuni*. Antimicrob Agents Chemother 52:2699–2708
Jernberg C, Löfmark S, Edlund C, Jansson JK (2007) Long-term ecological impacts of antibiotic administration on the human intestinal microbiota. ISME J 1:56–66
Johansson MEV, Phillipson M, Petersson J, et al (2008) The inner of the two Muc2 mucin-dependent mucus layers in colon is devoid of bacteria. Proc Natl Acad Sci USA 105:15064–15069
Johansson MEV, Larsson JMH, Hansson GC (2011) The two mucus layers of colon are organized by the MUC2 mucin, whereas the outer layer is a legislator of host-microbial interactions. Proc Natl Acad Sci USA 108 Suppl 1:4659–4665
Johansson MEV, Jakobsson HE, Holmén-Larsson J et al (2015) Normalization of host intestinal mucus layers requires long-term microbial colonization. Cell Host Microbe 18:582–592
Johnning A, Kristiansson E, Angelin M et al (2015) Quinolone resistance mutations in the faecal microbiota of Swedish travellers to India. BMC Microbiol 15:235
Just I, Richter HP, Prepens U et al (1994) Probing the action of *Clostridium difficile* toxin B in Xenopus laevis oocytes. J Cell Sci 107 (Pt 6):1653–1659
Kelly CR, de Leon L, Jasutkar N (2012) Fecal microbiota transplantation for relapsing *Clostridium difficile* infection in 26 patients: methodology and results. J Clin Gastroenterol 46:145–149

Khanna S, Gupta A, Baddour LM, Pardi DS (2015) Epidemiology, outcomes, and predictors of mortality in hospitalized adults with *Clostridium difficile* infection. Intern Emerg Med 1–9

Kinnebrew MA, Ubeda C, Zenewicz LA et al (2010) Bacterial flagellin stimulates Toll-like receptor 5-dependent defense against vancomycin-resistant *Enterococcus* infection. J Infect Dis 201:534–543

Korpela K, Salonen A, Virta LJ et al (2016) Intestinal microbiome is related to lifetime antibiotic use in Finnish pre-school children. Nat Commun 7:10410.0

Landy J, Al-Hassi HO, McLaughlin SD et al (2011) Review article: faecal transplantation therapy for gastrointestinal disease. Aliment Pharmacol Ther 34:409–415

Larentis DZ, Rosa RG, Santos Dos RP, Goldani LZ (2015) Outcomes and risk factors associated with *Clostridium difficile* diarrhea in hospitalized adult patients. Gastroenterol Res Pract 2015:346341–6

Lau HY, Huffnagle GB, Moore TA (2008) Host and microbiota factors that control *Klebsiella pneumoniae* mucosal colonization in mice. Microbes Infect 10:1283–1290

Lautenbach E, Marsicano R, Tolomeo P et al (2009) Epidemiology of antimicrobial resistance among gram-negative organisms recovered from patients in a multistate network of long-term care facilities. Infect Control Hosp Epidemiol 30:790–793

Lawley TD, Clare S, Walker AW et al (2012) Targeted restoration of the intestinal microbiota with a simple, defined bacteriotherapy resolves relapsing *Clostridium difficile* disease in mice. PLoS Pathog 8:e1002995

Le Lay C, Dridi L, Bergeron MG et al (2016) Nisin is an effective inhibitor of *Clostridium difficile* vegetative cells and spore germination. J Med Microbiol 65:169–175

Lessa FC, Mu Y, Bamberg WM et al (2015) Burden of *Clostridium difficile* infection in the United States. N Engl J Med 372:825–834

Levy SB, FitzGerald GB, Macone AB (1976) Spread of antibiotic-resistant plasmids from chicken to chicken and from chicken to man. Nature 260:40–42

Lewis BB, Buffie CG, Carter RA et al (2015) Loss of microbiota-mediated colonization resistance to *Clostridium difficile* infection with oral vancomycin compared with metronidazole. J Infect Dis jiv256

Ley RE, Hamady M, Lozupone C et al (2008) Evolution of mammals and their gut microbes. Science 320:1647–1651. doi:10.1126/science.1155725

Li H, Limenitakis JP, Fuhrer T et al (2015) The outer mucus layer hosts a distinct intestinal microbial niche. Nat Commun 6:8292

Loo VG, Poirier L, Miller MA et al (2005) A predominantly clonal multi-institutional outbreak of *Clostridium difficile*-associated diarrhea with high morbidity and mortality. N Engl J Med 353:2442–2449

Lozupone CA, Stombaugh JI, Gordon JI et al (2012) Diversity, stability and resilience of the human gut microbiota. Nature 489:220–230

Machado AMD, Sommer MOA (2014) Human intestinal cells modulate conjugational transfer of multidrug resistance plasmids between clinical *Escherichia coli* isolates. PLoS ONE 9: e100739–5

Mathur S, Singh R (2005) Antibiotic resistance in food lactic acid bacteria—a review. Int J Food Microbiol 105:281–295

Maurice CF, Haiser HJ, Turnbaugh PJ (2013) Xenobiotics shape the physiology and gene expression of the active human gut microbiome. Cell 152:39–50

Michod RE, Wojciechowski MF, Hoelzer MA (1988) DNA repair and the evolution of transformation in the bacterium *Bacillus subtilis*. Genetics 118:31–39

Milazzo I, Speciale A, Musumeci R et al (2006) Identification and antibiotic susceptibility of bacterial isolates from probiotic products available in Italy. New Microbiol 29:281–291. doi:10.1016/S0168-1605(02)00162-9

Mullany P, Pallen M, Wilks M et al (1996) A group II intron in a conjugative transposon from the gram-positive bacterium, *Clostridium difficile*. Gene 174:145–150

Mullany P, Allan E, Roberts AP (2015) Mobile genetic elements in *Clostridium difficile* and their role in genome function. Res Microbiol 166:361–367

Neuhauser MM, Weinstein RA, Rzdman R (2003) Antibiotic resistance among gram-negative bacilli in us intensive care units: implications for fluoroquinolone use. JAMA 289:885–888

Niess JH, Brand S, Gu X et al (2005) CX3CR1-mediated dendritic cell access to the intestinal lumen and bacterial clearance. Science 307:254–258

Nigro SJ, Holt KE, Pickard D, Hall RM (2015) Carbapenem and amikacin resistance on a large conjugative *Acinetobacter baumannii* plasmid. J Antimicrob Chemother 70:1259–1261

Nordmann P, Naas T, Poirel L (2011) Global spread of carbapenemase-producing enterobacteriaceae. Emerging Infect Dis 17:1791–1798

O'Connor JR, Johnson S, Gerding DN (2009) *Clostridium difficile* infection caused by the epidemic BI/NAP1/027 strain. Gastroenterology 136:1913–1924

Oethinger M, Kern WV, Jellen-Ritter AS et al (2000) Ineffectiveness of topoisomerase mutations in mediating clinically significant fluoroquinolone resistance in *Escherichia coli* in the absence of the AcrAB efflux pump. Antimicrob Agents Chemother 44:10–13

Orenstein R, Dubberke E, Hardi R et al (2016) Safety and durability of RBX2660 (Microbiota Suspension) for recurrent *Clostridium difficile* infection: results of the PUNCH CD study. Clin Infect Dis 62:596–602

Paltansing S, Vlot JA, Kraakman MEM et al (2013) Extended-spectrum β-lactamase-producing enterobacteriaceae among travelers from the Netherlands. Emerging Infect Dis 19:1206–1213

Pantosti A, Del Grosso M, Tagliabue S et al (1999) Decrease of vancomycin-resistant enterococci in poultry meat after avoparcin ban. Lancet 354:741–742

Paramsothy S, Borody TJ, Lin E et al (2015) Donor recruitment for fecal microbiota transplantation. Inflamm Bowel Dis 21:1600–1606

Petersson J, Schreiber O, Hansson GC et al (2011) Importance and regulation of the colonic mucus barrier in a mouse model of colitis. Am J Physiol Gastrointest Liver Physiol 300:G327–33

Petrof EO, Gloor GB, Vanner SJ et al (2013) Stool substitute transplant therapy for the eradication of *Clostridium difficile* infection: "RePOOPulating" the gut. Microbiome 1:3

Pham TAN, Clare S, Goulding D et al (2014) Epithelial IL-22RA1-mediated fucosylation promotes intestinal colonization resistance to an opportunistic pathogen. Cell Host Microbe 16:504–516

Pickard JM, Maurice CF, Kinnebrew MA et al (2014) Rapid fucosylation of intestinal epithelium sustains host-commensal symbiosis in sickness. Nature 514:638–641

Pokharel BM, Koirala J, Dahal RK et al (2006) Multidrug-resistant and extended-spectrum beta-lactamase (ESBL)-producing *Salmonella enterica* (serotypes Typhi and Paratyphi A) from blood isolates in Nepal: surveillance of resistance and a search for newer alternatives. Int J Infect Dis 10:434–438

Ramirez MS, Traglia GM, Lin DL et al (2014) Plasmid-mediated antibiotic resistance and virulence in gram-negatives: the *Klebsiella pneumoniae* paradigm. Microbiol Spectr 2:1–15

Raymond F, Ouameur AA, Déraspe M et al (2016) The initial state of the human gut microbiome determines its reshaping by antibiotics. ISME J 10:707–720

Redelings MD, Sorvillo F, Mascola L (2007) Increase in *Clostridium difficile*-related mortality rates, United States, 1999–2004. Emerging Infect Dis 13:1417–1419

Reeves AE, Theriot CM, Bergin IL et al (2011) The interplay between microbiome dynamics and pathogen dynamics in a murine model of *Clostridium difficile* infection. Gut Microbes 2:145–158

Reeves AE, Koenigsknecht MJ, Bergin IL, Young VB (2012) Suppression of *Clostridium difficile* in the gastrointestinal tracts of germfree mice inoculated with a murine isolate from the family Lachnospiraceae. Infect Immun 80:3786–3794

Ruppe E, Armand-Lefevre L, Estellat C et al (2014) Acquisition of carbapenemase-producing enterobacteriaceae by healthy travellers to India, France, February 2012 to March 2013. Eurosurveillance 19:20768–4

Rutten NBMM, Rijkers GT, Meijssen CB et al (2015) Intestinal microbiota composition after antibiotic treatment in early life: the INCA study. BMC Pediatr 15:204

Saari A, Virta LJ, Sankilampi U et al (2015) Antibiotic exposure in infancy and risk of being overweight in the first 24 months of life. Pediatrics 135:617–626

Saleeby J, Ducea M, Clemens-Knott D (2003) Production and loss of high-density batholithic root, southern Sierra Nevada, California. Tectonics 22:n/a–n/a

Salyers AA, Gupta A, Wang Y (2004) Human intestinal bacteria as reservoirs for antibiotic resistance genes. Trends Microbiol 12:412–416

Schmieger H, Schicklmaier P (1999) Transduction of multiple drug resistance of *Salmonella enterica* serovar typhimurium DT104. FEMS Microbiol Lett 170:251–256

Searle LEJ, Best A, Nunez A et al (2009) A mixture containing galactooligosaccharide, produced by the enzymic activity of Bifidobacterium bifidum, reduces *Salmonella enterica* serovar typhimurium infection in mice. J Med Microbiol 58:37–48

Sebaihia M, Wren BW, Mullany P et al (2006) The multidrug-resistant human pathogen *Clostridium difficile* has a highly mobile, mosaic genome. Nat Genet 38:779–786

Sekirov I, Tam NM, Jogova M et al (2008) Antibiotic-induced perturbations of the intestinal microbiota alter host susceptibility to enteric infection. Infect Immun 76:4726–4736

Sghir A, Gramet G, Suau A et al (2000) Quantification of bacterial groups within human fecal flora by oligonucleotide probe hybridization. Appl Environ Microbiol 66:2263–2266

Sievert DM, Ricks P, Edwards JR et al (2013) Antimicrobial-resistant pathogens associated with healthcare-associated infections: summary of data reported to the National Healthcare Safety Network at the Centers for Disease Control and Prevention, 2009–2010. Infect Control Hosp Epidemiol 34:1–14

Singh R, Schroeder CM, Meng J et al (2005) Identification of antimicrobial resistance and class 1 integrons in Shiga toxin-producing *Escherichia coli* recovered from humans and food animals. J Antimicrob Chemother 56:216–219

Sommer MOA, Church GM, Dantas G et al (2009) Functional characterization of the antibiotic resistance reservoir in the human microflora. Science 325:1128–1131

Sorg JA, Sonenshein AL (2008) Bile salts and glycine as cogerminants for *Clostridium difficile* spores. J Bacteriol 190:2505–2512

Stecher B, Denzler R, Maier L et al (2012) Gut inflammation can boost horizontal gene transfer between pathogenic and commensal enterobacteriaceae. Proc Natl Acad Sci USA 109:1269–1274

Talham GL, Jiang HQ, Bos NA, Cebra JJ (1999) Segmented filamentous bacteria are potent stimuli of a physiologically normal state of the murine gut mucosal immune system. Infect Immun 67:1992–2000

Tanaka S, Kobayashi T, Songjinda P, et al (2009) Influence of antibiotic exposure in the early postnatal period on the development of intestinal microbiota. FEMS Immunol Med Microbiol 56:80–87

Taylor DE, Chau AS (1997) Cloning and nucleotide sequence of the gyrA gene from Campylobacter fetus subsp. fetus ATCC 27374 and characterization of ciprofloxacin-resistant laboratory and clinical isolates. Antimicrob Agents Chemother 41:665–671

Threlfall EJ (2000) Epidemic *Salmonella* Typhimurium DT 104—a truly international multiresistant clone. J Antimicrob Chemother 46:7–10

Tolmasky ME, Chamorro RM, Crosa JH, Marini PM (1988) Transposon-Mediated Amikacin resistance in *Klebsiella pneumoniae*. Antimicrob Agents Chemother 32:1416–1420

Tvede M, Rask-Madsen J (1989) Bacteriotherapy for chronic relapsing *Clostridium difficile* diarrhoea in six patients. Lancet 1:1156–1160

Ubeda C, Taur Y, Jenq RR et al (2010) Vancomycin-resistant *Enterococcus* domination of intestinal microbiota is enabled by antibiotic treatment in mice and precedes bloodstream invasion in humans. J Clin Invest 120:4332–4341

Ubeda C, Bucci V, Caballero S et al (2013) Intestinal microbiota containing Barnesiella species cures vancomycin-resistant *Enterococcus faecium* colonization. Infect Immun 81:965–973

Uttley AH, Collins CH, Naidoo J, George RC (1988) Vancomycin-resistant enterococci. Lancet 1:57–58

Vaishnava S, Behrendt CL, Ismail AS et al (2008) Paneth cells directly sense gut commensals and maintain homeostasis at the intestinal host-microbial interface. Proc Natl Acad Sci USA 105:20858–20863

Wani KA, Thakur MA, Fayaz AS et al (2009) Extended spectrum B-Lactamase mediated resistance in *Escherichia coli* in a tertiary care hospital. Int J Health Sci 3:155–163

Webber MA, Piddock LJV (2003) The importance of efflux pumps in bacterial antibiotic resistance. J Antimicrob Chemother 51:9–11

Wenzel RP (2002) The antibiotic paradox: how the misuse of antibiotics destroys their curative powers (Book review), second edn. By Stuart B. Levy. 353 pp., illustrated. Cambridge, Mass., Perseus Publishing, 2002. $17.50. 0-7382-0440-4. N Engl J Med 347:1213–1213

Willing BP, Vacharaksa A, Croxen M et al (2011) Altering host resistance to infections through microbial transplantation. PLoS ONE 6:e26988

Wilson KH (1983) Efficiency of various bile salt preparations for stimulation of *Clostridium difficile* spore germination. J Clin Microbiol 18:1017–1019

Wlodarska M, Willing B, Keeney KM et al (2011) Antibiotic treatment alters the colonic mucus layer and predisposes the host to exacerbated *Citrobacter rodentium*-induced colitis. Infect Immun 79:1536–1545

Woloj M, Tolmasky ME, Roberts MC, Crosa JH (1986) Plasmid-encoded amikacin resistance in multiresistant strains of *Klebsiella pneumoniae* isolated from neonates with meningitis. Antimicrob Agents Chemother 29:315–319

Zaura E, Brandt BW, Teixeira de Mattos MJ et al (2015) Same exposure but two radically different responses to antibiotics: resilience of the salivary microbiome versus long-term microbial shifts in feces. MBio 6:e01693–15

Zhang L, Kinkelaar D, Huang Y et al (2011) Acquired antibiotic resistance: are we born with it? Appl Environ Microbiol 77:7134–7141

Zoch B, Karch A, Dreesman J et al (2015) Feasibility of a birth cohort study dedicated to assessing acute infections using symptom diaries and parental collection of biomaterials. BMC Infect Dis 15:436

Anti-virulence Strategies to Target Bacterial Infections

Sabrina Mühlen and Petra Dersch

Abstract Resistance of important bacterial pathogens to common antimicrobial therapies and the emergence of multidrug-resistant bacteria are increasing at an alarming rate and constitute one of our greatest challenges in the combat of bacterial infection and accompanied diseases. The current shortage of effective drugs, lack of successful prevention measures and only a few new antibiotics in the clinical pipeline demand the development of novel treatment options and alternative antimicrobial therapies. Our increasing understanding of bacterial virulence strategies and the induced molecular pathways of the infectious disease provides novel opportunities to target and interfere with crucial pathogenicity factors or virulence-associated traits of the bacteria while bypassing the evolutionary pressure on the bacterium to develop resistance. In the past decade, numerous new bacterial targets for anti-virulence therapies have been identified, and structure-based tailoring of intervention strategies and screening assays for small-molecule inhibitors of such pathways were successfully established. In this chapter, we will take a closer look at the bacterial virulence-related factors and processes that present promising targets for anti-virulence therapies, recently discovered inhibitory substances and their promises and discuss the challenges, and problems that have to be faced.

Contents

S. Mühlen · P. Dersch (✉)
Helmholtz-Zentrum für Infektionsforschung and Deutsches Zentrum für Infektionsforschung, Inhoffenstr. 7, 38124 Brunswick, Germany
e-mail: petra.dersch@helmholtz-hzi.de

Current Topics in Microbiology and Immunology (2016) 398: 147–184
DOI 10.1007/82_2015_490

Published Online: 28 January 2016

1 Introduction—Challenges in Fighting Bacterial Infections

Although prevention and treatment of infectious diseases have improved over the last decades due to the widespread use of vaccines and anti-infectives and the development of infection control measurements, bacterial infections are still a major cause of morbidity and mortality worldwide. In particular, global spreading of antibiotic resistance genes and their acquisition by clinically relevant bacterial pathogens constitute a serious public health problem. In the European Union alone, about 3 million healthcare-associated infections were reported in 2004 leading to an approximate 50,000 deaths (McHugh et al. 2010). A report published by the Centre for Disease Control and Prevention (CDC) in 2013 estimates that more than 2 million infections and 23,000 deaths annually are caused by antibiotic-resistant bacteria in the USA alone and lists the top 18 drug-resistant pathogens considered a threat in the USA (Control 2013). These pathogens are categorized into three threat classes, urgent, serious and concerning, and include the clinically most relevant pathogens *Enterococcus faecium*, *Staphylococcus aureus*, *Acinetobacter baumannii*, *Pseudomonas aeruginosa*, and *Enterobacter* species summarized as ESKAPE pathogens (Rice 2008) to emphasize that they efficiently "escape" the effects of common antibacterial drugs. This list also includes *Clostridium difficile*, *Neisseria gonorrhoeae*, carbapenem- and Extended Spectrum β-Lactam-resistant *Enterobacteriaceae* (ESBL), drug-resistant *Campylobacter*, *Salmonella* and *Shigella*, and *Streptococcus pneumoniae* (Control 2013).

At a time, when antibiotic and multidrug resistance is becoming increasingly common in the clinic, new treatments are urgently needed to circumvent high mortality rates due to untreatable infections. Increasing effort is currently being put into the discovery and development of new resistance-breaking antibiotics, and some novel antimicrobial agents are currently in the preclinical and clinical pipeline (Hesterkamp 2015). However, we do not yet know whether this classical approach will succeed to identify new agents with activities against these pan-resistant pathogens in the foreseeable future. Recent work on the occurrence of antibiotic resistance determinants in bacterial populations further demonstrated that non-pathogenic bacteria can be identified in the environment that are already resistant to recently developed antimicrobial drugs (D'Costa et al. 2006; Martinez et al. 2009). For instance, many antibiotic-producing microorganisms encode resistance genes to the antibiotics they synthesize for self-protection (Hopwood 2007). In addition, many of the acquired antibiotic resistance genes are carried on mobile genetic elements such as transposons, integrons, and plasmids, which are frequently transferred to other bacteria of the same or related species by horizontal gene transfer. Evidence exists that the transfer of resistance determinants between commensal bacteria and pathogens, in particular in the intestinal tract, is extensive, leading to the rapid spread of resistance in bacterial populations and communities (Aarestrup 2005; Allen et al. 2010; Salyers et al. 2004). Moreover, an alarmingly high prevalence of antibiotic-resistant bacterial strains has been reported both in domestic and wild animals and in the environment. Prophylactic use of antibiotics in agricultural settings and in feed/water in the animal food production has significantly enhanced the evolution and global spread of antibiotic resistances (Allen et al. 2010; Berendonk et al. 2015; Canton 2009). These findings are daunting as bacteria can rapidly catch up or are even on par with the development of new chemical entities and the danger that we re-enter an apocalyptic preantibiotic era is looming on the horizon.

Consequently, novel intervention strategies are required to respond to current antimicrobial resistance and anticipate evolving resistance mechanisms. One compelling approach to antibiotic therapy is the development of anti-virulence strategies, by which only virulence-associated, but not survival/fitness-relevant traits are targeted. In other words, the anti-infective drug interferes with pathogenicity mechanisms, in particular properties of the bacteria that cause disease. This targeted intervention effectively disarms the pathogen and enables its clearance by the host immune system, but contrary to common antibiotic therapies, it is not bacteriostatic (inhibiting bacterial growth) or bacteriocidal (killing bacteria). Active agents acting in this fashion, the so-called **patho-** or **virulence blockers**, alleviate the pressure on the pathogen to develop resistance by solely affecting pathogens expressing the targeted pathogenicity factor (Baron 2010; Beckham and Roe 2014; Cegelski et al. 2008; Clatworthy et al. 2007; Escaich 2008; Lee and Boucher 2015; Lynch and Wiener-Kronish 2008; O'Connell et al. 2013; Rasko and Sperandio 2010; Zambelloni et al. 2015). Consequently, they neither damage nor modify the composition of the natural host microbiota, a process, which is increasingly recognized to facilitate the development, progress, and persistence of chronic inflammatory diseases and other morbidities such as diabetes,

2 Potential Virulence Targets and Strategies

Recent efforts in the development of anti-virulence therapies are directed to target various factors or mechanisms of pathogens that are crucial to initiate an infection and cause disease. This includes interference with various pathogenicity factors promoting cell adhesion, cell invasion, intracellular replication and damage of host tissues, biofilm formation and maintenance (Fig. 1), stress adaptation and metabolic functions important to adapt to the different host environments, mechanisms to evade or overcome the host immune defense, and control systems regulating the expression of virulence-relevant genes.

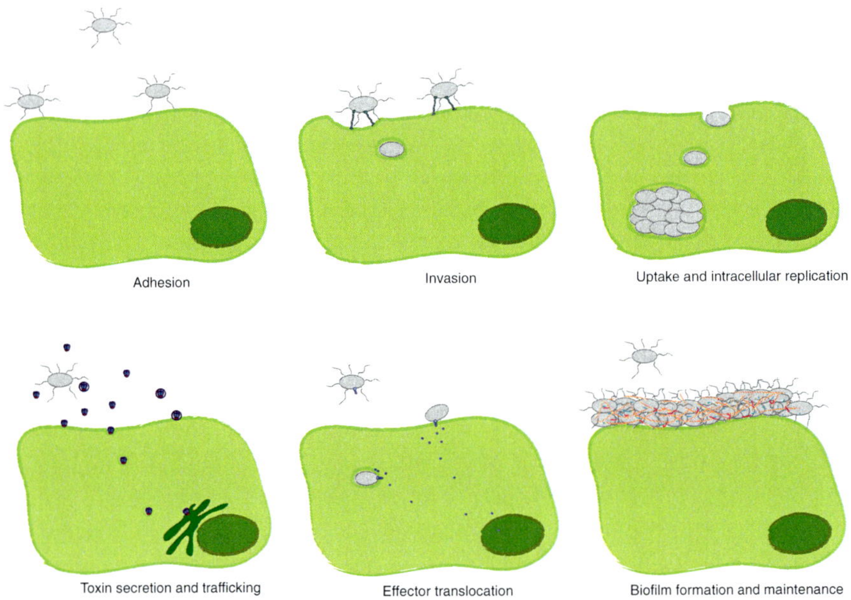

Fig. 1 Schematic representation of the infection processes that can be targeted by anti-virulence drugs. Bacterial pathogens coordinate highly complex and organized temporal and spatial events to colonize and to disseminate to different sites within the host. They produce special pathogenicity factors to adhere to, invade, persist, and in some cases replicate within host cells. Many pathogens are also able to form and replicate with a biofilm-like bacterial community on and/or within cells. Knowledge about the molecular mechanisms will allow the interference with colonization by the (**i**) inhibition of adhesin biosynthesis and function and (**ii**) inhibition of invasion, persistence, and proliferation in cells or outside cells in distinct host tissues

The development of novel technologies, such as in vivo transcriptome analysis by RNA-Seq (dual RNA-Seq) and transposon-directed insertion site sequencing (TraDIS), now enables us to identify in vivo active genes and gene functions that are crucial for the survival of pathogens in certain tissues (Chaudhuri et al. 2013; Langridge et al. 2009; Westermann et al. 2012). The discovery of the bottlenecks of an infection will allow us to develop anti-infectives that particularly target fundamental virulence mechanisms (the Achilles heel) of a pathogen. Particularly, approaches that (i) target classical virulence factors, such as adhesins/invasins, (ii) inhibit pathogen-induced host signaling disruption by toxins, effectors, and immune modulators, (iii) manipulate microbial signal transduction and regulation, or (iv) interfere with functions required for bacterial survival and/or persistence during the infection appear promising for the development of new therapeutics against infections (Fig. 2).

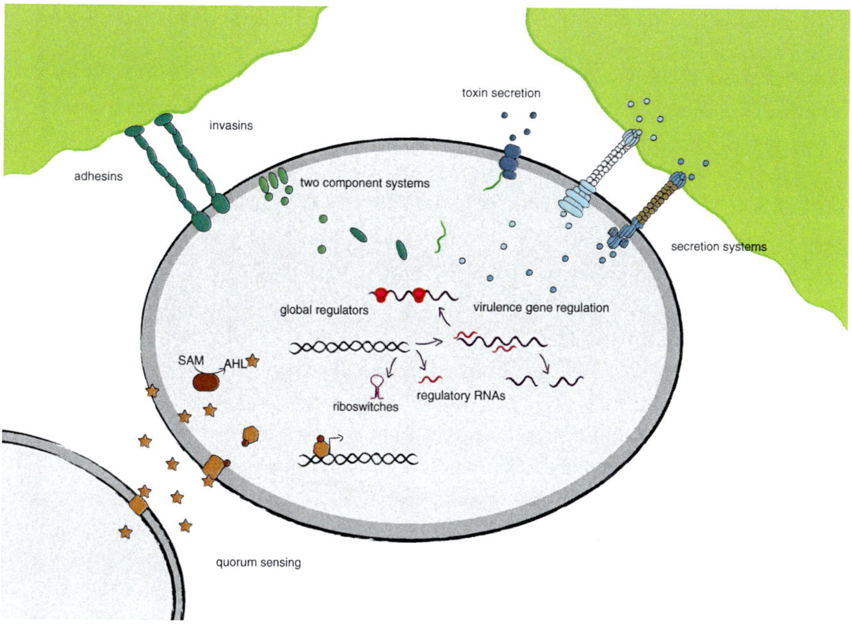

Fig. 2 Schematic overview of the targets of current anti-virulence strategies designed for Gram-negative bacterial pathogens. Pathoblockers can target toxin function, quorum and environmental parameter sensing, gene regulation by global or specific regulators, sensory or regulatory RNAs, cell adhesion and invasion-promoting surface structures, and bacterial secretion systems (i.e., T3SS, T4SS)

2.1 Denying Access: Adherence and Invasion Inhibitors

Upon entering the host, bacterial pathogens must travel to their respective site of infection in order to initialize the disease process. Once the bacteria reach the site of infection, bacterial cell surface structures and appendages such as **pili/fimbriae** and **afimbrial adhesins** detect and interact tightly with specific host cell receptors to adhere to the host cell (Thanassi et al. 2012). Cell attachment enables bacteria to withstand host mechanical and immunological clearance and is crucial for the initiation of an infection. Furthermore, adhesion is essential for the pathogen to get into close proximity with the host cell surface, a process required for the formation and activation of secretion systems (Thanassi et al. 2012). Agents targeting bacterial adherence, e.g., by the inhibition of **pili/fimbriae or adhesin formation** would not only deny access to host tissues, but would also promote rapid clearance of the bacteria and avoid the release and translocation of tissue-damaging factors (e.g., toxins) (Krachler and Orth 2013). As the adhesion structures are further specific to the pathogen and will mediate only attachment to host cells that express the corresponding receptor, they are potential targets for new anti-infectives.

Pili or fimbriae are hairlike, multi-subunit bacterial cell surface protrusions that facilitate bacterial adhesion and aid colonization. They are important virulence factors in a wide range of bacterial pathogens, including *E. coli*, *Salmonella*, *Yersinia*, *Haemophilus*, *Pseudomonas*, and *Klebsiella* (Krachler and Orth 2013). Many pili structures are assembled in the bacterial cell via the chaperone–usher pathway. In this pathway, the pilin subunits reach the periplasm via the Sec pathway, where they interact with the chaperone, which aids the folding of the pilin subunit's immunoglobulin fold (Li et al. 2004). The chaperone–pilin complex traffics to bacterial outer membrane usher proteins, where the pilus is formed by the incorporation of the pilin subunit into the growing pilin fiber resulting in the liberation of the chaperone (Barnhart et al. 2003).

Bicyclic 2-pyridones and N-substituted amino acid derivatives were discovered as potential pilicides (inhibitors of pili formation/biogenesis) that target conserved regions on the chaperone and competitively inhibit the binding of the chaperone to the pilin subunits. They were shown to inhibit biogenesis of type 1 and P pili of uropathogenic *E. coli* (UPEC), the major cause of urinary tract infections, which decrease UPEC binding to bladder cells by 90 % (Pinkner et al. 2006). Furthermore, they also inhibit hemagglutination and biofilm formation in *E. coli* strains (Pinkner et al. 2006). While none of the substances affect bacterial growth, pilicides may have broad-spectrum activity as both the chaperone structure and the chaperone–usher pathway are highly conserved among bacteria.

The interaction of pili or fimbriae with host cell surface receptors is often mediated through binding of the terminal pilin subunits and usually involves specific sugar or peptide moieties at the receptor (Ofek et al. 2003). Therefore, competitive inhibition of bacterial binding by tailoring anti-adhesive compounds to the specific receptor can be a promising strategy for anti-adhesion therapies. For instance, FimH or PapG are bacterial lectins that are located at the tip of type 1 or P pili of UPEC and they are one of the bacterium's main virulence factors, responsible for colonization, invasion of host bladder epithelial cells, and biofilm formation (Wright and Hultgren 2006). FimH recognizes mannosylated receptors on the host cell surface. Initial studies showed that monovalent mannose derivatives display rather weak inhibitory effects (Firon et al. 1987), but recent use of multivalent compounds with increased binding activity is promising success (Han et al. 2010). Biphenyl mannosides were shown to be 200,000-fold more potent than the originally tested monovalent mannose. The substances are orally available and demonstrate an overall low toxicity (Han et al. 2012; Hartmann et al. 2012). Additionally, mouse models indicate that biphenyl mannosides decrease colonization levels significantly (Klein et al. 2010). Alternatively, **adhesin binding** to glycosylated receptor proteins can be inhibited by mucins. Mucins are glycoproteins of the mucus that mimic the glycosylation pattern found on host cell receptors. Purified Muc1, a mucin derived from cow milk that is highly glycosylated, can selectively inhibit binding of Gram-negative pathogens such as *E. coli* and *Salmonella* (Parker et al. 2010). It can also limit infection with *Helicobacter pylori* (Linden et al. 2009). Muc1 does, however, have a less pronounced effect on Gram-positive bacteria (Parker et al. 2010). Another very effective option to block cell adhesion by a particular bacterial adhesin is the application of synthetic peptides that mimic the epitope or use of monoclonal antibodies raised against the binding epitope that interacts with the host cell receptor. This strategy has been successfully applied to the streptococcal antigen (SA) I/II, an adhesin of *Streptococcus mutants* responsible for bacterial binding to host salivary receptors. A synthetic peptide and an antibody directed against a binding epitope of SA I/II have been shown to prevent colonization of the oral cavity by *S. mutans* (Lehner et al. 1985; Ma et al. 1989).

Another option to inhibit bacterial adhesion is interference with the biogenesis and presentation of the host receptor. Many bacterial adhesins and toxins bind to host glycosphingolipids. Blocking of the ceramide-specific glycosyltransferase that catalyzes the formation of a glycosphingolipid precursor has been used to successfully diminish the amount of colonization of UPEC in cell culture and mouse studies (Svensson et al. 2003). Also, depletion of glycosphingolipids using enzyme replacement therapy was successfully used to treat systemic salmonellosis (Margalit et al. 2002).

2.2 *Fighting Biofilms and Chronic Infections*

More than 80 % of all microbial infectious diseases, which are difficult to treat, involve the formation of biofilms (Romling and Balsalobre 2012). Therefore, **biofilms** and **quorum sensing** implicated in the control of biofilm formation constitute other important treatment targets.

2.2.1 Interference with Cell–Cell Communication—Quorum Sensing

Quorum sensing was discovered in the early 1970s as a means of cell–cell communication between bacteria of a species. It allows the cells to react to environmental changes/stresses as a group by coordinating gene expression according to the local cell population density (LaSarre and Federle 2013). Quorum sensing is a three-step process, in which the bacteria produce small signal molecules, so-called autoinducers that are released from the cell and detected by membrane-bound or cytoplasmic receptors (Ng and Bassler 2009). When bacterial numbers are low, the concentration of autoinducers is also low and the molecules dissipate in the environment. If, however, the concentration of bacteria reaches a threshold level, they come into contact with the released autoinducer molecules, inferring that they have neighbors. This, in turn, induces the formation of biofilms, antibiotic resistance, and expression of multiple pathogenicity factors (Deep et al. 2011). By regulating genes that are mainly associated with virulence and persistence, quorum sensing poses an important target for anti-virulence therapies. Different strategies identified for its inhibition include the **blockage of signal production**, interference with or **degradation of signal molecules** (signal dissemination), and **disruption of the**

signal reception and response (Hentzer and Givskov 2003; LaSarre and Federle 2013; Lu et al. 2014a, b).

One important strategy to interfere with quorum sensing is to prevent the production of autoinducers. Gram-negative bacteria generate autoinducer molecules such as acyl-homoserine lactone (AHL) from S-adenosyl methionine (SAM) by proteins homologous to LuxI of *Vibrio fischeri*. This knowledge was exploited for the generation of quorum sensing inhibitors. For instance, analogues of SAM (e.g., S-adenosyl-homocysteine) were shown to inhibit AHL synthesis in *P. aeruginosa* (Parsek et al. 1999), and analogues of AHLs resulted in the inhibition of the expression of quorum sensing-regulated genes (Smith et al. 2003b) and, subsequently, biofilm formation (Smith et al. 2003a). Furthermore, substances blocking regulators of AHL synthesis were identified (Park et al. 2015; Soheili et al. 2015; Lu et al. 2014a, b). Gram-positive bacteria such as *S. aureus* produce autoinducing peptides that require cleaving for activation and recognition. Administration of inhibitory autoinducer peptides to mice during the initial stages of *S. aureus* infection was shown to inhibit *S. aureus*-induced abscess formation (Wright et al. 2005). In addition, diverse small-molecule inhibitors of biofilm formation have been identified. These include hydnocarpin-type flavonolignans and streptorubin B for *S. aureus* and cyclosporine and valspodar for *Streptococcus* biofilm inhibition (Aggarwal et al. 2015; Suzuki et al. 2015; Vimberg et al. 2015).

Several species of bacteria also produce enzymes that are capable of inactivating AHL molecules. *Bacillus subtilis* produces an acyl-homoserine lactonase, which hydrolyzes the lactone ring of AHL molecules, resulting in the ablation of AHL function (Dong et al. 2000). Interestingly, tobacco plants expressing this AHL-lactonase show an enhanced resistance to *Erwinia* infection suggesting at a broad spectrum function of these enzymes (Dong et al. 2001, 2002). Moreover, AHL-acylases, which cleave AHL after the N-terminal acyl, inactivating AHL molecules in a substrate-specific manner have been identified in different strains of *Pseudomonas* and *Ralstonia* (LaSarre and Federle 2013) and are also of interest as anti-quorum sensing agents.

Another possibility is the blockage of AHL signaling by competitive inhibitors that prevent AHL binding to its LuxR-type receptor (Hentzer and Givskov 2003). Several AHL analogues with modified acyl side chains were identified with agonistic activities (Chhabra et al. 1993; Kline et al. 1999; Smith et al. 2003b), and halogenated furanones of marine algae were shown to modulate quorum sensing regulators at a post-transcriptional level (Hentzer and Givskov 2003; Hentzer et al. 2002; Manefield et al. 1999).

Cyclic-di-GMP (c-di-GMP) is a widely used second messenger that plays a crucial role in bacterial biofilm formation as it stimulates the biosynthesis of adhesins and exopolysaccharide matrix components and prevents bacterial motility. C-di-GMP is synthesized by diguanylate cyclase (DGC), which is considered to be an attractive drug target. Recent research led to the identification of a catechol-containing sulfonohydrazide compound that inhibits the DGC PleD (Fernicola et al. 2015).

2.2.2 Prevention and Resolution of Biofilms

Biofilms form once bacteria sense a sufficiently high population density in their vicinity. This alters their gene expression and induces the secretion of a mix of polysaccharides, proteins, and extracellular DNA. These secreted substances interact to form an extracellular matrix for the pathogens. Once the biofilm has been formed and bacteria are embedded in the matrix, the targeting of these bacteria becomes difficult. The formation of a biofilm increases bacterial resistance to exogenous stresses such as antibiotics, UV damage, acidity, metal toxicity, and to host immune clearance and phagocytosis (Costerton et al. 1999; Hall-Stoodley et al. 2004; Romling and Balsalobre 2012). An additional disadvantage of biofilm formation in terms of treatment is that growth in biofilms aids the occurrence of processes that lead to the acquisition of inheritable resistance traits, such as horizontal gene transfer and adaptive mutations (Madsen et al. 2012). Growth within biofilms also raises the probability of bacterial persistence and antibiotic tolerance as bacteria within the biofilm alter their metabolism (Lewis 2005, 2008). This may lead to resistance to bacterial clearance, enabling the bacteria to regrowth within the host when the conditions become more hospitable. To target biofilm formation to inhibit persistence and recurrent infection, substances are being investigated that can inhibit secretion of biofilm components and biofilm matrix formation or destroy or resolve existing biofilm matrices.

Strategies to target bacteria within the context of a mature biofilm or to prevent biofilms from forming include the induction of bacterial motility, inhibition of host cell adhesion (see also 2.1), and/or the initiation of cell dispersal. Under natural circumstances, molecules inducing cell dispersal are produced by bacteria within the biofilm, allowing some cells to detach and infect new cells or surfaces. *B. subtilis* produces D-amino acids that result in the disruption of amyloid fibers, which link bacteria within the biofilm (Kolodkin-Gal et al. 2010). Exogenous treatment of biofilms with low levels of D-amino acids results in the disruption of mature biofilms and inhibition of biofilm formation by *B. subtilis* (Kolodkin Gal et al. 2010). Norspermidine is another molecule produced by bacteria and plants that targets polysaccharides present in the matrix. It disperses biofilms formed by *B. subtilis* and was further shown to inhibit the formation of biofilms by *E. coli* and *S. aureus*. Administration of a combination of D-amino acids and norspermidine enhances their activity against mature biofilms (Kolodkin-Gal et al. 2012). Moreover, nitric oxide or the addition of the nitric oxide donor sodium nitroprusside (SNP) induces the dispersion of *P. aeruginosa* biofilms, and cotreatment of biofilms with SNP and antibiotics increases the efficacy of the antibiotics (Barraud et al. 2006). Nitric oxide was suggested to reduce biofilms by stimulating c-di-GMP-degrading phosphodiesterases, decreasing local c-di-GMP levels (Barraud et al. 2009). Furthermore, DNase I can degrade the extracellular DNA present in the biofilm matrix (Okshevsky et al. 2015; Qin et al. 2007), interfering with its formation and stability as shown for *Bordetella pertussis* (Conover et al. 2011), *Listeria monocytogenes* (Harmsen et al. 2010), and *Campylobacter jejuni* (Brown et al. 2015). A bacterial glycoside hydrolase, named Dispersin B, has been isolated from *Actinobacillus actinomycetemcomitans*, which was shown to disrupt mature *Actinobacillus* biofilms (Kaplan et al. 2004) and to inhibit *S. aureus* biofilms when added exogenously to the cells (Izano et al. 2008). Furthermore, the use of lytic bacteriophages to treat bacteria within a biofilm has been investigated, and it was shown that using a phage engineered to express Dispersin B could lead to complete dispersal of *E. coli* biofilms (Lu and Collins 2007) targeting even potential persister cells.

2.3 Interference with Global Virulence Control (Regulation of Virulence Gene Expression)

Bacteria tightly control the production of energy-consuming pathogenicity factors and virulence-associated traits to avoid unnecessary energy expenses and optimize their biological fitness. Recent advances in our understanding of virulence regulation have identified many control circuits and networks implicating many different often conserved sensory and regulatory components acting at the transcriptional or post-transcriptional level that could be targeted. This includes bacterial sensory and signal transduction molecules, global and specific transcriptional regulators, and RNA-based regulatory mechanisms. A major advantage of **targeting signal transduction** and **regulatory mechanism** is that these control systems are specific for bacteria and not present in eukaryotic host cells.

2.3.1 Two-Component Systems Involved in Virulence

Environmental cues, which are important for the pathogenesis and the biological fitness of bacterial pathogens during infection, are sensed by ubiquitous, highly conserved two-component systems. They constitute a membrane-bound histidine sensor kinase, which activates a corresponding cytoplasmic response regulator by phosphorylation. In particular, the **two-component systems** EnvZ/OmpR, RcsB/RcsC, PhoP/PhoQ, BarA/SirA, CpxR/CpxS, AgrC/AgrA, and QseC/QseB of bacterial pathogens have been well characterized and shown to control complex gene networks important for virulence in response to temperature, osmolarity, nutrients, secondary metabolites, and ions (Altier et al. 2000; Arya and Princy 2013; Clarke 2010; Forst and Roberts 1994; Groisman and Mouslim 2006; Vogt and Raivio 2012; Weigel and Demuth 2015). Moreover, studies exist that inhibitors of bacterial two-component systems worked in animal models and blocked pathogenesis of important pathogens (Rasko et al. 2008; Stephenson et al. 2000; Wilke et al. 2015; Worthington et al. 2013). One prominent example is the small-molecule inhibitor savirin, which reduces the expression of AgrCA-regulated genes in *S. aureus*, affecting its virulence, but not its survival and has no impact on the commensal *Staphylococcus epidermidis* (Sully et al. 2014). Another promising approach has identified a small-molecule LED209, which prevents the autophosphorylation of the sensor kinase QseC without influencing bacterial growth or inducing cell cytotoxicity. The QseC kinase contributes to virulence in a number of Gram-negative pathogens and has been studied extensively in enterohemorrhagic *E. coli*. Here, QseC phosphorylates three transcription factors, QseB (regulates flagella and motility genes by binding the master regulator *flhDC*), QseF (activates Stx production), and KdpE (binds to *ler*, the regulator of the main EHEC pathogenicity island LEE). Furthermore, QseC knockout strains of *Salmonella* and *Francisella* were shown to be attenuated in animal models. Strikingly, the inhibitor LED209 prevents the expression of the EHEC pathogenicity island LEE and Shiga toxin 2 without triggering an SOS response in EHEC, a reaction, which has been shown to result in the activation of Shiga toxin expression. On the contrary, a decrease in Stx2 expression could be observed (Curtis et al. 2014; Rasko et al. 2008).

2.3.2 Global Transcriptional Regulators of Virulence

A growing number and range of **global transcriptional regulators** of virulence have been identified which are highly conserved among bacteria, but specific for prokaryotic gene expression control, and which adjust coexpression of host-adapted metabolic processes, adaptation to host stresses, and virulence factors. Targeting of these regulators is an attractive concept as rapid adjustment to continuously changing environments during the course of an infection is a prerequisite for successful persistence in the host and the development of the infectious disease. Among the most promising drug targets are transcriptional regulators, such as the cAMP repressor protein (Crp) of Gram-negative bacteria, the equivalent catabolite control protein (CcpA) of Gram-positive bacteria, the carbon storage regulator (CsrA/RsmA, see also below), and AraC-type activators (e.g., RhaR). They are produced by both Gram-negative and Gram-positive bacteria which all adjust expression of important virulence relevant processes (e.g., toxin production) in response to available nutrients in the infected tissues (Brautaset et al. 2009; Bruckner and Titgemeyer 2002; Deutscher et al. 2005; Heroven et al. 2012; Romeo et al. 2013; Skredenske et al. 2013; Vakulskas et al. 2015). The strength of this approach is supported by a recent study demonstrating that the phenylpropanoid anethole, which influences the virulence regulatory cascade by overproduction of Crp suppressed toxigenic *V. cholerae*-mediated fluid accumulation in ligate ileum of rabbits (Zahid et al. 2015).

2.3.3 Regulatory and Sensory RNAs

Novel deep-sequencing-based strategies have discovered an unprecedented level of complexity of transcriptional networks by the identification of **small *trans*-acting regulatory RNAs** (sRNAs), **antisense RNAs**, and sensory RNAs, such as **RNA thermometers** and **riboswitches**. Many of these RNA-based control elements were shown to influence the expression of virulence-relevant processes (Oliva et al. 2015; Papenfort and Vogel 2014). Several regulatory RNAs (e.g., Qrr RNAs, CsrB/C, and RsmZ/Y RNAs) have redundant functions and are used to fine-tune the regulation of pathogenicity, stress adaptation, and/or metabolic genes by sequestration of regulatory proteins, hindrance of translation, and/or control of RNA degradation (Feng et al. 2015; Heroven et al. 2012; Vakulskas et al. 2015). Redundancy and distinct regulatory mechanisms hamper the development of anti-regulatory RNA inhibitors. However, mechanistic insights further revealed that the function of many regulatory RNAs is governed by highly conserved, global RNA-binding proteins, such as the RNA chaperone Hfq and CsrA/RsmA (Heroven et al. 2012; Lucchetti-Miganeh et al. 2008; Oliva et al. 2015; Papenfort and Vogel 2014; Vakulskas et al. 2015). Hfq is known as the central mediator of sRNA-based gene regulation in bacteria as it establishes dynamic interactions of a wide range of RNA molecules and manipulates translation and degradation of many mRNAs important for pathogenesis (Chao and Vogel 2010; Vogel and Luisi 2011). In fact, *hfq* knockout derivatives of many important pathogens (e.g., *Francisella*, *Neisseria*, *Legionella*, *Salmonella*, *Yersinia*, and *Listeria*) showed a severe growth defect and were drastically attenuated in animal infection models (Oliva et al. 2015). Another well-characterized RNA-binding protein implicated in the control of multiple regulatory RNAs and a large set of virulence-linked traits is the CsrA/RsmA protein. This RNA-binding regulator is highly conserved among bacteria, but it is not produced in archaea and eukaryotes. It predominantly controls RNA translation and degradation by binding to A(N)GGA motifs within the 5′-untranslated region of the mRNA targets (Duss et al. 2014). The CsrA/RsmA-controlled network is very versatile and includes target mRNAs implicated in the control of cell morphology, motility, biofilm formation, multiple stress responses, and crucial virulence factors/regulators (i.e., secretion systems and secreted effectors, adhesins, and invasins). Accordingly, *csrA/rsmA* mutants of many pathogens are avirulent or strongly attenuated (Heroven et al. 2012; Lucchetti-Miganeh et al. 2008; Vakulskas et al. 2015). Furthermore, multiple regulatory RNAs of pathogenic *Enterobacteriaceae* implicated in the control of virulence functions are controlled by the global transcriptional regulator Crp (see above) in response to the available nutrients in the medium. Strikingly, transcriptional profiling using RNA-Seq recently revealed Crp as a master regulator of 50 % of all identified small RNAs in *Y. pseudotuberculosis* which are reprogrammed by Crp in response to temperature (Nuss et al. 2015). The major impact of Hfq, CsrA/RsmA, and Crp on the expression of the virulence phenotype of many pathogens makes these global regulators also to promising targets for anti-virulence strategies.

RNA riboswitches and thermometers represent another type of regulatory RNA elements. They are predominantly located in the 5′-untranslated regions of target mRNAs and comprise complex RNA structures (e.g., stem loops, which include the ribosome-binding site and/or the start codon in base-pairing of the stem structure) that sense and react to thermal or biochemical signals by conformational changes, which mainly affect translation of the downstream gene(s) (Kortmann and Narberhaus 2012). Prominent examples are the RNA thermometers controlling the expression of major virulence regulators, such as PrfA of *Listeria monocytogenes* (Johansson et al. 2002), LcrF/VirF of *Yersinia* (Böhme et al. 2012), and the cholera toxin regulator ToxT of *Vibrio cholerae* (Weber et al. 2014). Additionally, three RNA thermosensors were described to be essential for *Neisseria meningitidis* resistance against immune killing (Loh et al. 2013).

Several virulence-related metabolic functions are also controlled by riboswitches —metabolite-binding mRNA or part of regulatory RNA structures (Mellin et al. 2014; Peselis and Serganov 2014; Serganov and Nudler 2013). Utilization of ethanolamine and propanediol, by-products of rhamnose and fucose fermentation by the intestinal microbiota, is controlled by riboswitches in response to vitamin B12 binding. This leads to the synthesis of short or longer regulatory RNAs, which differentially modulate transcription of the *pdu* and *eut* genes (Mellin et al. 2013, 2014; Toledo-Arana et al. 2009). Moreover, *trans*-regulatory riboswitches can function as small regulatory RNAs to link metabolic and virulence control. Several potential S-adenosylmethionine (SAM)-binding riboswitches were identified in enteric pathogens and two of them—SreA and SreB—were shown to bind the 5′-untranslated region of the *prfA* transcript encoding the master virulence regulator of *L. monocytogenes* (Toledo-Arana et al. 2009). Rapidly increasing numbers of crucial RNA thermometers and riboswitches involved in virulence control may warrant the design of potential RNA-based inhibitors, e.g., RNA fragments that interfere with these crucial RNA elements or show perfect complementarities to the ribosomal binding site and the start codon of crucial virulence regulators.

2.4 *Preventing Host Damage and Development of the Disease*

Toxin-producing pathogens exhibit the severest effect on their host. The most serious clinical symptoms associated with infectious diseases are the results of severe tissue damage, cellular malfunction, or destruction caused by **bacterial exotoxins** such as botulinum, cholera, diphtheria, anthrax, tetanus, and Shiga toxins (Henkel et al. 2010; Schmitt et al. 1999). A deletion of the toxin gene(s) generally disarms the bacteria and results in avirulence without harming their overall biological fitness. This makes these virulence factors ideal targets for new inhibitors, and multiple approaches are currently being followed to prevent toxin-mediated damages of the host.

2.4.1 Targeting Exotoxin Trafficking or Function

Exotoxins are bacterial virulence factors that are actively released into the surrounding environment from the bacterium during its growth, commonly via bacterial type II secretion systems (Henkel et al. 2010) and/or outer membrane vesicles

(Kulp and Kuehn 2010; Kunsmann et al. 2015). As toxins remain outside the cell for a period of time prior to binding of their specific target cells, the toxins themselves make for good therapeutic targets, for example for **competitive inhibitors** or **neutralizing antibodies**. Once released into the environment, the toxins traffic through the host until they recognize and bind to specific host cell receptors. Interaction of the toxin with its receptor induces the uptake of the receptor from the cell surface by endocytosis. Once inside the host cell cytosol, the trafficking and destination of toxins varies according to their specificity. Some toxins are activated at neutral pH, while others require the acidification of the endosome for activation or trafficking to a particular compartment (Henkel et al. 2010; Schmitt et al. 1999). Knowledge of the route of action of a specific toxin now enables us to develop new **toxin-trafficking inhibitors**.

For many exotoxins, screening assays have been designed to identify small-molecule compounds that inhibit the action of the particular exotoxin. For instance, several potent inhibitors have been identified for the *Bacillus anthracis*-encoded anthrax toxin, a multimeric toxin consisting of three proteins, the protective antigen (PA), the edema factor (EF), and the lethal factor (LF) (Montecucco et al. 2004; Nestorovich and Bezrukov 2014; Tonello et al. 2002). The PA binds to the host cell surface, where it gets cleaved and forms multimeric channels in the host cell membrane through which the EF and the LF can enter the cell. The EF, an adenylate cyclase, raises the level of cAMP within the host cell, leading to edema (Baldari et al. 2006; Dell'Aica et al. 2004). The LF, a zinc metalloprotease, which has been shown to be critical for infection, disrupts host MAP kinase signaling pathways. Initial screening of a library of known zinc metalloprotease inhibitors identified a potent sulfonamide derivative, anthrax LF inhibitor 40 ((2R)-2-[(4-fluoro-3-methylphenyl) sulfonylamino]-N-hydroxy-2-(tetrahydro-2H-pyran-4-yl)acetamide), which was further shown to be effective in several animal models (Xiong et al. 2006). The compound binds competitively within the active site of LF and was found to be efficient in prophylactic therapy as well as for therapeutic treatment when used in combination with classic antibiotics (Shoop et al. 2005). Other screening assays using small-molecule libraries and application of a mixture-based peptide library approach identified additional small-molecule inhibitors and peptide analogues, which show competitive inhibition of anthrax LF (Bannwarth et al. 2012; Goldman et al. 2006; Kim et al. 2011; Panchal et al. 2004; Shoop et al. 2005; Turk et al. 2004). Furthermore, some polyphenols such as catechin gallate (CG) and epigallocatechin-3-gallate (EGCG) abundantly identified in green tea were found to exert a strong inhibition of the LF proteolytic activity (Dell'Aica et al. 2004). Several non-competitive/exosite-targeting inhibitors that prevent LF function were also identified (Bannwarth et al. 2012; Kuzmic et al. 2006). Similarly, antitoxins targeting the EF enzymatic activity have been identified of which the most potent inhibitors interact with the catalytic site of the protein (Nestorovich and Bezrukov 2014).

As the PA subunit of the anthrax toxin is responsible for LF and EF delivery and represents the major antigen for toxin-neutralizing antibodies, it has been the most important target for preventive and therapeutic measures. Several attempts have been made to (i) prevent PA binding to its host cell receptors (ATR/TEM8, CMG2) by blocking the receptor-binding domain of PA or the cell receptors or by the design of soluble polyvalent peptide analogues which compete with the natural receptors for PA binding (Cryan et al. 2013; Cryan and Rogers 2011; Rogers et al. 2012; Scobie et al. 2005), (ii) to prevent endocytosis of the toxin by blocking pH-dependent cell entry and endosomal trafficking, or (iii) to prevent PA-promoted translocation of the LF and the EF. The most potent inhibitors include liposome-functionalized multiple copies of the AWPLSQLDHSYN peptide that binds and neutralizes the LF (Basha et al. 2006), dominant negative derivatives of the PA that coassemble with wild-type PA, and the small-molecule inhibitor (3-aminopropylthio)-β-cyclodextrin, which disrupts proper PA channel formation and/or blocks its activity to translocate the LF and the EF (Karginov et al. 2006; Nestorovich and Bezrukov 2014).

Shiga toxin and Shiga-like toxin are found in *Shigella* and Shiga toxin-producing *E. coli* strains (EHEC). In EHEC pathogenesis, Shiga toxin has been identified as the main virulence factor responsible for bloody diarrhea, destruction of red blood cells and platelets, and the development of hemolytic uremic syndrome (HUS) resulting in severe kidney and neurological damage (Greener 2000; Kaplan et al. 1998; Tarr et al. 2005). Shiga toxin is an AB_5 toxin, of which the B subunit promotes binding to Gb3 glycolipid receptors most commonly found on kidney cells, but also on thrombocytes and neuronal cells (Boyd and Lingwood 1989; Kaplan et al. 1998). Several groups have designed strategies to interfere with receptor binding of the Shiga toxin. Sugars mimicking the Gb3 receptor have been designed in previous studies with varying success at neutralizing the free toxin (Kitov et al. 2000; Nishikawa et al. 2005; Trachtman et al. 2003). Despite toxin binding and removal, clinical symptoms were not significantly reduced. In another approach, glycan-encapsulated gold nanoparticles have been used to display ligands for Shiga toxin in an attempt to bind the free toxin. The nanoparticles were able to limit Vero cell cytotoxicity in response to Shiga toxins 1 and 2, but were unable to neutralize certain Shiga toxin 2 variants (Kulkarni et al. 2010). Furthermore, C-9, an inhibitor of glucosylceramide synthase, was applied to downregulate Gb3 expression and prevented the cytotoxic effect of Stx2 on Vero cells (Silberstein et al. 2011). Using a rat model of infection, C-9 protected animals against Shiga toxin 2-associated disease and both prophylactic and therapeutic treatment of rats with C-9 decreased expression of Gb3 receptors and the development of a disease phenotype (Kulkarni et al. 2010).

2.4.2 Inhibition of Exotoxin Synthesis

The two main virulence factors of *V. cholerae* are cholera toxin (CT) and toxin coregulated pilus (TCP). CT is another AB_5 toxin of which the catalytically active A subunit activates host cell G proteins upon uptake into the host cell. This in turn leads to the activation of adenylate cyclase, increasing the concentration of cAMP in intestinal epithelial cells, inducing strong secretory diarrhea (Field 2003).

The specific transcription factor ToxT is known to directly activate the expression of both cholera toxin (*ctxAB*) and *tcp* genes. The small-molecule virstatin was found to interfere with the homodimerization of the ToxT N-termini and thus blocks CT and TCP production resulting in reduced colonization of *V. cholerae* in mice (Hung et al. 2005; Shakhnovich et al. 2007). Unfortunately, virstatin-resistant *toxT* mutants containing a single amino acid substitution in the N-terminus have already been isolated (Hung et al. 2005).

2.4.3 Antibody-Mediated Exotoxin Neutralization

A very effective strategy to block toxin function and prevent the deleterious effect of very aggressive exotoxins is the treatment with antibodies that specifically bind and neutralize the toxin. One example for a highly potent neurotoxin that can be efficiently treated with neutralizing antibodies is the *Clostridium botulinum*-produced botulinum toxin. In the case of infections of adults, the US Centers for Disease Control supply an antitoxin that contains horse antibodies raised against type A, B, and/or E strains of the neurotoxin. For the treatment of children, the FDA approved a drug containing anti-botulinum toxin antibodies produced from human [marketed as human botulism immune globulin (BabyBIG)] (Arnon et al. 2006). Due to the high cost of the antibodies, a new equine alternative of the antibody is being tested (Vanella de Cuetos et al. 2011).

In addition, monoclonal antibodies binding directly to the PA subunit of the anthrax toxin and thus preventing its interaction with host cells were developed, which were shown to protect rats and chimpanzees against *B. anthracis* infections (Chen et al. 2011).

Furthermore, antibodies targeting Shiga toxin have been isolated from rabbits immunized with a fusion protein consisting of an epitope of the B subunit of heat-labile toxin (LT) of enterotoxic *E. coli* (ETEC) and the A subunit of Shiga toxin (Stx) fused to the fimbrial protein FaeG. The produced antibodies were able to inhibit the adhesion of *E. coli* to enterocytes and neutralize Shiga toxin (Zhang and Zhang 2010). Similarly, a fusion protein of two different Shiga toxin antigens as well as the bacterial cell surface protein intimin was used to immunize mice and shown to result in the production of anti-Stx and anti-intimin antibodies. Mice immunized with this fusion protein were immune to lethal doses of EHEC (Gu et al. 2011).

2.4.4 Targeting of Secretion Systems

Secretion systems are used by the bacterium to translocate virulence factors (effectors) directly into the host cytosol. Several types of secretion systems have a strong association with disease. The secretion systems most innately connected to pathogenesis are the type III (T3SS) and type IV (T4SS) secretion systems. T3SSs are needle-like structures with a high similarity to flagella, and T4SSs are evolutionary related to bacterial conjugation systems. Both secretion systems are highly conserved in structure between the pathogens that employ them. They span the bacterial inner and outer membrane and connect to the host cell. Here, they form a pore in the host cell membrane enabling the pathogen to translocate virulence proteins directly into the host cell cytosol in which they can interfere with cell signaling pathways in favor of bacterial persistence (Chandran 2013; Galan and Wolf-Watz 2006; Hueck 1998; Trokter et al. 2014; Waksman and Orlova 2014).

The genes for T3SSs and T4SSs are only found in pathogenic bacteria and are usually encoded within mobile genetic regions that are associated with virulence (pathogenicity islands). T3SSs are found in more than 25 species of Gram-negative pathogens including pathogenic *Chlamydia*, *E. coli*, *Salmonella*, *Shigella*, *Yersinia*, and *Pseudomonas* (Coburn et al. 2007; Galan and Wolf-Watz 2006; Schroeder and Hilbi 2008). T4SSs are associated with virulence in pathogens such as *Legionella*, *Bartonella*, *Helicobacter*, *Coxiella*, and *Brucella* (Nagai and Kubori 2011; Voth et al. 2012). As the structures and functions of the secretion systems themselves show a high conservation between the different strains of bacteria, so does their synthesis and assembly. Furthermore, the proteins that make up the secretion systems are exposed on the bacterial cell surface, making them accessible. Taken everything into consideration, a number of different potential targets present themselves: synthesis of needle components, assembly of the secretion system, interaction with the host cell, and secretion/translocation of the substrates. Due to the high similarity of the secretion systems, it is likely that inhibitors can be found which target not only one but also several different pathogens at once. Furthermore, as only pathogenic bacteria express these types of secretion systems, non-pathogenic bacteria will not be targeted. Additionally, as secretion system inhibition does not influence the overall survival of the bacterium, the selective pressure to develop resistance is low.

The design and use of special secretion test assays, i.e., fusion of the β-lactamase gene to the translocation signal of effectors secreted by T3SSs, GFP-labeled chaperone, and tagged effector labeling, allowed the identification of different natural compounds and chemical inhibitors that block the function of T3SSs of different important pathogens (Baron 2010; Izore et al. 2011; Keyser et al. 2008; Marshall and Finlay 2014; McShan and De Guzman 2015; Pan et al. 2009; Tsou et al. 2013). The glycolipids caminoside A, B, and C isolated from the marine sponge *Caminus sphaeroconia* were the first T3SS inhibitors. They block the secretion of the effector EspB, a protein that makes up part of the translocon of the *E. coli* T3SS, by varying degrees. Caminoside B appears to be the most potent inhibitor of this class; however, the cellular target remains unknown (Linington et al. 2002). The kirromycin derivative aurodox, isolated from *Streptomyces*, inhibited T3SS-mediated hemolysis of red blood cells in response to incubation with enteropathogenic *E. coli* (EPEC). It was further shown to decrease the secretion of the effectors EspB, EspF, and Map without affecting bacterial growth. Furthermore, aurodox protected mice against infection with a lethal dose of *Citrobacter rodentium*, the mouse homologue of EPEC. The mode of action of aurodox is still unknown; however, it was suggested that it might interact with transcriptional regulators (Kimura et al. 2011). Guadinomines A and B are *Streptomyces*-produced natural compounds with a strong activity against the T3SS of EPEC as shown by the inhibition of T3S-induced hemolysis. The mode of action of these compounds is also unknown (Duncan et al. 2014; Iwatsuki et al. 2008). The insecticidal, actinomycete-derived respiratory chain inhibitor piericidin A1 and its closely related derivative Mer-A2026B were also shown to act against T3SS of *Y. pseudotuberculosis*, inhibiting the secretion and translocation of the *Yersinia* virulence proteins (Yops) into host cells (Duncan et al. 2014). A series of thiazolidinone inhibitors were discovered to inhibit both the *Salmonella* and the *Yersinia* T3SSs. However, they seem to also block T2SS and type IV pili, most likely due to an inhibition of the common outer membrane protein secretin (Felise et al. 2008). The polyphenol (-)-hopeaphenol is a natural product that also decreases Yop secretion in *Y. pseudotuberculosis* without affecting Yop expression. The compound is also active against *Pseudomonas*, but the exact target of (-)-hopeaphenol is still unknown (Zetterstrom et al. 2013). The salicylidene acylhydrazides (SAHs) are by far the best-studied chemical substances targeting T3SSs. They are effective against the T3SS of a range of pathogens including *Salmonella* Typhimurium (Hudson et al. 2007; Negrea et al. 2007), *Y. pseudotuberculosis* (Nordfelth et al. 2005), *Chlamydia* (Bailey et al. 2007; Muschiol et al. 2006; Slepenkin et al. 2007; Wolf et al. 2006), *E. coli* (Tree et al. 2009), and *Shigella* (Veenendaal et al. 2009). The mode of action seems to be via the inhibition of needle subunit secretion or assembly, but recent analysis further demonstrated that SAHs significantly repressed the expression of main regulators of the T3SS machinery and modulate the function or activity of several protein targets (Layton et al. 2010; Tree et al. 2009; Wang et al. 2011). An involvement in iron chelation (Layton et al. 2010; Slepenkin et al. 2007) and a role for the metabolism (Wang et al. 2011) are also debated. In addition to these inhibitors, several other virulence blockers have been identified that inhibit transcription factors important for the expression of the T3SS in different pathogens. For instance, N-hydroxybenzimidazole derivatives block T3SS expression in *Pseudomonas* and were found to prevent T3SS expression of *Yersinia* by the inhibition of the MarA-type transcriptional activator LcrF/VirF of *Yersinia* and ExsA of *Pseudomonas* (Bowser et al. 2007; Garrity-Ryan et al. 2010; Harmon et al. 2010; Kim et al. 2009). Furthermore, salicylanilides were identified as inhibitors of LcrF expression (Kauppi et al. 2003), and the small-molecule inhibitor SE-1 inhibits the master T3SS regulator VirF of *Shigella* (Koppolu et al. 2013).

Another very promising approach includes the use of antibodies directed against the tip complex of the secretion systems, e.g., PcrV for the *Pseudomonas* T3SS. A PcrV-specific antibody of KaloBios Pharmaceuticals is in clinical phase II and already showed that it is non-immunogenic and safe in pharmacokinetic studies (Francois et al. 2012).

The *Brucella abortus* protein VirB8 is an essential component of the T4SS and indispensable for its assembly. B8I-2 was identified to inhibit the dimerization of VirB8 as well as the interaction with other VirB proteins (Paschos et al. 2011). The compound markedly reduced *virB* transcription and subsequently VirB protein levels, but it also strongly reduced the intracellular survival of *Brucella* in macrophages (Paschos et al. 2011). Another study identified three substances to inhibit VirB11, the crucial ATPase of the T4SS of *H. pylori*, which were named CHIR-1, CHIR-2, and CHIR-3. All were shown to inhibit the secretion of CagA into cells in vitro which resulted in a strong reduction of *H. pylori* colonization in mice (Hilleringmann et al. 2006).

2.4.5 Sortases

Gram-positive bacteria use enzymes named sortases to present proteins such as pilins or glycoproteins on their surface. In addition, pathogens such as *S. aureus* use these proteins to display their virulence factors (Cascioferro et al. 2014). This presentation is essential, as sortase A (*srtA*) mutant strains are impaired in their ability to cause infection in the mouse model (Jonsson et al. 2002). Sortase A mutants in *Streptococcus suis* and *L. monocytogenes* also show a marked reduction in pathogenicity (Bierne et al. 2002; Vanier et al. 2008). The conservation, essentiality, and widespread use of sortase by pathogens suggest that compounds, which inhibit their activity, will function as potent anti-infective agents (Cascioferro et al. 2015). The surface proteins anchored to the cell wall by *S. aureus* SrtA include virulence factors that play key roles in the infection process by promoting nutrient acquisition from the host, bacterial adhesion, and immune evasion. Among the first sortase inhibitors were peptidomimetic molecules, the small-molecule inhibitor diarylacrylonitriles, and flavonols such as morin (Cascioferro et al. 2015). More recently, the compound (2-(2,3-dihydro-1*H*-perimidin-2-yl)-phenoxy)-acetic acid was found to inhibit SrtA without affecting bacterial growth (Chan et al. 2013; Zhang et al. 2014). It also protects mice from bacteremia (Zhang et al. 2014). Furthermore, aryl beta-amino(ethyl) ketone (AAEK) was identified as a sortase inhibitor in a high-throughput screen and selected for further investigation on the basis of its marked effect on sortases of staphylococci and bacilli (Maresso et al. 2007).

2.5 *Anti-resistance Drugs*

The development and use of anti-resistance drugs that are administered together with known antibiotics, to circumvent acquired resistance mechanisms of bacterial pathogens, opens a door to extending the life span of known antibiotics.

2.5.1 Resistance Against Antibiotics

β-lactamase Inhibitors

β-lactam antibiotics have been widely used for almost 80 years. They are bactericidal compounds that act by inhibiting cell wall-synthesizing enzymes found only in bacteria (Kong et al. 2010). The continuous use of this family of antibiotics has given rise to the extensive proliferation of bacterial β-lactamases, which hydrolyze the β-lactam ring of the antibiotic, rendering the bacteria resistant to antibiotics of the carbapenem family (Palzkill 2013). A combination therapy of β-lactamase inhibitors as an adjuvant to suppress resistance with a β-lactam antibiotic increases the efficacy and the spectrum of the antibiotic. Successful combinations of antibiotics and β-lactamase inhibitors include clavulanic acid, sulbactam or tazobactam, and penicillins (Carlier et al. 2014; Totir et al. 2007). Avibactam, a non-beta-lactam bicyclic diazabicyclooctane with no antibiotic activity, forms reversible bonds with several beta-lactamases including *K. pneumoniae* carbapenemase and ESBL- or AmpC-overexpressing strains (Castanheira et al. 2014; Coleman et al. 2014). Similar modes of action have been described for Merck MK-7655, a piperidine analogue used with imipenem (Blizzard et al. 2014).

Efflux Pump Inhibitors

Overexpression of efflux pumps to expel toxic compounds such as antibiotics from the cell is a common resistance strategy found in bacteria. With the added advantage that uptake of antibiotics into Gram-negative bacteria is slowed down by decreased permeability of the bacterial outer membrane, the bacteria are less susceptible to efflux pump substrates even those with poor affinity, as the rate of efflux usually exceeds that of influx (Nikaido and Pages 2012). Discovering compounds that selectively inhibit the bacterial efflux pumps may lead to the development of therapies that could restore sensitivity to antibiotics. Several bacterial efflux pumps have been studied in detail and would make for promising therapeutic targets.

Resistance–nodulation–division (RND) efflux pumps are encoded by Gram-negative *Enterobacteriaceae* and *Pseudomonas*. They transport a variety of toxic substances from the bacterial cell, including antibiotics such as fluoroquinolones, β-lactams, tetracyclines, and oxazolidines, mediating intrinsic resistance of the bacteria to these substances (Nikaido and Pages 2012). Phenylalanyl arginyl β-naphthylamide (PAβN) was discovered in a high-throughput screen for molecules that sensitize efflux pump-overexpressing *P. aeruginosa* to levofloxacin (Renau et al. 1999). PAβN inhibits the efflux pumps of several bacterial pathogens such as *E. coli*, *S. Typhimurium*, *K. pneumoniae*, and *Campylobacter* sp. with divergent efficacy. It was shown that while PAβN sensitizes *Pseudomonas* to antibiotics such as levofloxacin, erythromycin, and chloramphenicol, it showed little effect for enhancing the susceptibility to carbenicillin, suggesting that its activity is strongly dependent on the antibiotic (Lomovskaya et al. 2001). It was therefore suggested that PAβN is a competitive inhibitor that binds within the substrate-binding pocket used by a specific set of antibiotics. PAβN was found to increase the permeability of the cellular membrane (Lomovskaya et al. 2001) in addition to its function as an efflux pump. Increased uptake into the bacterial cell increases its potency. Naphtylpiperazines (NMPs) were identified as potentiators of levofloxacin in RND efflux pump-overexpressing *E. coli* (Bohnert and Kern 2005). Addition of NMP to cells increases the intracellular concentration of levofloxacin. It further affects the susceptibility of the bacteria to other antibiotic substances such as rifampicin and chloramphenicol. NMP is able to reverse multidrug resistance in clinical *E. coli* isolates (Kern et al. 2006) and shows some effects against multidrug-resistant *K. pneumoniae* (Schumacher et al. 2006) and *A. baumannii* (Pannek et al. 2006). However, NMP is ineffective in increasing antibiotic susceptibility of *P. aeruginosa* (Coban et al. 2009). Notably, antisense peptide nucleic acids (PNAs) are synthetic homologues of nucleic acids that bind complementary DNA and RNA sequences with very high specificity (Paulasova and Pellestor 2004). They can be used as antisense peptides to specifically inhibit the gene expression of efflux pump genes. This approach was employed to decrease the expression of the *C. jejuni* RND efflux pump CmeABC, resulting in a significant increase of susceptibility to antibiotic treatment (Jeon and Zhang 2009; Mu et al. 2013). NorA efflux pumps are found in Gram-positive pathogens such as *S. aureus*. They confer resistance to antibiotics including those of the family of fluoroquinolones. Capsaicin, an alkaloid, has recently been shown to potently reduce the resistance of *S. aureus* to ciprofloxacin and inhibit the efflux of ethidium bromide in vitro (Kalia et al. 2012).

2.5.2 Environmental Stress Resistance

Depletion of Iron

Pathogenic bacteria are exposed to multiple host-associated stresses directed to prevent proliferation and eliminate the invaders. Among the most important stresses is the depletion of iron from the blood and lymph systems by specific iron-complexing molecules. To overcome this problem, bacteria synthesize a plethora of iron-chelating siderophores for iron uptake, which are essential for virulence (Miethke and Marahiel 2007). Several attempts were made to identify compounds that prevent siderophore synthesis and transport. A salicylsulfamoyl adenosine and other nucleoside bisubstrate analogues were found to prevent the early step of siderophore synthesis of *Yersinia* and *Mycobacterium* (Ferreras et al. 2005; Miethke and Marahiel 2007; Neres et al. 2008).

2.5.3 Resistance Against Reactive Oxygen and Nitrogen Species (ROS, NOS)

In a similar manner, virulence factors can be targeted which are essential to promote resistance against reactive oxygen or nitrogen species used by innate immune cells to destroy bacteria. Enzymes that have been successfully targeted by inhibitors are staphyloxanthin, a *S. aureus* pigment with antioxidant activity, which is blocked by phosphosulfonate, and the *Mycobacterium tuberculosis* factor DlaT essential to resist NOS intermediates which are targeted by rhodanine analogues (Escaich 2010).

2.5.4 Interference with Bacterial Host Resistance Mechanisms

Components of the innate immunity, such as complement factors, antimicrobial peptides (e.g., defensins), and professional phagocytes (e.g., neutrophils, macrophages) circulating in the blood, lymph system, and tissues, are able to efficiently eliminate invading pathogens. Lipopolysaccharides (LPSs) of the outer membrane of Gram-negative bacteria are required for the resistance to complement and cationic antimicrobial peptides (CAMPs). Modification of LPS to decrease surface charges (e.g., by the addition of aminoarabinose or heptoses) and inhibition of several enzymes of LPS biosynthesis have been identified as treatment strategies (Desroy et al. 2013; Escaich 2010). Bacterial resistance to CAMPs is linked to an increase in D-alanylation of the lipoteichoic acids in the cell wall of staphylococci and streptococci (Peschel et al. 1999; Saar-Dover et al. 2012). Furthermore, inhibitors of D-alanylation enzymes have been identified, some of which were also shown to reduce bacteremia (Escaich 2010; Santa Maria et al. 2014).

3 Challenges and Problems Associated with Virulence Blockers

The large variety of newly identified natural compounds, structural analogues, mimetic peptides, and antibodies that act as inhibitors of crucial virulence traits over the last ten years has demonstrated that anti-virulence strategies can be successfully applied to combat bacterial infections. However, there are still many challenges and problems for the development of anti-virulence drugs, which will have to be addressed in the near future.

Recent development of target-based high-throughput screening or rational drug design with chemical and natural compound libraries or structural analogues has furthered the field and facilitated the identification of inhibitors. However, for many inhibitors, the demonstration that inhibition of a particular virulence strategy leads to an inhibition of the bacterial infection in vitro and the validation of inhibitor activities in animal models is still missing.

Another important issue concerns the mode of action of many compounds. Although several effective inhibitors of certain virulence traits (e.g., T3SS-mediated translocation of bacterial effectors) have been identified, the precise molecular mechanism of the inhibition and the exact targets are often unclear. Moreover, some of the identified inhibitor classes (i.e., SAHs) seem to have multiple molecular targets, which contribute to the inhibitory effect of the compounds, and other inhibitors are active on more than one species.

One major challenge in the development of successful anti-virulence strategies is the general redundancy of many crucial virulence mechanisms of pathogens. Bacterial pathogens usually expose multiple adhesive surface structures with distinct cell receptor specificities, which allow binding and colonization of different tissues during the infection cycle. Furthermore, alternative infection routes are used by several pathogens to disseminate into deeper tissues. Hence, the development of a successful universal adherence inhibitor is rather difficult. Nonetheless, certain adhesion structures have been identified as hallmark requirements for specific pathogens, which are promising targets for drug discovery and development. Other examples are bacterial effector proteins, which are injected into host cells by T3SS or T4SSs to manipulate host cells and circumvent immune responses. Many bacteria that cannot evade detection by the receptors encode multiple effector proteins, which are able to specifically modify the host innate immune response in their favor. Pathogenic *Yersinia* species encode 5 effector proteins YopH, YopE, YopJ/YopP, YopT, and YopM (Bliska et al. 2013), and in EPEC and enterohemorrhagic *E. coli* (EHEC), at least seven translocated effector proteins (NleB1, NleC, NleD, NleE, NleF, NleH2, and Tir) are known to modulate different aspects of the proinflammatory and apoptotic cell signaling pathways (Wong et al. 2011). In *Legionalla pneumophila*, which encodes a staggering amount of at least 300 effector proteins, the redundancy of effector function is consequently much higher (Pearson et al. 2015). In addition, many host adapted metabolic processes, ion/nutrient uptake systems, and regulatory circuits are redundant, and elimination of one is almost fully compensated by the upregulation or activation of (an) alternative pathway(s).

Crucial pathogenicity traits are often only expressed at distinct sites and at certain time points when they are required during the infection to prevent immune responses and balance their energy budget to optimize their fitness. Hence, detailed knowledge about the tempo-spatial expression pattern and the role of the potential targets is required for the design/development of effective anti-virulence drugs.

Over the past years, an increasing number of reports demonstrated that certain crucial virulence traits are only expressed by a certain ratio of the bacterial population at any time during the infection (Burton et al. 2014; Claudi et al. 2014; Diard et al. 2013; Helaine et al. 2014; Manina et al. 2015; Putrins et al. 2015; Sturm et al. 2011). The presence of genetically identical but phenotypically heterogeneous subpopulations is advantageous for a pathogen to survive within a fluctuating environment with varying nutrient, ion, and stress conditions. It allows the bacteria to prepare themselves for uncertainties (bet-hedging) and to divide a biological task into different subtasks executed by the different subpopulations (division of labor) (Ackermann 2015; Avery 2006; Smits et al. 2006). Furthermore, certain pathogen subsets reside in diverse tissue microenvironments and biofilms as a response to local conditions and molecular interactions, and this has disparate consequences on the expression of virulence-relevant traits. Differential expression of certain virulence traits in individual subpopulations can result in a failure to control the infection foci by identified inhibitors or in selective targeting of pathogen subsets.

Even if the selective pressure of virulence blockers is less than that induced by antibiotics, the appearance of resistance against anti-virulence drugs may result in pathogens that have an advantage over the remaining population. This allows for (involuntary) selection for these better-adapted pathogens over time. Many crucial virulence factors are located on mobilizable DNA elements, such as plasmids and phages. Rapid exchange of mutated gene variants of a targeted virulence trait by horizontal DNA transfer could result in the rapid evolution of drug-resistant bacterial populations. Application of combinational therapies may be a good option to suppress these better-adapted pathogens at the moment they appear.

4 Conclusions and Outlook

The present review takes a close look at current anti-virulence strategies and identified classes of novel virulence blockers and illustrates the promising advances made in our attempts to develop alternatives to antibiotic therapies. To fully exploit this strategy, it is imperative that we improve our understanding of the molecular mechanisms and the consequences of host–pathogen interactions, as many crucial virulence-associated processes remain unclear. Future research in this field will also be aided by the increasing number of crystal structures of crucial pathogenicity factors allowing directed drug design for high-throughput screening assays. This will not only facilitate structure–function analyses and optimization of identified inhibitors, but also allow an exploitation of other less well-characterized adhesion structures and secretion systems (T6SS, T7SS).

The loss of interest in antimicrobial drug development by large pharmaceutical companies and the urgent need for alternatives have triggered science funding agencies around the world to support projects that advance the development of alternatives to antibiotics. This generated a boost for academic researchers and small biotechnology enterprises to pursue different strategies with the goal to find alternative solutions for future applications. However, the following pipeline for the development of novel antibiotics is often not available. There is an urgent task to provide test systems to study pharmacokinetics/dynamics parameters and evaluate the efficacy in animal models. In particular, the specificity, safety, and tolerability of the compounds need to be assessed, an appropriate formulation for delivery must be developed, and setups must be installed that help transfer successful virulence blockers into broad-scale clinical trials. A major task will be to convince the pharmaceutical industry to participate and support this mission early on.

Another important aspect is that an effective application of virulence blockers requires a rapid and precise diagnosis of infectious agents in the clinics. This will include a more detailed profiling of the pathogen and its virulence traits and demands a more patient-specific, personalized analysis of the responsible disease agent. Such advancements are not only attractive due to their potential to combat resistant pathogens, but also help to improve the use of standard antibiotics and offer the possibility to use both antibiotics and virulence blockers in a synergistic therapy approach to minimize the selection of resistant variants.

Acknowledgements Work of P. Dersch is supported by the German Research Foundation and the German Center of Infection Research (DZIF, TTU-GI).

References

Aarestrup FM (2005) Veterinary drug usage and antimicrobial resistance in bacteria of animal origin. Basic Clin Pharmacol Toxicol 96:271–281

Ackermann M (2015) A functional perspective on phenotypic heterogeneity in microorganisms. Nat Rev Microbiol 13:497–508

Aggarwal C, Jimenez JC, Lee H, Chlipala GE, Ratia K, Federle MJ (2015) Identification of quorum-sensing inhibitors disrupting signaling between Rgg and short hydrophobic peptides in Streptococci. MBio 6:e00393–e00315

Allen HK, Donato J, Wang HH, Cloud-Hansen KA, Davies J, Handelsman J (2010) Call of the wild: antibiotic resistance genes in natural environments. Nat Rev Microbiol 8:251–259

Altier C, Suyemoto M, Ruiz AI, Burnham KD, Maurer R (2000) Characterization of two novel regulatory genes affecting *Salmonella* invasion gene expression. Mol Microbiol 35:635–646

Arnon SS, Schechter R, Maslanka SE, Jewell NP, Hatheway CL (2006) Human botulism immune globulin for the treatment of infant botulism. N Engl J Med 354:462–471

Arya R, Princy SA (2013) An insight into pleiotropic regulators Agr and Sar: molecular probes paving the new way for antivirulent therapy. Future Microbiol 8:1339–1353

Avery SV (2006) Microbial cell individuality and the underlying sources of heterogeneity. Nat Rev Microbiol 4:577–587

Bailey L, Gylfe A, Sundin C, Muschiol S, Elofsson M, Nordstrom P, Henriques-Normark B, Lugert R, Waldenstrom A, Wolf-Watz H et al (2007) Small molecule inhibitors of type III secretion in *Yersinia* block the *Chlamydia pneumoniae* infection cycle. FEBS Lett 581:587–595

Baldari CT, Tonello F, Paccani SR, Montecucco C (2006) Anthrax toxins: a paradigm of bacterial immune suppression. Trends Immunol 27:434–440

Bannwarth L, Goldberg AB, Chen C, Turk BE (2012) Identification of exosite-targeting inhibitors of anthrax lethal factor by high-throughput screening. Chem Biol 19:875–882

Barnhart MM, Sauer FG, Pinkner JS, Hultgren SJ (2003) Chaperone-subunit-usher interactions required for donor strand exchange during bacterial pilus assembly. J Bacteriol 185:2723–2730

Baron C (2010) Antivirulence drugs to target bacterial secretion systems. Curr Opin Microbiol 13:100–105

Barraud N, Hassett DJ, Hwang SH, Rice SA, Kjelleberg S, Webb JS (2006) Involvement of nitric oxide in biofilm dispersal of *Pseudomonas aeruginosa*. J Bacteriol 188:7344–7353

Barraud N, Schleheck D, Klebensberger J, Webb JS, Hassett DJ, Rice SA, Kjelleberg S (2009) Nitric oxide signaling in *Pseudomonas aeruginosa* biofilms mediates phosphodiesterase activity, decreased cyclic di-GMP levels, and enhanced dispersal. J Bacteriol 191:7333–7342

Basha S, Rai P, Poon V, Saraph A, Gujraty K, Go MY, Sadacharan S, Frost M, Mogridge J, Kane RS (2006) Polyvalent inhibitors of anthrax toxin that target host receptors. Proc Natl Acad Sci USA 103:13509–13513

Beckham KS, Roe AJ (2014) From screen to target: insights and approaches for the development of anti-virulence compounds. Front Cell Infect Microbiol 4:139

Berendonk TU, Manaia CM, Merlin C, Fatta-Kassinos D, Cytryn E, Walsh F, Burgmann H, Sorum H, Norstrom M, Pons MN et al (2015) Tackling antibiotic resistance: the environmental framework. Nat Rev Microbiol 13:310–317

Bierne H, Mazmanian SK, Trost M, Pucciarelli MG, Liu G, Dehoux P, Jansch L, Garcia-del Portillo F, Schneewind O, Cossart P et al (2002) Inactivation of the *srtA* gene in *Listeria monocytogenes* inhibits anchoring of surface proteins and affects virulence. Mol Microbiol 43:869–881

Bliska JB, Wang X, Viboud GI, Brodsky IE (2013) Modulation of innate immune responses by *Yersinia* type III secretion system translocators and effectors. Cell Microbiol 15:1622–1631

Blizzard TA, Chen H, Kim S, Wu J, Bodner R, Gude C, Imbriglio J, Young K, Park YW, Ogawa A et al (2014) Discovery of MK-7655, a beta-lactamase inhibitor for combination with Primaxin(R). Bioorg Med Chem Lett 24:780–785

Böhme K, Steinmann R, Kortmann J, Seekircher S, Heroven AK, Berger E, Pisano F, Thiermann T, Wolf-Watz H, Narberhaus F et al (2012) Concerted actions of a thermo-labile regulator and a unique intergenic RNA thermosensor control *Yersinia* virulence. PLoS Pathog 8:e1002518

Bohnert JA, Kern WV (2005) Selected arylpiperazines are capable of reversing multidrug resistance in *Escherichia coli* overexpressing RND efflux pumps. Antimicrob Agents Chemother 49:849–852

Bowser TE, Bartlett VJ, Grier MC, Verma AK, Warchol T, Levy SB, Alekshun MN (2007) Novel anti-infection agents: small-molecule inhibitors of bacterial transcription factors. Bioorg Med Chem Lett 17:5652–5655

Boyd B, Lingwood C (1989) Verotoxin receptor glycolipid in human renal tissue. Nephron 51:207–210

Brautaset T, Lale R, Valla S (2009) Positively regulated bacterial expression systems. Microb Biotechnol 2:15–30

Brown HL, Hanman K, Reuter M, Betts RP, van Vliet AH (2015) *Campylobacter jejuni* biofilms contain extracellular DNA and are sensitive to DNase I treatment. Front Microbiol 6:699

Bruckner R, Titgemeyer F (2002) Carbon catabolite repression in bacteria: choice of the carbon source and autoregulatory limitation of sugar utilization. FEMS Microbiol Lett 209:141–148

Burton NA, Schurmann N, Casse O, Steeb AK, Claudi B, Zankl J, Schmidt A, Bumann D (2014) Disparate impact of oxidative host defenses determines the fate of *Salmonella* during systemic infection in mice. Cell Host Microbe 15:72–83

Canton R (2009) Antibiotic resistance genes from the environment: a perspective through newly identified antibiotic resistance mechanisms in the clinical setting. Clin Microbiol Infect 15(Suppl 1):20–25

Carlier M, Carrette S, Stove V, Verstraete AG, De Waele JJ (2014) Does consistent piperacillin dosing result in consistent therapeutic concentrations in critically ill patients? A longitudinal study over an entire antibiotic course. Int J Antimicrob Agents 43:470–473

Cascioferro S, Totsika M, Schillaci D (2014) Sortase A: an ideal target for anti-virulence drug development. Microb Pathog 77:105–112

Cascioferro S, Raffa D, Maggio B, Raimondi MV, Schillaci D, Daidone G (2015) Sortase A inhibitors: recent advances and future perspectives. J Med Chem

Castanheira M, Williams G, Jones RN, Sader HS (2014) Activity of ceftaroline-avibactam tested against contemporary *Enterobacteriaceae* isolates carrying beta-lactamases prevalent in the United States. Microb Drug Resist 20:436–440

Cegelski L, Marshall GR, Eldridge GR, Hultgren SJ (2008) The biology and future prospects of antivirulence therapies. Nat Rev Microbiol 6:17–27

Chan AH, Wereszczynski J, Amer BR, Yi SW, Jung ME, McCammon JA, Clubb RT (2013) Discovery of *Staphylococcus aureus* sortase A inhibitors using virtual screening and the relaxed complex scheme. Chem Biol Drug Des 82:418–428

Chandran V (2013) Type IV secretion machinery: molecular architecture and function. Biochem Soc Trans 41:17–28

Chao Y, Vogel J (2010) The role of Hfq in bacterial pathogens. Curr Opin Microbiol 13:24–33

Chaudhuri RR, Morgan E, Peters SE, Pleasance SJ, Hudson DL, Davies HM, Wang J, van Diemen PM, Buckley AM, Bowen AJ et al (2013) Comprehensive assignment of roles for *Salmonella typhimurium* genes in intestinal colonization of food-producing animals. PLoS Genet 9:e1003456

Chen Z, Moayeri M, Purcell R (2011) Monoclonal antibody therapies against anthrax. Toxins (Basel) 3:1004–1019

Chhabra SR, Stead P, Bainton NJ, Salmond GP, Stewart GS, Williams P, Bycroft BW (1993) Autoregulation of carbapenem biosynthesis in *Erwinia carotovora* by analogues of N-(3-oxohexanoyl)-L-homoserine lactone. J Antibiot (Tokyo) 46:441–454

Clarke DJ (2010) The Rcs phosphorelay: more than just a two-component pathway. Future Microbiol 5:1173–1184

Clatworthy AE, Pierson E, Hung DT (2007) Targeting virulence: a new paradigm for antimicrobial therapy. Nat Chem Biol 3:541–548

Claudi B, Sprote P, Chirkova A, Personnic N, Zankl J, Schurmann N, Schmidt A, Bumann D (2014) Phenotypic variation of *Salmonella* in host tissues delays eradication by antimicrobial chemotherapy. Cell 158:722–733

Coban AY, Tanriverdi Cayci Y, Erturan Z, Durupinar B (2009) Effects of efflux pump inhibitors phenyl-arginine-beta-naphthylamide and 1-(1-naphthylmethyl)-piperazine on the antimicrobial susceptibility of *Pseudomonas aeruginosa* isolates from cystic fibrosis patients. J Chemother 21:592–594

Coburn B, Sekirov I, Finlay BB (2007) Type III secretion systems and disease. Clin Microbiol Rev 20:535–549

Coleman K, Levasseur P, Girard AM, Borgonovi M, Miossec C, Merdjan H, Drusano G, Shlaes D, Nichols WW (2014) Activities of ceftazidime and avibactam against beta-lactamase-producing *Enterobacteriaceae* in a hollow-fiber pharmacodynamic model. Antimicrob Agents Chemother 58:3366–3372

Conover MS, Mishra M, Deora R (2011) Extracellular DNA is essential for maintaining *Bordetella* biofilm integrity on abiotic surfaces and in the upper respiratory tract of mice. PLoS ONE 6: e16861

Control CfD (2013) Antibiotic resistance threats in the United States

Costerton JW, Stewart PS, Greenberg EP (1999) Bacterial biofilms: a common cause of persistent infections. Science 284:1318–1322

Cryan LM, Rogers MS (2011) Targeting the anthrax receptors, TEM-8 and CMG-2, for anti-angiogenic therapy. Front Biosci (Landmark Ed) 16:1574–1588

Cryan LM, Habeshian KA, Caldwell TP, Morris MT, Ackroyd PC, Christensen KA, Rogers MS (2013) Identification of small molecules that inhibit the interaction of TEM8 with anthrax protective antigen using a FRET assay. J Biomol Screen 18:714–725

Curtis MM, Russell R, Moreira CG, Adebesin AM, Wang C, Williams NS, Taussig R, Stewart D, Zimmern P, Lu B et al (2014) QseC inhibitors as an antivirulence approach for Gram-negative pathogens. MBio 5:e02165

D'Costa VM, McGrann KM, Hughes DW, Wright GD (2006) Sampling the antibiotic resistome. Science 311:374–377

Deep A, Chaudhary U, Gupta V (2011) Quorum sensing and bacterial pathogenicity: from molecules to disease. J Lab Phys 3:4–11

Dell'Aica I, Dona M, Tonello F, Piris A, Mock M, Montecucco C, Garbisa S (2004) Potent inhibitors of anthrax lethal factor from green tea. EMBO Rep 5:418–422

Desroy N, Denis A, Oliveira C, Atamanyuk D, Briet S, Faivre F, LeFralliec G, Bonvin Y, Oxoby M, Escaich S et al (2013) Novel HldE-K inhibitors leading to attenuated Gram negative bacterial virulence. J Med Chem 56:1418–1430
Deutscher J, Herro R, Bourand A, Mijakovic I, Poncet S (2005) P-Ser-HPr—a link between carbon metabolism and the virulence of some pathogenic bacteria. Biochim Biophys Acta 1754:118–125
Diard M, Garcia V, Maier L, Remus-Emsermann MN, Regoes RR, Ackermann M, Hardt WD (2013) Stabilization of cooperative virulence by the expression of an avirulent phenotype. Nature 494:353–356
Dong YH, Xu JL, Li XZ, Zhang LH (2000) AiiA, an enzyme that inactivates the acylhomoserine lactone quorum-sensing signal and attenuates the virulence of *Erwinia carotovora*. Proc Natl Acad Sci USA 97:3526–3531
Dong YH, Wang LH, Xu JL, Zhang HB, Zhang XF, Zhang LH (2001) Quenching quorum-sensing-dependent bacterial infection by an N-acyl homoserine lactonase. Nature 411:813–817
Dong YH, Gusti AR, Zhang Q, Xu JL, Zhang LH (2002) Identification of quorum-quenching N-acyl homoserine lactonases from *Bacillus* species. Appl Environ Microbiol 68:1754–1759
Duncan MC, Wong WR, Dupzyk AJ, Bray WM, Linington RG, Auerbuch V (2014) An NF-kappaB-based high-throughput screen identifies piericidins as inhibitors of the *Yersinia pseudotuberculosis* type III secretion system. Antimicrob Agents Chemother 58:1118–1126
Duss O, Michel E, Yulikov M, Schubert M, Jeschke G, Allain FH (2014) Structural basis of the non-coding RNA RsmZ acting as a protein sponge. Nature 509:588–592
Escaich S (2008) Antivirulence as a new antibacterial approach for chemotherapy. Curr Opin Chem Biol 12:400–408
Escaich S (2010) Novel agents to inhibit microbial virulence and pathogenicity. Expert Opin Ther Pat 20:1401–1418
Felise HB, Nguyen HV, Pfuetzner RA, Barry KC, Jackson SR, Blanc MP, Bronstein PA, Kline T, Miller SI (2008) An inhibitor of gram-negative bacterial virulence protein secretion. Cell Host Microbe 4:325–336
Feng L, Rutherford ST, Papenfort K, Bagert JD, van Kessel JC, Tirrell DA, Wingreen NS, Bassler BL (2015) A *qrr* noncoding RNA deploys four different regulatory mechanisms to optimize quorum-sensing dynamics. Cell 160:228–240
Fernicola S, Paiardini A, Giardina G, Rampioni G, Leoni L, Cutruzzola F, Rinaldo S (2015) In silico discovery and in vitro validation of catechol-containing sulfonohydrazide compounds as potent inhibitors of the diguanylate cyclase PleD. J Bacteriol
Ferreras JA, Ryu JS, Di Lello F, Tan DS, Quadri LE (2005) Small-molecule inhibition of siderophore biosynthesis in *Mycobacterium tuberculosis* and *Yersinia pestis*. Nat Chem Biol 1:29–32
Field M (2003) Intestinal ion transport and the pathophysiology of diarrhea. J Clin Invest 111:931–943
Firon N, Ashkenazi S, Mirelman D, Ofek I, Sharon N (1987) Aromatic alpha-glycosides of mannose are powerful inhibitors of the adherence of type 1 fimbriated *Escherichia coli* to yeast and intestinal epithelial cells. Infect Immun 55:472–476
Forst SA, Roberts DL (1994) Signal transduction by the EnvZ-OmpR phosphotransfer system in bacteria. Res Microbiol 145:363–373
Francois B, Luyt CE, Dugard A, Wolff M, Diehl JL, Jaber S, Forel JM, Garot D, Kipnis E, Mebazaa A et al (2012) Safety and pharmacokinetics of an anti-PcrV PEGylated monoclonal antibody fragment in mechanically ventilated patients colonized with Pseudomonas aeruginosa: a randomized, double-blind, placebo-controlled trial. Crit Care Med 40:2320–2326
Galan JE, Wolf-Watz H (2006) Protein delivery into eukaryotic cells by type III secretion machines. Nature 444:567–573
Garrity-Ryan LK, Kim OK, Balada-Llasat JM, Bartlett VJ, Verma AK, Fisher ML, Castillo C, Songsungthong W, Tanaka SK, Levy SB et al (2010) Small molecule inhibitors of LcrF, a

Yersinia pseudotuberculosis transcription factor, attenuate virulence and limit infection in a murine pneumonia model. Infect Immun 78:4683–4690

Goldman ME, Cregar L, Nguyen D, Simo O, O'Malley S, Humphreys T (2006) Cationic polyamines inhibit anthrax lethal factor protease. BMC Pharmacol 6:8

Greener M (2000) How *Escherichia coli* kills. Mol Med Today 6:411

Groisman EA, Mouslim C (2006) Sensing by bacterial regulatory systems in host and non-host environments. Nat Rev Microbiol 4:705–709

Gu J, Ning Y, Wang H, Xiao D, Tang B, Luo P, Cheng Y, Jiang M, Li N, Zou Q et al (2011) Vaccination of attenuated EIS-producing *Salmonella* induces protective immunity against enterohemorrhagic *Escherichia coli* in mice. Vaccine 29:7395–7403

Hall-Stoodley L, Costerton JW, Stoodley P (2004) Bacterial biofilms: from the natural environment to infectious diseases. Nat Rev Microbiol 2:95–108

Han Z, Pinkner JS, Ford B, Obermann R, Nolan W, Wildman SA, Hobbs D, Ellenberger T, Cusumano CK, Hultgren SJ et al (2010) Structure-based drug design and optimization of mannoside bacterial FimH antagonists. J Med Chem 53:4779–4792

Han Z, Pinkner JS, Ford B, Chorell E, Crowley JM, Cusumano CK, Campbell S, Henderson JP, Hultgren SJ, Janetka JW (2012) Lead optimization studies on FimH antagonists: discovery of potent and orally bio-available ortho-substituted biphenyl mannosides. J Med Chem 55:3945–3959

Harmon DE, Davis AJ, Castillo C, Mecsas J (2010) Identification and characterization of small-molecule inhibitors of Yop translocation in *Yersinia pseudotuberculosis*. Antimicrob Agents Chemother 54:3241–3254

Harmsen M, Lappann M, Knochel S, Molin S (2010) Role of extracellular DNA during biofilm formation by *Listeria monocytogenes*. Appl Environ Microbiol 76:2271–2279

Hartmann M, Papavlassopoulos H, Chandrasekaran V, Grabosch C, Beiroth F, Lindhorst TK, Rohl C (2012) Inhibition of bacterial adhesion to live human cells: activity and cytotoxicity of synthetic mannosides. FEBS Lett 586:1459–1465

Helaine S, Cheverton AM, Watson KG, Faure LM, Matthews SA, Holden DW (2014) Internalization of *Salmonella* by macrophages induces formation of nonreplicating persisters. Science 343:204–208

Henkel JS, Baldwin MR, Barbieri JT (2010) Toxins from bacteria. EXS 100:1–29

Hentzer M, Givskov M (2003) Pharmacological inhibition of quorum sensing for the treatment of chronic bacterial infections. J Clin Invest 112:1300–1307

Hentzer M, Riedel K, Rasmussen TB, Heydorn A, Andersen JB, Parsek MR, Rice SA, Eberl L, Molin S, Hoiby N et al (2002) Inhibition of quorum sensing in *Pseudomonas aeruginosa* biofilm bacteria by a halogenated furanone compound. Microbiology 148:87–102

Heroven AK, Böhme K, Dersch P (2012) The Csr/Rsm system of *Yersinia* and related pathogens: a post-transcriptional strategy for managing virulence. RNA Biol 9:379–391

Hesterkamp T (2015) Antibiotics clinical development and pipeline. Curr Top Microbiol Immunol

Hilleringmann M, Pansegrau W, Doyle M, Kaufman S, MacKichan ML, Gianfaldoni C, Ruggiero P, Covacci A (2006) Inhibitors of *Helicobacter pylori* ATPase Cagalpha block CagA transport and *cag* virulence. Microbiology 152:2919–2930

Hopwood DA (2007) How do antibiotic-producing bacteria ensure their self-resistance before antibiotic biosynthesis incapacitates them? Mol Microbiol 63:937–940

Hudson DL, Layton AN, Field TR, Bowen AJ, Wolf-Watz H, Elofsson M, Stevens MP, Galyov EE (2007) Inhibition of type III secretion in *Salmonella enterica* serovar Typhimurium by small-molecule inhibitors. Antimicrob Agents Chemother 51:2631–2635

Hueck CJ (1998) Type III protein secretion systems in bacterial pathogens of animals and plants. Microbiol Mol Biol Rev 62:379–433

Hung DT, Shakhnovich EA, Pierson E, Mekalanos JJ (2005) Small-molecule inhibitor of *Vibrio cholerae* virulence and intestinal colonization. Science 310:670–674

Iwatsuki M, Uchida R, Yoshijima H, Ui H, Shiomi K, Kim YP, Hirose T, Sunazuka T, Abe A, Tomoda H et al (2008) Guadinomines, type III secretion system inhibitors, produced by

Streptomyces sp. K01-0509. II: physico-chemical properties and structure elucidation. J Antibiot (Tokyo) 61:230–236

Izano EA, Amarante MA, Kher WB, Kaplan JB (2008) Differential roles of poly-N-acetylglucosamine surface polysaccharide and extracellular DNA in *Staphylococcus aureus* and *Staphylococcus epidermidis* biofilms. Appl Environ Microbiol 74:470–476

Izore T, Job V, Dessen A (2011) Biogenesis, regulation, and targeting of the type III secretion system. Structure 19:603–612

Jeon B, Zhang Q (2009) Sensitization of *Campylobacter jejuni* to fluoroquinolone and macrolide antibiotics by antisense inhibition of the CmeABC multidrug efflux transporter. J Antimicrob Chemother 63:946–948

Johansson J, Mandin P, Renzoni A, Chiaruttini C, Springer M, Cossart P (2002) An RNA thermosensor controls expression of virulence genes in *Listeria monocytogenes*. Cell 110:551–561

Jonsson IM, Mazmanian SK, Schneewind O, Verdrengh M, Bremell T, Tarkowski A (2002) On the role of *Staphylococcus aureus* sortase and sortase-catalyzed surface protein anchoring in murine septic arthritis. J Infect Dis 185:1417–1424

Kalia NP, Mahajan P, Mehra R, Nargotra A, Sharma JP, Koul S, Khan IA (2012) Capsaicin, a novel inhibitor of the NorA efflux pump, reduces the intracellular invasion of *Staphylococcus aureus*. J Antimicrob Chemother 67:2401–2408

Kaplan BS, Meyers KE, Schulman SL (1998) The pathogenesis and treatment of hemolytic uremic syndrome. J Am Soc Nephrol 9:1126–1133

Kaplan JB, Velliyagounder K, Ragunath C, Rohde H, Mack D, Knobloch JK, Ramasubbu N (2004) Genes involved in the synthesis and degradation of matrix polysaccharide in *Actinobacillus actinomycetemcomitans* and *Actinobacillus pleuropneumoniae* biofilms. J Bacteriol 186:8213–8220

Karginov VA, Yohannes A, Robinson TM, Fahmi NE, Alibek K, Hecht SM (2006) Beta-cyclodextrin derivatives that inhibit anthrax lethal toxin. Bioorg Med Chem 14:33–40

Kauppi AM, Nordfelth R, Uvell H, Wolf-Watz H, Elofsson M (2003) Targeting bacterial virulence: inhibitors of type III secretion in *Yersinia*. Chem Biol 10:241–249

Kern WV, Steinke P, Schumacher A, Schuster S, von Baum H, Bohnert JA (2006) Effect of 1-(1-naphthylmethyl)-piperazine, a novel putative efflux pump inhibitor, on antimicrobial drug susceptibility in clinical isolates of *Escherichia coli*. J Antimicrob Chemother 57:339–343

Keyser P, Elofsson M, Rosell S, Wolf-Watz H (2008) Virulence blockers as alternatives to antibiotics: type III secretion inhibitors against Gram-negative bacteria. J Intern Med 264:17–29

Kim OK, Garrity-Ryan LK, Bartlett VJ, Grier MC, Verma AK, Medjanis G, Donatelli JE, Macone AB, Tanaka SK, Levy SB et al (2009) N-hydroxybenzimidazole inhibitors of the transcription factor LcrF in *Yersinia*: novel antivirulence agents. J Med Chem 52:5626–5634

Kim S, Jiao GS, Moayeri M, Crown D, Cregar-Hernandez L, McKasson L, Margosiak SA, Leppla SH, Johnson AT (2011) Antidotes to anthrax lethal factor intoxication. Part 2: structural modifications leading to improved in vivo efficacy. Bioorg Med Chem Lett 21:2030–2033

Kimura K, Iwatsuki M, Nagai T, Matsumoto A, Takahashi Y, Shiomi K, Omura S, Abe A (2011) A small-molecule inhibitor of the bacterial type III secretion system protects against in vivo infection with *Citrobacter rodentium*. J Antibiot (Tokyo) 64:197–203

Kitov PI, Sadowska JM, Mulvey G, Armstrong GD, Ling H, Pannu NS, Read RJ, Bundle DR (2000) Shiga-like toxins are neutralized by tailored multivalent carbohydrate ligands. Nature 403:669–672

Klein T, Abgottspon D, Wittwer M, Rabbani S, Herold J, Jiang X, Kleeb S, Luthi C, Scharenberg M, Bezencon J et al (2010) FimH antagonists for the oral treatment of urinary tract infections: from design and synthesis to in vitro and in vivo evaluation. J Med Chem 53:8627–8641

Kline T, Bowman J, Iglewski BH, de Kievit T, Kakai Y, Passador L (1999) Novel synthetic analogs of the *Pseudomonas* autoinducer. Bioorg Med Chem Lett 9:3447–3452

Kolodkin-Gal I, Romero D, Cao S, Clardy J, Kolter R, Losick R (2010) D-amino acids trigger biofilm disassembly. Science 328:627–629

Kolodkin-Gal I, Cao S, Chai L, Bottcher T, Kolter R, Clardy J, Losick R (2012) A self-produced trigger for biofilm disassembly that targets exopolysaccharide. Cell 149:684–692

Kong KF, Schneper L, Mathee K (2010) Beta-lactam antibiotics: from antibiosis to resistance and bacteriology. APMIS 118:1–36

Koppolu V, Osaka I, Skredenske JM, Kettle B, Hefty PS, Li J, Egan SM (2013) Small-molecule inhibitor of the *Shigella flexneri* master virulence regulator VirF. Infect Immun 81:4220–4231

Kortmann J, Narberhaus F (2012) Bacterial RNA thermometers: molecular zippers and switches. Nat Rev Microbiol 10:255–265

Krachler AM, Orth K (2013) Targeting the bacteria-host interface: strategies in anti-adhesion therapy. Virulence 4:284–294

Kulkarni AA, Fuller C, Korman H, Weiss AA, Iyer SS (2010) Glycan encapsulated gold nanoparticles selectively inhibit shiga toxins 1 and 2. Bioconjug Chem 21:1486–1493

Kulp A, Kuehn MJ (2010) Biological functions and biogenesis of secreted bacterial outer membrane vesicles. Annu Rev Microbiol 64:163–184

Kunsmann L, Ruter C, Bauwens A, Greune L, Gluder M, Kemper B, Fruth A, Wai SN, He X, Lloubes R et al (2015) Virulence from vesicles: novel mechanisms of host cell injury by *Escherichia coli* O104: H4 outbreak strain. Sci Rep 5:13252

Kuzmic P, Cregar L, Millis SZ, Goldman M (2006) Mixed-type noncompetitive inhibition of anthrax lethal factor protease by aminoglycosides. FEBS J 273:3054–3062

Langridge GC, Phan MD, Turner DJ, Perkins TT, Parts L, Haase J, Charles I, Maskell DJ, Peters SE, Dougan G et al (2009) Simultaneous assay of every *Salmonella* Typhi gene using one million transposon mutants. Genome Res 19:2308–2316

LaSarre B, Federle MJ (2013) Exploiting quorum sensing to confuse bacterial pathogens. Microbiol Mol Biol Rev 77:73–111

Layton AN, Hudson DL, Thompson A, Hinton JC, Stevens JM, Galyov EE, Stevens MP (2010) Salicylidene acylhydrazide-mediated inhibition of type III secretion system-1 in *Salmonella enterica* serovar Typhimurium is associated with iron restriction and can be reversed by free iron. FEMS Microbiol Lett 302:114–122

Lee B, Boucher HW (2015) Targeting antimicrobial-resistant bacterial respiratory tract pathogens: it is time to 'get smart'. Curr Opin Pulm Med 21:293–303

Lehner T, Caldwell J, Smith R (1985) Local passive immunization by monoclonal antibodies against streptococcal antigen I/II in the prevention of dental caries. Infect Immun 50:796–799

Lewis K (2005) Persister cells and the riddle of biofilm survival. Biochemistry (Mosc) 70:267–274

Lewis K (2008) Multidrug tolerance of biofilms and persister cells. Curr Top Microbiol Immunol 322:107–131

Li H, Qian L, Chen Z, Thibault D, Liu G, Liu T, Thanassi DG (2004) The outer membrane usher forms a twin-pore secretion complex. J Mol Biol 344:1397–1407

Linden SK, Sheng YH, Every AL, Miles KM, Skoog EC, Florin TH, Sutton P, McGuckin MA (2009) MUC1 limits *Helicobacter pylori* infection both by steric hindrance and by acting as a releasable decoy. PLoS Pathog 5:e1000617

Linington RG, Robertson M, Gauthier A, Finlay BB, van Soest R, Andersen RJ (2002) Caminoside A, an antimicrobial glycolipid isolated from the marine sponge *Caminus sphaeroconia*. Org Lett 4:4089–4092

Loh E, Kugelberg E, Tracy A, Zhang Q, Gollan B, Ewles H, Chalmers R, Pelicic V, Tang CM (2013) Temperature triggers immune evasion by *Neisseria meningitidis*. Nature 502:237–240

Lomovskaya O, Warren MS, Lee A, Galazzo J, Fronko R, Lee M, Blais J, Cho D, Chamberland S, Renau T et al (2001) Identification and characterization of inhibitors of multidrug resistance efflux pumps in *Pseudomonas aeruginosa*: novel agents for combination therapy. Antimicrob Agents Chemother 45:105–116

Lu TK, Collins JJ (2007) Dispersing biofilms with engineered enzymatic bacteriophage. Proc Natl Acad Sci USA 104:11197–11202

Lu C, Maurer CK, Kirsch B, Steinbach A, Hartmann RW (2014a) Overcoming the unexpected functional inversion of a PqsR antagonist in *Pseudomonas aeruginosa*: an in vivo potent antivirulence agent targeting pqs quorum sensing. Angew Chem Int Ed Engl 53:1109–1112

Lu C, Kirsch B, Maurer CK, de Jong JC, Braunshausen A, Steinbach A, Hartmann RW (2014b) Optimization of anti-virulence PqsR antagonists regarding aqueous solubility and biological properties resulting in new insights in structure-activity relationships. Eur J Med Chem 79:173–183

Lucchetti-Miganeh C, Burrowes E, Baysse C, Ermel G (2008) The post-transcriptional regulator CsrA plays a central role in the adaptation of bacterial pathogens to different stages of infection in animal hosts. Microbiology 154:16–29

Lynch SV, Wiener-Kronish JP (2008) Novel strategies to combat bacterial virulence. Curr Opin Crit Care 14:593–599

Ma JK, Hunjan M, Smith R, Lehner T (1989) Specificity of monoclonal antibodies in local passive immunization against *Streptococcus mutans*. Clin Exp Immunol 77:331–337

Madsen JS, Burmolle M, Hansen LH, Sorensen SJ (2012) The interconnection between biofilm formation and horizontal gene transfer. FEMS Immunol Med Microbiol 65:183–195

Manefield M, de Nys R, Kumar N, Read R, Givskov M, Steinberg P, Kjelleberg S (1999) Evidence that halogenated furanones from *Delisea pulchra* inhibit acylated homoserine lactone (AHL)-mediated gene expression by displacing the AHL signal from its receptor protein. Microbiology 145(2):283–291

Manina G, Dhar N, McKinney JD (2015) Stress and host immunity amplify *Mycobacterium tuberculosis* phenotypic heterogeneity and induce nongrowing metabolically active forms. Cell Host Microbe 17:32–46

Maresso AW, Wu R, Kern JW, Zhang R, Janik D, Missiakas DM, Duban ME, Joachimiak A, Schneewind O (2007) Activation of inhibitors by sortase triggers irreversible modification of the active site. J Biol Chem 282:23129–23139

Margalit M, Ash N, Zimran A, Halkin H (2002) Enzyme replacement therapy in the management of longstanding skeletal and soft tissue salmonella infection in a patient with Gaucher's disease. Postgrad Med J 78:564–565

Marshall NC, Finlay BB (2014) Targeting the type III secretion system to treat bacterial infections. Expert Opin Ther Targets 18:137–152

Martinez JL, Fajardo A, Garmendia L, Hernandez A, Linares JF, Martinez-Solano L, Sanchez MB (2009) A global view of antibiotic resistance. FEMS Microbiol Rev 33:44–65

McHugh SM, Hill AD, Humphreys H (2010) Preventing healthcare-associated infection through education: have surgeons been overlooked? Surgeon 8:96–100

McShan AC, De Guzman RN (2015) The bacterial type III secretion system as a target for developing new antibiotics. Chem Biol Drug Des 85:30–42

Mellin JR, Tiensuu T, Becavin C, Gouin E, Johansson J, Cossart P (2013) A riboswitch-regulated antisense RNA in *Listeria monocytogenes*. Proc Natl Acad Sci USA 110:13132–13137

Mellin JR, Koutero M, Dar D, Nahori MA, Sorek R, Cossart P (2014) Riboswitches. Sequestration of a two-component response regulator by a riboswitch-regulated noncoding RNA. Science 345:940–943

Miethke M, Marahiel MA (2007) Siderophore-based iron acquisition and pathogen control. Microbiol Mol Biol Rev 71:413–451

Montecucco C, Tonello F, Zanotti G (2004) Stop the killer: how to inhibit the anthrax lethal factor metalloprotease. Trends Biochem Sci 29:282–285

Mu Y, Shen Z, Jeon B, Dai L, Zhang Q (2013) Synergistic effects of anti-CmeA and anti-CmeB peptide nucleic acids on sensitizing *Campylobacter jejuni* to antibiotics. Antimicrob Agents Chemother 57:4575–4577

Muschiol S, Bailey L, Gylfe A, Sundin C, Hultenby K, Bergstrom S, Elofsson M, Wolf-Watz H, Normark S, Henriques-Normark B (2006) A small-molecule inhibitor of type III secretion inhibits different stages of the infectious cycle of *Chlamydia trachomatis*. Proc Natl Acad Sci USA 103:14566–14571

Nagai H, Kubori T (2011) Type IVB secretion systems of *Legionella* and other Gram-negative bacteria. Front Microbiol 2:136

Negrea A, Bjur E, Ygberg SE, Elofsson M, Wolf-Watz H, Rhen M (2007) Salicylidene acylhydrazides that affect type III protein secretion in *Salmonella enterica* serovar typhimurium. Antimicrob Agents Chemother 51:2867–2876

Neres J, Labello NP, Somu RV, Boshoff HI, Wilson DJ, Vannada J, Chen L, Barry CE 3rd, Bennett EM, Aldrich CC (2008) Inhibition of siderophore biosynthesis in *Mycobacterium tuberculosis* with nucleoside bisubstrate analogues: structure-activity relationships of the nucleobase domain of 5′-O-[N-(salicyl)sulfamoyl]adenosine. J Med Chem 51:5349–5370

Nestorovich EM, Bezrukov SM (2014) Designing inhibitors of anthrax toxin. Expert Opin Drug Discov 9:299–318

Ng WL, Bassler BL (2009) Bacterial quorum-sensing network architectures. Annu Rev Genet 43:197–222

Nikaido H, Pages JM (2012) Broad-specificity efflux pumps and their role in multidrug resistance of Gram-negative bacteria. FEMS Microbiol Rev 36:340–363

Nishikawa K, Matsuoka K, Watanabe M, Igai K, Hino K, Hatano K, Yamada A, Abe N, Terunuma D, Kuzuhara H et al (2005) Identification of the optimal structure required for a Shiga toxin neutralizer with oriented carbohydrates to function in the circulation. J Infect Dis 191:2097–2105

Nordfelth R, Kauppi AM, Norberg HA, Wolf-Watz H, Elofsson M (2005) Small-molecule inhibitors specifically targeting type III secretion. Infect Immun 73:3104–3114

Nuss AM, Heroven AK, Waldmann B, Reinkensmeier J, Jarek M, Beckstette M, Dersch P (2015) Trans-criptomic profiling of *Yersinia pseudotuberculosis* reveals reprogramming of the Crp regulon by temperature and uncovers Crp as a master regulator of small RNAs. PLoS Genet 11: e1005087

O'Connell KM, Hodgkinson JT, Sore HF, Welch M, Salmond GP, Spring DR (2013) Combating multidrug-resistant bacteria: current strategies for the discovery of novel antibacterials. Angew Chem Int Ed Engl 52:10706–10733

Ofek I, Hasty DL, Sharon N (2003) Anti-adhesion therapy of bacterial diseases: prospects and problems. FEMS Immunol Med Microbiol 38:181–191

Okshevsky M, Regina VR, Meyer RL (2015) Extracellular DNA as a target for biofilm control. Curr Opin Biotechnol 33:73–80

Oliva G, Sahr T, Buchrieser C (2015) Small RNAs, 5′ UTR elements and RNA-binding proteins in intracellular bacteria: impact on metabolism and virulence. FEMS Microbiol Rev 39:331–349

Palzkill T (2013) Metallo-beta-lactamase structure and function. Ann N Y Acad Sci 1277:91–104

Pan NJ, Brady MJ, Leong JM, Goguen JD (2009) Targeting type III secretion in *Yersinia pestis*. Antimicrob Agents Chemother 53:385–392

Panchal RG, Hermone AR, Nguyen TL, Wong TY, Schwarzenbacher R, Schmidt J, Lane D, McGrath C, Turk BE, Burnett J et al (2004) Identification of small molecule inhibitors of anthrax lethal factor. Nat Struct Mol Biol 11:67–72

Pannek S, Higgins PG, Steinke P, Jonas D, Akova M, Bohnert JA, Seifert H, Kern WV (2006) Multidrug efflux inhibition in *Acinetobacter baumannii*: comparison between 1-(1-naphthylmethyl)-piperazine and phenyl-arginine-beta-naphthylamide. J Antimicrob Chemother 57:970–974

Papenfort K, Vogel J (2014) Small RNA functions in carbon metabolism and virulence of enteric pathogens. Front Cell Infect Microbiol 4:91

Park S, Kim HS, Ok K, Kim Y, Park HD, Byun Y (2015) Design, synthesis and biological evaluation of 4-(alkyloxy)-6-methyl-2H-pyran-2-one derivatives as quorum sensing inhibitors. Bioorg Med Chem Lett 25:2913–2917

Parker P, Sando L, Pearson R, Kongsuwan K, Tellam RL, Smith S (2010) Bovine Muc1 inhibits binding of enteric bacteria to Caco-2 cells. Glycoconj J 27:89–97

Parsek MR, Val DL, Hanzelka BL, Cronan JE Jr, Greenberg EP (1999) Acyl homoserine-lactone quorum-sensing signal generation. Proc Natl Acad Sci USA 96:4360–4365

Paschos A, den Hartigh A, Smith MA, Atluri VL, Sivanesan D, Tsolis RM, Baron C (2011) An in vivo high-throughput screening approach targeting the type IV secretion system component VirB8 identified inhibitors of Brucella abortus 2308 proliferation. Infect Immun 79:1033–1043

Paulasova P, Pellestor F (2004) The peptide nucleic acids (PNAs): a new generation of probes for genetic and cytogenetic analyses. Ann Genet 47:349–358

Pearson JS, Zhang Y, Newton HJ, Hartland EL (2015) Post-modern pathogens: surprising activities of translocated effectors from *E. coli* and *Legionella*. Curr Opin Microbiol 23:73–79

Peschel A, Otto M, Jack RW, Kalbacher H, Jung G, Gotz F (1999) Inactivation of the *dlt* operon in *Staphylococcus aureus* confers sensitivity to defensins, protegrins, and other antimicrobial peptides. J Biol Chem 274:8405–8410

Peselis A, Serganov A (2014) Themes and variations in riboswitch structure and function. Biochim Biophys Acta 1839:908–918

Pinkner JS, Remaut H, Buelens F, Miller E, Aberg V, Pemberton N, Hedenstrom M, Larsson A, Seed P, Waksman G et al (2006) Rationally designed small compounds inhibit pilus biogenesis in uropathogenic bacteria. Proc Natl Acad Sci USA 103:17897–17902

Putrins M, Kogermann K, Lukk E, Lippus M, Varik V, Tenson T (2015) Phenotypic heterogeneity enables uropathogenic *Escherichia coli* to evade killing by antibiotics and serum complement. Infect Immun 83:1056–1067

Qin Z, Ou Y, Yang L, Zhu Y, Tolker-Nielsen T, Molin S, Qu D (2007) Role of autolysin-mediated DNA release in biofilm formation of *Staphylococcus epidermidis*. Microbiology 153:2083–2092

Rasko DA, Sperandio V (2010) Anti-virulence strategies to combat bacteria-mediated disease. Nat Rev Drug Discov 9:117–128

Rasko DA, Moreira CG, de Li R, Reading NC, Ritchie JM, Waldor MK, Williams N, Taussig R, Wei S, Roth M et al (2008) Targeting QseC signaling and virulence for antibiotic development. Science 321:1078–1080

Renau TE, Leger R, Flamme EM, Sangalang J, She MW, Yen R, Gannon CL, Griffith D, Chamberland S, Lomovskaya O et al (1999) Inhibitors of efflux pumps in *Pseudomonas aeruginosa* potentiate the activity of the fluoroquinolone antibacterial levofloxacin. J Med Chem 42:4928–4931

Rice LB (2008) Federal funding for the study of antimicrobial resistance in nosocomial pathogens: no ESKAPE. J Infect Dis 197:1079–1081

Rogers MS, Cryan LM, Habeshian KA, Bazinet L, Caldwell TP, Ackroyd PC, Christensen KA (2012) A FRET-based high throughput screening assay to identify inhibitors of anthrax protective antigen binding to capillary morphogenesis gene 2 protein. PLoS ONE 7:e39911

Romeo T, Vakulskas CA, Babitzke P (2013) Post-transcriptional regulation on a global scale: form and function of Csr/Rsm systems. Environ Microbiol 15:313–324

Romling U, Balsalobre C (2012) Biofilm infections, their resilience to therapy and innovative treatment strategies. J Intern Med 272:541–561

Saar-Dover R, Bitler A, Nezer R, Shmuel-Galia L, Firon A, Shimoni E, Trieu-Cuot P, Shai Y (2012) D-alanylation of lipoteichoic acids confers resistance to cationic peptides in group B *streptococcus* by increasing the cell wall density. PLoS Pathog 8:e1002891

Salyers AA, Gupta A, Wang Y (2004) Human intestinal bacteria as reservoirs for antibiotic resistance genes. Trends Microbiol 12:412–416

Santa Maria JP Jr, Sadaka A, Moussa SH, Brown S, Zhang YJ, Rubin EJ, Gilmore MS, Walker S (2014) Compound-gene interaction mapping reveals distinct roles for *Staphylococcus aureus* teichoic acids. Proc Natl Acad Sci USA 111:12510–12515

Schmitt CK, Meysick KC, O'Brien AD (1999) Bacterial toxins: friends or foes? Emerg Infect Dis 5:224–234

Schroeder GN, Hilbi H (2008) Molecular pathogenesis of *Shigella* spp.: controlling host cell signaling, invasion, and death by type III secretion. Clin Microbiol Rev 21:134–156

Schumacher A, Steinke P, Bohnert JA, Akova M, Jonas D, Kern WV (2006) Effect of 1-(1-naphthylmethyl)-piperazine, a novel putative efflux pump inhibitor, on antimicrobial drug susceptibility in clinical isolates of *Enterobacteriaceae* other than *Escherichia coli*. J Antimicrob Chemother 57:344–348

Scobie HM, Thomas D, Marlett JM, Destito G, Wigelsworth DJ, Collier RJ, Young JA, Manchester M (2005) A soluble receptor decoy protects rats against anthrax lethal toxin challenge. J Infect Dis 192:1047–1051

Serganov A, Nudler E (2013) A decade of riboswitches. Cell 152:17–24

Shakhnovich EA, Hung DT, Pierson E, Lee K, Mekalanos JJ (2007) Virstatin inhibits dimerization of the transcriptional activator ToxT. Proc Natl Acad Sci USA 104:2372–2377

Shoop WL, Xiong Y, Wiltsie J, Woods A, Guo J, Pivnichny JV, Felcetto T, Michael BF, Bansal A, Cummings RT et al (2005) Anthrax lethal factor inhibition. Proc Natl Acad Sci USA 102:7958–7963

Silberstein C, Lucero MS, Zotta E, Copeland DP, Lingyun L, Repetto HA, Ibarra C (2011) A glucosylceramide synthase inhibitor protects rats against the cytotoxic effects of shiga toxin 2. Pediatr Res 69:390–394

Skredenske JM, Koppolu V, Kolin A, Deng J, Kettle B, Taylor B, Egan SM (2013) Identification of a small-molecule inhibitor of bacterial AraC family activators. J Biomol Screen 18:588–598

Slepenkin A, Enquist PA, Hagglund U, de la Maza LM, Elofsson M, Peterson EM (2007) Reversal of the antichlamydial activity of putative type III secretion inhibitors by iron. Infect Immun 75:3478–3489

Smith KM, Bu Y, Suga H (2003a) Induction and inhibition of *Pseudomonas aeruginosa* quorum sensing by synthetic autoinducer analogs. Chem Biol 10:81–89

Smith KM, Bu Y, Suga H (2003b) Library screening for synthetic agonists and antagonists of a *Pseudomonas aeruginosa* autoinducer. Chem Biol 10:563–571

Smits WK, Kuipers OP, Veening JW (2006) Phenotypic variation in bacteria: the role of feedback regulation. Nat Rev Microbiol 4:259–271

Soheili V, Bazzaz BS, Abdollahpour N, Hadizadeh F (2015) Investigation of *Pseudomonas aeruginosa* quorum-sensing signaling system for identifying multiple inhibitors using molecular docking and structural analysis methodology. Microb Pathog 89:73–78

Stephenson K, Yamaguchi Y, Hoch JA (2000) The mechanism of action of inhibitors of bacterial two-component signal transduction systems. J Biol Chem 275:38900–38904

Sturm A, Heinemann M, Arnoldini M, Benecke A, Ackermann M, Benz M, Dormann J, Hardt WD (2011) The cost of virulence: retarded growth of *Salmonella* Typhimurium cells expressing type III secretion system 1. PLoS Pathog 7:e1002143

Sully EK, Malachowa N, Elmore BO, Alexander SM, Femling JK, Gray BM, DeLeo FR, Otto M, Cheung AL, Edwards BS et al (2014) Selective chemical inhibition of *agr* quorum sensing in *Staphylococcus aureus* promotes host defense with minimal impact on resistance. PLoS Pathog 10:e1004174

Suzuki N, Ohtaguro N, Yoshida Y, Hirai M, Matsuo H, Yamada Y, Imamura N, Tsuchiya T (2015) A compound inhibits biofilm formation of *Staphylococcus aureus* from *Streptomyces*. Biol Pharm Bull 38:889–892

Svensson M, Frendeus B, Butters T, Platt F, Dwek R, Svanborg C (2003) Glycolipid depletion in antimicrobial therapy. Mol Microbiol 47:453–461

Tarr PI, Gordon CA, Chandler WL (2005) Shiga-toxin-producing *Escherichia coli* and haemolytic uraemic syndrome. Lancet 365:1073–1086

Thanassi DG, Bliska JB, Christie PJ (2012) Surface organelles assembled by secretion systems of Gram-negative bacteria: diversity in structure and function. FEMS Microbiol Rev 36:1046–1082

Toledo-Arana A, Dussurget O, Nikitas G, Sesto N, Guet-Revillet H, Balestrino D, Loh E, Gripenland J, Tiensuu T, Vaitkevicius K et al (2009) The *Listeria* transcriptional landscape from saprophytism to virulence. Nature 459:950–956

Tonello F, Seveso M, Marin O, Mock M, Montecucco C (2002) Screening inhibitors of anthrax lethal factor. Nature 418:386

Totir MA, Helfand MS, Carey MP, Sheri A, Buynak JD, Bonomo RA, Carey PR (2007) Sulbactam forms only minimal amounts of irreversible acrylate-enzyme with SHV-1 beta-lactamase. Biochemistry 46:8980–8987

Trachtman H, Cnaan A, Christen E, Gibbs K, Zhao S, Acheson DW, Weiss R, Kaskel FJ, Spitzer A, Hirschman GH et al (2003) Effect of an oral Shiga toxin-binding agent on diarrhea-associated hemolytic uremic syndrome in children: a randomized controlled trial. JAMA 290:1337–1344

Tree JJ, Wang D, McInally C, Mahajan A, Layton A, Houghton I, Elofsson M, Stevens MP, Gally DL, Roe AJ (2009) Characterization of the effects of salicylidene acylhydrazide compounds on type III secretion in *Escherichia coli* O157: H7. Infect Immun 77:4209–4220

Trokter M, Felisberto-Rodrigues C, Christie PJ, Waksman G (2014) Recent advances in the structural and molecular biology of type IV secretion systems. Curr Opin Struct Biol 27:16–23

Tsou LK, Dossa PD, Hang HC (2013) Small molecules aimed at type III secretion systems to inhibit bacterial virulence. Medchemcomm 4:68–79

Turk BE, Wong TY, Schwarzenbacher R, Jarrell ET, Leppla SH, Collier RJ, Liddington RC, Cantley LC (2004) The structural basis for substrate and inhibitor selectivity of the anthrax lethal factor. Nat Struct Mol Biol 11:60–66

Vakulskas CA, Potts AH, Babitzke P, Ahmer BM, Romeo T (2015) Regulation of bacterial virulence by Csr (Rsm) systems. Microbiol Mol Biol Rev 79:193–224

Vanella de Cuetos EE, Fernandez RA, Bianco MI, Sartori OJ, Piovano ML, Luquez C, de Jong LI (2011) Equine botulinum antitoxin for the treatment of infant botulism. Clin Vaccine Immunol 18:1845–1849

Vanier G, Sekizaki T, Dominguez-Punaro MC, Esgleas M, Osaki M, Takamatsu D, Segura M, Gottschalk M (2008) Disruption of *srtA* gene in *Streptococcus suis* results in decreased interactions with endothelial cells and extracellular matrix proteins. Vet Microbiol 127:417–424

Veenendaal AK, Sundin C, Blocker AJ (2009) Small-molecule type III secretion system inhibitors block assembly of the *Shigella* type III secreton. J Bacteriol 191:563–570

Vimberg V, Kuzma M, Stodulkova E, Novak P, Bednarova L, Sulc M, Gazak R (2015) Hydnocarpin-type Flavonolignans: semisynthesis and inhibitory effects on *Staphylococcus aureus* biofilm formation. J Nat Prod 78:2095–2103

Vogel J, Luisi BF (2011) Hfq and its constellation of RNA. Nat Rev Microbiol 9:578–589

Vogt SL, Raivio TL (2012) Just scratching the surface: an expanding view of the Cpx envelope stress response. FEMS Microbiol Lett 326:2–11

Voth DE, Broederdorf LJ, Graham JG (2012) Bacterial type IV secretion systems: versatile virulence machines. Future Microbiol 7:241–257

Waksman G, Orlova EV (2014) Structural organisation of the type IV secretion systems. Curr Opin Microbiol 17:24–31

Wang D, Zetterstrom CE, Gabrielsen M, Beckham KS, Tree JJ, Macdonald SE, Byron O, Mitchell TJ, Gally DL, Herzyk P et al (2011) Identification of bacterial target proteins for the salicylidene acylhydrazide class of virulence-blocking compounds. J Biol Chem 286:29922–29931

Weber GG, Kortmann J, Narberhaus F, Klose KE (2014) RNA thermometer controls temperature-dependent virulence factor expression in *Vibrio cholerae*. Proc Natl Acad Sci USA 111:14241–14246

Weigel WA, Demuth DR (2015) QseBC, a two component bacterial adrenergic receptor and global regulator of virulence in *Enterobacteriaceae* and *Pasteurellaceae*. Mol Oral Microbiol. doi:10.1111/omi.12138. [Epub ahead of print]

Westermann AJ, Gorski SA, Vogel J (2012) Dual RNA-seq of pathogen and host. Nat Rev Microbiol 10:618–630

Wilke KE, Francis S, Carlson EE (2015) Inactivation of multiple bacterial histidine kinases by targeting the ATP-binding domain. ACS Chem Biol 10:328–335

Wolf K, Betts HJ, Chellas-Gery B, Hower S, Linton CN, Fields KA (2006) Treatment of *Chlamydia trachomatis* with a small molecule inhibitor of the *Yersinia* type III secretion system disrupts progression of the chlamydial developmental cycle. Mol Microbiol 61:1543–1555

Wong AR, Pearson JS, Bright MD, Munera D, Robinson KS, Lee SF, Frankel G, Hartland EL (2011) Enteropathogenic and enterohaemorrhagic *Escherichia coli*: even more subversive elements. Mol Microbiol 80:1420–1438

Worthington RJ, Blackledge MS, Melander C (2013) Small-molecule inhibition of bacterial two-component systems to combat antibiotic resistance and virulence. Future Med Chem 5:1265–1284

Wright KJ, Hultgren SJ (2006) Sticky fibers and uropathogenesis: bacterial adhesins in the urinary tract. Future Microbiol 1:75–87

Wright JS 3rd, Jin R, Novick RP (2005) Transient interference with staphylococcal quorum sensing blocks abscess formation. Proc Natl Acad Sci USA 102:1691–1696

Xiong Y, Wiltsie J, Woods A, Guo J, Pivnichny JV, Tang W, Bansal A, Cummings RT, Cunningham BR, Friedlander AM et al (2006) The discovery of a potent and selective lethal factor inhibitor for adjunct therapy of anthrax infection. Bioorg Med Chem Lett 16:964–968

Zahid MS, Awasthi SP, Asakura M, Chatterjee S, Hinenoya A, Faruque SM, Yamasaki S (2015) Suppression of virulence of toxigenic *Vibrio cholerae* by anethole through the cyclic AMP (cAMP)-cAMP receptor protein signaling system. PLoS ONE 10:e0137529

Zambelloni R, Marquez R, Roe AJ (2015) Development of antivirulence compounds: a biochemical review. Chem Biol Drug Des 85:43–55

Zetterstrom CE, Hasselgren J, Salin O, Davis RA, Quinn RJ, Sundin C, Elofsson M (2013) The resveratrol tetramer (-)-hopeaphenol inhibits type III secretion in the gram-negative pathogens *Yersinia pseudotuber-culosis* and *Pseudomonas aeruginosa*. PLoS ONE 8:e81969

Zhang C, Zhang W (2010) *Escherichia coli* K88ac fimbriae expressing heat-labile and heat-stable (STa) toxin epitopes elicit antibodies that neutralize cholera toxin and STa toxin and inhibit adherence of K88ac fimbrial *E. coli*. Clin Vaccine Immunol 17:1859–1867

Zhang J, Liu H, Zhu K, Gong S, Dramsi S, Wang YT, Li J, Chen F, Zhang R, Zhou L et al (2014) Antiinfective therapy with a small molecule inhibitor of *Staphylococcus aureus* sortase. Proc Natl Acad Sci USA 111:13517–13522

Strategies to Block Bacterial Pathogenesis by Interference with Motility and Chemotaxis

Marc Erhardt

Abstract Infections by motile, pathogenic bacteria, such as *Campylobacter* species, *Clostridium* species, *Escherichia coli*, *Helicobacter pylori*, *Listeria monocytogenes*, *Neisseria gonorrhoeae*, *Pseudomonas aeruginosa*, *Salmonella* species, *Vibrio cholerae*, and *Yersinia* species, represent a severe economic and health problem worldwide. Of special importance in this context is the increasing emergence and spread of multidrug-resistant bacteria. Due to the shortage of effective antibiotics for the treatment of infections caused by multidrug-resistant, pathogenic bacteria, the targeting of novel, virulence-relevant factors constitutes a promising, alternative approach. Bacteria have evolved distinct motility structures for movement across surfaces and in aqueous environments. In this review, I will focus on the bacterial flagellum, the associated chemosensory system, and the type-IV pilus as motility devices, which are crucial for bacterial pathogens to reach a preferred site of infection, facilitate biofilm formation, and adhere to surfaces or host cells. Thus, those nanomachines constitute potential targets for the development of novel anti-infectives that are urgently needed at a time of spreading antibiotic resistance. Both bacterial flagella and type-IV pili (T4P) are intricate macromolecular complexes made of dozens of different proteins and their motility function relies on the correct spatial and temporal assembly of various substructures. Specific type-III and type IV secretion systems power the export of substrate proteins of the bacterial flagellum and type-IV pilus, respectively, and are homologous to virulence-associated type-III and type-II secretion systems. Accordingly, bacterial flagella and T4P represent attractive targets for novel antivirulence drugs interfering with synthesis, assembly, and function of these motility structures.

M. Erhardt (✉)
Helmholtz Centre for Infection Research, Inhoffenstr. 7, 38124 Braunschweig, Germany
e-mail: marc.erhardt@helmholtz-hzi.de

Current Topics in Microbiology and Immunology (2016) 398: 185–206
DOI 10.1007/82_2016_493

Published Online: 15 March 2016

Contents

1 Introduction—Bacterial Motility and Chemotaxis Systems

Prokaryotic organisms have evolved different mechanisms for movement through liquid or across solid environments, which are categorized into swimming, swarming, gliding, or twitching motility. Many bacterial pathogens utilize various forms of motility during some stage of the infection cycle in order to reach the site of infection, facilitate pathogen–host interactions, and promote biofilm formation.

The different forms of motility are mediated by a diverse set of motility structures—extracellular appendages as well as intracellular motors—in addition to complex chemosensory systems, which together allow for active and directed movement.

1.1 Flagella-Mediated Motility and Chemotactic Behavior

The best-studied motility structure is the flagellum of Gram-positive and Gram-negative bacteria (Fig. 1a). Flagella are rigid motility organelles, and rotation of flagellar filaments provides the bacteria with propulsion forces for chemotactic movement through liquids (swimming) and highly viscous environments or surfaces (swarming) (Berg and Anderson 1973; Chevance and Hughes 2008). For many pathogens, flagella play also a crucial role in bacterial pathogenesis by promoting bacteria–host interactions and biofilm formation.

The term "swimming motility" is used for movement through aqueous environments by the means of flagella and has been extensively studied in the peritrichously flagellated enteric bacteria *E. coli* and *Salmonella enterica serovar* Typhimurium (Chevance and Hughes 2008).

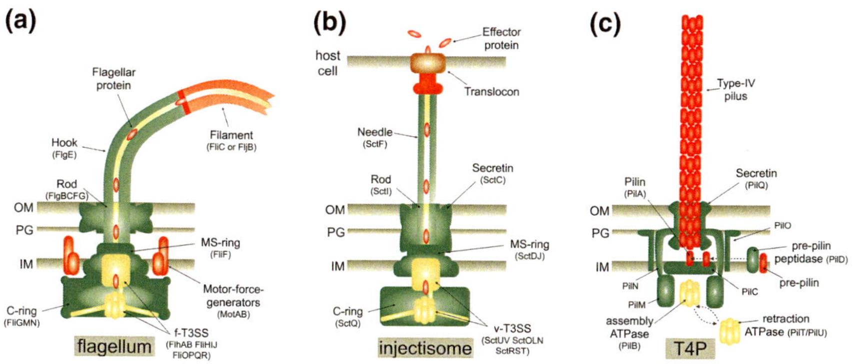

Fig. 1 Schematic drawings of the bacterial flagellum, the virulence-associated injectisome, and T4P. **a** Schematic overview of the bacterial flagellum. The structure of the flagellum consists of three parts: (i) a basal body with a f-T3SS within the inner membrane ring; (ii) a flexible hook acting as a universal joint to (iii) the rigid filament. **b** Schematic overview of the virulence-associated injectisome. Many components of the flagellum and injectisome are structurally and/or functionally related, including the T3SS. **c** Schematic overview of the T4P complex consisting of a flexible pilus (PilA), a channel in the outer membrane (secretin, PilQ), a membrane complex (PilC), an assembly platform (PilM, PilN, and PilO), and ATPase proteins driving pilus assembly (PilB) or retraction (PilT, PilU). The prepilin peptidase PilD processes PilA prior to export. *OM* outer membrane; *PG* peptidoglycan (cell wall); *IM* inner membrane

Many other bacteria move also through liquid environments using a wide variety of polar, lateral, or—in case of spirochetes—periplasmic flagella. In addition, *Archaea* swim using flagella-like organelles, but the so-called archaellum evolved convergently to the flagellum of eubacteria.

Movement across solid surfaces—termed “swarming motility”—is also powered by bacterial flagella (Harshey 2003; Kearns 2010). The transition from a planktonic to a surface-associated form of motility is frequently associated with changes in cell morphology and number of flagella per cell or even mediated by another, distinct set of flagella, such as in the case of *Vibrio parahaemolyticus* (Jarrell and McBride 2008; Merino et al. 2006).

The different forms of flagella-mediated motility allow for directed movement in concert with sensing of environmental signals by a battery of chemoreceptors. This chemotactic behavior enables motile bacteria to move toward favorable chemicals (attractants) and away from harmful substances (repellents) and thus facilitates survival in rapidly changing environments. The information about attractant and repellent concentrations is sensed by membrane-bound chemoreceptors and forwarded via a two-component phosphorelay system to the flagellar motor (Hazelbauer et al. 2008). Depending on the phosphorylation state of the response regulator, the forwarded signal determines the rotational direction of the flagellar motor. Counterclockwise (CCW) rotation of all flagella of a cell allows the formation of a flagellar bundle at one cell pole that functions as a propeller and pushes the cell forward. In the absence of attractant or in the presence of repellent, the response regulator is phosphorylated and binds to the flagellar motor to change rotational direction. The change in rotational direction toward clockwise of one or more flagella disrupts the flagellar bundle and causes the cell to tumble. A random reorientation of the cell pole during the tumble event results in a change in direction on the subsequent run cycle. The flagellar motor rotates in the presence of attractant or the absence of repellent in a CCW direction, and thus, the system is biased toward the run phase. For a directed chemotactic movement, a sequence of run and tumble events result in a random walk in the direction of a chemical gradient of attractant (Macnab and Koshland 1972; Tsang et al. 1973; Brown and Berg 1974). Importantly, such chemotactic behavior plays a crucial role in virulence of many pathogenic bacteria, including *P. aeruginosa* and *H. pylori* (Williams et al. 2007; Garvis et al. 2009).

1.2 Structure and Function of the Bacterial Flagellum

The flagellum of *E. coli* and *S.* Typhimurium is a sophisticated nanomachine made of more than 30 different proteins and composed of three main parts: (i) a basal body that is embedded in the cytoplasmic membrane and traverses the periplasm and cell wall up to the outer membrane (the motor); (ii) a long external filament (the propeller); and (iii) a flexible, curved structure that connects the basal body and the rigid filament (the hook) (Chevance and Hughes 2008) (Fig. 1a). Most components of the flagellum self-assemble outside the cytoplasm in a tightly structured process. The self-assembly of the flagellum relies on multiple mechanisms to achieve the accurate size and subunit composition, including spatiotemporal coordination of gene expression, hierarchical sorting and export of substrates, and length control of substructures.

The structural subunits of the flagellum need to be exported through a narrow channel within the flagellar structure. This protein secretion process depends on a flagella-specific type-III secretion system (f-T3SS) embedded within the basal body structure that uses energy derived from the proton-motive force (pmf) and ATP hydrolysis for the protein secretion process (Paul et al. 2008; Minamino and Namba 2008). Structurally and functionally closely related to the f-T3SS is the virulence-associated type-III secretion system (v-T3SS) of the injectisome devices employed by many Gram-negative pathogens to invade host cells (Abby and Rocha 2012) (Fig. 1b).

The assembly of a flagellum initiates with the formation of the inner membrane (MS-ring) and cytoplasmic C-ring components (Zhao et al. 1996a, b). The C-ring interacts with phosphorylated CheY response regulator of the chemosensory system to switch rotational direction of the flagellar motor dependent on the input of environmental stimuli (Toker and Macnab 1997) and with both the MS-ring and the stator proteins to generate the torque for flagellar rotation (Francis et al. 1992; Lloyd et al. 1996; Kubori et al. 1997). The membrane-embedded components of the f-T3SS assemble within the central pore of the MS-ring. The associated, flagella-specific cytoplasmic ATPase interacts with both inner membrane components of the f-T3SS and the C-ring, presumably to energize export substrate unfolding and/or chaperone release (Aizawa 1996; Hara et al. 2012; Ibuki et al. 2013; Abrusci et al. 2013). Pmf-dependent secretion of substrates occurs through a narrow 2.5-nm-wide channel, indicating that substrate proteins are translocated in an unfolded or partially folded state (Yonekura et al. 2003; Paul et al. 2008; Minamino and Namba 2008). While being essential under physiological conditions, the requirement of ATPase activity for the export process can be overcome by an increase in pmf or enhanced substrate protein levels, indicating that flagellar protein export has evolved for maximal speed and efficiency (Erhardt et al. 2014).

Assembly of the flagellum is dependent on self-assembly of subunits at the growing tip of the structure after traversing the narrow secretion channel (Chevance and Hughes 2008). The f-T3SS secretes most extracytoplasmic components of the flagellum, starting with components of the rod that spans the periplasmic space. After the formation of the PL-rings in the periplasmic space/outer membrane, the extracellular hook structure assembles. A molecular ruler protein controls the final length of the hook structure (Erhardt et al. 2010, 2011) and induces a switch in secretion specificity from early (rod–hook-type) substrates to late (filament-type) secretion substrates upon measuring the final hook length (Ferris and Minamino 2006; Minamino et al. 2006). Only after the switch in secretion specificity, the f-T3SS commences export of late substrates. The class of late secretion substrates includes the anti-σ28 factor, FlgM, and thereby couples the assembly state of the flagellum to gene expression (Hughes et al. 1993). The filament subunits are also only made and secreted after a completed hook basal body complex exists, onto top of which they can assembly to complete a functional flagellum (Chevance and Hughes 2008). Needed for pmf-dependent rotation of the flagellum are the stator protein complexes, which assemble in the inner membrane, attach non-covalently to the peptidoglycan layer, and interact with the C-ring for torque generation (Manson et al. 1977; Chun and Parkinson 1988; Blair and Berg 1990).

1.3 Twitching Motility and Gliding Motility

Twitching motility and gliding motility are other forms of motility across solid surfaces, which are independent of bacterial flagella, but are mediated instead by type-IV pili (T4P) and the less characterized gliding machinery.

'Twitching motility' is a form of movement across solid surfaces powered by extension, attachment, and retraction of T4P in diverse bacteria, including *Myxococcus xanthus*, cyanobacteria, and pathogenic bacteria such as *N. gonorrhoeae*, *P. aeruginosa*, *Clostridium perfringens*, and other *Clostridium* species (Mattick 2002; Varga et al. 2006; Jarrell and McBride 2008). The retraction of pili can be in a smooth or jerky motion, which originated the term "twitching motility." Motility mediated by a smooth retraction of pili has originally been described as "gliding motility"; however, there are several examples of bacteria that move in a gliding motion across surfaces in the absence of T4P. Therefore, the term "gliding motility" is now used to describe the movement across solid surfaces by distinct and less characterized motility structures. This form of motility is mediated by pmf-powered motors that are embedded in the cell envelope and propel cell surface adhesins (Nakane et al. 2013). Interestingly, the T4P-independent form of gliding motility is characteristic for many Bacteroidetes species (McBride 2001; McBride and Zhu 2013).

The T4P machinery is also implicated in translocation of virulence effector proteins across the bacterial cell envelope and/or into host cells in many pathogens, such as *Agrobacterium tumefaciens*, *Brucella*, *H. pylori*, and *Legionella pneumophila*. T4P further mediate DNA uptake by the process of transformation, as well as conjugative DNA transfer, and thus contribute to the spread of antibiotic resistances (Fronzes et al. 2009).

It is also important to note that the type-II secretion system used for secretion of virulence factors by many bacterial pathogens, including pathogenic *E. coli*, *Klebsiella*, *P. aeruginosa*, and *V. cholerae*, is evolutionary and functionally closely related to the T4P machinery (Peabody et al. 2003; Johnson et al. 2006; Fronzes et al. 2009).

2 The Importance of Bacterial Motility and Chemotaxis in Virulence

Flagella-mediated motility plays an important role for *Salmonella* and other bacterial pathogens at various stages of the infection. In case of *S.* Typhimurium, flagella-mediated movement in the intestinal lumen is required for efficient colonization and induction of colitis in the murine gastroenteritis model by allowing the pathogen to benefit from the increased nutrient availability in the inflamed intestine (Stecher et al. 2004, 2008). The chemotactic behavior is also an important virulence factor for colonization of other hosts, including chicken and rabbit (Allen-Vercoe and Woodward 1999; Marchetti et al. 2004). In addition, a majority of *Salmonella* clinical isolates displayed increased motility if compared to a laboratory strain (Martins et al. 2013). *Salmonella* Typhi is known to require motility for virulence (Ottemann and Miller 1997), and non-motile mutants of *S.* Typhimurium have a reduced efficiency to invade cultured epithelial cells in vitro, unless the bacteria are brought into contact with eukaryotic cells by centrifugation (Finlay and Falkow 1989). Flagellum-mediated motility is also important for pathogenicity of *Campylobacter jejuni* (Black et al. 1988), *Yersinia enterocolitica* virulence and biofilm formation (Young et al. 2000; Kim et al. 2008), mucus colonization by *V. cholerae* (Krukonis and DiRita 2003), and wound infection by *P. aeruginosa* (Arora et al. 2005).

Flagella-mediated motility does not only play an important role in reaching a preferred site of infection, but has also other functions during pathogenicity, in particular in adherence, biofilm formation, and immune system modulation (Rossez et al. 2015).

In this context, flagellin—the major component of the bacterial flagellum—is a prime target for recognition by the immune system. Therefore, interactions between invading bacteria and its cognate host lead to the development of protective immunity against the pathogens. Initially, the innate immune system deals with an infection by alarming the cellular and molecular defense mechanisms of the host and facilitating the transition to adaptive immunity (Iwasaki and Medzhitov 2010). Infection is sensed by discriminating "non-self" from "self"; therefore, the innate immune system is equipped with a series of germline-encoded receptors recognizing the conserved pathogen-associated molecular patterns (PAMPs) that are present in microorganisms but not in host cells. Among these pattern recognition receptors (PRRs), the Toll-like receptor (TLR) family is best characterized and able to recognize lipids, lipoproteins, proteins, and nucleic acids derived from a wide range of microorganisms (Kawai and Akira 2010). In this context, TLR5 recognizes conserved regions in the flagellar protein FliC (flagellin) (Andersen-Nissen et al. 2005).

In addition, another set of receptors exists in the intracellular compartment that senses a microbial threat from the inside. Among these receptors, the NOD-like protein NLRC4 is able to recognize flagellin, which has been secreted directly into host cells by *Salmonella* (Miao et al. 2006). This interaction leads to the formation of the proinflammatory multiprotein inflammasome complex and results in the activation of caspase-1 and the subsequent maturation of IL-1β and IL-18 (Mariathasan and Monack 2007). Caspase-1 activation can also lead to inflammatory cell death called pyroptosis, which contributes to the control of bacterial infection (Miao et al. 2006). Finally, it was demonstrated that NLRC4 discriminates between pathogenic *Salmonella* and commensal microbiota to promote immunity against infectious agents but not against the intestinal microflora (Franchi et al. 2012). Thus, components of the flagellum are not only central to bacterial movement and host cell invasion, but are also essential for immune recognition and control of bacterial infection.

In addition to the chemotactic behavior used for a directed movement toward the site of infection and modulation of the immune system response, flagellar motility contributes to pathogenesis by promoting adherence to host cell tissue and biofilm formation (Haiko and Westerlund-Wikstrom 2013; Rossez et al. 2015). Flagella are needed for adherence, colonization, and subsequent invasion of host cell tissue for several bacterial pathogens. For instance, invasion of Caco-2 cells by *L. monocytogenes* or *S.* Enteritidis is dependent on the presence of flagella (Dons et al. 2004; van Asten et al. 2004), flagella purified from enteropathogenic *E. coli* (EPEC) bind directly to HeLa cells (Giron et al. 2002), and flagellin of enterotoxigenic *E. coli* (ETEC) interacts with the two-partner secretion adhesins EtpA to mediate adherence and colonization (Roy et al. 2009).

The presence of flagella and the initial interaction of flagella with surface structures are crucial for the formation of biofilms and to maintain microcolony structures via physical interactions (Pratt and Kolter 1998). In addition, flagella are able to reach into crevices and thereby allow bacteria to access and probe surface structures during the biofilm formation by enabling attachment to unfavorable surfaces (Friedlander et al. 2013).

3 Strategies to Interfere with Assembly and Function of Motility Structures

Motility structures of bacteria are complex, multicomponent nanomachines. Assembly of both bacterial flagella and T4P are processes of great complexity, which are dependent on the correct spatial and temporal expression and export of a large number of building blocks. The directed movement in various environments depends on signal transduction by a complex chemotaxis machinery. Accordingly, a number of crucial mechanisms are attractive targets to inhibit the ability of bacteria to move in a directed manner by interfering with chemotaxis, synthesis, assembly, and function of various motility structures.

3.1 General Strategies to Inhibit Bacterial Motility and Chemotaxis

High-throughput screens for compounds that generally inhibit motility or chemotaxis functions could be based on reporting swimming, swarming, or twitching motility phenotypes of various pathogenic bacteria of interest. Screens for compounds that specifically inhibit swarming motility have been performed for *P. aeruginosa*, and hits included branched-chain fatty acids, hydroxyindoles, and naphthalene derivatives (Inoue et al. 2008; Oura et al. 2015). Quinazoline-2,4-diamino analogs were identified as inhibitors of the Na^+-driven flagellar motor of *V. cholerae* in another high-throughput screen for compounds interfering with motility (Rasmussen et al. 2011).

Compounds that affect chemotaxis, but not assembly or function of the motility structures, could be identified by microscopic analyses of moving bacteria in the presence of inhibitor to analyze the chemotactic behavior. In this respect, it would also be feasible to design screens that would directly target the chemosensory system. For instance, compounds could be designed that block the membrane-bound chemoreceptors at the periplasmic substrate binding domains and thereby mimic either attractant- or repellent-bound states. The resulting inhibition of the signal transduction cascade would lead to a loss of chemotactic movement by locking the flagellar motor in either counterclockwise or clockwise rotation, respectively.

Such broad-range approaches to inhibit bacterial motility might yield interesting compounds; however, it is reasonable to assume that many hits would generally target cellular functions and would not be specific for assembly or function of the motility structures. Accordingly, I will describe in the following paragraphs strategies that are designed to specifically interfere with the assembly of motility structures on the level of (i) regulation of gene expression; (ii) posttranslational regulation; and (iii) activity of protein export systems.

3.2 Interfering with Flagellar Assembly at the Level of Gene Regulation

The correct spatiotemporal assembly of the flagellum is a complex process. In *S.* Typhimurium and *E. coli*, this process involves the temporal regulation of more than 60 genes organized into a transcriptional hierarchy of three promoter classes (Kutsukake et al. 1990; Chevance and Hughes 2008). Assembly of the flagellar structure constitutes a significant metabolic burden for the bacterium, and thus, a myriad of negative and positive regulation has been described.

On top of the transcriptional hierarchy, many environmental stimuli are integrated on the level of the flagellar master regulatory operon, *flhDC*, which is expressed from a σ70-dependent flagellar class 1 promoter (Yanagihara et al. 1999). The $FlhD_4C_2$ heteromultimeric complex (FlhDC) directs σ70-RNA polymerase (RNAP) to transcribe from flagellar class 2 promoters (Liu and Matsumura 1994; Wang et al. 2006). Gene products transcribed from class 2 promoters assemble the flagellar hook basal body (HBB) complex including the f-T3SS components. Other class 2 gene products include a flagella-specific, alternative σ factor, σ28 (encoded by *fliA*), and its cognate anti-σ factor, FlgM. Transcription from flagellar class 3 promoters is specific for σ28-RNAP and occurs only after a functional HBB structure has been assembled (Ohnishi et al. 1992; Chadsey et al. 1998; Karlinsey et al. 2000). Gene products transcribed from class 3 promoters include structural components of the filament, motor force generators, and components of the chemotaxis machinery.

Many environmental signals are integrated on the level of *flhDC* gene expression, translation, and FlhDC complex stability that ultimately decide the commencement of flagellar biosynthesis. Regulation occurs in *Salmonella* on the level of *flhDC* gene transcription by a myriad of transcription factors, including the global regulator cAMP-CRP (Komeda et al. 1976; Yanagihara et al. 1999; Mouslim and Hughes 2014), nucleoid-associated proteins Fur and Fis (Stojiljkovic et al. 1994; Kelly et al. 2004), the virulence regulator HilD (Singer et al. 2014), the phosphorelay system RcsCDB (Francez-Charlot et al. 2003; Wang et al. 2007), the EnvZ–OmpR two-component system (Shin and Park 1995), and many more (Erhardt and Dersch 2015).

As described above, negative regulation of the *flhDC* master regulatory operon occurs by action of the conserved two-component regulatory systems RcsCDB and EnvZ–OmpR. Accordingly, screening assays for small molecules could be devised that activate the respective membrane-bound sensor kinases, which would result in the activation of the corresponding response regulators and repression of flagellar synthesis. The Rcs phosphorelay system might be an attractive candidate for such a screen since RcsB is a potent repressor of *flhDC* operon transcription, and in addition, the Rcs system is broadly conserved in many Enterobacteriaceae and required for virulence (Mouslim et al. 2004).

An alternative approach to the activation of repressors could be the inhibition of global activators of flagellar gene expression. Expression of *flhDC* is dependent on the activity of cAMP-CRP and the presence of the nucleoid-associated proteins Fis and Fur. Interestingly, *crp* mutants of *Salmonella* and *Yersinia* have previously been shown to be strongly attenuated (Zhang et al. 1997; Heroven et al. 2012).

Specific for *Salmonella* is the activation of flagellar gene expression by a major regulatory protein of virulence genes, HilD. Inhibition of the transcriptional activity of HilD would have a twofold effect on virulence of *Salmonella*, affecting both expression of *flhDC* and simultaneously inhibiting virulence gene expression.

3.3 Inhibition of Posttranslational Mechanisms Regulating Flagellar Assembly

Posttranscriptional regulation occurs at the level of FlhDC protein translation and FlhDC complex stability. CsrA protects *flhDC* mRNA from RNase E cleavage by binding to the 5′ segment of the *flhDC* transcript (Wei et al. 2001; Yakhnin et al. 2013). Proteolysis of $FlhD_4C_2$ complex is mediated by the adapter proteins FliT and YdiV, which target FlhDC to ClpXP-dependent degradation (Takaya et al. 2012; Sato et al. 2014).

As noted, several regulatory mechanisms that affect protein translation, protein activity, or turnover are important for flagellar assembly. In particular, the interaction of the anti-σ factor FlgM with the flagella-specific σ28 factor provides a mechanism to couple expression of late flagellar genes to the assembly state of the flagellum. Interestingly, an inactivation of *flgM* results in the overexpression of flagellin due to a premature release of σ28 and thereby attenuates *Salmonella* virulence (Schmitt et al. 1994, 1996; Yang et al. 2012). Accordingly, a screen could be designed to disrupt the tight interaction of the anti-σ factor FlgM with its target, σ28. A compound specifically affecting the FlgM-σ28 interaction would result in upregulation of flagella and presumably rapid clearance of flagellin-overproducing bacteria by the immune system.

Further posttranslational regulation mechanisms at the level of FlhDC protein translation and FlhDC complex stability could be used for the development of screens affecting flagellar assembly. Such screens could target the RNA-binding activity of CsrA, which protects *flhDC* mRNA from degradation (Wei et al. 2001; Yakhnin et al. 2013), or modulate the activity of adapter proteins such as YdiV or FliT that target the FlhDC master regulatory complex to proteolytic degradation

(Takaya et al. 2012; Sato et al. 2014). For instance, compounds could be designed that disrupt the FliT–FliD interaction, thereby freeing FliT to initiate ClpXP-dependent proteolytic degradation of FlhDC.

3.4 Strategies to Interfere with Protein Export of Bacterial Flagella and Type-IV Pili (T4P)

3.4.1 The Type-III Secretion System (T3SS)

The T3SS is conserved in both the bacterial flagellum and injectisome needle complex and is an essential virulence factor of many enterobacterial pathogens. Thus, the T3SS constitutes an excellent target for the development of novel broad-spectrum antivirulence drugs (see also concurrent paper on antivirulence strategies; Mühlen and Dersch 2016).

Several screenings for T3SS inhibitors have been performed in *Y. pseudotuberculosis* (Kauppi et al. 2003; Nordfelth et al. 2005), enteropathogenic *E. coli* (Gauthier et al. 2005), and *S.* Typhimurium (Felise et al. 2008). Initial screens reported transcription of late v-T3SS genes or secretion of effector proteins into the culture supernatant as readout (Kauppi et al. 2003; Gauthier et al. 2005; Nordfelth et al. 2005). Any compound that affected the function and assembly of the v-T3SS at some stage would have a negative effect on reporter gene expression or secretion of the reporter substrate. Hits included acylated salicylaldehyde hydrazones, a halogenated salicylaldehyde molecule, and salicylidene anilines. While also inhibiting intracellular replication of *Chlamydia trachomatis* (Muschiol et al. 2006), this promising class of compounds was recently shown to have global cellular effects and does not specifically target the T3SS export machinery (Martinez-Argudo et al. 2013).

Another class of broad-range inhibitors of T3SS function was identified in a screen for compounds that inhibited secretion of an effector protein in *S.* Typhimurium (Felise et al. 2008). This class of molecules—thiazolidinones—was further shown to have broad-range activity against both the v-T3SS and the type-II secretion system of *P. aeruginosa* and *Francisella novicida*, indicating that these compounds might not directly target the type-III protein secretion function.

Accordingly, novel approaches for reporter systems are required to screen and identify compounds that specifically target the protein translocation step of bacterial T3SS independent of the assembly of any distal structural components or changes in gene expression. Based on the fact that the components of the T3SS are conserved and functionally interchangeable, such a class of compounds would be expected to generally inhibit all T3SS of enterobacterial pathogens, including the f-T3SS of the bacterial flagellum and v-T3SS of injectisome devices.

The f-T3SS would be uniquely suited to develop such a novel in vivo screening system since assembly and function of the f-T3SS does not require any extracytoplasmic structures. Beta-lactamase fused to a f-T3SS secretion substrate is exported into the periplasm in a strain deleted for the proximal rod components and confers ampicillin resistance dependent on the presence of a functional f-T3SS (Lee and Hughes 2006). Therefore, a high-throughput screen could be designed that would use the f-T3SS-dependent secretion of a reporter protein into the periplasmic space in reporter strain missing the rod subunits as readout, thereby enabling to screen for f-T3SS inhibitors independent on the assembly of any other structural components.

An in vitro screen for inhibitors that specifically target the protein export process could also be devised using purified T3SS components. For instance, it might proof feasible to purify functional f-T3SS export apparatus complexes or reconstitute functional T3SS in vitro in artificial lipid bilayers. Such in vitro systems to monitor T3SS-dependent protein translocation could be used to screen libraries of small molecules for specific compounds that inhibit the pmf-dependent function of the export apparatus. Another in vitro screen could be based on the ATP hydrolysis activity of the T3SS-associated ATPase, which also is an important component for T3SS function. Several flagella- and virulence T3SS-associated ATPases of *Salmonella* (InvC, FliI), *Yersinia* (YscN), and *E. coli* (EscN) can be recombinantly produced and retain function (Claret et al. 2003; Akeda and Galan 2005; Zarivach et al. 2007; Swietnicki et al. 2011).

Structure-based Approach to Design-Specific T3SS Inhibitors

Several structures of T3SS export apparatus components have been elucidated and could be used to design and screen for specific inhibitors against conserved, essential mechanisms of the protein translocation step.

In particular, several structures of T3SS-associated ATPases are known. ATP hydrolysis by the T3SS-associated ATPase is required for the protein translocation function of the export apparatus. Structures of the injectisome v-T3SS-associated ATPase EscN of *E. coli* (PDB 2OBL) and the f-T3SS ATPase FliI of *Salmonella* (PDB 2DPY) have been determined (Imada et al. 2007; Zarivach et al. 2007). Structure-based, computer-aided approaches could be employed to design small-molecule inhibitors that specifically bind to the active site of T3SS ATPases, thereby inhibiting the type-III protein secretion process.

Interfering with Substrate Specificity Switching

The T3SS export apparatus undergoes an essential switch in substrate specificity that is dependent on a posttranslational, autocatalytic cleavage event of the cytoplasmic domain of a conserved membrane protein of the export gate. Late secretion substrates, which include the filament protein of bacterial flagella and translocated effector proteins, are recognized and secreted by the T3SS only after the switch in secretion specificity.

Structures of the cytoplasmic domain of the T3SS membrane protein that is responsible for the substrate specificity switch are known for several enterobacterial species: the *E. coli* EPEC injectisome component EscU (PDB 3BZO), the *Salmonella* injectisome protein SpaS (PDB 3C01), the *Yersinia* injectisome component YscU (PDB 2JLI), and the *Salmonella* flagellar T3SS protein FlhB

(PDB 3B0Z) (Zarivach et al. 2008; Lountos et al. 2009; Meshcheryakov et al. 2013). The cyclization of a conserved Asn residue is essential for the autocleavage event, and a structure-based approach could be used to design molecules inhibiting the cyclization event or the subsequent conformational change of the cytoplasmic domain.

In case of the f-T3SS of *Salmonella*, activation of late substrate gene expression is dependent on the substrate specificity switch of the FlhB component of the export gate. Thus, an in vivo screen could monitor expression of GFP or other reporter proteins from a late substrate promoter. Complementary, for the injectisome v-T3SS of *Salmonella*, an ELISA-based assay could be designed that monitors secretion of a late-type substrate into the culture supernatant.

3.4.2 The Type-IV Secretion System (T4SS)

T4P consist of several-micrometer-long flexible filaments and a complex membrane-embedded protein export machinery used for pili assembly and retraction (Fig. 1c).

A core set of 12–14 proteins is needed for assembly and function of a T4P machine (Jarrell and McBride 2008). Recently, the in situ structure of a T4P machinery has been determined using electron cryotomography and revealed the true complexity of this nanomachine (Gold et al. 2015). The assembly and disassembly of a pilus made of the pilin protein is energized by different cytoplasmic ATPases. The ATPase PilB (PilF in *Neisseria*) is needed for the addition of new pilin subunits to the base during pilus extension and the ATPases PilT and PilU (PilT2) for removal of subunits during retraction (Burrows 2005; Salzer et al. 2014). Rapid retraction of the pilus enables adherence and is needed for motility (Hahn 1997; Merz et al. 2000). A large oligomeric outer membrane secretin complex functions as a central dynamic component of T4P and allows the passage of the pilus (Gold et al. 2015). The remaining components form the pilus assembly platform and a central inner membrane complex (Karuppiah et al. 2013).

Targeting the assembly and function of T4SS with specifically inhibiting compounds would contribute to control bacterial pathogenicity on several levels: (i) T4P are needed for twitching motility of many bacterial pathogens; (ii) T4SS are used to secrete effector proteins in host cells; and (iii) T4SS mediate DNA uptake from and release into the extracellular milieu and the conjugative transfer of plasmid DNA or transposons into recipient cells.

Several screens have been performed to identify inhibitors of the assembly of an essential membrane component of T4SS or activity of the associated ATPase (Hilleringmann et al. 2006; Paschos et al. 2011). To specifically inhibit the twitching motility function of T4P, the ATPase activity of the different ATPase proteins that are needed for pilus extension or retraction, respectively, could be an attractive target for high-throughput screens.

Based on the recently published in situ structure of a T4P system, structure-based approaches could be devised to design compounds that affect assembly or function of the complex macromolecular machinery, such as the dynamic movement of the secretin component. An alternative approach could target the activity of transcription factors needed for expression of T4SSs.

Finally, the option to inhibit DNA uptake or DNA transfer via T4SSs has not been extensively explored. Atypical unsaturated fatty acids were previously identified as potential inhibitors interfering with conjugative transfer of plasmid DNA between bacteria (Fernandez-Lopez et al. 2005). However, it remains unclear whether those compounds act specifically or unspecifically.

4 Conclusions

The increasing emergence and spread of multidrug-resistant bacteria are cause of great concern and require the development of novel antibiotics against pathogenic bacteria. Many bacterial pathogens use motility to reach a preferred site of infection, initiate biofilm formation, or adhere to host cells. The function of the best-studied motility organelles, the bacterial flagellum, and the type-IV pilus depends on the correct spatial and temporal assembly of a complex macromolecular structure. Thus, these motility devices and the corresponding protein secretion systems constitute attractive targets for the development of specific inhibitors that could be used as antivirulence drugs. Inhibition of bacterial motility does not directly reduce viability of the bacterial pathogens during infection, and therefore, it can be assumed that resistance mechanisms will not or will only slowly evolve. Further, the protein export systems needed for assembly of bacterial flagella (f-T3SS) and T4P (T4SS) are homologous to the related v-T3SS of injectisome devices or type-II secretion systems, respectively. Thus, a compound that would interfere with the motility function by inhibiting the protein secretion process might also simultaneously target the related virulence-associated secretion system. Importantly, many structures of components of the bacterial flagellum, the T3SS, and T4P are known, and thus, the development of high-throughput screens using purified components of the secretion systems might be an attractive approach, which has not been extensively explored in the past.

An important bottleneck for the development and optimization of novel antivirulence drugs is the identification of the molecular target of the compound. A specific, cellular target has not yet been identified for previously published inhibitors of the type-III secretion process. Exploiting the bacterial motility and chemotaxis machinery for the development of novel antivirulence compounds has the important additional advantage that any inhibition of motility provides at the same time the opportunity to positively select for restored motility using genetic screens. The isolation of spontaneous or induced mutations that restore some motility function in the presence of an inhibiting compound would enable target identification. Knowledge about the cellular target(s) would greatly facilitate further modification and optimization of inhibitors of bacterial motility and allow for the development of such inhibiting compounds into novel, highly potent antivirulence drugs.

Acknowledgements Marc Erhardt is supported by the Helmholtz Association Young Investigator grant VH-NG-932 and the People Programme (Marie Curie Actions) of the European Union's Seventh Framework Programme (grant 334030).

References

Abby SS, Rocha EP (2012) The non-flagellar type III secretion system evolved from the bacterial flagellum and diversified into host-cell adapted systems. PLoS Genet 8(9):e1002983. doi:10.1371/journal.pgen.1002983

Abrusci P, Vergara-Irigaray M, Johnson S, Beeby MD, Hendrixson DR, Roversi P, Friede ME, Deane JE, Jensen GJ, Tang CM, Lea SM (2013) Architecture of the major component of the type III secretion system export apparatus. Nat Struct Mol Biol 20(1):99–104. doi:10.1038/nsmb.2452

Aizawa SI (1996) Flagellar assembly in *Salmonella typhimurium*. Mol Microbiol 19(1):1–5

Akeda Y, Galan JE (2005) Chaperone release and unfolding of substrates in type III secretion. Nature 437(7060):911–915. doi:10.1038/nature03992

Allen-Vercoe E, Woodward MJ (1999) The role of flagella, but not fimbriae, in the adherence of *Salmonella enterica* serotype Enteritidis to chick gut explant. J Med Microbiol 48(8):771–780

Andersen-Nissen E, Smith KD, Strobe KL, Barrett SL, Cookson BT, Logan SM, Aderem A (2005) Evasion of Toll-like receptor 5 by flagellated bacteria. Proc Natl Acad Sci USA 102(26):9247–9252. doi:10.1073/pnas.0502040102

Arora SK, Neely AN, Blair B, Lory S, Ramphal R (2005) Role of motility and flagellin glycosylation in the pathogenesis of *Pseudomonas aeruginosa* burn wound infections. Infect Immun 73(7):4395–4398. doi:10.1128/IAI.73.7.4395-4398.2005

Berg HC, Anderson RA (1973) Bacteria swim by rotating their flagellar filaments. Nature 245 (5425):380–382

Black RE, Levine MM, Clements ML, Hughes TP, Blaser MJ (1988) Experimental *Campylobacter jejuni* infection in humans. J Infect Dis 157(3):472–479

Blair DF, Berg HC (1990) The MotA protein of *E. coli* is a proton-conducting component of the flagellar motor. Cell 60(3):439–449

Brown DA, Berg HC (1974) Temporal stimulation of chemotaxis in *Escherichia coli*. Proc Natl Acad Sci USA 71(4):1388–1392

Burrows LL (2005) Weapons of mass retraction. Mol Microbiol 57(4):878–888. doi:10.1111/j.1365-2958.2005.04703.x

Chadsey MS, Karlinsey JE, Hughes KT (1998) The flagellar anti-sigma factor FlgM actively dissociates *Salmonella typhimurium* sigma28 RNA polymerase holoenzyme. Genes Dev 12 (19):3123–3136

Chevance FF, Hughes KT (2008) Coordinating assembly of a bacterial macromolecular machine. Nat Rev Microbiol 6(6):455–465. doi:10.1038/nrmicro1887

Chun SY, Parkinson JS (1988) Bacterial motility: membrane topology of the *Escherichia coli* MotB protein. Science 239(4837):276–278

Claret L, Calder SR, Higgins M, Hughes C (2003) Oligomerization and activation of the FliI ATPase central to bacterial flagellum assembly. Mol Microbiol 48(5):1349–1355

Dons L, Eriksson E, Jin Y, Rottenberg ME, Kristensson K, Larsen CN, Bresciani J, Olsen JE (2004) Role of flagellin and the two-component CheA/CheY system of Listeria monocytogenes in host cell invasion and virulence. Infect Immun 72(6):3237–3244. doi:10.1128/IAI.72.6.3237-3244.2004

Erhardt M, Dersch P (2015) Regulatory principles governing *Salmonella* and *Yersinia* virulence. Front Microbiol 6:949. doi:10.3389/fmicb.2015.00949

Erhardt M, Hirano T, Su Y, Paul K, Wee DH, Mizuno S, Aizawa S, Hughes KT (2010) The role of the FliK molecular ruler in hook-length control in *Salmonella enterica*. Mol Microbiol 75 (5):1272–1284. doi:10.1111/j.1365-2958.2010.07050.x

Erhardt M, Singer HM, Wee DH, Keener JP, Hughes KT (2011) An infrequent molecular ruler controls flagellar hook length in *Salmonella enterica*. EMBO J 30(14):2948–2961. doi:10.1038/emboj.2011.185

Erhardt M, Mertens ME, Fabiani FD, Hughes KT (2014) ATPase-independent type-III protein secretion in *Salmonella enterica*. PLoS Genet 10(11):e1004800. doi:10.1371/journal.pgen.1004800

Felise HB, Nguyen HV, Pfuetzner RA, Barry KC, Jackson SR, Blanc MP, Bronstein PA, Kline T, Miller SI (2008) An inhibitor of gram-negative bacterial virulence protein secretion. Cell Host Microbe 4(4):325–336. doi:10.1016/j.chom.2008.08.001

Fernandez-Lopez R, Machon C, Longshaw CM, Martin S, Molin S, Zechner EL, Espinosa M, Lanka E, de la Cruz F (2005) Unsaturated fatty acids are inhibitors of bacterial conjugation. Microbiology 151(Pt 11):3517–3526. doi:10.1099/mic.0.28216-0

Ferris HU, Minamino T (2006) Flipping the switch: bringing order to flagellar assembly. Trends Microbiol 14(12):519–526. doi:10.1016/j.tim.2006.10.006

Finlay BB, Falkow S (1989) *Salmonella* as an intracellular parasite. Mol Microbiol 3(12):1833–1841

Francez-Charlot A, Laugel B, Van Gemert A, Dubarry N, Wiorowski F, Castanie-Cornet MP, Gutierrez C, Cam K (2003) RcsCDB His-Asp phosphorelay system negatively regulates the *flhDC* operon in *Escherichia coli*. Mol Microbiol 49(3):823–832

Franchi L, Kamada N, Nakamura Y, Burberry A, Kuffa P, Suzuki S, Shaw MH, Kim YG, Nunez G (2012) NLRC4-driven production of IL-1beta discriminates between pathogenic and commensal bacteria and promotes host intestinal defense. Nat Immunol 13(5):449–456. doi:10.1038/ni.2263

Francis NR, Irikura VM, Yamaguchi S, DeRosier DJ, Macnab RM (1992) Localization of the *Salmonella typhimurium* flagellar switch protein FliG to the cytoplasmic M-ring face of the basal body. Proc Natl Acad Sci USA 89(14):6304–6308

Friedlander RS, Vlamakis H, Kim P, Khan M, Kolter R, Aizenberg J (2013) Bacterial flagella explore microscale hummocks and hollows to increase adhesion. Proc Natl Acad Sci USA 110 (14):5624–5629. doi:10.1073/pnas.1219662110

Fronzes R, Christie PJ, Waksman G (2009) The structural biology of type IV secretion systems. Nat Rev Microbiol 7(10):703–714. doi:10.1038/nrmicro2218

Garvis S, Munder A, Ball G, de Bentzmann S, Wiehlmann L, Ewbank JJ, Tummler B, Filloux A (2009) *Caenorhabditis elegans* semi-automated liquid screen reveals a specialized role for the chemotaxis gene *cheB2* in *Pseudomonas aeruginosa* virulence. PLoS Pathog 5(8):e1000540. doi:10.1371/journal.ppat.1000540

Gauthier A, Robertson ML, Lowden M, Ibarra JA, Puente JL, Finlay BB (2005) Transcriptional inhibitor of virulence factors in enteropathogenic *Escherichia coli*. Antimicrob Agents Chemother 49(10):4101–4109. doi:10.1128/AAC.49.10.4101-4109.2005

Giron JA, Torres AG, Freer E, Kaper JB (2002) The flagella of enteropathogenic *Escherichia coli* mediate adherence to epithelial cells. Mol Microbiol 44(2):361–379

Gold VA, Salzer R, Averhoff B, Kuhlbrandt W (2015) Structure of a type IV pilus machinery in the open and closed state. eLife 4. doi:10.7554/eLife.07380

Hahn HP (1997) The type-4 pilus is the major virulence-associated adhesin of *Pseudomonas aeruginosa*–a review. Gene 192(1):99–108

Haiko J, Westerlund-Wikstrom B (2013) The role of the bacterial flagellum in adhesion and virulence. Biology 2(4):1242–1267. doi:10.3390/biology2041242

Hara N, Morimoto YV, Kawamoto A, Namba K, Minamino T (2012) Interaction of the extreme N-terminal region of FliH with FlhA is required for efficient bacterial flagellar protein export. J Bacteriol 194(19):5353–5360. doi:10.1128/JB.01028-12

Harshey RM (2003) Bacterial motility on a surface: many ways to a common goal. Annu Rev Microbiol 57:249–273. doi:10.1146/annurev.micro.57.030502.091014

Hazelbauer GL, Falke JJ, Parkinson JS (2008) Bacterial chemoreceptors: high-performance signaling in networked arrays. Trends Biochem Sci 33(1):9–19. doi:10.1016/j.tibs.2007.09.014

Heroven AK, Sest M, Pisano F, Scheb-Wetzel M, Steinmann R, Bohme K, Klein J, Munch R, Schomburg D, Dersch P (2012) Crp induces switching of the CsrB and CsrC RNAs in *Yersinia pseudotuberculosis* and links nutritional status to virulence. Front Cell Infect Microbiol 2:158. doi:10.3389/fcimb.2012.00158

Hilleringmann M, Pansegrau W, Doyle M, Kaufman S, MacKichan ML, Gianfaldoni C, Ruggiero P, Covacci A (2006) Inhibitors of *Helicobacter pylori* ATPase Cagalpha block CagA transport and *cag* virulence. Microbiology 152(Pt 10):2919–2930. doi:10.1099/mic.0.28984-0

Hughes KT, Gillen KL, Semon MJ, Karlinsey JE (1993) Sensing structural intermediates in bacterial flagellar assembly by export of a negative regulator. Science 262(5137):1277–1280

Ibuki T, Uchida Y, Hironaka Y, Namba K, Imada K, Minamino T (2013) Interaction between FliJ and FlhA, components of the bacterial flagellar type III export apparatus. J Bacteriol 195 (3):466–473. doi:10.1128/JB.01711-12

Imada K, Minamino T, Tahara A, Namba K (2007) Structural similarity between the flagellar type III ATPase FliI and F1-ATPase subunits. Proc Natl Acad Sci USA 104(2):485–490. doi:10.1073/pnas.0608090104

Inoue T, Shingaki R, Fukui K (2008) Inhibition of swarming motility of *Pseudomonas aeruginosa* by branched-chain fatty acids. FEMS Microbiol Lett 281(1):81–86. doi:10.1111/j.1574-6968.2008.01089.x

Iwasaki A, Medzhitov R (2010) Regulation of adaptive immunity by the innate immune system. Science 327(5963):291–295. doi:10.1126/science.1183021

Jarrell KF, McBride MJ (2008) The surprisingly diverse ways that prokaryotes move. Nat Rev Microbiol 6(6):466–476. doi:10.1038/nrmicro1900

Johnson TL, Abendroth J, Hol WG, Sandkvist M (2006) Type II secretion: from structure to function. FEMS Microbiol Lett 255(2):175–186. doi:10.1111/j.1574-6968.2006.00102.x

Karlinsey JE, Tanaka S, Bettenworth V, Yamaguchi S, Boos W, Aizawa SI, Hughes KT (2000) Completion of the hook-basal body complex of the *Salmonella typhimurium* flagellum is coupled to FlgM secretion and *fliC* transcription. Mol Microbiol 37(5):1220–1231

Karuppiah V, Collins RF, Thistlethwaite A, Gao Y, Derrick JP (2013) Structure and assembly of an inner membrane platform for initiation of type IV pilus biogenesis. Proc Natl Acad Sci USA 110(48):E4638–E4647. doi:10.1073/pnas.1312313110

Kauppi AM, Nordfelth R, Uvell H, Wolf-Watz H, Elofsson M (2003) Targeting bacterial virulence: inhibitors of type III secretion in *Yersinia*. Chem Biol 10(3):241–249

Kawai T, Akira S (2010) The role of pattern-recognition receptors in innate immunity: update on Toll-like receptors. Nat Immunol 11(5):373–384. doi:10.1038/ni.1863

Kearns DB (2010) A field guide to bacterial swarming motility. Nat Rev Microbiol 8(9):634–644. doi:10.1038/nrmicro2405

Kelly A, Goldberg MD, Carroll RK, Danino V, Hinton JC, Dorman CJ (2004) A global role for Fis in the transcriptional control of metabolism and type III secretion in *Salmonella enterica* serovar Typhimurium. Microbiology 150(Pt 7):2037–2053. doi:10.1099/mic.0.27209-0

Kim TJ, Young BM, Young GM (2008) Effect of flagellar mutations on *Yersinia enterocolitica* biofilm formation. Appl Environ Microbiol 74(17):5466–5474. doi:10.1128/AEM.00222-08

Komeda Y, Suzuki H, Ishidsu JI, Iino T (1976) The role of cAMP in flagellation of *Salmonella typhimurium*. Mol Gen Genet MGG 142(4):289–298

Krukonis ES, DiRita VJ (2003) From motility to virulence: sensing and responding to environmental signals in *Vibrio cholerae*. Curr Opin Microbiol 6(2):186–190

Kubori T, Yamaguchi S, Aizawa S (1997) Assembly of the switch complex onto the MS ring complex of *Salmonella typhimurium* does not require any other flagellar proteins. J Bacteriol 179(3):813–817

Kutsukake K, Ohya Y, Iino T (1990) Transcriptional analysis of the flagellar regulon of *Salmonella typhimurium*. J Bacteriol 172(2):741–747

Lee HJ, Hughes KT (2006) Posttranscriptional control of the *Salmonella enterica* flagellar hook protein FlgE. J Bacteriol 188(9):3308–3316. doi:10.1128/JB.188.9.3308-3316.2006

Liu X, Matsumura P (1994) The FlhD/FlhC complex, a transcriptional activator of the *Escherichia coli* flagellar class II operons. J Bacteriol 176(23):7345–7351

Lloyd SA, Tang H, Wang X, Billings S, Blair DF (1996) Torque generation in the flagellar motor of *Escherichia coli*: evidence of a direct role for FliG but not for FliM or FliN. J Bacteriol 178 (1):223–231

Lountos GT, Austin BP, Nallamsetty S, Waugh DS (2009) Atomic resolution structure of the cytoplasmic domain of *Yersinia pestis* YscU, a regulatory switch involved in type III secretion. Protein Sci Publ Protein Soc 18(2):467–474. doi:10.1002/pro.56

Macnab RM, Koshland DE Jr (1972) The gradient-sensing mechanism in bacterial chemotaxis. Proc Natl Acad Sci USA 69(9):2509–2512

Manson MD, Tedesco P, Berg HC, Harold FM, Van der Drift C (1977) A protonmotive force drives bacterial flagella. Proc Natl Acad Sci USA 74(7):3060–3064

Marchetti M, Sirard JC, Sansonetti P, Pringault E, Kerneis S (2004) Interaction of pathogenic bacteria with rabbit appendix M cells: bacterial motility is a key feature in vivo. Microbes Infect/Inst Pasteur 6(6):521–528. doi:10.1016/j.micinf.2004.02.009

Mariathasan S, Monack DM (2007) Inflammasome adaptors and sensors: intracellular regulators of infection and inflammation. Nat Rev Immunol 7(1):31–40. doi:10.1038/nri1997

Martinez-Argudo I, Veenendaal AK, Liu X, Roehrich AD, Ronessen MC, Franzoni G, van Rietschoten KN, Morimoto YV, Saijo-Hamano Y, Avison MB, Studholme DJ, Namba K, Minamino T, Blocker AJ (2013) Isolation of *Salmonella* mutants resistant to the inhibitory effect of Salicylidene acylhydrazides on flagella-mediated motility. PLoS ONE 8(1):e52179. doi:10.1371/journal.pone.0052179

Martins M, McCusker MP, McCabe EM, O'Leary D, Duffy G, Fanning S (2013) Evidence of metabolic switching and implications for food safety from the phenome(s) of *Salmonella enterica* serovar Typhimurium DT104 cultured at selected points across the pork production food chain. Appl Environ Microbiol 79(18):5437–5449. doi:10.1128/AEM.01041-13

Mattick JS (2002) Type IV pili and twitching motility. Annu Rev Microbiol 56:289–314. doi:10.1146/annurev.micro.56.012302.160938

McBride MJ (2001) Bacterial gliding motility: multiple mechanisms for cell movement over surfaces. Annu Rev Microbiol 55:49–75. doi:10.1146/annurev.micro.55.1.49

McBride MJ, Zhu Y (2013) Gliding motility and Por secretion system genes are widespread among members of the phylum bacteroidetes. J Bacteriol 195(2):270–278. doi:10.1128/JB.01962-12

Merino S, Shaw JG, Tomas JM (2006) Bacterial lateral flagella: an inducible flagella system. FEMS Microbiol Lett 263(2):127–135. doi:10.1111/j.1574-6968.2006.00403.x

Merz AJ, So M, Sheetz MP (2000) Pilus retraction powers bacterial twitching motility. Nature 407 (6800):98–102. doi:10.1038/35024105

Meshcheryakov VA, Kitao A, Matsunami H, Samatey FA (2013) Inhibition of a type III secretion system by the deletion of a short loop in one of its membrane proteins. Acta Crystallogr D Biol Crystallogr 69(Pt 5):812–820. doi:10.1107/S0907444913002102

Miao EA, Alpuche-Aranda CM, Dors M, Clark AE, Bader MW, Miller SI, Aderem A (2006) Cytoplasmic flagellin activates caspase-1 and secretion of interleukin 1beta via Ipaf. Nat Immunol 7(6):569–575. doi:10.1038/ni1344

Minamino T, Namba K (2008) Distinct roles of the FliI ATPase and proton motive force in bacterial flagellar protein export. Nature 451(7177):485–488. doi:10.1038/nature06449

Minamino T, Ferris HU, Moriya N, Kihara M, Namba K (2006) Two parts of the T3S4 domain of the hook-length control protein FliK are essential for the substrate specificity switching of the flagellar type III export apparatus. J Mol Biol 362(5):1148–1158. doi:10.1016/j.jmb.2006.08.004

Mouslim C, Hughes KT (2014) The effect of cell growth phase on the regulatory cross-talk between flagellar and Spi1 virulence gene expression. PLoS Pathog 10(3):e1003987. doi:10.1371/journal.ppat.1003987

Mouslim C, Delgado M, Groisman EA (2004) Activation of the RcsC/YojN/RcsB phosphorelay system attenuates *Salmonella* virulence. Mol Microbiol 54(2):386–395. doi:10.1111/j.1365-2958.2004.04293.x

Mühlen S, Dersch P (2016) Anti-virulence strategies to target bacterial infections. Curr Top Microbiol Immunol. doi:10.1007/82_2015_490

Muschiol S, Bailey L, Gylfe A, Sundin C, Hultenby K, Bergstrom S, Elofsson M, Wolf-Watz H, Normark S, Henriques-Normark B (2006) A small-molecule inhibitor of type III secretion inhibits different stages of the infectious cycle of *Chlamydia trachomatis*. Proc Natl Acad Sci USA 103(39):14566–14571. doi:10.1073/pnas.0606412103

Nakane D, Sato K, Wada H, McBride MJ, Nakayama K (2013) Helical flow of surface protein required for bacterial gliding motility. Proc Natl Acad Sci USA 110(27):11145–11150. doi:10.1073/pnas.1219753110

Nordfelth R, Kauppi AM, Norberg HA, Wolf-Watz H, Elofsson M (2005) Small-molecule inhibitors specifically targeting type III secretion. Infect Immun 73(5):3104–3114. doi:10.1128/IAI.73.5.3104-3114.2005

Ohnishi K, Kutsukake K, Suzuki H, Lino T (1992) A novel transcriptional regulation mechanism in the flagellar regulon of *Salmonella typhimurium*: an antisigma factor inhibits the activity of the flagellum-specific sigma factor, sigma F. Mol Microbiol 6(21):3149–3157

Ottemann KM, Miller JF (1997) Roles for motility in bacterial-host interactions. Mol Microbiol 24(6):1109–1117

Oura H, Tashiro Y, Toyofuku M, Ueda K, Kiyokawa T, Ito S, Takahashi Y, Lee S, Nojiri H, Nakajima-Kambe T, Uchiyama H, Futamata H, Nomura N (2015) Inhibition of *Pseudomonas aeruginosa* swarming motility by 1-naphthol and other bicyclic compounds bearing hydroxyl groups. Appl Environ Microbiol 81(8):2808–2818. doi:10.1128/AEM.04220-14

Paschos A, den Hartigh A, Smith MA, Atluri VL, Sivanesan D, Tsolis RM, Baron C (2011) An in vivo high-throughput screening approach targeting the type IV secretion system component VirB8 identified inhibitors of *Brucella abortus* 2308 proliferation. Infect Immun 79(3):1033–1043. doi:10.1128/IAI.00993-10

Paul K, Erhardt M, Hirano T, Blair DF, Hughes KT (2008) Energy source of flagellar type III secretion. Nature 451(7177):489–492. doi:10.1038/nature06497

Peabody CR, Chung YJ, Yen MR, Vidal-Ingigliardi D, Pugsley AP, Saier MH Jr (2003) Type II protein secretion and its relationship to bacterial type IV pili and archaeal flagella. Microbiology 149(Pt 11):3051–3072

Pratt LA, Kolter R (1998) Genetic analysis of *Escherichia coli* biofilm formation: roles of flagella, motility, chemotaxis and type I pili. Mol Microbiol 30(2):285–293

Rasmussen L, White EL, Pathak A, Ayala JC, Wang H, Wu JH, Benitez JA, Silva AJ (2011) A high-throughput screening assay for inhibitors of bacterial motility identifies a novel inhibitor of the Na^+-driven flagellar motor and virulence gene expression in *Vibrio cholerae*. Antimicrob Agents Chemother 55(9):4134–4143. doi:10.1128/AAC.00482-11

Rossez Y, Wolfson EB, Holmes A, Gally DL, Holden NJ (2015) Bacterial flagella: twist and stick, or dodge across the kingdoms. PLoS Pathog 11(1):e1004483. doi:10.1371/journal.ppat.1004483

Roy K, Hilliard GM, Hamilton DJ, Luo J, Ostmann MM, Fleckenstein JM (2009) Enterotoxigenic *Escherichia coli* EtpA mediates adhesion between flagella and host cells. Nature 457(7229):594–598. doi:10.1038/nature07568

Salzer R, Joos F, Averhoff B (2014) Type IV pilus biogenesis, twitching motility, and DNA uptake in *Thermus thermophilus*: discrete roles of antagonistic ATPases PilF, PilT1, and PilT2. Appl Environ Microbiol 80(2):644–652. doi:10.1128/AEM.03218-13

Sato Y, Takaya A, Mouslim C, Hughes KT, Yamamoto T (2014) FliT selectively enhances proteolysis of FlhC subunit in FlhD4C2 complex by an ATP-dependent protease, ClpXP. J Biol Chem 289(47):33001–33011. doi:10.1074/jbc.M114.593749

Schmitt CK, Darnell SC, Tesh VL, Stocker BA, O'Brien AD (1994) Mutation of *flgM* attenuates virulence of *Salmonella typhimurium*, and mutation of *fliA* represses the attenuated phenotype. J Bacteriol 176(2):368–377

Schmitt CK, Darnell SC, O'Brien AD (1996) The attenuated phenotype of a *Salmonella typhimurium flgM* mutant is related to expression of FliC flagellin. J Bacteriol 178(10):2911–2915

Shin S, Park C (1995) Modulation of flagellar expression in *Escherichia coli* by acetyl phosphate and the osmoregulator OmpR. J Bacteriol 177(16):4696–4702

Singer HM, Kuhne C, Deditius JA, Hughes KT, Erhardt M (2014) The *Salmonella* Spi1 virulence regulatory protein HilD directly activates transcription of the flagellar master operon *flhDC*. J Bacteriol 196(7):1448–1457. doi:10.1128/JB.01438-13

Stecher B, Hapfelmeier S, Muller C, Kremer M, Stallmach T, Hardt WD (2004) Flagella and chemotaxis are required for efficient induction of *Salmonella enterica* serovar Typhimurium colitis in streptomycin-pretreated mice. Infect Immun 72(7):4138–4150. doi:10.1128/IAI.72.7.4138-4150.2004

Stecher B, Barthel M, Schlumberger MC, Haberli L, Rabsch W, Kremer M, Hardt WD (2008) Motility allows *S.* Typhimurium to benefit from the mucosal defence. Cell Microbiol 10 (5):1166–1180. doi:10.1111/j.1462-5822.2008.01118.x

Stojiljkovic I, Baumler AJ, Hantke K (1994) Fur regulon in gram-negative bacteria. Identification and characterization of new iron-regulated *Escherichia coli* genes by a *fur* titration assay. J Mol Biol 236(2):531–545. doi:10.1006/jmbi.1994.1163

Swietnicki W, Carmany D, Retford M, Guelta M, Dorsey R, Bozue J, Lee MS, Olson MA (2011) Identification of small-molecule inhibitors of *Yersinia pestis* Type III secretion system YscN ATPase. PLoS ONE 6(5):e19716. doi:10.1371/journal.pone.0019716

Takaya A, Erhardt M, Karata K, Winterberg K, Yamamoto T, Hughes KT (2012) YdiV: a dual function protein that targets FlhDC for ClpXP-dependent degradation by promoting release of DNA-bound FlhDC complex. Mol Microbiol 83(6):1268–1284. doi:10.1111/j.1365-2958.2012.08007.x

Toker AS, Macnab RM (1997) Distinct regions of bacterial flagellar switch protein FliM interact with FliG, FliN and CheY. J Mol Biol 273(3):623–634. doi:10.1006/jmbi.1997.1335

Tsang N, Macnab R, Koshland DE Jr (1973) Common mechanism for repellents and attractants in bacterial chemotaxis. Science 181(4094):60–63

van Asten FJ, Hendriks HG, Koninkx JF, van Dijk JE (2004) Flagella-mediated bacterial motility accelerates but is not required for *Salmonella* serotype Enteritidis invasion of differentiated Caco-2 cells. Int J Med Microbiol IJMM 294(6):395–399. doi:10.1016/j.ijmm.2004.07.012

Varga JJ, Nguyen V, O'Brien DK, Rodgers K, Walker RA, Melville SB (2006) Type IV pili-dependent gliding motility in the Gram-positive pathogen *Clostridium perfringens* and other *Clostridia*. Mol Microbiol 62(3):680–694. doi:10.1111/j.1365-2958.2006.05414.x

Wang S, Fleming RT, Westbrook EM, Matsumura P, McKay DB (2006) Structure of the *Escherichia coli* FlhDC complex, a prokaryotic heteromeric regulator of transcription. J Mol Biol 355(4):798–808. doi:10.1016/j.jmb.2005.11.020

Wang Q, Zhao Y, McClelland M, Harshey RM (2007) The RcsCDB signaling system and swarming motility in *Salmonella enterica* serovar typhimurium: dual regulation of flagellar and SPI-2 virulence genes. J Bacteriol 189(23):8447–8457. doi:10.1128/JB.01198-07

Wei BL, Brun-Zinkernagel AM, Simecka JW, Pruss BM, Babitzke P, Romeo T (2001) Positive regulation of motility and *flhDC* expression by the RNA-binding protein CsrA of *Escherichia coli*. Mol Microbiol 40(1):245–256

Williams SM, Chen YT, Andermann TM, Carter JE, McGee DJ, Ottemann KM (2007) *Helicobacter pylori* chemotaxis modulates inflammation and bacterium-gastric epithelium interactions in infected mice. Infect Immun 75(8):3747–3757. doi:10.1128/IAI.00082-07

Yakhnin AV, Baker CS, Vakulskas CA, Yakhnin H, Berezin I, Romeo T, Babitzke P (2013) CsrA activates *flhDC* expression by protecting *flhDC* mRNA from RNase E-mediated cleavage. Mol Microbiol 87(4):851–866. doi:10.1111/mmi.12136

Yanagihara S, Iyoda S, Ohnishi K, Iino T, Kutsukake K (1999) Structure and transcriptional control of the flagellar master operon of *Salmonella typhimurium*. Genes Genet Syst 74 (3):105–111

Yang X, Thornburg T, Suo Z, Jun S, Robison A, Li J, Lim T, Cao L, Hoyt T, Avci R, Pascual DW (2012) Flagella overexpression attenuates *Salmonella* pathogenesis. PLoS ONE 7(10):e46828. doi:10.1371/journal.pone.0046828

Yonekura K, Maki-Yonekura S, Namba K (2003) Complete atomic model of the bacterial flagellar filament by electron cryomicroscopy. Nature 424(6949):643–650. doi:10.1038/nature01830

Young GM, Badger JL, Miller VL (2000) Motility is required to initiate host cell invasion by *Yersinia enterocolitica*. Infect Immun 68(7):4323–4326

Zarivach R, Vuckovic M, Deng W, Finlay BB, Strynadka NC (2007) Structural analysis of a prototypical ATPase from the type III secretion system. Nat Struct Mol Biol 14(2):131–137. doi:10.1038/nsmb1196

Zarivach R, Deng W, Vuckovic M, Felise HB, Nguyen HV, Miller SI, Finlay BB, Strynadka NC (2008) Structural analysis of the essential self-cleaving type III secretion proteins EscU and SpaS. Nature 453(7191):124–127. doi:10.1038/nature06832

Zhang X, Kelly SM, Bollen WS, Curtiss R 3rd (1997) Characterization and immunogenicity of *Salmonella typhimurium* SL1344 and UK-1 delta *crp* and delta *cdt* deletion mutants. Infect Immun 65(12):5381–5387

Zhao R, Amsler CD, Matsumura P, Khan S (1996a) FliG and FliM distribution in the *Salmonella typhimurium* cell and flagellar basal bodies. J Bacteriol 178(1):258–265

Zhao R, Pathak N, Jaffe H, Reese TS, Khan S (1996b) FliN is a major structural protein of the C-ring in the *Salmonella typhimurium* flagellar basal body. J Mol Biol 261(2):195–208. doi:10.1006/jmbi.1996.0452

New Horizons in the Development of Novel Needle-Free Immunization Strategies to Increase Vaccination Efficacy

Kai Schulze, Thomas Ebensen, Peggy Riese, Blair Prochnow, Claus-Michael Lehr and Carlos A. Guzmán

Abstract The young twenty-first century has already brought several medical advances, such as a functional artificial human liver created from stem cells, improved antiviral (e.g., against HIV) and cancer (e.g., against breast cancer) therapies, interventions controlling cardiovascular diseases, and development of new and optimized vaccines (e.g., HPV vaccine). However, despite this substantial progress and the achievements of the last century, humans still suffer considerably from diseases, especially from infectious diseases. Thus, almost one-fourth of all deaths worldwide are caused directly or indirectly by infectious agents. Although vaccination has led to the control of many diseases, including smallpox, diphtheria, and tetanus, emerging diseases are still not completely contained. Furthermore, pathogens such as *Bordetella pertussis* undergo alterations making adaptation of the

K. Schulze (✉) · T. Ebensen (✉) · P. Riese · B. Prochnow · C.A. Guzmán
Department of Vaccinology and Applied Microbiology, Helmholtz Centre for Infection Research (HZI), Braunschweig, Germany
e-mail: Kai.Schulze@helmholtz-hzi.de

T. Ebensen
e-mail: Thomas.Ebensen@helmholtz-hzi.de

P. Riese
e-mail: Peggy.Riese@helmholtz-hzi.de

B. Prochnow
e-mail: Blair.Prochnow@helmholtz-hzi.de

C.A. Guzmán
e-mail: Carlos.Guzman@helmholtz-hzi.de

C.-M. Lehr
Department of Drug Delivery, Helmholtz Centre for Infection Research (HZI), Helmholtz Institute for Pharmaceutical Research Saarland (HIPS), Braunschweig, Germany
e-mail: Claus-Michael.Lehr@helmholtz-hzi.de

C.-M. Lehr
Department of Pharmacy, Helmholtz Centre for Infection Research (HZI), Saarland University, Saarbrücken, Germany

Current Topics in Microbiology and Immunology (2016) 398: 207–234
DOI 10.1007/82_2016_495

Published Online: 28 June 2016

respective vaccine necessary. Moreover, insufficient implementation of vaccination campaigns leads to re-emergence of diseases which were believed to be already under control (e.g., poliomyelitis). Therefore, novel vaccination strategies need to be developed in order to meet the current challenges including lack of compliance, safety issues, and logistic constraints. In this context, mucosal and transdermal approaches constitute promising noninvasive vaccination strategies able to match these demands.

Abbreviation

αGalCer	Alpha galactosylceramide
APCs	Antigen-presenting cells
c-di-AMP	Bis-(3′,5′)-cyclic dimeric adenosine monophosphate
CNS	Central nervous system
CT	Cholera toxin
CTB	Cholera toxin B subunit
CTL	Cytotoxic T lymphocyte
cVDPV	Circulating vaccine-derived polioviruses
DCs	Dendritic cells
EMA	European Medicines Agency
ETEC	Enterotoxigenic *Escherichia coli*
FDA	Food and Drug Administration
GRAS	Generally regarded as safe
HIV	Human immunodeficiency virus
i.d.	Intradermal
i.m.	Intramuscular
IMSG NPs	Inverse micellar sugar glass nanoparticles
i.n.	Intranasal
i.t.	Intratracheal
i.v.	Intravenous
i.vag.	Intravaginal
LAIV	Live attenuated influenza virus vaccines
LCs	Langerhans cells
LPS	Lipopolysaccharide
LT	Heat-labile toxin
MALP-2	TLR2/6-binding macrophage-activating lipopeptide-2
MCTs	Medium-chain triglycerides
MPL	Monophosphoryl lipid A
NALT	Nasal-associated lymphoid tissue
PAMPs	Pathogen-associated molecular patterns
PEMs	Polyelectrolyte multiple layers
PLA	Polylactic acid
PLA NPs	Polylactic acid nanoparticles
PLGA	Polylactic-co-glycolic acid

PLGA NPs	Polylactic-co-glycolic acid nanoparticles
PRRs	Pattern recognition receptors
SC	Stratum corneum
s.c.	Subcutaneous
s.l.	Sublingual
STING	Stimulator of interferon genes
TCV	Transcutaneous vaccination
t.f.	Transfollicular
TLR	Toll-like receptor

Contents

1 Introduction

Infectious diseases remain a major worldwide health problem. Vaccination is probably the most valuable intervention tool, since it helps to prevent the onset of a subsequent disease (Riese et al. 2013, 2014). However, for many infectious diseases causing disability and death, vaccines are still not available. Furthermore, the performance of some of the existing vaccines can be further improved in terms of efficacy, safety, and cost. During the last decade, significant progress has been made in the generation of efficient vaccines based on well-defined antigens, which exhibit an enhanced safety profile. However, irrespective of the type of vaccine (e.g., inactivated, attenuated, subunit), they have to be administered by an appropriate route, since this can affect the overall safety, immunogenicity, efficacy, and logistic feasibility. In this context, the injection/application technique, inductive site reached by the antigen included in the vaccine formulation, and type of adjuvant, as well as age, sex, and race/ethnicity of the vaccinee, have a strong impact on the tolerability and immunogenicity of specific vaccines. Today, most of the vaccines are administered by parenteral route, *i.e.*, intramuscular (i.m.), subcutaneous (s.c.), or intradermal (i.d). However, there are many disadvantages associated with conventional injection, which in turn hamper broad acceptance and access to vaccination, in particular in developing countries. In this context, easy-to-use vaccination methods would be able to overcome certain issues associated with injectable vaccinations. Thus, there is increasing interest in the discovery of novel approaches and technologies to develop noninvasive and needle-free vaccines. Easy-to-implement needle-free vaccination methods encompass the administration of vaccine formulations by either the mucosal or transdermal route by using different delivery systems and/or application devices.

1.1 Needle-Free Vaccines

Vaccines are usually administered using syringes. This approach is associated with numerous disadvantages, such as lack of patient comfort that leads to reduced compliance, requirement of trained personnel, and potential safety risks (e.g., needle-stick accidents, needle reuse, and wound infection). In order to overcome these limitations, various needle-free methods have been developed, which either focus on mucosal or on transcutaneous vaccination (TCV) (Fig. 1).

Mucosal approaches aim to deliver vaccines to the mucosa of eyes, nose, lung, vagina, gut, and rectum, whereas TCV targets the immune cells within the skin (Mitragotri 2005). One major advantage of mucosal vaccination is the stimulation of both systemic and mucosal immune responses, as compared to needle-based parenteral vaccination. The mucosa is the major port of entry for infectious agents. Thus, stimulation of efficient mucosal immune responses not only protects against disease but also promotes protection already against infection (i.e., colonization). This in turn minimizes the risk of horizontal transfer to susceptible contacts and disease spread, especially of mucosally transmitted infections, such as respiratory, gastrointestinal, and sexually transmitted infections (Anjuere et al. 2012; Ruffin et al. 2012; Yu and Vajdy 2010). Besides the mucosa, also the skin represents a particularly attractive target for noninvasive vaccination, since it constitutes both an easily accessible administration site and a natural infection route (via insect bites, superficial or deeper wounds, etc.). Like the mucosa, the skin is equipped with a rich pool of immune cells able to initiate adaptive immune responses, the so-called antigen-presenting cells (APCs), which are capable of antigen detection, uptake, processing, and finally presentation to naïve T cells (Levin et al. 2015). Depending on the particular subset of APCs the vaccine is interacting with, the stimulation of different types of immune responses is favored.

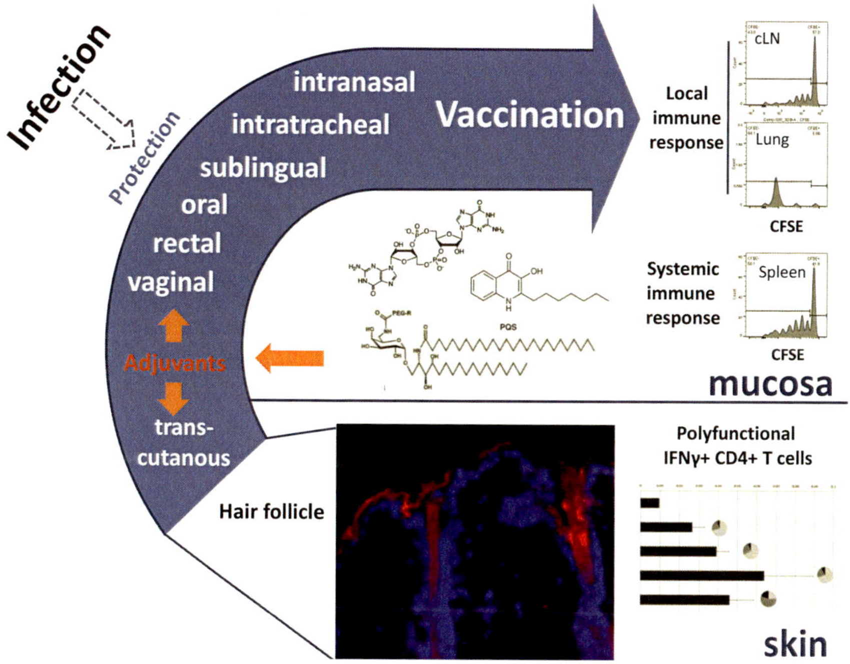

Fig. 1 New vaccination strategies across the mucosal and skin barriers using novel mucosal adjuvants and needle-free devices result in the elicitation of protective responses at both local and systemic levels. *cLN* cervical lymph node, Pie charts indicate mono- (*light grey*), bi- (*grey*) and tri-functional (*black*) IFNg+ CD4+ T cells

2 Mucosal Vaccines

Mucosal vaccines offer several physiological and practical advantages over parenteral immunization. Among others, the most important ones are (i) easy administration logistics (i.e., decreased costs due to lack of requirement for qualified personal), (ii) needle-free, (iii) high compliance, and (iv) suitability for mass vaccination (Allison 1997; Lamichhane et al. 2014; Wang and Wang 2015). However, despite these benefits, only few mucosal vaccines have been approved for human use, such as the live attenuated oral vaccine against polioviruses and the live attenuated intranasal vaccine against the influenza virus. Challenges associated with the development of an effective mucosal vaccine are the efficient antigen delivery across the mucosal barrier and the need for a strong safety profile. For instance, the use of well-defined antigens for the generation of subunit vaccines has led to products with an improved safety profile but at the same time resulted in usually only poorly immunogenic vaccines. Today, several strategies have been developed to overcome the poor immunogenicity of mucosal vaccines. One of the major approaches focuses on the coadministration of antigens with mucosal adjuvants, which generate a local microenvironment conductive toward antigen processing and presentation (Vajdy et al. 2004). Unfortunately, there is no mucosal adjuvant available for human use yet. The most promising candidates tested so far (e.g., derivatives of bacterial toxins) were associated with concerning neurological side effects (Mutsch et al. 2004). Furthermore, vaccine formulations can be (i) cleared by non-specific mechanisms (e.g., ciliary activity, mucous entrapment, and peristalsis), (ii) degraded by enzymes, and (iii) affected by extreme pH, before having the chance to interact with immune cells of the host (Bennett et al. 2009; Davitt and Lavelle 2015; des Rieux et al. 2006). To overcome these hurdles, there are attempts focusing on the use of nanoparticle-based delivery systems, which not only protect antigens against degradation but can also target vaccine antigens to the inductive site, thereby promoting the elicitation of robust responses at both local and systemic levels (Honma et al. 2004). Accordingly, there is still an urgent need for new and safe strategies to stimulate mucosal immunity.

2.1 Mucosal Administration Routes

As mentioned before, almost all pathogens enter the body through mucosal surfaces by ingestion, inhalation, or sexual contact. In this regard, vaccines mimicking the natural primary exposure to a pathogen that induces the production of a strong protective immune response are at the base of vaccinology (Dlugonska and Grzybowski 2012). Mucosal vaccines under investigation make use of various routes of immunization including intranasal (i.n.), sublingual (s.l.), intratracheal (i.t.), oral, rectal, or intravaginal (i.vag.). All the different routes induce antigen-specific immune responses both at local mucosal sites (portal of pathogen entry) but also at systemic level (Stevceva and Strober 2004). Interestingly, despite a certain degree of compartmentalization within the mucosal immune system, vaccination at a certain inductive site can also promote local responses at distant mucosal territories (Holmgren and Czerkinsky 2005).

Mucosal vaccines can be intrinsically more powerful than their systemic counterparts. As an example, i.n. administration of influenza vaccines showed distinct features compared with the parenteral administration (i.e., intramuscular) such as (i) providing protection at the pathogen's entry site, (ii) higher levels of mucosal antibodies, (iii) cross-protection, and (iv) needle-free application (Rose et al. 2012; Srivastava et al. 2015). However, this kind of administration strategy poses the risk of antigen or adjuvant delivery into the central nervous system (CNS) when A-B moiety toxins or their derivatives were used as adjuvants. An alternative approach circumventing this problem focuses on the s.l. route. Whereas oral, i.n., and i.t. vaccine delivery strategies have been described extensively in the past, the s.l. route is now receiving considerably more attention (Kraan et al. 2014; Pedersen and Cox 2012). Recent studies demonstrated the potential of the s.l. route for delivering vaccines capable of inducing both mucosal and systemic immune responses. It was shown that similar to administration via i.n. route, s.l. immunization is also able to stimulate efficient immune responses at both local and distinct mucosal areas, such as in the vaginal mucosa. Thus, s.l. immunization was able to induce not only an effective protection against a lethal challenge with influenza virus (H1N1) but also against genital papillomavirus (Kweon 2011). This is of great interest, since in many countries more than 100 million new cases of sexually transmitted bacterial infections, such as *Chlamydia*, are estimated annually. These acute infections translate into significant downstream healthcare costs, particularly for women (Hafner et al. 2014). In this regard, mucosal vaccination provides a promising tool to increase the overall immunization rate by enhancing public acceptance due to painless application (Caetano et al. 2014; van den Dobbelsteen and van Rees 1995).

2.2 *Mucosal Delivery System*

Recent advances in our understanding of mucosal immunity as well as the identification of correlates of protective immunity against specific mucosal pathogens have set the agenda for developing efficient delivery strategies (Woodrow et al. 2012). In this context, polymeric nanocarriers have been developed that are able to encapsulate vaccine components (i.e., protection of sensitive payload), including mucosal adjuvants, resulting in an improved antigen delivery to the targeted immune cells of the mucosal immune system and the subsequent stimulation of efficient immune responses (Chadwick et al. 2009; Sharma et al. 2015). This kind of nanocarrier-based delivery system offers the versatility of being formulated with multiple adjuvants and antigenic cargo which can be tailored effectively to prime immune responses across the mucosal barrier (Jia et al. 2015). Effective and stable vaccine delivery systems allowing painless mucosal vaccination even by self-administration can encompass biocompatible materials in the form of a matrix or formulated micro/nanoparticles, liposomes, polylactic acid (PLA)/polylactic-co-glycolic acid (PLGA)/chitosan nanoparticles, hydrogels, polyelectrolyte multiple layers (PEMs), and virosomes (Wang and Wang 2015). For example, chitosan polymer-based nanoparticles efficiently stimulated mucosal secretory IgA and IgG antibodies, pro inflammatory TNF-α and IL-6 cytokines, and IFN-γ production in nasal-associated lymphoid tissue (NALT) by the stimulation of Toll-like receptor (TLR)2- or TLR5-mediated innate responses, thus making them a very attractive antigen delivery system for mucosal vaccines (Baaten et al. 2010; Bolhassani et al. 2014; Read et al. 2005). Another particle-based approach focuses on the mucosal delivery of dry powder vaccine formulations given to the respiratory tract by i.t. route. These dry powder vaccines offer the potential to provide a needle-free and cold chain-independent vaccination strategy for the induction of protective immunity against either systemic or mucosal pathogens (Wang et al. 2012).

An important role in protection and cross-protection against a variety of infectious agents plays the secretory IgA antibodies which are mainly stimulated when vaccines are given by mucosal routes (Brandtzaeg 2007). However, only a few

Table 1 Mucosal vaccines under clinical development or licensed to the vaccine market (2005–2015)

Vaccine	Mucosal route	Vaccine market
FluMist® [live attenuated influenza virus vaccines (LAIV)]	Intranasal	USA (children and adults)
Fluenz™ [live attenuated influenza virus vaccines (LAIV)]	Intranasal	Europe (children)
NasoVac® (live attenuated influenza trivalent intranasal vaccine)	Intranasal	India
Polio vaccine (monovalent, bivalent, and trivalent)	Oral	Poliomyelitis endemic countries
Vivotif® (live attenuated vaccine against typhoid fever)	Oral	27 countries worldwide (children over 5 years and adults)
Dukoral® (cholera vaccine)	Oral	Over 60 countries worldwide (children over 2 years and adults)
RotaTeq® (pentavalent) *Rotarix*® (monovalent) (live attenuated vaccine against rotavirus)	Oral	USA/Europe (children from 6 to 24/32 weeks old)

mucosal vaccines have been approved for human use (see Table 1), and more basic work is needed in areas of delivery systems and adjuvants to design or further refine effective mucosal vaccines.

2.3 *Mucosal Adjuvants*

Whereas the enhancement of the immunogenicity of antigens by adjuvants for parenteral vaccines started already more than 80 years ago (Rappuoli et al. 2011), the exploration of potent mucosal adjuvants is just emerging. This research is driven by the necessity to increase the immune responses to antigens that are poorly immunogenic at the mucosal sites. Although mucosal vaccines are simpler to administer and show a reduced risk of transmitting infections when compared with injection strategies, the number of mucosal vaccines developed over the past decade and licensed today (Table 1) is small. More specifically, oral vaccines have been developed against cholera, typhoid fever, polio, and rotavirus, and one nasal vaccine exists which protects against influenza (FluMist®).

Mucosal adjuvants often behave like pathogen-associated molecular patterns (PAMPs) that are recognized by innate immune system pattern recognition receptors (PRR) and act on APCs by altering their cytokine production potential. However, the chemical nature of the adjuvants and their mode of action are extremely variable. Potential mechanisms by which they exert their biological activity are the (i) "depot" effect, (ii) antigen targeting to APC, (iii) improvement of antigen processing and presentation, and (iv) immune activation or modulation through the upregulated expression of cellular mediators. Today, mucosal adjuvants mainly consist of TLR agonists (e.g., Ampligen® poly(I:C_{12}U), a synthetic analogue of poly(I:C), lipopolysaccharide (LPS), or (CpG-ODN) that enhance adaptive immune responses by promoting APC activation and maturation with enhanced expression of MHC class II and costimulatory molecules (Savelkoul et al. 2015).

However, it is also important to consider that their biological activities may not only favor the elicitation of adaptive responses, but also the appearance of unwanted side effects. A strong stimulation may result in local or general inflammation and tissue destruction, resulting in distress for the vaccinees. On the other hand, the stimulation of vigorous responses is insufficient to protect against a specific pathogen. It is essential to stimulate an adequate type of response to promote efficient clearance, but simultaneously prevent immune pathological reactions which can in turn lead to a more severe course of infection. Thus, the identification of adjuvants able to promote predictable responses is a major issue in vaccinology. Unfortunately, most of the molecules identified until now exhibiting these properties are also associated with strong reactogenicity in humans. Therefore, the choice of a specific adjuvant for a formulation reflects a compromise between the required immune modulatory effect and an acceptable level of mild side effects (e.g., fever), which might be needed to induce a protective immune response. For example, the use of live attenuated vaccines, such as the H1N1 influenza strain A/17/CA/2009/38, induced effective protective immunity, but also mild-to-moderate adverse effects (Phonrat et al. 2013).

2.4 *Development of Mucosal Adjuvants for Human Vaccine Market*

Nowadays, there are only a few parenteral adjuvants that are included in licensed human and veterinary vaccines. These are Alum (Petrovsky 2006; Schmidt et al. 2007), MF59 (Ott et al. 1995; Podda and Del Giudice 2003), monophosphoryl lipid A (MPL) (De Vendittis et al. 1981; Kundi 2007; Patel and Salapatek 2006; Persing et al. 2002), a squalene-based adjuvant system 03 (AS03®), and a combination of MPL and Alum (AS04®) (Savelkoul et al. 2015). Although severe side effects are rare, even Alum has the potential to cause sterile abscesses, eosinophilia, and myofascitis and exhibits a potential role in neurodegenerative diseases (Petrovsky and Aguilar 2004). Alum induces strong antibody and Th2 responses, whereas the availability of adjuvants promoting Th1 and cytotoxic T lymphocyte (CTL) responses is still a major need. Interestingly, the combination of Alum with IL-2 is able to redirect Th2-dominated responses into a Th1-dominant response (Holmgren et al. 2005). MF59 promotes both humoral and cellular immune responses (Mesa and Fernandez 2004; O'Hagan et al. 1997; Ott et al. 1995; Podda and Del Giudice 2003). Such squalene or squalane emulsions are efficient adjuvants which can be stabilized by microfluidization, so that the emulsions can be frozen or kept for years at room temperature, allowing also their sterilization by terminal filtration, thereby enabling vaccination strategies in the third world where a cold chain is sometimes problematic (Ott et al. 1995). Antigens are added after emulsification, so denaturation of conformational epitopes can be avoided (Hisert et al. 2005; O'Hagan 1998). Clinical trials of MF59-adjuvanted vaccines which were performed in different age groups (i.e., newborns and elderly) have demonstrated their safety and immunogenicity (Kuroda et al. 2004). However, the strong adjuvant activity of MF59 demonstrated for parenteral administration is completely absent when tested by the intranasal route, suggesting that MF59 is not efficient for mucosal use (Boyce et al. 2000). Furthermore, phase IV studies are mandatory, since the use of oil-based adjuvants may be associated with a higher risk of autoimmunity, which is defined as "adjuvant disease" (Israeli et al. 2009).

Newly developed mucosal adjuvants are still in preclinical or clinical development. One reason for this delay, at least in part, is the potential toxicity of these molecules, representing important safety issues (O'Hagan and Rappuoli 2004). The lack of an effective mucosal adjuvant is even more significant as none of the licensed adjuvants revealed a strong enhancement in immune response when given by mucosal administration route (Newsted et al. 2015). The most commonly used mucosal adjuvants in the preclinical studies are (a) ganglioside receptor-binding toxins and their derivatives (cholera toxin [CT], heat-labile toxin from *E. coli* [LT]), (b) surface immunoglobulin-binding complex CTA1-DD, (c) TLR4-binding lipopolysaccharide, (d) TLR2-binding muramyl dipeptide, (e) Mannose receptor-binding mannan, (f) Dectin-1-binding glucans, (g) TLR9-binding CpG-oligodeoxynucleotides, (h) chemokines, (i) APC-targeting ISCOMATRIX and ISCOM, (j) TLR2/6-binding macrophage-activating lipopeptide-2 (MALP-2) (Becker et al. 2006; Borsutzky et al. 2003; Link et al. 2004; Rharbaoui et al. 2002), (k) CD1d-binding alpha galactosylceramide (αGalCer) (Ebensen et al. 2007a, b), (l) STING-binding cyclic dinucleotides (c-di-AMP, c-di-GMP, cGAMP) (Ebensen et al. 2011; Sanchez et al. 2014; Skrnjug et al. 2014), and (m) an adjuvant combination including a cationic formulation (CAF01) containing a cationic liposome as delivery system and a synthetic mycobacterial cord factor as adjuvant (Agger et al. 2008; ClinicalTrials.gov 2012; Cox et al. 2006). In addition, also pro-inflammatory cytokines, such as IL-1α, IL-4, IL-6, IL-12, IL-13, IL-15, and IL-18, are broadly used as mucosal adjuvants (Ranasinghe et al. 2014). These molecules activate innate immunity and polarize the adaptive immunity by inducing vaccine-specific cellular (e.g., mucosal CD8+ CTL) and humoral (e.g., IgA antibody production) immune responses (Rhee et al. 2012; Wang and Meng 2015). CT and LT are powerful mucosal adjuvants when coadministered with soluble antigens. Another type of mucosal adjuvant is represented by synthetic oligodeoxynucleotides containing CpG motifs linked to cholera toxin B subunit (CTB) and has been found to effectively stimulate both innate and adaptive mucosal immune responses (Holmgren et al. 2005). However, their use in humans is hindered by their extremely high toxicity (e.g., diarrhea, fever, weight loss). During the past years, a new generation of LT and CT mutants (e.g., LTK63 and LTR72) with reduced toxicity, but retained strong adjuvanticity at the mucosal level, was developed (Beignon et al. 2002; Pizza et al. 2000, 2001). However, they were shown to be associated with the appearance of concerning side effects in human trials (e.g., Bell's palsy). Another interesting molecule showing mucosal adjuvant properties is chitosan and its derivatives, such as the thermo-sensitive chitosan system (Xia et al. 2015).

2.5 *Approved Mucosal Vaccines*

Currently, all intranasal-delivered vaccines approved for human use are based on live attenuated influenza virus vaccines (LAIV). The US Food and Drug Administration (FDA)-approved quadrivalent vaccine FluMist® for children and adults (2–49 years of age) consists of the influenza A strains H1N1 and H3N2 and two influenza B strains. The European equivalents approved by the European Medicines Agency (EMA) for children (2–17 years of age) are represented by the trivalent Fluenz™ and the quadrivalent Fluenz™ Tetra (Carter and Curran 2011). NasoVac®, a trivalent intranasal vaccine licensed for India and under the approval process by the EMA, contains live attenuated strains of H1N1 and H3N2 and influenza B (Dhere et al. 2011). The currently existing live attenuated intranasal vaccines are cold adapted but heat sensitive, thereby allowing the viral replication in the nostrils but preventing viral spread to the lungs or downstream. The local replication results in a prolonged antigen presentation by APCs, thus inducing efficient protective immune responses as indicated by increased levels of locally secreted IgA and T cell-derived IFNγ (He et al. 2006). All intranasal influenza vaccines are administered annually and contain the seasonal adapted antigens. Nevertheless, in comparison with inactivated influenza vaccine, the LAIV displayed only poor performance in terms of efficacy (relative efficacy 52 %), as shown in a study by Ambrose et al. (Ambrose et al. 2014).

Oral vaccines are mainly intended to prevent gastrointestinal infections. The first vaccine delivered by the oral route was developed against poliomyelitis, a paralytic disease due to neuron destruction in the CNS, and approved in 1961. The trivalent oral polio vaccine generated by Albert Sabin consists of a mixture of three live, attenuated poliovirus strains. Following oral vaccination, the virus persists in the gastrointestinal tract for 4–6 weeks, thereby inducing antipolio-specific IgA antibodies in the serum and the mucosa. The generation of neutralizing antibodies, which results in reduced viral shedding and transmission following re-exposure, is considered as a correlate of protection. Currently, the oral vaccine is mainly used in developing countries due to its high efficacy, low production costs, and easy administration. However, in very rare cases of immune deficiencies, the live attenuated vaccine can cause paralysis. Therefore, the oral polio vaccine is not suitable for use in immune-compromised individuals. Since 1998, the oral polio vaccine is no longer recommended in Europe and was substituted by an intramuscularly delivered inactivated polio vaccine generated by Jonas Salk in 1955.

Thus, under conditions of insufficient population immunity, viruses can circulate in the population (circulating vaccine-derived polioviruses; cVDPV) and due to genetic mutation can even cause polio epidemics (e.g., Hispaniola, Philippines). As the number of wild-type virus-caused polio cases becomes rare, the risk of cVDPV epidemics increases. In order to prevent cVDPV epidemics, inactivated polio vaccines have been implemented (Minor 2009). Although the inactivated vaccine cannot induce paralytic polio and is therefore considered to be safe, it cannot elicit mucosal immunity at the site of viral replication. In addition to the trivalent oral polio vaccine, a monovalent oral polio vaccine and a bivalent oral polio vaccine were licensed in 2005 and 2009, respectively. Both vaccines are mainly used for mass vaccination campaigns in poliomyelitis endemic countries. Clinical trials evaluating the seroconversion rates and neutralizing antibody titers demonstrated superiority of the monovalent and bivalent vaccines over the trivalent vaccine (el-Sayed et al. 2008, Sutter et al. 2010).

Another orally delivered vaccine, Vivotif® (PaxVax Berna, Switzerland), is a live attenuated vaccine against typhoid fever, a disease caused by the highly virulent and invasive enteric bacterium *Salmonella enterica* serovar Typhi. The vaccine was generated by chemical mutagenesis of the wild-type strain Ty21, and it is delivered as lyophilized live bacteria in an enteric coated capsule. Vivotif® was approved in 1989 by the FDA and is indicated for children over 5 years and adults. It is currently used in over 27 countries. Beside the capsule vaccine Vivotif®, a liquid formulation of the Ty21a vaccine was licensed in 1997. In two clinical trials, the liquid formulation was found to be superior in terms of protection 3 years after vaccination and equally protective 7 years after administration when compared to the capsule formulation. However, this vaccine formulation was not well accepted by travelers from industrialized countries, and therefore, its manufacturing was discontinued. The oral Ty21a vaccine was demonstrated to elicit antigen-specific serum IgG and mucosal IgA against the *O*-polysaccharide as well as Th1 and CTL. Further, studies comparing typhoid vaccines delivered by the oral or parenteral route revealed the generation of circulating antibody-secreting cells bearing the gut mucosa homing marker $\alpha 4\beta 7$ in the oral vaccinated individuals but not in the parenteral group. Unfortunately, Vivotif® requires the administration of 3–4 doses, one every other day, to generate efficient immunogenicity. Thus, a single-dose mucosal vaccine against typhoid fever is highly desired.

Different types of orally delivered cholera vaccines are available on the international market. Cholera, a severe diarrheic disease, is caused by the waterborne bacterium *Vibrio cholerae* which colonizes the epithelial gut layer where it secretes the CT. Of the 200 serogroups, only 2 (O1 and O139) can cause the disease. The most commonly used oral cholera vaccine is Dukoral™ (SBL Vaccin), which was developed in Sweden and licensed in 1991 for children over 2 years and adults. Currently, it is used internationally in over 60 countries, mainly for travelers to endemic regions. In contrast to most of the other mucosal vaccines which are based on live attenuation, Dukoral™ consists of the inactivated *V. cholerae* serotypes O1 and O139 and the recombinant non-toxic B subunit. Dukoral™ was demonstrated to induce intestinal antibacterial and antitoxin IgA antibodies. Interestingly, Dukoral™ can induce 86 % cross-protection against clinically severe episodes of enterotoxigenic *Escherichia coli* (ETEC) infection, most probably due to the structural and functional similarities of the cholera toxin B to the heat-labile toxin of ETEC. The immunization schedule for children older than 5 years and adults consists of two oral doses 7 days apart. Another inactivated oral cholera vaccine was licensed in 2009, named mORCVAX in Vietnam where it is only intended for local use and called Shanchol™ in India for international application. It contains the serotypes O1 and O139 but not the cholera non-toxic subunit B. This vaccine was shown to induce longer term protection in children under 5 years against cholera, but not ETEC as compared to Dukoral™. The third internationally approved oral cholera vaccine CVD 103-HgR (Orochol, Bern, SSVI) consists of the live attenuated O1 strain with a deletion of the cholera toxin gene. Despite showing an efficacy of approximately 72 % in clinical trials in healthy individuals from developed countries, in endemic areas, e.g., Indonesia, the vaccine CVD 103-HgR failed to induce sufficient protection. Currently, this vaccine is not available on the market.

Oral vaccines also exist against rotavirus, the infectious agent of severe gastroenteritis in infants and young children. Of the two currently licensed, the live pentavalent vaccine RotaTeq® (Merck & Co, Inc) was approved in 2006 by the EMA and FDA. RotaTeq® contains five human-bovine monoreassortant rotaviruses encoding G1, G2, G3, G4, and P1A [8] outer capsid proteins of human rotaviruses on a bovine rotavirus background. The rotavirus G1 serotype is the most prevalent strain worldwide, and the genotype P [8] is shared by most of the common circulating rotavirus strains. Given as a three-dose vaccine to infants between the ages of 6–32 weeks, RotaTeq® was demonstrated to be protective against the G1, G3, G4, and P1A [8] rotavirus serotypes. Placebo-controlled phase III studies revealed a primary efficacy of 68 % against any grade of severity and approximately 98 % against severe rotavirus-induced gastroenteritis in the first two rotavirus seasons after vaccination. The hospitalization rate was reduced by 94.5 % (Vesikari et al. 2009). The live attenuated monovalent vaccine Rotarix® (GSK) represents the second oral vaccine against rotavirus infection and was approved in 2008 by the FDA. Rotarix® is administered in a two-dose regimen to children from 6 to 24 weeks of age. The vaccine is based on the human 89-12 strain which belongs to the G1 serotype and P [8] genotype. Rotarix® was shown to be effective against the G1, G3, G4, and G9 strains. In clinical trials including phase I, II, and III performed in different countries, Rotarix® conferred 85–96 % protection against severe gastroenteritis and reduced hospitalization by 40–75 % (O'Ryan and Linhares 2009; Ruiz-Palacios et al. 2006). Both live attenuated rotavirus vaccine strains are suggested to replicate in the small intestine, thereby leading to the generation of intestinal and serum IgA as well as neutralizing antibodies, which might mediate long-term immunity (Franco et al. 2006; Vesikari et al. 2006). However, the mechanism of protection is still not completely understood.

3 Transcutaneous Vaccines

3.1 Transdermal and Transfollicular Vaccines

In the context of noninvasive vaccination strategies meeting the current challenges in vaccine delivery (e.g., lack of compliance, safety issues, and logistic constraints), TCV seems to be a valid alternative to mucosal approaches (Karande and Mitragotri 2010). The skin has been used for vaccination purposes from the very beginning when Jenner was implementing his vaccinations against smallpox in 1796 (Stern and Markel 2005). While Jenner was using a lancet to immunize by s.c. inoculation, nowadays needle-free jet injectors and bifurcated needles for i.d. vaccine delivery are used (Kendall 2010; Pattanayak et al. 1970). Thus, the skin has been known for centuries to be an attractive site for immunization, and recent developments in needle-free systems have renewed the interest in its exploitation as inductive site. TCV offers an attractive approach for the development of highly accepted and needle- or pain-free vaccines, which are not only safe but also effective due to the presence of abundant professional APCs, such as dendritic cells (DCs) and Langerhans cells (LCs), in different layers of the skin (Bangert et al. 2011). Thus, several studies reported on the stimulation of efficient immune responses encompassing strong humoral as well as cellular responses (Frenck et al. 2011; Icardi et al. 2012; Kim et al. 2010). In some cases, even low vaccine doses could be used, which would be especially important for pandemics in which vaccines become scarce goods and dose reduction is necessary (Bramson et al. 2003; Chen et al. 2004).

However, the main challenge for TCV is to enhance the transport of antigens across the stratum corneum (SC) barrier (Kaurav et al. 2016). To this end, reversible barrier disruption methods are often applied, such as chemical permeation enhancers, abrasion (e.g., tape stripping), electroporation, microneedles, laser-assisted generation of micropores, PowderJect, and gene gun (Bal et al. 2010). The disruption of the skin usually lasts for a significant time and also promotes an activation of the innate immune system that results in enhanced immune responses (Karande and Mitragotri 2010; Mittal et al. 2013; Rancan et al. 2013). When using polylactic acid nanoparticles (PLA NPs) as antigen delivery system, activation of LCs and subsequently CD8+ effector cells has been observed following TCV via cyanoacrylate-stripped skin (Liard et al. 2011; Rancan et al. 2013). Nevertheless, barrier disruption considerably increases the risk of infection, making these methods suboptimal for mass vaccination campaigns in countries with critical hygienic conditions or vaccination of immune-compromised individuals, elderly, young children, or people with poor wound healing (Kugelberg et al. 2005; Wanke et al. 2013). Thus, even though some of these methods are needle-free or pain-free, they are, to a certain extent, still invasive. Furthermore, for many of these approaches, such as gene gun, PowderJect, laser-assisted generation of micropores, and ultrasound, special devices are needed and make necessary application by trained personal. Recently, an alternative pathway has been reported targeting the skin immune system via the hair follicles without skin disruption by using nanoparticles. This transfollicular (t.f.) route is very attractive for vaccine delivery as LCs and DCs accumulate around the hair follicles and antigens are cleared only slowly by hair growth and sebum production. Thus, hair follicles represent an excellent antigen reservoir. Furthermore, antigens entering the follicles (e.g., glycoproteins and pro-inflammatory fatty acids) were shown to (i) very rapidly cross the sebum-filled follicle, (ii) be ingested by the present APCs, and (iii) promote immune responses (Mahe et al. 2009). While LCs are mainly present in the epidermis and responsible for the stimulation of Th2, Th17, and Treg cells, different subsets of DCs are found in the dermis. The main DC population is characterized by the expression of CD11b but not CD103 and langerin. These cells are non-migratory residing in the skin, and they play a major role in the stimulation of $CD4^+$ T cells including not only Th1 and Th2 but also Treg cells (Heath and Carbone 2013; McLachlan et al. 2009). The other DC subset is characterized by the expression of CD103 and langerin as well as Clec9A and seems to be important with regard to the stimulation of cytotoxic CD8+ T cell responses (Bedoui et al. 2009; Caminschi et al. 2008; Heath and Carbone 2013; McLachlan et al. 2009; Shortman and Heath 2010). However, antigen delivery into the hair follicles without disrupting the skin barrier is difficult, since most of the material remains on the skin surface (Mittal et al. 2015a, b). Furthermore, pure antigens are usually only poorly immunogenic, and therefore, they are not properly recognized by the innate immune cells of the skin. In this context, nano- or microsized carriers, such as polylactic-co-glycolic acid nanoparticles (PLGA NPs) or inverse micellar sugar glass nanoparticles (IMSG NPs), designed for the delivery of vaccine antigens and adjuvants to the hair follicles have been shown not only to increase the amount of antigen reaching deeper areas of the hair follicle (i.e., the infundibulum) by a factor of three, but also to significantly increase the strength of the stimulated antigen-specific immune responses without the necessity of prior skin disruption (Mittal et al. 2013, 2015a, b). Thus, beside the stimulation of strong humoral immune responses, also efficient cellular responses have been obtained following t.f. immunization, as indicated by the stimulation of antigen-specific multifunctional T cells (Mittal et al. 2015a, b). There is a general consensus that efficient immunity to infection depends to a substantial part on the functional properties of antigen-specific T cells. In this regard, there is increasing evidence that T cells which produce several key cytokines at one time are more potent in their effector functions as compared to single producers (Almeida et al. 2007; Darrah et al. 2007, 2010; Graw and Regoes 2014; Kannanganat et al. 2007, Seder et al. 2008a, b). While IFNγ mediates the effector function of T cells, TNFα enhances their proliferation capacity and IL-2 is needed for maintaining the immune response (reviewed in Seder et al. 2008a, b). T cells which produce both IFNγ and TNFα were shown to mediate more efficient killing, as compared with cells expressing either cytokine alone, whereas those expressing IL-2 and TNFα seem to serve as a reservoir of memory T cells (reviewed in Seder et al. 2008a, b).

Although the use of NPs as a carrier system for vaccine antigens already increases the efficacy of vaccination via the t.f. route in a way that makes it unnecessary to destroy the skin barrier, incorporation of a suitable adjuvant seems to be required in order to promote strong immune responses. The adjuvants used so far encompass TLR agonists (e.g., CpG, Poly I:C), A/B moiety toxins or their derivatives (e.g., CT, LT), STING agonists (e.g., c-di-AMP), and others (e.g., polyphosphazenes, saponins) (reviewed in Engelke et al. 2015). For example, TCV with a CTL epitope using the imidazoquinoline derivative imiquimod as adjuvant induces potent CTL responses (Stein et al. 2014). Strong Th2 responses have also been observed by using bacterial toxins as adjuvant applied on shaved skin (Glenn et al. 2007; McKenzie et al. 2007). However, often it is not clear from the studies whether the skin of the treated animals was actually intact. For example, immunizations were either performed only shortly after the shaving when microtrauma might have not been healed or the skin was hydrated prior to immunization, thereby increasing its permeability (Glenn et al. 2007; Mondoulet et al. 2012). In contrast, t.f. vaccination with antigen and c-di-AMP-loaded NPs 48 h after hair depilation, when skin integrity reached basal level again, resulted in strong humoral and cellular responses (Mittal et al. 2015a, b). The c-di-AMP originates from prokaryotes, in which it acts as second messenger (Schaap 2013). Several studies have shown that c-di-AMP exhibits strong adjuvant properties regardless of the administration route (Ebensen et al. 2007a, b, 2011; Sanchez et al. 2014). The transmembrane protein STING (stimulator of IFN genes) was identified as an innate sensor of cyclic dinucleotides. Interaction with STING activates the TBK1-IRF3 signaling pathway, subsequently resulting in the production of type I IFN and TNFα (Blaauboer et al. 2014; Burdette et al. 2011; McWhirter et al. 2009; Shu et al. 2012).

Nevertheless, generation of antigen-loaded NPs still has some limitations. The use of chemical solvents and physical stresses during the production process can negatively affect antigen and/or adjuvant stability. Also, the involvement of complex multistep generation processes leads to increased production costs (De Geest et al. 2012). Some of those problems can be overcome by using, for example, simplified approaches based on inverse micelles. Micelles can be prepared using pharmaceutical excipients, such as medium-chain triglycerides (MCTs) and phospholipids (e.g., lectins), which are generally regarded as safe (GRAS listed), biocompatible, and biodegradable. MCTs are broadly used for parenteral nutrition or as a readily digestible energy source for neonates. The production process of these kinds of micelles is gentler compared to the formation of antigen-loaded PLGA NPs, resulting in an improved integrity of the encapsulated antigen as compared to PLGA NPs. This in turn secures the stimulation of efficient immune responses tailored to the specific needs (Mittal et al. 2015a, b). However, although the encapsulation efficiency of these NPs is about 100 %, the amount of antigen and adjuvant needed for NP formulation is still considerable. Therefore, new optimized particle preparation techniques need to be developed (e.g., "hydrophobic ion pairing—(HIP–) and reverse micelle-hybridized o/w emulsion technique") in order to minimize the amount of antigen and adjuvant needed (reviewed in Sah and Sah 2015). Besides optimizing the NP formulation process, the follicular uptake of the particles as well as the subsequent interaction with dermal APCs needs to be improved when the SC barrier remains intact. Aljuffali and coworkers developed nanoparticles formed from sebum-derived lipids, i.e., squalene and fatty esters, resulting in an up to sevenfold higher follicular uptake compared to the non-particulate drug formulation (Aljuffali et al. 2014). Another approach aims at improved targeting of the antigen-loaded NPs to the dermal APCs by conjugation of DC-directed antibodies (e.g., anti-Clec9A, DEC205) or DC receptor–ligands, such as chitosan or mannose (Demoulins et al. 2014; He et al. 2007; Thomann-Harwood et al. 2013; Tsuji et al. 2011). Although DC-targeting seems to be sufficient for stimulating humoral immune responses, elicitation of strong cellular immunity, especially CTL responses, seems to require the incorporation of an adjuvant (Badillo-Godinez et al. 2015; He et al. 2007; Sancho et al. 2008; van Kooyk et al. 2013).

In any case, besides putting enormous efforts to develop efficient particle-based carrier systems with increased loading, encapsulating and targeting efficacies, immunogenicity, and antigenicity of the used particles as well as the incorporated adjuvants need to be determined. Thus, particle- and/or adjuvant-specific immune responses could result in decreased vaccine efficacy due to rapid clearance of the carrier system and will hamper its repeated use for different vaccinations (Ilinskaya and Dobrovolskaia 2016).

Taken together, a variety of novel approaches are currently under development to allow for minimal-invasive, needle-free vaccination against infectious diseases. In this context, NP-based immunization via the t.f. route constitutes a promising approach to deliver antigens through the intact skin.

3.2 Approved Transcutaneous Vaccine

The first needle-free intramuscular-delivered vaccine is available on the market. The seasonal trivalent inactivated influenza vaccine Afluria® (CSL Biotherapies), consisting of the influenza A strains H1N1 and H3N2 and the B strain, which is administered with a needle and syringe, was first approved in 2007 for children and adults. In 2014, the FDA approved the delivery of Afluria® (CSL Biotherapies) with the PharmaJet® Stratis® needle-free injection system for persons aged 18–64 years. The PharmaJet® injector works with a precise fluid stream penetrating the skin in about one-tenth of a second, thereby preventing potential needle-stick injuries and cross-contamination. In a randomized phase IV trial in healthy adults, it was clearly demonstrated that Afluria® delivered by jet injection met the criteria for non-inferiority to that of the needle and syringe groups with regard to immunogenicity, as determined by the strain-specific geometric mean titer ratios and seroconversion rates (McAllister et al. 2014). The majority of the study participants provided a strong positive feedback and a preference for jet injection for their next vaccination. The high acceptance in the public accompanied by the prevention of needle-stick-induced side effects renders needle-free vaccination a promising tool for future vaccines. The following section will introduce some of the needle-free vaccine candidates currently in the clinical development pipeline.

3.3 Needle-Free Transcutaneous Vaccine Candidates Under Clinical Development

Several needle-free transcutaneous vaccine candidates against influenza are currently under investigation in clinical trials encompassing phase I–IV (e.g., ClinicalTrials.gov: NCT00386542, NCT00694213, NCT00987350) aimed to evaluate the safety and efficacy of influenza vaccines that are delivered by needle-free injector systems as compared to needle syringe administration. Beside vaccines against influenza, a relatively low number of clinical trials study the effect of needle-free administration against other infections.

The phase I study CUT*HIVAC001 is focused on assessing the safety and immunogenicity of a GTU®-multi-HIV B clade DNA vaccine using different invasive delivery strategies and one transcutaneous needle-free approach. The experimental HIV vaccine is administered to healthy volunteers being at low risk of HIV infection in a three-dose regimen over 12 weeks. For the needle-free approach, the hairs of the volunteers are stripped away and a water-based solution containing the vaccine is placed on the skin surface. The HIV DNA is intended to enter the body through the hair follicles, thereby inducing anti-HIV immunity. This needle-free approach will be compared with intramuscular and intradermal injections as well as with an intramuscular approach amplified by electroporation, especially with regard to the generation of HIV-specific CD8 T cell responses measured 2 weeks after the last vaccination. Another phase I study performed in healthy adults compares the immune response and side effects of an experimental adenovirus-based HIV vaccine (VRC-HIVADV014-00-VP) given by needle injection or by use of a needle-free device called the Biojector 2000™. The single-dose vaccine comprises four adenoviral vectors coding for the HIV-1 gag/pol polyprotein from clade B and HIV-1 env glycoproteins from clades A, B, and C. Blood samples are collected at weeks 2, 4, 12, and 24 and analyzed for HIV-specific humoral and cellular immune responses as well as for adenovirus serotype 5 antibodies. An exploratory phase I b study (ClinicalTrials.gov, NCT01765426) also evaluates the safety and immunogenicity of a recombinant live attenuated tetravalent dengue vaccine (DENVax), comparing intradermal administration by needle with the needle-free PharmaJet injector. The vaccine is administered to healthy adults in a two-dose regimen at day 0 and 90. Blood samples collected at days 0, 120, and 270 after vaccination will be used to measure viremia encompassing all four dengue serotypes, the generation of neutralizing antibodies and the induction of cellular immunity. A phase IV study (ClinicalTrials.gov, NCT02409095) is aimed at assessing and comparing the immunogenicity and reactogenicity of the DTP-HB-Hib vaccine (Serum Institute of

India Ltd) against diphtheria, tetanus, pertussis, hepatitis B, and haemophilus influenza type B administered either with a syringe jet injector or syringe needle in children aged 6–8 weeks. Serum samples are taken at day 28 after administration of the 3rd dose and will be used to look for seroconversion against each vaccine component.

The usage of hydrogel patches represents another promising needle-free alternative for TCV. In a clinical phase I study, a TCV formulation consisting of a hydrogel patch which contains tetanus toxoid and diphtheria toxoid was evaluated for safety and efficacy. Vaccine delivery by hydrogel patches resulted in increased levels of anti-tetanus toxoid and anti-diphtheria toxoid antibodies in most of the volunteers measured at days 60 and 365 after application. A second vaccination of the non-responders could in part increase the antibody titers, thereby indicating that a repeated TCV can enhance vaccine efficacy. A passive challenge experiment using sera of vaccinated volunteers to protect mice against a lethal challenge with tetanus toxoid displayed the generation of neutralizing antibodies suggesting that TCV induces protective immunity against bacterial toxins. The analysis of the skin revealed increased water content in the SC, which suggests that antigen penetration is associated with changes in the skin condition (Hirobe et al. 2012).

A phase III placebo-controlled field trail addressed the safety and efficacy of a skin patch vaccine against ETEC which contains the LT in Europeans' travelers. Healthy volunteers aged between 18 and 64 years received 2 skin patch vaccine doses 14 days apart prior to traveling (ClinicalTrials.gov, number NCT00993681). The study revealed an antigen-specific induction of serum IgG but a vaccine efficacy of only 34.6 %, as defined by the occurrence of moderate or severe ETEC-induced diarrhea. Furthermore, 93 % of the participants of the vaccine patch group displayed local adverse events including erythema, rash, pruritus, hyperpigmentation, hypopigmentation, pain, and edema, whereas in the placebo group (i.e., same patch and application without the LT protein), 56 % showed these kinds of adverse effects (Behrens et al. 2014). These findings further underline the importance of developing TCV strategies with improved tolerability, while avoiding methods associated with dermatitis. Interestingly, a similar previously performed phase II study in healthy US participants demonstrated a vaccine efficacy of 75 % without induction of side effects (ClinicalTrials.gov, number NCT00516659, Frech et al. 2008) Nevertheless, this study was smaller with efficacy results based on 59 participants in contrast to about 2000 in the study mentioned above.

A clinical phase I/II study using a vaccine against measles evaluated the safety and efficacy of a skin patch delivery as compared to the subcutaneous route. Here, healthy volunteers were vaccinated with the live attenuated measles vaccine ROUVAX®. This study clearly showed an elevated induction of measles-specific salivary IgA compared to subcutaneous vaccination and a slight increase of IFNγ-producing T cells. However, this study also revealed that in contrast to s.c. vaccination, TCV failed to efficiently stimulate measles-specific neutralizing serum antibody titers, most probably due to inefficient vaccine uptake by skin DCs (Clinical Trials Registry, ISRCTN88861431, Etchart et al. 2007).

Another phase I study focused on vaccine targeting to human hair follicle ducts. This trial investigated the delivery of a conventional licensed tetanus/influenza vaccine (TETAGRIP®) by TCV as compared to i.m. immunization in healthy adults and HIV exposed individuals. TCV displayed superiority regarding the generation of influenza-specific CD8 T cells in both healthy and HIV-infected volunteers, whereas no differences were observed with respect to the induction of influenza-specific CD4 T cells. Similar to other studies, only i.m. immunization resulted in the induction of neutralizing antibodies (Clinicaltrials.gov NCT00261001, Combadiere et al. 2010).

4 Conclusions and Outlook

Vaccination remains one of the most effective tools for preventing infectious diseases. However, potential rare adverse events together with a decreased perception of the real threat posed by infectious agents result in a certain degree of vaccination fatigue among the general public. Thus, switching from injectable vaccines to noninvasive needle-free vaccination strategies would not only increase vaccine safety, especially in terms of mass vaccination campaigns in countries with low hygiene standards, but also enhance the acceptance by the public due to painless application, stilling the fear associated with needle injection. In this context, mucosal as well as needle-free TCV approaches constitute promising strategies able to stimulate not only systemic but already mucosal immunity toward protection against infectious diseases. However, further effort needs to be taken in order to improve vaccine performance following needle-free application. Thus, delivery systems need to be developed allowing efficient transport of vaccine antigens across the different mucosal and skin barriers. Furthermore, specific targeting of vaccine antigens to those APCs activating the appropriate arms of the immune system would further improve vaccine efficacy. Increased understanding of inflammation processes, and innate and adaptive immune responses, has resulted in a better understanding of the underlying principles for the mechanisms of adjuvant actions and provides the base for the design of new vaccines with improved performance. The development of new noninvasive vaccines could prove a windfall to vaccine developers aiming at safer vaccines to improve public health. Vaccination strategies might emerge as the defining trend in the foreseeable future allowing not only immune enhancement along with a simultaneous dose sparing but also self-administration, even in elderly patients with less responsive immune systems.

Acknowledgement This work was in part supported by the BMBF-funded project PeTrA (Project Number 13N11455).

References

Agger EM, Rosenkrands I, Hansen J, Brahimi K, Vandahl BS, Aagaard C et al (2008) Cationic liposomes formulated with synthetic mycobacterial cordfactor (CAF01): a versatile adjuvant for vaccines with different immunological requirements. PLoS One 3(9):e3116

Aljuffali IA, Sung CT, Shen FM, Huang CT, Fang JY (2014) Squarticles as a lipid nanocarrier for delivering diphencyprone and minoxidil to hair follicles and human dermal papilla cells. AAPS J 16(1):140–150

Allison AC (1997) Immunological adjuvants and their modes of action. Arch Immunol Ther Exp (Warsz) 45(2–3):141–147

Almeida JR, Price DA, Papagno L, Arkoub ZA, Sauce D, Bornstein E et al (2007) Superior control of HIV-1 replication by CD8+ T cells is reflected by their avidity, polyfunctionality, and clonal turnover. J Exp Med 204(10):2473–2485

Ambrose CS, Wu X, Caspard H, Belshe RB (2014) Efficacy of live attenuated *influenza* vaccine against *influenza* illness in children as a function of illness severity. Vaccine 32(43):5546–5548

Anjuere F, Bekri S, Bihl F, Braud VM, Cuburu N, Czerkinsky C et al (2012) B cell and T cell immunity in the female genital tract: potential of distinct mucosal routes of vaccination and role of tissue-associated dendritic cells and natural killer cells. Clin Microbiol Infect 18 (Suppl 5):117–122

Baaten BJ, Clarke B, Strong P, Hou S (2010) Nasal mucosal administration of chitin microparticles boosts innate immunity against *influenza A virus* in the local pulmonary tissue. Vaccine 28(25):4130–4137

Badillo-Godinez O, Gutierrez-Xicotencatl L, Plett-Torres T, Pedroza-Saavedra A, Gonzalez-Jaimes A, Chihu-Amparan L et al (2015) Targeting of rotavirus VP6 to DEC-205 induces protection against the infection in mice. Vaccine 33(35):4228–4237

Bal SM, Ding Z, van Riet E, Jiskoot W, Bouwstra JA (2010) Advances in transcutaneous vaccine delivery: do all ways lead to Rome? J Control Release 148(3):266–282

Bangert C, Brunner PM, Stingl G (2011) Immune functions of the skin. Clin Dermatol 29(4):360–376

Becker PD, Fiorentini S, Link C, Tosti G, Ebensen T, Caruso A et al (2006) The HIV-1 matrix protein p17 can be efficiently delivered by intranasal route in mice using the TLR 2/6 agonist MALP-2 as mucosal adjuvant. Vaccine 24(25):5269–5276

Bedoui S, Whitney PG, Waithman J, Eidsmo L, Wakim L, Caminschi I et al (2009) Cross-presentation of viral and self antigens by skin-derived CD103+ dendritic cells. Nat Immunol 10(5):488–495

Behrens RH, Cramer JP, Jelinek T, Shaw H, von Sonnenburg F, Wilbraham D et al (2014) Efficacy and safety of a patch vaccine containing heat-labile toxin from Escherichia coli against travellers' diarrhoea: a phase 3, randomised, double-blind, placebo-controlled field trial in travellers from Europe to Mexico and Guatemala. Lancet Infect Dis 14(3):197–204

Beignon AS, Briand JP, Rappuoli R, Muller S, Partidos CD (2002) The LTR72 mutant of heat-labile enterotoxin of *Escherichia coli* enhances the ability of peptide antigens to elicit CD4 (+) T cells and secrete gamma interferon after coapplication onto bare skin. Infect Immun 70 (6):3012–3019

Bennett E, Mullen AB, Ferro VA (2009) Translational modifications to improve vaccine efficacy in an oral influenza vaccine. Methods 49(4):322–327

Blaauboer SM, Gabrielle VD, Jin L (2014) MPYS/STING-mediated TNF-alpha, not type I IFN, is essential for the mucosal adjuvant activity of (3'-5')-cyclic-di-guanosine-monophosphate in vivo. J Immunol 192(1):492–502

Bolhassani A, Javanzad S, Saleh T, Hashemi M, Aghasadeghi MR, Sadat SM (2014) Polymeric nanoparticles: potent vectors for vaccine delivery targeting cancer and infectious diseases. Hum Vaccines Immunotherapeutics 10(2):321–332

Borsutzky S, Fiorelli V, Ebensen T, Tripiciano A, Rharbaoui F, Scoglio A et al (2003) Efficient mucosal delivery of the HIV-1 Tat protein using the synthetic lipopeptide MALP-2 as adjuvant. Eur J Immunol 33(6):1548–1556

Boyce TG, Hsu HH, Sannella EC, Coleman-Dockery SD, Baylis E, Zhu Y et al (2000) Safety and immunogenicity of adjuvanted and unadjuvanted subunit *influenza* vaccines administered intranasally to healthy adults. Vaccine 19(2–3):217–226

Bramson J, Dayball K, Evelegh C, Wan YH, Page D, Smith A (2003) Enabling topical immunization via microporation: a novel method for pain-free and needle-free delivery of adenovirus-based vaccines. Gene Ther 10(3):251–260

Brandtzaeg P (2007) Induction of secretory immunity and memory at mucosal surfaces. Vaccine 25(30):5467–5484

Burdette DL, Monroe KM, Sotelo-Troha K, Iwig JS, Eckert B, Hyodo M et al (2011) STING is a direct innate immune sensor of cyclic di-GMP. Nature 478(7370):515–518

Caetano LA, Almeida AJ, Goncalves LM (2014) Approaches to tuberculosis mucosal vaccine development using nanoparticles and microparticles: a review. J Biomed Nanotechnol 10 (9):2295–2316

Caminschi I, Proietto AI, Ahmet F, Kitsoulis S, Shin Teh J, Lo JC et al (2008) The dendritic cell subtype-restricted C-type lectin Clec9A is a target for vaccine enhancement. Blood 112 (8):3264–3273

Carter NJ, Curran MP (2011) Live attenuated *influenza* vaccine (FluMist(R); Fluenz): a review of its use in the prevention of seasonal influenza in children and adults. Drugs 71(12):1591–1622

Chadwick S, Kriegel C, Amiji M (2009) Delivery strategies to enhance mucosal vaccination. Expert Opin Biol Ther 9(4):427–440

Chen D, Burger M, Chu Q, Endres R, Zuleger C, Dean H et al (2004) Epidermal powder immunization: cellular and molecular mechanisms for enhancing vaccine immunogenicity. Virus Res 103(1–2):147–153

ClinicalTrials.gov (2012) A service of the U.S. National Institutes of Health. Trial on the safety of a new liposomal adjuvant system, CAF01, when given with the tuberculosis subunit vaccine Ag85B-ESAT-6 as two injections with two months interval to healthy adult volunteers. Available from: http://clinicaltrials.gov/ct2/show/NCT00922363

Combadiere B, Vogt A, Mahe B, Costagliola D, Hadam S, Bonduelle O et al (2010) Preferential amplification of CD8 effector-T cells after transcutaneous application of an inactivated *influenza* vaccine: a randomized phase I trial. PLoS One 5(5):e10818

Cox E, Verdonck F, Vanrompay D, Goddeeris B (2006) Adjuvants modulating mucosal immune responses or directing systemic responses towards the mucosa. Vet Res 37(3):511–539

Darrah PA, Hegde ST, Patel DT, Lindsay RW, Chen L, Roederer M et al (2010) IL-10 production differentially influences the magnitude, quality, and protective capacity of Th1 responses depending on the vaccine platform. J Exp Med 207(7):1421–1433

Darrah PA, Patel DT, De Luca PM, Lindsay RW, Davey DF, Flynn BJ et al (2007) Multifunctional TH1 cells define a correlate of vaccine-mediated protection against *Leishmania major*. Nat Med 13(7):843–850

Davitt CJ, Lavelle EC (2015) Delivery strategies to enhance oral vaccination against enteric infections. Adv Drug Deliv Rev 91:52–69

De Geest BG, Willart MA, Lambrecht BN, Pollard C, Vervaet C, Remon JP et al (2012) Surface-engineered polyelectrolyte multilayer capsules: synthetic vaccines mimicking microbial structure and function. Angew Chem Int Edit 51(16):3862–3866

De Vendittis E, Palumbo G, Parlato G, Bocchini V (1981) A fluorimetric method for the estimation of the critical micelle concentration of surfactants. Anal Biochem 115(2):278–286

Demoulins T, Milona P, McCullough KC (2014) Alginate-coated chitosan nanogels differentially modulate class-A and class-B CpG-ODN targeting of dendritic cells and intracellular delivery. Nanomed Nanotechnol Biol Med 10(8):1739–1749

des Rieux A, Fievez V, Garinot M, Schneider YJ, Preat V (2006) Nanoparticles as potential oral delivery systems of proteins and vaccines: a mechanistic approach. J Control Release 116 (1):1–27

Dhere R, Yeolekar L, Kulkarni P, Menon R, Vaidya V, Ganguly M et al (2011) A pandemic *influenza* vaccine in India: from strain to sale within 12 months. Vaccine 29(Suppl 1):A16–A21

Dlugonska H, Grzybowski M (2012) Mucosal vaccination–an old but still vital strategy. Ann Parasitol 58(1):1–8

Ebensen T, Libanova R, Schulze K, Yevsa T, Morr M, Guzman CA (2011) Bis-(3',5')-cyclic dimeric adenosine monophosphate: strong Th1/Th2/Th17 promoting mucosal adjuvant. Vaccine 29(32):5210–5220

Ebensen T, Link C, Riese P, Schulze K, Morr M, Guzman CA (2007a) A pegylated derivative of alpha-galactosylceramide exhibits improved biological properties. J Immunol 179(4):2065–2073

Ebensen T, Schulze K, Riese P, Morr M, Guzman CA (2007b) The bacterial second messenger cdiGMP exhibits promising activity as a mucosal adjuvant. Clin Vaccine Immunol CVI 14 (8):952–958

el-Sayed N, el-Gamal Y, Abbassy AA, Seoud I, Salama M, Kandeel A et al (2008) Monovalent type 1 oral poliovirus vaccine in newborns. N Engl J Med 359(16):1655–1665

Engelke L, Winter G, Hook S, Engert J (2015) Recent insights into cutaneous immunization: how to vaccinate via the skin. Vaccine 33(37):4663–4674

Etchart N, Hennino A, Friede M, Dahel K, Dupouy M, Goujon-Henry C et al (2007) Safety and efficacy of transcutaneous vaccination using a patch with the live-attenuated measles vaccine in humans. Vaccine 25(39–40):6891–6899

Franco MA, Angel J, Greenberg HB (2006) Immunity and correlates of protection for rotavirus vaccines. Vaccine 24(15):2718–2731

Frech SA, Dupont HL, Bourgeois AL, McKenzie R, Belkind-Gerson J, Figueroa JF et al (2008) Use of a patch containing heat-labile toxin from *Escherichia coli* against travellers' diarrhoea: a phase II, randomised, double-blind, placebo-controlled field trial. Lancet 371(9629): 2019–2025

Frenck RW Jr, Belshe R, Brady RC, Winokur PL, Campbell JD, Treanor J et al (2011) Comparison of the immunogenicity and safety of a split-virion, inactivated, trivalent *influenza* vaccine (Fluzone(R)) administered by intradermal and intramuscular route in healthy adults. Vaccine 29(34):5666–5674

Glenn GM, Flyer DC, Ellingsworth LR, Frech SA, Frerichs DM, Seid RC et al (2007) Transcutaneous immunization with heat-labile enterotoxin: development of a needle-free vaccine patch. Expert Rev Vaccines 6(5):809–819

Graw F, Regoes RR (2014) Predicting the impact of CD8+ T cell polyfunctionality on HIV disease progression. J Virol 88(17):10134–10145

Hafner LM, Wilson DP, Timms P (2014) Development status and future prospects for a vaccine against *Chlamydia trachomatis* infection. Vaccine 32(14):1563–1571

He LZ, Crocker A, Lee J, Mendoza-Ramirez J, Wang XT, Vitale LA et al (2007) Antigenic targeting of the human mannose receptor induces tumor immunity. J Immunol 178(10):6259–6267

He XS, Holmes TH, Zhang C, Mahmood K, Kemble GW, Lewis DB et al (2006) Cellular immune responses in children and adults receiving inactivated or live attenuated *influenza* vaccines. J Virol 80(23):11756–11766

Heath WR, Carbone FR (2013) The skin-resident and migratory immune system in steady state and memory: innate lymphocytes, dendritic cells and T cells. Nat Immunol 14(10):978–985

Hirobe S, Matsuo K, Quan YS, Kamiyama F, Morito H, Asada H et al (2012) Clinical study of transcutaneous vaccination using a hydrogel patch for tetanus and diphtheria. Vaccine 30 (10):1847–1854

Hisert KB, MacCoss M, Shiloh MU, Darwin KH, Singh S, Jones RA et al (2005) A glutamate-alanine-leucine (EAL) domain protein of *Salmonella* controls bacterial survival in mice, antioxidant defence and killing of macrophages: role of cyclic diGMP. Mol Microbiol 56 (5):1234–1245

Holmgren J, Adamsson J, Anjuere F, Clemens J, Czerkinsky C, Eriksson K et al (2005) Mucosal adjuvants and anti-infection and anti-immunopathology vaccines based on cholera toxin, cholera toxin B subunit and CpG DNA. Immunol Lett 97(2):181–188

Holmgren J, Czerkinsky C (2005) Mucosal immunity and vaccines. Nat Med 11(4 Suppl):S45–S53

Honma K, Abraham JL, Chiyotani K, De Vuyst P, Dumortier P, Gibbs AR et al (2004) Proposed criteria for mixed-dust pneumoconiosis: definition, descriptions, and guidelines for pathologic diagnosis and clinical correlation. Hum Pathol 35(12):1515–1523

Icardi G, Orsi A, Ceravolo A, Ansaldi F (2012) Current evidence on intradermal *influenza* vaccines administered by Soluvia licensed micro injection system. Hum Vaccines immunotherapeutics 8(1):67–75

Ilinskaya AN, Dobrovolskaia MA (2016) Understanding the immunogenicity and antigenicity of nanomaterials: Past, present and future. Toxicol Appl Pharmacol 299:70–77

Israeli E, Agmon-Levin N, Blank M, Shoenfeld Y (2009) Adjuvants and autoimmunity. Lupus 18 (13):1217–1225

Jia Y, Krishnan L, Omri A (2015) Nasal and pulmonary vaccine delivery using particulate carriers. Expert Opin Drug Deliv 12(6):993–1008

Kannanganat S, Ibegbu C, Chennareddi L, Robinson HL, Amara RR (2007) Multiple-cytokine-producing antiviral CD4 T cells are functionally superior to single-cytokine-producing cells. J Virol 81(16):8468–8476

Karande P, Mitragotri S (2010) Transcutaneous immunization: an overview of advantages, disease targets, vaccines, and delivery technologies. Annu Rev Chem Biomol 1:175–201

Kaurav M, Minz S, Sahu K, Kumar M, Madan J, Pandey RS (2016) Nanoparticulate mediated transcutaneous immunization: Myth or reality. Nanomed Nanotechnol Biol Med 12:1063–1081

Kendall MA (2010) Needle-free vaccine injection. Handb Exp Pharmacol 197:193–219

Kim YC, Quan FS, Yoo DG, Compans RW, Kang SM, Prausnitz MR (2010) Enhanced memory responses to seasonal H1N1 *influenza* vaccination of the skin with the use of vaccine-coated microneedles. J Infect Dis 201(2):190–198

Kraan H, Vrieling H, Czerkinsky C, Jiskoot W, Kersten G, Amorij JP (2014) Buccal and sublingual vaccine delivery. J Control Release 190:580–592

Kugelberg E, Norstrom T, Petersen TK, Duvold T, Andersson DI, Hughes D (2005) Establishment of a superficial skin infection model in mice by using *Staphylococcus aureus* and *Streptococcus pyogenes*. Antimicrob Agents Chemother 49(8):3435–3441

Kundi M (2007) New hepatitis B vaccine formulated with an improved adjuvant system. Expert Rev Vaccines 6(2):133–140

Kuroda Y, Nacionales DC, Akaogi J, Reeves WH, Satoh M (2004) Autoimmunity induced by adjuvant hydrocarbon oil components of vaccine. Biomed Pharmacother 58(5):325–337

Kweon MN (2011) Sublingual mucosa: a new vaccination route for systemic and mucosal immunity. Cytokine 54(1):1–5

Lamichhane A, Azegamia T, Kiyonoa H (2014) The mucosal immune system for vaccine development. Vaccine 32(49):6711–6723

Levin C, Perrin H, Combadiere B (2015) Tailored immunity by skin antigen-presenting cells. Hum Vaccines immunotherapeutics 11(1):27–36

Liard C, Munier S, Arias M, Joulin-Giet A, Bonduelle O, Duffy D et al (2011) Targeting of HIV-p24 particle-based vaccine into differential skin layers induces distinct arms of the immune responses. Vaccine 29(37):6379–6391

Link C, Gavioli R, Ebensen T, Canella A, Reinhard E, Guzman CA (2004) The Toll-like receptor ligand MALP-2 stimulates dendritic cell maturation and modulates proteasome composition and activity. Eur J Immunol 34(3):899–907

Mahe B, Vogt A, Liard C, Duffy D, Abadie V, Bonduelle O et al (2009) Nanoparticle-based targeting of vaccine compounds to skin antigen-presenting cells by hair follicles and their transport in mice. J Invest Dermatol 129(5):1156–1164

McAllister L, Anderson J, Werth K, Cho I, Copeland K, Le Cam Bouveret N et al (2014) Needle-free jet injection for administration of *influenza* vaccine: a randomised non-inferiority trial. Lancet 384(9944):674–681

McKenzie R, Bourgeois AL, Frech SA, Flyer DC, Bloom A, Kazempour K et al (2007) Transcutaneous immunization with the heat-labile toxin (LT) of enterotoxigenic *Escherichia coli* (ETEC): protective efficacy in a double-blind, placebo-controlled challenge study. Vaccine 25(18):3684–3691

McLachlan JB, Catron DM, Moon JJ, Jenkins MK (2009) Dendritic cell antigen presentation drives simultaneous cytokine production by effector and regulatory T cells in inflamed skin. Immunity 30(2):277–288

McWhirter SM, Barbalat R, Monroe KM, Fontana MF, Hyodo M, Joncker NT et al (2009) A host type I interferon response is induced by cytosolic sensing of the bacterial second messenger cyclic-di-GMP. J Exp Med 206(9):1899–1911

Mesa C, Fernandez LE (2004) Challenges facing adjuvants for cancer immunotherapy. Immunol Cell Biol 82(6):644–650

Minor P (2009) Vaccine-derived *poliovirus* (VDPV): impact on poliomyelitis eradication. Vaccine 27(20):2649–2652

Mitragotri S (2005) Immunization without needles. Nat Rev Immunol 5(12):905–916

Mittal A, Raber AS, Schaefer UF, Weissmann S, Ebensen T, Schulze K et al (2013) Non-invasive delivery of nanoparticles to hair follicles: a perspective for transcutaneous immunization. Vaccine 31(34):3442–3451

Mittal A, Schulze K, Ebensen T, Weissmann S, Hansen S, Guzman CA et al (2015a) Inverse micellar sugar glass (IMSG) nanoparticles for transfollicular vaccination. J Controlled Release (official journal of the Controlled Release Society) 206:140–152

Mittal A, Schulze K, Ebensen T, Weissmann S, Hansen S, Lehr CM et al (2015b) Efficient nanoparticle-mediated needle-free transcutaneous vaccination via hair follicles requires adjuvantation. Nanomed Nanotechnol Biol Med 11(1):147–154

Mondoulet L, Dioszeghy V, Puteaux E, Ligouis M, Dhelft V, Letourneur F et al (2012) Intact skin and not stripped skin is crucial for the safety and efficacy of peanut epicutaneous immunotherapy (EPIT) in mice. Clin Transl Allergy 2(1):22

Mutsch M, Zhou W, Rhodes P, Bopp M, Chen RT, Linder T et al (2004) Use of the inactivated intranasal influenza vaccine and the risk of Bell's palsy in Switzerland. N Engl J Med 350 (9):896–903

Newsted D, Fallahi F, Golshani A, Azizi A (2015) Advances and challenges in mucosal adjuvant technology. Vaccine 33(21):2399–2405

O'Hagan DT (1998) Recent advances in vaccine adjuvants for systemic and mucosal administration. J Pharm Pharmacol 50(1):1–10

O'Hagan DT, Ott GS, Van Nest G (1997) Recent advances in vaccine adjuvants: the development of MF59 emulsion and polymeric microparticles. Mol Med Today 3(2):69–75

O'Hagan DT, Rappuoli R (2004) The safety of vaccines. Drug Discov Today 9(19):846–854

O'Ryan M, Linhares AC (2009) Update on Rotarix: an oral human rotavirus vaccine. Expert Rev Vaccines 8(12):1627–1641

Ott G, Barchfeld GL, Chernoff D, Radhakrishnan R, van Hoogevest P, Van Nest G (1995) MF59. Design and evaluation of a safe and potent adjuvant for human vaccines. Pharm Biotechnol 6:277–296

Patel P, Salapatek AM (2006) Pollinex Quattro: a novel and well-tolerated, ultra short-course allergy vaccine. Expert Rev vaccines 5(5):617–629

Pattanayak S, Arora DD, Sehgal CL, Raghavan NG, Topa PK, Subrahmanyam YK (1970) Comparative studies of *smallpox* vaccination by the bifurcated needle and rotary lancet techniques. Bull World Health Organ 42(2):305–310

Pedersen G, Cox R (2012) The mucosal vaccine quandary: intranasal vs. sublingual immunization against *influenza*. Hum Vaccines Immunotherapeutics 8(5):689–693

Persing DH, Coler RN, Lacy MJ, Johnson DA, Baldridge JR, Hershberg RM et al (2002) Taking toll: lipid A mimetics as adjuvants and immunomodulators. Trends Microbiol 10(10 Suppl): S32–S37

Petrovsky N (2006) Novel human polysaccharide adjuvants with dual Th1 and Th2 potentiating activity. Vaccine 24(Suppl 2):S2-26-29

Petrovsky N, Aguilar JC (2004) Vaccine adjuvants: current state and future trends. Immunol Cell Biol 82(5):488–496

Phonrat B, Pitisuttithum P, Chamnanchanunt S, Puthavathana P, Ngaosuwankul N, Louisirirotchanakul S, et al. (2013) Safety and immune responses following administration of H1N1 live attenuated influenza vaccine in Thais. Vaccine 31(11):1503–1509

Pizza M, Giuliani MM, Fontana MR, Douce G, Dougan G, Rappuoli R (2000) LTK63 and LTR72, two mucosal adjuvants ready for clinical trials. Int J Med Microbiol 290(4–5):455–461

Pizza M, Giuliani MM, Fontana MR, Monaci E, Douce G, Dougan G et al (2001) Mucosal vaccines: non toxic derivatives of LT and CT as mucosal adjuvants. Vaccine 19(17–19):2534–2541

Podda A, Del Giudice G (2003) MF59-adjuvanted vaccines: increased immunogenicity with an optimal safety profile. Expert Rev Vaccines 2(2):197–203

Ranasinghe C, Trivedi S, Wijesundara DK, Jackson RJ (2014) IL-4 and IL-13 receptors: roles in immunity and powerful vaccine adjuvants. Cytokine Growth Factor Rev 25(4):437–442

Rancan F, Amselgruber S, Hadam S, Munier S, Pavot V, Verrier B et al (2013) Particle-based transcutaneous administration of HIV-1 p24 protein to human skin explants and targeting of epidermal antigen presenting cells. J Control Release 176:115–122

Rappuoli R, Mandl CW, Black S, De Gregorio E (2011) Vaccines for the twenty-first century society. Nat Rev Immunol 11(12):865–872

Read RC, Naylor SC, Potter CW, Bond J, Jabbal-Gill I, Fisher A et al (2005) Effective nasal *influenza* vaccine delivery using chitosan. Vaccine 23(35):4367–4374

Rharbaoui F, Drabner B, Borsutzky S, Winckler U, Morr M, Ensoli B et al (2002) The Mycoplasma-derived lipopeptide MALP-2 is a potent mucosal adjuvant. Eur J Immunol 32 (10):2857–2865

Rhee JH, Lee SE, Kim SY (2012) Mucosal vaccine adjuvants update. Clin Exp Vaccine Res 1 (1):50–63

Riese P, Sakthivel P, Trittel S, Guzmán CA (2014) Intranasal formulations: promising strategy to deliver vaccines. Expert Opin Drug Deliv 11(10):1619–1634

Riese P, Schulze K, Ebensen T, Prochnow B, Guzman CA (2013) Vaccine adjuvants: key tools for innovative vaccine design. Curr Top Med Chem 13(20):2562–2580

Rose MA, Zielen S, Baumann U (2012) Mucosal immunity and nasal *influenza* vaccination. Expert Rev Vaccines 11(5):595–607

Ruffin N, Borggren M, Euler Z, Fiorino F, Grupping K, Hallengard D et al (2012) Rational design of HIV vaccines and microbicides: report of the EUROPRISE annual conference 2011. J Transl Med 10:144

Ruiz-Palacios GM, Perez-Schael I, Velazquez FR, Abate H, Breuer T, Clemens SC et al (2006) Safety and efficacy of an attenuated vaccine against severe *rotavirus* gastroenteritis. N Engl J Med 354(1):11–22

Sah E, Sah H (2015) Recent trends in preparation of poly(lactide-co-glycolide) nanoparticles by mixing polymeric organic solution with antisolvent. J Nanomaterials 2015:22

Sanchez MV, Ebensen T, Schulze K, Cargnelutti D, Blazejewska P, Scodeller EA et al (2014) Intranasal delivery of influenza rNP adjuvanted with c-di-AMP induces strong humoral and cellular immune responses and provides protection against virus challenge. PLoS One 9(8): e104824

Sancho D, Mourao-Sa D, Joffre OP, Schulz O, Rogers NC, Pennington DJ et al (2008) Tumor therapy in mice via antigen targeting to a novel, DC-restricted C-type lectin. J Clin Investig 118 (6):2098–2110

Savelkoul HF, Ferro VA, Strioga MM, Schijns VE (2015) Choice and design of adjuvants for parenteral and mucosal vaccines. Vaccines (Basel) 3(1):148–171

Schaap P (2013) Cyclic di-nucleotide signaling enters the eukaryote domain. IUBMB Life 65 (11):897–903
Schmidt CS, Morrow WJ, Sheikh NA (2007) Smart adjuvants. Expert Rev Vaccines 6(3):391–400
Seder RA, Darrah PA, Roederer M (2008a) T-cell quality in memory and protection: implications for vaccine design. Nat Rev Immunol 8(4):247–258
Seder RA, Darrah PA, Roederer M (2008) T-cell quality in memory and protection: implications for vaccine design. Nat Rev Immunol 8(6):247
Sharma R, Agrawal U, Mody N, Vyas SP (2015) Polymer nanotechnology based approaches in mucosal vaccine delivery: challenges and opportunities. Biotechnol Adv 33(1):64–79
Shortman K, Heath WR (2010) The CD8+ dendritic cell subset. Immunol Rev 234(1):18–31
Shu C, Yi G, Watts T, Kao CC, Li P (2012) Structure of STING bound to cyclic di-GMP reveals the mechanism of cyclic dinucleotide recognition by the immune system. Nat Struct Mol Biol 19(7):722–724
Skrnjug I, Rueckert C, Libanova R, Lienenklaus S, Weiss S, Guzman CA (2014) The mucosal adjuvant cyclic di-AMP exerts immune stimulatory effects on dendritic cells and macrophages. PLoS One 9(4):e95728
Srivastava A, Gowda DV, Madhunapantula SV, Shinde CG, Iyer M (2015) Mucosal vaccines: a paradigm shift in the development of mucosal adjuvants and delivery vehicles. APMIS 123 (4):275–288
Stein P, Gogoll K, Tenzer S, Schild H, Stevanovic S, Langguth P et al (2014) Efficacy of imiquimod-based transcutaneous immunization using a nano-dispersed emulsion gel formulation. PLoS One 9(7):e102664
Stern AM, Markel H (2005) The history of vaccines and immunization: familiar patterns, new challenges. Health Aff 24(3):611–621
Stevceva L, Strober W (2004) Mucosal HIV vaccines: where are we now? Curr HIV Res 2(1):1–10
Sutter RW, John TJ, Jain H, Agarkhedkar S, Ramanan PV, Verma H et al (2010) Immunogenicity of bivalent types 1 and 3 oral poliovirus vaccine: a randomised, double-blind, controlled trial. Lancet 376(9753):1682–1688
Thomann-Harwood LJ, Kaeuper P, Rossi N, Milona P, Herrmann B, McCullough KC (2013) Nanogel vaccines targeting dendritic cells: contributions of the surface decoration and vaccine cargo on cell targeting and activation. J Control Release 166(2):95–105
Tsuji T, Matsuzaki J, Kelly MP, Ramakrishna V, Vitale L, He LZ et al (2011) Antibody-targeted NY-ESO-1 to mannose receptor or DEC-205 in vitro elicits dual human CD8 + and CD4 + T cell responses with broad antigen specificity. J Immunol 186(2):1218–1227
Vajdy M, Srivastava I, Polo J, Donnelly J, O'Hagan D, Singh M (2004) Mucosal adjuvants and delivery systems for protein-, DNA- and RNA-based vaccines. Immunol Cell Biol 82(6):617–627
van den Dobbelsteen GP, van Rees EP (1995) Mucosal immune responses to pneumococcal polysaccharides: implications for vaccination. Trends Microbiol 3(4):155–159
van Kooyk Y, Unger WW, Fehres CM, Kalay H, Garcia-Vallejo JJ (2013) Glycan-based DC-SIGN targeting vaccines to enhance antigen cross-presentation. Mol Immunol 55(2):143–145
Vesikari T, Clark HF, Offit PA, Dallas MJ, DiStefano DJ, Goveia MG et al (2006) Effects of the potency and composition of the multivalent human-bovine (WC3) reassortant *rotavirus* vaccine on efficacy, safety and immunogenicity in healthy infants. Vaccine 24(22):4821–4829
Vesikari T, Itzler R, Karvonen A, Korhonen T, Van Damme P, Behre U et al (2009) RotaTeq, a pentavalent *rotavirus* vaccine: efficacy and safety among infants in Europe. Vaccine 28 (2):345–351
Wang SH, Thompson AL, Hickey AJ, Staats HF (2012) Dry powder vaccines for mucosal administration: critical factors in manufacture and delivery. Curr Top Microbiol Immunol 354:121–156
Wang T, Wang N (2015) Biocompatible mater constructed microneedle arrays as a novel vaccine adjuvant-delivery system for cutaneous and mucosal vaccination. Curr Pharm Des 21:5245–5255

Wang X, Meng D (2015) Innate endogenous adjuvants prime to desirable immune responses via mucosal routes. Protein Cell 6(3):170–184

Wanke I, Skabytska Y, Kraft B, Peschel A, Biedermann T, Schittek B (2013) *Staphylococcus aureus* skin colonization is promoted by barrier disruption and leads to local inflammation. Exp Dermatol 22(2):153–155

Woodrow KA, Bennett KM, Lo DD (2012) Mucosal vaccine design and delivery. Annu Rev Biomed Eng 14:17–46

Xia Y, Fan Q, Hao D, Wu J, Ma G, Su Z (2015) Chitosan-based mucosal adjuvants: sunrise on the ocean. Vaccine 44:5997–6010

Yu M, Vajdy M (2010) Mucosal HIV transmission and vaccination strategies through oral compared with vaginal and rectal routes. Expert Opin Biol Ther 10(8):1181–1195

Part II
Natural Compound Research and Anti-infective Development

History of Antibiotics Research

Kathrin I. Mohr

Abstract For thousands of years people were delivered helplessly to various kinds of infections, which often reached epidemic proportions and have cost the lives of millions of people. This is precisely the age since mankind has been thinking of infectious diseases and the question of their causes. However, due to a lack of knowledge, the search for strategies to fight, heal, and prevent the spread of communicable diseases was unsuccessful for a long time. It was not until the discovery of the healing effects of (antibiotic producing) molds, the first microscopic observations of microorganisms in the seventeenth century, the refutation of the abiogenesis theory, and the dissolution of the question "What is the nature of infectious diseases?" that the first milestones within the history of antibiotics research were set. Then new discoveries accelerated rapidly: Bacteria could be isolated and cultured and were identified as possible agents of diseases as well as producers of bioactive metabolites. At the same time the first synthetic antibiotics were developed and shortly thereafter, thousands of synthetic substances as well as millions of soil borne bacteria and fungi were screened for bioactivity within numerous microbial laboratories of pharmaceutical companies. New antibiotic classes with different targets were discovered as on assembly line production. With the beginning of the twentieth century, many of the diseases which reached epidemic proportions at the time—e.g., cholera, syphilis, plague, tuberculosis, or typhoid fever, just to name a few, could be combatted with new discovered antibiotics. It should be considered that hundred years ago the market launch of new antibiotics was significantly faster and less complicated than today (where it takes 10–12 years in average between the discovery of a new antibiotic until the launch). After the first euphoria it was quickly realized that bacteria are able to develop, acquire, and spread numerous resistance mechanisms. Whenever a new antibiotic reached the market it did not take long until scientists observed the first resistant germs. Since the marketing of the first antibiotic there is a neck-on-neck race between scientists who discover natural or develop semisynthetic and synthetic

K.I. Mohr (✉)
Department Microbial Drugs and German Center for Infection Research,
Helmholtz-Centre for Infection Research, Inhoffenstraße 7, 38124 Braunschweig, Germany
e-mail: Kathrin.mohr@helmholtz-hzi.de

Current Topics in Microbiology and Immunology (2016) 398: 237–272
DOI 10.1007/82_2016_499

Published Online: 01 October 2016

bioactive molecules and bacteria, which have developed resistance mechanisms. The emphasis of this chapter is to give an overview of the history of antibiotics research. The situation within the pre-antibiotic era as well as in the early antibiotic era will be described until the Golden Age of Antibiotics will conclude this time travel. The most important antibiotic classes, information about their discovery, activity spectrum, mode of action, resistance mechanisms, and current application will be presented.

Contents

1 Definition of the Term Antibiotic

In 1947, S. A. Waksman defined the term "antibiotic" as follows: "An antibiotic is a chemical substance, produced by micro-organisms, which has the capacity to inhibit the growth of and even to destroy bacteria and other micro-organisms" (Waksman 1947). Today, "antibiotic" has multiple meanings: (I) an organic chemical of natural or synthetic origin that inhibits or kills pathogenic bacteria; (II) any antimicrobial substance, or, (III) in the Waksman tradition, limited to antimicrobial substances of microbial origin. Anyway, all of these definitions coexist in literature and for a detailed discussion of "What is an antibiotic?" I refer to Bentley and Bennett (2003).

Within this review the term antibiotic is used as any antimicrobial substance, independent if it is of natural, semisynthetic, or synthetic origin.

2 From the Early Beginning to the Golden Age of Antibiotics Research

2.1 *The Pre-antibiotic Era*

The discovery and subsequent large-scale production of antibiotics in the early twentieth century was one of the most important achievements in the history of medicine. This new wonder drugs, in addition to comprehensive knowledge about pathogens and improved hygiene measures, took away the fear of many infectious diseases and extremely increased quality and expectancy of life. While in 1900 contagious diseases were the most common cause of death in the USA, in 2000 only a negligible percentage died of infectious diseases (US National Center for Health Statistics). However, nowadays in many countries of the Third World the situation is comparable to 1900, whereas in industrial countries infectious diseases are mainly a problem of immune-suppressed people (HIV or cancer patients) or patients infected with multi-resistant pathogens. In the days when nothing was known about the path of infections, infection prevention, antibiotics, and vaccination, for thousands of years mankind was tortured by huge epidemics like syphilis, smallpox, malaria, typhus, yellow fever, leprosy, tuberculosis, Spanish Influenza, cholera, and plague, to name just a few.

The disastrous situation of people in cases of pandemics during the pre-antibiotic era should briefly be described exemplarily for plague: Plague is caused by the bacterium *Yersinia pestis*, which was named in honor of its discoverer Alexandre Émile Yersin (1863–1943). Yersin was a Swiss-born bacteriologist and as a member of the French Colonial Health Service he was sent to Hong Kong in 1894 to investigate the outbreak of bubonic plague (Hawgood 2008). He was the first who isolated the causative agent of pestilence from dissected buboes. Overall, *Y. pestis* was responsible for at least three pandemics in history: the Justinian plague which caused almost 100 million deaths, the "Black Death"—period in the fourteenth century during which tens of millions of people in Europe died (Haensch et al. 2010), and the outbreak between 1895 and 1930 with about 12 million victims, mostly in India (Kool 2005). Plague has not been eradicated yet and without antibiotic therapy the mortality rate is between 50 and 90 % (Lippi and Conti 2002). Furthermore nowadays pest is still a great public health and infection control threat. An aerosol of the agent could be used by terrorists as a biological weapon, which would cause the human transmissible, contagious pneumonic plague (Inglesby et al. 2000), in comparison to bubonic plague which is exclusively transmitted by bites of infected rodent fleas.

However, in the Middle Ages, bacteria and viruses as pathogenic agents, prophylactic measures as vaccines or even effective drugs as antibiotics were completely unknown. Bad air, miasms, inconvenient planetary constellations, or an imbalance between the four elements fire, earth, water and air and the four body juices blood, mucus, yellow bile, and black bile (Hippocratic Doctrine of the Four Juices; Hart 2001) have been held responsible of any kind of illnesses. Concerning serious infection diseases, doctors and physicians were unable to provide valuable assistance for suffering patients. In the case of plague infections, for example, they recommended energy sapping bloodlettings, diets with herbs, sometimes mixed with meat of vipers, open north windows, and if everything else fails, praying. Christian clerics and scholars of the occident interpreted such catastrophes as a "sign of God's wrath" and a "deserved punishment of mankind's sins." While the Black Death raged most furiously around the year 1350, the desperation of people was such big that curious phenomena like the "flagellant movements" became very popular: Increasing numbers of followers joined desperate penitent religious groups which swarmed through the land and whose members whipped themselves to bleed. In this way they hoped to avert God's anger from humankind (Würth 2012). Nonetheless, physicians' advices for post exposure prophylaxis like wearing masks, avoiding close contact to, or separation of, infected persons were quiet useful from today's infectious biology point of view. The government of Milano had even ordered walling of pest-infected persons in their houses, but probably these extreme measures were exceptional cases (Bulst 1977).

Thousands of years ago, since at least 1500 before Christ, at least healing effects of mushrooms, beer yeast, and molds are known to be valuable in treatment of infected wounds. These microorganisms were gladly used for medical applications, even if the practitioners were not able to explain the phenomenon of the effect (Duckett 1999). It was reported about ancient physicians in Egypt, Persian, and Greek, for example, that they treated patients with compresses and tonics made from herbs, molds, and organic compounds for more than 1000 years ago. Also bioactive compounds of plants were medically used as seen in the example of the first systematic chemotherapy, reported in the seventeenth century: Indigenous people of South America used a powder made of the cinchona tree bark (*Cinchona pubescens* (**a**, Fig. 1) or *C. officinalis*) against fever and later on for prevention and treatment of malaria, which has been introduced to South America by Spanish conquistadors. Later on (in 1820), the most active agent quinine, an alkaloid with antipyretic activity, was isolated by the French chemists Pierre-Joseph Pelletier and Joseph-Bienaimé Caventou (Rezende 2006) (**1**, Fig. 1) and synthesized by the American chemists Woodward and Doering in 1945. As reported by Jäger (2011), today quinine and other antimalaria agents in the bark's extract of *Cinchona* spp. are still of medical importance, because the evolution of *Plasmodium falciparum* strains which are resistant to synthetic drugs (mainly chloroquine) led to rise in the malaria incidence in Asia and Africa.

However, without the discovery and description of bacteria (and later on also viruses, both are agents of many infectious diseases), the successful story of antibiotics could not have been written. Although it has long been assumed that

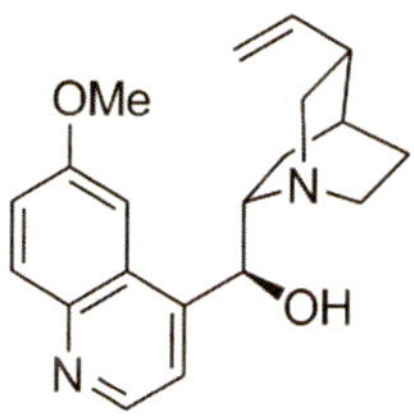

1

Fig. 1 Chemical structure of quinine *1* and photograph of *Cinchona pubescens* (**a**) (photo kindly provided by Dr. Heinke Jäger)

creatures exist which are indistinguishable with the naked eye, the first important milestone in the history of microbiology was set by Antonie van Leeuwenhoek (1632–1723), a tradesman and "amateur" scientist. Using his handcrafted microscopes he could visualize microstructures of various objects. Without any university education, he was the first who observed, described, and drew bacteria, sperms, and erythrocytes. His illustrations impressed with precision and attention to detail and gave a hitherto unknown insight to micro objects.

Also in the seventeenth century, the healing effects of molds were described by John Parkington, an English apothecary. He recommended in his book *Theatrum Botanicum* (Zimdahl 2015) to use these microorganisms for treatment of infections. Then, for about 200 years, there were no major advances with regard to "infection research." However, in the nineteenth century, more and more scientists and physicians became interested in molds with antibacterial activity: Theodor Billroth (1829–1894), a German-Austrian surgeon, performed experiments with bacterial cultures and moulds to research their role in "accidental wound diseases." He described, that occasionally when *Penicillium* grew in a culture, bacteria failed to grow. He observed the antibiotic activity of the fungi, but he could not draw the right conclusions, because he suggested that *Penicillium* could have "modified the medium" so as to render it unsuitable for bacterial growth (Majno and Joris 1979). In addition, Billroth searched for the causes of sepsis. Therefore he investigated microbial blood and pus preparations and described "*Coccobacteria septica*," later called staphylococci and streptococci (Klein et al. 2012). In 1870 Sir John Scott Burdon-Sanderson (1828–1905), an English physiologist, also mentioned that culture fluid covered with mould inhibits bacterial growth (Williams 2013). At this time bacteria were already well known and described as contaminants in milk, wine, and silkworms by Louis Pasteur, as well as in blood by Billroth, but their role as agents of human diseases had not yet been proven. In 1875, another English physician, John Tyndall (1820–1893), conducted contamination experiments with air-exposed tubes of broth to investigate contamination by bacteria from the air. He also confirmed the aforementioned effect that in tubes with *Penicillium* growth bacteria could not be observed. Tyndall concluded that there was a battle between the bacteria and the mould, and "in every case where the mould was thick and coherent, the bacteria died or became dormant and fell to the bottom as a sediment" (Tyndall 1877). He observed that *Penicillium* was able to inhibit the growth of bacteria falling into the tubes from the air. However, as mentioned above in Tyndall's times, the importance of bacteria as human pathogens was still not proven and therefore he also failed to assess the dimension of this observation (Friedman 1998). A few years later Joseph Lister (1827–1912), an English surgeon, also suspected bacteria to be responsible for infections. He was pioneering the development of antiseptic surgery by using chemical antiseptics, such as phenol, to kill bacteria on operating equipment and on wounds (De la Bédoyère 2005). He also cultivated a mould, described as *Penicillium glaucum*, together with bacteria in urine and observed that the bacteria did not grow in presence of the mold (Rubin 2007). In contrast to his predecessors, he postulated an antimicrobial activity and proved this theory in 1884, when he cured an abscess of a nurse with *P. glaucum* soaked tissues. Unfortunately Lister did not publish his results and was not able to produce sufficient amounts of extract or even isolate the active substance, later known as penicillin.

In the second part of the nineteenth century two main questions tasked the scientists: "Does spontaneous generation (abiogenesis) exist?" and "What is the nature of infectious diseases?" (Madigan et al. 2000). According to the theory of abiogenesis, bacteria, which could be observed under the microscope in numerous amounts from samples of rotten, but not from fresh food, develop spontaneously from dead material—but exclusively after contact with fresh air. Therefore the supporters of abiogenesis agreed that air is necessary for microbial development, but they could not explain the underlying cause. The myth was definitely refuted by the French chemist Louis Pasteur (1822–1895), who performed experiments to prove that without contamination microbial growth is not possible. He showed that in a sterile broth microbes could not develop spontaneously, but only after air contact. And he further detected that the same microbes, which grow in large amounts in air-exposed broth, can be found in small amounts in the air. Preliminary work had been carried out by the Italian priest Lazzaro Spallanzani (1729–1799) who already proved in the middle of the eighteenth century that microbial growth did not develop in boiled and afterwards hermetically sealed broth (Klein et al. 2012). As a consequence of the refuted germ theory numerous efficient sterilization techniques have been developed (Madigan et al. 2000). Furthermore, Pasteur discovered the principles of vaccination (anthrax, fowl cholera, rabies), lactic acid and alcoholic fermentation, and pasteurization.

The challenge to elucidate the "nature of infectious diseases" was met by Robert Koch (1843–1910). Long before Koch's time, priest physicians in ancient India and China already presumed that pathogens could be transmitted between living beings. They tried to prevent smallpox infections by immunization of healthy persons with smallpox material from ill persons (immunization by variolation; Klein et al. 2012). In the sixteenth century, people also assumed that "something" could be transmitted from ill to healthy persons which therefore came down with the same illness. Consequently some diseases were classified as contagious, because they are able to spread within a population. Ignaz Semmelweis (1818–1865), an Austrian-Hungarian gynecologist found out that puerperal fever could be transmitted from deceased women in childbed to healthy pregnant woman. As the transmission vehicle he identified unwashed hands of physicians who carried out vaginal examination of birthing mothers after they autopsied women who died on puerperal fever. Thereupon Semmelweis ordered a consequently disinfection of hands with chlorine water before vaginal examinations and thus secured a place in history as a pioneer of infection prophylaxis and later on called "Mother's Savior" (Klein et al. 2012). Joseph Lister, who also provided some evidence that bacteria could be responsible for infections, successfully introduced antiseptic measures in surgery. After a comprehensive literature research, Jakob Henle (1809–1885) came to the conclusion that during infection a specific infection material is transferred from ill to healthy persons (Klein et al. 2012). Koch was the first who successfully proved this theory and the correlation between a causative agent (*Bacillus anthracis*) and a disease (anthrax; Madigan et al. 2000). Later, in 1877, his colleague Pasteur observed that cultures of the anthrax bacilli, when contaminated with molds, became inhibited. Further great achievements of Koch included the development of solid media for bacterial cultivation and separation, the implementation of microphotography and, in 1876 the discovery of *B. anthracis* spores. Hereby he could describe the previously incompletely explained infection chain of anthrax. Using this example, he postulated four criteria to establish a causative relationship between a specific microbe and a disease (Henle Koch postulates), which are still valid. Koch also identified the specific causative agents of cholera (*Vibrio cholerae*) and tuberculosis (*Mycobacterium tuberculosis*). When he started his work on tuberculosis in 1881, consumption was responsible for one-seventh of the reported deaths (Madigan et al. 2000). He succeeded in both, staining *M. tuberculosis* in tissue and cultivating the demanding germ, but staining and cultivating was challenging: The mycobacterial surface is rich in lipids (mycolic acid), acid resistant and therefore not accessible to traditional staining techniques. Koch invented a dye (alkaline methylene blue and Bismarck-brown) which served as a precursor of the Ziehl–Neelsen dye, a standard dye in many microbiological laboratories today. For the isolation and enrichment of mycobacteria, Koch used coagulated blood serum and thus created the basis for realistic experiments of antibiotics evaluation in vitro where the test conditions should mimic the situation in vivo as much as possible. Next, he infected guinea pigs with pure cultures of *M. tuberculosis* and proved that his formerly established four postulates could be fulfilled. As appreciation of his work about tuberculosis he was awarded the 1905 Novel Prize.

Now, where bacteria as the possible agents of diseases were identified, the discovery and isolation of the first antibiotic was just a blink away. About 30 years before Alexander Fleming falsely passed into history as the discoverer of penicillin (see also Karwehl and Stadler 2016), Ernest Duchesne (1874–1912), a French medical officer, noticed the antibacterial activity of molds (Duckett 1999), as so many before him. However, he went one step further and tried to identify the specific cause of the inhibition. During his work in a military hospital he observed Arabian hostlers, who stored their saddles in a dark, moisture chamber to support mould growth. They explained that saddles treated by this method supported wound healing of saddle sores. As mentioned above, the wound healing effect of molds was not new. Duchesne started a couple of precise experiments. He infected guinea pigs with *Escherichia coli* (formerly named "*Bacillus coli communis*") or *Salmonella typhi* (formerly named "Eberth's bacilli"), the latter of which is the agent of typhoid fever, and treated them with a suspension of the isolated mould. To his surprise all medicated animals survived. He also investigated interactions between bacteria and antibiotic producing fungi and was the first who discussed a therapeutic application of these antibiotic producing microorganisms. Duchesne described his results in the thesis "Contribution to the study of vital competition between microorganisms: antagonism between moulds and microbes" (Duchesne 1897). His thesis was unfortunately not accepted by the Institute Pasteur. Later on, in 1940, his work was confirmed by Chain, Florey, and coworkers, who identified the antibiotic produced by a *Penicillium* sp. and its effect against Gram-positive and Gram-negative bacteria (Chain et al. 1940; see chapter "Beta lactam antibiotics: penicillin, ampicillin, methicillin" elsewhere in this review). Last but not least Ernest Duchesne was honored posthumously by the Académie nationale de Médecine in 1949, four years after Fleming and colleagues had received the Nobel Prize for the rediscovery of the antibiotic effect of penicillin.

2.2 The Early Antibiotic Era

Anyway, the first antibiotic was described by the Italian medical scientist Bartolomeo Gosio (1863–1944). In 1893 Gosio explored the question why many people of Southern Europe and Southern USA, who lived in impoverished conditions, came down with pellagra. The diet among this social class mainly consisted of corn, which was suspected to be fungi contaminated. It was assumed that the consumption of this contaminated corn causes pellagra (Sydenstricker 1958; Bentley 2000; Nord 2010). The clinical symptoms of pellagra include photosensitive dermatitis, diarrhea, dementia, and death. Gosio isolated and cultivated *Penicillium brevi-compactum* from the corn and obtained a crystalline product within a filtrate. He found out that this substance showed antibiotic activity against *Bacillus anthracis*, the causative agent of anthrax. Therefore Gosio was the first one ever described a natural antibiotic (Gosio 1893; Gosio 1896). Gosio's discovery and therefore a possible use for medical application was forgotten until two

American scientists, Carl Alsberg and Otis Fisher Black resynthesized the substance in 1912 and named it mycophenolic acid (Bentley 2000; for more information and structure of mycophenolic acid see Fig. 1 in Karwehl and Stadler 2016). However, in 1937 the cause of pellagra was identified as a nicotinic acid deficiency by Conrad Elvehjem (Burris et al. 1990).

2.3 *Salvarsan*

The activity of the first synthetic antibiotic, the arsenic derivative arsphenamine, (Dioxy-diamino-arsenobenzol-dihydrochloride; Ehrlich and Bertheim 1912), trade name Salvarsan, was discovered by Paul Ehrlich (1854–1915), a German physician in 1909 in cooperation with the chemist Alfred Bertheim (1879–1914; synthesis) and the bacteriologist Sahachiro Hata (1873–1938; biological testing). Syphilis is a chronic, venereal lethal disease caused by the spirochaete *Treponema pallidum*, which was discovered by Fritz Schaudinn (1871–1906) and Erich Hoffmann (1868–1959) in 1905. At Ehrlich's time syphilis was a rampant, worldwide affliction. It is assumed that *T. pallidum* was introduced to Spain in 1493 by returnees of Columbus' discovery trip. They imported the pathogen from Haiti to Spain from where syphilis was later on spread rapidly fast by mercenaries from Napoli throughout Europe (Berger et al. 2012). Initially, this disease was treated with guaiac-bark, a cure adopted from the Indios. Later on, till the end of the nineteenth century, syphilis was usually treated with inorganic mercury salts, which has often resulted in fatal heavy metal poisoning with a simultaneous low efficacy. Around the year 1900 Ehrlich searched for dyes with specific affinity to bacteria but not to human or animal tissue. He argued that certain chemoreceptors on parasites, microorganisms, and cancer cells would be different from analogous structures in host tissues (Drews 2000). The idea was to find toxic substances, like arsenic, with these discriminatory properties for medical purposes against infectious diseases ("magic bullet"; Madigan et al. 2000). He tested a large number of aniline dyes and found out that methylene blue stains parasitic protozoa like plasmodia. Afterwards he administered methylene blue to two patients with milder symptoms of malaria and noted that the fever attacks disappeared after a few days of methylene blue administration and at the latest after eight days *Plasmodium falciparum* could not be detected in the patient's blood anymore (Guttmann and Ehrlich 1891). Even if this experiment had no statistical significance, herewith Ehrlich can be given credit of having laid the foundation stone of modern chemotherapy. He was the first one to report that a synthetic drug could be used successfully to treat a specific disease, although the dye showed much less effect than the traditional used quinine.

In connection with his research about syphilis, which was still incurable at his time, Ehrlich took special note of a study published by Thomas (1905), who described aminophenyl arsenic acid, commercial name atoxyl, to be effective in the treatment of sleeping sickness. This disease is caused by the parasitic protozoan *Trypanosoma brucei* and was the greatest cause of death in Africa at that time. Atoxyl

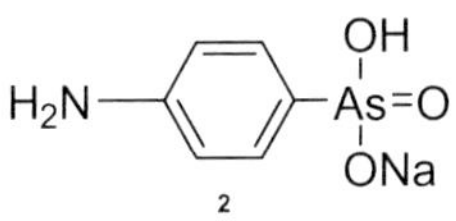

Fig. 2 Atoxyl; sodium;(4-aminophenyl)-hydroxyarsinate *2*

has been synthesized and published by Pierre Béchamp (1816–1908) in 1863 (Béchamp 1863). Ehrlich started an extensive screen of several hundred organoarsenic derivatives of atoxyl which were tested by his colleague Hata in vivo in syphilis-infected rabbits (Ehrlich and Hata 1910; **2**, Fig. 2).

In 1909 compound no. 606 turned out to be the big hit: the arsenic derivative arsphenamine was also called "Ehrlich 606" and later on sold as Salvarsan (Fig. 3) (Ehrlich 1910). Salvarsan consists of about 30 % arsenic, showed "parasitotropic" properties but lacked "organotropic" properties (Bosch and Rosich 2008). With Salvarsan the scientists pioneered in 1910 the field of targeted therapies, the breakthrough treatment for *Treponema*, other spirochetes and also trypanosomiasis. This new drug selectively binds the pathogens and the arsenate kills them—the perfect "magic bullet." Salvarsan is generally viewed as the first modern chemotherapeutic agent because it was discovered as a result of a rational screening program (Bennett 2015). From November 1910, the pharmaceutical company Hoechst (Frankfurt) was producing 12,000–14,000 ampoules Salvarsan per day for clinical trials and therapy worldwide. By treatment with Salvarsan two-thirds of syphilis-infected people could be cured up to 1928, although it had serious side effects like hypersensitivity problems due to arsenic poisoning.

Ehrlich continued to evaluate further synthesized compounds and in 1914 an enhanced version of Salvarsan, neoarsphenamine, marked as Neosalvarsan, was developed (**4**, Fig. 4; Williams 2009). Increased solubility and less toxicity due to a lower arsenical content (19 %) were the main advantages of Neosalvarsan. Unfortunately this improved product was also associated with side effects like nausea and vomiting. A further problem of both compounds was the need to store

Fig. 3 Information sheet of Salvarsan (kindly provided by Prof. Dr. Christoph Friedrich, Bildarchiv des Instituts für Geschichte der Pharmazie der Universität Marburg)

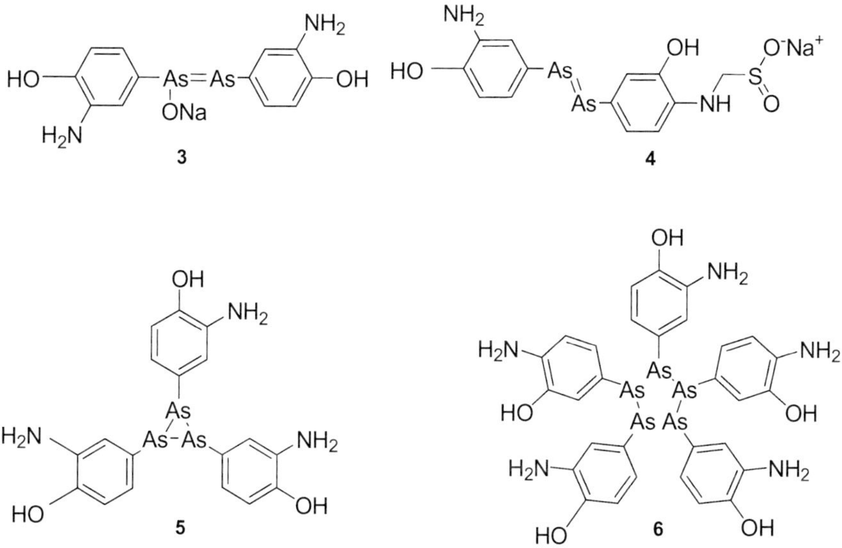

Fig. 4 Salvarsan, arsphenamine; structure as assumed by Ehrlich with an As-As double bond *3*; Neosalvarsan *4*, *5* and *6* mixture of the trimer and pentamer of Salvarsan as elucidated by Lloyd et al. (2005)

them in sealed vials under nitrogen atmosphere to prevent oxidation (Bosch and Rosich 2008). However, Salvarsan/Neosalvarsan was the drug of choice until penicillin became available in 1940. Ehrlich assigned the structure of Salvarsan (**3**, Fig. 4) with an As = As double bond. In 2005, an extensive mass spectral analysis by Lloyd and coworkers identified the structure of Salvarsan to have As–As single bonds and is a mixture consisting of cyclo-As_3 and cyclo-As_5 species (**5** and **6**, Fig. 4; Lloyd et al. 2005). Neosalvarsan also had an overall narrow spectrum activity (Aminov 2010).

Already in 1908 Paul Ehrlich was awarded the Nobel Prize in Physiology or Medicine together with Ilya Ilyich Mechnikov "*in recognition of their work on immunity*." The mode of action of Salvarsan is still unknown.

3 The Golden Age of Antibiotic Discovery: From Sulfa Drugs to Quinolones

3.1 Sulfa Drugs: Prontosil

The discovery of antibiotic activity of the synthetic sulfa drug class by Paul Gerhard Domagk (1895–1964) in the 1930th marked a further milestone in the history of antibiotics. Domagk, a German physician, already practiced in treating war wounds, was tasked with the identification of antibacterial azo dyes. These dyes were tested in large scale in vitro and in vivo (mice, rats, rabbits, guinea pigs) in his institute at Bayer. At Domagk's time Bayer was a part of IG Farben, a huge German chemical trust of six companies: BASF, Bayer, Hoechst, Agfa, Cassella, and Kalle founded in winter 1925/26 (Spoerer and Streb 2013). The screening program was inspired by Ehrlich's success to combat syphilis, as described above. After World War I, research and development concerning antibiotics have been intensified substantially as a result of the horrendous battle injuries suffered by the armies of the war, which were frequently magnified by subsequent bacterial infection of the wounds. However, also poorly nourished civilian populations were highly susceptible to bacterial infections and birthing women were endangered acutely to die from childbed fever (Bentley 2009). Initial comprehensive screening efforts of hundreds of dyes, synthesized by chemists at the Hoechst Company failed to produce a useful antibacterial agent (Fig. 5).

Probably the main problem of these studies was the lack of reliable tests for antibacterial activity (Rubin 2007). Domagk developed an ingenious method to test the survival rate of mice previously infected with a highly virulent strain of hemolytic *Streptococcus*. He confirmed the reliability of his test system by using known antibacterial substances and those without activity. On the basis of preliminary studies a large number of sulfonamide-containing dyes were synthesized by the chemists Fritz Mietzsch (1896–1956) and Josef Klarer (1898–1953) at Bayer and subsequently tested by Domagk. In 1932 a red colored dye turned out to be highly effective in protecting mice from a lethal dose of *Streptococci*. For this substance, sulfamidochrysoidine, Domagk proclaimed a 100 % success rate, as assessed by the mouse protection assay, when administered prior to a challenge with the potentially lethal microorganisms (Rubin 2007). Interestingly, this dye, trade name Prontosil (**7**, Fig. 6), was highly active in infected mice but nearly ineffective in vitro (Domagk 1935). Later it was found out by Tréfouël and colleagues that sulfamidochrysoidine is a precursor and is degraded by enzymes within the body to the active drug sulfanilamide (**8**, Fig. 6; Tréfouël 1935). Sulfanilamide was already described by the Austrian chemist Paul Gelmo (1879–1961) in 1908

Fig. 5 Examples of some azo dyes (from *left* to *right*): Sudan IV (staining of lipids), *Evans Blue* (dye for immunofluorescence), *methyl red* (pH-indicator), *Sudan black* (staining of lipids), *Sudan green*, and basic aniline dye *methylene blue* (stains parasitic protozoa like plasmodia)

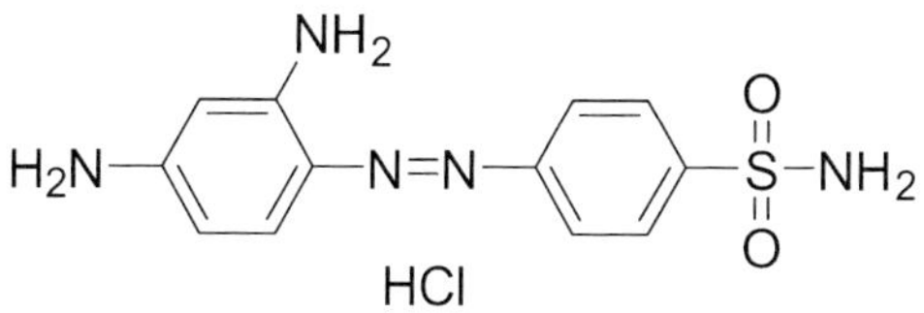

7

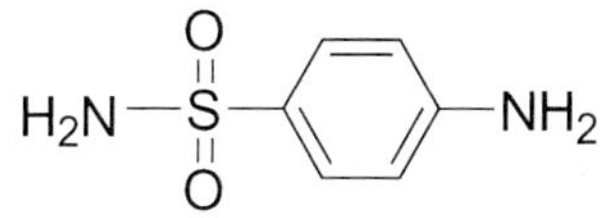

8

Fig. 6 Prodrug Prontosil; 4-[(2,4-diaminophenyl)diazenyl] benzenesulfonamide) *7*; Active form sulfanilamide (4-aminobenzenesulfonamide) *8*

(patent registration by Heinrich Hörlein, Bayer AG, in 1909). He had synthesized and characterized the substance for his doctoral dissertation, but unfortunately the potent therapeutic properties of sulfanilamide went unrecognized until the above-mentioned rediscovery.

Prontosil was marketed from 1935 on, and immediately deaths as a result of pneumonia, childbed fever and meningitis decreased drastically. With Prontosil an effective drug against Gram-positive cocci (mainly *Streptococcus*) was discovered and Domagk awarded the 1939 Nobel Prize for the discovery of the antibiotic effect of Prontosil.

The target of sulfonamides, and the basis for their selectivity, is the enzyme dihydropteroate synthase (DHPS) in the folic acid pathway. Mammalian cells are not dependent on endogenous synthesis of folic acid and generally lack DHPS (Sköld 2000); therefore sulfonamides have a bacteriostatic effect on many Gram-positive and on some Gram-negative bacteria, too. As a result of the enormously profitable marketing of Prontosil, the search and development of new sulfa drugs boomed in the late 1930 and many thousands of derivatives with sulfanilamide structure have been created in the following years. Today sulfonamides are infrequently used, in part due to widespread resistance (Sköld 2000, which were first observed in 1942 (Lewis 2013). Resistances against sulfanilamide-derivatives are commonly based on pathogen's capability to use external folic acid sources. Nowadays, sulfa drugs are mainly given in combination with trimethoprim, which inhibits a later folic acid synthesis step (dihydrofolate reductase inhibitor). In combination with trimethoprim sulfa drugs show a bactericide effect.

In 1940, despite the discovery of quinine for malaria, Salvarsan for syphilis, and the sulfa drugs for mainly Gram-positive cocci-infections, most agents of infectious diseases could still not be treated.

3.2 *Beta Lactam Antibiotics: Penicillin, Ampicillin, Methicillin*

"It was noticed that around a large colony of a contaminating mold the staphylococcus colonies became transparent and were obviously undergoing lysis." This observation, published by Alexander Fleming (1881–1955) in 1929, is often called the "birth of the antibiotic era" (Fleming 1929). However, Fleming was neither able to produce appreciable amounts nor to elucidate the structure of penicillin and therefore large-scale production and structure elucidation took a further 10 years. Howard Walter Florey (1898–1968) and Ernst Boris Chain (1906–1979) elucidated the structure of penicillin in 1939 (for structure of penicillin G, the first penicillin used in therapy see Karwehl and Stadler 2016). In March 1940 Chain isolated a small quantity of a substance that contained only 1 % of the active compound. Subsequently Norman Heatley (1911–2004) enhanced the purification process and since 1939 large-scale production of penicillin was possible. It is worth mentioning that the antibacterial properties of mold had been known from ancient times, and that many scientists before Fleming had come upon the similar observations regarding the antimicrobial activity of *Penicillium*, as already described in detail by Duchesne in 1897 (see previous section of this review). The introduction of penicillin in therapy in 1941 was the next big milestone of this successful story (Chain et al. 1940) (**c**, Fig. 7). In 1945 the Nobel Prize in Physiology or Medicine was awarded Fleming, Chain and Florey "*for the discovery of penicillin and its curative effect in various infectious diseases*" and at the end of the Second World War penicillin was available for US military and civil population. It was the first antibiotic capable of killing Gram-positive bacteria including the pathogens that caused gonorrhea, syphilis, and puerperal infections.

A great contribution to the history of antibiotic research was Fleming's development of a new screening method. He revolutionized the search for bioactive compounds from microorganisms by spreading soil or soil dilutions on agar plates, inoculated with pathogenic bacteria. Subsequently he searched for inhibition zones in lawns of these pathogens. This procedure saved time, money and required much less resources than any testing in animal disease models. As a result, this method became widely used in mass screenings for antibiotic producing microorganisms by

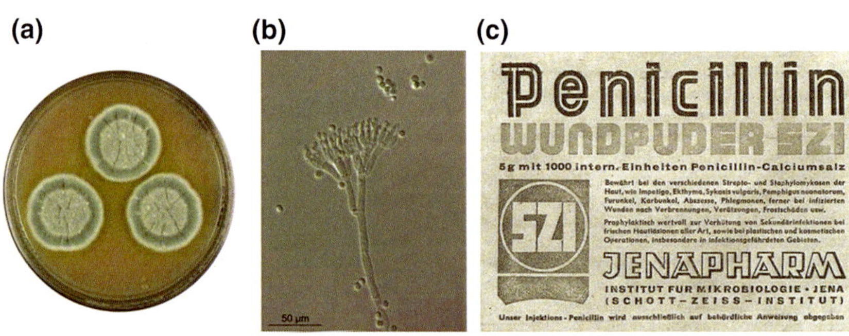

Fig. 7 *Penicillium chrysogenum* Q176 on agar **a**; Light microscopic image of a conidiophore of *P. chrysogenum* **b** (both pictures are kindly provided by Prof. Dr. Ulrich Kück and Dr. Julia Böhm, Ruhr-University Bochum); Label of Penicillin containing antiseptic powder **c** (kindly provided by Prof. Dr. Christoph Friedrich, Bildarchiv des Instituts für Geschichte der Pharmazie der Universität Marburg)

many researchers in academia and industry (Aminov 2010). As also mentioned by Karwehl and Stadler (2016), modern taxonomic studies revealed that the name of the fungus from which penicillin was first obtained ("*Penicillium notatum*" and later "*P. chrysogenum*" Fig. 7a and b), is actually *P. rubens* (Houbraken et al. 2011).

Penicillin and derivatives belong to the class of β-lactam antibiotics, which further include cephalosporins, carbapenems, and monobactams. Characteristic for this group is a β-lactam ring within the molecular structure. Most β-lactam antibiotics are potent inhibitors of cell wall biosynthesis. The first β-lactam antibiotic penicillin G was mainly effective against Gram-positive bacteria, because the cell wall of most Gram-negatives is impermeable for this agent. Subsequently developed semisynthetic penicillins (methicillin, oxacillin, ampicillin, carbenicillin) show a broad-spectrum activity and also inhibit several Gram-negative pathogens. Many bacteria, like staphylococci or *E. coli* for example, are able to produce β-lactamases; enzymes which destroy the β-lactam ring and therefore render the antibiotic ineffective. Today, β-lactam antibiotics are often given in combination with β-lactamase inhibitors (clavulanic acid, sulbactam). As cited by Aminov (2010), Rollo and coworkers investigated possible resistance emergence under laboratory conditions in 1952 and predicted that: "Syphilis has now been treated with arsenicals for about 40 years without any indications of an increased incidence of arsenic-resistant infections, and this work gives grounds for hoping that the widespread use of penicillin will equally not result in an increasing incidence of infections resistant to penicillin" (Rollo et al. 1952). While this is true for *Treponema pallidium* as reported by Cha et al. (2004), this is not the case for many other pathogenic bacteria, including the Enterobacteriaceae, which have become resistant not only to the original penicillin but also to semisynthetic penicillins, cephalosporins, and newer carbapenemes (Kumarasamy et al. 2010). Fleming was among the first who cautioned about the potential resistance to penicillin and already noted in 1929, "that the growth of *E. coli* and a number of other bacteria belonging to the coli-typhoid group was not inhibited by penicillin." He explained this observation that the dosage was too little or too shortly given. A few years later, Abraham and Chain made an extract of *E. coli* and found a substance that destroyed the growth-inhibiting property of penicillin. They published these results in 1940 (Abraham and Chain 1940) and in 1945 the first resistances against penicillin in therapy were already observed (Lewis 2013). This was the beginning of an armament race with uncertain outcomes between scientists, who develop new antibiotics and microorganism, which already have, develop, or acquire and, still worse, spread diverse resistant mechanisms by conjugation, transformation, and transduction of the encoding genes. Usually resistance develops within two to three years after the introduction of a new antibiotic treatment (Davies 2006). This is nowhere more apparent than in the steady evolution of beta-lactamases by point mutation under the selective pressure of successive introductions of new beta-lactamase-resistant penicillins, cephalosporins, carbapenems and monobactams (Jacoby and Bush 2005). Between 2010 and 2014 two novel cephalosporins (ceftaroline fosamil; ceftobiprole) for treatment of acute bacterial skin/skin structure infections and community-acquired pneumonia as well as a cephalosporin-beta-lactamase inhibitor combination (ceftolozane-tazobactam), reached the market (Hesterkamp 2015).

The semisynthetic penicillin-derivate ampicillin (para-aminobenzyl penicillin), was first described by Brewer and Jonson (1953) and later on patented by the US chemist John Clark Sheehan. Ampicillin is a broad-spectrum antibiotic. It shows the same activity like penicillin but in addition ampicillin acts against a couple of Gram-negative pathogens like *Enterococcus faecalis*, *Escherichia coli*, *Haemophilus influenza*, and *Proteus mirabilis*. Ampicillin is still used in therapy.

Methicillin (celbenin), also a semisynthetic penicillin-derivate, was developed by Beecham, a British pharmaceutical company in 1959 (later SmithKlineBeecham [SKB] and now Glaxo). It has a narrow spectrum activity but was, together with oxacillin, the first beta-lactamase-resistant substance of this antibiotic class and therefore it seemed to be a very promising drug in the early 1960. Methicillin was used in therapy against beta-lactamase-producing pathogens such as *Staphylococcus aureus* that would otherwise be resistant to most penicillins. However, already in 1961 first resistances against methicillin were reported from staphylococci (Barber 1961) and today methicillin is no longer in clinical use. The resistance of **m**ethicillin-**r**esistant ***S. aureus*** strains (MRSA) is based on a modified penicillin-binding protein (PBP2a). Because in the past resistance tests have been conducted with methicillin as lead-antibiotic, the designation "methicillin-resistant" was deduced for strains which show acquired resistance against all beta-lactam-antibiotics (penicillins, cephalosporins, carbapenems). Later on oxacillin was used as test antibiotic and led to **o**xacillin-**r**esistant ***S. aureus*** strains (ORSA).

In 1945, an Italian physician, Guiseppe Brotzu, isolated the ascomycete *Acremonium chrysogenum* from Sardinian coastal seawater. Brotzu detected an antibiotic effects of extracts generated from this fungus and later, in 1955, the structure of the active compound, cephalosporin C, was elucidated (Newton and Abraham 1955). For more information of the history of cephalosporins see Bloemendal and Kück (2014) and Karwehl and Stadler (2016).

3.3 The Discovery of Soil Inhabiting Bacteria as Reliable Sources for New Antibiotics

The discovery and successful marketing of penicillin as well as World War II, which suggested the need for new agents to control infectious diseases and epidemics certain to arise, has led to intensive search for new synthetic and natural antibiotics around the 1940. One main focus was set on the isolation and screen of microorganisms from the environment for antimicrobial activity. Many drug companies established departments of microbiology and fermentation units, and there were only a few large Pharma companies that did not participate in the search for new antibiotics (Drews 2000).

3.4 Tyrothricin/Gramicidin

In 1939, the American microbiologist René Dubos (1901–1982) isolated tyrothricin, a mixture of linear and cyclic polypeptide antibiotics, from the soil bacterium *Bacillus brevis*. He found out that tyrothricin decomposed the capsule of *Pneumococcus* bacteria, which were the major cause of pneumonia since the late nineteenth century down to the present day. The main constituent of tyrothricin is alkaline tyrocidine, in addition to lipophilic gramicidins A, B, and C. Tyrothricin was the first commercially available antibiotic (Dubos and Hotchkiss 1942; Dubos et al. 1942), but the current application is limited to skin infections and infections of mouth and pharynx.

Also from *B. brevis* Rollin D. Hotchkiss and Dubos isolated gramicidin D (Hotchkiss and Dubos 1940 a, b), a heterogeneous mixture of antibiotic compounds (gramicidin A, B, and C) which are synthesized via the nonribosomal pathway by nonribosomal peptide-synthetases. Gramicidin D (D from **D**ubos) is a membrane channel forming linear pentadecapeptide antibiotic, which is active against Gram-positive bacteria. In 1944 Georgyi F. Gause and Mariya G. Brazhnikova detected the antibacterial gramicidin S (**S**oviet Gramicidin) also in supernatants of *B. brevis*. This cyclic decapeptide was produced by a strain from Russian soil (Gause and Brazhnikova 1944).

In the early 1940 the biochemist Selman Abraham Waksman (1888–1973), along with Albert Schatz (1922–2005) and Elizabeth Bugie (1920–2001) also searched for soil borne microorganisms that, among other test germs, would kill or inhibit the growth of penicillin-resistant bacteria. The main focus was set on drugs to combat *M. tuberculosis,* the causative agent of tuberculosis. *M. tuberculosis* is insensitive against Salvarsan, Prontosil and penicillins and was responsible for the death of every fourth adult in Europe and USA around 1900. Waksman and colleagues made the pioneering discovery that in particular bacteria of the genus *Streptomyces* (group actinobacteria) are highly promising candidates for the production of new antibiotics. He and his students screened soil bacteria for bioactivity and searched for growth inhibition zones surrounding single colonies of isolated soil microbes on agar plates, based on Fleming's method. Subsequently, they tested the inhibition on specifically targeted pathogenic bacteria.

3.5 Actinomycin

The first promising substance isolated under this initial screening program was found in 1940: Actinomycin, a polypeptide, was also the first antibiotic isolated from an actinobacterium: *Streptomyces antibioticus* subsp. *antibioticus* (formerly named *Actinomyces antibioticus*) (Fig. 8, a–c) (Waksman and Woodruff 1940).

The substance is active against a broad range of bacteria and even showed promise of attacking a tuberculosis strain, but it proved too toxic for antibacterial

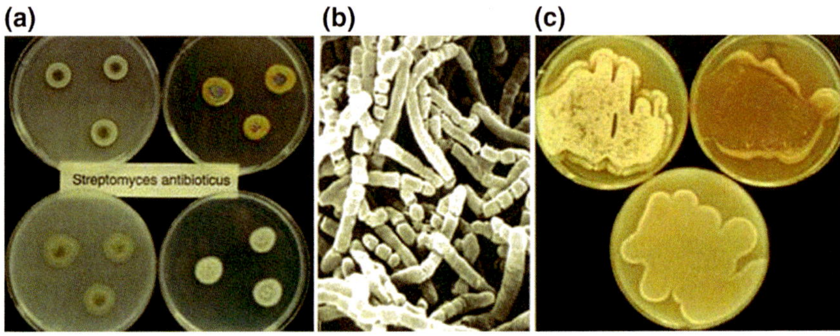

Fig. 8 *Streptomyces antibioticus* subsp. *antibioticus* agar plates media ISP 5, ISP 2, ISP 3, ISP 4 **a**; Spore chain morphology and spore surface in SEM C × 7500 **b**. *S. antibioticus* subsp. *antibioticus* agar plates media 5006, 5265, 5315 **c**. Pictures were kindly provided by PD Dr. Joachim Wink

therapeutic use in humans (Waksman and Woodruff 1941). Actinomycin D was also the first antibiotic with anticancer activity and is nowadays still in use in antitumor therapy. It binds to DNA duplexes, thereby interfering with the action of enzymes engaged in replication and transcription. Based on its mode of action, actinomycin has also become an important tool in molecular and cell biology (Hollstein 1973).

3.6 Bacitracin

Bacitracin, a broad-spectrum polypeptide antibiotic complex, inhibits mainly Gram-positive pathogens like streptococci and staphylococci and the anaerobic germ *Clostridium difficile* by preventing the bacterial cell wall synthesis. It was originally isolated from *Bacillus subtilis* and *Bacillus licheniformis* and described by Johnson et al. (1945). It inhibits peptidoglycan synthesis in Gram-positive bacteria by binding to a lipid pyrophosphate carrier that transports cell wall precursors to the growing cell wall (Husain 2004). Commercial bacitracin is a mixture of different related polypeptides with the main compound bacitracin B. Due to its systemic toxicity, bacitracin is mainly used local against wound infections, burn injuries and skin grafts. Four major bacitracin resistance mechanisms have been detected until 2012 and are described in detail in Charlebois et al. (2012).

3.7 Aminoglycosides: Streptomycin, Neomycin, Kanamycin, Gentamicin

The next promising hit within the screening program was the first described antibiotic of the aminoglycoside class: Streptomycin. This compound was isolated

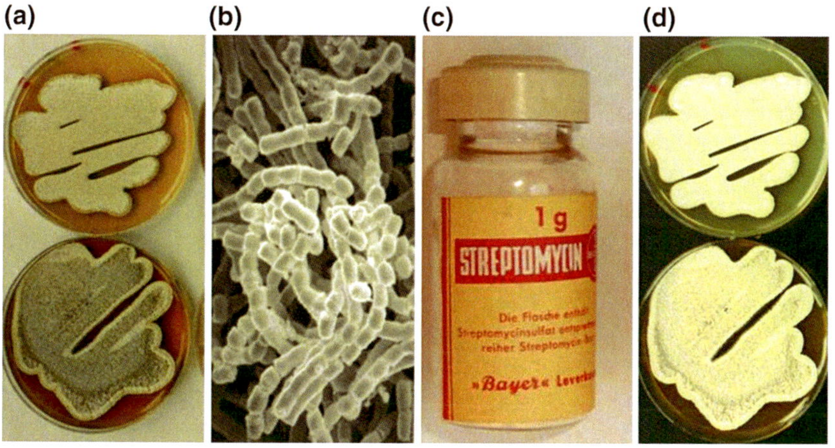

Fig. 9 *Streptomyces anulatus* subsp. *griseus* agar plates media ISP 1 + 3 **a**; Spore chains and spore surface (SEM × 7500) **b** Streptomycin ampoule **c**. This image was kindly provided by Prof. Dr. Christoph Friedrich, Bildarchiv des Instituts für Geschichte der Pharmazie der Universität Marburg; *Streptomyces anulatus* subsp. *griseus* agar plates media ISP 2 + 4 **d**. Pictures 1, 2, and 4 were kindly provided by PD Dr. Joachim Wink

by Schatz et al. (1944) from *Streptomyces anulatus* subsp. *griseus* (previously named *S. griseus*) (Fig. 9a–d). Streptomycin and many other aminoglycosides are high potent broad-spectrum antibiotics and act primarily by impairing bacterial protein synthesis through binding to the 30S ribosomal subunit. This antibiotic class is still in clinical use, albeit mostly in combination therapy or in topical application because of their insufficient pharmaceutical properties and severe side effects. Streptomycin was also the first anti-tuberculoticum used in therapy (1946) and furthermore it showed activity against several other diseases that could not be cured by penicillin. The history of aminoglycosides thereafter is marked by the successive introduction of a series of milestone compounds (kanamycin, gentamicin, and tobramycin) which definitively established the usefulness of this class of antibiotics for the treatment of Gram-negative bacillary infections (Mingeot Leclercq et al. 1999; Shaw et al. 1993).

For the co-discovery of streptomycin Waksman was awarded the Nobel Prize in 1952. Resistances against aminoglycosides are based on alteration of the ribosomal binding sites (streptomycin only), decreased uptake and/or accumulation of the drug in bacteria and the bacterial expression of enzymes which modify the antibiotic and thereby inactivate it (Davies and Wright 1997). Today, out of the various microorganisms screened, species of *Streptomyces* are still the most important producers of naturally occurring antibiotics, producing approximately two-thirds of all known antibiotics (Baltz 1998; Weber et al. 2003). However, no new drugs have been obtained from these approaches since the mid-1980s and it is even very difficult to find novel carbon skeletons in *Streptomyces*, suggesting that the genus has been exhaustively explored.

In the 1940, a large number of antibiotics with activities against Gram-negatives, including rickettsiae, and Gram-positives, including mycobacteria, were isolated from various species of the genus *Streptomyces*. Therefore the focus was set on the actinomycetes as potential producers of antimicrobial agents that might possess promising chemotherapeutic properties. The following characteristics of new antibiotics were particularly important: (1) High activity against Gram-negative bacteria and mycobacteria; (2) antibiotic action against streptomycin-resistant bacteria; (3) low toxicity to animals; (4) other desirable properties, such as activity against rickettsiae, viruses, tumors and phages (fide Waksman et al. 1949[1]).

Within this scope, neomycin, a new aminoglycoside broad-spectrum antibiotic active against streptomycin-resistant bacteria, including *M. tuberculosis* was isolated from *Streptomyces fradiae* and described by Waksman and Lechevalier (1949). Neomycin shows broad-spectrum activity as other antibiotics of the aminoglycoside group. Resistance against neomycin is conferred by aminoglycoside 3′-phosphotransferase genes, which inactivate neomycin by phosphorylation. There is a cross-resistance of neomycin to kanamycin and partially also to gentamicin.

In the 1950s another important broad-spectrum aminoglycoside antibiotic, kanamycin, a mixture of derivatives A, B and C, was isolated from *Streptomyces kanamyceticus*, a strain cultivated from Japanese soil (Takeuchi et al. 1957). The antibacterial spectrum includes mycobacteria, many Gram-positive and most Gram-negative pathogens. However, due to a couple of serious side effects, today kanamycin is mainly used for local therapy of eye infections (in the form of eye drops) and as a reserve antibiotic. Resistance mechanisms are as described for neomycin.

Gentamicin (genticin), a mixture of similar aminoglycosides, is produced by *Micromonospora purpurea* and was discovered in 1963 in the laboratories of the Schering Corporation, Bloomington, New Jersey (No authors listed, 1967). Gentamicin has a similar activity spectrum to related antibiotics such as neomycin and kanamycin (exerting both a bacteriostatic and bactericidal effect), but a rather greater activity: Almost all enterobacteria, including species of *Aerobacter*, *Escherichia*, *Klebsiella*, *Salmonella*, *Shigella*, *Proteus*, and also some *Pseudomonas aeruginosa*-strains are sensitive to gentamicin. Among Gram-positives, staphylococci are the most sensitives. Due to strong side effects (nephrotoxic reactions, ototoxicity) nowadays gentamicin is only used as an emergency antibiotic.

Back to the 1940s, the screening of fermentation broths of actinomycetes yielded a variety of antibiotics that were relatively quickly developed for clinical use. The antibiotic activity was detected, as described above, by agar diffusion assays in which fermentation samples were applied to filter paper discs that were placed on an agar plate inoculated with a bacterial test germ (Silver 2012). After this screening

[1]This deviates from the modern definition of antibiotics, but is a literal citation of Waksman's concept.

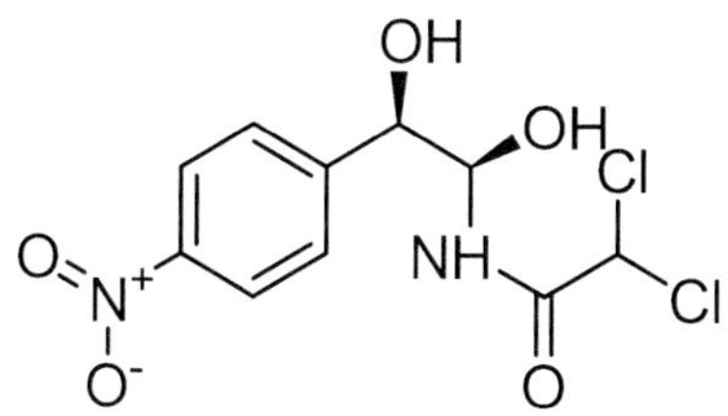

9

Fig. 10 Structure of chloramphenicol *9*

method had been successfully established in many laboratories, one hit followed another.

3.8 Chloramphenicol

A new antibiotic with broad-spectrum activity against various Gram-positive and Gram-negative pathogens, anaerobes, spirochetes, rickettsiae, chlamydiae and mycoplasma, was chloramphenicol (**9**, Fig. 10), isolated by John Ehrlich (Ehrlich et al. 1947) and colleagues from *Streptomyces venezuelae* (Ehrlich et al. 1948). The biological activity of chloramphenicol was described by Smith et al. (1947) and Gottlieb et al. (1948). The drug was introduced into clinical practice in 1949 (trade name Chloromycetin). Chloramphenicol also became the first antibiotic accessible by total synthesis (Controulis et al. 1949).

Chloramphenicol had been the only antibiotic available that was consistently active against *Salmonella* species including *Salmonella typhi* but due to its high toxicity, today the compound is given only in cases of specific indication. Chloramphenicol acts as inhibitor of the protein synthesis in bacteria and binds reversibly to the 50S subunit. Here the transfer of amino acids to growing peptide chains is prevented by suppression of peptidyl transferase activity. Thus the drug inhibits peptide bond formation, ultimately leading to inhibition of protein formation.

3.9 Tetracyclines

Another important antibiotic class discovered from *Streptomyces* species are the tetracyclines. The first tetracyclines, aureomycin (chlortetracycline), discovered by Duggar (1948) and terramycin (oxytetracycline: **10**, Fig. 11), described by Finlay et al. (1950) have been isolated from the fermenter broths of *Streptomyces aureofaciens* and *S. rimosus*, respectively. Tetracycline molecules act bacteriostatic by inhibition of protein biosynthesis and comprise a linear fused tetracyclic nucleus to which a variety of functional groups are attached.

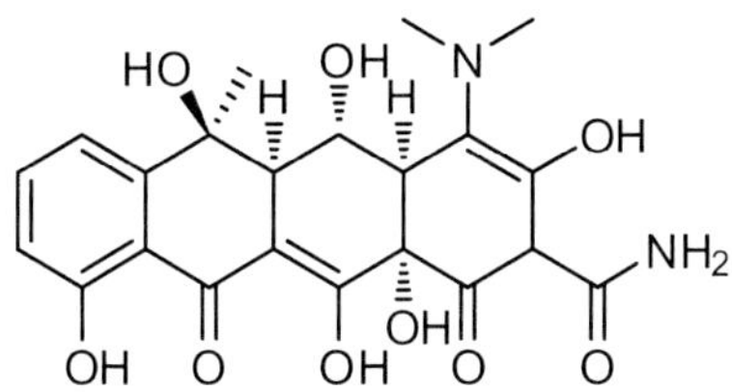

10

Fig. 11 Structure of terramycin *10*

Commercially applied tetracycline molecules bind to the 30S ribosomal subunit and block the entry of the amino-acyl tRNA into the A site of the ribosome. Therefore they are broad-spectrum antibiotics and show activity against a wide range of Gram-positive and Gram-negative germs including pathogens such as chlamydiae, mycoplasmas, rickettsiae, and protozoan parasites.

The absence of major adverse side effects has led to extensive use of tetracyclines in human and animal infection therapy, but the emergence of microbial resistance has limited their effectiveness (Chopra and Roberts 2001). As referred by Chopra and Roberts (2001), Levy (1984) reported that until the mid-1950, the majority of commensal and pathogenic bacteria was sensitive against tetracyclines. Hughes and Datta (1983) found out that among 433 different strains of Enterobacteriaceae, collected between 1917 and 1954, only 2 % were resistant to tetracyclines. Four different mechanisms of tetracycline resistance have been identified so far: drug inactivation, active efflux, ribosomal protection, and alteration of the target. As the majority of tetracycline resistance genes are associated with mobile plasmids, transposons, conjugative transposons, and integrons, these genes can spread from species to species into a wide range of genera by conjugation (Chopra and Roberts 2001). The first tetracycline-resistant bacterium, *Shigella dysenteriae*, was isolated in 1953 (Falkow 1975), and just two years later the first multiple-drug resistant *Shigella* was isolated. Multi-drug resistances which include resistance against tetracyclines has been identified in an increasing number of Gram-negative and Gram-positive species. Goldstein et al. (1994) found out that at the beginning of the 1990s approximately 90 % of methicillin-resistant *S. aureus*, 70 % of *Streptococcus agalactiae*, 70 % of multiple-drug resistant *Enterococcus faecalis*, and 60 % of the multiple-drug resistant *Streptococcus pneumoniae* strains were also resistant against tetracycline. Until the 1970s, a couple of naturally tetracyclines as well as those from semisynthetic approaches reached the market (Chopra and Roberts 2001). Aureomycin, for example, has been synthesized since 1959. Despite the widespread of tetracycline resistances, this antibiotic class is still of great interest for antibiotic therapy. Today the focus is set on new derivatives which are under examination for potential introduction as clinical agents to circumvent existing tetracycline resistance mechanisms (McMurry and Levy 2000; Tymiak et al. 1993) and on tetracycline efflux pump inhibitors that could be used in conjunction with older tetracyclines to restore their activity (Nelson et al. 1993; Nelson and Levy 1999).

Another broad-spectrum tetracycline, chelocardin, is produced by *Amycolatopsis sulphurea* (former *Nocardia sulphurea*) and was first described by Oliver et al. (1962). The structure was elucidated by Mitscher et al. (1970). Chelocardins are regarded as structurally atypical tetracyclines with antibacterial activity (Oliva et al. 1992). The exact mode of action is still unknown. Like other existing natural product scaffolds, chelocardins have not been developed because their suboptimal pharmacological properties could not be addressed at the time. Lešnik et al. (2015) demonstrated that reviving such compounds through the application of biosynthetic engineering can deliver novel drug candidates. They introduced the carboxamido moiety of tetracyclines (an important structural feature for their bioactivity) into the chelocardins, and generated a broad-spectrum antibiotic lead with significantly improved activity, including all Gram-negative pathogens of the ESKAPE panel (*Enterococcus faecium*, *S. aureus*, *Klebsiella pneumoniae*, *Acinetobacter baumannii*, *P. aeruginosa*, *Enterobacter* spp.). As the molecular target site of chelocardin seems to be different from that of the regular tetracyclins, there is some hope that this molecule can soon be turned into a novel broad-spectrum antibiotic with which we can combat the newly arising multi-resistant Gram-negative bacteria.

3.10 Macrolides/Ketolides

The important class of macrolides was discovered by James M. McGuire, who isolated and described the first lead compound erythromycin (Ilotycin) from *Streptomyces erythreus*. The strain was isolated from a soil sample collected on a small Philippine island in 1952 (McGuire et al. 1952). Macrolides are macrocyclic polyketides that typically consist of a 12–16 chained macrocyclic lactone ring with a glycosidic linked amino sugar. Macrolide antibiotics bind to the 50S subunit of the bacterial ribosome and inhibit the translocase and therefore inhibit the protein synthesis. They act primarily bacteriostatic, mainly against metabolically active bacteria. The pathogen spectrum of macrolides includes Gram-positive cocci (staphylococci, streptococci) and rods, Gram-negative pathogens like *Legionella pneumophila*, chlamydia, *Bordetella pertussis*, spirochaetes, *Haemophilus influenza,* and cell wall lacking mycoplasmas.

Three resistance mechanisms against macrolide antibiotics have been reported: (I) mutation or modification of the target (methylation of 23S rRNA of the 50S subunit); (II) efflux of exclusively 14- and/or 15-membered macrolides by increasing the number of efflux pumps, and (III) enzymatic inactivation by plasmid-coding esterases, as described for staphylococci and streptococci (Höck 2012b). To date erythromycin is still the standard antibiotic against infections of the respiratory tract and of throat-nose-ear sections, against pertussis and for patients with allergy to penicillin.

To remove the acid instability of erythromycin, semisynthetic macrolides of the second generation have been developed (clarithromycin, azithromycin). When resistances against first- and second-generation macrolides appeared, the ketolides, an enhancement of macrolides, have been designed (Chellat et al. 2016). Ketolides are semisynthetic derivatives of the macrolide erythromycin A (Zhanel et al. 2002) and were first mentioned in literature by Griesgraber et al. (1996). The mode of action of ketolides is very similar to erythromycin and ketolides exhibit good activity against Gram-positive and some Gram-negative aerobes. They also have excellent activity against drug resistant *S. pneumoniae*, including some macrolide-resistant strains (Zhanel et al. 2002). Actually, telithromycin is the only ketolide that has so far made it to the market.

3.11 Glycopeptides

Also in the 1950s the glycopeptides (glycosylated, cyclic or polycyclic, nonribosomal peptides) produced by a diverse group of soil actinomycetes have been discovered. Glycopeptides inhibit a late stage in bacterial cell wall peptidoglycan synthesis of exclusively Gram-positive bacteria. The first described glycopeptide, vancomycin was described by McCormick et al. (1955) from *Amycolatopsis orientalis*, formerly *Streptomyces orientalis*. The producer was isolated from a soil sample collected in Borneo during a research program carried out by Eli Lilly. Vancomycin became available for clinical use as an anti-staphylococcal agent after the approval of the U.S. Food and Drug Administration (FDA) in 1958 (Jovetic et al. 2010), although its early use was somewhat limited by side effects.

The second glycopeptide antibiotic in clinical use is teicoplanin, which belongs to the same family as vancomycin. It was isolated from the fermentation broth of *Actinoplanes teichomyceticus* (Somma et al. 1984). Teicoplanin is a mixture of five components of very similar polarity including teichomycin, published by Parenti et al. (1978). Today, vancomycin and teicoplanin are used to treat serious Gram-positive bacterial infections that are resistant to other antibiotics like β-lactams (Kahne et al. 2005) and methicillin-resistant *S. aureus* infections. They serve as reserve antibiotics in the treatment of serious infections of oxacillin-resistant staphylococci and ampicillin-resistant enterococci as well as against enterocolitis caused by toxin producing *Clostridium difficile*.

The frequency of resistance to glycopeptide antibiotics has increased significantly over the past decades and multiple genera, including enterococci and staphylococci, have developed resistance to these drugs (Ferber 2003). Some bacteria show natural, intrinsic resistance against vancomycin and teicoplanin, for example enterococcal species and non-enterococcal organisms like *Leuconostoc*, *Pediococcus*, *Lactobacillus*, and *Erysipelothrix* (Nelson 1999).

Acquired resistances have manifested itself largely through the expression of genes encoding proteins that reprogram cell wall biosynthesis and, thus, evade the action of vancomycin and teicoplanin (Binda et al. 2014). Reduced sensitivity (only against vancomycin), resistance by overproduction of peptidoglycan, or change of the target by acquisition of resistance genes via horizontal gene transfer are possible resistance mechanisms (Höck 2012a).

Another glycopeptide antibiotic, ristocetin (ristomycin), was first described by Grundy et al. (1956/1957) from *Amycolatopsis lurida*. It was originally used to treat Gram-positive pathogenic infections in humans, particularly staphylococcal infections, but its use was soon discontinued, however, due to toxic side effects related to its ability to cause thrombocytopenia and platelet agglutination (Weiss et al. 1973).

Actually three new lipoglycopeptides, i.e., telavancin, dalbavancin, and oritavancin have been registered for infections due to *S. aureus* including MRSA (Hesterkamp 2015).

3.12 Streptogramins

Streptogramins, also discovered from soil inhabiting streptomycetes, were first mentioned in literature in 1953 by Charney et al. Streptogramins are natural cyclic peptides of different unique classes. Each member of the class is a combination of at least three structurally unrelated molecules (group A and group B). Group A streptogramins are polyunsaturated macrolactones, group B streptogramins are cyclic hexadepsipeptides (Khosla et al. 1999). Group A and group B streptogramins alone act bacteriostatic, together the effect is bactericidal. Both groups bind to bacterial ribosomes and inhibit the translation of messenger RNA at the elongation step (DiGiambattista et al. 1989). Their combination generates bactericidal activities and reduces the possibility of emergence of resistant strains. However, resistances against streptogramins have been described: The substances can be conferred by modification of the drug target (methylation of the 23S ribosomal RNA), resulting in resistance to group B but not group A streptogramins or by active efflux, which has been described in *Staphylococcus epidermidis*, affecting inter alia group B streptogramins (DiGiambattista et al. 1989). Cross-resistance between streptogramins, macrolides and lincosamides is common. A combination product of later developed derivatives showed strong activity against streptococci, staphylococci and pneumococci. Due to several side effects of streptogramins and better investigated monopreparations like linezolid and daptomycin, today streptogramins are no longer relevant in antibiotic therapy.

3.13 Lincosamide

The antibiotic class of lincosamides, including the natural product lincomycin (**11**, Fig. 12) and the semisynthetic analogue clindamycin (**12**, Fig. 12), was first characterized in the 1960s and is now used for treatment of a broad spectrum of infections. Lincosamides act bacteriostatically by binding to the 23S rRNA of the 50S subunit. They mimic the intermediate formed in the initial phase of the elongation cycle, thereby inhibiting the protein synthesis.

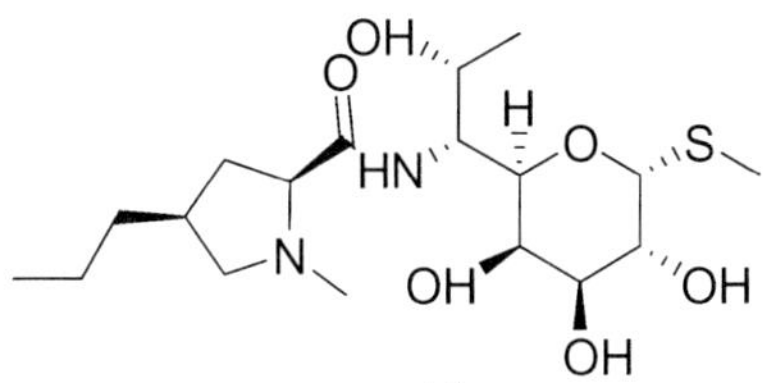

11

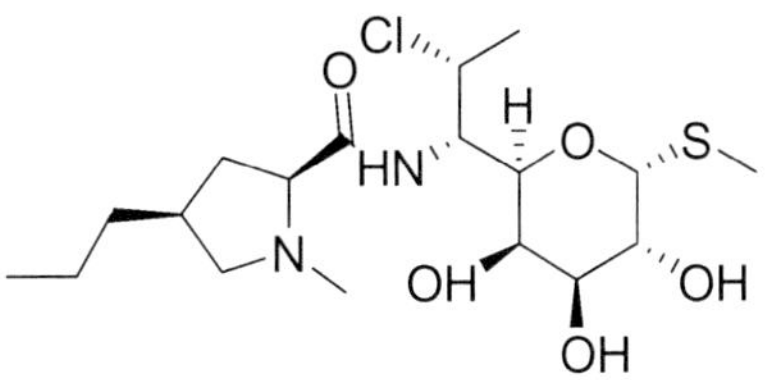

12

Fig. 12 Structures of lincomycin *11* and clindamycin *12*

Lincomycin is produced by *Streptomyces lincolnensis*. The producer was originally isolated from a soil sample collected in Lincoln, Nebraska (Mason et al. 1962). Also already in 1962 it was found that the in vitro activity of lincomycin against Gram-positives was very similar to the macrolide erythromycin (Lewis et al. 1962). Macrolides and lincosamides are chemically distinct but share a similar mode of action: inhibition is limited to Gram-positive cocci (mainly staphylococci and streptococci) and bacilli, to Gram-negative cocci, and intracellular bacteria (chlamydia and rickettsia species). Gram-negative bacilli are generally resistant, with some exceptions (Leclercq 2002).

The later developed semisynthetic clindamycin is four to eight times more active than lincomycin against most Gram-positive organisms (Meyers et al. 1969), and also acts against anaerobic germs and toxoplasmas. Today, lincosamides make up an important class of antibiotics used against a wide range of pathogens, including MRSA (Morar et al. 2009).

3.14 Ansamycins—Rifamycin, Rifampicin

Rifamycins, substances of the ansamycin family, are a group of antibiotics which are naturally produced or chemically synthesized. The first rifamycin (rifomycin; **13**, Fig. 13) was discovered and isolated in Italy in 1957 from *Streptomyces mediterranei* (Sensi et al. 1960) which has been renamed to *Nocardia mediterranei*, later to *Amycolatopsis mediterranei*, and finally to *Amycolatopsis rifamycinica*. The five rifamycin components are designated A, B, C, D, and E, whereby rifamycins

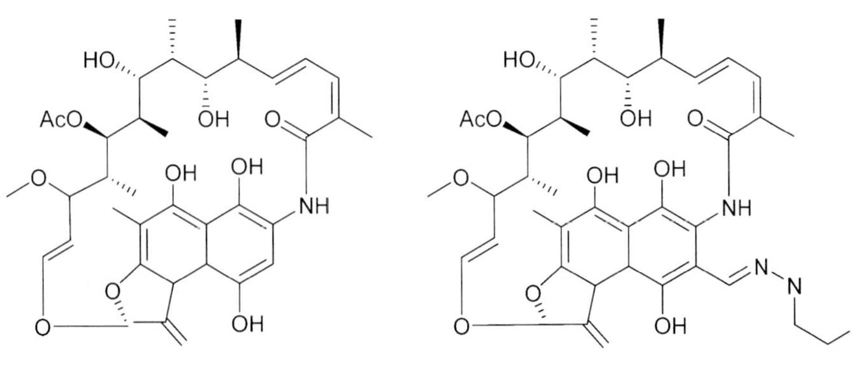

Fig. 13 Structures of rifamycin B *13* and rifampicin *14*

O, S, and SV are derivatives of the inactive B component, and AG and X are derivatives of the O component. Rifamycin B, which chemical structure has been elucidated by Oppolzer et al. (1964), is the precursor of clinically used antibiotics that are effective against *M. tuberculosis*, the pathogenic agent of tuberculosis and *M. leprae*, the agent of leprosy. Greater potency was achieved through substitutions as exemplified by the semisynthetic derivatives rifamide, rifaximin, rifapentine, rifampicin (rifampin; **14**, Fig. 13), and rifabutin (Carter 2009).

Rifampicin inhibits the RNA synthesis and acts bactericide on extra- and intracellular bacteria. In addition to mycobacteria, also staphylococci (including MRSA-strains), streptococci (including penicillin-resistant pneumococci), enterococci, Neisseria, Legionella, and other bacteria can be combated with rifampicin (Bange and Fille 2012). Due to resistance development caused by mutations of the RNA-polymerase gene (rpoB), normally a combination of different anti-tuberculosis drugs is given in combination: within the first two months: isoniazid, rifampicin, pyrazinamide and ethambutol; Up to the third month (stabilization phase) for at least four months: rifampicin and isoniazid (Bange et al. 2012). If a patient is infected with multi-drug resistant Mycobacteria (MDR-TB), the WHO recommends second-line or reserve drugs like kanamycin or linezolid, for example (WHO 2009).

3.15 (Fluoro) Quinolones: Nalidixic Acid, Ciprofloxacin

The synthetic quinolone antibiotics are arguably one of the most important classes of anti-infective agents. Their mode of action was new at the time of their discovery. They act bactericidal by inhibiting the bacterial enzyme DNA gyrase (topoisomerase II). DNA gyrase catalyses changes in the topology of DNA and can interconvert relaxed and supercoiled forms, introduce and remove catenanes and knots. Therefore these enzymes are essential in all bacteria to survival, but absent in higher eukaryotes, making them an attractive target for antibiotics (Collin et al. 2011).

When gyrase was discovered in *E. coli* in 1976 (Gellert et al. 1976), the two most important classes of gyrase inhibitors had already been described.

The first clinical quinolone, nalidixic acid (**15**, Fig. 14), was discovered accidentally as a by-product of the synthesis of the important antimalarial drug chloroquine. Nalidixic acid was patented by the Sterling Drug Company in 1963

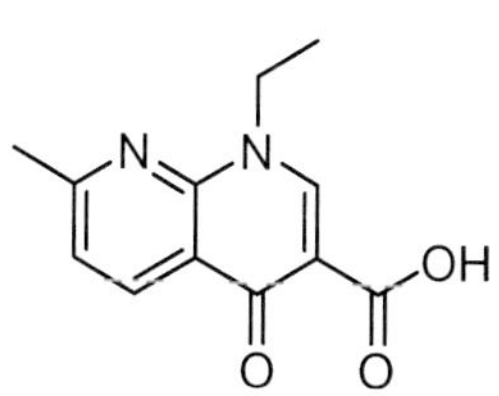

15

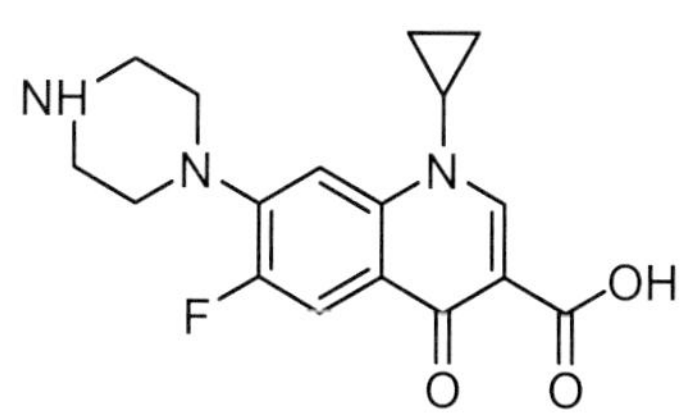

16

Fig. 14 Structure of nalidixic acid *15* and ciprofloxacin *16*

(Nielsen et al.). Already in 1962 the antibacterial properties of nalidixic acid were described (Lesher et al. 1962; Bisacchi 2015). In 1964 the first clinical reports and launch by Sterling as a first-in-class antibacterial agent followed. Based on nalidixic acid the fluoroquinolones were developed. Within the search for more active derivatives of nalidixic acid, the chemist K. Grohe (Bayer Company) found a way to produce highly effective antibiotics (cycloaracylation):

The top product was the broad-spectrum fluoroquinolone ciprofloxacin (**16**, Fig. 14) (Grohe and Heitzer 1987). In comparison to nalidixic acid ciprofloxacin showed an enhanced activity and antibacterial spectrum. Out of more than 20.000 tested bacterial strains, 98.2 % were sensitive against this new drug. The substance was patented in 1981 (Grohe et al. 1981) and introduced to the market as *ciprobay* in 1987 (Neufeldt 2003). After their discovery, fluoroquinolones became one of the principal clinical weapons in the fight against bacterial infections. Up to now more than 10,000 quinolones have been synthesized.

At the beginning resistances against quinolones developed slowly, in comparison to other antibiotic classes, because quinolones act on two different targets, i.e., DNA gyrase and topoisomerase IV. However, their usefulness is now endangered by the rapid emergence of resistance which typically arises as a result of alterations in these target enzymes and of changes in drug entry and efflux (Jacoby 2005).

The only other class of gyrase inhibitors which has been introduced into clinical use are the aminocoumarins, with the most prominent drug novobiocin, produced by *Streptomyces spheroides* (later reclassified to *Streptomyces niveus*), and discovered in the 1950 (Heide 2014; Harris et al. 1955). Within the last two years, two promising DNA gyrase inhibitors reached the medical pipeline: Baumann et al. (2014) discovered three cystobactamids, which have been isolated from the soil borne myxobacterium *Cystobacter* sp.. Cystobactamids were highly active against various Gram-positive pathogens and one cystobactamid, designated 919-2, also efficiently inhibits the growth of *E. coli* and *A. baumannii*, confirming that the compounds can penetrate the outer membrane of Gram-negative bacteria (Baumann et al. 2014).

Albicidin, a peptide antibiotic with remarkable antibacterial activity against various Gram-positive and Gram-negative microorganisms, is naturally produced by the sugarcane pathogenic bacterium *Xanthomonas albilineans* and can also be produced through heterologous expression. However, the yield of albicidin by natural production as well as by heterologous expression was too low for providing sufficient material for bioactivity profiling and structure–activity studies. With the total synthesis of albicidin Kretz et al. (2015) succeeded in sufficient production of this promising substance, which is already known since the 1980s (Birch and Patil 1985). As the potential of quinolones is largely exhausted as a template for new type II topoisomerase inhibitors, cystobactamids and albicidin offer promising alternatives to generate novel antibiotics using medicinal chemistry and biosynthetic engineering.

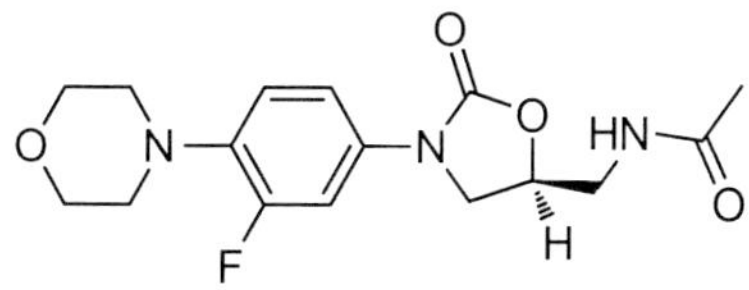

17

Fig. 15 Structure of linezolid *17*

3.16 Oxazolidinones

A further clinical important class of antibiotics was detected after the "Golden Age of Antibiotics" and should also be mentioned within this chapter: The oxazolidinones, a new class of synthetic compounds against Gram-positives, were first described in 1978 in a patent (DuPont), primary as antibiotics against plant pathogens. In 1987 oxazolidinones were introduced to clinical tests in human medicine. The first commercial substance of this class, linezolid (**17**, Fig. 15), was introduced to the US market in 2000 (Stahlmann and Riecke 2004). As the name suggests, the most important structure element of oxazolidinones is the five member heterocyclic oxazolidinone ring with an acetamidomethyl group. Linezolid, like vancomycin, is a reserve antibiotic. While most of the widely known antibiotics (macrolides, chloramphenicol) inhibit bacterial protein synthesis at the peptide chain elongation stage, linezolid acts earlier by potent interaction with the 50S ribosomal subunit and therefore hinders the protein biosynthesis at the first step. Linezolid is active against all Gram-positives, aerobic as well as anaerobic growing germs and also against methicillin resistant *S. aureus* vancomycin resistant enterococci and acts mainly bacteriostatic. Gram-negatives, with the exception of *Pasteurella*, are resistant. Resistance to other protein synthesis inhibitors does not affect oxazolidinone activity; however rare development of oxazolidinone resistance cases, associated with 23S rRNA alterations during treatment, have been reported (Pandit et al. 2012).

A second-generation oxazolidinone, tedizolid, has been registered for acute bacterial skin and skin structure infections by MRSA (Hesterkamp 2015).

Between 1940 and 1970 there was an explosion of antibiotic discoveries and their use in human medicine. Since 1970 the situation began to change and there has been a rapid drop in the introduction of new antibiotics and a sad departure by many large pharmaceutical companies from antibiotics as commercial products. In a comprehensive review Newman and Cragg (2016) summarized all new chemical entities and medical indications by source of compound between 1981 and 2014 and showed that natural products/natural product structures continue to play a dominant role in the discovery of leads for the development of drugs for the treatment of human diseases. But the Golden Age of antibiotics is a distant memory and the pipeline is almost empty (Cole 2014). Therefore, the actual situation is alarming: Dangerous rise in resistance to present antibiotics by pathogenic microbes, especially those nosocomial bacteria from the ESKAPE panel present an urgent need for the discovery of new antibiotics and chemical modifications of known antibiotic scaffolds.

For a current overview about "Natural products as sources of new drugs from 1981 to 2014" see Newman and Cragg (2016). For more information about "The clinical development and pipeline of antibiotics" see Hesterkamp (2015) and for "Strategies for the discovery and development of new antibiotics from natural products—three case studies" see Herrmann et al. (2016), the last two articles are within this EBook.

Acknowledgments I would like to thank Prof. Dr. Christoph Friedrich, Bildarchiv des Instituts für Geschichte der Pharmazie der Universität Marburg (Germany), Prof. Dr. Ulrich Kück and Dr. Julia Böhm, Ruhr-University Bochum, Dr. Heinke Jäger, Restoration Ecologist Charles Darwin Foundation Puerta Ayora, Galápagos (Ecuador), and PD Dr. Joachim Wink for providing the nice pictures and Dr. Rolf Jansen for examination of the chemical structures.

References

Abraham EP, Chain E (1940) An enzyme from bacteria able to destroy penicillin. Nature 146:837

Aminov RI (2010) A brief history of the antibiotic era: lessons learned and challenges for the future. Front Microbiol 1:134

Baltz RH (1998) Genetic manipulation of antibiotic producing *Streptomyces*. Trends Microbiol 2:76–83

Bange F-C, Fille M (2012) Antimykobakterielle therapie chapter 105. In: Suerbaum S, Hahn H, Burchard G-D, Kaufmann SHE, Schulz TF (eds) Medizinische Mikrobiologie und Infektiologie 7 Auflage Springer, Berlin

Bange F-C, Hahn H, Kaufmann SHE, Ulrichs T (2012) Mykobakterien. Chapter 41. In: Suerbaum S, Hahn H, Burchard G-D, Kaufmann SHE, Schulz TF (eds) Medizinische Mikrobiologie und Infektiologie 7 Auflage. Springer, Berlin

Barber M (1961) Methicillin-resistant staphylococci. J Clin Pathol 14:385–393

Baumann S, Herrmann J, Raju R, Steinmetz H, Mohr KI, Hüttel S, Harmrolfs K, Stadler M, Müller R (2014) Cystobactamids: myxobacterial topoisomerase inhibitors exhibiting potent antibacterial activity. Angew Chem Int Ed 53:1–6

Béchamp MA (1863) De l'action de la chaleur sur l'arseniate d'analine et de la formation d'un anilide de l'acide arsenique. Compt Rend 56:1172–1175

Bennett JW (2015) What is an antibiotic? In: Sánchez S, Demain AL (eds) Antibiotics-current innovations and future trends. Caister Academic Press

Bentley R (2000) Mycophenolic acid: a one hundred year odyssey from antibiotic to immunosuppressant. Chem Rev 100:3801–3826

Bentley R, Bennett JW (2003) What is an antibiotic? Revisited. Adv Appl Microbiol 52:303–331

Bentley R (2009) Different roads to discovery; Prontosil (hence sulfa drugs) and penicillin (hence b-lactams). J Ind Microbiol Biotechnol 36:775–786

Berger A, Fingerle V, Sing A (2012) Treponemen. Chapter 43. In: Suerbaum S, Hahn H, Burchard G-D, Kaufmann SHE, Schulz TF (2012) Medizinische Mikrobiologie und Infektiologie. 7. Auflage. Springer, Berlin

Binda E, Marinelli F, Marcone GL (2014) Old and new glycopeptide antibiotics: action and resistance. Antibiotics 3:572–594

Birch RG, Patil SS (1985) Preliminary characterization of an antibiotic produced by *Xanthomonas albilineans* which inhibits DNA synthesis in *Escherichia coli*. J Gen Microbiol 131:1069–1075

Bisacchi GS (2015) Origins of the Quinolone class of antibacterials: an expanded "Discovery Story". J Med Chem 58:4874–4882

Bloemendal S, Kück U (2014) Cephalosporins. In: Martín JF, García-Estrada C, Zeilinger S (eds) Biosynthesis and molecular genetics of fungal secondary metabolites. Springer, New York, pp 43–64

Bosch F, Rosich L (2008) The contributions of Paul Ehrlich to pharmacology: a tribute on the occasion of the centenary of his Nobel Prize. Pharmacology 82:171–179

Brewer GA, Johnson MJ (1953) Activity and properties of para-aminobenzyl penicillin. Appl Microbiol 1:163–166

Bulst N (1977) Der Schwarze Tod. Demographische, wirtschaftliche und kulturgeschichtliche Aspekte der Pestkatastrophe von 1347–1352. Bilanz der neueren Forschung. Antrittsvorlesung an der Universität Heidelberg

Burris RH, Baumann A, Potter VR (1990) Conrad Arnold Elvehjem. Biographical memoir, Nat Acad Sc

Carter GT (2009) Natural products in drug discovery. In: Stromgaard K, Krogsgaard-Larsen P, Madsen U (eds.) Textbook of drug design and discovery, 4th edn. CRC Press Boca Raton, pp 89–106

Cha JY, Ishiwata A, Mobashery S (2004) A novel β-lactamase activity from a penicillin-binding protein of *Treponema pallidum* and why syphilis is still treatable with penicillin. J Biol Chem 279:14917–14921

Chain E, Florey HW, Gardner AD, Heatley NG, Jennings MA, Orr-Ewing J, Sanders AG (1940) Penicillin as a chemotherapeutic agent. Lancet ii: 226–228

Charlebois A, Jalbert LA, Harel J, Masson L, Archambault M (2012) Characterization of genes encoding for acquired bacitracin resistance. *Clostridium perfringens* PloS One 7(9). doi:10.1371

Charney J, Fisher WP, Curran C, Machlowitz RA, Tytell AA (1953) Streptogramin, a new antibiotic Antibiot Chemother 3: 1283–1286

Chellat MF, Raguz L, Riedl R (2016) Antibiotikaresistenzen gezielt überwinden. Angew Chem 128:2–32

Chopra I, Roberts M (2001) Tetracycline antibiotics: mode of action, applications, molecular biology, and epidemiology of bacterial resistance. Microbiol Mol Biol Rev 65:232–260

Cole ST (2014) Who will develop new antibacterial agents? Phil Trans R Soc B 369:20130430

Collin F, Karkare S, Maxwell A (2011) Exploiting bacterial DNA gyrase as a drug target: current state and perspectives. Appl Microbiol Biotechnol 92:479–497

Controulis J, Rebstock MC, Crooks HM (1949) Chloramphenicol (Chloromycetin). V. Synthesis. J Am Chem Soc 71:2463–2468

Davies J, Wright GD (1997) Bacterial resistance to aminoglycoside antibiotics. Trends Microbiol 5:234–240

Davies J (2006) Where have all the antibiotics gone? Can J Infect Dis Med Microbiol 17:287–290

De la Bédoyère G (2005) The discovery of penicillin (milestones in modern science). Evans Brothers Ltd

Di Giambattista M, Chinali G, Cocito C (1989) The molecular basis of the inhibitory activities of type A and type B synergimycins and related antibiotics on ribosomes. J Antimicrob Chemother 24:485–507

Domagk G (1935) Ein Beitrag zur Chemotherapie der bakteriellen Infektionen. Deutsch Med Wochenschrift 61:250–253

Drews J (2000) Drug discovery: A historical perspective. Science 287:1960–1964

Dubos RJ, Hotchkiss RD (1942) Origin, nature and property of gramicidin and tyrocidine. Trans Studies Coll Phys Phila 4(10):11–19

Dubos RJ, Hotchkiss RD, Coburn AF (1942) The effect of gramicidin and tyrocidine on bacterial metabolism. J Biol Chem 146:421–426

Duchesne E (1897) Contribution à l'etude de la concurrence vitale chez les microorganismes. Antagonisme entre les moisissures et le microbes. Thèse. Lyon: Faculté de Médecine et de Pharmacie de Lyon

Duckett S (1999) Ernest Duchesne and the concept of fungal antibiotic therapy. Lancet 11 (354):2068–2071

Duggar BM (1948) Aureomycin; a product of the continuing search for new antibiotics. Ann N Y Acad Sci 30:177–181

Ehrlich P (1910) Die Behandlung der Syphilis mit dem Ehrlichschen Präparat 606. Deutsche medizinische Wochenschrift: 1893–1896

Ehrlich P, Hata S (1910) Die Experimentelle Chemotherapie der Spirillosen. Springer, Berlin

Ehrlich P, Bertheim A (1912): Über das salzsaure 3.3-Diamino-4.4-dioxy-arsenobenzol und seine nächsten Verwandten. Berichte der deutschen chemischen Gesellschaft Bd 45(1):756–766

Ehrlich J, Bartz QR, Smith RM, Joslyn DA, Burkholder PR (1947) Chloromycetin, a new antibiotic from a soil actinomycete. Science 31:417

Ehrlich J, Gottlieb D, Burkholder PR, Anderson LE, Pridham TG (1948) *Streptomyces venezuelae*, N. sp., The source of chloromycetin. J Bacteriol 56:467–477

Falkow S (1975) Infectious multiple drug resistance. Pion Ltd London United Kingdom

Ferber D (2003) Triple-threat microbe gained powers from another bug. Science 302:1488

Finlay AC, Hobby GL, P'an SY, Regna PP, Routien JB, Seeley DB, Shull GM, Sobin BA, Vinson IAJW, Kane JH (1950) Terramycin, a new antibiotic. Science 27:85

Fleming A (1929) On the antibacterial action of cultures of a *Penicillium*, with special reference to their use in the isolation of *B. influenzae*. Br J Exp Pathol 10(3):226–236

Gause GF, Brazhnikova MG (1944) Gramicidin S and its use in the treatment of infected wounds. Nature 154:703

Gellert M, Mizuuchi K, O'Dea MH, Nash HA (1976) DNA gyrase: an enzymethat introduces superhelical turns into DNA. Proc Natl Acad Sci USA 73:3872–3876

Gelmo P (1908) Über Sulfamide der p-Amidobenzolsulfonsäure. J prak Chem 77:369–382

Goldstein FW, Kitzis MD, Acar JF (1994) N, N-Dimethylglycylamido derivative of minocycline and 6-demethyl-6-deoxytetracycline, two new glycylcyclines highly effective against tetracycline-resistant Gram positive cocci. Antimicrob Agents Chemother 38: 2218–2220

Gosio B (1893) Contributo all'etiologia della pellagra; ricerche chimiche e batteriologische sulle alterazioni del mais. Giornale della Reale Accademia di Medicina di Torino 61:484–487

Gosio B (1896) Ricerdre batteriologische chirmide sulle alterazioni del mais. Rivista d'Ingienne e Sanita Publica 7:825–868

Gottlieb D, Bhattacharyya PK, Anderson HW, Carter HE (1948) Some properties of an antibiotic from a species of Streptomyces. J Bact 55:409–417

Griesgraber G, Or YS, Chu DTW, Nilius AM, Johnson PM, Flamm RK, Henry RF, Plattner JJ (1996) 3-Keto-11, 12-carbazate Derivatives of 6-*O*-Methylerythromycin A synthesis and in vitro activity. J Antibiot 49:465–477

Grohe K, Heitzer H (1987) Cycloaracylierung von Enaminen, I Synthese von 4-Chinolon-3-carbonsauren. Lieb Ann Chem 1:29–37

Grohe K, Zeiler HJ, Metzger K (1981) 1-cyclopropyl-6-fluor-1,4-dihydro-4-oxo-7- piperazino-chinolin-3-carbonsaeuren, Verfahren zu ihrer Herstellung sowie diese enthaltende antibakterielle Mittel. DE 3142854:A1

Grundy WE, Sinclair AC, Theriault RJ, Goldstein AW, Rickher CJ, Warren HB Jr, Oliver TJ, Sylvester JC (1956–1957) Ristocetin, microbiologic properties. Antibiot Annu 687–692

Guttmann P, Ehrlich P (1891) Über die Wirkung des Methylenblau bei Malaria. Berl Klin Wochenschr 28:953–956

Haensch S, Bianucci R, Signoli M et al (2010) Distinct clones of yersinia pestis caused the black death. Ed Nora J Besansky PLoS Pathogens 6(10):e1001134

Hart GD (2001) Descriptions of blood and blood disorders before the advent of laboratory studies. Histl Rev Brit J Haem 115:719–728

Harris DA, Reagan MA, Ruger M, Wallick H, Woodruff HB (1955) Discovery and antimicrobial properties of cathomycin, a new antibiotic produced by *Streptomyces spheroides* n. sp. Antibiot Annu 3:909–917

Hawgood BJ (2008) Alexandre Yersin (1863–1943): discoverer of the plague bacillus, explorer and agronomist. J Med Biog 16:167–172

Heide L (2014) New aminocoumarin antibiotics as gyrase inhibitors. Int J Med Microbiol 304:31–36

Herrmann J, Lukežič T, Kling A, Baumann S, Hüttel S, Petković H, Müller R (2016) Strategies for the discovery and development of new antibiotics from natural products: three case studies. Curr Top Microbiol Immunol Springer EBook

Hesterkamp T (2015) Antibiotics clinical development and pipeline. Curr Top Microbiol Immunol. (Springer EBook). doi:10.1007/82_2015_451

Höck M (2012a) Glycopeptidantibiotika. Chapter 98. Suerbaum S, Hahn H, Burchard G-D, Kaufmann SHE, Schulz TF (eds) Medizinische Mikrobiologie und Infektiologie 7 Auflage. Springer, Berlin

Höck M (2012b) Makrolide. Chapter 102. Suerbaum S, Hahn H, Burchard G-D, Kaufmann SHE, Schulz TF (eds) Medizinische Mikrobiologie und Infektiologie 7. Auflage. Springer, Berlin

Hollstein U (1973) Actinomycin. chemistry and mechanism of action. Chem Rev 6:625–652

Hotchkiss RD, Dubos RJ (1940a) Fractionation of bactericidal agent from cultures of a soil bacillus. J Biol Chem 132:791–792

Hotchkiss RD, Dubos RJ (1940b) Chemical properties of bactericidal substances isolated from cultures of a soil bacillus. J Biol Chem 132:793–794

Houbraken J, Frisvad JC, Samson RA (2011) Fleming's penicillin producing strain is not *Penicillium chrysogenum* but *P. rubens*. IMA Fungus 2(1):87–95

Hughes VM, Datta N (1983) Conjugative plasmids in bacteria of the 'pre-antibiotic' era. Nature 302:725–726

Husain MA (2004) Bacitracin, glycopeptide antibiotics, and the polymyxins. In: Craig CR, Stitzel RE (eds) Modern pharmacology with clinical applications 48:552

Inglesby TV, Dennis DT, Henderson DA et al (2000) Plague as a biological weapon: medical and public health management. Working Group on Civilian Biodefense JAMA 283:2281–2290

Jacoby GA (2005) Mechanisms of resistance to quinolones. Clin Inf Dis 41:120–126

Jacoby GA, Bush K (2005) Beta-lactam resistance in the 21st century. In: White DG, Alekshun MN, McDermott PF (eds) Frontiers in antimicrobial resistance. ASM Press, Washington, DC, pp 53–65

Jäger H (2011) *Cinchona pubescens*. In: Roloff, A., Weisgerber, H., Lang, U., Stimm, B. (Hrsg.): Enzyklopädie der Holzgewächse, Wiley-VCH, Weinheim, 58:1–14

Johnson BA, Anker H, Meleney FL (1945) Bacitracin: A new antibiotic produced by a member of the *B. subtilis* group. Science 12:376–377

Jovetic S, Zhu Y, Marcone GL, Marinelli F, Tramper J (2010) β-Lactam and glycopeptide antibiotics: first and last line of defense? Trends Biotechnol 28:596–604

Kahne, D Leimkuhler C, Lu W, Walsh C (2005) Glycopeptide and lipoglycopeptide antibiotics. Chem Rev 105:425–448

Karwehl S, Stadler M (2016) Exploitation of fungal biodiversity for discovery of novel antibiotics. In: Curr Top Microbiol Immunol, in press. doi:10.1007/82_2016_496

Khosla R, Verma DD, Kapur A, Aruna RV, Khanna N (1999) Streptogramins: a new class of antibiotics. Indian J Med Sci 53:111–119

Klein P, Falke D, Hahn H (2012) Ursprung der medizinischen Mikrobiologie. Chapter 2. In: Suerbaum S, Hahn H, Burchard G-D, Kaufmann SHE, Schulz TF (eds) Medizinische Mikrobiologie und Infektiologie. 7. Auflage. Springer, Berlin

Kool JL (2005) Risk of person-to-person transmission of pneumonic plague. Clin Infect Dis 40 (8):1166–1172

Kretz J, Kerwat D, Schubert V, Grätz S, Pesic A, Semsary S, Cociancich S, Royer M, Süssmuth RD (2015) Total synthesis of albicidin: A lead structure from Xanthomonas albilineans for potent antibacterial gyrase inhibitors. Angew Chem Int Ed 54:1969–1973

Kumarasamy KK, Toleman MA, Walsh TR, Bagaria J, Butt F, Balakrishnan R, Chaudhary U, Doumith M, Giske CG, Irfan S, Krishnan P, Kumar AV, Maharjan S, Mushtaq S, Noorie T, Paterson DL, Pearson A, Perry C, Pike R, Rao B, Ray U, Sarma JB, Sharma M, Sheridan E, Thirunarayan MA, Turton J, Upadhyay S, Warner M, Welfare W, Livermore DM, Woodford N (2010) Emergence of a new antibiotic resistance mechanism in India, Pakistan, and the UK: a molecular, biological, and epidemiological study. Lancet Infect Dis 10:597–602

Leclercq R (2002) Mechanisms of resistance to macrolides and lincosamides: nature of the resistance elements and their clinical implications. Clin Infect Dis 34:482–492

Lesher GY, Froelich EJ, Gurett MD, Baily JH, Brundage RP (1962) 1,8-Naphthyridine derivatives. A new class of chemotherapeutic agents. J Med Pharm Chem 91:1063–1065

Lešnik U, Lukežič T, Podgoršek A, Horvat J, Polak T, Šala M, Jenko B, Harmrolfs K, Ocampo-Sosa A, Martínez-Martínez L, Herron PR, Fujs S, Kosec G, Hunter IS, Müller R (2015) Construction of a new class of tetracycline lead structures with potent antibacterial activity through biosynthetic engineering. Angew Chem Int Ed 54:3937–3940

Levy SB (1984) Resistance to the tetracyclines. In Bryan LE (ed) Antimicrobial drug resistance. Academic Press, pp 191–240

Lewis K (2013) Platforms for antibiotic discovery. Nat Rev Drug Discovery 12:371–387

Lewis C, Clapp HW, Grady JE (1962) In vitro and in vivo evaluation of lincomycin, a new antibiotic. Antimicrob Agents Chemother 2:570

Lippi D, Conti AA (2002) Plague, policy, saints and terrorists: a historical survey. J Infect 44 (4):226–228

Lloyd NC, Morgan HW, Nicholson BK, Ronimus RS (2005) The composition of Ehrlich's Salvarsan: resolution of a century-old debate. Angew Chem Int Ed 44:941–944

Madigan MT, Martinko JM, Parker J (2000) Brock Mikrobiologie. Spektrum Akademischer Verlag

Majno G, Joris I (1979) Billroth and Penicillium. Rev Infect Dis 5:880–884

Mason DJ, Dietz A, Deboer C (1962) Lincomycin-a new antibiotic. I. Discovery and biological properties AAC 2:554–559

McCormick MH, McGuire JM, Pittenger GE, Pittenger RC, Stark WM (1955–1956) Vancomycin, a new antibiotic. I. Chemical and biologic properties. Antibiot Annu 3:606–611

McGuire JM, Bunch RL, Anderson RC, Boaz HE, Flynn EH, Powell HM, Smith JW (1952) Ilotycin, a new antibiotic. Antibiot Chemother 2:281–283

McMurry LM, Levy SB (2000) Tetracycline resistance in Gram positive bacteria. 660–677. In: Fischetti VA, Novick RP, Ferretti JJ, Portnoy DA, and Rood JI (eds) Gram positive pathogens. American Society for Microbiology, Washington DC

Meyers BR, Kaplan K, Weinstein L (1969) Micro- biological and pharmacological behaviour of 7-chloro- lincomycin. Appl Microbiol 17:653

Mingeot-Leclercq M-P, Glupczynski Y, Tulkens PM (1999) Aminoglycosides: activity and resistance. AAC 43:727–737

Mitscher LA, Juvarkar JV, Rosenbrook W, Andres WW, Schenck JR, Egan RS (1970) Structure of chelocardin, a novel tetracycline antibiotic. J Am Chem Soc 92:6070–6071

Morar M, Bhullar K, Hughes DW, Junop M, Wright GD (2009) Structure and mechanism of the lincosamide antibiotic adenylyltransferase LinB. Structure 17:1649–1659

Nelson ML, Park BH, Andrew JS, Georgian VA, Thomas BC, Levy SB (1993) Inhibition of the tetracycline efflux antiport protein by 13-thio-substituted 5-hydroxy-6-deoxytetracyclines. J Med Chem 36:370–377

Nelson ML, Levy SB (1999) Reversal of tetracycline resistance mediated by different bacterial tetracycline resistance determinants by an inhibitor of the Tet(B) antiport protein. Antimicrob Agents Chemother 43:1719–1724

Nelson RRS (1999) Intrinsically vancomycin-resistant Gram-positive organisms: clinical relevance and implications for infection control. J Hosp Inf 42:275–282

Neufeldt S (2003) Chronologie Chemie: Entdecker und Entdeckungen Ed. Neufeldt S. 3. Auflage Wiley VCH

Newman DJ, Cragg GM (2016) Natural products as sources of new drugs from 1981 to 2014. J Nat Prod 79:629–661

Newton GGF, Abraham EP (1955) Cephalosporin C, a new antibiotic containing sulphur and D-aminoadipic acid. Nature 175:548

Nielsen ED, Hamilton PB, David Rosi D, Peruzzotti GP (1963) Microbiological oxidation of 7-methyl-1, 8-naphthyridines to 7-hydroxymethyl-1, 8-naphthyridines. US 3317401 A Sterling Drug Inc

No authors listed (1967) Gentamicin. Br Med J 21:158–159

Nord C (2010) Mycophenolic Acid—The road from an early antibiotic to a modern drug. Biologiskt aktiva naturprodukter I läkemedelsutvecklingen, pp 1–11

Oliva B, Gordon G, McNicholas P, Ellestad G, Chopra I (1992) Evidence that tetracycline analogs whose primary target is not the bacterial ribosome cause lysis of *Escherichia coli*. Antimicrob Agents Chemother 36:913–919

Oliver TJ, Ptokop JF, Bower RR, Otto RH (1962) Chelocardin, a new broad spectrum antibiotic. I. Discovery and biological properties. Antimicrob Agents Chemother 583–591

Oppolzer W, Prelog V, Sensi P (1964) Experientia 20:336–339

Pandit N, Singla RK, Shrivastava B (2012) Current updates on oxazolidinone and its significance. Int J Med Chem. doi:10.1155/2012/159285

Parenti F, Beretta G, Berti M, Arioli V (1978) Teicomycins, new antibiotics from *Actinoplanes teicomyceticus* nov. sp. I. Description of the producer strain, fermentation studies and biological properties. J Antibiot 31:276–283

Rezende L (2006) Chronology of science. Checkmark, New York

Rollo IM, Williamson J, Plackett RL (1952) Acquired resistance to penicillin and to neoarsphenamine in *Spirochaeta recurrentis*. Br J Pharmacol Chemother 7:33–41

Rubin RP (2007) A brief history of great discoveries in pharmacology. In celebration of the centennial anniversary of the founding of the american society of pharmacology and experimental therapeutics. Pharmacol Rev 59:289–359

Schatz A, Bugie E, Waksman S (1944) Streptomycin: A substance exhibiting antibiotic activity against gram positive and gram negative bacteria. Proc Exp Biol Med 55:66–69

Sensi P, Greco AM, Ballotta R (1960) Rifomycins. I. Isolation and properties of rifomycin B and rifomycin complex. Antibiot Annu 262–270

Shaw KJ, Rather PN, Hare RS, Miller GH (1993) Molecular genetics of aminoglycoside resistance genes and familial relationships of the aminoglycoside-modifying enzymes. Microbiol Rev 57:138–163

Silver LL (2012) Chapter 2 Rational Approaches to antibacterial discovery: pre-genomic directed and phenotypic screening, 33–75. In: Dougherty TJ, Pucci MJ (eds) Antibiotic discovery and development, Springer, Berlin

Sköld O (2000) Sulfonamide resistance: mechanisms and trends. Drug Resist Updat 3(3):155–160

Smith RM, Joslyn DA, Gruhzit OM, McLean W, Penner MA, Ehrlich J (1947) Chloromycetin: biological studies

Somma S, Gastaldo L, Corti A (1984) Teicoplanin, a new antibiotic from *Actinoplanes teichomyceticus* nov. sp. Antimicrob Agents Chemother 26:917–923

Spoerer M, Streb J (2013) Neue deutsche Wirtschaftsgeschichte des 20. Walter de Gruyter, Jahrhunderts

Stahlmann R, Riecke K (2004) Die Infektiologie. 5.3.14.1 Editors: Adam D, Doerr HW, Link H, Lode H (Hrsg.), Springer

Sydenstricker VP (1958) The History of Pellagra. Its Recognition as a Disorder of Nutrition and Its Conquest. Am J Clin Nutr 6(4):409–414

Takeuchi T, Hikiji T, Nitta K, Yamazaki S, Abe S, Takayama H, Umezawa H (1957) Biological studies on kanamycin. J Antibiot (Tokyo) 10:107–114

Thomas HW (1905) Some experiments in the treatment of trypanosomiasis. Brit Med J 1140

Tréfouël J and T, Nitti F, Bovet D (1935) Activité du *p*-aminophénylsulfamide sur l'infection streptococcique expérimentale de la souris et du lapin. C R Soc Biol 120(23):756

Tymiak AA, Aklonis C, Bolgar MS et al (1993) Novel tetracycline glycosides active against tetracycline-resistant bacteria. J Org Chem 58:535–537

Tyndall J (1877) The optical deportment of the atmosphere in relation to the phenomena of putrefaction and infection. Philos Trans R Soc Lond 166:27–74

Waksman SA, Woodruff HB (1940) Bacteriostatic and bacteriocidal substances produced by soil actinomycetes. Proc Soc Exp Biol 45:609–614

Waksman S, Woodruff HB (1941) Actinomyces Antibioticus, a new soil organism antagonistic to pathogenic and non-pathogenic bacteria. J Bacteriol 42:231–249

Waksman SA (1947) What is an antibiotic or an antibiotic substance? Mycologia 39:565–569
Waksman SA, Lechevalier HA, Harris DA (1949) Neomycin-production and antibiotic properties. J Clin Invest 28:934–939
Waksman SA, Lechevalier HA (1949) Neomycin, a new antibiotic active against streptomycin-resistant bacteria, including tuberculosis organisms. Science 25:305–307
Weber T, Welzel K, Pelzer S, Vente A, Wohlleben W (2003) Exploiting the genetic potential of polyketide producing streptomycetes. J Biotechnol 106:221–232
Weiss HJ, Rogers J, Brand H (1973) Defective ristocetin-induced platelet aggregation in von Willebrand's disease and its correction by factor VIII. J Clin Invest 52:2697–2707
WHO (2009) Treatment of tuberculosis: guidelines for national programmes
Williams KJ (2009) The introduction of 'chemotherapy' using arsphenamine—The first magic bullet. J R Soc Med 102(8):343–348
Williams C (2013) Medicinal plants in Australia Vol 4: An antipodean apothecary. Rosenberg Publishing
Woodward RB, Doering WE (1945) The total synthesis of quinine. J. Amer Chem Soc 67:860–874
Würth I (2012) Geißler in Thüringen: Die Entstehung einer spätmittelalterlichen Häresie (Hallische Beiträge zur Geschichte des Mittelalters und der Frühen Neuzeit, Band 10
Zhanel GG, Walters M, Noreddin A, Vercaigne LM, Wierzbowski A, Embil JM, Gin AS, Douthwaite S, Hoban DJ (2002) The ketolides: a critical review. Drugs 62:1771–1804
Zimdahl RL (2015) Six chemicals that changed agriculture. 1st edn. Academic Press Elsevier

Actinobacteria and Myxobacteria—Two of the Most Important Bacterial Resources for Novel Antibiotics

Wiebke Landwehr, Corinna Wolf and Joachim Wink

Abstract Bacteria have been by far the most promising resource for antibiotics in the past decades and will in all undoubtedly remain an important resource of innovative bioactive natural products in the future. Actinobacteria have been screened for many years, whereas the Myxobacteria have been underestimated in the past. Even though Actinobacteria belong to the Gram-positive and Myxobacteria to the Gram-negative bacteria both groups have a number of similar characters, as they both have huge genomes with in some cases more than 10kB and a high GC content and they both can differentiate and have often cell cycles including the formation of spores. Actinobacteria have been used for the antibiotic research for many years, hence it is often discussed whether this resource has now been exhaustively exploited but most of the screening programs from pharmaceutical companies were basing on the cultivation mainly of members of the genus *Streptomyces* or Streptomyces like strains (e.g., some *Saccharopolyspora, Amycolatopsis* or *Actinomadura* species) by use of standard methods so that many of the so called "neglected" Actinobacteria were overlooked the whole time. The present review gives an overview on the state of the art regarding new bioactive compounds with a focus on the marine habitats. Furthermore, the evaluation of Myxobacteria in our ongoing search for novel anti-infectives is highlighted.

Contents

W. Landwehr · C. Wolf · J. Wink (✉)
Helmholtz Centre for Infection Research, Microbial Strain Collection, Inhoffenstrasse 7, 38124 Braunschweig, Germany
e-mail: Joachim.wink@helmholtz-hzi.de

Current Topics in Microbiology and Immunology (2016) 398: 273–302
DOI 10.1007/82_2016_503

Published Online: 22 September 2016

1 Marine Actinobacteria

1.1 Introduction

Over the last years, the commercial natural product research came back into focus because "the pipeline for new antibiotics is running dangerously low" (Fenical and Jensen 2006). While the rate of newly discovered antibiotics from soil inhabiting Actinobacteria decreased, the rate of re-isolation increased (Fenical et al. 1999; Lam 2006). However, to further explore this promising source of novel bioactive secondary metabolites, new strains had to be isolated with alternative methods or in unexplored environments (Lam 2006). Therefore, not only isolation techniques, but also the sampling sites had to be altered. With 70 % of the earth surface and a microbial abundance of 10^6 per mL in sea water and 10^9 per mL in ocean bottom, oceans are the world's biggest environment (Fenical and Jensen 2006). However, not only sea water and sediments are microbial rich environments, also marine organisms, flora and fauna like sponges harbor abundant communities (Ward and Bora 2006).

1.2 Marine Actinobacteria?

Over a long time, it was not clear whether truly "marine" Actinobacteria really exist because of the fact that there has been a lot of re-isolation of terrestrial strains and known compounds (Moore et al. 2005). The explanation for this assumption was the wash-in from terrestrial spores into the sea (Goodfellow and Haynes 1984) and the sampling problems of marine samples which were taken mostly close to the coast (Fenical and Jensen 2006). However, in the year 1984, the first marine Actinobacterium was found: *Rhodococcus marinonscens* (Helmke and Weyland 1984). 7 Years later, in 1991, the first marine Actinobacteria genus "*Salinispora* spp." which obligately requires seawater for growth (Jensen et al. 1991) was published. But even with the application of DNA sequence-based methods and the corresponding ability to analyze the relationships between this genus and their terrestrial relatives, the first seawater-obligate Actinobacteria genus "*Salinospora*" (grammatically incorrect; corrected to *Salinispora*) was described in 2005 (Mincer et al. 2005) and its two species *Salinispora tropica* and *Salinispora arenicola* were published (Maldonado et al. 2005a). Moreover, with the help of the type strains of these species, which are actively growing in sediment samples, the metabolic activity in the natural marine environment was demonstrated (Mincer et al. 2005). In addition to this finding, Fenical and Jensen (2006) also detected uncommon secondary metabolites produced by *Salinispora* strains and were inspired to further search for new groups of marine Actinobacteria. With the help of 16S analysis of the phylogenetic diversity as well as new cultivation approaches, Stach and Bull (2005) demonstrated that deep-sea sediments contained more than 1300 different actinobacterial operational taxonomical units which led to the assumption that there is a great opportunity to find novel species and genera. Fenical and Jensen (2006) cultivated diverse strains within six Actinobacteria families and many of them seemed to represent new taxa like *Salinispora* and *Mariniphilus* (Fig. 1). In addition to the taxonomic findings and therefore the demonstration of the existence of marine Actinobacteria, these strains turned out to be "excellent producers for secondary metabolites".

1.3 Where Can One Find Marine Actinobacteria?

The geographical origin of the new actinobacterial producer strains, published compounds and bioactivities of 67 % of marine natural products (up to 2003) was restricted to Japan, the Mediterranean as well as the Western Pacific Ocean (Blunt et al. 2007; Bull and Starch, 2007). However, because of the growing focus on the research on marine natural products, a dramatic rise of the published data from the China Sea was observed just one year later (Bull and Starch 2007). Marine Actinobacteria are present in diverse marine habitats, which are widespread over the ocean. These habitats are influenced by numerous geographical as well as physical parameters like temperature and salinity. Furthermore, they underlie geochemical impacts and ocean currents. But also ecosystems like salt marshes, wetlands, estuaries, continental shelves as well as the open ocean and the deep sea are habitats for specialized marine Actinobacteria (Ward and Bora 2006). The marine habitat starts with the sea surface microlayer, followed by the water column, from a few millimeters below the surface to more than 10,000 m depth, down to the sea floor with the micro- and macro-fauna and-flora, which were used as host for epibiosis and symbiosis, as well as the sea subfloor and deep biosphere. Within the habitat of the sea floor, varying sediments of varying geology, mineral nodule fields, carbonate mounds, cold seeps, hydrocarbon seeps, saturated brines, and hydrothermal vents were observed (Ward and Bora 2006).

The sea surface microlayer is an environment which is to date poorly characterized. However, some studies showed the existence of Actinobacteria and Proteobacteria within this habitat (Ward and Bora 2006). In the water column, together with the β- and δ-Proteobacteria, Firmicutes, Cytophaga-Flavobacter-Bacteriodes (CFGs) and Chlorobia, Actinobacteria belong to the mid-range of abundance. α- and γ-Proteobacteria dominated this habitat. Interestingly, theses compositions of strain collectives were also found in coastal and pelagic waters, despite differences in isolation techniques used and scale (Ward and Bora 2006).

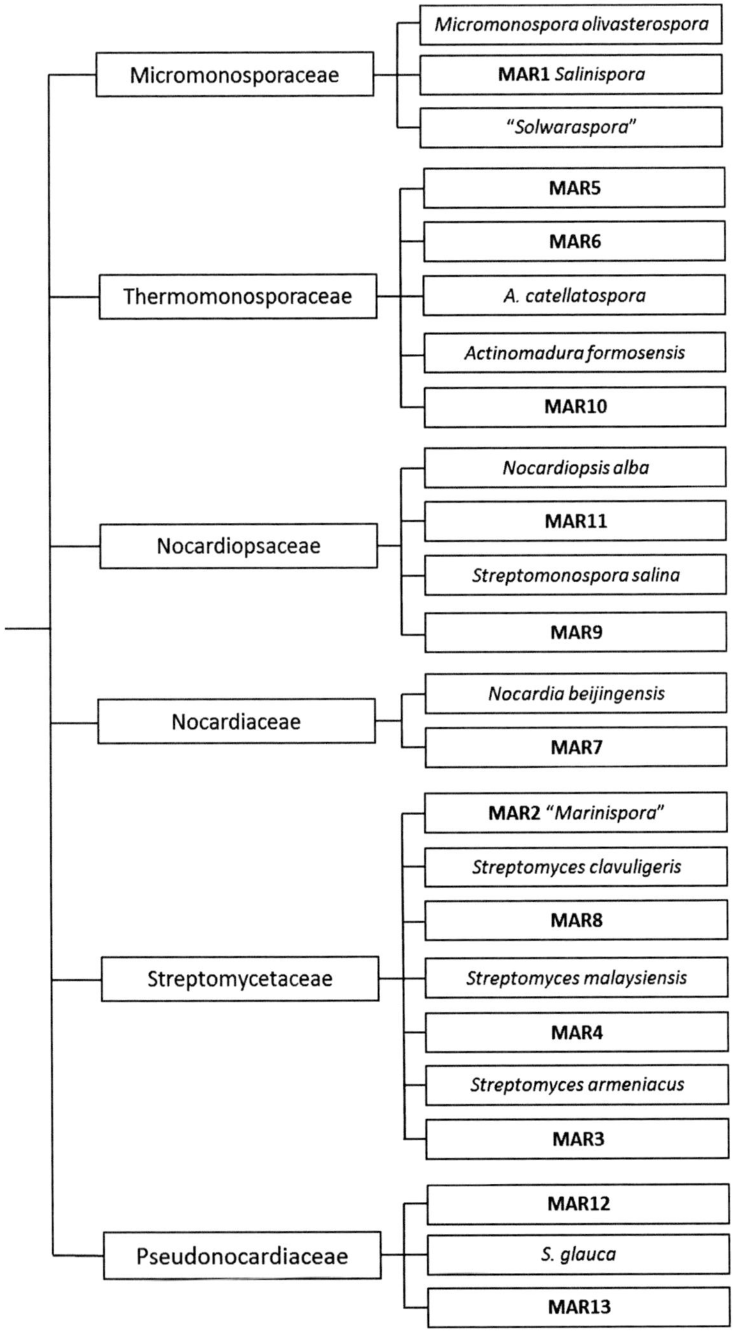

Fig. 1 Tree illustrating the phylogenetic relationships of 13 groups of marine-derived actinobacteria within six different families. The taxonomic status of the MAR groups is not really clear whereas it is known that they include numerous of new species. However, the MAR1 group was described as the genus *Salinispora* and the MAR2 group as the genus *Marinispora* (modified after Fenical and Jensen 2006)

Marine sediments are the best known source for the isolation of Actinobacteria from marine habitats. Studies on the 16S rRNA, to detect the phylogenetic diversity of Actinobacteria in marine sediments, showed that some deep-sea sediment contained up to 13,000 different actinobacterial operational taxonomic units which are forecast to belong to a large number of novel species and genera (Stach and Bull 2005). Also cultivation approaches of a range of depths and sediments (Maldonado et al. 2005b; Kim et al. 2004; Takami et al. 1997) illustrate the diversity and new insights in Actinobacteria classification (Maldonado et al. 2005a) and biogeography. These insights were used as inspirations for the isolation and recognition of novel marine Actinobacteria (Magarvey et al. 2004; Ward and Bora 2006). Furthermore, molecular studies on three different structural types of stromatolites ("organisms that have been present in the fossil record from greater than 3.5 billion years ago," cf. Ward and Bora 2006) ensured the presence 6–9 % of actinobacterial clones in the their complex mat community (Papineau et al. 2005).

However, there are not only the upper centimeters of marine sediments harboring Actinobacteria but also the below sea subfloor and the deep biosphere were found to be a natural habitat for these bacteria. Up to a depth of 800 m, a minor fraction of Actinobacteria were isolated (Stach et al. 2003; Inagaki et al. 2003) and molecular studies with specific primers showed a high actinobacterial diversity in clone libraries (Stach et al. 2003).

Moreover, marine Actinobacteria even live in marine organisms like many free swimming as well as sessile vertebrates and invertebrates. These organisms are mostly known to produce bioactive metabolites but these substances are often produced by symbiotic living microorganisms. The pufferfish, for example, was for a long time supposed to be the producer of the potent neurotoxin tetrodotoxin (TTX). However, to date this substance is known to be produced by numerous of marine organisms. Additionally, the production could also be correlated to many taxa of marine bacteria which live in symbiotic relationships with these organisms. In case of the pufferfish, a TTX-producing Actinobacteria (closely related to *Nocardiopsis dassonvillei*) and some TTX producing Bacillus strains were isolated out of its ovaries which are known to harbor high levels of TTX (Wu et al. 2005; Ward and Bora 2006). With the help of a sodium channel blocker assay using a mouse neuroblastoma cell culture bioassay, toxicity levels of 0.1–1.6 MU/g bacteria cells were observed [MU: amount of toxin which was expressed, calculated from the observed cell ratio (relationship between survival cell ratio and authentic TTX amount)] (Wu et al. 2005; Ward and Bora 2006). Furthermore, some TTX producing Actinobacteria like *Micrococcus* spp. and *Streptomyces* spp. were isolated out of deep sea and marine sediment (Do et al. 1990, 1991) which support the assumption that the symbiotic living organisms are responsible for the TTX production. However, to date the biosynthesis of TTX in bacteria or other organisms has not been described (Chau et al. 2011).

Furthermore, marine invertebrates such as sponges were described as natural habitat for marine Actinobacteria. They are known over a long time to be a prolific source of bioactive substances. Because of the reason that they are sessile organisms, they use the bioactive metabolites as a kind of chemical defence (Hill 2004; Ward and Bora 2006). However, up to 35 % of sponge biomass is comprised of microorganisms, which build an abundant microbial community (Hentschel et al. 2012; Webster and Taylor 2012; Steinert et al. 2014). Sponge-associated bacteria have been also frequently described as producers of bioactive natural products (Blunt et al. 2011, 2012, 2013). These communities include amongst other Actinobacteria also *Salinispora* and related strains (Hentschel et al. 2002).

1.4 Isolation of Marine Actinobacteria

To date, the most common sources for the isolation of marine Actinobacteria are sediments and sponges. However, as mentioned above, Actinobacteria were found almost everywhere in the marine environment. Because of the large amount of bacteria living in the marine sediment (10^9 bacteria per mL, Fenical and Jensen 2006), the cells have to be separated using diverse dilution or stamp techniques.

The common method for the isolation of marine Actinobacteria is very similar to the methods used to isolate terrestrial ones. On the one hand, sea mud can be directly spread over the agar media (Okami and Okazaki 1972) or the sediment samples may be diluted and treated with diverse methods as described below.

Some sediment samples were mixed with sterile sea water (Jensen et al. 1991; Mincer et al. 2002) or used as dried samples (Takizawa et al. 1993) before they were heated up between 6 and 60 min (Mincer et al. 2002; Jensen et al. 1991, Takizawa et al. 1993) at about 50 °C. This heat-shock treatment should dispatch most non-spore forming bacteria to provide the slow growing Actinobacteria a selective benefit for growth. Such measures had to be taken to avoid that, as described in the concurrent paper of Karwehl and Stadler (2016) for fungi, the fast-growing strains in soil samples would inadvertently overgrow the more interesting, hitherto unexploited ones. Afterwards, the samples were diluted in several dilution steps and plated on different types of nutrient rich agar plates. Pathom-Aree et al. (2006) used sterile saline solution (Ringer's solution) to pre-incubate the sediment samples for 30 min before the dilution and the following inoculation on a range of different media and an incubation temperature of 55 °C for the isolation of new thermophilic Actinobacteria taxa.

Another dilution technique was described by Mincer et al. (2002) for the isolation of the first *Salinispora* species. In this study, in addition to the heat-shock dilution series approach, the wet sediment was air dried and afterwards pressed into a sterile form plug. This plug was used as a stamp to inoculate the agar plates by stamping the plug several times in a circular fashion onto the plate to cause a dilution of the sediment. This approach was used with and without a previously described heat-shock treatment.

For the isolation of "novel marine-derived Actinobacteria taxa" Magarvey et al. (2004) used a modification of the medium Stan21, which is normally used to isolate Myxobacteria from soil samples (Shimkets et al. 2004). Therefore, the yeast extract was eliminated and the distilled water was replaced by artificial sea water. After the media preparation and the addition of cycloheximide, filter paper disks were placed on the agar plates and the wet sediment was spotted on the surface of the cellulose. The plates were incubated afterwards for 30–90 days at 30 °C in a humidified chamber. With this method they isolated some unknown marine Actinobacteria which cluster between the *Salinispora* clade and the genus *Micromonospora.*

One possibility for the isolation of marine Actinobacteria out of seawater is a simple filter technique. Okami and Okazaki (1972) described a method by which the seawater was centrifuged and filtered through a 0.3 µl pore size filter. The concentrated microorganisms on the filter were afterwards suspended in filtered sea water and plated on different isolation media.

To isolate Actinobacteria out of sponges several approaches are described. In classical isolation techniques, the sponge samples were directly stamped on isolation media or, in the dilution series method, sponge samples are crashed with a mortar, placed into sea water and after sedimentation, plated in different dilutions on specific sponge isolation agar (Mantalvo et al. 2005; Jiang et al. 2007). However, there are also some modifications of this method. Abdelmohsen et al. (2010) added the supernatant of crashed sponges (sponge extract) into the isolation media to rebuilt the natural environment. A further approach is the addition of both, "aqueous extract" like described before and "organic extract" where the sponge tissue was extracted with the help of hexane, dichlormethane and methanol. Finally, both studies demonstrated an enhanced number of novel isolated strains (Webster and Hill 2001; Selvin et al. 2004; Abdelmohsen et al. 2010). However, for all of these approaches, the sponges had to be harvested. To keep the sponges alive, Steinert et al. (2014) constructed a so called diffusion growth chamber (DGC). This chamber was built out of two combined centrifuge microfilter sections and was inoculated with different types of media. The media compositions differed in the amount of nutrients. Furthermore, every medium includes a small amount of sponge homogenate which was prepared out of a homogenized sponge sample in sterile sea water. After inoculation, the DGCs were directly inserted in the living sponge and retaken after 4 weeks of incubation (Fig. 2). Subsequently, one part of the inoculated media was plated on isolation media and the other part was used to inoculate new DGCs which were inserted into the same sponges. In this study, the authors showed that they were able to detect and cultivate more bacteria than using the classical direct plating method while the sponges stayed in their natural environment.

Most of the isolation media contained fungicidal agents like cycloheximide and nystatin to reduce fungal contamination as well as rifampicin and nalidixic acid to dispatch fast-growing Gram-negative bacteria. All plates were incubated between 2 and 6 weeks between room temperature and 28 °C.

1.5 Bioactive Substances Produced by Marine Actinobacteria

While we concentrate in this review on the bioactive secondary metabolites like antibiotics, cytotoxic agents and fungicides, marine Actinobacteria also produce melanins, enzymes, enzyme inhibitors, single cell proteins as well as probiotics which can be used for example in aquaculture (Manivasagan et al. 2013).

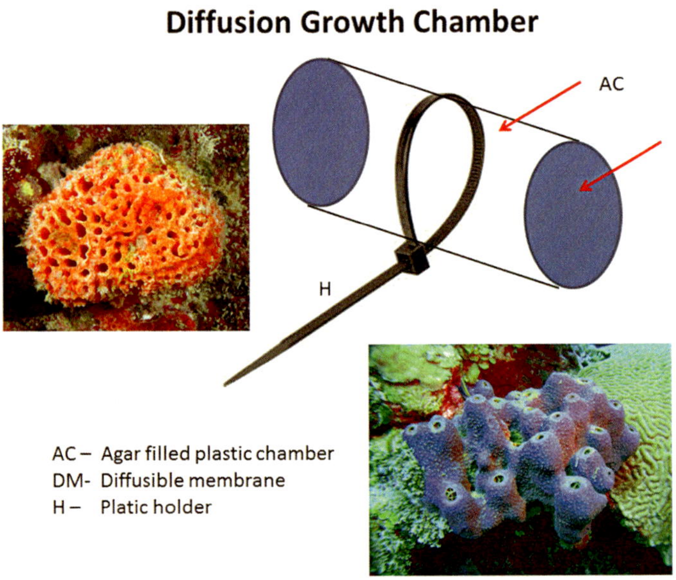

Fig. 2 Diffusion growth chamber (DGC) for in vivo cultivation of sponge-associated bacteria. Build-up of DGC out of two combined centrifuge microfilter sections with diffusible membranes (DM), inoculated with different types of media within the chamber (AC) and the plastic holder (H) for fixing within the sponge

Novel metabolites isolated from marine Actinobacteria include the antitumor agents chromomycins B, A2 and A3 from *Streptomyces coelicolor* (Lu et al. 2012), the antibacterial frigocyclinone from *Streptomyces griseus* (Bruntner et al. 2005), the antifungal and anticancer agents, daryamides from *Streptomyces* sp. (Sivakumar et al. 2007) and further ones as listed in Table 1. However, it becomes apparent that most of the producer strains were classified as *Streptomyces* species, which are common terrestrial bacteria. To date, "real new structures" were only isolated from the two genera of genuine marine Actinobacteria *Salinispora* and *Marinospora* (Manivasagan et al. 2013). Salinosporamides A and B (Fig. 3) are both produced by *Salinispora tropica*. These compounds are β-lactone-γ-lactams produced by a mixed PKS/NRPS biosynthesis. Salinosporamide A is an orally active proteasome inhibitor that induced apoptosis in multiple myeloma cells with mechanisms distinct from the commercial proteasome inhibitor anticancer drug Bortezomib (Chauhan et al. 2005). As NPI-0052, salinosporamide A entered as the first compound isolated out of an obligate marine organism, the clinical studies in multiple phase I trials for solid tumors, lymphoma, and multiple myeloma (http://www.nereuspharm.com/NPI-0052.shtml). The second compound, salinisporamide B, differs only in a lack of chorine and the subsequently less activity by a factor of 500 (Manivasagan et al. 2013). However, the terrestrial *Streptomyces* strain JS360 is the producer strain of the cinnabaramides A-G which are structurally close related to salinosporamide A (Stadler et al. 2007). These substances are also described as

Table 1 Examples of secondary metabolites produced by marine actinobacteria

Compound	Biological activity	Species/Isolate	Reference
1,8-Dihydroxy-2-ethyl-3 methylanthraquinone	Anticancer	*Streptomyces* sp.	Huang et al. (2006)
1-Hydroxy-1-norresistomycin	Antibacterial; anticancer	*Streptomyces chinaensis*	Gorajana et al. (2005) and Kock et al. (2005)
2-Allyloxyphenol	Antimycrobial; food preservative; oral desinfectant	*Streptomyces* sp.	Arumugam et al. (2010)
Abyssomycin C	Antibacterial (inhibition of PABA biosynthesis)	*Verrucosispora maris*	Bister et al. (2004)
Albidopyrone	Anticancer	*Streptomyces* sp.	Hohmann et al. (2009a)
Antracyclines	Anticancer	*Streptomyces galileus*	Fujii and Ebizuka (1997)
Arenicolides A-C	Mild anticancer activity	*Salinispora arenicola*	Jensen et al. (2007)
Arenimycin	Antibacterial; anticancer	*Salinispora arenicola*	Asolkar et al. (2006)
Aureoverticillactam	Anticancer	*Streptomyces aureoverticillatus*	Mitchell et al. (2004)
Avermectin	Antiparasitic	*Streptomyces avermitilis*	Burg et al. (1979)
Bafilomycin	ATPase-inhibitor of microorganisms, plant and animal cells	*Streptomyces griseus*, *Streptomyces halstedii*	Werner et al. (1984) and Frändberg et al. (2000)
Bisanthraquinone	Antibacterial	*Streptomyces* sp.	Socha et al. (2006)
Butenolides	Anticancer	*Streptoverticillium luteoverticillatum*	Li et al. (2006)
Carboxamycin	Antibacterial; anticancer	*Streptomyces* sp.	Hohmann et al. (2009b)
Chinikomycns	Anticancer	*Streptomyces* sp.	Li et al. (2005)
Chlaramphenicol	Antibacterial, inhibitor of protein biosynthesis	*Streptomyces venuzoelae*	Bewick et al. (1976)
Chlorodihydroquinones	Antibacterial; anticancer	*Novel Actinobacteria*	Soria-Mercado et al. (2005)
Cyanospraside A	Unknown	*Salinispora pacifica*	Jensen et al. (2007)
Cyclomarines	Antiinflammatory	*Streptomyces* sp., *Salinispora arenicola*	Schultz et al. (2008)

(continued)

Table 1 (continued)

Compound	Biological activity	Species/Isolate	Reference
Daryamides	Antifungal; anticancer	*Streptomyces* sp.	Sivakumar et al. (2007)
Dermacozines	Anticancer, radical scavenging	*Dermacoccus* sp.	Abdel-Mageed et al. (2010)
Diazepinomicin	Anticancer	*Micromonospora* sp.	Charan et al. (2004)
Enterocin	Bacteriostatic	*Streptomyces maritimus*	Piel et al. (2000) Max-Planck Inst., Jena
Essramycin	Antibacterial	*Streptomyces* sp.	El-Gendy et al. (2008)
Frigocyclinone	Antibacterial	*Streptomyces griseus*	Bruntner et al. (2005)
Glaciapyrroles	Antibacterial	*Streptomyces* sp.	Macherla et al. (2005)
Hygromycin	Antimicrobial, immunosuppressive	*Streptomyces hygroscopicus*	Omura et al. (1987), Uyeda et al. (2001)
Lajollamycin	Antibacterial	*Streptomyces nodosus*	Manam et al. (2005)
Lincomycin	Antibacterial, inhibitor of protein biosynthesis	*Streptomyces lincolnensis*	Peschke et al. (2006)
Lynamicins	Antibacterial	*Marinispora* sp.	McArthur et al. (2008)
Mansouramycins	Anticancer	*Streptomyces* sp.	Hawas et al. (2009)
Marinomycin A-D	Antimicrobial, anticancer	*Marinispora*	Kwon et al. (2006)
Mechercharmycins	Anticancer	*Thermoactinomyces* sp.	Kanoh et al. (2005)
Mitomycin C	Anticancer, binds to double stranded DNA	*Streptomyces lavendulae*	Mao et al. (1999)
ML-449	Anticancer	*Streptomyces* sp.	Jørgensen et al. (2010)
Pacificanones A and B	Antibacterial	*Salinispora pacifica*	Oh et al. (2008)
Piericidins	Anticancer	*Streptomyces* sp.	Hayakawa et al. (2007)
Proximicins	Antibacterial; anticancer	*Verrucosispora* sp.	Fiedler et al. (2008)
Rapamycin	Immunosuppressive, antifungal	*Streptomyces hygroscopicus*	Vezina et al. (1975)

(continued)

Table 1 (continued)

Compound	Biological activity	Species/Isolate	Reference
Resistoflavin methyl ester	Antibacterial; anti-oxidative	*Streptomyces* sp.	Kock et al. (2005)
Salinamides	Antiinflammatory	*Streptomyces* sp.	Moore et al. (1999)
Saliniketal	Cancer cemoprevention	*Salinispora arenicola*	Jensen et al. (2007)
Salinispyrone	Unknown	*Salinispora pacifica*	Jensen et al. (2007)
Salinispyrone A&B	Mild anticancer activity	*Salinispora pacifica*	Oh et al. (2008)
Salinosproramide A	Anticancer; antimalarial	*Salinispora tropica*	Jensen et al. (2007) and Prudhomme et al. (2008)
Salinosporamide B and C	Anticancer	*Salinispora tropica*	Williams et al. (2005)
Sesquiterpene	Unknown	*Streptomyces* sp.	Wu et al. (2006)
Staurosporine	Antitumor; phycotoxicity	*Streptomyces* sp.	Wu et al. (2006)
Streptokordin	Antitumor	*Streptomyces* sp.	Jeong et al. (2006)
Streptomycin	Antimicrobial	*Streptomyces griseus*	Egan et al. (1998)
Streptozotocin	Diabetogenic	*Streptomyces achromogenes*	Herr et al. (1967)
Tetracyclines	Antimicrobial	*Streptomyces achromogenes, Streptomyces rimosus*	Saleh et al. (1985) and Hansen et al. (2001)
Thiocoraline	Anticancer	*Micromonospora* spp.	Perez Baz et al. (1997)
Tirandamycins	Antibacterial	*Streptomyces* sp.	Carlson et al. (2009)
TP-1161	Antibacterial (inhibition of protein synthesis)	*Nocardiopsis* sp.	Engelhardt et al. (2010)
Valinomycin	Isophor, toxic for Pro- and Eukaryotes	*Streptomyces griseus*	Andersson et al. (1998)
ZHD-0501	Anticancer	*Actinomadura* sp.	Han et al. (2003)
Elaiomycins B and C	Anticancer	*Streptomyces* sp. BK 190	Helaly et al. (2011)
N-(2-hydroxyphenyl)-2phenazinamine (NHP)	Anticancer, antifungal	*Nocardia dassonvillei*	Gao et al. (2012)

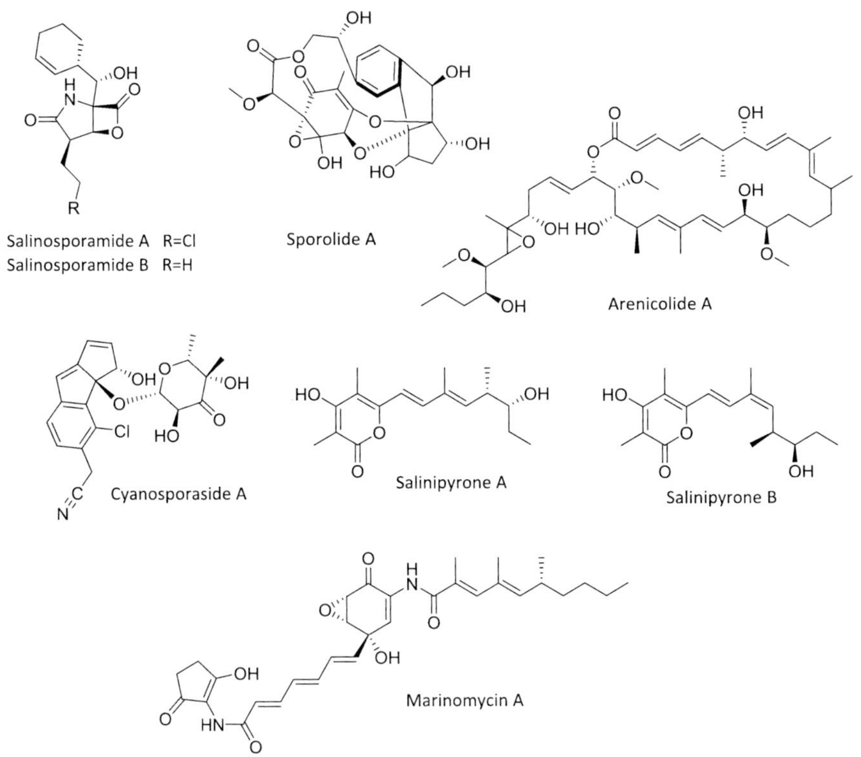

Fig. 3 Examples of new bioactive compounds and their corresponding structures produced by novel genera of marine actinobacteria

strong proteasome inhibitors. However, Rachid et al. (2011) detected a significantly weaker cytotoxic effect then caused by salinosporamide A. Other unprecedented substances produced by *Salinispora tropica* are sporolide A, arenicolide A, cyanosporaside A, and salinipyrone A (Fig. 3). Even though, the salinipyrones A and B (Fig. 3) were first isolated out of the obligate marine Actinobacterium *Salinispora pacifica* (Oh et al. 2008). Both substance did not show antibiotic activity against drug resistance human pathogens, however a moderate cytotoxic activity was detected (Jensen et al. 2007; Manivasagan et al. 2013).

Marinomycin A (Fig. 3), produced by a *Marinophilus* strain (Kwon et al. 2006), is a new polyene macrolide with a high toxicity to tumor cells and antibiotic effect against vancomycin-resistant *Enterococcus faecium* and methicillin-resistant *Staphylococcus aureus*.

In conclusion, the discovery of the new marine Actinobacteria genera *Salinispora* and *Marinophilus* could be directly correlated to the new and unprecedented compounds with new structures and partially new modes of action. Like described before, molecular studies indicate a great potential for the isolation of novel genera of marine Actinobacteria. However, new isolation techniques will have to be established and unexplored environments have to be sampled to avoid re-isolations of bacteria out of known genera producing common compounds.

2 Myxobacteria—The Underestimated Bacterial Resource

2.1 History of Myxobacteria

The first myxobacterium was discovered in 1809 by the German botanist H.F. Link and named *Polyangium vitellinum*, but erroneously it was characterized as fungus because of the characteristic fungi-like life cycle (Link 1809). It took many years, until 1892, until Roland Thaxter identified these organisms as bacteria (Thaxter 1892). Actually, the order of Myxococcales consists of 55 species including 28 genera (Fig. 4) and differentiates from other Gram-negative prokaryotes by the mutuality to have a special life cycle.

Myxobacteria belong to the δ-subgroup of proteobacteria and therefore are Gram-negative. The vegetative cells are rod-shaped with the ability to glide over

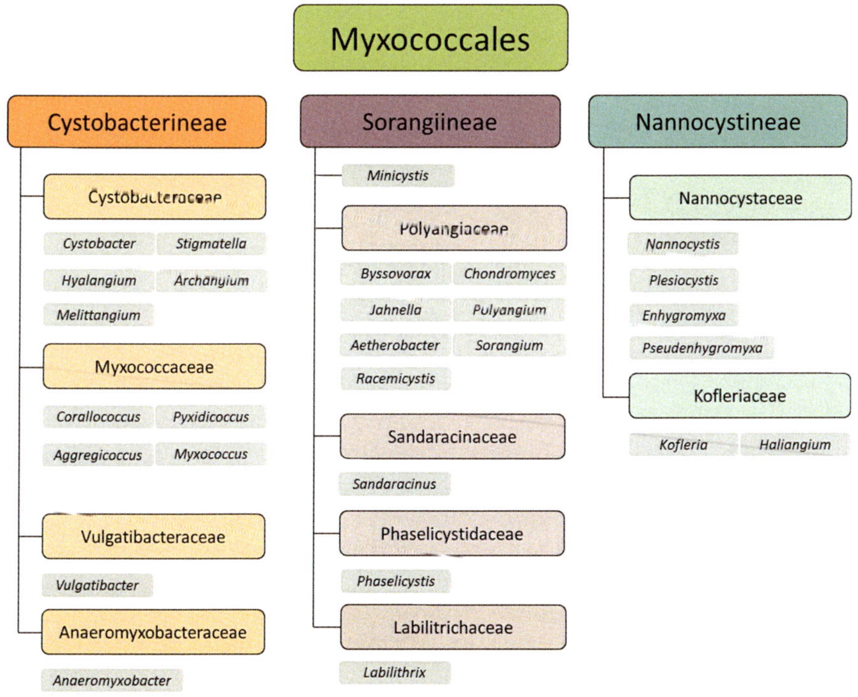

Fig. 4 Current taxonomy of the order Myxococcales

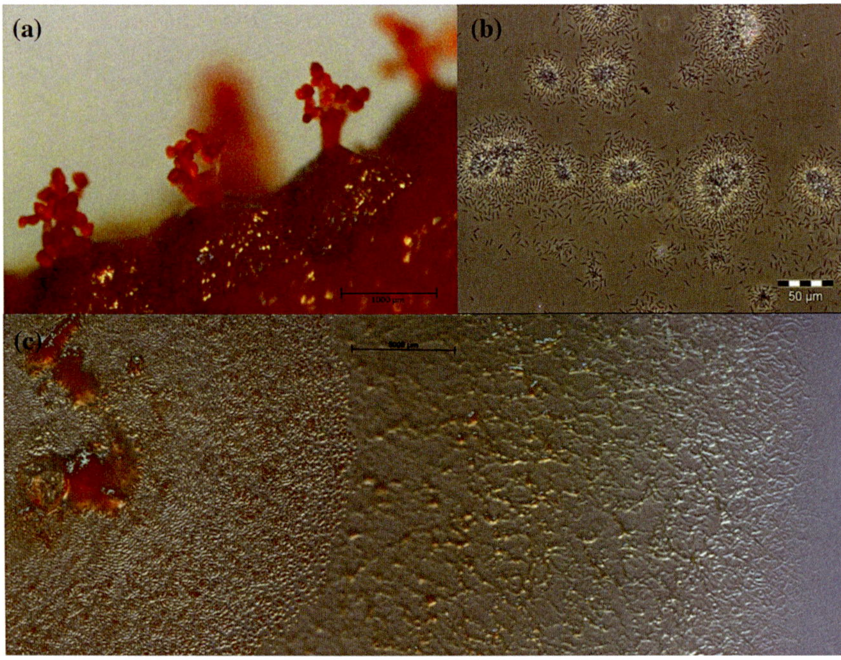

Fig. 5 **a** Fruiting bodies of *Stigmatella aurantiaca* (bar = 1000 µm); **b** Cells of *Myxococcus virescens* (bar = 50 µm); **c** Swarming of *Corallococcus coralloides* (bar = 5000 µm)

solid surfaces and form some kind of multicellular, species-specific "fruiting-bodies" under starvation conditions without access to a sufficient nutrient storage (Reichenbach et al. 1988) (Fig. 5). Those fruiting bodies can comprise up to 10^5 individuals and show a wide range of differences between the genera and species referring their height, shape and color, which mostly varies from yellow, orange or red until brown or even black (Reichenbach 1983; Garcia and Müller 2014a, b, c, d, e).

Within the fruiting bodies, the vegetative cells transform to short, so called myxospores, often enclosed in sporangioles. Because of desiccation resistance, myxospores are able to survive in unfavorable environmental conditions for many years (Reichenbach et al. 2006a, b). Myxobacteria can be found in large populations on many substrates in nature all over the world, e.g., soil, rotting wood, and other habitats. Due to their nutritional requirements, myxobacteria can be divided into two ecological groups that are also in agreement with their phylogeny. Predators use other bacteria or yeasts as food source, and cellulose decomposers, belonging to the genera *Sorangium* and *Byssovorax*, decompose organic materials by producing different types of lytic exoenzymes (Reichenbach et al. 1988; Reichenbach et al. 2006a, b).

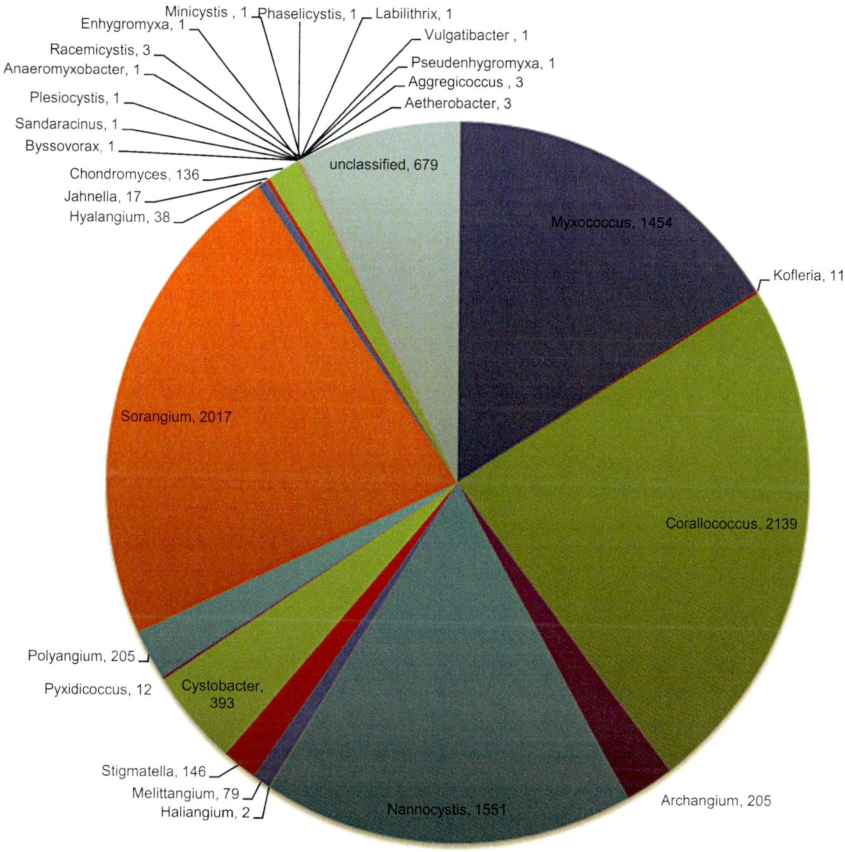

Fig. 6 Number of strains at the HZI

Another important and interesting feature of myxobacteria is their ability to produce a broad range of structurally diverse secondary metabolites, many of them with bioactivity. The working group Microbial Strain Collection at the Helmholtz-Center for Infection Research (HZI) in Braunschweig, has the largest collection of myxobacteria worldwide with more than 9000 strains, including all validly described type strains (Fig. 6).

The major part of the collection is represented by species of the frequently occurring genera *Corallococcus*, *Myxococcus*, *Nannocystis*, and *Sorangium* (Gerth et al. 2003). *Chondromyces*, *Myxococcus*, and *Sorangium* are producers of most of the interesting natural products, consequently they became enriched in the collection (Fig. 7).

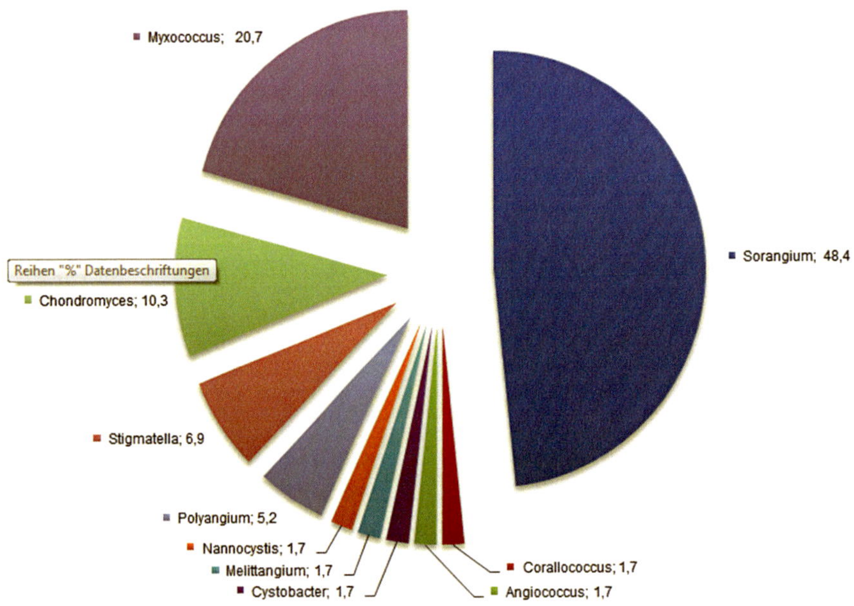

Fig. 7 Percentage of genera in the producers of our novel secondary metabolites from myxobacteria (modified according to Gerth et al. 2003)

2.2 *Isolation Methods*

To find new bioactive substances, it seems reasonable to exploit new producer organisms and to insert these novel species and genera in the latest taxonomic classification (Müller and Wink 2014). It is necessary to use basic routine methods (Dawid 2000; Shimkets et al. 2006) as well as new approaches (Garcia et al. 2009; Mohr et al. 2015) for the isolation and cultivation of potential and interesting new producer strains.

In general, myxobacteria are characterized as mesophilic soil organisms which prefer a temperature of 30° and a neutral pH. Nevertheless, also acidophilic (Brockman and Boyd 1963), alkaliphilic (Hook 1977) and psychrophilic (Dawid et al. 1988) species were found in the last decades, demonstrating the enormous adaptability of myxobacteria even to extreme uncomfortable habitats. Myxobacteria can also be found in marine, saline habitats, for example *Haliangium ochraceum* and *Haliangium tepidum* (Iizuka et al. 1998). To isolate new myxobacterial strains from different biotopes, environmental samples (e.g., soil, sand, dead wood, bark, leaves, compost) were treated by a standardized procedure. For soil samples, an amount of 50–150 g of sample is collected in 1–5 cm depth below the surface in sterile vials. Before the enrichment on different plates, the sample has to be air dried and characterized concerning different parameters like color, size, grain, content, and acidity (Dawid 2000). Commonly, different types of plates are used for the enrichment of a wide range especially of myxobacteria species referring to the methods of Reichenbach et al. (2006). Living *E. coli* cells on water agar are suitable as the feed organism for predatory myxobacteria (Shimkets et al. 2006). Many soil bacteria are able to degrade dead, but not living, microorganisms. Furthermore, most soil bacteria grow faster than myxobacteria. Using living bait organisms, myxobacteria have an advantage in comparison to other, not or less swarming, competitors (Garcia et al. 2009). To enrich and isolate specifically cellulose degrading myxobacteria, mineral salt agar Stan 21 is used. Sterile filter paper is placed on the top of the agar and serves as food source (Shimkets et al. 2006). The plates are normally incubated at 30° over a period of several weeks, until the visually striking myxobacteria can be seen with the naked eye. The following isolation, purification, and identification of new strains are carried out on VY/2 plates, containing yeast as nutrient source (Dawid 2000). It is supposed, that the preparation of the medium, combining the important characteristics like in the natural habitats of myxobacteria, leads to high advantage finding novel strains. For example, novel strains were isolated by growing at room temperature (23°) as well as under white light or sun light exposure and on acidic or saline agar plates (Mohr et al. 2015; Garcia et al. 2009). The identification of the facultative anaerobic genus *Anaeromyxobacter* (Sanford et al. 2002) leads to the assumption, that different air conditions tried within the isolation procedures can reveal facultative anaerobic and microaerophilic species (Garcia et al. 2009). Culture-independent methods, like clone bank analyzes, revealed that there is a high number of new potential producer species (Mohr et al. 2015). The development of new cultivation methods is important to find the best growth conditions for all these different types of uncultivated myxobacteria.

The procedure of the standardized screening method, developed at the HZI, starts with the cultivation of a new isolated strain in different liquid complete media (each 100 ml) with different C and N sources. The bacteria are cultivated at 30 °C and shaking (180 rpm) 7–14 days. Myxobacteria secrete the secondary metabolites out of the cells into their environment, in this case into the medium. To bind the metabolites and thereby preventing a feedback inhibition or a degradation of the metabolite by a producer strain, XAD16 adsorber resin is added to the culture. After sieving, the resin, and the adsorbed compounds are extracted with acetone, evaporated and finally eluted with methanol, resulting in a raw extract of a 1:100 concentration which is used for further analyzes (Reichenbach and Höfle 1993). However, it is necessary to find the best growth conditions and to optimize the specific production rate for each producer strain, as well as an economic justifiable fermentation process has to be established.

To investigate the effect of a crude extract from a new potential myxobacterial producer strain against different pathogens with clinical importance and to find the active principle, the working group Microbial Strain Collection at the HZI use a dereplication system, which is a combination of a biological activity assay and a chemical screening with liquid chromatography-high resolution mass spectrometry

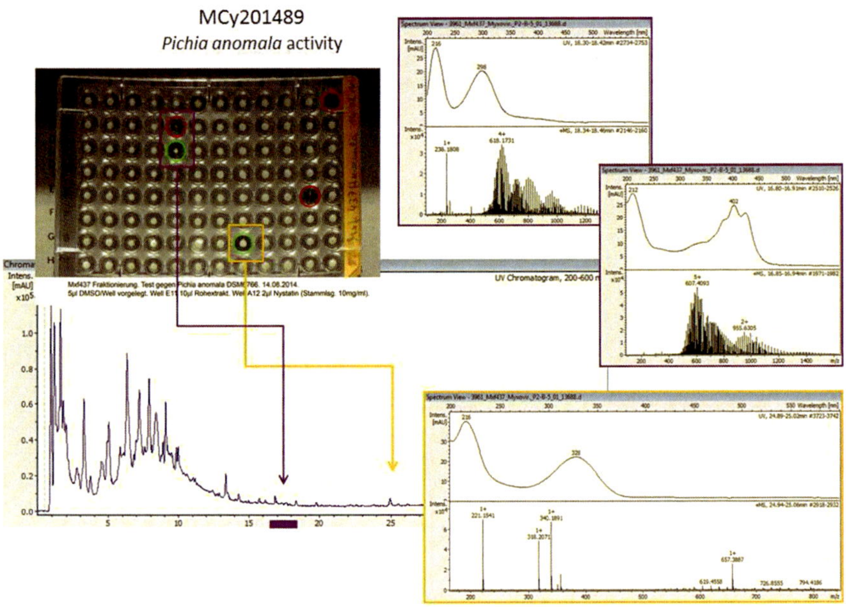

Fig. 8 Combination of activity assay and HPLC-MS

(LC/HRMS). First, the minimum inhibitory concentration (MIC) of the extract is determined by the use of a microtiter plate based serial dilution assay containing different Gram-positive and Gram-negative bacteria as well as fungi to assess selectivity. After semipreparative high performance liquid chromatography (HPLC), the inhibitory effect of the fractionated extract is examined against the affected organisms and further analyzed by ultra-high resolution-time of flight mass spectrometry (UHR-TOF-MS) regarding molecular masses and molecular formulae (Fig. 8). These obtained molecular features are matched against known myxobacterial compounds from the in-house database (Myxobase) or Dictionary of Natural Products (Taylor and Francis Group 2016). The strategy consisting of the use of the different methods above gives possibilities to find and identify new bioactive substances from myxobacteria.

2.3 Genetical Potential

The genomes of myxobacteria belong to the largest ever found in bacteria. The genome of the *Sorangium* strain So ce56, for example, consists of more than 13,000 base pairs and includes seventeen secondary metabolite gene clusters (Schneiker et al. 2007) and the genome of *M. xanthus* comprises at least around 9000 base pairs (Goldman et al. 2006). It is suggested that this large genome size is connected to the extraordinary and complex life cycle of myxobacteria. In contrast to other bacteria, they need a much more complex genetic constitution to demonstrate their special social behavior including swarming and culminating fruiting bodies. Within these enormously large genomes, there probably exists a high capacity concerning to the unlimited number of unknown and promising secondary metabolites (Wenzel and Müller 2009).

It is hardly possible to figure out the potential of a strain to produce bioactive substances only by phenotypic and biochemic analyzes, because some metabolites cannot be detected due to the applied extraction and detection methods (Wenzel and Müller 2009). Therefore, our research is focusing on the discovery of biosynthetic pathways and the connected gene clusters of novel, potentially anti-infective natural compounds as well as the full genome sequences of scientific interesting bacteria strains. With this information, it might become easier to find possibilities for enhancing the production of known metabolites as well as activating unused "silent" genes for novel substances (Müller and Wink 2014). The genome mining approach is based on the genome sequences, which are used to predict synthesized compounds with the help of bioinformatics analyzes. The recovered genetic information can lead to specific inducing of the supposed biosynthetic gene clusters that encode for new bioactive products (Müller and Wink 2014).

2.4 Pharmaceutically Important Secondary Metabolites

Myxobacteria are a rich source of novel and unique secondary metabolites, mainly polyketides and nonribosomal polypetides. Many of these metabolites show antibiotic activities and are urgently needed as new drugs for a broad range of applications. These skills probably evolved because of the natural competition between different kinds of organisms in varying habitats. For example, the cellulose degrading members of the genus Sorangium have to combat other cellulose degraders sharing the same habitat like fungi which also use wood (cellulose) as nutrient source, whereas proteolytic myxobacteria need to stand up to other degraders of decaying organic material (Gerth et al. 2003). The existence of a link between the production of biological active secondary metabolites and microbial predation is supposed, because about 20 % of the known myxobacterial compounds show antibiotic effects. The importance of secondary metabolites derived from natural producers should not be underestimated, because they are the source of almost 50 % of most important medications for humans (Demain 1999).

The scientific interest for myxobacteria increased already in the year 1947, when it was shown that a strain of *Myxococcus virescens* has a significant inhibitory effect on the growth of *Staphylococcus aureus* (Oxford 1947). In the following years many research groups all over the world tried to find the active compounds by optimizing the growth conditions. It was believed for a long time that it is very difficult to cultivate myxobacteria in liquid medium and that these organisms are somewhat problematic with regard to their axenic growth in general (Reichenbach and Höfle 1993). The breakthrough in discovering antibiotic substances from myxobacteria happened in 1977 by elucidation of the complete chemical structure of the potent antifungal secondary metabolite ambruticin, produced by *Sorangium cellulosum* (Connor et al. 1977; Ringel et al. 1977). Next to the use as antibiotic or antifungal drug, biologically active secondary metabolites can also be used as antiparasitic, antiviral and antitumor drugs in human and veterinary medicine and also as insecticides, acaricides, and herbicides. They can act, e.g., as inhibitors of carboxylases, polymerases, or mitochondrial respiration as well as inhibitors of eukaryotic protein synthesis (Weissmann and Müller 2009). Furthermore, also antidiabetic, antimalarial, antihypertensive, antihypercholesterolemic, insuline-sensitizing, and immunoregulatory characteristics can be attributed to microbial products (Grabley and Thiericke 1999; Schreurs et al. 2009; Berod et al. 2014). A molecule with biological effects can also work as a model for a synthetic production to get a higher output under better economic conditions (Reichenbach and Höfle 1993). More than 100 new and important myxobacterial core structures have been discovered (Garcia et al. 2009), some of them are summarized in Table 2. Another important ability of myxobacteria is the production of polyunsaturated fatty acids (PUFAs) like eicosaoentaenoic acid (EPA) and docosahexaenoic acid (DHA), which play an important role in food industry and for pharmaceutical applications (Garcia et al. 2011; Gemperlein et al. 2016).

Table 2 Important compounds and their biological activity found in myxobacteria

Compound	Activity	Mode of action	Species	Reference
Ambruticin	Antifungal	Interfere with high-osmolarity glycerol (HOG) signaling pathway	*S. cellulosum*	Ringel et al. (1977); Connor et al. (1977) Vetcher et al. (2013)
Aurachins	Antibacterial	Block NADH oxidation	*S. aurantiaca*	Kunze et al. (1987)
Chrondramide	Antifungal/ cytostatic	Interfere with actine polymerisation	*C. crocatus*	Kunze et al. (1995)
Crocacin	Antibacterial	Inhibits electron transport	*C. crocatus*	Kunze et al. (1994)
Cystobactamids	Antibacterial	Inhibit type II topoisomerase	*Cystobacter sp.*	Baumann et al. (2014)
Cystothiazol	Antifungal/ cytostatic	Inhibits submitochondrial NADH oxidation	*C. fuscus*	Ojika et al. (1998)
Disciformycins	Antibacterial	n/a	*P. fallax*	Surup et al. (2014)
Epothilones	Cytotoxic	Inhibition of microtubule function	*S. cellulosum*	Gerth et al. (1996)
Etnangien	Antibacterial	Inhibits nucleic acid polymerases	*S. cellulosum*	Irschik et al. (2007)

(continued)

Table 2 (continued)

Compound	Activity	Mode of action	Species	Reference
Melithiazols	Antibacterial	Inhibit NADH oxidation	*M. lichenicola, A. gephyra, M. stipitatus*	Sasse et al. (1999)
Myxothiazol	Antifungal	Inhibits electron transport	*M. fulvus*	Gerth et al. (1980)
Myxovalargin	Antibacterial	Inhibits protein synthesis and damages cell membranes	*M. fulvus*	Irschik et al. (1983) Irschik and Reichenbach (1985)
Myxovirescin	Antibacterial	Inhibition of signal peptidase	*M. virescens*	Gerth et al. (1982)
Rhizopodin	Cytostatic	Alteration of protein phosphorilation	*M. stipitatus*	Sasse et al. (1993)
Ripostatin	Antibacterial	Inhibits RNA polymerase	*S. cellulosum*	Irschik et al. (1995)
Sorangicin	Antibacterial	Inhibits RNA polymerase	*S. cellulosum*	Irschik et al. (1987)
Soraphens	Antifungal, antiviral, Cancerocidal, Immunoregulatory, Insulin-sensitizing	Inhibit acetyl-CoA carboxylase	*S. cellulosum*	Gerth et al. (1994) Schreurs et al. (2009) Martinez et al. (2013) Berod et al. (2014) Corominas-Faja et al. (2014) Koutsoudakis et al. (2015)
Stigmatellin	Antibacterial	Inhibits electron transport	*S. aurantiaca*	Kunze et al. (1984)

Furthermore, the natural compounds can also work as a basic structure for chemical modifications which can lead to an exploitation in the pharmaceutical area. Epothilones A and B, for example, were originally found in a *Sorangium cellulosum* strain and are distinguished by antifungal and cytotoxic activity (Gerth et al. 1996). Meanwhile, there are modified versions of these molecules already used in the active treatment against different types of cancer or are present in clinical trials. Until epothilones were discovered, the common medication for advanced and early-stage breast cancer was taxanes, anthracyclines, and capecetabine, but these agents are often subjected to a multidrug-resistance, which has a natural origin in the patient. It is suggested that there is an overexpression of efflux pumps and other proteins serving as efflux pumps, so that the anticancer agents can be removed very easy out of the targeted cancer cell (Egerton 2008). In fact, this resistance limits dramatically the chances of success within a therapy (Burger et al. 2003). As an effective alternative to the chemotherapies with taxanes and

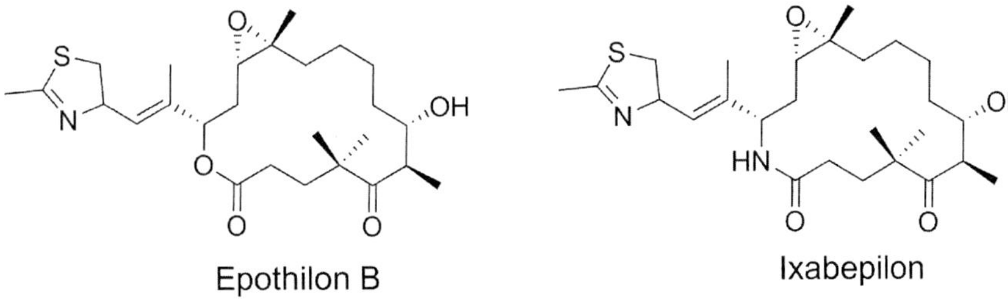

Fig. 9 Chemical structures of epothilon B and ixabepilon

anthracyclines, the semi-synthetic epothilone derivative ixabepilone (Ixempra®) was developed in October 2007 for monotherapie of different stages of breast cancer (Fig. 9) (Reichenbach and Höfle 2008). In contrast to the other available pharmaceuticals referring to this disease, ixabepilone is just low affected by multidrug-resistant mechanisms and consequently offers a chance to eliminate the tumor cells effectively (Pivot et al. 2007).

References

Abdel-Mageed WM, Milne BF, Wagner M, Schumacher M, Sandor P, Pathom-Aree W, Goodfellow M, Bull AT, Horikoshi K, Ebel R, Diederich M, Fiedler HP, Jaspars M (2010) Dermacozines, a new phenazine family from deep-sea dermacocci isolated from a Mariana Trench sediment. Org Biomol Chem 8:2352–2362

Abdelmohsen UR, Pimentel-Elardo SM, Hanora A, Radwan M, Abou-El-Ela SH, Ahmed S, Hentschel U (2010) Isolation, phylogenetic analysis and anti-infective activity screening of marine sponge-associated actinomycetes. Mar Drugs 8:399–412

Andersson M, Mikkola R, Kroppenstedt R, Rainey F, Peltola J, Helin J et al (1998) The mitochondrial toxin produced by *Streptomyces griseus* strains isolated from an indoor environment is valinomycin. Appl Environ Microbiol 64:4767–4773

Arumugam M, Mitra A, Jaisankar P, Dasgupta S, Sen T, Gachhui R et al (2010) Isolation of an unusual metabolite 2-allyloxyphenol from a marine actinobacterium, its biological activities and applications. Appl Microbiol Biotechnol 86:109–117

Asolkar RN, Jensen PR, Kauffman CA, Fenical W, Daryamides AC (2006) Weakly cytotoxic polyketides from a marine-derived actinomycete of the genus *Streptomyces* strain CNQ-085. J Nat Prod 69:1756–1759

Baumann S, Herrmann J, Raju R, Steinmetz H, Mohr KI, Hüttel S, Harmrolfs K, Stadler M, Müller R (2014) Cystobactamids: myxobacterial topoisomerase inhibitors exhibiting potent antibacterial activity. Angew Chem Int Ed Engl 53:14605–14609

Berod L, Friedrich C, Nandan A, Freitag J, Hagemann S, Harmrolfs K, Sandouk A, Hesse C, Castro CN, Bähre H, Tschirner SK, Gorinski N, Gohmert M, Mayer CT, Huehn J, Ponimaskin E, Abraham WR, Müller R, Lochner M, Sparwasser T (2014) De novo fatty acid synthesis controls the fate between regulatory T and T helper 17 cells. Nat Med 20:1327–1333

Bewick M, Williams S, Veltkamp C (1976) Growth and ultrastructure of *Streptomyces venezuelae* during chloramphenicol production. Microbios 16:191–199

Bister B, Bischoff D, Ströbele M, Riedlinger J, Reicke A, Wolter F, Bull AT, Zähner H, Fiedler HP, Süssmuth RD (2004) Abyssomicin C-A—polycyclic antibiotic from a marine

Verrucosispora strain as an inhibitor of the p-aminobenzoic acid/tetrahydrofolate biosynthesis pathway. Angew Chem Int Ed Engl 43:2574–2576

Blunt JW, Copp BR, Hu WP, Munro MH, Northcote PT, Prinsep MR (2007) Marine natural products. Nat Prod Rep 24:31–86

Blunt JW, Copp BR, Munro MHG, Northcote PT, Prinsep MR (2011) Marine natural products. Nat Prod Rep 28:196–268

Blunt JW, Copp BR, Keyzers RA, Munro MHG, Prinsep MR (2012) Marine natural products. Nat Prod Rep 29:144–222

Blunt JW, Copp BR, Keyzers RA, Munro MHG, Prinsep MR (2013) Marine natural products. Nat Prod Rep 30:237–323

Brockman ER, Boyd WL (1963) Myxobacteria from soils of the Alaskan and Canadian Arctic. J Bacteriol 86:605–606

Bruntner C, Binder T, Pathom-Aree W, Goodfellow M, Bull AT, Potterat O et al (2005) Frigocyclinone, a novel angucyclinone antibiotic produced by a *Streptomyces griseus* strain from Antarctica. J Antibiot 58:346–349

Bull AT, Starch JEM (2007) Marine actinobacteria: new opportunities for natural product search and discovery. Trends in Microbiol 15:491–499

Burg RW, Miller BM, Baker EE, Birnbaum J, Currie SA, Hartman R et al (1997) Avermectins, new family of potent anthelmintic agents: producing organism and fermentation. Antimicrob Agents Chemother 15:361–367

Burger H, Foekens JA, Look MP, Meijer-van Gelder ME, Klijn JG, Wiemer EA, Stoter G, Nooter K (2003) RNA expression of breast cancer resistance protein, lung resistance-related protein, multidrug resistance-associated proteins 1 and 2, and multidrug resistance gene 1 in breast cancer: correlation with chemotherapeutic response. Clin Cancer Res 9:827–836

Carlson JC, Li S, Burr DA, Sherman DH (2009) Isolation and characterization of tirandamycins from a marine-derived *Streptomyces* sp. J Nat Prod 72:2076–2079

Charan RD, Schlingmann G, Janso J, Bernan V, Feng X, Carter GT (2004) Diazepinomicin, a new antimicrobial alkaloid from a marine *Micromonospora* sp. J Nat Prod 67:1431–1433

Chau R, Kalaitzis JA, Neilan BA (2011) On the origins and biosynthesis of tetrodotoxin. Aqua Toxicol 104:61–72

Chauhan D, Catley L, Li G, Podar K, Hideshima T, Velankar M et al (2005) A novel orally active proteasome inhibitor induces apoptosis in multiple myeloma cells with mechanisms distinct from Bortezomib. Cancer Cell 8:407–419

Connor DT, Greenough RC, Strandtmann M (1977) W-7783, a unique antifungal antibiotic. J Org Chem 42:3664–3669

Corominas-Faja B, Cuyàs E, Gumuzio J, Bosch-Barrera J, Leis O, Martin ÁG, Menendez JA (2014) Chemical inhibition of acetyl-CoA carboxylase suppresses self-renewal growth of cancer stem cells. Oncotarget 5:8306–8316

Dawid W (1978) Fruchtkörperbildene Myxobakterien in Böden Brasiliens. Z Allg Mikrobiol 34:333–335

Dawid W, Gallikowski CA, Hirsch P (1988) 3.8 Psychrophilic myxobacteria from antarctic soils. Polarforschung 58:271–278 (Alfred Wegener Institute for Polar and Marine Research & German Society of Polar Research, Bremerhaven)

Demain A (1999) Pharmaceutically active secondary metabolites of microorganisms. Appl Microbiol Biotechnol 52:455–463

Do H, Kogure K, Simidu U (1990) Identification of deep-sea-sediment bacteria which produce tetrodotoxin. Appl Environ Microbiol 56:1162

Do H, Kogure K, Imada C, Noguchi T, Ohwada K, Simidu U (1991) Tetrodotoxin production of actinomycetes isolated from marine sediment. J Appl Microbiol 70:464–468

Egan S, Wiener P, Kallifidas D, Wellington EMH (1998) Transfer of streptomycin biosynthesis gene clusters within streptomycetes isolated from soil. Appl Environ Microbiol 64:5061–5063

Egerton N (2008) Ixabepilone (ixempra), a therapeutic option for locally advanced or metastatic breast cancer. P T 33:523–531

El-Gendy MM, Shaaban M, Shaaban KA, El-Bondkly AM, Laatsch H (2008) Essramycin: a first triazolopyrimidine antibiotic isolated from nature. J Antibiot 61:149–157

Engelhardt K, Degnes KF, Kemmler M, Bredholt H, Fjaervik E, Klinkenberg G, Sletta H, Ellingsen TE, Zotchev SB (2010) Production of a new thiopeptide antibiotic, TP-1161, by a marine *Nocardiopsis* species. Appl Environ Microbiol 76:4969–4976

Fenical W, Jensen P (2006) Developing a new resource for drug discovery: marine actinomycete bacteria. Nature Chem Biol 2:666–673

Fenical W, Baden D, Burg M, de Goyet CV, Grimes JD, Katz M, Marcus NH, Pomponi S, Rhines P, Tester P, Vena J (1999) Marine derived pharmaceuticals and related bioactive compounds. In: Fenical W (ed) From monsoons to microbes: understanding the ocean's role in human health, National Academies Press, pp 71–86

Fiedler HP, Bruntner C, Riedlinger J, Bull AT, Knutsen G, Goodfellow M, Jones A, Maldonado L, Pathom-Aree W, Beil W, Schneider K, Keller S, Sussmuth RD (2008) Proximicin A, B and C, novel aminofuran antibiotic and anticancer compounds isolated from marine strains of the actinomycete *Verrucosispora*. J Antibiot 61:158–163

Frändberg E, Petersson C, Lundgren LN, Schnürer J (2000) *Streptomyces halstedii* K122 produces the antifungal compounds bafilomycin B1 and C1. Can J Microbiol 46:753–758

Fujii I, Ebizuka Y (1997) Anthracycline biosynthesis in *Streptomyces galilaeus*. Chem Rev 97:2511–2524

Gao X, Lu Y, Xing Y, Ma Y, Lu J, Bao W, Wang Y, Xi T (2012) A novel anticancer and antifungus phenazine derivative from a marine actinomycete BM-17. Microbiol Res 167:616–622

Garcia RO, Müller R (2014a) The family haliangiaceae. In: Rosenberg E, DeLong EF, Lory S, Stackebrandt E, Thompson F (eds) The prokaryotes deltaproteobacteria and epsilonproteobacteria, Springer, pp 173–181

Garcia RO, Müller R (2014b) The family myxococcaceae. In: Rosenberg E, DeLong EF, Lory S, Stackebrandt E, Thompson F (eds) The prokaryotes deltaproteobacteria and epsilonproteobacteria, Springer, pp 192–212

Garcia RO, Müller R (2014c) The family nannocystaceae. In: Rosenberg E, DeLong EF, Lory S, Stackebrandt E, Thompson F (eds) The prokaryotes deltaproteobacteria and epsilonproteobacteria, Springer, pp 213–229

Garcia RO, Müller R (2014d) The family phaselicastaceae. In: Rosenberg E, DeLong EF, Lory S, Stackebrandt E, Thompson F (eds) The prokaryotes deltaproteobacteria and epsilonproteobacteria, Springer, pp 239–245

Garcia RO, Müller R (2014e) The family polyangiaceae In: Rosenberg E, DeLong EF, Lory S, Stackebrandt E, Thompson F (eds) The prokaryotes deltaproteobacteria and epsilonproteobacteria, Springer, pp 247–279

Garcia RO, Krug D, Müller R (2009) Discovering natural products from myxobacteria with emphasis on rare producer strains in combination with improved analytical methods. Methods Enzymol 458:59–91

Garcia R, Pistorius D, Stadler M, Müller R (2011) Fatty acid-related phylogeny of myxobacteria as an approach to discover polyunsaturated omega-3/6 fatty acids. J Bacteriol 139:1930–1942

Gemperlein K, Zipf G, Bernauer HS, Müller R, Wenzel SC (2016) Metabolic engineering of *Pseudomonas putida* for production of docosahexaenoic acid based on a myxobacterial PUFA synthase. Metab 33:98–108

Gerth K, Irschik H, Reichenbach H, Trowitzsch W (1980) Myxothiazol, an antibiotic from *Myxococcus fulvus* (myxobacterales). I. Cultivation, isolation, physico-chemical and biological properties. J Antibiot 33:1474–1479 Tokyo

Gerth K, Pradella S, Perlova O, Beyer S, Müller R (2003) Myxobacteria: proficient producers of novel natural products with various biological activities–past and future biotechnological aspects with the focus on the genus *Sorangium*. J Biotechnol 106:233–253

Gerth K, Ischik H, Reichenbach H, Trowitzsch WPG (1982) The myxovirescins, a family of antibiotics from *Myxococcus virescens* (myxobacterales). J Antibiot 35:1454–1459

Gerth K, Bedorf N, Irschik H, Höfle G, Reichenbach H (1994) The soraphens: a family of novel antifungal compounds from *Sorangium cellulosum* (Myxobacteria). I. Soraphen A1 alpha: fermentation, isolation, biological properties. J Antibiot 47:23–31

Gerth K, Bedorf N, Irschik H, Höfle G, Reichenbach H (1996) Epothilons A and B: antifungal and cytotoxic compounds from *Sorangium cellulosum* (myxobacteria) production, physico-chemical and biological properties. J Antibiot 49:560–563

Goldman BS, Nierman WC, Kaiser D, Slater SC, Durkin AS, Eisen JA, Ronning CM, Barbazuk WB, Blanchard M, Field C, Halling C, Hinkle G, Iartchuk O, Kim HS, Mackenzie C, Madupu R, Miller N, Shvartsbeyn A, Sullivan SA, Vaudin M, Wiegand R, Kaplan HB (2006) Evolution of sensory complexity recorded in a myxobacterial genome. Proc Natl Atlac Sci USA 103:15200–15205

Goodfellow M, Haynes JA (1984) Actinomycetes in marine sediments. In: Ortiz-Ortiz L, Bojalil LF, Yakoleff V (eds) Biological, biochemical, and biomedical aspects of actinomycetes. Academic Press, New York, pp 453–472

Gorajana A, Kurada BV, Peela S, Jangam P, Vinjamuri S, Poluri E et al (2005) 1-Hydroxy-1-norresistomycin, a new cytotoxic compound from a marine actinomycete, *Streptomyces chibaensis* AUBN1/7. J Antibiot 58:526–529

Grabley S, Thiericke R (1999) The impact of natural products on drug discovery. Drug discovery from nature, Springer, pp 3–37

Han SK, Nedashkovskaya OI, Mikhailov VV, Kim SB, Bae KS (2003) *Salinibacterium amurskyense* gen. nov., sp. nov., a novel genus of the family microbacteriaceae from the marine environment. Int J Syst Evol Microbiol 53:2061–2066

Hansen LH, Ferrari B, Sørensen AH, Veal D, Sørensen S (2001) Detection of oxytetracycline production by *Streptomyces rimosus* in soil microcosmos by combining whole cell biosensors and flow cytometry. Appl Environ Microbiol 67:239–244

Hawas UW, Shaaban M, Shaaban KA, Speitling M, Maier A, Kelter G, Fiebig HH, Meiners M, Helmke E, Laatsch H (2009) Mansouramycins A-D, cytotoxic isoquinolinequinones from a marine streptomycete. J Nat Prod 72:2120–2124

Hayakawa Y, Shirasaki S, Shiba S, Kawasaki T, Matsuo Y, Adachi K et al (2007) Piericidins C7 and C8, new cytotoxic antibiotics produced by a marine *Streptomyces* sp. J Antibiot 60:196–200

Helaly SE, Pesic A, Fiedler HP, Süßmuth RD (2011) Elaiomycins B and C: Alkylhydrazide antibiotics from *Streptomyces* sp. BK 190. Org Lett 13:1052–1055

Helmke E, Weyland H (1984) *Rhodococcus marinonascens* sp. nov., an actinomycete from the sea. Int J Syst Bacteriol 34:127–138

Hentschel U, Hopke J, Horn M, Friedrich AB, Wagner M, Hacker J, Moore BS (2002) Molecular evidence for a uniform microbial community in sponges from different oceans. Appl Environ Microbiol 68:4431–4440

Hentschel U, Piel J, Degnan SM, Taylor MW (2012) Genomic insights into marine sponge microbiome. Nat Rev Microbiol 10:641–654

Herr RR, Jahnke HK, Argoudelis AD (1967) Structure of streptozotocin. J Am Chem Soc 89:4808–4809

Hill RT (2004) Microbes from marine sponges: a treasure trove of biodiversity for natural products discovery. In: Bull AT (ed) Microbial diversity and bioprospecting, ASM Press, pp. 225–231

Hohmann C, Schneider K, Bruntner C, Brown R, Jones AL, Goodfellow M, Krämer M, Imhoff JF, Nicholson G, Fiedler HP, Süssmuth RD (2009a) Albidopyrone, a new alpha-pyrone-containing metabolite from marine-derived *Streptomyces* sp. NTK 227. J Antibiot 62:75–79

Hohmann C, Schneider K, Bruntner C, Irran E, Nicholson G, Bull AT et al (2009b) Caboxamycin, a new antibiotic of the benzoxazole family produced by the deep-sea strain *Streptomyces* sp. NTK 937. J Antibiot 62:99–104

Hook LA (1977) Distribution of myxobacters in aquatic habitats of an alkaline bog. Appl Environ Microbiol 34:333–335

Huang YF, Tian L, Fu HW, Hua HM, Pei YH (2006) One new anthraquinone from marine *Streptomyces* sp. FX-58. Nat Prod Res 20:1207–1210

Iizuka T, Jojima Y, Fudou R, Yamanaka S (1998) Isolation of myxobacteria from the marine environment. FEMS Microbiol Lett 169:317–322

Inagaki F, Suzuki M, Takai K, Oida H, Sakamoto T, Aoki K, Nealson KH, Horikoshi K (2003) Microbial communities associated with geological horizons in coastal subseafloor sediments from the Sea of Okhotsk. Appl Environ Microbiol 69:7224–7235

Irschik H, Reichenbach H (1985) The mechanism of action of myxovalargin A, a peptide antibiotic from *Myxococcus fulvus*. J Antibiot 38:1237–1245 Tokyo

Irschik H, Gerth K, Kemmer T, Steinmetz H, Reichenbach H (1983) The myxovalargins, new peptide antibiotics from *Myxococcus fulvus* (Myxobacterales). I. Cultivation, isolation, and some chemical and biological properties. J Antibiot 36:6–12 Tokyo

Irschik H, Jansen R, Gerth K, Höfle G, Reichenbach H (1987) The sorangicins, novel and powerful inhibitors of eubacterial RNA polymerase isolated from myxobacteria. J Antibiot 0:7–13 (Tokyo)

Irschik H, Augustiniak H, Gerth K, Höfle G, Reichenbach H (1995) The ripostatins, novel inhibitors of eubacterial RNA polymerase isolated from myxobacteria. J Antibiot 48:787–792 Tokyo

Irschik H, Schummer D, Höfle G, Reichenbach H, Steinmetz H, Jansen R (2007) Etnangien, a macrolide-polyene antibiotic from *Sorangium cellulosum* that inhibits nucleic acid polymerases. J Nat Prod 70:1060–1063

Jensen PR, Dwight R, Fenical W (1991) Distribution of actinomycetes in near-shore tropical marine sediments. Appl Environ Microbiol 57:1102–1108

Jensen PR, Williams PG, Oh DC, Zeigler L, Fenical W (2007) Species-specific secondary metabolite production in marine actinomycetes of the genus *Salinispora*. Appl Environ Microbiol 73:1146–1152

Jeong SY, Shin HJ, Kim TS, Lee HS, Park S, Kim HM (2006) Streptokordin, a new cytotoxic compound of the methylpyridine class from a marine-derived *Streptomyces* sp. KORDI-3238. J Antibiot 59:234–240

Jiang S, Sun W, Chen M, Dai S, Zhang L, Liu Y, Lee KJ, Li X (2007) Diversity of culturable actinobacteria isolated from marine sponge *Haliclona* sp. Antonie Van Leeuwenhoek 92:405–416

Jørgensen H, Degnes KF, Dikiy A, Fjaervik E, Klinkenberg G, Zotchev SB (2010) Insights into the evolution of macrolactam biosynthesis through cloning and comparative analysis of the biosynthetic gene cluster for a novel macrocyclic lactam, ML-449. Appl Environ Microbiol 76:283–293

Kanoh K, Matsuo Y, Adachi K, Imagawa H, Nishizawa M, Shizuri Y (2005) Mechercharmycins A and B, cytotoxic substances from marine-derived Thermoactinomyces sp. YM3-251. J Antibiot 58:289–292

Karwehl S, Stadler M (2016) Exploitation of fungal biodiversity for discovery of novel antibiotics. Curr Top Microbiol Immunol, in press (doi: 10.1007/82_2016_496)

Kim SB, Oh HM, Kang H, Park SS, Chun J (2004) Remarkable bacterial diversity in the tidal flat sediment as revealed by 16S rDNA analysis. J Microbiol Biotechnol 14:205–211

Kock I, Maskey RP, Biabani MAF, Helmke E, Laatsch H (2005) 1-Hydroxy-1-norresistomycin and resistoflavin methyl ether: new antibiotics from marine-derived streptomycetes. J Antibiot 58:530–535

Koutsoudakis G, Romero-Brey I, Berger C, Pérez-Vilaró G, Monteiro Perin P, Vondran FW, Kalesse M, Harmrolfs K, Müller R, Martinez JP, Pietschmann T, Bartenschlager R, Brönstrup M, Meyerhans A, Díez J (2015) Soraphen A: a broad-spectrum antiviral natural product with potent anti-hepatitis C virus activity. J Hepatol 63:813–821

Kunze B, Kemmer T, Höfle G, Reichenbach H (1984) Stigmatellin, a new antibiotic from *Stigmatella aurantiaca* (Myxobacterales). I. Production, physico-chemical and biological properties. J Antibiot 37:454–461 Tokyo

Kunze B, Höfle G, Reichenbach H (1987) The aurachins, new quinoline antibiotics from myxobacteria: production, physico-chemical and biological properties. J Antibiot 40:258–265 Tokyo

Kunze B, Jansen R, Höfle G, Reichenbach H (1994) Crocacin, a new electron transport inhibitor from *Chondromyces crocatus* (myxobacteria). Production, isolation, physico-chemical and biological properties. J Antibiot 47:881–886 Tokyo

Kunze B, Jansen R, Sasse F, Höfle G, Reichenbach H (1995) Chondramides A approximately D, new antifungal and cytostatic depsipeptides from *Chondromyces crocatus* (myxobacteria). Production, physico-chemical and biological properties. J Antibiot 48:1262–1266 Tokyo

Kwon HC et al (2006) Marinomycins A-D, antitumor-antibiotics of a new structure class from a marine actinomycete of the recently discovered genus "*Marinospora*". J Am Chem Soc 128:1622–1632

Lam KS (2006) Discovery of novel metabolites from marine actinomycetes. Curr Opin Microbiol 9:245–251

Li DH, Zhu TJ, Liu HB, Fang YC, Gu QQ, Zhu WM (2006) Four butenolides are novel cytotoxic compounds isolated from the marine-derived bacterium, *Streptoverticillium luteoverticillatum* 11014. Arch Pharmacal Res 29:624–626

Li F, Maskey RP, Qin S, Sattler I, Fiebig HH, Maier A et al (2005) Chinikomycins A and B: isolation structure elucidation, and biological activity of novel antibiotics from a marine *Streptomyces* sp. isolate M045. J Nat Prod 68:349–353

Link HF (1809) Observations in Ordines plantarum naturales. Dissertatio prima, complectens Anandrarum ordines Epiphytas, Mucedines Gastomycos et Fungos. Der Geselllschaft Naturforschender Freunde zu Berlin Magazin für die neuesten Entdeckungen in der gesamten Naturkunde 3:1–42

Lu J, Ma Y, Liang J, Xing Y, Xi T, Lu Y (2012) Aureolic acids from a marine-derived *Streptomyces* sp. WBF16. Microbiol Res 167:590–595

Macherla VR, Liu J, Bellows C, Teisan S, Nicholson B, Lam KS et al (2005) Glaciapyrroles A, B and C, pyrrolosesquiterpenes from a Streptomyces sp. isolated from an Alaskan marine sediment. J Nat Prod 68:780–783

Magarvey NA, Keller JM, Bernan V, Dworkin M, Sherman DH (2004) Isolation and characterization of novel marine-derived actinomycete taxa rich in bioactive metabolites. Appl Environ Microbiol 70:7520–7529

Maldonado LA, Fenical W, Jensen PR, Kauffman CA, Mincer TJ, Bull AT, Ward AC, Goodfellow M (2005a) *Salinispora arenicola* gen. nov., sp nov and *Salinispora tropica* sp nov., obligate marine actinomycetes belonging to the family micromonosporaceae. Int J Syst Evol Microbiol 55:1759–1766

Maldonado LA, Starch JEM, Pathom-Aree W, Ward AC, Bill AT, Goodfellow M (2005b) The diversity of cultivable actinobacteria in geographically widespread marine sediments. Antonie Van Leeuwenhoek 87:11–18

Manam RR, Teisan S, White DJ, Nishino T, Grodberg J, Neuteboom STC et al (2005) Lajollamycin, a nitro-tetraene spiro-b-lactone-g-lactam antibiotic from the marine actinomycete *Streptomyces nodosus*. J Nat Prod 68:240–243

Mao Y, Varoglu M, Sherman DH (1999) Molecular characterization and analysis of the biosynthetic gene cluster for the antitumor antibiotic mitomycin C from *Streptomyces lavendulae* NRRL 2564. Chem Biol 6:251–263

Martinez JP, Hinkelmann B, Fleta-Soriano E, Steinmetz H, Jansen R, Diez J, Frank R, Sasse F, Meyerhans A (2013) Identification of myxobacteria-derived HIV inhibitors by a high-throughput two-step infectivity assay. Microb Cell Fact 12:85

Manivasagan P, Venkatesan J, Sivakumar K, Kim S-K (2013) Marine actinobacterial metabolites: Current status and future perspectives. Microbiol Res 168:311–332

Mantalvo NF, Mohamed NM, Enticknap JJ, Hill RT (2005) Novel actinobacteria from marine sponges. Antonie Van Leeuwenhoek 87:29–36

McArthur KA, Mitchell SS, Tsueng G, Rheingold A, White DJ, Grodberg J, Lam KS, Potts BC (2008) Lynamicins A-E, chlorinated bisindole pyrrole antibiotics from a novel marine actinomycete. J Nat Prod 71:1732–1737

Mincer TJ, Jensen PR, Kauffman CA, Fenical W (2002) Widespread and persistent populations of a major new marine actinomycete taxon in ocean sediments. Appl Environ Microbiol 68:5005–5011

Mincer TJ, Fenical W, Jensen PR (2005) Culture-dependent and culture-independent diversity within the obligate marine actinomycete genus *Salinispora*. Appl Environ Microbiol 71:7019–7028

Mitchell SS, Nicholson B, Teisan S, Lam KS, Potts BC (2004) Aureoverticillactam, a novel 22-atom macrocyclic lactam from the marine actinomycete *Streptomyces aureoverticillatus*. J Nat Prod 67:1400–1402

Mohr KI, Stechling M, Wink J, Wilharm E, Stadler M (2015) Comparison of myxobacterial diversity and evaluation of isolation success in two niches: Kiritimati Island and German compost. Microbiologyopen 5:268–278

Moore BS, Trischman JA, Seng D, Kho D, Jensen PR, Fenical W (1999) Salinamides, antiinflammatory depsipeptides from a marine streptomycete. J Org Chem 64:1145–1150

Moore BS, Kalaitzis JA, Xiang L (2005) Exploiting marine actinomycete biosynthetic pathways for drug discovery. Antonie Van Leeuwenhoek 87:49–57

Müller R, Wink J (2014) Future potential for anti-infectives from bacteria—how to exploit biodiversity and genomic potential. Int J Med Microbiol 304:3–13

Oh DC, Gontang EA, Kauffman CA, Jensen PR, Fenical W (2008) Salinipyrones and pacificanones, mixed-precursor polyketides from the marine actinomycete salinispora pacifica. J Nat Prod 71:570–575

Ojika M, Suzuki Y, Tsukamoto A, Sakagami Y, Fudou R, Yoshimura T, Yamanaka S (1998) Cystothiazoles A and B, new bithiazole-type antibiotics from the myxobacterium *Cystobacter fuscus*. J Antibiot 51:275–281 Tokyo

Okami Y, Okazaki T (1972) Studies on marine microorganisms. J Antibiot 25:456–460

Omura S, Nakagawa A, Fujimoto T, Saito K, Otoguro K, Walsh JC et al (1987) Hygromycin A, an antitreponemal substance. I. Screening method and therapeutic effect for *Treponema hyodysenteriae*-caused infection in CF-1 mice. J Antibiot 40:1619

Oxford AE (1947) Observations concerning the growth and metabolic activities of myxococci in a simple protein-free liquid medium. J Bacteriol 53:129–138

Papineau D, Walker JJ, Mojzsis SJ, Pace NR (2005) Composition and structure of microbial communities from stromatolites of Hamelin Pool in Shark Bay, Western Australia. Appl Environ Microbiol 71:4822–4832

Pathom-Aree W, Stach JEM, Ward AC, Horikoshi K, Bull AT, Goodfellow M (2006) Diversity of actinomycetes isolated from challenger deep sediment (10,898 m) from the Mariana Trench. Extremophiles 10:181–189

Perez Baz J, Cañedo LM, Fernández Puentes JL, Silva Elipe MV (1997) Thiocoraline, a novel depsipeptide with antitumor activity produced by a marine *Micromonospora*. II. Physico-chemical properties and structure determination. J Antibiot 50:738–741

Peschke U, Schmidt H, Zhang HZ, Piepersberg W (2006) Molecular characterization of the lincomycin-production gene cluster of *Streptomyces lincolnensis* 78–11. Mol Microbiol 16:1137–1156

Piel J, Hertweck C, Shipley PR, Hunt DM, Newman MS, Moore BS (2000) Cloning, sequencing and analysis of the enterocin biosynthesis gene cluster from the marine isolate '*Streptomyces maritimus*': evidence for the derailment of an aromatic polyketide synthase. Chem Biol 7:943–955

Pivot X, Dufresne A, Villanueva C (2007) Efficacy and safety of ixabepilone, a novel epothilone analogue. Clin Breast Canc 7:543–549

Prudhomme J, McDaniel E, Ponts N, Bertani S, Fenical W, Jensen P et al (2008) Marine actinomycetes: a new source of compounds against the human malaria parasite. PLoS ONE 3:2335

Rachid S, Huo L, Herrmann J, Stadler M, Köpcke B, Bitzer J, Müller R (2011) Mining the cinnabaramide biosynthetic pathway to generate novel proteasome inhibitors. ChemBioChem 12:922–931

Reichenbach H (1983) A simple method for the purification of myxobacteria. J Microbiol Methods 1:77–79

Reichenbach H, Höfle G (1993) Biologically active secondary metabolites from myxobacteria. Biotech Adv 11:219–277

Reichenbach H, Gerth K, Irschik H, Kunze B, Höfle G (1988) Myxobacteria: a source of new antibiotics. Trends Biotechnol 6:115–121

Reichenbach H, Lang E, Schumann P, Spröer C (2006a) *Byssovorax cruenta* gen. nov., sp. nov., nom. rev., an cellulose-degrading myxobacterium: rediscovery of '*Myxococcus cruentus*' Thaxter 1897. Int J Syst Evol Microbiol 56:2357–2363

Reichenbach H, Dworkin M, Shimkets LJ (2006) The myxobacteria In: Dworkin M, Falkow S, Rosenberg E, Schleifer K.-H. Stackebrandt E (eds) The prokaryotes,Springer, Berlin, vol 7, pp. 31–115

Reichenbach H, Höfle G (2008) Discovery and development of the epothilones: a novel class of antineoplastic drugs. Drugs R D 9:1–10

Ringel SM, Greenough RC, Roemer S (1977) Ambruticin (W7783), a new antifungal antibiotic. J Antibiot 30:371–375

Saleh EA, Mahmoud SAZ, El-Haddad ME, Abdel-Fatah MK (1985) Purification and identification of *Streptomyces aureofaciens* ID13 antibiotic. Zentralblatt für Mikrobiologie 140:325–332

Sanford RA, Cole JR, Tiedje JM (2002) Characterization and description of *Anaeromyxobacter dehalogenans* gen. nov., sp. nov., an aryl-halorespiring facultative anaerobic myxobacterium. Appl Environ Microbiol 68:893–900

Sasse F, Steinmetz H, Höfle G, Reichenbach H (1993) Rhizopodin, a new compound from *Myxococcus stipitatus* (myxobacteria) causes formation of rhizopodia-like structures in animal cell cultures. Production, isolation, physico-chemical and biological properties. J Antibiot 46:741–748

Sasse F, Böhlendorf B, Herrmann M, Kunze B, Forche E, Steinmetz H, Höfle G, Reichenbach H (1999) Melithiazols, new beta-methoxyacrylate inhibitors of the respiratory chain isolated from myxobacteria. Production, isolation, physico-chemical and biological properties. J Antibiot 52:721–729

Selvin J, Joseph S, Asha KRT, Manjusha WA, Sangeetha VS, Jayaseema DM, Antony MC, Denslin Vinitha AJ (2004) Antibacterial potential of antagonistic *Streptomyces* sp. isolated from marine sponge *Dendrilla nigra*. FEMS Microbiol Ecol 50:117–122

Schneiker S, Perlova O, Kaiser O, Gerth K, Alici A, Altmeyer MO, Bartels D, Bekel T, Beyer S, Bode E, Bode HB, Bolten CJ, Choudhuri JV, Doss S, Elnakady YA, Frank B, Gaigalat L, Goesmann A, Groeger C, Gross F, Jelsbak L, Jelsbak L, Kalinowski J, Kegler C, Knauber T, Konietzny S, Kopp M, Krause L, Krug D, Linke B, Mahmud T, Martinez-Arias R, McHardy AC, Merai M, Meyer F, Mormann S, Muñoz-Dorado J, Perez J, Pradella S, Rachid S, Raddatz G, Rosenau F, Rückert C, Sasse F, Scharfe M, Schuster SC, Suen G, Treuner-Lange A, Velicer GJ, Vorhölter FJ, Weissman KJ, Welch RD, Wenzel SC, Whitworth DE, Wilhelm S, Wittmann C, Blöcker H, Pühler A, Müller R (2007) Complete genome sequence of the myxobacterium *Sorangium cellulosum*. Nat Biotechnol 25:1281–1289

Schreurs M, Van Dijk TH, Gerding A, Havinga R, Reijngoud DJ, Kuipers F (2009) Soraphen, an inhibitor of the acetyl-CoA carboxylase system, improves peripheral insulin sensitivity in mice fed a high-fat diet. Diabetes Obes Metab 11:987–991

Schultz AW, Oh DC, Carney JR, Williamson RT, Udwary DW, Jensen PR, Gould SJ, Fenical W, Moore BS (2008) Biosynthesis and structures of cyclomarins and cyclomarazines, prenylated cyclic peptides of marine actinobacterial origin. J Am Chem Soc 130:4507–4516

Shimkets LJ, Dworkin M, Reichbach H (2004) The myxobacteria, 3rd edn. Springer, New York

Sivakumar K, Sahu MK, Thangaradjou T, Kannan L (2007) Research on marine actinobacteria in India. Ind J Microbiol 47:186–196

Socha AM, LaPlante KL, Rowley DC (2006) New bisanthraquinone antibiotics and semisynthetic derivatives with potent activity against clinical *Staphylococcus aureus* and *Enterococcus faecium* isolates. Bioorganic Med Chem 14:8446–8454

Soria-Mercado IE, Prieto-Davo A, Jensen PR, Fenical W (2005) Antibiotic terpenoid chloro-dihydroquinones from a new marine actinomycete. J Nat Prod 68:904–910

Stach JEM, Bull AT (2005) Estimating and comparing the diversity of marine actinobacteria. Antonie Van Leeuwenhoek 87:3–9

Stach JEM, Maldonado LA, Masson DG, Ward AC, Goodfellow M, Bull AT (2003) Statistical approaches to estimating bacterial diversity in marine sediments. Appl Environ Microbiol 69:6189–6200

Stadler M, Bitzer J, Mayer-Bartschmid A, Müller H, Benet-Buchholz J, Gantner F, Tichy HV, Reinemer P, Bacon KB (2007) Cinnabaramides A-G: analogues of lactacystin and salinosporamide from a terrestrial streptomycete. J Nat Prod 70:246–252

Steinert G, Whitfield S, Taylor MW, Thoms C, Schupp PJ (2014) Application of diffusion growth chambers for the cultivation of marine sponge-associated bacteria. Mar Biotechnol 16:594–603 Springer

Surup F, Viehrig K, Mohr KI, Herrmann J, Jansen R, Müller R (2014) Disciformycins A and B: 12-membered macrolide glycoside antibiotics from the myxobacterium *Pyxidicoccus fallax* active against multiresistant staphylococci. Angew Chem Int Ed 53:13588–13591

Takami H, Inoue A, Fuji F, Horikoshi K (1997) Microbial flora in the deepest sea mud of the Mariana Trench. FEMS Microbiol Lett 152:279–285

Takizawa M, Colwell RR, Hill RT (1993) Isolation and diversity of actinomycetes in the Chesapeake Bay. Appl Environm Micobiol 59:997–1002

Taylor and Francis Group (2016) Dictionary of natural products (online)

Thaxter R (1892) On the myxobacteriaceae, a new order of schizomycetes. Bot Gaz 17:389–406

Uyeda M, Mizukami M, Yokomizo K, Suzuki K, Pentalenolactone I (2001) Hygromycin A. immunosuppressants produced by *Streptomyces filipinensis* and *Streptomyces hygroscopicus*. Biosci Biotechnol Biochem 65:1252–1254

Vetcher L, Menzella HG, Kudo T, Motoyama T, Katz L (2013) The antifungal polyketide ambruticin targets the HOG pathway. Antimicrob Agents Chemother 51:3734–3736

Vezina C, Kudelski A, Sehgal S (1975) Rapamycin (AY-22,989), a new antifungal antibiotic. I. Taxonomy of the producing streptomycete and isolation of the active principle. J Antibiot 28:721–726

Ward AC, Bora N (2006) Diversity and biogeography of marine actinobacteria. Curr Opin Microbiol 9:279–286

Webster NS, Hill RT (2001) The culturable microbial community of the great barrier reef sponge Rhopaloeides odorabile. Appl Environ Microbiol 138:843–851

Webster NS, Taylor MW (2012) Marine sponges and their microbial symbionts: love and other relationships. Environ Microbiol 14:335–346

Weissmann KJ, Müller R (2009) A brief tour of myxobacterial secondary metabolism. Bioorg Med Chem 17:2121–2136

Wenzel SC, Müller R (2009) The impact of genomics on the exploitation of the myxobacterial secondary metabolome. Nat Prod Rep 26:1385–1407

Werner G, Hagenmaier H, Drautz H, Baumgartner A, Zähner H (1984) Metabolic products of microorganisms. bafilomycins, a new group of macrolide antibiotics. Production, isolation, chemical structure and biological activity. J Antibiot 37:110–117

Williams PG, Buchanan GO, Feling RH, Kauffman CA, Jensen PR, Fenical W (2005) New cytotoxic Salinosporamides from the marine actinomycete Salinispora tropica. J Org Chem 70:6196–6203

Wu SJ, Fotso S, Li F, Qin S, Kelter T, Fiebig HH et al (2006) 39-N-carboxamidostaurosporine and selina-4(14),7(11)-diene-8,9-diol, new metabolites from a marine *Streptomyces* sp. J Antibiot 59:331–337

Wu Z, Xie L, Xia G, Zhang J, Nie Y, Hu J, Wang S, Zhang R (2005) A new tetrodotoxin-producing actinomycete, *Nocardiopsis dassonvillei*, isolated from the ovaries of pufferfish *Fugu rubripes*. Taxoicon 45:851–859

Exploitation of Fungal Biodiversity for Discovery of Novel Antibiotics

Sabrina Karwehl and Marc Stadler

Abstract Fungi were among the first sources for antibiotics. The discovery and development of the penicillin-type and cephalosporin-type β-lactams and their synthetic versions were transformative in emergence of the modern pharmaceutical industry. They remain some of the most important antibiotics, even 70 years after their discovery. Meanwhile, thousands of fungal metabolites have been discovered, yet these metabolites have only contributed a few additional compounds that have entered clinical development. Substantial expansion in fungal biodiversity assessment along with the availability of modern "-OMICS" technology and revolutionary developments in fungal biotechnology have been made in the last 15 years subsequent to the exit of most of the big Pharma companies from the field of novel antibiotics discovery. Therefore, the timing seems opportune to revisit these fascinating chemically rich organisms as a reservoir of small-molecule templates for lead discovery. This review will describe ongoing interdisciplinary scenarios in which specialists in fungal biology collaborate with chemists, pharmacologists and biochemical and process engineers in order to reveal and make new antibiotics. The utility of a pre-selection process based on phylogenetic data and distribution of secondary metabolite encoding gene cluster will be highlighted. Examples of novel bioactive metabolites from fungi derived from special ecological groups and new phylogenetic lineages will also be discussed.

Contents

S. Karwehl · M. Stadler (✉)
Helmholtz-Zentrum für Infektionsforschung, Abt. Mikrobielle Wirkstoffe,
Deutsches Zentrum für Infektionsforschung, Inhoffenstraße 7,
38124 Braunschweig, Germany
e-mail: marc.stadler@helmholtz-hzi.de

Current Topics in Microbiology and Immunology (2016) 398: 303–338
DOI 10.1007/82_2016_496

Published Online: 09 July 2016

1 Introduction

Fungi have been utilised by mankind since the onset of human civilisation. Records of edible, poisonous and medicinal mushrooms date back to the prehistory (Wasson 1968; Wasson et al. 2008). The Sumerians and Egyptians domesticated the wild yeasts, *Saccharomyces cerevisiae* and related species, to brew beer. The same organisms have also been used for millennia to prepare bread, only to name two of the manifold biotechnological applications for fungal organisms. However, it was not until around the end of the nineteenth century, that scientists began to recognise and formally investigate the fascinating biology of these organisms. Yet, today, we are still far away from understanding their ecology and their importance in the global ecosystems. The slow development of fungal biology as discipline unto itself is to a great part due to the fact that the fungi have traditionally been treated under various other disciplines, i.e. botany (as "cryptogams" among the "lower" plants) on the one hand and microbiology (as "microorganisms") on the other. Likewise, the easily tractable genetic systems of fungi, such as *Neurospora crassa* and *Aspergillus nidulans,* have endeared them as model organisms in disciplines of genetics, physiology and cell biology. Mycology as an independent discipline, combining the historical knowledge of all aforementioned areas and embarking on the modern -OMICS techniques, is rather young, yet mycologists could and should become very actively involved in tackling the challenges of the human civilisation in the twenty-first century (Hawksworth 2012; Rambold et al. 2013). These challenges also encompass the fight against newly arising infectious diseases and in particular, the multi-resistant bacterial pathogens. In this chapter, we will briefly review the history of antibiotics research on fungi and its current state of the art. We also provide a future perspective on how fungi and their secondary metabolites could be beneficially employed in antimicrobial therapy and prophylaxis.

2 History

The historical part of this article is rather brief, since these aspects have been treated at length in other recent reviews, to which the reader is referred (e.g. Clardy et al. 2009; Kirst 2013). We will only point out some aspects that are not common knowledge even in the community of infection biologists and drug researchers.

Although penicillin G (**1**, Fig. 1) was the first natural product that gave rise to the development of a therapeutically useful antibiotic (Lax 2005), even earlier,

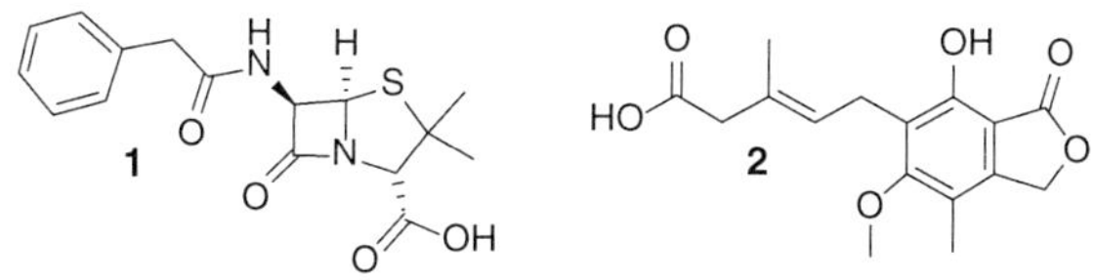

Fig. 1 Chemical structures of penicillin G (**1**) and mycophenolic acid (**2**)

mycophenolic acid (MPA) (**2**) was discovered. In 1893, Bartolomeo Gosio discovered and isolated MPA from a *Penicillium brevicompactum* strain that he found to inhibit the growth of *Bacillus anthracis*. Later on, MPA was distinguished as the first purified natural product antibiotic. The story around MPA is discussed in detail by Bentley (2000). Today, after being dismissed as an antibiotic because of an inherent toxicity and the elucidation of its mode of action, MPA was used to develop mycophenolate mofetil, marketed under the names CellCept® (Hoffmann LaRoche) and Myfortic® (Novartis). Beyond its impact as the precursor of this important new class of immunosuppressant drugs, MPA illustrates the current transformation underway in modern natural products research. The complete biosynthetic pathway has been characterised in two producing fungi: *P. brevicompactum* (Hansen et al. 2011a, b) and *Penicillium roqueforti* (Del-Cid et al. 2016).

Many other fungal metabolites were first found in screening programmes aimed at novel antibiotics discovery but were later found to possess other biological properties that led to recognition of their natural functions as plant and animal virulent factors and to practical applications in other fields of therapy.

The rapid development of therapeutic methods and industrial scale processes for penicillin during the 1940s was arguably the most significant milestone in the history of drug discovery research (Lax 2005). Substantial increases in the human lifespan can be traced back to the introduction of the β-lactam antibiotics penicillin and cephalosporin and their modern derivatives. The penicillin and cephalosporin biosynthetic gene clusters (BGCs) are among the best characterised of all NRPSs. Recently, the penicillin gene cluster from *Penicillium chrysogenum* was reconstituted as a single synthetic three-ORF polycistronic gene, and the construct was placed under the control of a single promoter (Ozcengiz and Demain 2013).

Making all of this possible in the first place was the groundbreaking work of the American mycologist Charles Thom. Thom was directly imparted in the Penicillin project by classifying Fleming's producer strain as *Penicillium notatum* Westling. But even before that he established identification and fermentation methods for *Penicillium* and *Aspergillus* which directly led to the successful production of Penicillin during World War II. His work on cheese moulds made him realise that different fungal strains depended upon unique environmental conditions for optimal growth. Being now aware of this, he was able to culture various *Aspergillus* and *Penicillium* strains to start his identification project. By culturing all strains on defined, reproducible media to describe and identify the various fungal strains he was able to "clean up the mess in *Penicillium* and *Aspergillus*" which were only inadequately and incoherently described at that time (Raper 1965). Modern taxonomic studies (Houbraken et al. 2011) recently found that the name of Fleming's fungus from which the compound was first obtained was actually *Penicillium rubens*. Nonetheless, the vast culture collection Thom had accumulated over the years, including over one thousand strains already in 1929, laid the foundation to find the best producer strain for penicillin production when it was most needed.

Only recently, Houbraken et al. (2012) have provided a clear picture on the distribution of penicillin among *Penicillium* species. This was only possible because the authors could rely on a data matrix that had arisen from over 25 years of intensive morphological, chemotaxonomic and molecular phylogenetic studies involving numerous scientists from all over the world. The advantage of combining classical and modern know-how on taxonomy to establish correlations between biological and chemical diversity within the fungal kingdom will be outlined further in Sect. 2.2.

Statistically, the case for investing antibacterial antibiotics from fungi is not especially strong. Besides penicillin, less than a handful of fungal-derived antibacterial antibiotics have made it to the market. Historically, the majority of antibiotics have come from secondary metabolites of actinobacteria. These "Actinomycetes" (which had for a long time been taxonomically classified as fungi), also produce β-lactam antibiotics, e.g. cephamycins, nocardicins (monobactams) and carbapenems. The cephalosporins (e.g. cephalosporin C (**3**), Fig. 2) were the second important type of fungal β-lactam antibiotic and were discovered about 15 years after the penicillin from the hypocrealean ascomycete *Acremonium chrysogenum* (previously known as "*Cephalosporium acremonium*") by Giuseppe Brotzu, a professor of Hygiene at the University of Cagliari, Sardinia, Italy. He formulated a hypothesis about why young people that were swimming at "Su Siccu" Bay, precisely at the site where the city sewer system drained into the sea, never contracted typhus. There were no outbreaks of typhoid fever cases related to bathing in these sewage-contaminated waters. Then, he decided to take a water sample and test its effect on a *Salmonella typhi* culture. Investigations by Abraham and Newton in England then led to the isolation of cephalosporins P, N and C from culture fluids of the Sardinian fungus—the cephalosporin family has given rise to several generations of derivatisations since their introduction. Remarkably, these semisynthetic derivatives have finally been optimised to become broad-spectrum antibiotics, and some of the third-generation cephalosporins (e.g. **4**–**5**, Fig. 2) are even effective against highly recalcitrant gram-negative pathogens such as *Pseudomonas aeruginosa*, while the natural cephalosporins as well as the penicillin are only active against gram-positive bacteria.

This historical example demonstrates that medicinal chemistry approaches can lead to substantial improvements of natural lead structures. Unfortunately, because of the discovery of so many excellent antibiotics during the Golden Ages of the 1950s to the 1970s, many other potential lead compounds were never allocated MedChem capacities for their optimisation and were developed, often because they initially only showed narrow spectrum effects, or had other undesirable properties. Clearly, it would be worthwhile to revisit some of these overlooked metabolites with new methods to elucidate their mechanism of action and modify their

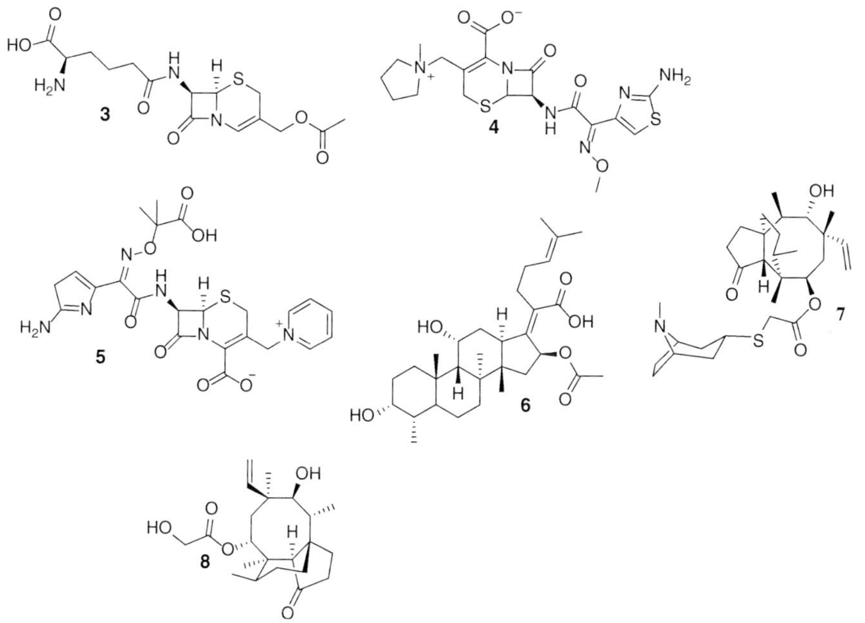

Fig. 2 Chemical structures of cephalosporin C (**3**), the "third-generation cephalosporins", cefepim (**4**) and ceftazidim (**5**), fusidic acid (**6**), retapamulin (**7**) and pleuromutilin (**8**)

chemistry, as has proved to be the cases in antibiotics that were passed over during the Golden Age and recently resurrected and brought to the market, e.g. fidaxomicin or tigecycline.

Despite the fact that fungi have been screened extensively since the inception of the era of antibiotics research, only a few other compounds derived from these organisms aside from the aforementioned β-lactams have reached the market, and they are not widely used as therapeutic agents. Fusidic acid (**6**) is a fusidane-type triterpenoid produced by a strain of *Acremonium fusidioides*. It was discovered at Leo Pharma, Denmark, during testing strains purchased from the Centraalbureau voor Schimmelcultures. The strain had been isolated from excrement of wild monkey by the Japanese mycologist, Keisuke Tubaki (Fernandes 2016). It inhibits bacterial protein synthesis by locking elongation factor-G to the ribosome after GTP hydrolysis. Fusidic acid has no known cross-resistance with any other class of antibiotic, and its clinical use outside the USA indicates a track record of safety for more than 40 years. Based on the potential new applications for resistant bacterial infections, Cempra Pharmaceutical has undertaken clinical trials to reintroduce and register an orally active formulation for use in the USA (Fernandes 2016; Fernandes and Pereira 2011).

One of the most recently launched classes of antibiotics traces its origins back to a diterpenoid compound that was discovered from cultures of basidiomycetes over sixty years ago: retapamulin (**7**) was approved as a drug in 2007 for use as a topical antibiotic and is a semisynthetic derivative of pleuromutilin (**8**), a characteristic secondary metabolite of the genus *Clitopilus*. Previously, other synthetic derivatives of pleuromutilin, valnemulin and tiamulin have been developed for treatment of infections in poultry and swine. A major reason why this compound was not developed earlier was the lack of a sustainable production process to produce the parent compound in adequate titres from the relatively slow-growing producer organism. The failure to develop sustainable production processes, for numerous fungal lead compounds likely has been a factor hindering progress in preclinical stages of development. During the Golden Age of antibiotics discovery, the researchers were accustomed to exploiting the "low hanging fruits" from *Streptomyces* and other bacteria, and fast-growing moulds could be handled more easily for scale-up than other secondary metabolite producing microbes.

2.1 Improved Workflow and Screening Methods

Figure 3 shows the general workflow that is being employed in natural products screening at our laboratory. The crucial steps, such as isolation, identification and preservation of organisms, screening fermentation, extract preparation, bioassay-guided fractionation and subsequent structure elucidation of active principles, were established during the 1950s and have continued to this day. Although the process may appear "old-fashioned" in the view of some scientists, funding agencies and industrial R&D managers, an efficient and comprehensive alternative has yet to materialise. This viewpoint, along with the dwindling interest in the pharmaceutical industry in the development of new antibiotics, has certainly contributed to the decline of natural products research since the early 1990s. Another reason is the fact that the entire workflow is labour- and capital-intensive. Consequently, many natural product facilities were sacrificed during re-organisations and mergers of the Big Pharma and Agrochemical companies. Indeed, research in various therapeutic categories, including antiparasitics, antivirals, neurology and cardiovascular, have relied on natural products during the early stages of lead discovery, resulting in the discovery of valuable leads and even blockbuster drugs such as ara A (**9**), ara C (**10**) and lovastatin (**11**, Fig. 4), respectively. Over time, the majority of drugs launched for these and other indications were based on synthetic compounds, and therefore, natural products were believed to have become expedient. However, for discovery of antibacterials, anticancer agents, and even antiparasitics, CNS-active and immunosuppressive agents, the majority of marketed and investigational drugs or at least some of the most important products continued to be derived from natural molecules or synthesised from natural product inspired templates (Newman and Cragg 2016). Among the antibacterial antibiotics, only two classes, sulphonamides and oxazolidonones, have no natural precedents (Peláez 2006). Even the fluoroquinolones, which are often regarded as "synthetics", since their discovery and optimisation exclusively relied on synthetic approaches, have natural counterparts featuring

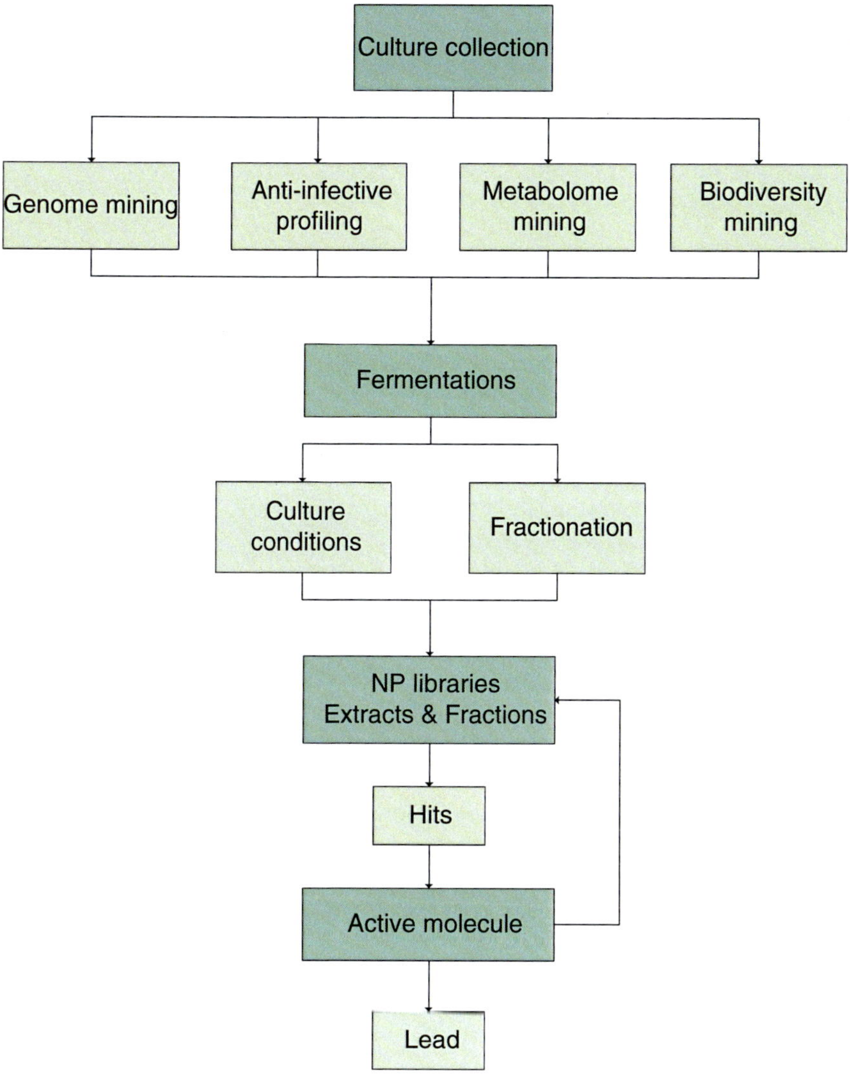

Fig. 3 The screening workflow for antibiotic natural product discovery

quinolone skeletons derived from plants, whose antibiotic activities had been known long before the first-generation quinolones were developed independently from the plant metabolites by de novo synthesis.

Many companies gave up their in-house natural products expertise long ago. As of recently, there seems to be a renewed interest in reviving the natural product discovery process by incorporating modern technologies to overcome the bottlenecks that have resulted in lack of success in the past 30 years. Measures to eliminate redundancies starting at the strain level are outlined below in Sect. 2.2,

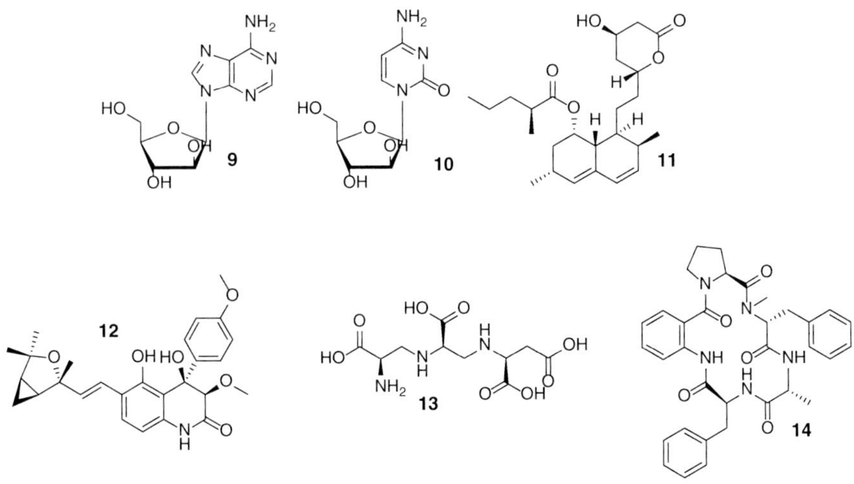

Fig. 4 Chemical structures of ara A (**9**), ara C (**10**), mevinolin (**11**), aspoquinoline A (**12**), aspergillomarasmine A (**13**) and avellanin C (**14**)

but it is even more important to provide high-quality extracts for screening samples. For instance, many studies of fungal metabolites from academic laboratories continue to use potato dextrose or malt extract media, which may be adequate for maintenance of fungal cultures, but are rather poor choices of media for secondary metabolite production, particularly in a resource-limited scheme where each strain will be tested on a single medium. These media will not favour good titres of NRPS-derived metabolites because of their poor protein content. In order to promote varied metabolite expression, it is recommended to employ fermentation in multiple media that are complementary with regard to their C/N ratio, inclusion of plant-derived nutrients, buffering agents and other parameters relating the composition. In the laboratory, the metabolite profiles may be highly dependent on the culture medium, a well-known phenomenon that has been formally designated OSMAC (One Strain—Many Active Compounds) hypothesis (Bode et al. 2002). Using this approach, Scherlach and Hertweck (2006) tried out 40 different culture conditions for *A. nidulans* to find the one condition that yielded four new secondary metabolites, aspoquinolines A (**12**)-D. The practice of testing each strain in multiple media was well established in many successful industrial screening programmes with fungi and actinomycetes where analysis of screening data statistics clearly demonstrated that important active fermentations were observed in only one medium among several that were tested (Yarbrough et al. 1993). Bills et al. (2008) took advantage of this well-known phenomenon and used nutritional arrays in miniaturised fermentations, in which a large number of culture media could be evaluated concurrently and quickly to recognise the optimal conditions for hit detection and scale-up. This methodology appears promising for optimisation once a certain lead compound has been discovered. A disadvantage of this method is that the screening samples are of limited volume, about 1 mL or less, and therefore, little extract will remain for follow-up biology and chemistry. More frequently, the goal will be to prepare a library that can be tested in multiple assays overtime. In this workflow, larger quantities of culture media are needed to prepare larger extract samples. The input and processing parameters of such a library must be carefully organised in a database that can record selectivity of assay data, as well as data on the detected constituents and other outcomes. In fact, scale-up of production from very small volumes into larger vessels such as Erlenmeyer flasks or even fermenters is not always a linear process, and occasionally may require challenging experimentation to achieve an effective scaled up process (Running and Bansal 2016). Since the penicillin discovery era, screening of microbial metabolites has relied on growing strains in combination with static and shaken liquid culture, and solid grain culture using glass vials, Erlenmeyer and Fernbach flasks, with or without baffles, and Roux and Glaxo bottles, among other types of vessels (Katzer et al. 2001; Weijman and van Eijk 1982; Running and Bansal 2016). On the other hand, it is well known that many fungi will preferentially produce certain interesting compounds in solid culture. Therefore, rationally solid fermentations of fungi have employed cooked grains, e.g. whole rice, corn or wheat, the FERMEX method, developed and described by Bills et al. (2012b), constitutes a very interesting alternative. The fungi are grown in roller bottles in medium ingredient enriched vermiculite and were shown to produce a higher diversity and larger amounts of bioactive compounds than in conventional liquid fermentation. It has a great advantage over other related methods that involve the propagation of the fungi on cellulose, rice grains or even agar media.

Since multiple active compounds are usually present in the same fermentation-derived extract, pre-fractionating the crude samples can be useful to separate and optimise concentrations of individual components prior to testing, but this procedure is laborious and in many cases, the allocated resources are insufficient. Furthermore, crude extracts are incompatible with certain screening readouts, so natural products can sometimes only be screened concurrently with synthetic libraries if they constitute pure compounds or fractions. An alternative may be to produce extracts where some of the "nuisance compounds" that often cause false-positive screening activities and may occur in fungal samples in extraordinarily large quantities, such as fatty acids, have been depleted from the extracts. Using solid phase extraction procedures or organic extractions to remove the hydrophilic constituents of the culture broths such as sugars amino acids and hydrophilic macromolecules has also become a customary method to enrich the medium-polar to non-polar small molecules that are most often preferred as candidates for hit-to-lead evaluation and are compatible with the follow-up workflow.

If extracts are being used for screening, a powerful dereplication methodology, best based on hyphenated HPLC (or UPLC) methodologies with mass spectrometric and diode array detectors (HPLC-DAD/MS), is indispensable for success. There are numerous reviews illustrating this approach (e.g. Bitzer et al. 2007; Lang et al. 2008; Wolfender et al. 2010; Allard et al. 2016), demonstrating the revolutionary developments regarding the increase of sensitivity, due to improvements and adaptations of instrumentation. While the initial techniques still afforded a rather complex sample preparation procedure, ambient ionisation mass spectrometry techniques have facilitated the detection of secondary metabolites substantially. Sica et al. (2015) have recently described an approach using the droplet–liquid microjunction–surface sampling probe (droplet–LMJ–SSP), coupled with UPLC–PDA–HRMS–MS/MS, which even allows for the detection of secondary metabolites in fungal cultures in situ. Such techniques can be highly valuable to study phenomena of chemical ecology and may also contribute to the discovery of novel metabolites, which, however, remain to be produced in larger quantities to allow for their complete structure elucidation and the evaluation of their biological activities.

Another important factor for success of new antibiotics discovery is the choice of the screening system. Initially, mainly cell-based assays used suspensions of pathogenic organisms and were tested for susceptibility of growth towards fungal and microbial strains, or extracts derived thereof. This procedure continues to be useful until today, since alternatives such as biochemically target-based cell-free assays have yielded few cell-penetrable novel compounds that were subsequently developed to marketed drugs; none from this approach have been derived from a fungus. In general, the high throughput screening campaigns started to dominate since the late 1980s into the 1990s in many pharmaceutical companies. However, they have not been especially successful for the discovery of novel antibacterial anti-infectives. One major reason was that numerous lead compounds that were found during screening (no matter whether derived from synthetic or natural product libraries) showed non-selective activities or toxicity or did not penetrate the bacterial cell envelope in second line assays. For natural products, the target-based assays were often not suitable to test extract libraries, since certain constituents of the matrix of fungal or bacterial crude extracts interfered with the readout. In general, screenings of pure metabolites were preferred for many reasons, but the preparation of libraries of pure compounds with fully elucidated structures is very costly and only exceptionally, e.g. preliminary chemistry around lead natural product, would it be an option. Nevertheless, the screening of such metabolites in newly established bioassays can lead to the rediscovery of new applications for previously known metabolites. This was recently demonstrated by a study on aspergillomarasmine A (**13**, Fig. 4) (King et al. 2014), a known fungal metabolite that might be able to overcome resistance by its interference with β-lactam degrading metallolactamases.

Notably, fungal metabolites have not been tested exhaustively in screening systems aimed at finding drug candidates to follow-up alternative strategies to classical antibiotic screening, some of which are being outlined in other contributions of the current project (Mühlen and Dersch 2016; Erhardt 2016). Aside from the pathoblocker approach suggested in these papers, numerous other strategies have been followed, and at least, it has been proven that new natural products can be found. For instance, Donald et al. (2009) and Goetz et al. (2010) have discovered fungal metabolites that showed significant activities in a chemical genetic screening based on a method that employs plasmid-encoded antisense RNA to increase susceptibility for each essential gene target in the bacterial cell, the so-called

Staphylococcus aureus fitness test. This screening method relied on testing of numerous strains of the pathogen, each of which was engineered for reduced expression of target genes essential for *S. aureus* growth. The technology was directed to find novel compounds with a novel mode of action; however, so far no "druggable" molecule seems to have resulted from this approach, possibly because the company (Merck) abandoned the programme before allowing enough time for it to fully explore potential lead molecules.

Inhibition of quorum sensing and other signalling factors that can ultimately disrupt biofilm formation of the pathogens in the tissue of the host is an alternative to the classical approach aimed at killing the pathogen, since biofilm inhibitors are not likely to cause resistance immediately. Igarashi et al. (2015) have recently published such a quorum sensing inhibitor named avellanin C (**14**) from a culture of the ascomycete genus *Hamigera*. Otherwise, reports on bioactive natural products that are able to target quorum sensing and biofilm formation in pathogenic bacteria are still rather rare. It might be worthwhile to follow such complementary approaches to discover novel candidate compounds for development as antibacterial agents. However, as demonstrated in the next sections, the quality of the libraries is also important, since the most innovative and intelligent screen may fail to deliver novel drug candidates if the compounds to be tested lack substantial chemical diversity and innovative potential.

2.2 Correlations Between Phylogeny and Metabolite Production

The early stages of natural product-based drug discovery had been dominated by empirical screening approaches, such as the mass cultivation of soil isolates and studies of their crude extracts in cell-based growth inhibition bioassays. Despite the undisputed success of this strategy, which has resulted in the discovery of many of the currently marketed antibiotics, especially from actinobacteria (see concurrent paper by Mohr 2016), problems with redundancies (production of the same compound by multiple isolates) and "frequent hitters" (reactive compounds with non-selective modes of action in biological systems) soon became evident. Both problems persisted even when organisms from habitats other than marine, plant-associated, etc. (see Sect. 2.3) were exploited simply because the gene clusters encoding many of the metabolites are widespread and their production in culture is facile.

Antibiosis is a widespread phenomenon among all microbial groups that are "talented" secondary metabolite producers (Bills et al. 2009a, b; Dreyfuss and Chapela 1994); hence, bioactive extracts are easy to find. So long as the secondary metabolite production was thought to be "strain-specific", it was therefore difficult to find a way to overcome this problem and redundant work could not be avoided as the number of screening isolates increased in industrial natural product screening scenarios.

Since the process of hit-to-lead evaluation, including the scale-up of production of selected strains and in particular the isolation of the active principles by bioassay-guided fractionation has always constituted the most laborious part of the natural product-based lead discovery process, the continuous rediscovery of known compounds contributed to discouragement and led many companies to abandon their R&D in natural products.

Taxonomy can help to some extent to overcome this problem because it has been known for a long time that secondary metabolite production and taxonomy may be closely aligned within narrow phylogenetic lineages in many organism groups. Frisvad et al. (2008) have reviewed the literature on this matter with respect to filamentous fungi and have given several examples where taxonomic groups can be delineated by the distribution of secondary metabolites. For instance, in lichenised Ascomycota and certain genera and families of mushrooms, pigments have been traditionally employed to discriminate species or genera (Gill and Steglich 1987). Larsen et al. (2005) have summarised manifold examples, above all from the economically important fungal genera *Aspergillus* and *Penicillium* (at that time even including *Talaromyces*), which demonstrated that the combination of concise morphotaxonomy and secondary metabolite profiles can lead to the recognition of novel metabolite producers even in this well-studied group of ascomycetes, which are classified in the family Trichocomaceae. Today, the taxonomy of this family is based on a polythetic concept, uniting data from phenotypic and genotypic studies. This work was facilitated by the fact that well-preserved viable cultures of most of the relevant species had been available, which is unfortunately not the case for most other fungal taxa. With regard to the utility of the aforementioned approach to natural product discovery, the study of de la Cruz et al. (2012) provided an example that inherent correlations between the production of certain metabolites and the taxonomy of the producing strains (which were chosen from *Emericella*, the sexual state of certain *Aspergillus* species including *A. nidulans*) exist, even when the fungi are subjected to a broad range of fermentation conditions that are also being employed in an industrial screening scenario. The authors used solid vermiculite media, adsorptive resin to capture the secondary metabolites and added epigenetic modifiers (see further below) to the cultures under study and identified several hitherto unknown producer strains of the antifungal agent, echinocandin B. However, they observed strain-specific differences in metabolite production, demonstrating that a pre-selection of strains based on taxonomy can never replace a thorough manipulative study of the respective strains, in order to reveal their full metabolic potential. Only a few strains were capable of producing the echinocandins regardless of the fermentation conditions. This example illustrates the value applying of many approaches to large sets of related strains, as opposed to testing strains under a few simple conditions that have previously failed to yield any new lead compounds.

Such case studies are extremely useful also for identification of novel congeners and novel producers of a certain metabolite class and can even help to solve taxonomic problems and result in the recognition of new species.

The astounding diversity of fungi is shown in the studies of Röhrich et al. (2014) and Schardl et al. (2013), which investigated peptaibiotics from *Trichoderma/Hypocrea* spp. and dynamics of alkaloid loci in Clavicipitaceae, respectively. These very different studies could show that the fungal secondary metabolome is not only dependent on the fermentation conditions used in antibiotic screening programmes but also greatly depends on the natural environment from which the fungi are collected. More recently, Bills et al. (2015) were able to show that a comparison of the genomes of closely related strains can reveal important differences and similarities in capacity for secondary metabolite biosynthesis. In this example, it was found that the genome of the newly described taxon, *Aspergillus mulundensis,* shared 45 % of the BGCs encoding for secondary metabolites with the related species, *A. nidulans*. This example previews "phylogenomic" approaches that can be employed widely in fungal taxonomy and phylogeny. This can be demonstrated, based on the echinocandin type of antimycotics (**15**–**17**, Fig. 5), which were discovered concurrently in several phylogenetically unrelated species of the Ascomycota.

Yue et al. (2015) have sequenced the genomes of several echinocandin producing strains and identified the echinocandin BGCs, and attempted a phylogenetic reconstruction of their evolution. Their results pointed towards a single evolutionary origin of the gene cluster in the producer strains belonging to the classes Eurotiomycetes versus Leotiomycetes, i.e. two lineages of Ascomycota that are believed to have diverged ca. 290–390 million years ago. Since many of the close relatives of the known echinocandin producers do not produce the compound, a horizontal gene transfer of the echinocandin NRPS appears plausible. A possible horizontal gene transfer of certain NRPSs from bacteria to fungi was further supported by a study of Bushley and Turgeon (2010) who analysed a comprehensive dataset of fungal NRPSs. However, many other cases of coincidental occurrence of the same, complex metabolite in seemingly unrelated fungi remain to be studied. The mechanisms by which such large portions of DNA can be transferred from one fungus to another in the natural environment also remain to be clarified. In any case, it is far more common to encounter similar complex metabolites in closely related species and genera, as demonstrated also by chemotaxonomic studies on other potent natural fungicides with interesting structural and biological properties, such as the parnafungins (e.g. **18**–**19**, Fig. 5) (Bills et al. 2009a, b) and the sordarins (e.g. **20**, Fig. 5) (Vicente et al. 2009). In both cases, the majority of producer strains were found to be taxonomically related. Nevertheless, there are numerous examples in the literature where the same compound was found from apparently unrelated taxa, and therefore, taxonomy and phylogeny can only serve as a criterion for pre-selection but should not be generally used as a criterion for exclusion of certain strains from the screening. The exceptions are taxa which are well known for the production of potent mycotoxins, such as trichothecenes. This applies for instance for the genus *Myrothecium* and part of the genus *Fusarium*, where these compounds are almost always responsible for the strong bioactivities observed in the crude extracts. In a study by Li et al. (2016) previously published data, generated during the last decades, about fungal natural products and their associated BGCs were curated and organised to reveal yet uncharacterised BGCs with the potential to

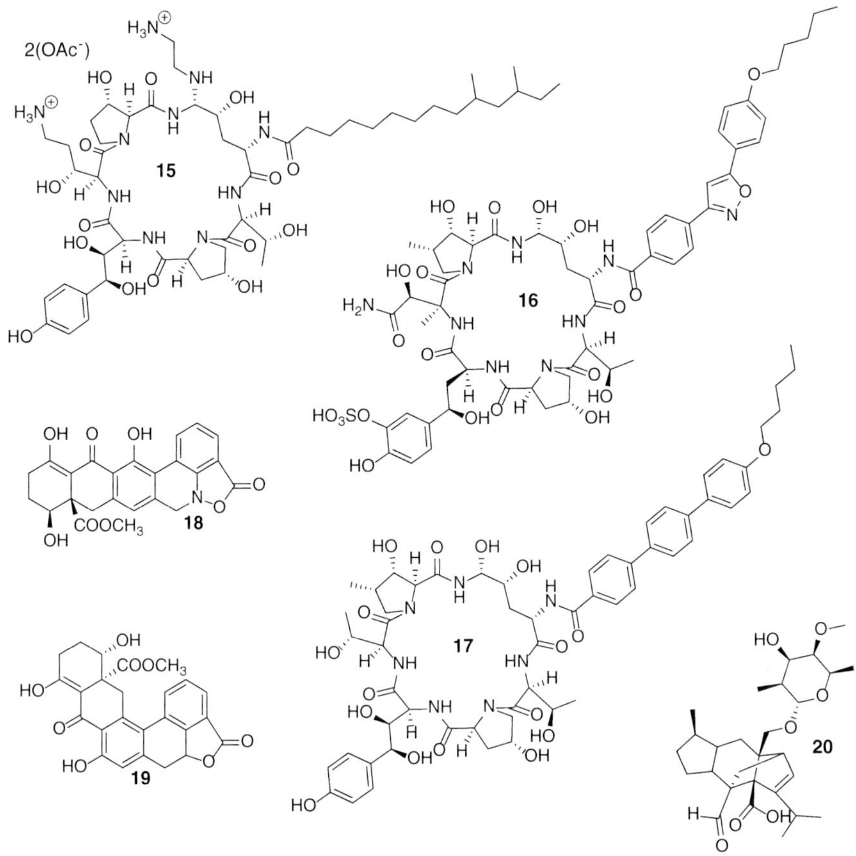

Fig. 5 Chemical structures of caspofungin (**15**), micafungin (**16**), anidulafungin (**17**), parnafungin A1 (**18**), parnafungin B1 (**19**) and sordarin (**20**)

encode the production of biologically active metabolites. The resulting information was incorporated into the Minimum Information about Biosynthetic Gene cluster (MIBiG) repository creating the largest set of linked fungal BGCs and natural products effectively reducing efforts for future genome mining and synthetic biology projects as natural product scientists are now able to find and characterise unknown BGCs much faster.

Our own work on the establishment of correlations between biological and chemical diversity has focused on the fungal family Xylariaceae. These fungi have long been regarded as saprotrophic wood destroyers (except for the genera treated further below in Sect. 2.3 and some additional smaller genera and species groups that are restricted to dung, marine environments or associated with termites). Recent studies, mostly relying on methods of molecular ecology, have revealed they are among the predominant endophytes of vascular plants and even marine algae. An endophytic stage is part of the life cycle in many of the known species (cf.

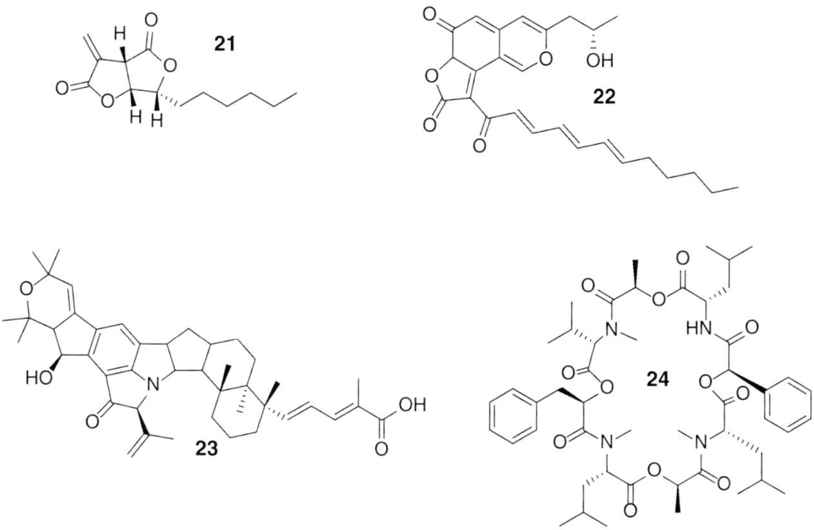

Fig. 6 Chemical structures of sporothriolide (**21**), lenormandin A (**22**), nodulisporic acid A (**23**) and PF-1022A (**24**)

Sect. 2.3). They are ubiquitous in all forested areas of the world and exhibit their highest diversity in the tropics, and numerous novel bioactive secondary metabolites are continuously being obtained from both their ascomata and mycelial cultures (Stadler 2011). Extensive work on the taxonomy and phylogeny of these fungi (e.g. Kuhnert et al. 2014a; Stadler et al. 2014) has led to a matrix of morphological, molecular phylogenetic and chemotaxonomic data that meanwhile allows for a concise identification of novel species, and these have very often yielded novel bioactive secondary metabolites. In addition, interesting correlations between the taxonomy and the production of secondary metabolites have become evident. For instance, recent studies on the genus *Hypoxylon* have revealed that all mycelial cultures studied of the pantropical species *Hypoxylon monticulosum* produce sporothriolides (e.g. **21**, Fig. 6), which are strong antifungal agents (Surup et al. 2014), while all stromata of *Hypoxylon lenormandii* and closely related species produce the lenormandins (e.g. **22**), which constitute a novel type of azaphilone pigments (Kuhnert et al. 2015). In some cases, the metabolite production is apparently species-specific. For example, so far only the stromata of the recently described species *Hypoxylon lechatii* were found to contain tetramic acid antibiotics of the vermelhotin type as major metabolites, while its closest relatives produce other metabolites instead (Kuhnert et al. 2014b). Last but not least, Bills et al. (2012a) established the identity of an endophytic Xylariaceae species that was found throughout the tropics with the newly discovered species, *Hypoxylon pulicicidum*, thereby elucidating the life cycle of this fungus. Nodulisporic acids (e.g. **23**), gave rise to a developmental candidate for invertebrate parasites and apparently constitute species specific marker metabolites.

In the related genus *Daldinia* and other, predominantly tropical Xylariaceae, such as *Entonaema*, *Phylacia*, *Rhopalostroma* and *Thamnomyces*, recent polyphasic taxonomic work has revealed strong correlations between the production of multiple PKS-derived molecules and concurrently obtained molecular phylogenetic data (Bitzer et al. 2008; Stadler et al. 2010). All these fungi, however, also produce specific compounds besides the common metabolites, of which many remain to be identified and tested for their biological activities. Notably, such chemotaxonomic studies need to be carried out under standardised conditions, where the full potential of the fungi under study cannot be revealed. Aside from the "phylogenetically significant" metabolites, every species may contain specific metabolites. However, "genetic barcoding" may give valuable information on how to sort out redundancies in large strain contingencies and focus on the right organisms during preparation of screening libraries and selection of hits for the laborious hit follow-up work, such as bioassay-guided fractionation (see Sect. 2.1). Molecular phylogeny is also of great utility in the assessment of redundancies among strain collectives of endophytes and other ecological groups, which are treated in the next chapter.

2.3 Ecology-driven Antibiotics Discovery

One important field of innovation in terms of secondary metabolite discovery that has been hitherto neglected is the chemical ecology of fungi and the evaluation of the potential of certain taxonomic and ecological groups that could hitherto not be cultured and remain unstudied for the production of antibiotics and other secondary metabolites. While some pioneers of antibiotics discovery believed that these compounds were just produced by coincidence or constitute waste products (Vining 1990), it is now well established that secondary metabolism is mediated by highly sophisticated biochemical mechanisms, and numerous genes are always involved in the biosynthesis of a given compound class and the regulation of its production. Strachan and Davies (2016) and Davies (2013) have summarised the "modern" view of many researchers that secondary metabolites are "molecular messengers" that mediate intraspecific and interspecific interactions, and the evidence that corroborates this hypothesis is steadily increasing. In this chapter, we will provide an outline on how ecological evidence can help to increase the probability of success in finding novel antibiotics.

The traditional antibiotics screening started with organisms from soils over 60 years ago, and a high percentage of the known fungal secondary metabolites have been obtained from soil isolates. Soils were a logical place to start because of convenience, high species diversity, and also because of the parallel phenomenal successes of Selman Waksman and his students in discovering new antibiotics from soil actinomycetes. Ecologically, soils were also thought to be fertile sources of antibiotic-producing microbes because of strong competition for nutrients and territory in this microbially rich habitat. Many of these fungi are easy to isolate, using simple plating techniques. With the help of selective media, the basic protocols can be modified to obtain a wider array organism groups according to nutrient requirements, temperature and pH optimum and other selective methods which are extensively described by Bills et al. (2004). The limitation in relying on soil-inhabiting fungi only was that certain genera always predominant in soils, and many of these have already been heavily exploited for several decades. The fast-growing strains in such soil samples will overgrow the more interesting, hitherto unexploited ones, unless special techniques such as dilution-to-extinction culturing, baiting with special substrata, or pre-pasteurisation of the soil are employed (see also concurrent paper by Landwehr et al. 2016 for similar problems with bacteria). The dilution-to-extinction technique disperses washed soil particles into microtiter plates at very high dilution rates so that the slow-growing strains can have a chance to grow (Collado et al. 2007). This technique requires some additional technical preparations, and of course, the characterisation of the resulting cultures in order to recognise redundancies is required as in any other direct isolation method. The morphological and molecular characterisation of large sets of strain remains work intensive (cf. Stadler and Hellwig 2004); hence, the pre-selection procedures described in Sect. 2.2 also come in handy during the recognition of redundancies in groups of fresh isolates of the ecological groups of fungi mentioned further below.

The most widely explored ecological group during the past years has been endophytes of vascular plants. These organisms have been included in industrial screening programmes since the 1970s (cf. Petrini et al. 1993); the methods were later adopted by many academic groups, who have been able to isolate hundreds of novel bioactive metabolites from their cultures (cf. Schulz et al. 2002; Aly et al. 2010). This is in part due to hyperbole arising from the detection of the anticancer agent, taxol in cultures of an endophyte that was derived from the host tree that produces the aforementioned anticancer agent. As reflected by Heinig et al. (2013), this observation has never resulted in a sustainable production process for taxol from cultures of a fungus that could compete with the processes based on plant cell cultures and extraction from the plant combined with semisynthesis. Even the frequently postulated horizontal transfer of biosynthesis genes between the host plant and its fungal endophytes remains highly controversial because a plausible mechanism has yet to be proven by molecular genetics and genomic methods. Nevertheless, endophytes are exciting organisms to study for both ecologists and "bioprospectors" and are far from being exhaustively exploited. Some groups have recently started to evaluate the possibility of obtaining additional plant metabolites from cultures of fungal endophytes (cf. Kusari et al. 2012). In many cases, the respective compounds were detected by HPLC-MS in the fungal cultures, but have not been isolated to purity in amounts that would allow for unambiguous proof of their identity using NMR spectroscopy. For discovery of novel antibacterial agents, these research activities are of secondary importance because plant metabolites thus far have played a minor role as lead structures for antibiotic agents. It should also be pointed out that the only metabolites from endophytic fungi that have ever entered late preclinical development (i.e. the antiparasitic agents PF-1022A **24**, Fig. 6; see

Scherkenbeck et al. 2002, and nodulisporic acid; see Bills et al. 2012a), constitute "typical" fungal metabolites (derived from NRPS or NRPS/terpenoid biosynthesis) that have never been obtained from a plant. Recently, SCY-078 a semisynthetic derivative of enfumafungin, an acidic triterpenoid β-1,3-D-glucan synthase inhibitor which was originally isolated from *Hormonema* sp. (Butler et al. 2014; Lepak et al. 2015, see also http://www.scynexis.com/pipeline/pipeline.php), is entering Phase II clinical trials.

On the other hand, lead compounds that have been isolated first from soil—or plant debris-inhabiting fungi are frequently the same as those found in endophytes. This is unsurprising, considering they often possess the same gene clusters and that most endophytic fungi only spend part of their life cycle in the host plant and are transferred horizontally from one host plant to another. It is thought that the propagules of the endophytes mostly enter the host plant via the stomata or the roots, in a similar manner as many phytopathogens (Schulz and Boyle 2005). The ascospore-based propagation seems to be particularly common in the Xylariaceae, a fungal family that contains numerous endophytic species. These fungi have thick-walled, highly melanised ascospores that are designed to persist in the environment for a long time (Stadler et al. 2014). Many fungal species may be present in numerous host plants as endophytes, but may cause extremely high damage when transferred to the "wrong" host; on the other hand, endophytes have also been demonstrated to protect the plant host and several species are being exploited for their potential as biocontrol agents or "biofertilisers" because they show antagonistic effects against invertebrate pests.

As pointed out by Rohlfs and Churchill (2011), fungi and invertebrates have manifold ecological interactions in which secondary metabolites are hypothesised to be involved. Indeed, as demonstrated by Pažoutová et al. (2013), endophytic fungi may even be closely associated with a certain insect species that acts as vector for the distribution of the fungus, resulting in highly interesting trilateral ecological relationships between plants, fungi and invertebrates. Since invertebrate-associated fungi constitute another "hotspot" of metabolic diversity within the fungal kingdom (Isaka et al. 2005; Gibson et al. 2014; Wang et al. 2016; Donzelli and Krasnoff 2016), it appears all the more worthwhile to further embark on such organisms. However, the involvement of fungi in such multilateral interactions has only recently been discovered, as modern methods of molecular ecology became available. The same is true for other ecological groups of fungi that form consortia on the phylloplane, such as the "sooty moulds". These fungi have adapted to an epiphytic lifestyle and probably use the sugar-rich excrements of aphids as major nutrient source (Chomnunti et al. 2014). Recent molecular phylogenetic work on sampling sites in warmer climates, where insects are also abundant, has revealed that these fungi show an unprecedented taxonomic diversity, and only a few species were already studied in-depth for the production of antibiotics, but are otherwise almost unexplored. Methiosetin (**25**, Fig. 7), the only antibiotic from such a fungus so far reported (Herath et al. 2012) constitutes a member of the "equisetin" type of decaline tetramic acid type. Many other members of this class, which differ in the modification of the decaline carbon skeleton and/or the tetramic acid moiety, have

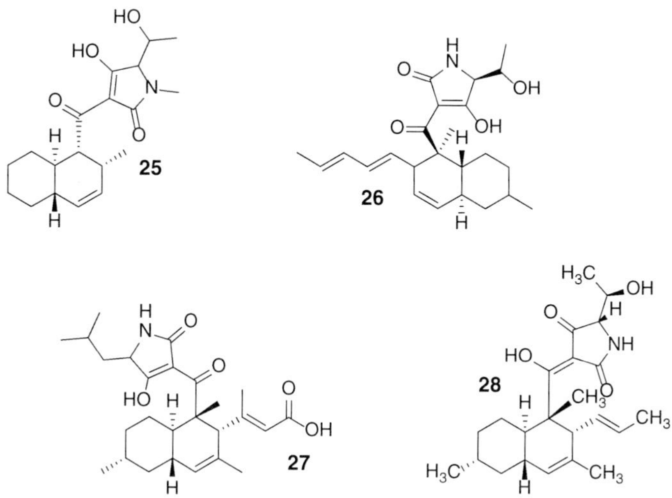

Fig. 7 Chemical structures of methiosetin (**25**), altersetin (**26**), ascosetin (**27**) and hymenosetin (**28**)

been reported from other fungi, including many endophytes. While some of them were reported to have pronounced antifungal or antiviral activities, at least three, namely altersetin (**26**, Hellwig et al. 2002), ascosetin (**27**, Ondeyka et al. 2014) and hymenosetin (**28**, Halecker et al. 2014), possess strong activities against gram-positive human pathogens, and in particular MRSA. The natural function of these tetramic acid antibiotics may be a means of non-specific defence, and like many other natural compounds, they may serve the producer organisms well in their natural habitat, yet they are not optimised for therapeutic use in humans. Chemical optimisation of these molecules appears feasible—the total synthesis of hymenosetin was accomplished shortly after the original publication (Kauhl et al. 2016). A medicinal chemistry programme to optimise this compound class, e.g. by broadening the activity spectrum and enhancing the selectivity of the antibacterial versus the concurrently observed cytotoxic effects, appears feasible. However, as with many other hit compounds that showed apparently non-selective activities, chances for concise drug development are much lower than with compounds that initially already show selectivity against bacteria and are devoid of cytotoxicity.

Aside from endophytes, marine-derived fungi have now also been screened intensively for several decades. As summarised by Debbab et al. (2012), even fungal endophytes of marine habitats (including algae as well as mangroves) already proved to be a rich source for novel compounds (e.g. **29–41**, Figs. 8 and 9). However, it can be concluded from these studies that many of the new compounds from "marine" fungal taxa merely constitute derivatives of molecules that had already been known from the terrestrial relatives of the respective fungi. In addition, 92 % of the fungal taxa that were subject of the respective studies were not strictly associated with marine habitats, while many of the truly marine fungi still remain to

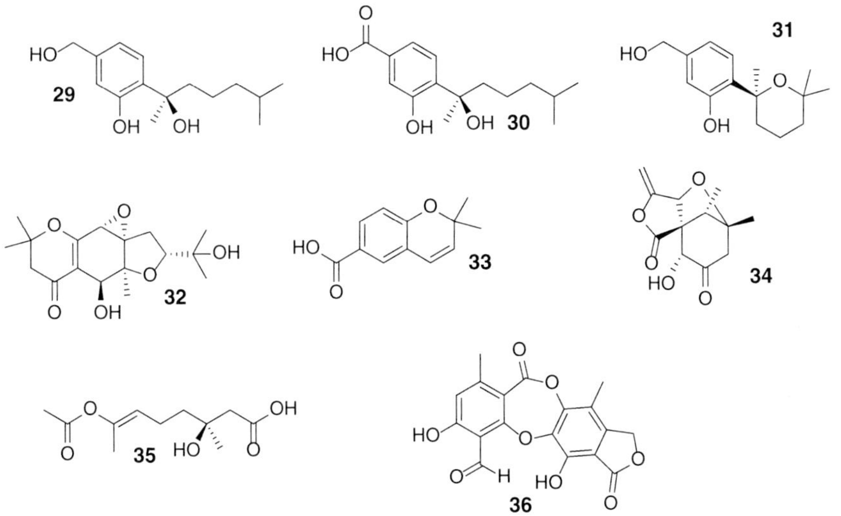

Fig. 8 Antibiotics from marine-derived endophytic fungi: chemical structures of (-)-sydonol (**29**, Li et al. 2012), (-)-sydonic acid (**30**, Li et al. 2012), (-)-5-(hydroxymethyl)-2-(2′,6′,6′-trimethyltetrahydro-2H-ypran-2-yl)phenol (**31**, Qin et al. 2011), pestalotheol G (**32**, Qin et al. 2011), anofinic acid (**33**, Sun et al. 2011 (The compounds reported by Sun et al. were obtained from a fungus named "Massrison", but this name does not actually exist in fungal taxonomy! Nevertheless, the paper was accepted by a chemical scientific journal.)), spiromassaritone (**34**, Sun et al. 2011), penicimonoterpene (**35**, Gao et al. 2011) and cordycepsidone (**36**, Varughese et al. 2012)

be exploited. In particular, strains that are obtained from marine sediments and even sponges and other sessile, marine animals are often identical to known terrestrial fungi, even including human pathogens such as *Aspergillus fumigatus*. Consequently, the metabolites reported from those are often related to those that are already known from the terrestrial counterparts. A thorough summary of the issues regarding the ecological origins of marine-derived fungi was recently published by Overy et al. (2014). The problem is that the cultivation and handling of the true marine fungi often requires special expertise and they are not easily obtained by simple plating methods from tissues of marine organisms, which will often just result in isolation of the more trivial "marine-derived" strains. The natural product community could advance to the knowledge of the active marine mycota through collaborations with mycologists, rather than to continue in their attempts to adopt outdated methods developed during the 1950s.

Coprophilous fungi are another ecological group that seem predestined for antibiotic production, due to their lifestyle where they dwell in highly competitive and transient habitats. They remain to be studied thoroughly for secondary metabolite production (Bills et al. 2013). Indeed, representatives of many phylogenetic lineages among the higher fungi contain genera that have apparently arisen from plant-associated fungi have specialised on dung as habitat. Most of these

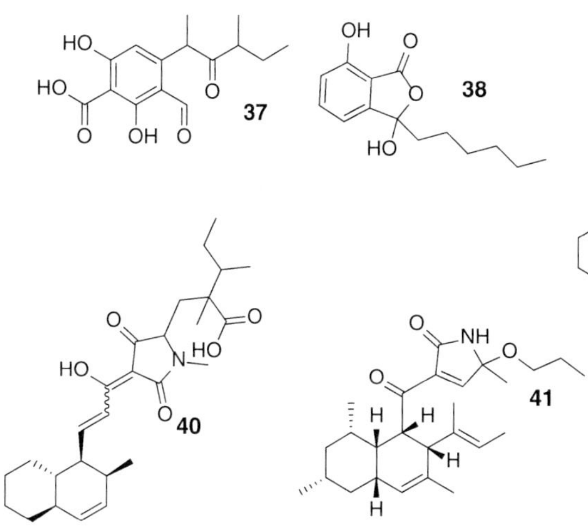

Fig. 9 Chemical structures of antibiotics from marine fungi: ascochital (**37**, Kusnick et al. 2002), ascosalipyrrolidinone (**38**, Osterhage et al. 2000), zopfiellamide A (**39**, Daferner et al. 2002), zopfiellamide B (**40**, Daferner et al. 2002) and corollosporine (**41**, Liberra et al. 1998)

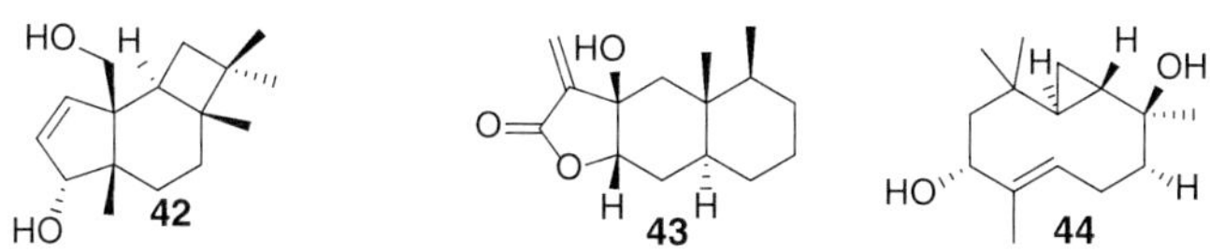

Fig. 10 Chemical structures of punctaporonin A (**42**), tulasnein (**43**), hypocoprin A (**44**)

genera remain to be studied thoroughly and systematically for secondary metabolite production. However, in those fungal groups that can be regarded as relatively well studied, some interesting correlations between metabolite production and taxonomy have already become evident. To give an example, the coprophilous genera of the Xylariaceae have yielded rather unique antibiotic polyketides and terpenoids, including the punctaporonins (**42**, Fig. 10), (Anderson et al. 1988), tulasnein (**43**) (Ridderbusch et al. 2004) and the recently discovered hypocoprin (**44**), (Jayanetti et al. 2015). These compounds have not yet been isolated from related genera, suggesting that the coprophilous Xylariaceae may have developed a complementary secondary metabolism to that of their plant-associated relatives.

Empirical work on the secondary metabolism of many fungal species has also revealed that different secondary metabolites often prevail during different stages of their life cycle, owing to differential expression of secondary metabolite biosynthesis genes. However, the differences in the composition of extracts from fruiting bodies and corresponding cultures of the same species may be even more striking. Stadler et al. (2006) have demonstrated drastic changes of secondary metabolite profiles in the xylariaceous ascomycete *Hypoxylon fragiforme*, even during production of stromata in the natural environment. Cytochalasins and numerous yet unidentified metabolites with prominent biological activities were preferentially produced in the growing stromata and the asexual state that spread on wood, while these compounds disappeared when the ascospores of the fungus became mature, giving rise to mitorubrin-type azaphilones. Even in the sclerotia of the well-studied genus *Aspergillus*, certain metabolites can be found that are not apparently produced in the vegetative mycelia (e.g. Whyte et al. 1996). Many other compounds from Ascomycota and Basidiomycota are apparently exclusively produced by the cultured mycelia or the fruiting bodies. For instance, hericenones are prevailing in basidiomata of the medicinal mushroom of the genus *Hericium*, while cyathane-type diterpenoids are commonly encountered in their mycelial cultures instead (Thongbai et al. 2015). These examples show that it is very important to study the entire life cycle of a given fungus in order to fully exploit its potential for secondary metabolite production. As their production is often correlated with morphological differentiation processes that accompany the switch from the asexual to the sexual state and vice versa, basic research in the activation of the corresponding morphogenetic genes might reveal new opportunities to improve the accessibility of hitherto unknown compounds that can so far not easily be obtained, e.g. from the tiny "fruitbodies" of many fungi. Epigenetics and dual culturing can also help to awaken silent biosynthetic genes, as will be further described in Sect. 2.5.

2.4 Culturing of Previously Uncultured Fungal Groups

The higher fungi (Dikarya) are a subkingdom of the fungi that includes the phyla Ascomycota and Basidiomycota. With the exception of pleuromutilin (**8**) (see 2.1), all other fungal antibiotics with antibacterial activity that have found their way into the developmental pipeline are derived from Ascomycota. In fact, they constitute a complementary source of secondary metabolites to the Ascomycota but have been explored far less intensively and exhaustively than their sister phylum (Stadler and Hoffmeister 2015). The reason for this is that the Basidiomycota contain many symbiotic fungi that have never been cultured or grow extremely slowly. Even the saprotrophic species are very difficult to handle upon scale-up and the slow-growing cultures may easily contaminate. Nevertheless, there are numerous examples demonstrating that it pays off to study these organisms because they may yield highly interesting and unique metabolites, in particular terpenoids (Qin et al. 2011). As mentioned above, pleuromutilin (**8**) was discovered several decades ago. In particular, since the late 1970s, numerous additional compounds have been discovered from Basidiomycota, and even in recent years, the number of bioactive metabolites is steadily increasing (e.g. **45–48**, Fig. 11) (De Silva et al. 2013, Schüffler and Anke 2014). It even pays off to re-evaluate species that had already been explored thoroughly in the past because the fermentation methods have improved and the amounts of pure compounds necessary for structure elucidation

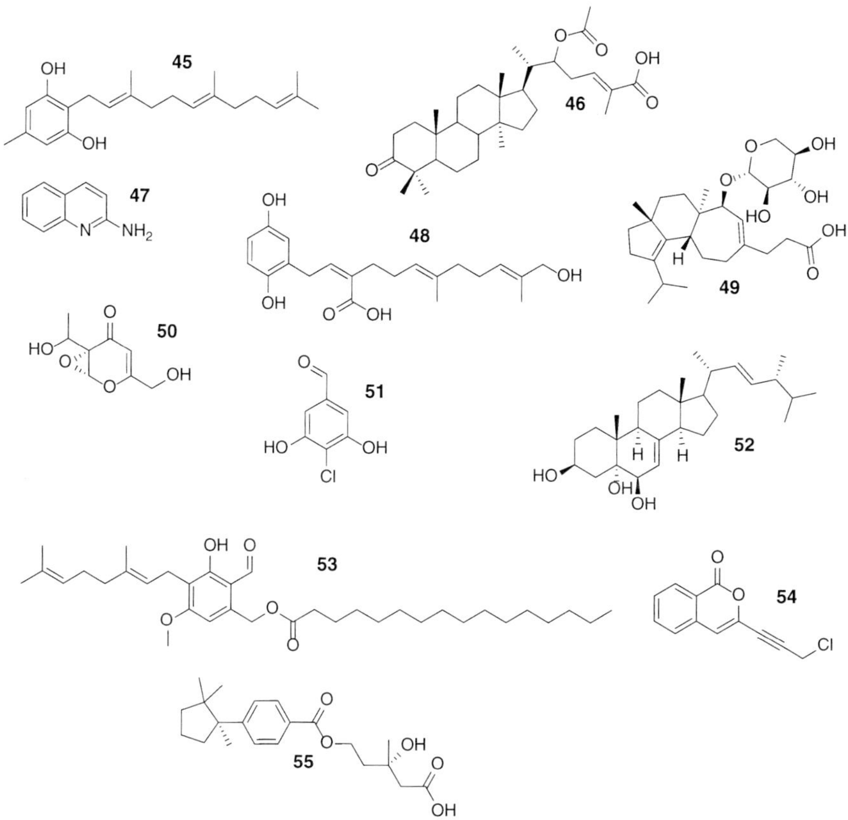

Fig. 11 Chemical structures of grifolin (**45**), astraodoric acid A (**46**), 2-aminoquinoline (**47**), ganomycin A (**48**), erinacine K (**49**), erinapyrone C (**50**), 4-chloro-3,5-dihydroxybenzaldehyde (**51**), 3β-5α-dihydroxy-6β-methoxyergosta-7,22-diene (**52**), hericene A (**53**), gymnopalyne A (**54**), deconin A (**55**)

have decreased drastically over the past years. Traditional Asian mushrooms, e.g. the genera *Ganoderma* (Richter et al. 2015) and *Hericium* (e.g. **49**–**53**) (Thongbai et al. 2015), have not only proven to be extremely rich in unique compounds but are even being grown at the ton scale and constitute valuable market products not only in Asia. Recent studies have revealed that especially the evaluation of tropical mushroom species can still be rewarding (e.g. **54**–**55**) (Thongbai et al. 2013; Surup et al. 2015; Mudalungu et al. 2016).

Another way to gain access to untapped metabolic diversity will be the identification of biosynthetic genes and gene clusters from slow-growing Basidiomycota and their transfer to faster growing heterologous host organisms that can be handled more easily in bioprocess development (Quin et al. 2014; Schmidt-Dannert et al. 2015a). Such work has recently started on the biosynthesis of some terpenoids and polyketides from *Armillaria* (Engels et al. 2011; Lackner et al. 2012), as well as on a pleuromutilin (**8**) producing *Clitopilus* species (Kilaru et al. 2009). A patent application from Sandoz AG described the pleuromutilin pathway which consists of six transcriptionally co-regulated genes, including a labdane-type diterpene synthase gene, a putative geranylgeranyl diphosphate synthase gene, three cytochrome P450 monooxygenase genes and a putative acyltransferase-encoding gene (Mitterbauer and Specht 2011). Notably, the aforementioned species are saprotrophs and grow reasonably well in the laboratory, so the respective compounds can also be obtained from their cultures. On the other hand, a recent molecular ecology study at the global scale (Tedersoo et al. 2014) has revealed an unprecedented biodiversity in soil, and the Agaricomycotina (mushroom forming fungi) constituted the majority of taxa detected. These species are presumably mycorrhizal symbionts that can hardly grow in culture, but even in the sequenced genomes of such organisms, numerous BGCs can be detected by means of bioinformatics. Hence, it could be feasible to develop special techniques for their isolation and produce small amounts of biomass for DNA extraction, followed by genome mining approaches assess their respective metabolites. The development of adequate expression systems for such antibiotics and other secondary metabolites from mycorrhizal symbionts will pose a challenge for the future, but it appears feasible, owing to the rapid progress in genomics and transcriptomics methodology (Schmidt-Dannert et al. 2015b).

2.5 Genome Mining and Epigenetic Activation of Silent Genes

In the past decade, the genome sequencing across a broad cross section of fungi has revealed a vast diversity of biosynthetic genes and gene clusters in filamentous fungi. For instance, the genome of an endophytic strain of *Pestalotiopsis fici* (Xylariales) has recently been found to contain 97 different loci that putatively encode for secondary metabolite biosynthesis (Wang et al. 2015). The first studies of this kind had been carried out on *Aspergillus* species as model organisms, which are far better studied than the majority of the Ascomycota (e.g. Keller et al. 2005; Bok et al. 2006; Schwab et al. 2007; Williams et al. 2008; Fisch et al. 2009; Henrikson et al. 2009; Brakhage and Shroeckh 2011; Sanchez et al. 2012). In general, these experiments were either based on the supplementation of molecules that act as modulators of DNA methylation or acetylation or on a knockout of the respective enzymes. In these studies, the secondary metabolism of the model organisms has changed drastically, as the constitutive metabolites were no longer produced and "silent" biosynthetic genes were activated instead. However, it should be kept in mind that only positive results will be publishable, and in our own experience, epigenetic modifiers often lead to growth inhibition or other phenomena that are not related to changes in the secondary metabolism of the producer strains.

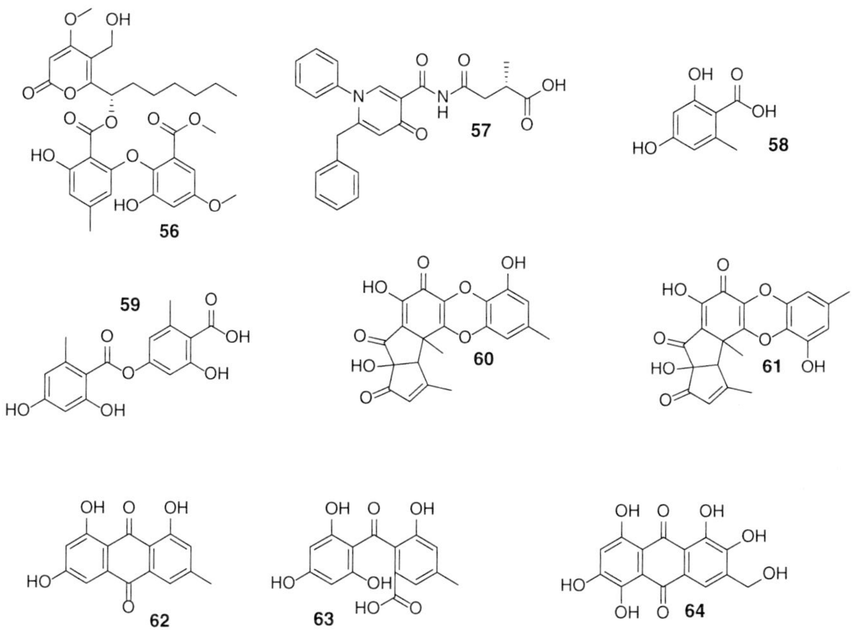

Fig. 12 Chemical structures of pestalofíciol T (**56**), nygerone A (**57**), orsellinic acid (**58**), lecanoric acid (**59**), F-9775 A (**60**), F-9775 B (**61**), emodin (**62**), monodictyphenone (**63**) and asperthecin (**64**)

More recently, different fungal organisms have been targeted, using similar approaches, which has already led to considerable success. The molecular mechanisms of these phenomena have been described in a topical review by Strauss and Reyes-Dominguez (2011), and for a concise general overview of the employed techniques, see Wiemann and Keller (2014). The epigenetic approach may be another option to fully exploit the biosynthetic potential of fungi, even though so far not resulted in the discovery of any novel antibiotic lead compounds.

Epigenetics encompass various methods resulting in the expression of BGCs which are silent under normal laboratory conditions. One of these methods which was used successfully in a recent study by Wu et al. (2016) is gene knockout. Working with *P. fici* Wu et al. were able to isolate 15 new secondary metabolites (e.g. pestaloficiol T (**56**), Fig. 12) by targeted gene disruption of the two genes PfCclA and PfHdaA. In another approach, a culture of *Aspergillus niger* was supplemented with suberoylanilide hydroxamic acid (SAHA) which resulted in the appearance of nygerone A (**57**, Fig. 12) (Henrikson et al. 2009). Furthermore, thinking about the fungi in their natural habitats and realising that they often co-inhabit ecosystems with other species brought on the idea to co-cultivate fungi with different bacteria. This was demonstrated with a co-culture of *A. nidulans* and *Streptomyces rapamycinicus* where physical contact between the two organisms resulted in the emergence of orsellinic acid (**58**), lecanoric acid (**59**) and dioxins

F9775 A (**60**) and B (**61**) (Schroeckh et al. 2009; Sanchez et al. 2010). The two dioxins were also found in a gene knockout approach in *A. nidulans* with the cclA gene as knockout. Additionally, the anthraquinone emodin (**62**) and its derivative monodictyphenone (**63**) were produced (Bok et al. 2009). Another anthraquinone derivative, asperthecin (**64**), was produced from *A. nidulans* when the sumoylation gene *sumO* was knocked out (Szewczyk et al. 2008).

Phytopathogens such as *Botrytis cinerea* and *Cochliobolus heterostrophus* have also been studied very well (Collado and Viaud 2016; Condon et al. 2013), and their results have certainly contributed a lot to our understanding of the functional genomics of the fungal metabolome. However, these fungi mainly produce phytotoxins, and the research on their secondary metabolism was to a great extent directed towards the role of these compounds as virulence factors.

However, as a plethora of fungal genome data have become available and even the bioinformatics and transcriptomics methods have been steadily improving, it should be rewarding to transfer the methodology and protocols established using a few model organism to a larger number of species comprising a broad variety of taxonomic groups. Since the aforementioned work on *Aspergillus* has been the subject on various topical reviews over the past years, we will focus here on exemplifying recent work dedicated to fungi that belong to the less studied fungal organisms.

Mao et al. (2015) have recently studied the mycoparasite *Calcarisporium arbuscula* (falsely referred to as an "endophyte", despite the fact that this fungus is associated with basidiomata of various mushrooms, rather than with a plant). In standard laboratory culture, wild-type strains of this species overproduce aurovertin-type mycotoxins that are polyketides and act as ATPase inhibitors. By inactivation of a histone deacetylase, the authors were able to suppress aurovertin biosynthesis and achieve pleiotropic activation and overexpression of an estimate of more than 75 % of the biosynthetic genes. They could finally obtain various metabolites that were not produced in the wild-type strain, including four novel natural products. Interestingly, those are derived from NRPS and terpenoid, rather than polyketide biosynthetic pathways as the "constitutional" metabolites (**65–69**, Fig. 13). These results point towards the existence of a regulation mechanism that may be involved in switching out undesired polyketides and elicitate the biosynthesis of different metabolic pathways. However, the corresponding molecular mechanisms remain to be elucidated, and this will afford the transfer of protocols that have so far only been elaborated, using model organisms such as *A. nidulans*.

Van der Molen et al. (2014) did not use the "classical" epigenetic modifiers such as inhibitors of acetylation or methylation but supplied instead the proteasome inhibitor bortezomib to the culture of a pleosporalean ascomycete that was not characterised to genus and species level. These experiments resulted in the de novo biosynthesis of a polyketide that had hitherto only been known from a plant. This example shows that it is possible to "awaken" silent BGCs also by adding elicitors other than epigenetic modifiers. A proteasome inhibitor, for instance, will in all likelihood systemically disturb the metabolism of the producing organism and therefore inadvertently expose it to stress. Like the aforementioned HDAC

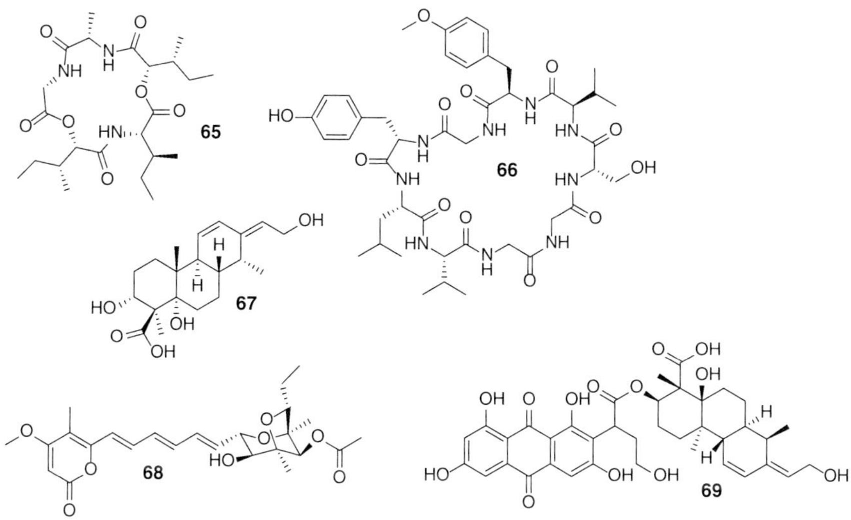

Fig. 13 Chemical structures of arbumycin (**65**), arbumelin (**66**), arbuscullic acid A (**67**), aurovertin (**68**) and arbuscullic acid B (**69**)

inhibitors, such compounds are, however, rather toxic to the producer organism and it is not always easy to find the optimal concentration that will result in overproduction of the novel metabolites. While it can be rather costly and even awkward to scale up such processes for mass production of the respective elicitors (some HDAC inhibitors are extremely expensive and have prohibitive costs of up to several thousand Euros per mg!), stable HDAC knockout strains may be more easily subjected to a more straightforward optimisation.

Almost nothing is known about the options to achieve epigenetic modification of secondary metabolite production in basidiomycetes, even though some studies have been dealing with experiments to elicitation of secondary metabolites in medicinal mushrooms by adding jasmonate, a well-known elicitor of secondary metabolite (phytoalexin) biosynthesis in plant cells (Gundlach et al. 1992). Even though a significant increase of production of the target compounds (which are cytotoxic triterpenoids) was observed, the genetic and biochemical background of this observation remains obscure. Currently, there are no data available to suggest that addition of epigenetic modifying agents to a broad cross section of fungal fermentations over time would significantly enhance the probability of discovering a new antibiotic. Thus far, it is unclear how epigenetic modifiers would be implemented in discovery libraries. The deliberate addition of bioactive metabolites to cultures would need to be carefully controlled and active extracts would need to be compared to untreated control extracts. Nor do any of these preliminary investigations address the question of scale-up of fermentation that would be dependent on addition of epigenetic agents.

In general, there are probably many ways to manipulate fungal secondary metabolite production in small scale, and such methods are another option to recover hitherto unknown compounds that were either not known from the respective (model) organisms or may even constitute novel chemical entities. The highly sensitive and sophisticated analytical methods that are now available and the miniaturised biological screening assays allow for identification of novel chemical entities on the basis of quantities of less than one milligram.

However, before straightforward development of marketable antibiotics can be envisaged, these compounds must also be made accessible in sufficient quantities to allow for a concise and straightforward optimisation, including biotechnological production, medicinal chemistry approaches and/or or total synthesis.

3 Conclusions and Outlook

This review has tried the spectrum efforts have been made to exploit fungi as potential sources of novel antibiotics and does not even take into account the large-scale screening programmes conducted during 1960-(ca.) 1995 when all Big Pharma companies were still on the hunt for natural antibiotics and invested substantial resources. Up to date, it was not possible to repeat the success of the beta-lactams and no single broad-spectrum antibiotic with market potential has ever been discovered from a fungal natural source.

However, the advent of molecular phylogeny and (phylo-)genomics, paired with highly sophisticated analytics and revolutionary developments in biotechnology as well as bioassay development and modern methods of medicinal chemistry, makes it possible to revisit these fascinating organisms, many of which certainly rely on antibiotics production in the course of their life cycle. Sadly, over the past decades, the number of taxonomists who know how to spot and discover these organisms in the field or to develop techniques to isolate specific ecological groups from hitherto unexplored habitats and know how to tell the notorious mycotoxin producers from the unique, unexplored taxa has been steadily decreasing for lack of funding. Blindly relying on new technology, however, does not always provide adequate solutions, and novel antibiotics discovery is no exception. For instance, over the past years, there have been some attempts involving substantial resources to exploit fungal secondary metabolome by using metagenomics approaches (retrieving the DNA fragments from environmental samples and aiming at heterologous expression). While this approach is workable (and has a proven success record!) in case of other biotechnological products like enzymes that are encoded by a single gene, it seems ridiculous, considering the size of the gene clusters encoding for PKS and NRPS, plus the fact that regulatory genes that switch secondary metabolite biosynthesis on and off are not normally located in these clusters.

Other modern techniques are more promising, but remain to be subjected to for further development and adaptation. Several "-OMICS" tools have first been developed based on certain model organisms, but are apparently not yet applicable even to their close relatives. Whole classes of widely untapped fungi, above all in the Basidiomycota, cannot even be handled in the laboratory, and there is also a lack of host systems for, e.g., heterologous expression of their secondary metabolites. Accordingly, more efforts should be taken to adapt the "standard" techniques that are commonly used in White Biotechnology to be able to cope with the challenge of secondary metabolite production in a broad range of genera and species. Tackling this challenge could result in a number of highly interesting scientific results, and ultimately lead to the discovery of novel antibiotics that can combat the currently arising human pathogens.

Acknowledgment We acknowledge the help of Gerald F. Bills, who spent a lot of his valuable time, critically proofreading a preliminary version of this paper, which has led to substantial improvements.

References

Allard P-M, Peresse T, Bisson J, Gindro K, Marcourt L, Pham VC, Roussi F, Litaudon M, Wolfender J-L (2016) Integration of molecular networking and in-silico MS/MS fragmentation for natural products dereplication. Anal Chem 88:3317–3323

Aly AH, Debbab A, Kjer J, Proksch P (2010) Fungal endophytes from higher plants: a prolific source of phytochemicals and other bioactive natural products. Fungal Divers 41:1–16

Anderson JR, Edwards RL, Poyser JP, Whalley AJS (1988) Metabolites of the higher fungi. Part 23. The punctaporonins. Novel bi-, tri-, and tetra-cyclic sesquiterpenes related to caryophyllene, from the fungus *Poronia punctata* (Linnaeus: Fries) Fries. J Chem Soc Perkin Trans 1 823–831

Bentley R (2000) Mycophenolic acid: a one hundred year odyssey from antibiotic to Immunosuppressant. Chem Rev 100:3801–3826

Bills GF, Christensen M, Powell M, Thorn G (2004) Saprobic fungi. In: Mueller GM, Bills GF, Foster M (eds) Biodiversity of fungi. Inventory and monitoring methods. Elsevier Academic Press, Burlington, pp 271–302

Bills GF, Platas G, Fillola A, Jiménez MR, Bur-Zimmerman J, Tormo JR, Peláez F (2008) Enhancement of antibiotic and secondary metabolite detection from filamentous fungi by growth on nutritional arrays. J Appl Bacteriol 104:1644–1658

Bills GF, Platas G, Overy DP, Collado J, Fillola A, Jiménez MR, Martín J, del Val AG, Vicente F, Tormo JR, Peláez F, Calati K, Harris G, Parish C, Xu D, Roemer T (2009a) Discovery of the parnafungins, antifungal metabolites that inhibit mRNA polyadenylation, from the *Fusarium larvarum* complex and other hypocrealean fungi. Mycologia 101:449–472

Bills GF, Martín J, Collado J, Platas G, Overy D, Tormo JR, Vicente F, Verkleij G, Crous P (2009b) Measuring the distribution and diversity of antibiosis and secondary metabolites in the filamentous fungi. SIMB News 59:133–147

Bills GF, González-Menéndez V, Martín J, Platas G, Fournier J, Peršoh D, Stadler M (2012a) *Hypoxylon pulicicidum* sp. nov. (Ascomycota, Xylariales), a pantropical Insecticide-producing endophyte. PLoS ONE 7(10):e46687

Bills GF, Dombrowski AW, Goetz MA (2012b) The "FERMEX" method for metabolite-enriched fungal extracts. Methods Mol Biol 944:79–96

Bills GF, Gloer JB, An Z (2013) Coprophilous fungi: antibiotic discovery and functions in an underexplored arena of microbial defensive mutualism. Curr Opin Microbiol 16:549–565

Bills GF, Yue Q, Chen L, Li Y, An Z, Frisvad JC (2015). *Aspergillus mulundensis* sp. nov., a new species for the fungus producing the antifungal echinocandin lipopeptides, mulundocandins. J Antibiot. doi:10.1038/ja.2015.105

Bitzer J, Köpcke B, Stadler M, Hellwig V, Ju YM, Seip S, Henkel T (2007) Accelerated dereplication of natural products, supported by reference libraries. Chimia 51:332–338

Bitzer J, Læssøe T, Fournier J, Kummer V, Decock C, Tichy HV, Piepenbring M, Peršoh D, Stadler M (2008) Affinities of *Phylacia* and the daldinoid Xylariaceae, inferred from chemotypes of cultures and ribosomal DNA sequences. Mycol Res 112:251–270

Brakhage AA, Schroeckh V (2011) Fungal secondary metabolites—strategies to activate silent gene clusters. Fungal Genet Biol 48:15–22

Bode HB, Bethe B, Höfs R, Zeeck A (2002) Big effects from small changes: possible ways to explore nature's chemical diversity. ChemBioChem 3:619–627

Bok JW, Chiang YM, Szewczyk E, Reyes-Dominguez Y, Davidson AD, Sanchez JF, Lo HC, Watanabe K, Strauss J, Oakley BR, Wang CCC, Keller NP (2009) Chromatin-level regulation of biosynthetic gene clusters. Nat Chem Biol 5:462–464

Bok JW, Hoffmeister D, Maggio-Hall LA, Murillo R, Glasner JD, Keller NP (2006) Genomic mining for *Aspergillus* natural products. Chem Biol 13:31–37

Bushley KE, Turgeon BG (2010) Phylogenomics reveals subfamilies of fungal nonribosomal peptide synthetases and their evolutionary relationships. BMC Evol Biol 10:26

Butler MS, Robertson AAB, Cooper MA (2014) Natural product and natural product derived drugs in clinical trials. Nat Prod Rep 31:1612–1661

Chomnunti P, Hongsanan S, Aguirre-Hudson B, Qing T, Peršoh D, Dhami MK, Alias SA, Xu J, Liu J, Stadler M, Hyde KD (2014) The sooty moulds. Fungal Divers 66:1–36

Clardy J, Fischbach M, Currie C (2009) The natural history of antibiotics. Curr Biol 19:R437–R441

Collado J, Platas G, Paulus B, Bills GF (2007) High-throughput culturing of fungi from plant litter by a dilution-to-extinction technique. FEMS Microbiol Ecol 60:521–533

Collado IG, Viaud M (2016) Secondary metabolism in *Botrytis cinerea*: combining genomic and metabolomic approaches. In *Botrytis*—the fungus, the pathogen and its management in agricultural Systems. Springer, Berlin, pp 291–313

Condon BJ, Leng Y, Wu D, Bushley KE, Ohm RA, Otillar R, Martin J, Schackwitz W, Grimwood J, MohdZainudin N, Xue C, Wang R, Dhillon B, Tu ZJ, Steffenson BJ, Salamov A, Sun H, Lowry S, LaButti K, Han J, Copeland A, Lindquist E, Lucas S, Barry K, Schmutz J, Baker S, Grigoriev IV, Zhong S, Turgeon BG (2013) Comparative genome structure, secondary metabolite and effector coding capacity across *Cochliobolus* pathogens. PLoS Genet 9(1):e1003233. doi:10.1371/journal.pgen.233

Daferner M, Anke T, Sterner O (2002) Zopfiellamides A and B, antimicrobial pyrrolidinone derivatives from the marine fungus *Zopfiella latipes*. Tetrahedron 58:7781–7784

Davies J (2013) Specialized microbial metabolites: functions and origins. J Antibiot 66:361–364

de la Cruz M, Martín J, González-Menéndez V, Pérez-Victoria I, Moreno C, Tormo JR, El Aouad N, Guarro J, Vicente F, Reyes F, Bills GF (2012) Chemical and physical modulation of antibiotic activity in *Emericella* species. Chem Biodivers 9:1095–1113

Debbab A, Aly AH, Proksch P (2012) Endophytes and associated marine derived fungi—ecological and chemical perspectives. Fungal Divers 57:45–83

De Silva DD, Rapior S, Sudarman E, Stadler M, Xu J, Aisyah Alias S, Hyde KD (2013) Bioactive metabolites from macrofungi: ethnopharmacology, biological activities and chemistry. Fungal Divers 61:1–40

Del-Cid A, Gil-Durán C, Vaca I, Rojas-Aedo JF, García-Rico RO, Levicán G, Chávez R (2016) Identification and functional analysis of the mycophenolic acid gene cluster of *Penicillium roqueforti*. PloS one 11(1). doi:10.1371/journal.pone.0147047

Donald RGK, Skwish S, Forsyth RA, Anderson JW, Zhong T, Burns C, Lee S, Meng X, LoCastro L, Jarantow LW, Martin J, Lee SH, Taylor I, Robbins D, Malone C, Wang LS, Zamudio CS, Youngman PJ, Phillips JW (2009) A *Staphylococcus aureus* fitness test platform for mechanism-based profiling of antibacterial compounds. Chem Biol 16:826–836

Donzelli BGG, Krasnoff SB (2016) Molecular genetics of secondary chemistry in *Metarhizium* fungi. Adv Genet 94:365–436

Dreyfuss MM, Chapela IH (1994) Potential of fungi in the discovery of novel, low-molecular weight pharmaceuticals. Biotechnol (Reading, Mass.) 26: 49–80

Engels B, Heinig U, Grothe T, Stadler M, Jennewein S (2011) Cloning and characterization of an *Armillaria gallica* cDNA encoding protoilludene synthase, which catalyzes the first committed step in the synthesis of antimicrobial melleolides. J Biol Chem 286:6871–6878

Erhardt M (2016) Strategies to block bacterial pathogenesis by interference with motility and chemotaxis. Curr Top Microbiol Immunol, in press. doi:10.1007/82_2016_493

Fernandes P, Pereira D (2011) Efforts to support the development of fusidic acid in the United States. Clin Infect Dis 52:S542–S546

Fernandes P (2016) Fusidic acid: a bacterial elongation factor inhibitor for the oral treatment of acute and chronic staphylococcal infections. Cold Spring Harb Perspect Med 6. doi:10.1101/cshperspect.a025437

Fisch KM, Gillaspy AF, Gipson M, Henrikson JC, Hoover AR, Jackson L, Najar FZ, Wägele H, Cichewicz RH (2009) Chemical induction of silent biosynthetic pathway transcription in *Aspergillus niger*. J Ind Microbiol Biotechnol 36:1199–1213

Frisvad JC, Andersen B, Thrane U (2008) The use of secondary metabolite profiling in fungal taxonomy. Mycol Res 112:231–240

Gao SS, Li XM, Li CH, Proksch P, Wang BG (2011) Penicisteroids A and B, antifungal and cytotoxic polyoxygenated steroids from the marine alga-derived endophytic fungus *Penicillium chrysogenum* QEN-24 S. Bioorg Med Chem Lett 21:2894–2897

Gibson DM, Donzelli BG, Krasnoff SB, Keyhani NO (2014) Discovering the secondary metabolite potential encoded within entomopathogenic fungi. Nat Prod Rep 31:1287–1305

Gill M, Steglich W (1987) Pigments of fungi (Macromycetes). Prog Chem Org Nat Prod 51:1–317

Goetz MA, Zhang C, Zink DL, Arocho M, Vicente F, Bills GF, Polishook J, Dorso K, Onishi R, Gill C, Hickey E, Lee S, Ball R, Skwish S, Donald RG, Phillips JW, Singh SB (2010) Coelomycin, a highly substituted 2,6-dioxo-pyrazine fungal metabolite antibacterial agent discovered by *Staphylococcus aureus* fitness test profiling. J Antibiot 63:512–518

Gundlach H, Müller MJ, Kutchan TM, Zenk MH (1992) Jasmonic acid is a signal transducer in elicitor-induced plant cell cultures. PNAS 89:2389–2393

Halecker S, Surup F, Kuhnert E, Mohr KI, Brock NL, Dickschat JS, Junker C, Schulz B, Stadler M (2014) Hymenosetin, a 3-decalinoyltetramic acid antibiotic from cultures of the ash dieback pathogen, *Hymenoscyphus pseudoalbidus*. Phytochemistry 100:86–91

Hansen BG, Genee HJ, Kaas CS, Nielsen JB, Regueira TB, Mortensen UH, Frisvad JC, Patil KR (2011a) A new class of IMP dehydrogenase with a role in self-resistance of mycophenolic acid producing fungi. BMC Microbiol 11:1–7

Hansen BG, Salomonsen B, Nielsen MT, Nielsen JB, Hansen NB, Nielsen KF, Regueira TB, Nielsen J, Patil KR, Mortensen UH (2011b) Versatile enzyme expression and characterization system for *Aspergillus nidulans*, with the *Penicillium brevicompactum* polyketide synthase gene from the mycophenolic acid gene cluster as a test case. Appl Environ Microbiol 77:3044–3051

Hawksworth DL (2012) Global species numbers of fungi: are tropical studies and molecular approaches contributing to a more robust estimate? Biodivers Conserv 21:2425–2433

Heinig U, Scholz S, Jennewein S (2013) Getting to the bottom of taxol biosynthesis in fungi. Fungal Divers 60:161–170

Hellwig V, Grothe T, Mayer-Bartschmid A, Endermann R, Geschke FU, Henkel T, Stadler M (2002) Altersetin, a new antibiotic from cultures of endophytic *Alternaria* spp. Taxonomy, fermentation, isolation, structure elucidation, and biological activities. J Antibiot 55:881–892

Herath K, Jayasuriya H, Zink DL, Sigmund J, Vicente F, de la Cruz M, Basilio A, Bills GF, Polishook JD, Donald R, Phillips J, Goetz M, Singh SB (2012) Isolation, structure elucidation, and antibacterial activity of methiosetin, a tetramic acid from a tropical sooty mold (*Capnodium* sp.). J Nat Prod 75:420–424

Henrikson JC, Hoover AR, Joyner PM, Cichewicz RH (2009) A chemical epigenetics approach for engineering the in situ biosynthesis of cryptic natural products from *Aspergillus niger*. Org Biomol Chem 7:435–438

Houbraken J, Frisvad JC, Samson RA (2011) Fleming's penicillin producing strain is not *Penicillium chrysogenum* but *P. rubens*. IMA Fungus 2:87–95

Houbraken J, Frisvad JC, Seifert KA, Overy DP, Tuthill DE, Valdez JG, Samson RA (2012) New penicillin-producing *Penicillium* species and an overview of section *Chrysogena*. Persoonia 29:78–100

Igarashi Y, Gohda F, Kadoshima T, Fukuda T, Hanafusa T, Shojima A, Nakayama J, Bills GF, Peterson S (2015) Avellanin C, an inhibitor of quorum-sensing signaling in *Staphylococcus aureus*, from *Hamigera ingelheimensis*. J Antibiot 68:707–710

Isaka M, Kittakoop P, Kirtikara K, Hywel-Jones NL, Thebtaranonth Y (2005) Bioactive substances from insect pathogenic fungi. Acc Chem Res 38:813–823

Jayanetti DR, Yue Q, Bills GF, Gloer JB (2015) Hypocoprins A–C: new sesquiterpenoids from the coprophilous fungus *Hypocopra rostrata*. J Nat Prod 78:396–401

Katzer W, Blackburn M, Charman K, Martin S, Penn J, Wrigley S (2001) Scale-up of filamentous organisms from tubes and shake-flasks into stirred vessels. Biochem Eng J 7:127–134

Kauhl U, Andernach L, Weck S, Sandjo LP, Jacob S, Thines E, Opatz T (2016) Total synthesis of (−)-hymenosetin. J Org Chem 81:215–228

Keller NP, Turner G, Bennett JW (2005) Fungal secondary metabolism—from biochemistry to genomics. Nat Rev Microbiol 3:937–947

Kilaru S, Collins CM, Hartley AJ, Bailey AM, Foster GD (2009) Establishing molecular tools for genetic manipulation of the pleuromutilin-producing fungus *Clitopilus passeckerianus*. Appl Environ Microbiol 75:7196–7204

King AM, Reid-Yu SA, Wang W, King DT, De Pascale G, Strynadka NC, Walsh TR, Coombes BK, Wright GD (2014) Aspergillomarasmine A overcomes metallo-β-lactamase antibiotic resistance. Nature 510:503–506

Kirst HA (2013) Developing new antibacterials through natural product research. Expert Opin Drug Discov 8:479–493

Kuhnert E, Fournier J, Peršoh D, Luangsa-ard JJD, Stadler M (2014a) New *Hypoxylon* species from Martinique and new evidence on the molecular phylogeny of *Hypoxylon* based on ITS rDNA and β-tubulin data. Fungal Divers 64:181–203

Kuhnert E, Heitkämper S, Fournier J, Surup F, Stadler M (2014b) Hypoxyvermelhotins A–C, novel pigments from *Hypoxylon lechatii* sp. nov. Fungal Biol 118:242–252

Kuhnert E, Surup F, Sir EB, Lambert C, Hyde KD, Hladki AI, Romero AI, Stadler M (2015) Lenormandins A—G, new azaphilones from *Hypoxylon lenormandii* and *Hypoxylon jaklitschii* sp. nov., recognised by chemotaxonomic data. Fungal Divers 71:165–184

Kusari S, Hertweck C, Spiteller M (2012) Chemical ecology of endophytic fungi: origins of secondary metabolites. Chem Biol 19:792–798

Kusnick C, Jansen R, Liberra K, Lindequist U (2002) Ascochital, a new metabolite from the marine ascomycete *Kirschsteiniothelia maritima*. Pharmazie 57:510–512

Lang G, Mayhudin NA, Mitova MI, Sun L, van der Sar S, Blunt JW, Laatsch H, Munro MH (2008) Evolving trends in the dereplication of natural product extracts: new methodology for rapid, small-scale investigation of natural product extracts. J Nat Prod 71:1595–1599

Lackner G, Misiek M, Braesel J, Hoffmeister D (2012) Genome mining reveals the evolutionary origin and biosynthetic potential of basidiomycete polyketide synthases. Fungal Genet Biol 49:996–1003

Landwehr W, Wolf C, Wink J (2016) Bacterial sources for new antibiotics (Concurrent contribution)

Larsen TO, Smedsgaard J, Nielsen KF, Hansen ME, Frisvad JC (2005) Phenotypic taxonomy and metabolite profiling in microbial drug discovery. Nat Prod Rep 22:672–695

Lax E (2005) The mold in Dr. Florey's coat: the story of the penicillin miracle. Henry Holt and Co. LLC, New York

Lepak AJ, Marchillo K, Andres DR (2015) Pharmacodynamic target evaluation of a novel oral glucan synthase inhibitor, SCY-078 (MK-3118), using an in vivo murine invasive candidiasis model. Antimicrob Agents Chemother 59:1265–1272

Li D, Xu Y, Shao CL, Yang RY, Zheng CJ, Chen YY, Fu XM, Qian PY, She ZG, de Voogd NJ, Wang CY (2012) Antibacterial bisabolane-type sesquiterpenoids from the sponge-derived fungus *Aspergillus* sp. Mar Drugs 10:234–241

Li YF, Tsai KJS, Harvey CJB, Li JJ, Ary BE, Berlew EE, Boehman BL, Findley DM, Friant AG, Gardner CA, Gould MP, Ha JH, Lilley BK, McKinstry EL, Nawal S, Parry RC, Rothschild KW, Silbert SD, Tentilucci MD, Thurston AM, Wai RB, Yoon Y, Aiyar RS, Medema MH, Hillenmeyer ME, Charkoudian LK (2016) Comprehensive curation and analysis of fungal biosynthetic gene clusters of published natural products. Fungal Genet Biol 89:18–28

Liberra K, Jansen R, Lindequist U (1998) Corollosporine, a new phthalide derivative from the marine fungus *Corollospora maritima*. Pharmazie 53:578–581

Mao XM, Xu W, Li D, Yin WB, Chooi YH, Li YQ, Tang Yi HuY (2015) Epigenetic genome mining of an endophytic fungus leads to the pleiotropic biosynthesis of natural products. Angew Chem Int Ed 127:7702–7706

Mitterbauer R, Specht T (2011) Polypeptide having diterpene synthase activity. US Patent US20130171701 A1

Mohr KI (2016) Introduction to antibiotics research—from the early beginning to the golden time of antibiotic research. Curr Top Microbiol Immunol (Concurrent contribution)

Mühlen S, Dersch P (2016) Anti-virulence strategies to target bacterial infections. Curr Top Microbiol Immunol, in press. doi:10.1007/82_2015_490

Mudalungu CM, Richter C, Wittstein K, Ali Abdalla M, Matasyoh JC, Stadler M, Süssmuth RD (2016) Laxitextines A and B, cyathane xylosides from the tropical fungus *Laxitextum incrustatum*. J Nat Prod 79:894–898

Newman DJ, Cragg GM (2016) Natural products as sources of new drugs from 1981 to 2014. J Nat Prod 79:629–661

Ondeyka JG, Smith SK, Zink DL, Vicente F, Basilio A, Bills GF, Polishook JD, Garlisi C, Mcguinness D, Smith E, Qiu H, Gill CJ, Donald RG, Phillips JW, Goetz MA, Singh SB (2014) Isolation, structure elucidation and antibacterial activity of a new tetramic acid, ascosetin. J Antibiot 67:527–531

Osterhage C, Kaminsky R, König GM, Wright AD (2000) Ascosalipyrrolidinone A, an antimicrobial alkaloid from the obligate marine fungus *Ascochyta salicorniae*. J Org Chem 65:6412–6417

Overy DP, Bayman P, Kerr RG, Bills GF (2014) An assessment of natural product discovery from marine (sensu strictu) and marine-derived fungi. Mycology 5:145–167

Ozcengiz G, Demain AL (2013) Recent advances in the biosynthesis of penicillins, cephalosporins and clavams and its regulation. Biotechnol Adv 31:287–311

Pažoutová S, Follert S, Bitzer J, Keck M, Surup F, Šrůtka P, Holuša J, Stadler M (2013) A new endophytic insect-associated *Daldinia* species, recognised from a comparison of secondary metabolite profiles and molecular phylogeny. Fungal Divers 60:107–123

Peláez F (2006) The historical delivery of antibiotics from microbial natural products—can history repeat? Biochem Pharmacol 71:981–990

Petrini O, Sieber TN, Toti L, Viret O (1993) Ecology, metabolite production, and substrate utilization in endophytic fungi. Nat Toxins 1:185–196

Qin S, Krohn K, Hussain H, Schulz B, Draeger S (2011) Pestalotheols E–H: antimicrobial metabolites from an endophytic fungus isolated from the tree *Arbutus unedo*. Eur J Org Chem 5163–5166

Quin MB, Flynn CM, Schmidt-Dannert C (2014) Traversing the fungal terpenome. Nat Prod Rep 10:1449–1473

Rambold G, Stadler M, Begerow D (2013) Mycology should be recognized as a field in biology at eye level with other major disciplines—a memorandum. Mycol Prog 12:455–463

Raper KB (1965) Charles Thom 1872–1956—biographical memoir. In: Raper KB (ed) National Academy of Science, Washington, DC, pp. 307–344

Richter C, Wittstein K, Kirk PM, Stadler M (2015) An assessment of the taxonomy and chemotaxonomy of *Ganoderma*. Fungal Divers 71:1–15

Ridderbusch DC1, Weber RW, Anke T, Sterner O (2004) Tulasnein and podospirone from the coprophilous xylariaceous fungus *Podosordaria tulasnei*. Z Naturforsch 59c:379–383

Röhrich CR, Jaklitsch WM, Voglmayr H, Iversen A, Vilcinskas A, Nielsen KF, Thrane U, von Döhren H, Brückner H, Degenkolb T (2014) Front line defenders of the ecological niche! Screening the structural diversity of peptaibiotics from saprotrophic and fungicolous *Trichoderma*/*Hypocrea* species. Fungal Divers 69:117–146

Rohlfs M, Churchill AC (2011) Fungal secondary metabolites as modulators of interactions with insects and other arthropods. Fungal Genet Biol 48:23–34

Running JA, Bansal K (2016) Oxygen transfer rates in shaken culture vessels from Fernbach flasks to microtiter plates. Biotechnol Bioeng 9999:1–7

Sanchez JF, Chiang YM, Szewczyk E, Davidson AD, Ahuja M, Oakley CE, Bok JW, Keller NP, Oakley BR, Wang CCC (2010) Molecular genetic analysis of the orsellinic acid/F9775 gene cluster of *Aspergillus nidulans*. Mol Biol Syst 6:587–593

Sanchez JF, Somoza AD, Keller NP, Wang CC (2012) Advances in *Aspergillus* secondary metabolite research in the post-genomic era. Nat Prod Rep 29:351–371

Schardl CL, Young CA, Hesse U, Amyotte SG, Andreeva K, Calie PJ, Fleetwood DJ, Haws DC, Moore N, Oeser B, Panaccione DG, Schweri KK, Voisey CR, Farman ML, Jaromczyk JW, Roe BA, O'Sullivan DM, Scott B, Tudzynski P, An Z, Arnaoudova EG, Bullock CT, Charlton ND, Chen L, Cox M, Dinkins RD, Florea S, Glenn AE, Gordon A, Güldener U, Harris DR, Hollin W, Jaromczyk J, Johnson RD, Khan AK, Leistner E, Leuchtmann A, Li C, Liu J, Liu J, Liu M, Mace W, Machado C, Nagabhyru P, Pan J, Schmid J, Sugawara K, Steiner U, Takach JE, Tanaka E, Webb JS, Wilson EV, Wiseman JL, Yoshida R, Zeng Z (2013) Plant-symbiotic fungi as chemical engineers: multi-genome analysis of the Clavicipitaceae reveals dynamics of alkaloid loci. PLOS Genet 9:e1003323

Scherkenbeck J, Jeschke P, Harder A (2002) PF1022A and related cyclodepsipeptides—a novel class of anthelmintics. Curr Topics Med Chem 2:759–777

Scherlach K, Hertweck J (2006) Discovery of aspoquinolones A–D, prenylated quinoline-2-ones from *Aspergillus nidulans*, motivated by genome mining. Org Biomol Chem 4:3517–3520

Schmidt-Dannert C (2015a) Biosynthesis of terpenoid natural products in fungi. Adv Biochem Eng Biotechnol 148:19–61

Schmidt-Dannert C (2015b) NextGen microbial natural products discovery. Microb Biotechnol 8:26–28

Schroeckh V, Scherlach K, Nutzmann HW, Shelest E, Schmidt-Heck W, Schuemann J, Martin K, Hertweck C, Brakhage AA (2009) Intimate bacterial-fungal interaction triggers biosynthesis of archetypal polyketides in *Aspergillus nidulans*. Proc Natl Acad Sci USA 106:14558–14563

Schüffler A, Anke T (2014) Fungal natural products in research and development. Nat Prod Rep 31:1425–1448

Schulz B, Boyle C (2005) The endophytic continuum. Mycol Res 109:661–686

Schulz B, Boyle C, Draeger S, Römmert AK, Krohn K (2002) Endophytic fungi: a source of novel biologically active secondary metabolites. Mycol Res 106:996–1004

Schwab EK, Bok JW, Tribus M, Galehr J, Graessle S, Keller NP (2007) Histone deacetylase activity regulates chemical diversity in *Aspergillus*. Eukaryot Cell 6:1656–1664

Sica VP, Raja HA, El-Elimat T, Kertesz V, Van Berkel GJ, Pearce CJ, Oberlies NH (2015) Dereplicating and spatial mapping of secondary metabolites from fungal cultures in situ. J Nat Prod 78:1926–1936

Stadler M (2011) Importance of secondary metabolites in the Xylariaceae as parameters for assessment of their taxonomy, phylogeny, and functional biodiversity. Curr Res Environ Appl Mycol 1:75–133

Stadler M, Hellwig V (2004) PCR-based data and secondary metabolites as chemotaxonomic markers in high-throughput screening for bioactive compounds from fungi. In: An Z (ed) Handbook of industrial mycology. Marcel Dekker, New York, pp 269–307

Stadler M, Quang DN, Tomita A, Hashimoto T, Asakawa Y (2006) Production of bioactive metabolites during stromatal ontogeny of *Hypoxylon fragiforme*. Mycol Res 110:811–820

Stadler M, Fournier J, Gardt S, Peršoh D (2010) The phylogenetic position of *Rhopalostroma* as inferred from a polythetic approach. Persoonia 25:11–21

Stadler M, Læssøe T, Fournier J, Decock C, Schmieschek B, Tichy HV, Peršoh D (2014) A polyphasic taxonomy of *Daldinia* (Xylariaceae). Stud Mycol 77:1–143

Stadler M, Hoffmeister D (2015) Fungal natural products—the mushroom perspective. Front Microbiol 6:127. doi:10.3389/fmicb.2015.00127

Strachan CR, Davies J (2016) Antibiotics and evolution: food for thought. J Ind Microbiol Biotechnol 43:149–153

Strauss J, Reyes-Dominguez Y (2011) Regulation of secondary metabolism by chromatin structure and epigenetic codes. Fungal Genet Biol 48:62–69

Surup F, Kuhnert E, Lehmann E, Heitkämper S, Hyde KD, Fournier J, Stadler M (2014) Sporothriolide derivatives as chemotaxonomic markers for *Hypoxylon monticulosum*. Mycol Int J Fungal Biol 5:110–119

Surup F, Thongbai B, Kuhnert E, Hyde KD, Stadler M (2015) Deconins A–E—unprecedented cuparenic and mevalonic or propionic acid conjugates from the basidomycete *Deconica* sp. 471. J Nat Prod 78:934–938

Sun ZL, Zhang M, Zhang J-F, Feng J (2011) Antifungal and cytotoxic activities of the secondary metabolites from endophytic fungus *Massrison* sp. Phytomedicine 18:859–862

Szewczyk E, Chiang YM, Oakley CE, Davidson AD, Wang CC, Oakley BR (2008) Identification and characterization of the asperthecin gene cluster of *Aspergillus nidulans*. Appl Environ Microbiol 74:7607–7612

Tedersoo L, Bahram M, Põlme S, Kõljalg U, Yorou NS, Wijesundera Ruiz LV, Vasco-Palacios AM, Thu PQ, Suija A, Smith ME, Sharp C, Saluveer E, Saitta A, Rosas M, Riit T, Ratkowski D, Pritsch K, Põldmaa K, Piepenbring M, Phosri C, Peterson M, Parts K, Pärtel K, Otsing E, Nouhra E, Njouonkou AL, Nilsson RH, Morgado LN, Mayor J, May TW, Majuakim L, Lodge DJ, Lee SS, Larsson K-H, Kohout P, Hosaka K, Hiiesalu I, Henkel TW, Harend H, Guo L, Greslebin A, Grelet G, Geml J, Gates G, Dunstan W, Dunk C, Drenkhan R, Dearnaley J, de Kesel A, Dang T, Chen X, Buegger F, Brearley FQ, Bonito G, Anslan S, Abell S, Abarenkov K (2014) Global diversity and geography of soil fungi. Science 346:1256688

Thongbai B, Rapior S, Wittstein K, Hyde KD, Stadler M (2015) *Hericium erinaceus*, an amazing medicinal mushroom. Mycol Prog 14:1–23

Thongbai B, Surup F, Mohr K, Kuhnert E, Hyde KD, Stadler M (2013) Gymnopalynes A and B—chloropropynyl-isocoumarin antibiotics from cultures of the basidiomycete *Gymnopus* sp. J Nat Prod 76:2141–2144

Van der Molen KM, Darveaux BA, Chen WL, Swanson SM, Pearce CJ, Oberlies NH (2014) Epigenetic manipulation of a filamentous fungus by the proteasome-inhibitor bortezomib induces the production of an additional secondary metabolite. RSC Adv 4:18329–18335

Varughese T, Rios N, Higginbotham S, Arnold AE, Coley PD, Kursar TA, Gerwick WH, Rios LC (2012) Antifungal depsidone metabolites from *Cordyceps dipterigena*, an endophytic fungus antagonistic to the phytopathogen *Gibberella fujikuroi*. Tetrahedron Lett 53:1624–1626

Vicente F, Basilio A, Platas G, Collado J, Bills GF, Del Val AG, Martín J, Tormo JR, Harris GH, Zink DL, Justice M, Kahn JN, Peláez F (2009) Distribution of the antifungal agents sordarins across filamentous fungi. Mycol Res 113:754–770

Vining LC (1990) Functions of secondary metabolites. Ann Rev Microbiol 44:395–427

Wang JB, St. Leger RJ, Wang C (2016) Advances in genomics of entomopathogenic fungi. Adv Genet 94:67–105 (Academic Press)

Wang X, Zhang X, Liu L, Xiang M, Wang W, Sun X, Che Y, Guo L, Liu G, Guo L, Wang C, Yin WB, Stadler M, Zhang X, Liu X (2015) Genomic and transcriptomic analysis of the endophytic fungus *Pestalotiopsis fici* reveals its lifestyle and high potential for synthesis of natural products. BMC Genom 16:28. doi:10.1186/s12864-014-1190-9

Wasson RG (1968) Soma—divine mushroom of immortality; Ethno-mycological Studie No. 1. Harcourt Brace Jovanovich, New York

Wasson RG, Hofmann A, Ruck CAP (2008) In: Forte R (ed) The road to Eleusis—unveiling the secret of the mysteries, 13th edn. North Atlantic Books, Berkeley

Weijman ACM, van Eijk GW (1982) Use of a single infusion bottle to culture and prepare potentially hazardous microorganisms for biochemical analysis. Culturing on solid media. Antonie Van Leeuwenhoek 48:457–459

Whyte AC, Gloer JB, Wicklow DT, Dowd PF (1996) Sclerotiamide: a new member of the paraherquamide class with potent antiinsectan activity from the sclerotia of *Aspergillus sclerotiorum*. J Nat Prod 59:1093–1095

Wiemann P, Keller NP (2014) Strategies for mining fungal natural products. J Ind Microbiol Biotechnol 41:301–313

Williams RB, Henrikson JC, Hoover AR, Lee AE, Cichewicz RA (2008) Epigenetic remodeling of the fungal secondary metabolome. Org Biomol Chem 6:1895–1897

Wolfender JL, Marti G, Queiroz EF (2010) Advances in techniques for profiling crude extracts and for the rapid identification of natural products: Dereplication, quality control and metabolomics. Curr Org Chem 14:1808–1832

Wu G, Zhou H, Zhang P, Wang X, Li W, Zhang W, Liu X, Liu H-W, Keller NP, An Z, Yin W-B (2016) Polyketide production of pestaloficiols and macrolide ficiolides revealed by manipulations of epigenetic regulators in an endophytic fungus. Org Lett 18:1832–1835

Yarbrough GC, Taylor DP, Rowlands RT, Crawford MS, Lasure LL (1993) Screening microbial metabolites for new drugs – theoretical and practical issues. J Antibiot 46:535–544

Yue Q, Chen L, Zhang X, Li K, Sun J, Liu X, An Z, Bills GF (2015) Evolution of chemical diversity in echinocandin lipopeptide antifungal metabolites. Eukaryot Cell 14:698–718

Strategies for the Discovery and Development of New Antibiotics from Natural Products: Three Case Studies

Jennifer Herrmann, Tadeja Lukežič, Angela Kling, Sascha Baumann, Stephan Hüttel, Hrvoje Petković and Rolf Müller

Abstract Natural products continue to be a predominant source for new anti-infective agents. Research at the Helmholtz Institute for Pharmaceutical Research Saarland (HIPS) and the Helmholtz Centre for Infection Research (HZI) is dedicated to the development of new lead structures against infectious diseases and, in particular, new antibiotics against hard-to-treat and multidrug-resistant bacterial pathogens. In this chapter, we introduce some of the concepts currently being employed in the field of antibiotic discovery. In particular, we will exemplarily illustrate three approaches: (1) Current sources for novel compounds are mainly soil-dwelling bacteria. In the course of our antimicrobial discovery program, a biodiverse collection of myxobacterial strains has been established and screened for antibiotic activities. Based on this effort, one successful example is presented in this chapter: Antibacterial cystobactamids were discovered and their molecular target, the DNA gyrase, was identified soon after the analysis of myxobacterial self-resistance making use of the information found in the respective biosynthesis gene cluster. (2) Besides our focus on novel natural products, we also apply strategies to further develop either neglected drugs or widely used antibiotics for which development of resistance in the clinical setting is an issue: Antimycobacterial griselimycins were first described in the 1960s but their

J. Herrmann · T. Lukežič · A. Kling · S. Baumann · R. Müller (✉)
Department of Microbial Natural Products, Helmholtz Centre for Infection Research (HZI) and Pharmaceutical Biotechnology, Helmholtz Institute for Pharmaceutical Research Saarland (HIPS), Saarland University, Campus E8 1, 66123 Saarbrücken, Germany
e-mail: rolf.mueller@helmholtz-hzi.de

J. Herrmann · T. Lukežič · A. Kling · S. Baumann · S. Hüttel · R. Müller
German Centre for Infection Research (DZIF), Partner Site Hannover-Braunschweig, Braunschweig, Germany

S. Hüttel
Department Microbial Drugs, Helmholtz Centre for Infection Research (HZI), Inhoffenstraße 7, 38124 Braunschweig, Germany

H. Petković
Department of Food Science and Technology, Biotechnical Faculty, University of Ljubljana, Jamnikarjeva 101, 1000 Ljubljana, Slovenia

Current Topics in Microbiology and Immunology (2016) 398: 339–364
DOI 10.1007/82_2016_498

Published Online: 26 August 2016

development and use in tuberculosis therapy was not further pursued. We show how a griselimycin derivative with improved pharmacokinetic properties and enhanced potency against *Mycobacterium tuberculosis* revealed and validated a novel target for antibacterial therapy, the DNA sliding clamp. (3) In a third approach, biosynthetic engineering was used to modify and optimize natural products regarding their pharmaceutical properties and their production scale: The atypical tetracycline chelocardin is a natural product scaffold that was modified to yield a more potent derivative exhibiting activity against multidrug-resistant pathogens. This was achieved by genetic engineering of the producer strain and the resulting compound is now subject to further optimization by medicinal chemistry approaches.

Contents

1 Introduction

Decades after the golden era of antibiotic discovery, it is becoming evident that microbes are gaining advantage again: Antibiotic resistance is rapidly emerging and spreading among clinically relevant pathogens. A few years after a new antibiotic is discovered and introduced into the clinic, resistant pathogenic strains evolve. In particular, the natural ability of bacteria to spread resistance determinants through horizontal gene transfer is constantly challenging us to develop new drugs to overcome their resistance (Arias and Murray 2009). To tackle this threat, promise lies in the continued and improved discovery of novel natural products, which have yielded most of the antibiotics that are in use today.

In contrast to earlier assumptions, the repertoire of natural product-based anti-infective molecules is by far not exhausted yet, as the capacity of single organisms for natural product biosynthesis has been largely underestimated (Wenzel and Müller 2009). This goes hand in hand with the discovery and "culturability" of new microorganism families capable of synthesizing novel bioactive secondary metabolites (Challinor and Bode 2015; Garcia et al. 2010). Antibiotic molecules with an innovative mechanism-of-action directly minimize the probability of a pre-existing resistance and depending on their specific cellular target may also prevent a rapid resistance development among pathogens. However, besides the search for novel antibiotic molecules by focusing on new targets or by exploring untapped sources of antibiotics (such as currently uncultured bacteria), the screening of existing libraries to repurpose drugs as antibiotics or the re-initiation of research on neglected antibiotics can be seen as alternative approaches.

The advantage of natural products over libraries of chemically synthesized compounds is in part based on the fact that natural products evolved over millions of years. Thus, they are believed to be structurally optimized to interact with biological macromolecules. The search for molecules with, e.g., antibacterial, antiviral, anticancer, or immunomodulatory activity should be faster and more fruitful through natural product libraries compared to chemical libraries, as natural product scaffolds are regarded as privileged structures with high structural diversity and numerous biological activities (Koehn and Carter 2005; Newman 2008). Screening of synthetic libraries in industrial—either target or whole cell-based—HT (high-throughput) campaigns only rarely resulted in the discovery of new anti-infective lead structures. This disappointing result was later on mainly contributed to the lack of chemical diversity and unfavorable pharmaceutical properties of the screened compounds (Payne et al. 2007) Although a significant number of inhibitors was found in vitro, most of them were difficult to develop as drugs since the cellular effectiveness of small molecule inhibitors is often hampered by the fact that the compounds do not reach their intracellular target structure as they cannot enter the bacterial cell and/or are prone to active export. Furthermore, available synthetic libraries are mainly designed to meet the needs for the discovery of drugs that interact with mammalian targets and, thus, often do not yield hits in anti-infective drug screening as this requires compounds with much different physicochemical properties to pass the bacterial cell envelope (Lewis 2013). By contrast, natural product libraries often yield higher-quality leads with a significantly higher hit rate as observed for synthetic compound collections (Li and Vederas 2009). Consequently, there is a renewed interest in natural products for drug discovery, especially for the development of new anti-infective compounds. However, within the hit-to-lead process natural products are often difficult to pursue which is mainly due to their usually restricted availability and chemical complexity, which hampers further derivatization to improve pharmaceutical properties.

There are different approaches to modify a natural product, through either genetic engineering or chemical synthesis. Many natural products with interesting biological activities are polyketides (e.g., erythromycin and tetracycline), nonribosomally made peptides (e.g., penicillin G and vancomycin), or hybrids thereof (e.g., FK506 and epothilone), synthesized by polyketide synthases (PKSs) or nonribosomal peptide synthases (NRPSs), respectively. Polyketide and nonribosomal peptides are both a result of a coordinated, multistep action of enzymes organized in assembly lines, where a small set of monomeric units is incorporated into a linear oligomer by iterative chemical condensation steps, forming C–C bonds in polyketides or C–N bonds in the case of nonribosomal peptides. Monomers utilized by PKSs are acyl-CoA thioesters, whereas NRPSs use proteinogenic and nonproteinogenic amino acids and other carboxylic acids (Fischbach and Walsh 2006). By manipulating enzymatic domains of multifunctional proteins, incorporation of different building blocks can be achieved, leading to alterations in the scaffold. Mutasynthesis (chemobiosynthesis), semisynthesis, and other approaches involving chemistry either provide alternative building blocks or allow modifications of the product after biosynthesis is completed (Goss et al. 2012). However, for many and sometimes very complex natural products a total synthesis could be achieved (Kadota et al. 2005; Nicolaou et al. 1997; Rahn and Kalesse 2008; Seiple et al. 2016), which makes the scaffold accessible for further derivatizations. The decision whether a biotechnological or chemical approach is pursued to improve the original natural product is mainly driven by economic considerations, i.e., if the total synthesis of a natural product can be achieved in a reasonable number of synthetic steps (usually less than ten) and good overall yields this often appears to be more efficient than producing the natural product and new derivatives thereof by means of biotechnological fermentations of genetically engineered strains.

At the Department of Microbial Natural Products (MINS) of the Helmholtz Institute for Pharmaceutical Research Saarland (HIPS), different strategies to discover and develop new antibiotics, produced by Actinomycetes and Myxobacteria, are employed. Due to reasons mentioned above, we primarily focus on natural compounds that show a favorable antimicrobial profile with a special emphasis on hard-to-treat Gram-positive and Gram-negative pathogens. One cornerstone is the initial screening for new bioactive metabolites by making use of biodiversity and underexploited microbial sources to discover novel producers (Müller and Wink 2014). Based on a unique and constantly growing collection of strains, genera, and families, new screening methods are being developed to search for metabolites with novel chemical scaffolds. During the last years, a comprehensive and diverse set of approximately 2300 myxobacterial strains has been (re)activated and underwent chemical and biological screening. Numerous bioactive compounds, such as antibacterial disciformycins (Surup et al. 2014), antiviral aetheramides (Plaza et al. 2012), or the antiproliferative nannocystin (Hoffmann et al. 2015), were recently isolated, confirming that especially new or little-studied genera of myxobacteria are a rich and viable source for natural products (Garcia et al. 2015; Weissman and Müller 2010; see case study 1: Cystobactamids). In the course of hit optimization, we aim at elucidating the mode-of-action (MoA). In particular, the identification of self-resistance determinants of the natural producer organisms and the genomic analysis of resistant pathogens can serve as reference point for the elucidation of the MoA and the resistance mechanism against the compound (Laxminarayan et al. 2013; Bielecki et al. 2012). We also aim at improving pharmaceutical properties, target-binding affinities, and the antimicrobial spectrum to ultimately generate analogs that can enter the hit-to-lead stage. Recent re-examination including application of modern state-of-the-art technologies showed that some of the neglected antibiotic compounds, which displayed very potent activity, but failed to progress through clinical evaluation due to different reasons, could also be very promising lead structures and should be developed further by means of in-depth mechanistic studies and improving pharmaceutical properties (see case study 2: Griselimycin). On the other hand, there are also groups of antibiotics that were used extensively in the past and resistance development already limits their use. However, their chemical scaffolds showed to be very effective and thus, it is worth to continuously modify these backbones to be "reused" as antibiotics. Chemical synthesis and semisynthesis approaches, to generate new analogs based on proven scaffolds, have been exhausted over the decades of drug development. Thus, new approaches of biosynthetic engineering and synthetic biology today offer unique opportunities for modifications of these proven chemical scaffolds (see case study 3: Chelocardin).

2 Cystobactamids: En Route to Novel Antibacterial Lead Structures

In our search for new compounds exhibiting activity against Gram-negative bacteria, a novel class of natural products from *Cystobacter* spp. came to our attention. Crude extracts prepared from small-scale cultivations of *Cystobacter* sp. Cbv34 efficiently inhibited the growth of several Gram-positive and Gram-negative bacteria. Following an LC-HRMS-assisted bioactivity-guided screening approach, we were able to identify cystobactamids 919-1, 919-2, and 507 as active components (Fig. 1), whose planar structures and stereochemistry were elucidated using various spectroscopic/spectrometric techniques (Baumann et al. 2014). The cystobactamid compound family has largely grown since its first description and the compounds as well as the corresponding biosynthetic gene clusters have been identified in additional myxobacterial genera (unpublished results).

2.1 *Chemical Structures of Cystobactamids*

A key feature of the cystobactamids is the complete novelty of their chemical structure. They can be regarded as peptides as their unique scaffold (as exemplified

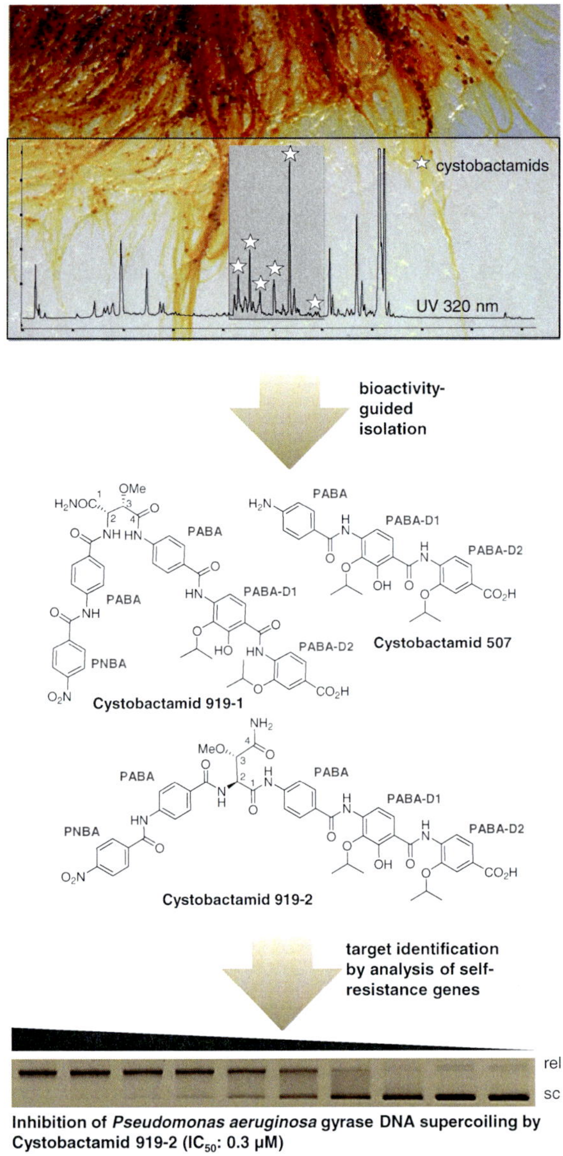

Fig. 1 Discovery and target identification of cystobactamids from *Cystobacter* sp. Cbv34 (*image source* Telkemeyer/HZI). PABA—*p*-amino-benzoic acid, PNBA—*p*-nitrobenzoic acid, rel—relaxed, sc—supercoiled

for cystobactamid 919-1 and 919-2) is characterized by six building blocks connected via amide bonds: five unusual aromatic moieties, namely *p*-nitrobenzoic acid, two unmodified *p*-amino-benzoic acid units (PABA), and two PABA derivatives with different oxidation patterns, which are connected via an *iso*-L-threo-methoxyasparagine unit (919-1) and yet another previously undescribed moiety in natural products, or an L-threo-methoxyasparagine unit (919-2) situated between the two PABA units. Cystobactamid 507 represents the tripeptidic eastern fragment of cystobactamids 919-1 and 919-2 (Fig. 1). Additionally, the cystobactamids show various special structural features like a rare nitro group (919-1 and 919-2), most likely originating from the oxidation of the amino group in one PABA moiety and the rather exotic isopropoxylation of the 3-hydroxy-4-aminobenzoic acid moieties also unprecedented in natural products. There is only one class of related natural products, the albicidins from the plant pathogen *Xanthomonas albilineans*, whose structure has been elucidated almost simultaneously to the cystobactamids (Kretz et al. 2015).

2.2 *Bioactivity of Cystobactamids*

When evaluating the antimicrobial activities of the purified compounds, it was found that cystobactamids 919-1 and 919-2 inhibited the growth of *Escherichia coli* at concentrations as low as 1 μg/mL. In particular 919-2 strongly inhibited growth of various pathogens belonging to the "ESKAPE" panel (Pendleton et al. 2013) including the Gram-negative pathogen *Acinetobacter baumannii* and Gram-positive *Enterococcus faecalis* and *Streptococcus pneumoniae* at concentrations comparable to or even exceeding the activity of ciprofloxacin (Cp), a clinically relevant second-generation fluoroquinolone antibiotic (Emmerson and Jones 2003). Importantly, none of the tested derivatives exhibited pronounced antifungal or cytotoxic activities up to concentrations of approximately 50–100 μg/mL. The activity of cystobactamids against Gram-negative pathogens is of particular interest since only very few new antibiotics are able to inhibit these often multidrug-resistant bacteria. Continuous efforts to further characterize and extend our knowledge on the cystobactamid scaffold recently enabled the discovery of new derivatives that are even active on an extended spectrum of Gram-negative pathogens, such as *Pseudomonas aeruginosa*.

2.3 *Biosynthesis Model*

Based on their peptidic chemical architecture, it appeared likely that the cystobactamids are products of a nonribosomal peptide synthetase (NRPS) (Strieker et al. 2010). Intriguingly, the major building block of these compounds is PABA, a moiety rarely incorporated in secondary metabolites and unprecedented in nonribosomal peptides (Walsh et al. 2012). As the production of this building block might be one of the bottlenecks of cystobactamid biosynthesis, it was hypothesized that the formation of both might be encoded in one biosynthetic cluster. Indeed, screening of the genome of *Cystobacter* sp. Cbv34 revealed an NRPS locus containing PABA biosynthetic enzymes (ADC lyase and anthranilate synthase). This gene cluster furthermore contains three large NRPS genes plus various genes encoding tailoring enzymes, transporters, and putative resistance genes. The biosynthesis exhibits a number of mostly unusual features including an *in trans* extender unit (iso-asparagine or asparagine) loading process, the isopropoxylation by subsequent SAM-dependent methylation of a phenolic hydroxyl-group followed by two further methylations and catalyzed by a radical SAM-dependent methyl transferase as well as the oxidation of an amino to a nitro group (for details see Baumann et al. 2014).

2.4 Target Identification and Mode-of-Action

The cystobactamid genetic locus encodes a putative self-resistance protein, CysO, belonging to the pentapeptide repeat protein family. Some of these proteins are known as resistance factors against topoisomerase poisons including fluoroquinolones (Qnr and MfpA), microcin B17 (McbG), and albicidin (AlbG) (Tran and Jacoby 2002; Montero et al. 2001; Garrido et al. 1988; Hashimi et al. 2007), suggesting that the cystobactamids might act as inhibitors of bacterial type IIa topoisomerases. Indeed, biochemical assays showed that the cystobactamids are able to inhibit the activity of the bacterial topoisomerases, such as gyrase and topoisomerase IV from *E. coli*. Cystobactamid 919-2 showed IC_{50} values for *E. coli* gyrase in the nM range, thus being equally potent as Cp. For the same enzyme, cystobactamids 919-1 and 507 show 100- to 1000-fold higher IC_{50} values (two to three-digit μM range). Presence of the first two PABA units and more importantly the configuration of the central amino acid (L-iso-asn vs. L-asn), and thus the overall shape of the molecule, seem to be mandatory to boost the activity from the μM to the nM range. Despite their high structural homology, cystobactamids show only moderate activities (mid- to high μM) on *E. coli* topoisomerase IV, indicating that the main target of the cystobactamids in *E. coli* is DNA gyrase.

Type IIa topoisomerase inhibitors can be generally divided into two classes: topoisomerase poisons such as the fluoroquinolone ciprofloxacin and competitive ATP-binding pocket inhibitors, such as the aminocoumarin natural product novobiocin (Vos et al. 2011). Biochemical and molecular biological experiments clearly identify the cystobactamids as topoisomerase poisons, as these compounds induce typical DNA double-strand breaks and are not inactivated by excess ATP. The cystobactamids can therefore be regarded as "natural quinolones," which represent one of the most successful classes of antibiotics. However, these findings also suggested that the primary binding site of the cystobactamids on gyrase might overlap with the quinolone-binding site, raising the question of possible cross-resistance induced by target mutation.

Target-based resistance to quinolones is reported to be mediated by mutations in *gyrA* and *parC* leading to alterations in the drug-binding sites. In GyrA, the quinolone-resistance determining region (QRDR) is located between amino acids 67 and 106, whereas S83 and D87 are most often involved (Gruger et al. 2004). The comparison of the MIC values of cystobactamids versus Cp using a quinolone-susceptible *E. coli* strain (0.5 and 0.02 μg/mL, respectively) and strains harboring the aforementioned mutations (isogenic mutants in the *gyrA* gene) indicated that the activity of cystobactamid 919-2 (the most potent derivative published to date) was only reduced 2- to 7-fold in the S83 and D87 mutants and a combination of both GyrA mutations had a marginally larger impact. By contrast, the activity of Cp was already diminished approximately by a factor of 30–60 for single GyrA mutants, whereas the combination of both mutations (S83L, D87G) resulted in clinical resistance (MIC ≥ 1 mg/L). These findings suggest an overlapping but not identical binding site of cystobactamids and quinolones on gyrase, as the impact of the quinolone-resistance mutations on cystobactamid activity is less pronounced. X-ray crystallographic studies to support this finding on the structural level are currently underway. We also investigated nontarget-based resistance, and importantly, some of the cystobactamid derivatives are not exported by major efflux pump systems such as *E. coli* AcrAB and *P. aeruginosa* MexAB and were still active on *Enterobacter* spp. and *E. coli* that harbor *qnr* genes. Thus, we conclude that it might indeed be possible to develop cystobactamids as novel gyrase inhibitors that do not show significant cross-resistance with fluoroquinolones.

2.5 *Biotechnological Production, Total Synthesis, and Outlook*

Cystobactamids have been isolated from *Cystobacter* sp. Cbv34 in low overall yields (<100 μg/L). Optimization of the fermentation process, as well as the isolation procedure, and the exploration of alternative myxobacterial producer strains finally yielded numerous additional natural derivatives of the cystobactamid family, some of which possess even more favorable biological activities against an extended spectrum of Gram-negative bacteria. These natural congeners, in turn, helped to get first insights into basic structure–activity relationship (SAR) of the cystobactamid family of compounds and, in addition, will guide the current chemical synthesis and derivatization program. The total synthesis of the smallest naturally occurring cystobactamid 507 has been published recently (Moreno et al. 2015) and total synthesis of full length cystobactamids is on its way. To enable further development of this promising new class of antibiotics, it appears crucial to gain access to chemical matter via organic synthesis. As soon as an efficient supply of the most active cystobactamids is assured, in vitro ADME (absorption, distribution, metabolism, excretion) and in vivo pharmacodynamic and pharmacokinetic studies using the compound class can be initiated, which will aid the development of a potential novel lead structure based on the natural cystobactamid scaffold either by total synthesis or using semisynthesis. A first proof-of-concept study could recently be initiated based on two natural congeners that were obtained in sufficient amounts in an improved biotechnological process (unpublished data).

3 Griselimycin: Rediscovery and Characterization of a Neglected Antibiotic Exhibiting an Innovative Mode-of-Action

A number of neglected antibiotics are promising natural product leads which were not advanced to clinical development due to various reasons. These include the perception after the 1970s that infectious diseases were almost eliminated and that the interest in the development of antibacterial drugs vanished (Laxminarayan et al. 2013; Rangel-Vega et al. 2015). Also the typically high attrition rate during drug development prevented some of these molecules to be brought forward. Today, at least some of the limitations of such molecules might be addressed by using state-of-the-art technology not available some decades ago. Griselimycins are an example for the rediscovery of neglected antibiotics which were subsequently optimized with respect to potency and pharmacokinetic characteristics and comprehensively characterized with respect to the mechanism-of-action and a possible resistance mechanism (Kling et al. 2015). Combined treatment of mouse models of TB with a griselimycin derivative and first-line anti-TB drugs showed the lack of cross-resistance and potential to significantly shorten the TB treatment period.

3.1 *Rediscovery and Optimization of Griselimycins*

Griselimycins are natural compounds produced by *Streptomyces* strains. Griselimycin was discovered in the 1960s and the natural derivative methylgriselimycin, which is naturally produced in only small amounts, was discovered in the 1970s (Terlain and Thomas 1971; Anonymous 1962). Griselimycins are highly active against drug-resistant *Mycobacterium tuberculosis* and the first human studies were promising but showed poor pharmacokinetic characteristics of griselimycin, particularly a poor stability (half-life) in plasma. Studies on griselimycin were not further pursued after rifampicin became available for tuberculosis treatment in the 1970s. Recently, griselimycin was rediscovered in search for compounds with antituberculosis activity and studies on griselimycins were reinitiated at Sanofi Company (Kling et al. 2015). In order to find griselimycin analogs with improved potency, metabolic stability, and exposure, structure–activity relationship (SAR) studies with a library of synthetic analogs were conducted. The synthetic analog cyclohexylgriselimycin was found to show improved potency and pharmacokinetic properties (such as a high oral bioavailability, a moderate total plasma clearance, and a large volume of distribution) and was chosen for in vivo proof-of-concept studies. The metabolic stability was improved for analogs with substituents at proline 8 of the molecule, such as methyl- and cyclohexylgriselimycin (cp. Fig. 3), indicating that proline 8 is the main site of metabolic degradation.

3.2 Self-resistance Mechanism Against Griselimycin

The examination of the griselimycin biosynthetic gene cluster in the producer *Streptomyces caelicus* revealed that this cluster contains an unusual gene (*griR*) which is not involved in the synthesis, transport, or regulation of griselimycin (Broenstrup et al. 2014; Fig. 2). The *griR* gene encodes an additional copy of the DNA polymerase III beta subunit (also called sliding clamp), which shows 51 % amino acid identity to the housekeeping sliding clamp of *S. caelicus*. The sliding clamp is known to tether DNA polymerase III to the DNA during DNA synthesis and thus provides processivity to the DNA polymerase. In the absence of the sliding clamp, the DNA polymerase III of *E. coli* synthesizes DNA at a rate of about 10 nucleotides (nt)/sec with a processivity of 10–20 base-pairs and in complex with the sliding clamp the polymerase incorporates 1–2 kilobase-pairs at a rate of about 350–500 nt/sec (Tanner et al. 2011). In addition, the sliding clamp acts as a switch for other enzymes and factors during replication and repair. The additional sliding clamp GriR was found to be the self-resistant determinant for griselimycin, since overexpression of *griR* in griselimycin sensitive *Streptomyces coelicolor* conferred resistance to griselimycin (Kling et al. 2015).

3.3 Mechanism-of-Action of Griselimycins

Griselimycins were found to exert bactericidal activity against *M. tuberculosis* in vitro and in vivo. They bind to the peptide-binding site of the mycobacterial sliding clamp (which is also the interaction site of other sliding clamp partner proteins including enzymes involved in DNA repair) and thus, inhibit DNA synthesis and most probably also DNA repair. The sliding clamp is a novel anti(myco)bacterial target that is essential for viability and does not share significant sequence similarity with the human sliding clamp equivalent (in fact, no binding of griselimycins to the human equivalent was observed). In addition, combination treatment of cyclohexylgriselimycin with first-line antituberculosis drugs in mouse models of tuberculosis showed improved activity (i.e., a shortened treatment duration) compared to the standard tuberculosis regimen (i.e., administration of rifampicin, isoniazid, and pyrazinamide for 2 months, followed by administration of rifampicin and isoniazid).

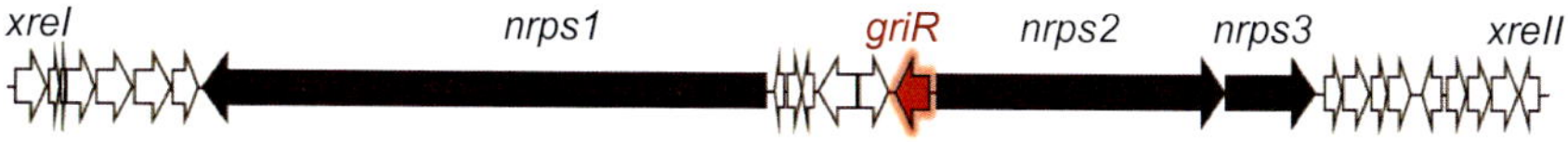

Fig. 2 Griselimycin self-resistance determinant. The griselimycin biosynthetic gene cluster of *Streptomyces caelicus* encodes 26 protein-encoding genes (Broenstrup et al. 2014), including genes encoding nonribosomal peptide synthetases (*nrps1*, *nrps2*, *nrps3*), genes encoding transcription factors (*xreI*, *xreII*), and a gene encoding an additional copy of the DNA polymerase sliding clamp (*griR*), which was found to mediate self-resistance (Kling et al. 2015)

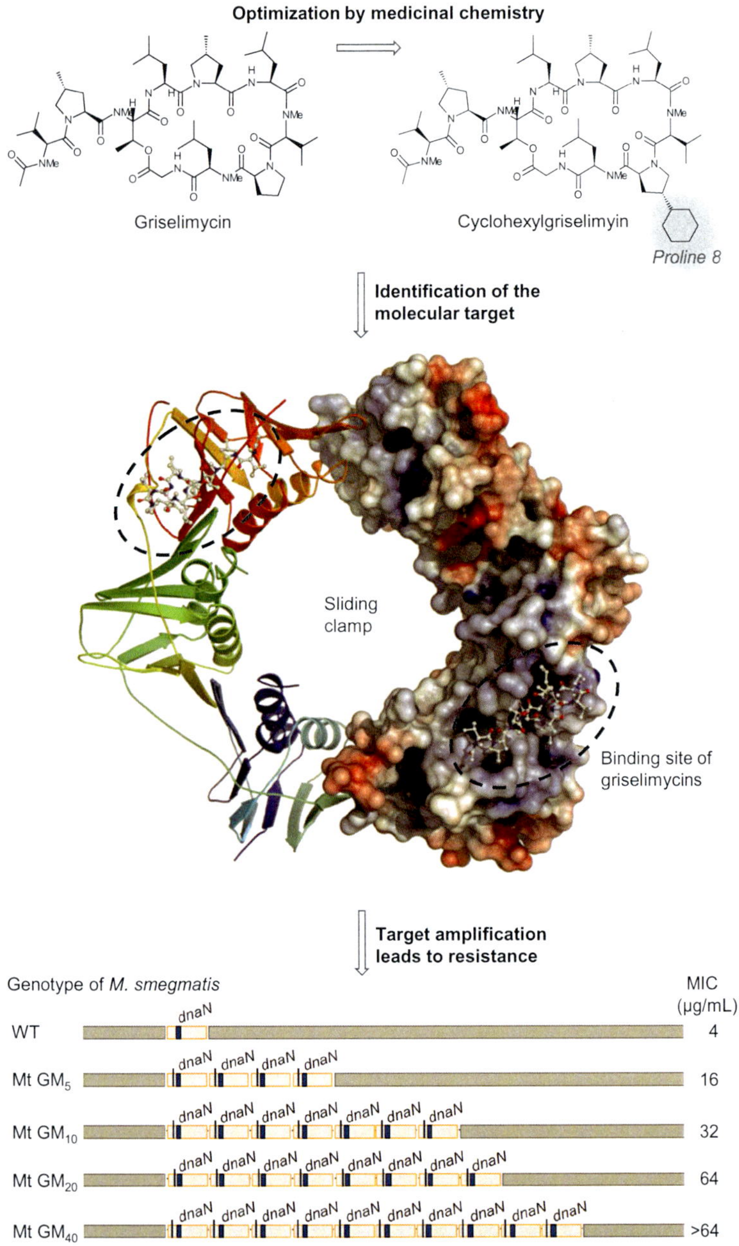

Optimization by medicinal chemistry
Griselimycin
Cyclohexylgriselimyin
Proline 8
Identification of the molecular target
Sliding clamp
Binding site of griselimycins
Target amplification leads to resistance
Genotype of M. smegmatis
MIC (µg/mL)
WT
Mt GM5
Mt GM10
Mt GM20
Mt GM40
dnaN
4
16
32
64
>64

◀ **Fig. 3** Optimization by medicinal chemistry and characterization of griselimycins. Cyclohexylgriselimycin is a synthetic analog that shows improved potency, metabolic stability, and pharmacokinetic parameters. Griselimycins bind to the peptide-binding site of the DNA polymerase sliding clamp and thereby inhibit replication and most probably DNA repair. Resistance to griselimycins is associated with amplification of the sliding clamp. Resistant *M. smegmatis* mutants were generated by exposure to stepwise increasing concentrations griselimycin (*subscript* indicates concentration of griselimycin (μg/mL) used for mutant selection; *GM*—griselimycin; *dnaN*—gene encoding the sliding clamp; *Mt*—M. smegmatis mutant; *WT*—M. smegmatis WT). Resistant *M. smegmatis* mutants contain an amplification of a chromosomal segment (indicated as *boxes*) that contains the gene encoding the sliding clamp (indicated as *bars within boxes*), the *ori* (indicated as *lines*), and several additional genes

3.4 *Resistance Mechanism Against Griselimycin in Mycobacteria*

To investigate a potential mechanism of resistance against griselimycin in mycobacteria, resistant *Mycobacterium smegmatis* mutants were generated in vitro by exposure to stepwise increasing concentrations of griselimycin and resistant *M. tuberculosis* mutants were selected in vivo from nude mice following subinhibitory monotherapy with cyclohexylgriselimycin. Although mycobacteria developed resistance against griselimycins, resistance was found to occur only with very low frequency of 5×10^{-10} (at the about twofold MIC of griselimycin in *M. smegmatis*) and was accompanied by a considerable fitness loss. Genome analysis of the resistant mutants revealed that resistance to griselimycins is associated with target amplification, as resistant mutants contain an amplification of a chromosomal segment that contains the gene encoding the sliding clamp, the origin of replication (*ori*), and several additional genes (Fig. 3). Griselimycin resistance turned out to be reversible, as *M. smegmatis* mutants grown in the absence of griselimycin showed an increased sensitivity for griselimycin and a decreased amplification of the gene encoding the molecular target.

3.5 *Griselimycins and Their Potential to Become a Tuberculosis Drug*

Further in vitro studies on the detailed mechanism-of-action of griselimycins are done to elucidate the role of other amplified genes found on the multiplied chromosomal segment, and to decipher other sliding clamp binding partners whose interaction with DnaN is hampered by griselimycin binding. Such data will provide a better understanding of how griselimycins inhibit DNA replication and repair.

Due to the very favorable properties—potent in vitro activity and in vivo efficacy, strong bactericidal effect, no cross-resistance with other TB drugs/new target, low frequency of resistance and reversible mutation, no cytochrome P450 induction or inhibition—the lead structure cyclohexylgriselimycin is currently codeveloped by the TB alliance and Sanofi within their lead optimization portfolio. Results from mouse models employing cyclohexylgriselimycin in combination with the first-line drugs rifampicin and pyrazinamide let us assume that it is possible to shorten the treatment period with such a regimen and by this, prevent resistance development. The ultimate goal is obviously an affordable, new drug regimen that can be applied for the treatment of multidrug-resistant TB including HIV coinfected patients.

4 Chelocardin: Generation of New Analogs by Genetic Engineering

Tetracyclines have been used as broad-spectrum antibiotics in human and veterinary medicine since their discovery in the late 1940s. Intensive use led to an increasing incidence of bacterial resistance to tetracyclines (Chopra and Roberts 2001). Interestingly, another group of tetracycline antibiotics was identified in the 1960s and 1970s (Chopra and Roberts 2001), so-called atypical tetracyclines, one of which is chelocardin, that are still active on tetracycline-resistant pathogens. We aimed at exploiting and further improving the chelocardin scaffold in order to obtain a lead structure with a more favorable antimicrobial spectrum and better pharmaceutical properties. With the idea to generate new analogs of known antibiotics, producing organisms can be genetically engineered to produce modified structures of target compounds; enzymes involved in the biosynthesis of natural products can be deleted, replaced, modified, or new heterologous genes introduced into the natural producer, for example, to generate alternatively glycosylated analogs of erythromycin and mithramycin (Mendez and Salas 2001), daptomycin analogs with different amino acid substitutions (Baltz 2009) or oxo-amphotericin analogs (Power et al. 2008). The availability of a large number of decoded biosynthetic gene clusters is making genetic engineering and combinatorial biosynthesis very attractive and importantly, realistic approaches. Novel scaffolds produced by biosynthetic engineering approaches can be further derivatized by means of semisynthesis (Fig. 4).

4.1 Generating New Natural Product Derivatives by Genetic Engineering

4.1.1 Manipulation of Multidomain Proteins

As we learn from nature, diverse natural products biosynthetic pathways evolved through both, mutations within genes and horizontal gene transfer between biosynthetic gene clusters. It is therefore assumed possible to manipulate polyketide or nonribosomal peptide assembly line logic through swapping of entire domains,

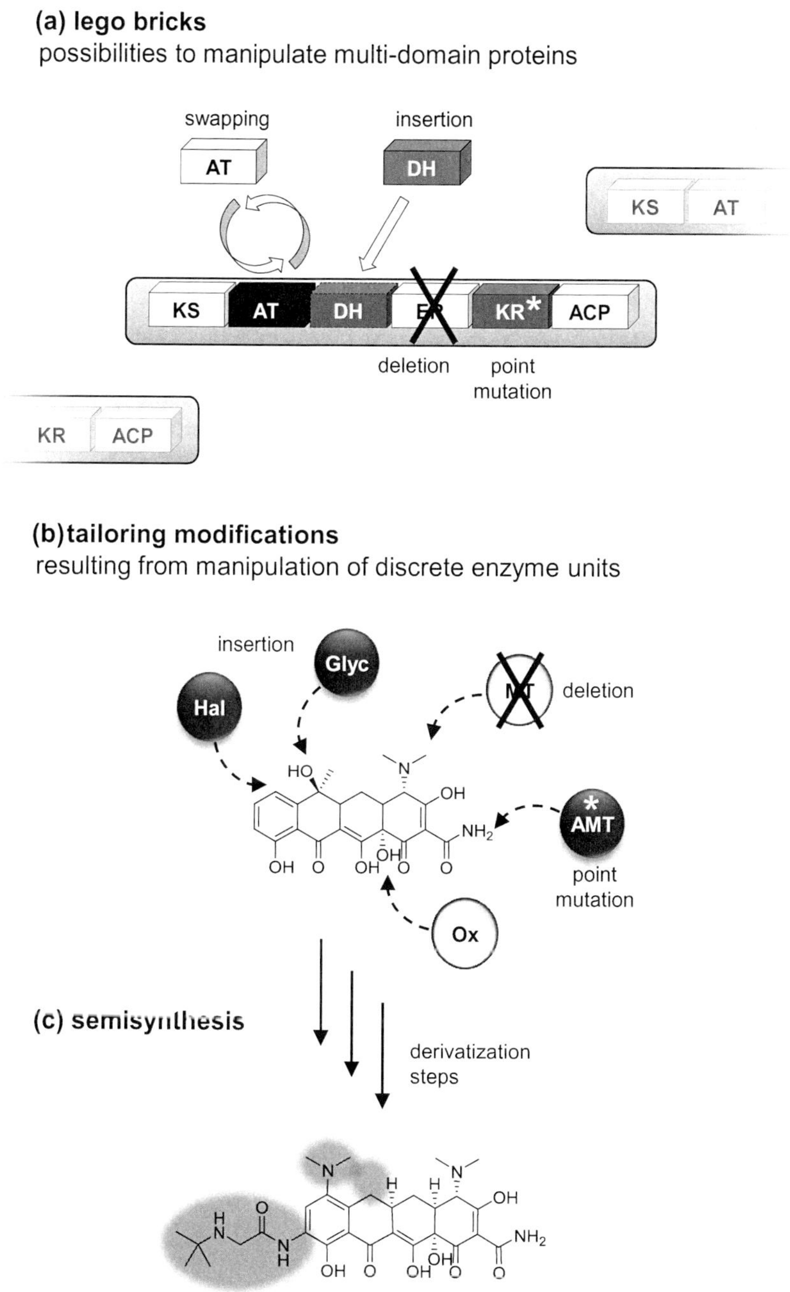

Fig. 4 Different approaches to modify a natural product through genetic engineering or semisynthetic chemistry: **a** "lego bricks" or manipulation of enzymatic domains of multifunctional proteins; **b** tailoring modifications as a result of a manipulation of discrete enzyme units; and **c** subsequent semisynthesis

modules, or subunits, where in fact exactly the modular organization of NRPS and type I PKS allows combinatorial biosynthesis. Introduction of changes in the assembly line should result in non-natural intermediates which due to relaxed substrate specificity of subsequent domains/enzymes often enables the processing of such noncognate substrates to eventually give rise to new analogs (Cummings et al. 2014; Sun et al. 2015; Wong and Khosla 2012). However, it needs to be pointed out that usually substrate recognition and specificity issues of engineered enzymatic complexes reduce the yield of the target analog. Engineered enzyme complexes are often entirely nonfunctional, thus resulting in nonproducing strains.

Therefore, gene substitutions, deletions, point mutations, or introduction of new functionalities can be used to introduce diversity into natural product biosynthesis. Different steps throughout biosynthesis can be modified: (1) manipulation of loading domains for starter unit selection and modules involved in selection and incorporation of extender units allows changing the choice of building blocks eventually introducing new moieties into the polyketide or peptide backbone; (2) changes in chain length are possible by module insertion or deletion; and (3) furthermore, it is possible to affect the extent of reduction and stereochemistry of β-keto groups, alkylation (methylation), β-branching, and cyclization by manipulating responsible domains of chosen modules (Cummings et al. 2014; Sun et al. 2015).

When swapping complete domains or modules, one has to take into account that the protein's overall structure might become disrupted and thus compromise its function, often leading to significantly reduced product yields or total stalling of the respective assembly line. As an alternative, it is possible to choose less invasive and therefore often more successful strategies to change the enzyme's function, leading to an altered structure of the natural product. This can be achieved by the introduction of amino acid substitutions, either by site-specific mutagenesis or by using directed evolution approaches. The latter generates a number of different enzyme variants and is sometimes described as most promising, as we still lack a lot of understanding of substrate specificity and catalysis on the molecular level (Sun et al. 2015; Williams 2013).

4.1.2 Manipulation of Discrete Enzyme Units

By manipulating discrete enzymes (for example tailoring enzymes), the polyketide or peptide scaffolds can also be modified. Besides gene deletion or point mutation, introduction of genes from other biosynthetic pathways can be used to either remove or add a new modification step, respectively, leading to new derivatives. Natural products can be decorated through glycosylation, halogenation, acyl transfer, hydroxylation, epoxidation, alkylation, transamination, and desaturation reactions (Cummings et al. 2014; Goss et al. 2012). As we have seen, manipulation of type I modular PKS and NRPS enables incorporation of various starter and extender units into the polyketide or peptide chain. On the contrary, iterative type II PKSs do not allow incorporation of alternative extender units; only malonyl-CoA can be used. Therefore, the only possibility to modify the polyketide chain is to incorporate a different starter unit or modification of late stages of the basic backbone biosynthesis (so-called post-PKS processing in polyketides), later resulting in a differently decorated final product, whereas the core structure still remains the same (described below on the example of chelocardin modification). Finally, by interchanging and combining heterologous genes from multiple species it is possible to generate combinatorial or hybrid pathways for the production of novel natural products, for example, different glycosylated analogs (Sun et al. 2015). Furthermore, it is possible to change the cyclization pattern of aromatic polyketides by inactivating or introducing cyclase genes. However, this approach might lead to a significantly changed scaffold, which is not a suitable substrate for downstream biosynthetic enzymes anymore.

4.1.3 Help from Chemistry

Apart from applying biosynthetic engineering to generate new drugs, chemical approaches are traditionally widely used to modify natural product scaffolds. Successfully combining these two approaches enables modifications at different parts of the molecule, ideally yielding an optimized drug with better activity and pharmaceutical properties. Abbreviations "CHEM" for (partial) chemical synthesis and "BIO" for biosynthesis or a biotransformation are used to describe these combinatorial approaches (Kirschning and Hahn 2012).

Chemical approaches can be used after the natural product is biosynthesized, extracted, and purified to introduce additional modifications (semisynthesis, BIO-CHEM) or at the very beginning by precursor-directed biosynthesis (CHEM-BIO) or mutasynthesis (chemobiosynthesis, CHEM-BIO), where alternative precursors are chemically synthesized and fed to the producer organism, which then incorporates the unnatural building block into the natural product. Another CHEM-BIO approach is chemical synthesis followed by enzymatic transformation using heterologously expressed late-stage enzymes or by subjecting it to whole-cell biotransformation (by fermentation process). However, a prerequisite for CHEM-BIO approaches is the natural flexibility of biosynthetic enzymes for accepting unnatural substrate analogs (Goss et al. 2012; Kirschning and Hahn 2012).

In the mutasynthesis approach, the biosynthetic enzymes required for the biosynthesis of specific precursors are inactivated, substituted, or modified and then the mutated producer strain culture is supplemented with the chemically prepared unnatural precursor (mutasynthon). It is worth mentioning that in contrast to enzyme inactivation, enzyme substitution or modification in order to broaden substrate specificity may result in downstream issues, which include also problems with separation of highly structurally similar derivatives. Successful mutasynthesis approaches again show how tolerant the downstream biosynthetic machinery can be toward nonnatural precursors (Goss et al. 2012; Kirschning and Hahn 2012; Williams 2013).

Furthermore, after the mutasynthesis step, semisynthetic modification (CHEM-BIO-CHEM) can be applied for further derivatization, or the order of chemical synthesis and biosynthesis can be altered to a BIO-BIO-CHEM or CHEM-BIO-BIO synthetic sequence. In the former approach, the target product of an engineered producer strain is fed to a second engineered strain, performing enzymatic transformation, followed by a semisynthetic modification. In the latter approach, chemical synthesis is followed by a two-step biotransformation (Kirschning and Hahn 2012).

4.2 Chelocardin Exerts Promising in Vitro Activity and in Vivo Efficacy

Chelocardin belongs to the group of so-called atypical tetracyclines and most likely acts on the bacterial membrane. A recent proteomics study revealed that sublethal concentrations of chelocardin induce changes in *Bacillus subtilis* that are specific for protein biosynthesis inhibitors (Stepanek et al. 2016). However, its exact mode-of-action remains unknown. In contrast to the bacteriostatic effect exerted by typical tetracyclines, the atypical tetracyclines act bactericidal (Chopra and Roberts 2001). Chelocardin showed activity against many multidrug-resistant pathogens, including some difficult-to-treat Gram-negative bacteria of the ESKAPE group. Importantly, it is also effective against tetracycline-resistant strains, except for *P. aeruginosa* (Oliver and Sinclair 1964; Proctor et al. 1978; Lešnik et al. 2015). Chelocardin was already tested in a small phase II clinical trial in Zagreb, Croatia, in 1977, where its effect on 12 patients suffering from urinary tract infections was studied. The results of the drug treatment were positive and were manifested by the disappearance of clinical symptoms and bacteriuria as well as the normalization of the relevant laboratory findings, while the drug tolerance was described as satisfactory. Previous experiments in rats and dogs have also shown a relative lack of toxicity (Molnar et al. 1977). However, chelocardin has not been developed further, because of its suboptimal pharmaceutical properties, an issue that could not be addressed at that time (Lešnik et al. 2015).

4.3 Engineering a Second-Generation Chelocardin

Chelocardin differs from typical tetracyclines in a few structural features, one of them being the C2 moiety (acetyl instead of carboxamido), a result of incorporation of a different starter unit, which can be concluded also from comparison of the respective biosynthetic gene clusters (Zhang et al. 2006; Lukežič et al. 2013). As the C2 carboxamido moiety was shown to be important for the antibacterial activity of typical tetracyclines, which act as translational inhibitors, the idea to introduce this moiety into the structure of chelocardin emerged. Following this rational approach, modification of chelocardin was achieved by genetic engineering of the chelocardin producer, which yielded a new chelocardin analog (2-carboxamido-2-deacetyl-chelocardin) indeed containing the C2 carboxamido moiety as found in typical tetracyclines such as oxytetracycline and chlortetracycline. Gratifyingly, the resulting compound was found to be a significantly more potent broad-spectrum antibiotic, active against various multidrug-resistant clinical pathogens. Intriguingly, the new chelocardin analog even gained activity against *P. aeruginosa* (Lešnik et al. 2015). Although the increased activity of the new chelocardin analog after modification is in agreement with the structure–activity relationships (SAR) found in typical tetracycline antibiotics (Nelson 1998), it is currently unclear how this modification relates to the interaction of chelocardin with its unknown target structure(s).

In order to achieve the modifications described above, genes encoding for amidotransferase and acyltransferase involved in the generation of the C2 carboxamido moiety in oxytetracycline biosynthesis were introduced into the chelocardin producer *Amycolatopsis sulphurea* (Lešnik et al. 2015). In oxytetracycline biosynthesis, the amidotransferase is the crucial enzyme responsible for incorporation of an amino group into malonate to generate a malonamate starter unit, whereas the acyltransferase only shows ancillary function. The function of acyltransferase homologs form the oxytetracycline gene cluster is not entirely understood; however, it is thought that the acyltransferase is actually acting as thiolase, thus removing the competing acetyl starter units from the minimal PKS system (ketosynthases alpha and beta, and acyl carrier protein) (Wang et al. 2011). By this, in oxytetracycline biosynthesis in *Streptomyces rimosus*, the acyltransferase helps to decrease the production of the less active acetate-primed oxytetracycline analog and increases the level of oxytetracycline (malonamate-primed) (Wang et al. 2011). Interestingly, no homologous acyltransferase gene is present in a second typical tetracycline biosynthetic gene cluster responsible for chlortetracycline biosynthesis. Presumably, the thiolase activity is not required for efficient malonamate-priming of chlortetracycline as the minimal PKS enzymes might have a much higher specificity for malonamate over the acetate starter unit (Pickens and Tang 2010). On the contrary, the chelocardin minimal PKS should exhibit a high preference for the acetate starter unit, as chelocardin is an acetate-primed polyketide. It was indeed discovered that the acyltransferase activity is almost indispensable for efficient biosynthesis (in terms of comparable yield to chelocardin in the wild-type producer) of the amidated chelocardin analog (CDCHD) in the system, where acetate is normally used for priming. When only the amidotransferase gene was expressed in *A. sulphurea*, CDCHD was biosynthesized in low yield (around 25 mg/L) and the main product was still unmodified chelocardin (around 500 mg/L), whereas expression of both amido- and acyltransferase led to the production of CDCHD in much higher yield (around 300 mg/L) and some chelocardin (85 mg/L); for comparison—production of chelocardin in wild-type bearing empty vector reached around 850 mg/L (see Table 1). Thus, in this setting the acyltransferase plays a much more important role than just priming of oxytetracycline: It is presumably essential for removing competing acetyl starter units to help increase the production of the malonamate-primed analog significantly. In conclusion, besides contributing to our understanding of the priming mechanism in tetracycline biosynthesis, a new tetracycline lead structure with potent antibacterial activity was generated, representing a platform for further development through genetic engineering and medicinal chemistry (Lešnik et al. 2015; Fig. 5).

Table 1 Approximate production yields of chelocardin and CDCHD in production strains expressing either only amidotransferase or both amido- and acyltransferase (compared to the control production strain)

	A. sulphurea + amidotransferase (mg/L)	*A. sulphurea* + amido- and acyltransferase (mg/L)	*A. sulphurea* + empty vector (mg/L)
Chelocardin production	500	85	850
CDCHD production	25	300	0

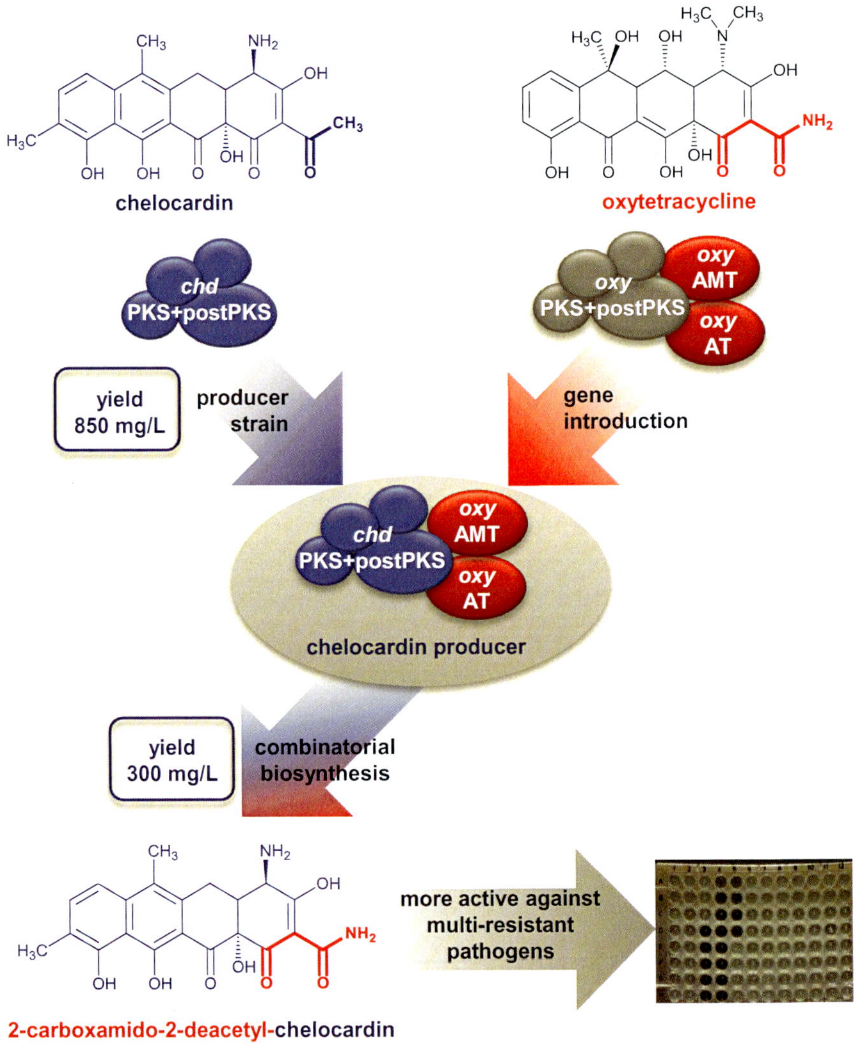

Fig. 5 Genetic engineering of the chelocardin producer toward generation of a new analog. *chd*—chelocardin; *oxy*—oxytetracycline; PKS—ketosynthases alpha and beta, and acyl carrier protein; post-PKS—tailoring enzymes; AMT—amidotransferase; AT—acyltransferase

4.4 Entering the Hit-to-Lead Stage with Chelocardins

The generation of 2-carboxamido-2-deactyl-chelocardin (CDCHD) through genetic engineering is regarded as a success because the antibacterial spectrum of this atypical tetracycline antibiotic could be significantly improved. In a next step, we aim at in-depth characterization of the scaffold by means of ADME (absorption, distribution, metabolism, and excretion) and PK/PD (pharmacokinetic/pharmacodynamics) studies to actually understand which pharmaceutical properties need to be improved in order to enable the definition of a clear target product profile, which will include a treatment for some tetracycline-resistant pathogens.

Once the most important sites for activity in the molecule have been probed in initial structure–activity relationship (SAR) studies with a concise set of semisynthetic CDCHD derivatives, a larger derivatization program can be initiated. Combining efforts to do analoging via semisynthesis and in parallel via genetic engineering appears to be most efficient and fruitful. As we have shown that early biosynthetic enzymes from chelocardin and oxytetracycline clusters are able to work together, and importantly, chelocardin late-stage (post-PKS) biosynthetic enzymes are able to accept the modified substrate, we hope that enzymes for the biosynthesis for other tetracyclic antibiotics would also be useful in combinatorial biosynthesis efforts to generate novel chelocardin analogs. In addition, semisynthesis can aid in modifying positions that are chemically easily accessible such as the C4 amino group and some positions on the western half of the molecule (Fig. 5).

5 Summary and Conclusions

Microorgansims continue to be an important source of new natural products. Within our discovery pipeline, we particularly explore the secondary metabolite profiles of myxobacterial strains belonging to new genera and families. In addition, samples are collected from biodiverse habitats to further increase the probability of finding new compounds with interesting biological activities as the novelty in biodiversity appears to be directly linked to chemical diversity. When whole cell-based activity screening is applied with crude extracts, it is assured that the active principle can pass the biological barriers of, e.g., a Gram-negative indicator strain, which is often not the case for hits from target-based assays where libraries of chemically synthesized compounds are screened. However, new natural products that are discovered within such a bioactivity-guided screening approach need to be studied further with regard to target identification and mode-of-action as well as optimization of their pharmaceutical properties. The former is often achieved by pull-down assays using tagged versions of the bioactive molecule but also the genome of the original producer strain can give important insights into either self-resistance mechanisms and/or modes-of-action. This strategy has been successfully applied for cystobactamids and for griselimycin. The myxobacterial secondary metabolite cystobactamid targets bacterial type IIa topoisomerases which was initially suggested from the analysis of its biosynthetic gene cluster in *Cystobacter* sp. Cbv34, which contains a resistance gene belonging to the pentapeptide repeat protein family that is often found to mediate resistance toward topoisomerase poisons. Further studies revealed novel cystobactamids as promising early lead structures as they possess a favorable antimicrobial profile against multidrug-resistant Gram-negative and Gram-positive pathogens and show only minor cross-resistance with widely applied fluoroquinolone antibiotics. In a similar manner, a gene encoding for a second copy of the DNA polymerase sliding clamp mediating resistance toward griselimycin was discovered within the biosynthesis gene cluster from the producer strain *S. caelicus*. Griselimycins were first described in the 1960s and were recently "rediscovered" because of their strong antimycobacterial activity. Generation of griselimycin-resistant *Mycobacterium* spp. finally leads to the discovery and validation of a new mycobacterial target, the DNA polymerase sliding clamp DnaN, and the description of a rather unusual resistance mechanism via target amplification in mycobacteria. Thus, neglected antibiotics from natural sources serve as highly interesting starting point for antibiotic discovery as their in-depth characterization may reveal new target structures and consequently, these antibiotics are also active on multidrug-resistant species found in the clinic. However, as in the case of griselimycins, the scaffold may be further optimized with regard to its pharmaceutical properties and in vivo efficacy by means of medicinal chemistry approaches. Changing the original scaffold can also lead to an extended activity spectrum as, e.g., in the case of chelocardin. This atypical tetracycline acts on currently undefined membrane-bound target(s) and was already shown to be effective in a human UTI (urinary tract infection) model. Chelocardin is bactericidal on a broad range of multidrug-resistant Gram-negative bacteria but lacks activity on *P. aeruginosa*, an important nosocomial pathogen. As chelocardin can be isolated in high yields from the producer strain *A. sulphurea* semisynthesis and genetic engineering approaches are feasible. In a first attempt, genes from the oxytetracycline producer were introduced into chelocardin producer strain to yield 2-carboxamido-2-deacetyl-chelocardin in high amounts. This new derivative indeed is active on *P. aeruginosa* and is a promising starting point for further optimization of the scaffold by chemical modifications to generate highly potent analogs with even more favorable properties.

In conclusion, novel or underexploited natural product scaffolds with potent antibacterial activity often exhibit new modes-of-action and, thus, might become the heavily demanded resistance-breaking antibiotics. In particular, compounds with activity against either multidrug-resistant Gram-negative bacteria or other hard-to-treat pathogens, such as *M. tuberculosis*, are worth to be studied in-depth. This refers not only to studies on their mechanism-of-action and efficacy, but also to pharmaceutical properties, which both can be significantly improved by means of either exploring more new natural analogs, medicinal chemistry, or genetic engineering approaches for the generation of new antibacterial lead structures.

References

Anonymous (1962) New antibiotic product, its preparation and compositions containing it. GB Patent 966,124

Arias CA, Murray BE (2009) Antibiotic-resistant bugs in the 21st century—a clinical super-challenge. N Engl J Med 360:439–443

Baltz RH (2009) Daptomycin: mechanisms of action and resistance, and biosynthetic engineering. Curr Opin Chem Biol 13(2):144–151

Baumann S, Herrmann J, Raju R, Steinmetz H, Mohr KI, Hüttel S, Harmrolfs K, Stadler M, Müller R (2014) Cystobactamids: myxobacterial topoisomerase inhibitors exhibiting potent antibacterial activity. Angew Chem Int Ed 53(52):14605–14609

Bielecki P, Lukat P, Hüsecken K, Dötsch A, Steinmetz H, Hartmann RW, Häussler S (2012) Mutation in Elongation Factor G Confers Resistance to the Antibiotic Argyrin in the Opportunistic Pathogen *Pseudomonas aeruginosa*. ChemBioChem 13(16):2339–2345

Broenstrup M, Koenig C, Toti L, Wink J, Leuschner W, Gassenhuber J, Müller R, Wenzel S, Binz T, Volz C (2014) Gene cluster for biosynthesis of griselimycin and methylgriselimycin. United States Patent US2014/0295457 A1

Challinor VL, Bode HB (2015) Bioactive natural products from novel microbial sources. Ann N Y Acad Sci 1354:82–97

Chopra I, Roberts M (2001) Tetracycline antibiotics: mode of action, applications, molecular biology, and epidemiology of bacterial resistance. Microbiol Mol Biol Rev 65(2):232–260

Cummings M, Breitling R, Takano E (2014) Steps towards the synthetic biology of polyketide biosynthesis. FEMS Microbiol Lett 351(2):116–125

Emmerson AM, Jones AM (2003) The quinolones: decades of development and use. J Antimicrob Chemother 51(Suppl 1):13–20

Fischbach MA, Walsh CT (2006) Assembly-line enzymology for polyketide and nonribosomal Peptide antibiotics: logic, machinery, and mechanisms. Chem Rev 106(8):3468–3496

Garcia R, Gerth K, Stadler M, Dogma IJ Jr, Müller R (2010) Expanded phylogeny of myxobacteria and evidence for cultivation of the 'unculturables'. Mol Phylogenet Evol 57 (2):878–887

Garcia R, Stadler M, Gemperlein K, Müller R (2015) Aetherobacter fasciculatus gen. nov., sp. nov. and Aetherobacter rufus gen. nov., sp. nov., two novel myxobacteria with promising biotechnological applications. Int J Syst Evol Microbiol. doi:10.1099/ijsem.0.000813

Garrido MC, Herrero M, Kolter R (1988) The export of the DNA replication inhibitor microcin B17 provides immunity for the host cell. EMBO J 7(6):1853–1862

Goss RJ, Shankar S, Fayad AA (2012) The generation of "unnatural" products: synthetic biology meets synthetic chemistry. Nat Prod Rep 29(8):870–889

Gruger T, Nitiss JL, Maxwell A, Zechiedrich EL, Heisig P, Seeber S, Pommier Y, Strumberg D (2004) A mutation in escherichia coli DNA gyrase conferring quinolone resistance results in sensitivity to drugs targeting eukaryotic topoisomerase II. Antimicrob Agents Chemother 48 (12):4495–4504

Hashimi SM, Wall MK, Smith AB, Maxwell A, Birch RG (2007) The phytotoxin albicidin is a novel inhibitor of DNA gyrase. Antimicrob Agents Chemother 51(1):181–187

Hoffmann H, Kogler H, Heyse W, Matter H, Caspers M, Schummer D, Klemke-Jahn C, Bauer A, Penarier G, Debussche L, Brönstrup M (2015) Discovery, structure elucidation, and biological characterization of nannocystin A, a macrocyclic myxobacterial metabolite with potent antiproliferative properties. Angew Chem Int Ed 54(35):10145–10148

Kadota I, Takamura H, Nishii H, Yamamoto Y (2005) Total synthesis of brevetoxin B. J Am Chem Soc 127(25):9246–9250

Kirschning A, Hahn F (2012) Merging chemical synthesis and biosynthesis: a new chapter in the total synthesis of natural products and natural product libraries. Angew Chem Int Ed 51 (17):4012–4022

Kling A, Lukat P, Almeida DV, Bauer A, Fontaine E, Sordello S, Zaburannyi N, Herrmann J, Wenzel SC, König C, Ammerman NC, Barrio MB, Borchers K, Bordon-Pallier F, Brönstrup M, Courtemanche G, Gerlitz M, Geslin M, Hammann P, Heinz DW, Hoffmann H, Klieber S, Kohlmann M, Kurz M, Lair C, Matter H, Nuermberger E, Tyagi S, Fraisse L, Grosset JH, Lagrange S, Müller R (2015) Targeting DnaN for tuberculosis therapy using novel griselimycins. Science 348(6239):1106–1112

Koehn FE, Carter GT (2005) The evolving role of natural products in drug discovery. Nat Rev Drug Discov 4:206–220

Kretz J, Kerwat D, Schubert V, Grätz S, Pesic A, Semsary S, Cociancich S, Royer M, Süssmuth RD (2015) Total synthesis of albicidin: a lead structure from xanthomonas albilineans for potent antibacterial gyrase inhibitors. Angew Chem Int Ed 54(6):1969–1973

Laxminarayan R, Duse A, Wattal C, Zaidi AKM, Wertheim HFL, Sumpradit N, Vlieghe E, Hara GL, Gould IM, Goossens H, Greko C, So AD, Bigdeli M, Tomson G, Woobhous W, Ombaka E, Peralta AQ, Qarnar FN, Mir F, Kariuki S, Bhutta ZA, Coates A, Bergstrom R, Wrigh GD, Brown ED, Cars O (2013) Antibiotic resistance-the need for global solutions. Lancet Infect Dis 13(12):1057–1098

Lešnik U, Lukežič T, Podgoršek A, Horvat J, Polak T, Šala M, Jenko B, Harmrolfs K, Ocampo-Sosa A, Martínez-Martínez L, Herron PR, Fujs Š, Kosec G, Hunter IS, Müller R, Petković H (2015) Construction of a new class of tetracycline lead structures with potent antibacterial activity through biosynthetic engineering. Angew Chem Int Ed Engl 54 (13):3937–3940

Lewis K (2013) Platforms for antibiotic discovery. Nat Rev Drug Discov 12(5):371–387

Li JWH, Vederas JC (2009) Drug discovery and natural products: end of an era or an endless frontier? Science 325(5937):161–165

Lukežič T, Lešnik U, Podgoršek A, Horvat J, Polak T, Šala M, Jenko B, Raspor P, Herron PR, Hunter IS, Petković H (2013) Identification of the chelocardin biosynthetic gene cluster from *Amycolatopsis sulphurea*: a platform for producing novel tetracycline antibiotics. Microbiology 159(12):2524–2532

Mendez C, Salas JA (2001) Altering the glycosylation pattern of bioactive compounds. Trends Biotechnol 19(11):449–456

Molnar V, Matković Z, Tambić T, Kozma C (1977) Clinico-pharmacological investigation of chelocardine in patients suffering from urinary tract infection (author's transl). Lijec Vjesn 99 (9):560–562

Montero C, Mateu G, Rodriguez R, Takiff H (2001) Intrinsic resistance of mycobacterium smegmatis to fluoroquinolones may be influenced by new pentapeptide protein MfpA. Antimicrob Agents Chemother 45(12):3387–3392

Moreno M, Elgaher WAM, Hermann J, Schläger N, Hamed MM, Baumann S, Müller R, Hartmann RW, Kirschning A (2015) Synthesis and biological evaluation of cystobactamid 507: a bacterial topoisomerase inhibitor from cystobacter sp. Synlett 26(09):1175–1178

Müller R, Wink J (2014) Future potential for anti-infectives from bacteria—how to exploit biodiversity and genomic potential. Int J Med Microbiol 304(1):3–13

Nelson ML (1998) Chemical and biological dynamics of tetracyclines. Adv Dent Res 12(2):5–11

Newman DJ (2008) Natural products as leads to potential drugs: an old process or the new hope for drug discovery? J Med Chem 51(9):2589–2599

Nicolaou KC, Winssinger N, Pastor J, Ninkovic S, Sarabia F, He Y, Vourloumis D, Yang Z, Li T, Giannakakou P, Hamel E (1997) Synthesis of epothilones A and B in solid and solution phase. Nature 387(6630):268–272

Oliver TJ, Sinclair AC (1964) Antibiotic M-319. United States Patent US3155582

Payne DJ, Gwynn MN, Holmes DJ, Pompliano DL (2007) Drugs for bad bugs: confronting the challenges of antibacterial discovery. Nat Rev Drug Discov 6:29–40

Pendleton JN, Gorman SP, Gilmore BF (2013) Clinical relevance of the ESKAPE pathogens. Expert Rev Anti Infect Ther 11(3):297–308

Pickens LB, Tang Y (2010) Oxytetracycline biosynthesis. J Biol Chem 285(36):27509–27515

Plaza A, Garcia R, Bifulco G, Martinez JP, Hüttel S, Sasse F, Meyerhans A, Stadler M, Müller R (2012) Aetheramides A and B, potent HIV-inhibitory depsipeptides from a myxobacterium of the new genus "aetherobacter". Org Lett 14(11):2854–2857

Power P, Dunne T, Murphy B, Lochlainn LN, Rai D, Borissow C, Rawlings B, Caffrey P (2008) Engineered synthesis of 7-oxo- and 15-deoxy-15-oxo-amphotericins: insights into structure-activity relationships in polyene antibiotics. Chem Biol 15(1):78–86

Proctor R, Craig W, Kunin C (1978) Cetocycline, tetracycline analog: in vitro studies of antimicrobial activity, serum binding, lipid solubility, and uptake by bacteria. Antimicrob Agents Chemother 13(4):598–604

Rahn N, Kalesse M (2008) The total synthesis of Chlorotonil A. Angew Chem Int Ed Engl 47 (3):2232–2247

Rangel-Vega A, Bernstein LR, Mandujano-Tinoco EA, Garcia-Cibtreras SJ, Contreras SJ, Garcia-Contreras R (2015) Drug repurposing as an alternative for the treatment of recalcitrant bacterial infections. Front Microbiol 6:1–8

Seiple IB, Zhang Z, Jakubec P, Langlois-Mercier A, Wright PM, Hog DT, Yabu K, Allu SR, Fukuzaki T, Carlsen PN, Kitamura Y, Zhou X, Condakes ML, Szczypiński FT, Green WD, Myers AG (2016) A platform for the discovery of new macrolide antibiotics. Nature 533 (7603):338–345

Stepanek JJ, Lukežič T, Teichert I, Petković H, Bandow JE (2016) Dual mechanism of action of the atypical tetracycline chelocardin. Biochim Biophys Acta 1864(6):645–654

Strieker M, Tanovic A, Marahiel MA (2010) Nonribosomal peptide synthetases: structures and dynamics. Curr Opin Struct Biol 20(2):234–240

Sun H, Liu Z, Zhao H, Ang EL (2015) Recent advances in combinatorial biosynthesis for drug discovery. Drug Des Devel Ther 9:823–833

Surup F, Viehrig K, Mohr KI, Herrmann J, Jansen R, Müller R (2014) Disciformycins A and B: 12-membered macrolide glycoside antibiotics from the myxobacterium *Pyxidicoccus fallax* active against multiresistant staphylococci. Angew Chem Int Ed 53(49):13588–13591

Tanner NA, Tolun G, Loparo JJ, Jergic S, Griffith JD, Dixon NE, van Oijen AM (2011) *E. coli* DNA replication in the absence of free β clamps. EMBO J 30(9):1830–1840

Terlain B, Thomas JP (1971) Structure of griselimycin, polypeptide antibiotic extracted from streptomyces cultures. Bull Soc Chim Fr 3:2349–2365

Tran JH, Jacoby GA (2002) Mechanism of plasmid-mediated quinolone resistance. Proc Natl Acad Sci USA 99(8):5638–5642

Vos SM, Tretter EM, Schmidt BH, Berger JM (2011) All tangled up: how cells direct, manage and exploit topoisomerase function. Nat Rev Mol Cell Biol 12(12):827–841

Walsh CT, Haynes ST, Ames BD (2012) Aminobenzoates as building blocks for natural product assembly lines. Nat Prod Rep 29(1):37–59

Wang P, Gao X, Chooi YH, Deng Z, Tang Y (2011) Genetic characterization of enzymes involved in the priming steps of oxytetracycline biosynthesis in Streptomyces rimosus. Microbiology-(UK) 157:2401–2409

Weissman KJ, Müller R (2010) Myxobacterial secondary metabolites: bioactivities and modes-of-action. Nat Prod Rep 27(9):1276–1295

Wenzel SC, Müller R (2009) The impact of genomics on the exploitation of the myxobacterial secondary metabolome. Nat Prod Rep 26(11):1385–1407

Williams GJ (2013) Engineering polyketide synthases and nonribosomal peptide synthetases. Curr Opin Struct Biol 23(4):603–612

Wong FT, Khosla C (2012) Combinatorial biosynthesis of polyketides—a perspective. Curr Opin Chem Biol 16(1–2):117–123

Zhang W, Ames BD, Tsai SC, Tang Y (2006) Engineered biosynthesis of a novel amidated polyketide, using the malonamyl-specific initiation module from the oxytetracycline polyketide synthase. Appl Environ Microbiol 72(4):2573–2580

New Structural Templates for Clinically Validated and Novel Targets in Antimicrobial Drug Research and Development

Philipp Klahn and Mark Brönstrup

Abstract The development of bacterial resistance against current antibiotic drugs necessitates a continuous renewal of the arsenal of efficacious drugs. This imperative has not been met by the output of antibiotic research and development of the past decades for various reasons, including the declining efforts of large pharma companies in this area. Moreover, the majority of novel antibiotics are chemical derivatives of existing structures that represent mostly step innovations, implying that the available chemical space may be exhausted. This review negates this impression by showcasing recent achievements in lead finding and optimization of antibiotics that have novel or unexplored chemical structures. Not surprisingly, many of the novel structural templates like teixobactins, lysocin, griselimycin, or the albicidin/cystobactamid pair were discovered from natural sources. Additional compounds were obtained from the screening of synthetic libraries and chemical synthesis, including the gyrase-inhibiting NTBI's and spiropyrimidinetrione, the tarocin and targocil inhibitors of wall teichoic acid synthesis, or the boronates and diazabicyclo[3.2.1]octane as novel β-lactamase inhibitors. A motif that is common to most clinically validated antibiotics is that they address hotspots in complex biosynthetic machineries, whose functioning is essential for the bacterial cell. Therefore, an introduction to the biological targets—cell wall synthesis, topoisomerases, the DNA sliding clamp, and membrane-bound electron transport—is given for each of the leads presented here.

P. Klahn (✉) · M. Brönstrup (✉)
Department of Chemical Biology, Helmholtz Centre for Infection Research,
Inhoffenstrasse 7, 38124 Braunschweig, Germany
e-mail: philipp.klahn@helmholtz-hzi.de

M. Brönstrup
e-mail: mark.broenstrup@helmholtz-hzi.de

Current Topics in Microbiology and Immunology (2016) 398: 365–418
DOI 10.1007/82_2016_501

Published Online: 22 September 2016

Contents

1 Introduction

The "golden era of antibiotic drug discovery" (Davies 2006) commenced with the breakthrough discovery of the sulfonamides and β-lactams in the 1930s and lasted o the mid 1960s, as the majority of the antibiotics used today had been discovered in that period. Since then, the speed of novel antibiotic discovery has become slower. This is expressed by a declining absolute number of launched antibiotics, and by the fact that only few of those addressed a new mode of action (Policy 2010; Boucher et al. 2013; O'Connell et al. 2013; Butler et al. 2013b). Indeed, most antibiotics launched in the past 30 years are chemical analogs of well-established drug classes. It has to be clearly stated that most of such analogs (so-called 'step-innovations') confer a medical benefit (Wright et al. 2014), e.g., through enlarged safety windows, improved ADMET properties, circumvented bacterial efflux or degradation systems (Brötz-Oesterhelt and Brunner 2008), and sometimes impressive shifts in antibacterial activity from a distinct activity against Gram-positive or Gram-negative bacteria to a broad-spectrum antibiotic activity, as demonstrated by the fourth and fifth generation of cephalosporins, advanced tetracycline, or fluoroquinolone analogs (Wright et al. 2014). However, the evolution of bacterial resistance against these analogs is often faster due to cross resistances (Coates et al. 2011).

Since 2000 five new classes of antibiotics have entered the market (Scheme 1): linezolid (2000), daptomycin (2003), retapamulin (2007), fidaxomicin (2010), and bedaquiline (2012) (Butler et al. 2013b).

These successes are a contribution to improve the innovation gap in antibiotic research and development, but they are clearly insufficient to cover the increasing medical need of a sustainable supply with effective antibiotics (O'Neill 2015). All of these five antibiotics only address infections with Gram-positive pathogens. Furthermore, retapamulin is confined to topical administrations, and fidaxomicin and bedaquiline are narrow spectrum agents that are only approved for treatments of *C. difficile* and MDR-TB, respectively. There remains a particularly strong, unmet need for novel antibiotics effective against Gram-negative pathogens.

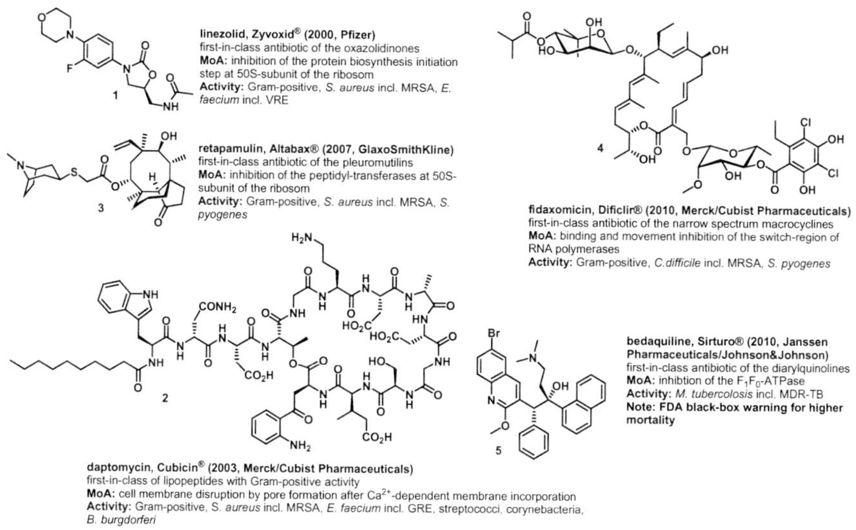

Scheme 1 Antibiotics for the treatment of Gram-positive infections launched since 2000 (*GRE* Glycopeptide-resistant Enterococci, *VRE* Vancomycin-resistant Enterococci, *MRSA* Methicillin-resistant *Staphylococcus aureus*)

Because the lack of innovation (Policy 2010; Boucher et al. 2013; O'Connell et al. 2013) in antibiotic drug research coincides with an increasing occurrence of infections with multi-drug-resistant pathogens (Boucher et al. 2009; Peterson 2009) associated with high mortality and morbidity (Pendleton et al. 2013), there are concerns that the increasing lack of effective therapeutics could develop into a serious threat for public health, or even into a fallback to a so-called pre-antibiotic era (Piddock 2012; Ventola 2015; Wright 2015b).

The anthropological reasons for this situation are manifold; they include the inadequate clinical use of existing antibiotics (Gilbert 2015; Sanchez and Demain 2015; Shiva 2015), extended misuse of antibiotics in intensive animal husbandry for food production (Bengtsson and Greko 2014; Littmann et al. 2015), and the economically driven exodus of big pharma companies from the antibiotics research field that contributed to the innovation gap mentioned above (Lowther 1979; Powers 2003; Projan 2003; Spellberg et al. 2004; Taubes 2008; Torres 2010; O'Connell et al. 2013).

Beyond these anthropological acceleration forces (Breu et al. 2001; Gillings 2013), we have to accept that bacterial resistance to antibiotics is not a side effect of modern drug therapy, but an inherent part of bacterial evolution to fight for their evolutionary niche with other bacteria and further organisms (Wright 2012; Wright and Poinar 2012; Rodríguez-Rojas et al. 2013).

It has been estimated that bacteria-producing antibacterial metabolites originated at least hundreds of millions of years ago (Baltz 2008; Wright and Poinar 2012). During the evolution of antibacterial metabolites, these bacteria had to intrinsically coevolve resistance mechanisms for self-preservation (Wright 2012; Wright and Poinar 2012; Perry et al. 2014), since in most cases they probably possess the same biological target for the drug. Therefore, it can be assumed that resistance mechanisms have existed for just as long as the corresponding antibacterial metabolites (O'Connell et al. 2013). This hypothesis is supported by a metagenomic analysis of DNA found in permafrost sediments, which was determined to be 30,000 years old and led to the discovery of genes encoding for resistance to β-lactams, tetracyclines, and glycopeptide antibacterials (D'Costa et al. 2011). This is most probably also true for purely synthetic antibiotics, as nature provides natural product antibiotics directed to nearly every known druggable target in bacteria (Lin et al. 1997; Keller et al. 2007; Johnston et al. 2016). Furthermore, we have to consider that every time an antibiotic is administered there is a significant influence on the resistome (Gillings 2013), which is defined as the collection of all genes in pathogenic and non-pathogenic bacteria that could contribute to a phenotype of antibiotic resistance (Frankel et al. 2006; Wright 2007, 2010; Forsberg et al. 2012; Gillings 2013; Nesme and Simonet 2015). This influence can lead to the evolution of resistance also in non-pathogenic bacteria in the patient or after clearance of the drug into the general environment, i.e., in organisms which have not been originally targeted by the treatment. The resistance genes can easily spread via horizontal gene transfer (HGT) (Syvanen 2012; Baltrus 2013; Polz et al. 2013) by mobile plasmids (Smillie et al. 2010; Harrison and Brockhurst 2012; Carattoli 2013), transposons (Casacuberta and Gonzalez 2013), or outer membrane vesicle (OMV) (Berleman and Auer 2013; Brown et al. 2015; Perez-Cruz et al. 2015; Schwechheimer and Kuehn 2015) transport through the whole pan-genome (Medini et al. 2005; Tettelin et al. 2008; Lapierre and Gogarten 2009) of the global microbiome (Whitman et al. 1998).

These facts imply that in pronounced contrast to other medical indications, the efficacy of antibacterial drugs deteriorates over time. Therefore, the identification of novel antimicrobials, especially with new modes of action (Fischbach and Walsh 2009; Wattal and Goel 2011), is a continuous, necessary task to keep a life-saving headway in the permanent race between bacterial evolution and the protection of human health (Rodríguez-Rojas et al. 2013). In addition, the way antibiotics are handled today should be seriously revised, since studies have shown that smart policies for the prudent use of antibiotics in the clinic and throughout agriculture can make a significant difference in the occurrence and the level of resistance (European Centre for Disease Prevention and Control: Annual Report of the European Antimicrobial Resistance Surveillance Network, EARS-Net 2012).

The threat of multi-drug-resistant pathogens has already been recognized and gained international political attention, leading to several national and international programs and initiatives for the research and development of novel antibiotics (Policy 2010; Boucher et al. 2013; Rex 2014; Bush 2015; Eichberg 2015). At this stage, most of the compounds that have recently been launched or that are currently undergoing phase II/III clinical trials are analogs of existing classes (Pucci and Bush 2013; Hesterkamp 2016).

As the marketed classes of antibiotics have been extensively reviewed before (Butler et al. 2013b; Zetts 2014; Paris 2015), we would like to focus this review on the feasibility to discover novel structural templates with antibiotic activity, and to advance such compounds to late-stage pre-clinical as well as clinical development (Boucher et al. 2013; O'Connell et al. 2013; Butler et al. 2013b; Brown et al. 2014; Xu et al. 2014a; Bush 2015; Paris 2015). For this purpose, a brief overview of existing druggable targets and common features among them is given, followed by a review of promising novel scaffolds that address existing as well as novel targets. The presented scaffolds result from a personal and non-comprehensive selection of compounds addressing essential bacterial machineries, i.e., DNA replication, cell wall synthesis, and membrane components.

2 Lessons Learned from Druggable Targets in Bacteria

In the over one hundred years of antibiotic research, numerous different bacterial targets and their corresponding interaction with antibiotics have been investigated (Tommasi et al. 2015). As in other indications, the majority of drug-like, optimized compounds did not reach the clinics, as they addressed non-valid targets hampered by an insufficient conservation in the microbial spectrum, unexpected metabolic bypass mechanisms, non-essentiality under in vivo conditions or unexpectedly rapid resistance formations (Payne et al. 2007). A special characteristic of antibiotic R&D is the dominance of natural products as the source of approved drugs (Koehn and Carter 2005; Bérdy 2012; Newman and Cragg 2012; Kirst 2013; Bauer and Brönstrup 2014; Butler et al. 2014; Schaefer 2014; Harvey et al. 2015). In fact, the far majority of antibiotics on the market are derived from secondary metabolites of bacteria (Gerth et al. 2003; Clardy et al. 2006; Diez et al. 2012; Schäberle et al. 2014; Elshahawi et al. 2015), marine microorganisms (Blunt et al. 2011), fungi (Schueffler and Anke 2014; Stadler and Hoffmeister 2015), and (in rare cases) plants (Gibbons 2004; Savoia 2012; Upadhyay et al. 2014). This is no surprise, considering that nature provides a huge diversity of structures, which have been evolutionary optimized for binding their biological targets and used by their producers to fight for their ecological niche against competing or harmful microbes. Even though most of the antibacterial natural products need further semi-synthetic optimization to fulfill the ADME (Absorption, Distribution, Metabolism, and Excretion) properties required for a drug applied to humans, nature is still the most effective source for novel antibacterial lead structures (Cragg and Newman 2013; Brown et al. 2014). Natural products have also often disclosed novel, relevant biological targets, which were subsequently addressed by synthetic compounds.

In Table 1, the most important approved antibiotic classes are summarized according to the metabolic pathway they address, their molecular targets, and their distinct mechanism of action. The targets they address share important common

Table 1 Bacterial targets and approved antimicrobial compounds addressing them (Brötz-Oesterhelt and Brunner 2008; Kohanski et al. 2010; Lewis 2013; Butler et al. 2013b; Wright et al. 2014; Bush 2015)

Targeted biological pathway	Molecular target	Class of antibiotics (year of discovery)	Mode of action/mechanism of action	First-in-class (year of introduction)	Approved representatives of the class	Origin	Activity
Peptidoglycan biosynthesis (PGS)	Penicillin binding proteins (PBPs), transpeptidases	Penicillins (Miller 2002; Zaffiri et al. 2012) (**1928**)	Inhibition of different transpeptidases	Penicillin (1938)	Ampicillin, oxacillin, methicillin, carbenicillin,	Natural	Gram+/–[a]
		Cephalosporins (Kalman and Barriere 1990; Zaffiri et al. 2012) (**1955**)		Cephalothin (1964)	Ceftobiprole, cefepime, ceftriaxone, cefazolin, cefoxitin,	Natural	Gram+/–[a]
		Carbapenems (Singh 2004; Nicolau 2008; Papp-Wallace et al. 2011; Hawkey and Livermore 2012) (**1976**)		Imipenem (1985)	Doripenem, meropenem, ertapenem	Natural	Gram+/–[a]
		Monobactams (Singh 2004) (**1978**)		Aztreonam (1986)	Tabtoxin, nocardicin A, tigemonam	Natural	Gram+/–[a]
	MurA1 and MurA2	Fosfomycins (Michalopoulos et al. 2011) (**1969**)	Blocking the formation of N-acetylmuramic acid	Fosfomycin (1971)	–	Natural	Gram+/–[a]
	Alanine–-racemases and D-Ala–D-Ala ligases	–	Competitive enzyme inhibition	D-cycloserine	–	Natural	*M. tuberculosis*
	Lipid II	Glycopeptides (Hammes and Neuhaus 1974; Watanakunakorn 1984; Williams and Bardsley 1999; Kahne et al. 2005) (**1953**)	Inhibition of peptidoglycan cross linkage by binding to Acyl-D-Ala-D-Ala dipeptide terminus of Lipid II	Vancomycin (1955)	Teicoplanin, bleomycin, telavancin	Natural	Gram+[a]
		Lantibiotics Type A ('t Hart et al. 2016) (**1928**)	Binding to Lipid II by recognition of the PP unit leading pore formation	Nisin[c] (2009)	–	Natural	Gram+
DNA replication and transcription	DNA gyrase (topoisomerase II) and topoisomerase IV	Fluoroquinolones and quinolones (Lipsky and Baker 1999; Oliphant and Green 2002; Aldred et al. 2014; Redgrave et al. 2014) (**1961**)	Formation of ternary complex with DNA and the GyrA subunit of topoisomerase II or IV	Nalidixic acid (1962)	Norfloxacin, moxiflocacin, gemifloxacin, ciprofloxacin	Synthetic	Gram+/–[a]
		Aminocoumarines (Hooper et al. 1982; Heide 2014) (**1955**)	Binding to the ATP binding site (GyrB subunit)	Novobiocin (1965)	Chlorobiocin, coumermycin A	Natural	Gram+
Folic acid metabolism	Dihydrofolate reductase (DHFR)	Benzyl diamino pyrimidines (Hawser et al. 2006; Sharma and Chauhan 2012) (**1962**)	Inhibition of DHFR (synergistic effects with sulfonamides)	Trimethoprim (1962)	–	Synthetic	Gram+/–[a]
	Dihydropteroate synthetase (DHPS)	Sulfonamides (Henry 1943; Maren 1976; Connor 1998) (**1932**)	Inhibition of DHPS (strong synergistic effects with benzyldiamino pyrimidines)	Prontosil (1932) sulfanilamid (1936)	Sulfamethylthiazol, sulfadiazine, sulfamethoxazole, sulfasalazine	Synthetic	Gram+

(continued)

Table 1 (continued)

Targeted biological pathway	Molecular target	Class of antibiotics (year of discovery)	Mode of action/mechanism of action	First-in-class (year of introduction)	Approved representatives of the class	Origin	Activity
Protein biosynthesis (PBS)	30S ribosomal subunit	Aminoglycosides (Davis 1987; Kotra et al. 2000) (**1943**)	Binding to the 16S rRNA near A site of ribosomal 30S subunit, disturbance of proofreading	Streptomycin (1946)	Gentamicin, tobramycin, kanamycin, neomycin	Natural	Gram+/−[a]
		Tetracyclines (Chopra and Roberts 2001; Thaker et al. 2010) (**1944**)	Binding to 16S rRNA at A site of the ribosomal 30S subunit, blocking of the attachment of charged aminoacyl-tRNA	Aureomycin (1948)	Doxycycline, minocycline, tigecycline evaracycline	Natural	Gram+/−[a]
	50S ribosomal subunit	Lincosamides (Giguère 2013) (**1963**)	Binding to domain V of 23S rRNA in PTC of ribosomal 50S subunit at A and P site, peptidyl transferase reaction blocked	Clindamycin (1967)	Pirlimycin, lincomycin	Natural	Gram+, anaerobic bacteria
	50S ribosomal subunit	Chloramphenicols (Yunis 1988) (**1946**)	Binding to A2451 and A2452 residues in the 23S rRNA of the 50S subunit	Chloramphenicol (1948)	Thiamphenicol	Natural	Gram+/−[a]
		Macrolides (Kannan and Mankin 2011) (**1948**)	Binding in the NPET close to PTC, blocking of elongation	Erythromycin (1951)	Clarithromycin, azithromycin, telithromycin, solithromycin	Natural	Gram+/−[a]
		Streptogramins (Giguère 2013; Mast and Wohlleben 2014) group A (**1963**) Marolactones	Binding to a tight pocket within the PTC (synergistic effects with group B)	Pristinamycin IIA (1998)	Dalfopristin	Natural	Gram+
		Streptogramins (Giguère 2013; Mast and Wohlleben 2014) group B (**1952**) cyclic hexapeptides	Binding to the 23S rRNA within NPET (synergistic effects with group A)	Virginiamycin S1	Quinupristin		
		Oxazolidinones (Bozdogan and Appelbaum 2004) (**1955**)	Binding to domain V of 23S rRNA in PTC of the ribosomal 50S subunit, initiation blocked	Linezolid (2000)	Radezolid, tedizolid	Synthetic	Gram+
		Pleuromutilins (Novak 2011; Giguère 2013) (**1950**)	Inhibition of peptidyl transferase at 50S ribosomal subunit	Repatamulin (2007)	–	Natural	Gram+
	EF-G elongation factor	Fusidanes (Godtfredsen et al. 1962; ZHao et al. 2013) (**1962**)	Binding to EF-G at the interface between domains G, II and IV, release of EF-G/GDP complex blocked	Fusidic acid (1968)	–	Natural	Gram+

(continued)

Table 1 (continued)

Targeted biological pathway	Molecular target	Class of antibiotics (year of discovery)	Mode of action/mechanism of action	First-in-class (year of introduction)	Approved representatives of the class	Origin	Activity
DNA transcription	RNA polymerase (RNAP)	Ansamycins (Floss and Yu 2005) (**1957**)	Binding to β-subunit of RNAP and blocking the path of the growing RNA chain	Rifampicin (1971)	Rifampin, rifapentine	Natural	Gram+, *M. tuberculosis*
		Macrocyclines (Mullane and Gorbach 2011; Scott 2013; Butler et al. 2013b; Zhanel et al. 2015) (**1948**)	Binding to the switch region of RNAP, movement inhibition	Fidaxomycin (2010)	–	Natural	Gram+
Cell membrane Biogenesis and functionality	Lipopolysaccharid (LPS) of outer membrane	Polypeptides (Yahav et al. 2011; Biswas et al. 2012) (**1959**)	Binding to LPS leads to electrostatic interaction and displacement of divalent cations	Colistin (1959)	Polymyxin B (1959)	Natural	Gram-
	Cell membrane	Lipopeptides (Anderson et al. 2012; Butler et al. 2013b; Epand et al. 2016) (**1986**)	Ca^{2+}-dependent incorporation to the lipid bilayer and oligomerization leads to pore formation	Daptomycin (2003)	–	Natural	Gram+
DNA integrity	Anaerobic electron transport proteins	Nitroimidazoles (Edwards 1993; Mital 2009) (**1955**)	Reactive nitroso radicals in vivo, (prodrugs)	Azomycin (1955)	Metronidazol, tinidazol, nimorazol, dimetridazol, megazol, ornidazol	Natural/synthetic	*T. vaginalis*, Anaerobic Gram+/–[a]
Enzyme functionality	Cysteine-containing proteins	Arsenicals (Zaffiri et al. 2012) (**1905**)	Oxidation to arsenic oxide in vivo, which crosslink sulfhydryl groups of proteins	Salvarsan (1911)	Neosalvarsan, solusalvarsan	Synthetic	Gram+/–[a]
Energy supply	F_1F_0-ATPase	Diarylquinolines (Cole and Riccardi 2011; Butler et al. 2013b) (**1997**)	Inhibition of ATP synthesis	Bedaquiline (2012)	–	Synthetic	*M. tubercolosis*[b]

[a]Broad spectrum activity
[b]Narrow spectrum activity
[c]FDA approved as additive for food use
PTC Peptidyl transferase catalytic centre, *NPET* Nascent peptide exit tunnel, *PP* Pyrophosphate

features, i.e., essentiality for the survival of the pathogen, an evolutionary conservation in a variety of pathogens, and a high evolutionary distance to the mammalian counterparts (Brötz-Oesterhelt and Brunner 2008). Antibiotics mainly applied in monotherapies such as β-lactams, glycopeptides, or fluoroquinolones show additional common characteristics: The most successful antibiotics all interact with large biological structures consisting of multi-enzyme complexes like the protein or cell wall biosynthesis machinery, whose impairment has drastic consequences for the bacterial cell that are hard to repair or compensate (Table 1) (Silver 2007; Brötz-Oesterhelt and Brunner 2008). In contrast, the few antibiotics targeting single enzymes such as sulfonamides and benzyl diaminopyrimidines have to be applied in combination therapies to avoid rapid resistance.

Furthermore, it is noticeable that there are fewer antibiotics available for the treatment of infections caused by Gram-negative pathogens compared to Gram-positive ones. The main reason for that issue is the difficulty for antibiotics to cross the outer membrane of Gram-negative bacteria (Zgurskaya et al. 2015). The surface of the outer membrane of the Gram-negative bilayer is covered with lipopolysaccharides (LPS), lipidic structures embedded and anchored into the bilayer that consist of three, covalently connected structural elements: A proximal hydrophobic lipid A, a core oligosaccharide region, and a distal O-antigen polysaccharide (Cohen 2011). It has been shown that LPS molecules interact with each other on the cell surface through divalent cations, thereby forming a permeability barrier (Nikaido 2003; Zhang et al. 2013). Due to the embedded cations, hydrophobic molecules have been shown to partition poorly into LPS and to permeate across the outer membrane bilayer with extremely low rates. Therefore, the LPS-containing bilayer serves as an efficient barrier for diffusion-mediated uptake of many chemical scaffolds known to be active against Gram-positive bacteria (Zhang et al. 2013). The uptake of most antibiotics into Gram-negative bacteria is mediated by porins, β-barrel-shaped proteins that act as a pore through which small, polar molecules can diffuse. Because the physicochemical constraints for porin diffusion are hardly met by existing compound collections designed for high affinity binding of canonical eukaryotic targets and for assuring oral bioavailability, attempts to discover novel lead structures active against Gram-negative pathogens have been by-and-large unsuccessful (Payne et al. 2007; Tommasi et al. 2015). Thus, the searches for relevant chemical matter (e.g., natural products preoptimized by microorganisms) or for alternative strategies [e.g., active transport machineries of Gram-negative pathogens (Ji et al. 2012; Górska et al. 2014; Mislin and Schalk 2014; Saha et al. 2013; Wang et al. 2014; Johnstone and Nolan 2015; Page 2013; Górska et al. 2014; Mislin and Schalk 2014; Xu et al. 2014b; Zgurskaya et al. 2015] need to be intensified significantly.

3 Lipid II and Lipid III—The Bottlenecks in Peptidoglycan and Wall Teichoic Acid Synthesis

Known, clinically validated targets represent the most promising choice for the development of novel antibiotics, in particular when novel chemical scaffolds that address different binding sites compared to existing drugs break resistance.

An excellent example for such a target is lipid II **6** (Scheme 2) (Breukink and de Kruijff 2006; de Kruijff et al. 2008), an amphiphilic, membrane-anchored peptidoglycan precursor molecule that is essential for cell membrane functionality.

The cell walls of all bacteria contain a layer of peptidoglycan, a biopolymer of the alternating amino sugars *N*-acetylglucosamine (D-NAc-Glu, NAG), and *N*-acetylmuramic acid (NAc-Mur, NAM), which is modified with pentapeptides of the sequence L-alanyl-γ-D-glutamyl-L-lysyl-D-alanyl-D-alanine (in the case of *S. aureus*) that is attached to the 3-hydroxy group of the NAc-Mur sugar (Schwartz et al. 2001; Breukink and de Kruijff 2006; de Kruijff et al. 2008). The cross linkage of the peptides by penicillin-binding proteins (PBPs) to macromolecules confers a structural rigidity and mechanical strength that prevents the bacterial protoplast to burst under the osmotic internal pressure.

The synthesis of lipid II as the precursor monomer of the peptidoglycan layer begins on the cytoplasmic site of the plasma membrane (Scheme 3, right side). *N*-Acetyl-muramic-acid-pentapeptide uridinyl pyrophosphate (UDP-NAc-Mur) is attached by the enzyme phosphor-MurNAc-pentapetide translocase (MraY) (Chung et al. 2013) to bactoprenol phosphate, which is embedded in the phospholipid membrane. The resulting membrane-anchored lipid I is then further glycosylated by the peripherally membrane-associated enzyme MurG (Mohammadi et al. 2007) leading to the formation of lipid II. Next, the FemXAB transpeptidases catalyze the coupling of a pentaglycine-moiety to the D-lysine of the pentapeptide unit and furthermore, the amidation of the γ-D-glutamate of the pentapeptide unit by the MurT/GatD two-enzyme complex leading to the corresponding D-*iso*glutamin (Münch et al. 2012; Zapun et al. 2013). This modified lipid II now contains the

Scheme 2 Structures of lipid II and binding sides of selected antibiotics

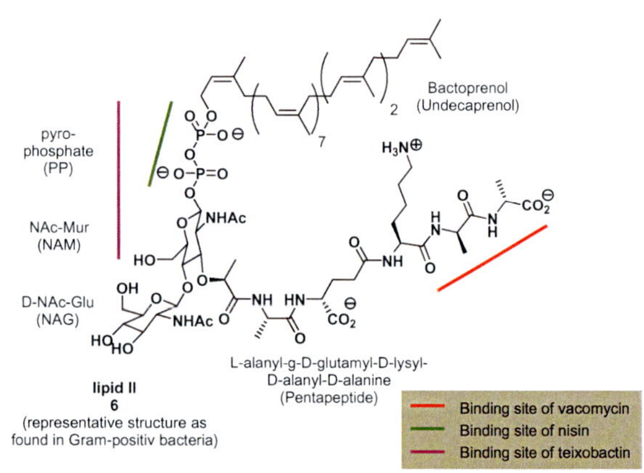

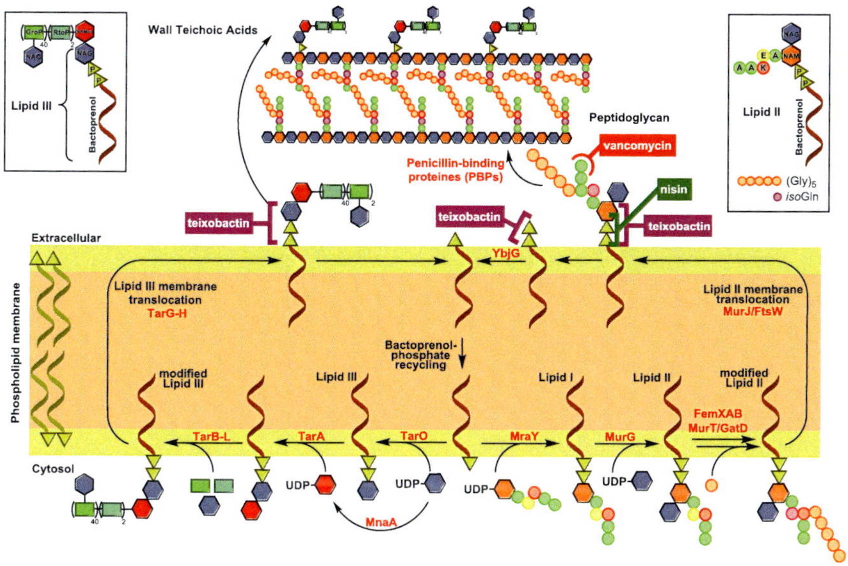

Scheme 3 Peptidoglycan and wall teichoic acid biosynthesis in Gram-positive bacteria, interactions with selected antibiotics

complete peptidoglycan monomer linked via a pyrophosphate to the bactoprenol membrane anchor, which is translocated through the phospholipid membrane by the MurJ flippase and presented on the extracellular membrane surface (Mohammadi et al. 2011; Butler et al. 2013a; Sham et al. 2014; Meeske et al. 2015).

On the extracellular membrane surface the peptidoglycan precursor is released, incorporated into the growing peptidoglycan layer and crosslinked (Zapun et al. 2013) with each other catalyzed by the PBPs (Sauvage et al. 2008).

The peptidoglycan layer in Gram-positive bacteria is generally around 20–80 nm thick and contains up to 20 sublayers with vast amounts of peptidoglycan subunits, which all have to be translocated through the membrane in the form of the membrane-anchored lipid II after assembly on the cytosolic membrane surface. Since supply with bactoprenol phosphate in bacteria is restricted (Storm and Strominger 1974; Kramer et al. 2004), the amount of lipid II that can be generated is limited. Thus, the lipid II cycle is a highly dynamic process that represents the bottleneck in bacterial cell wall synthesis. Therefore, the lipid II cycle is an ideal target for antibiotics (Breukink and de Kruijff 2006).

The first approved antibiotic targeting lipid II was the glycopeptide vancomycin **7** (Scheme 4). Glycopeptide antibiotics are active against a broad variety of Gram-positive pathogens and act by binding to the D-Ala-D-Ala-dipeptide terminus in lipid II on the extracellular membrane surface via several hydrogen bonds and ionic interactions (Jia et al. 2013; Münch et al. 2015).

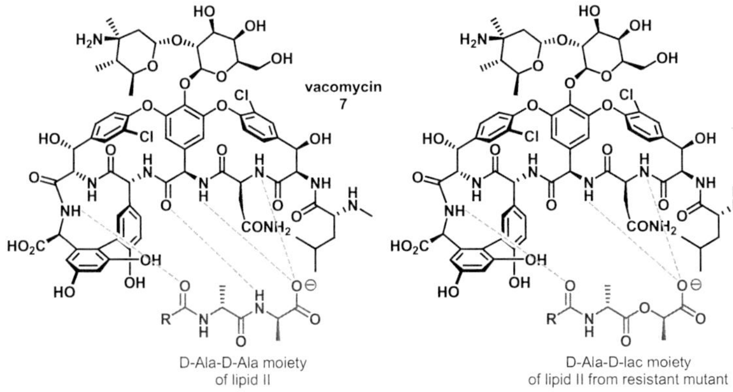

Scheme 4 Vancomycin and its binding mode to the D-Ala-D-Ala moiety of lipid II

Vancomycin has been for a long time the antibiotic of last resort against Gram-positive pathogens (Williams and Bardsley 1999). Due to the difficulty for bacteria to circumvent binding to an essential metabolite (rather than a protein target), the first significant levels of resistance to vancomycin took almost three decades to occur (Boneca and Chiosis 2003). However, nowadays there is an increasing prevalence of resistance against vancomycin (Williams and Bardsley 1999), which is caused in most cases by the exchange of the terminal D-alanine of the lipid II structure by a D-lactate (Williams and Bardsley 1999; Kahne et al. 2005), which results in the loss of a single hydrogen bond to the D-Ala residue, and a 100-fold drop in binding affinity (Mccomas et al. 2003).

Another well-investigated antibiotic interacting with lipid II is the peptide 1 antibiotic nisin **8** (Scheme 5). Nisin, obtained by fermentation of *Lactobacillus lactis* on natural substrates such as milk or dextrose, is widely used as a safe food preservation additive (Cleveland et al. 2001), since orally administered nisin is completely degraded by digesting enzymes in the stomach.

Nisin binds to the pyrophosphate moiety of lipid II (Scheme 5) forming a phosphate cage that prevents the release of the peptidoglycan subunits and leads to the formation of a pore in the membrane via inter-membrane assembly of a nisin–lipid II 8:4 complex (Hsu et al. 2004; Bauer and Dicks 2005; Martin and Breukink 2007).

The first isoprene unit and the pyrophosphate in lipid II and the N-terminal part of nisin have been identified as determinants for the nisin–lipid II interaction (Chan et al. 1989; Kuipers et al. 1993; Brötz et al. 1998; Wiedemann et al. 2001). Additionally, several other lipid II interacting compounds are currently under preclinical and clinical investigation such as ramoplanin, mannopeptimycin, lysobactin (Lee et al. 2016b), and plusbacin A_3 (Breukink and de Kruijff 2006).

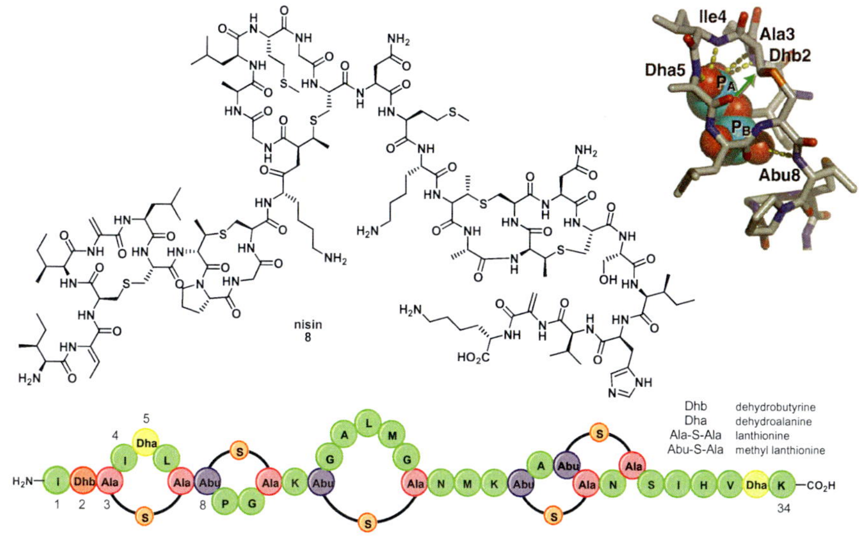

Scheme 5 Structure and amino acid sequence of nisin and its binding mode to lipid II

	Pathogen	Teixobactin MIC µg/mL
Gram-positive	*S. aureus (MRSA)*	0.25
	E. faecalis (VRE)	0.5
	E. faecium (VRE)	0.5
	S. pneumoniae[a]	<0.03
	S. pyogenes	0.06
	S. agalactiae	0.12
	B. anthracis	<0.06
	C. difficile	0.005
	P. acnes	0.08
	M. tuberculosis (H37Rv)	0.125
	M. catarrhalis	2
Gram-negative	*H. influenza*	4
	E. coli	25
	E. coli (asmB1)	2.5
	P. aeruginosa	>32
	K. pneumoniae	>32

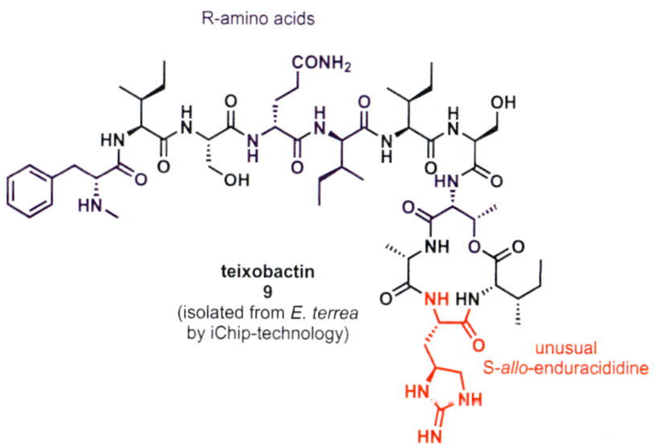

Scheme 6 Structure and antibacterial activity of teixobactin

Recently, a novel lipid II interacting cyclodepsipeptide has been isolated from the Gram-negative β-proteobacterium *Eleftheria terrae* (Ling et al. 2015). Teixobactin **9** (Scheme 6) is an undecapeptide bearing four R-amino acids and seven S-amino acids including the unusual S-*allo*-enduracididine. The N-terminal R-phenylalanine is mono methylated, and the C-terminus of the molecule is cyclized to the hydroxyl function of a R-threonine side chain in the sequence, forming a 13-membered depsipeptide ring.

Teixobactin was discovered within a screen of so far uncultured soil bacteria by in situ cultivation in diffusion chambers using the so-called iChip technology (Nichols et al. 2010; Piddock 2015). The compound showed highly potent antibacterial activity against a broad variety of Gram-positive pathogens including several multi-drug-resistant strains such as methicillin-resistant *S. aureus* (MRSA) and vancomycin-resistant *E. faecalis* (VRE). Teixobactin was also highly active against *C. difficile* (MIC = 0.005 μg/mL) and *M. tuberculosis* (MIC = 0.125 μg/mL). The compound was ineffective against Gram-negative bacteria with the exception of *E. coli* (asmB1), a mutant having a defective outer membrane barrier.

In vitro experiments have shown that teixobactin bound under formation of stable 2:1 drug:target complexes to lipid I and lipid II by interaction with the pyrophosphate sugar moiety (Schemes 2 and 3) (Ling et al. 2015). Furthermore, at higher concentrations teixobactin was able to completely inhibit the YbjG-catalyzed mono-dephosphorylation of bactoprenol pyrophosphate, which is an essential step of the bactoprenol phosphate recycling with in the lipid II cycle of peptidoglycan biosynthesis (Scheme 3).

In vivo the binding to lipid II is believed to be the primary effect of teixobactin, since lipid I is not exposed to the extracellular site. Nevertheless, the exact nature of the sugar in the binding motif seems to be less important, since it has been demonstrated that teixobactin efficiently bound lipid III **10** (Scheme 7), bearing a NAc-Glu instead of NAc-Mur attached to the pyrophosphate.

Lipid III is a precursor molecule in the synthesis of wall teichoic acids (WTA). Wall teichoic acids are glycol-phosphopolyol biopolymers that are attached to the NAc-Mur moieties on the surface of the peptidoglycan layer of Gram-positive bacteria (Pereira et al. 2008; Brown et al. 2013), rendering the bacterial surface strongly anionic (Scheme 3).

WTA biosynthesis inhibition is not per se lethal for Gram-positive bacteria. Early-stage WTA inhibition can lead to cells showing enhanced sensitivity to certain drugs (see next chapter) (Sewell and Brown 2014; Lee et al. 2016a). In contrast, late-stage inhibition of membrane-bound WTA precursors is lethal due to accumulation of toxic intermediates and the depletion of cellular pools of

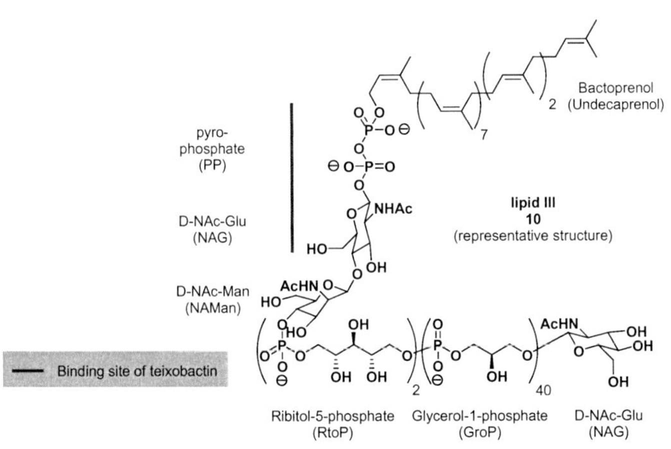

Scheme 7 Structures of lipid II and binding side of teixobactin

bactoprenol phosphate (D'Elia et al. 2006b; Swoboda et al. 2011; Brown et al. 2013; Wang et al. 2013), which is essential for peptidoglycan biosynthesis as discussed above. Furthermore, it is known that teichoic acids anchor autolysins to prevent uncontrolled hydrolysis of peptidoglycan (Bierbaum and Sahl 1985; Dubée et al. 2011). Thus, inhibition of WTA synthesis leads to enhanced concentrations of free autolysins, which significantly contributes to the outstanding antibacterial activity of teixobactin. With its binding to lipid II and lipid III, teixobactin is targeting two non-protein lipid structures that are bottlenecks of cell wall synthesis in Gram-positive bacteria (Wright 2015a). It is therefore no surprise that the escape strategies for the pathogens are limited. Whether a resistance mechanism already exists is not known, but so far it was not possible to obtain teixobactin-resistant mutants of *S. aureus* or *M. tuberculosis* at 4x MIC or in serial passaging experiments (Ling et al. 2015). As the teixobactin producer is a Gram-negative bacterium, it is intrinsically resistant against re-entry of the drug due to the impermeability of the outer membrane, and therefore does not encode a resistance mechanism that could be horizontally transferred to Gram-positive species (Kåhrström 2015).

Preliminary pharmacokinetic and pharmacodynamics data from *in vivo* studies in mice were promising (100 % survival rate in a mouse efficacy study). The compound retained its potency in blood serum, showed overall good stability and low toxicity to mammalian cells. Therefore, teixobactin could become a valuable candidate to fight back the antibiotic crisis (Scheme 8).

Not surprisingly, the compound has already motivated medicinal chemists to synthesize the first analogs (Jad et al. 2015; Parmar et al. 2016). Two research groups independently published the simple solid-supported synthesis of the teixobactin analog **11** (Scheme 7), where the unusual S-*allo*-enduracididine is substituted by the structurally related, simpler arginine. In a second analog (**12**), the four R-amino acids were additionally substituted by S-amino acids. Antibacterial tests demonstrated that substitution of S-*allo*-enduracididine leads to a 10-fold drop in activity for analog **11**. Furthermore, the R-configuration of several amino acids in the teixobactin sequence seems to be essential for the activity, since analog **12** leads to a 640-fold drop in activity compared to natural teixobactin. Nevertheless, more detailed structure–activity relationship (SAR) studies are needed to determine the exact binding interactions and to further optimize this emerging natural lead structure.

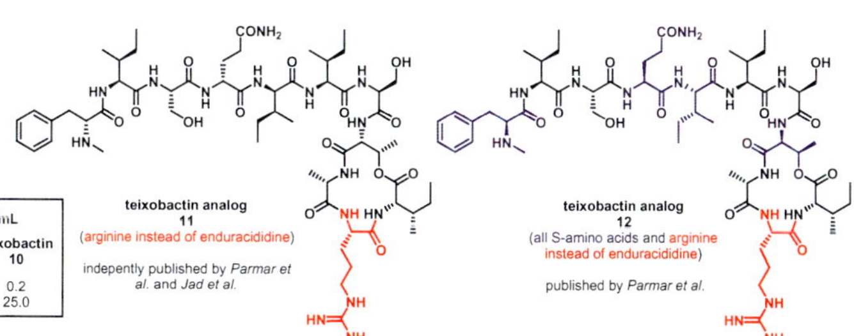

Gram-positive pathogen	MIC in ng/mL 11	12	teixobactin 10
S. aureus (25923)	1.6	128	0.2
E. coli (25922)	51.3	GAW‡	25.0

‡ Grown in all wells.

Scheme 8 Recently published analogs of teixobactin and their antibacterial activity

4 Early and Late-Stage WTA Biosynthesis Inhibition—Restoring the Efficacy of β-Lactams Against Gram-Positive Pathogens

The predominant resistance mechanisms toward β-lactams in Gram-negative and Gram-positive pathogens are different. Resistance in Gram-negative pathogens normally is achieved via two general mechanisms. One escape strategy is the production of degradation enzymes such as β-lactamases, rendering the antibiotic inactive due to hydrolysis of the β-lactam pharmacophore (Bush 2010; Rawat and Nair 2010). Alternatively, several Gram-negative pathogens overexpress membrane-embedded efflux pumps, which actively transport drugs out of the bacteria (Amaral et al. 2014; Blair et al. 2014). Therefore, considerable efforts have been made to restore the antibacterial activity of β-lactams against Gram-negative pathogens through inhibition of β-lactamases or of efflux pumps. While marketed β-lactamase inhibitors utilize a β-lactam motif themselves for suicide inhibition, chemists have recently identified and optimized novel structural templates: 1,6-diazabicyclo[3.2.1]octane-2-carboxamide-based, reversible β-lactamase inhibitors such as avibactam **13**, relebactam **14,** and OPO595 **15** (Scheme 9) have been advanced to clinical trials as additive in combination therapies with different β-lactam antibiotics (Hirsch et al. 2012; Livermore et al. 2013; Lapuebla et al. 2015a; Livermore et al. 2015; Paris 2015; Toussaint and Gallagher 2015). Avibactam has already been approved for combination therapy with ceftazidime and is in clinical trials for combination therapy with several other β-lactams (Zetts 2014; Paris 2015). Furthermore, RPX7009 **16**, a novel boronic acid-based β-lactamase inhibitor, is currently in phase III clinical trials in combination therapy with meropenem. Notably, some congeners of the boronic acid-based class of compounds also cover metallo-β-lactamases like NDM-1 in addition to serine-β-lactamases (Hecker et al. 2015; Paris 2015; Lapuebla et al. 2015b). The

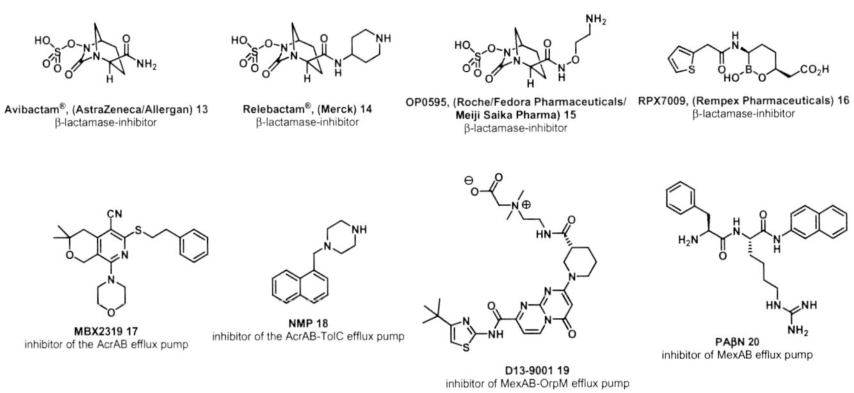

Scheme 9 Structures of advanced β-lactamase inhibitors and efflux pump inhibitors

significant progress achieved on this target class contributes to alleviate the upcoming antibiotic crisis in short to medium term.

Also, several promising efflux pump inhibitors for the AcrAB-TolC and MexAB-OrpM efflux pumps have been discovered (Scheme 9) (Askoura et al. 2011; Tegos et al. 2011; Opperman and Nguyen 2015; Venter et al. 2015), but so far none of them has entered the clinic.

The predominant resistance mechanism against β-lactams in clinically important Gram-positive pathogens is different, thus requiring a different strategy: It has been shown that resistance in MRSA is achieved by the horizontally acquired penicillin binding protein PBP2A, which has a considerably lower affinity for β-lactam antibiotics compared to the original, bifunctional transglycosylase/transpeptidase PBP2 (Pinho and Errington 2005; Lim and Strynadka 2002). β-Lactams acylate the active site for transpeptidation on PBP2 and thereby inhibit the autonomous localization to the division septum. In MRSA, PBP2A can take over the transpeptidase activity and additionally, act as a structural scaffold to recruit acylated PBP2 to the division septum to just exert its essential transglycosylation function (Pinho et al. 2001; Pinho and Errington 2005). This PBP2/PBP2A cooperation leads to the formation of peptidoglycan strands with low level of crosslinking (Leski and Tomasz 2005; Atilano et al. 2010), which are substrates for PBP4, a secondary transpeptidase that catalyzes the formation of highly crosslinked peptidoglycan (Wyke et al. 1981; Atilano et al. 2010).

It has been demonstrated that the correct localization of PBP4 at the division septum, which is crucial for the resistance mechanism, depends on the presence of wall teichoic acid (WTA) polymers (Scheme 10) (Atilano et al. 2010; Brown et al. 2012). As WTAs are also responsible for the anchoring of autolysins of the peptidoglycan as discussed above, simultaneous inhibition of WTA during the presence of β-lactams leads to depleted levels of crosslinked peptidoglycan and to non-viable bacteria, thus circumventing the resistance mechanism toward β-lactams and restoring their antibacterial activity against Gram-positive bacteria (Maki et al. 1994; Swoboda et al. 2011; Farha et al. 2013; Wang et al. 2013; Sewell and Brown 2014; Lee et al. 2016a).

WTA biosynthesis, as summarized earlier in Scheme 3, is categorized into two distinct groups: The non-essential, WTA early-stage genes tarO, tarA, and mnaA, responsible for attachment of *N*-acetylglucosamine (NAG, D-NAc-Glu) and *N*-acetylmuramic acid (NAc-Mur, NAM) to membrane-anchored bactoprenol phosphate, and the conditionally essential WTA late-stage genes tarB, tarD, tarF, tarI, tarJ, tarL, tarG, and tarH, responsible for the further assembly of the WTA biopolymers and finally for translocation through the phospholipid membrane to the extracellular environment (Lazarevic and Karamata 1995; Meredith et al. 2008; Pereira et al. 2008; Brown et al. 2013).

As mentioned earlier, the inhibition of late-stage wall teichoic acid biosynthesis steps or deletions of the encoding genes leads to antibacterial effects due to accumulation of toxic intermediates and the depletion of cellular pools of bactoprenol phosphate (D'Elia et al. 2006b; Swoboda et al. 2011; Brown et al. 2013; Wang

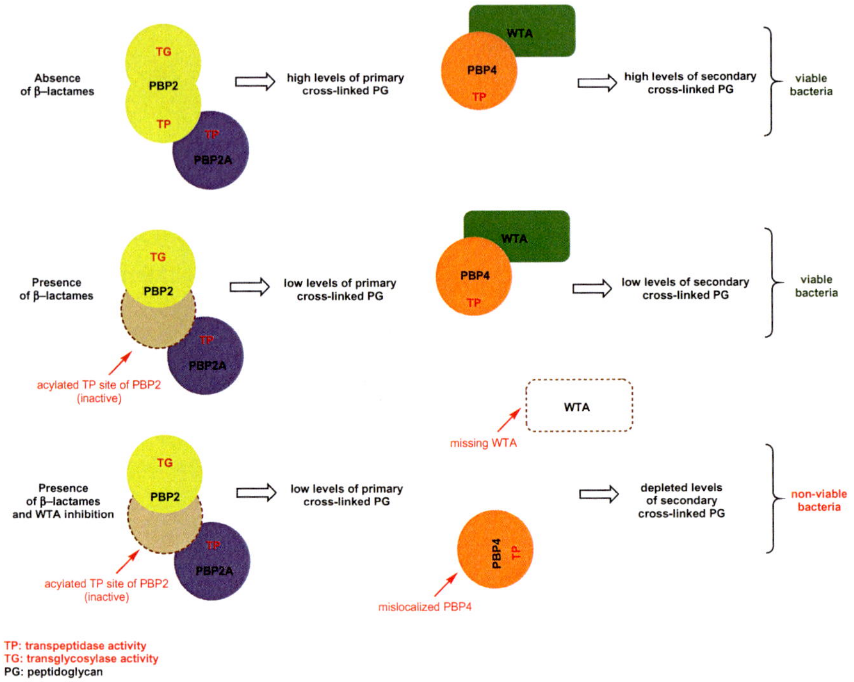

Scheme 10 Model of β-lactam resistance in MRSA and role of WTA. Adapted from Sewell and Brown (2014)

et al. 2013). In contrast, the inhibition of early-stage biosynthesis steps catalyzed by TarO and TarA or deletion of their corresponding genes leads to viable bacteria.

Paradoxically, it was demonstrated that the essentiality of late-stage WTA biosynthesis steps can be suppressed by the concomitant inactivation of early-stage WTA biosynthesis steps. This phenomenon has been referred to as the essential gene paradox (Scheme 11) (D'Elia et al. 2006a, b, 2009a, b; Farha et al. 2013; Sewell and Brown 2014; Lee et al. 2016a). Probably, concomitant inhibition of early-stage steps avoids depletion of the bactoprenol phosphate pool, which seems to be the lethal factor in late-stage WTA inhibition due to its significant effect on peptidoglycan biosynthesis. Bacteria under such early- and late-stage inhibitory pressure are synthetic-viable, but show significantly higher drug sensitivity. The concomitant early-stage inhibition of WTA synthesis, which represents also the major resistance mechanism toward late-stage WTA inhibitors, results in decreased fitness of the bacteria.

The first reported late-stage WTA biosynthesis inhibitor is targocil **21** (Scheme 12), which is a selective TarG inhibitor identified by the high throughput screening (HTS) of a library of 55000 small molecules against wild-type *S. aureus* and a tarO-deletion mutant (Lee et al. 2010; Swoboda et al. 2010). Targocil has been proven to induce significant cell wall stress (Campbell et al. 2012) and

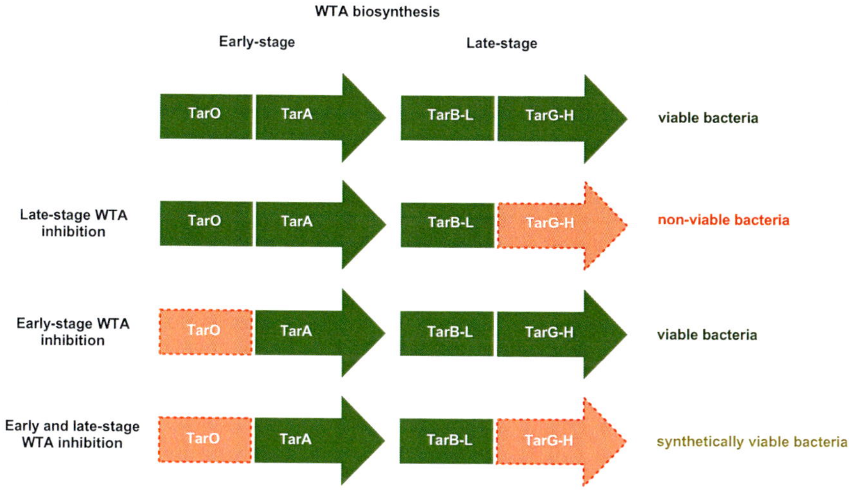

Scheme 11 Schematic presentation of the essential gene paradox in WTA biosynthesis. Adapted from Sewell and Brown (2014)

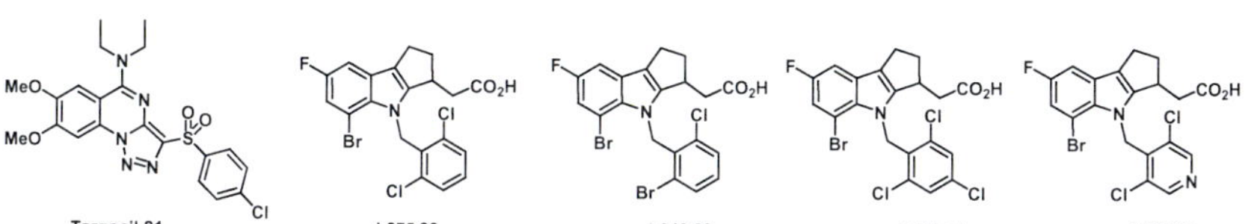

Targocil 21 L275 22 L640 23 L541 24 L638 25

Gram-positive pathogen and conditions	MIC µg/mL 21	22	23	25	26	27
MRSA (COL)	1	1	1	2	4	8
MRSA (COL) + 2 µg/mL tunicamycin A	>32	8	4	16	64	>32
MRSA (COL) + 4 µg/mL imipenem	1	0.5	0.5	1	2	4
MSSA (RN4220)	0.5	1	1	1	2	4

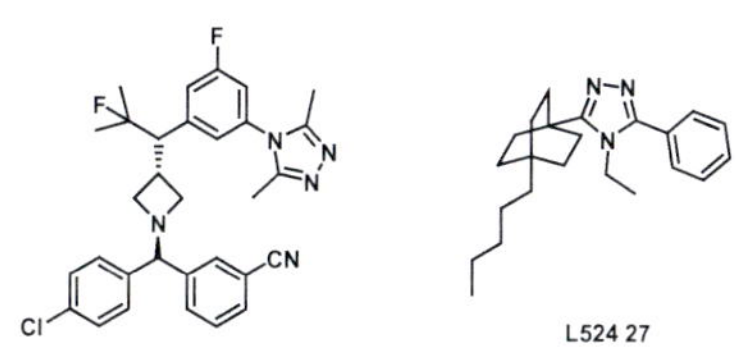

L555 26 L524 27

Scheme 12 Structures of reported TarG inhibitors and their antibacterial activities

displayed a submicromolar MIC against MRSA. Targocil treatment came with a high frequency of resistance selection (FOR 7×10^{-7} at $8 \times$ MIC (Lee et al. 2010)), which was substantially reduced in co-application with the β-lactam oxacillin, suggesting that the mechanism of resistance is an early-stage WTA biosynthesis gene mutation.

In another screening of a focused library of 20000 small molecules bioactive against *S. aureus* performed by researchers from Merck, six additional TarG inhibitors were identified (Wang et al. 2013). Interestingly, none of these inhibitors was structurally related to targocil. Four of them, L275 **22**, L640 **23**, L541 **24**, and L638 **25**, contain a tricyclic 1,2,3,4-tetrahydrocyclopenta[b]indole core. Compound

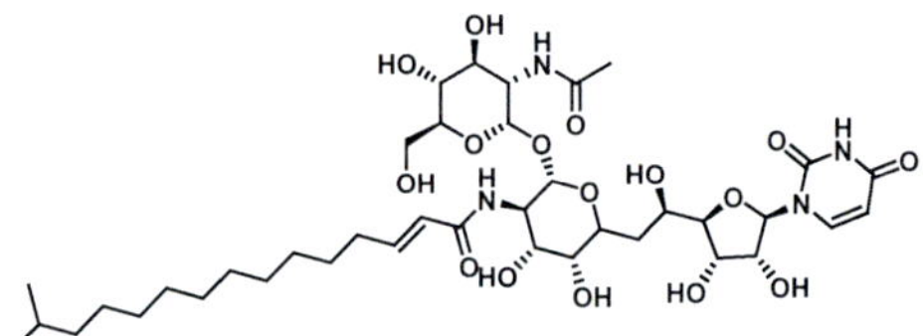

tunicamycin A 28
inhibitor of UDP-HexNAc/polyprenol-P-HexNAc-1-P family of enzymes, incl. TarO
(toxic for eucaryotic cells)

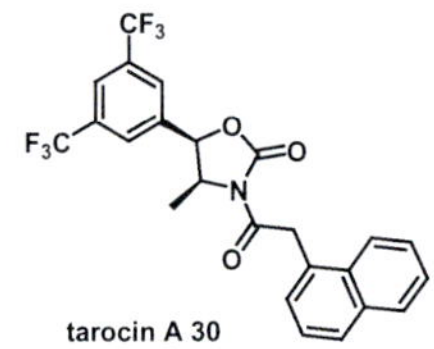

tarocin A 30
selective TarO inhibitor

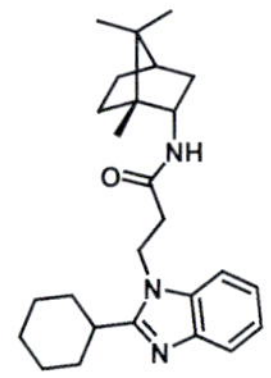

tarocin B 31
selective TarO inhibitor

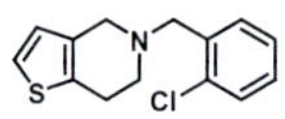

ticlopidine, Ticlid® (Sanofi) 29
adenosin diphosphate receptor inhibitor
(approved as antiplatelet drug)
and TarO inhibitor

in vitro ***S. aureus* strain**[a]	cefuroxime MIC (μg/mL)	cefuroxime FIC[b] (μg/mL)	29 MIC (μg/mL)	29 FIC[b] (μg/mL)
MRSA (USA600)	>1024	0.008	>256	0.032
MRSA (USA100/800/NY)	1024	0.125	>256	0.125
MRSA (USA200/EMRSA16)	1024	0.032	>256	0.063
MRSA (EMRSA15)	512	0.063	>256	0.063

[a] hospital-associated isolated. [b] FIC: Fractional inhibitory concentration FIC=[X]/MICX, where [X] is the lowest inhibitory concentration of drug in the presence of a co-drug

Phenotypic assay [μM]	**28**	**30**	**31**
MRSA (COL)	32	>200	>200
TarG inhibitor suppression (EC_{50})	<0.1	26	6
Phage K resistance (EC_{50})	0.1	14	3
Imipenem potentiation (SIC)	<0.1	3	3
Intrinsic bioactivity [μM]			
MSSA (RNA4220)	32	>200	>200
MRSA (COL)	32	>200	>200
MRSE (CLB27217)	16	>200	>200
C. albicans (MY1055)	1.19	>200	>200
H. influenzae (ATCC 49247)	0.595	>200	>200
S. pneumoniae (M86357)	>2.38	>200	>200
S. pyogenes (CL10253)	1.19	>200	>200
M. catarrhalis (ATCC 25238)	>2.38	>200	>200
HeLa (IC_{50})	0.197	>100	>100

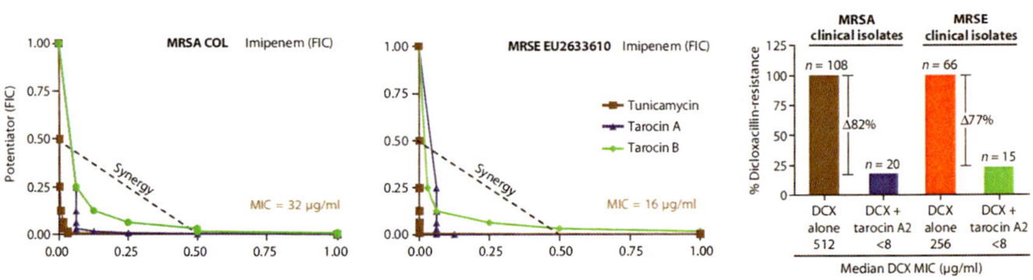

Scheme 13 Structures of reported TarO inhibitors and their antibacterial activities. Adapted from Lee et al. (2016a) and Farha et al. (2013)

L555 **26** is an active member of a large group of screened 4 *N*-aryl-4H-1,2,4-triazoles all depicting two chiral centers. Compound L524 **27** is a 3C-aryl-4H-1,2,4-triazol bearing a bicycle[2.2.2]octane motif. All these hits had submicromolar antibacterial activity against MRSA. Furthermore, combination experiments with the β-lactam imipenem showed clear synergistic effects. In contrast, concomitant inhibition of the early-stage WTA biosynthesis enzyme TarO by addition of the specific inhibitor tunicamycin A **28** (see Scheme 13) led to suppression of TarG inhibition and increased MICs. Since the TarG inhibitory effect of L275, L638, L640, and L555 could not be fully suppressed, it is likely that these compounds display an off-target associated non-specific cellular toxicity or even address secondary targets (Wang et al. 2013). The frequencies of resistance development for compounds L275 **22** (FOR <3.6×10^{-8} at 8 × MIC) and L638 **25** (FOR 1.9×10^{-10} at 8 × MIC) were significantly lower than for targocil.

Beyond tunicamycin A **28**, which displayed selective inhibitory activity for TarO in bacteria, but was cytotoxic for eukaryotic cells due to a promiscuous inhibition of UDP-HexNAc:polyprenol-P HexNAc-1-P family of enzymes including GlcNAc phosphotransferase (Swoboda et al. 2011), the first early-stage WTA biosynthesis inhibitor reported was ticlopidine **29**. Ticlopidine is an adenosine diphosphate receptor inhibitor approved as the antiplatelet drug Tilicid®, which has been identified as a TarO inhibitor in a screening of a library of 2080 previously approved drugs (Farha et al. 2013). Ticlopidine was demonstrated to have strong synergistic antibacterial activity against MRSA in combination with the β-lactam cefuroxime both in vitro and in vivo, while displaying no antibacterial activity when applied alone. In addition, ticlopidine as an approved drug has already optimized ADMET properties, a known side effect profile and established manufacturing processes. Nevertheless, it remains to be clarified whether the adenosine diphosphate receptor inhibition, the primary effect of ticlopidine, can be tolerated under the conditions of antibiotic therapy regimens. Recently, researchers at Merck reported a screen of 2.8 million synthetic small molecules for inhibitors of early-stage WTA biosynthesis in MRSA. Tarocin A **30** and tarocin B **31** (Scheme 13) were identified as compounds that could suppress TarG inhibition with IC_{50}'s of 26 and 6 μM, respectively (Lee et al. 2016a). The compounds were subsequently proven to selectively inhibit TarO. Interestingly, none of them is structurally related to ticlopidine: While tarocin A has an oxazolidin-2-one core, tarocin B consists of a benzimidazol scaffold bearing a bicyclo[2.2.1]heptane side chain.

Both compounds lack any intrinsic growth inhibitory activity against MRSA, MRSE, MSSA, or several other pathogens. In contrast to tunicamycin A, tarocin A and B show no cytotoxic activity against HeLa cells.

Furthermore, it could be demonstrated that both compounds as well as derivatives with improved water solubility (tarocin A1, tarocinA2, structures not shown) were able to restore β-lactam efficacy against MRSA and MRSE, as strong synergistic effects with imipenem and dicloxacillin in vitro and in vivo were observed. For example, panels of 108 different dicloxacillin-resistant MRSA strains and 66 dicloxacillin-resistant MRSE strains had significantly higher susceptibilities against a combination of dicloxacillin and tarocin A2 versus dicloxacillin alone (Lee et al. 2016a). In summary, the inhibition of WTA synthesis represents a novel and innovative strategy to restore the efficacy of β-lactam antibiotics in Gram-positive pathogens; the clinical proof of this concept has yet to be established, though.

5 DNA Gyrase Inhibition—Blocking the Relaxation of Supercoiled Bacterial DNA

Bacterial topoisomerases constitute a target class that has been successfully addressed in antibiotic therapy. Topoisomerases are enzymes controlling the topological state of DNA within bacterial cells and are therefore an integral part of essential processes such as DNA replication and transcription (Collin et al. 2011; Ehmann and Lahiri 2014). Topoisomerases in general are found in eukaryotic and prokaryotic cells and can be divided into two main types, type I and type II (Schoeffler and Berger 2008). Type I topoisomerases catalyze the relaxation of supercoiled DNA through the transient break of one DNA strand, while type II topoisomerases induce the relaxation through a transient double-strand breakage in an ATP hydrolysis-depending sequence (Liu et al. 1980). All topoisomerases are able to relax supercoiled DNA, but only the DNA gyrase (a type II topoisomerase in bacteria) can also introduce negative supercoils (Bates and Maxwell 2007; Nöllmann et al. 2007; Schoeffler and Berger 2008). There are significant differences between eukaryotic and prokaryotic type II topoisomerases. While eukaryotic type II topoisomerases are homodimers of large single-subunit enzymes, the prokaryotic type II topoisomerases, such as DNA gyrase (or topoisomerase II) and the structurally closely related topoisomerase IV, are A_2B_2 complexes of the subunits GyrA and GyrB (Scheme 14) (Champoux 2001; Wang 2002).

The DNA gyrase has three interfaces that exist in a closed or open conformation: The N-terminal domain of GyrB (N-gate), the GyrA-GyrB-DNA interface presenting tyrosine residues in the active cleaving site (DNA gate) and the C-terminal area of coiled coils (C-gate). The details of the complex mechanism of supercoiling by DNA gyrase are still under investigation (Gubaev and Klostermeier 2014). Nevertheless, the so-called two-gate model (Roca and Wang 1992, 1994; Roca et al. 1996) (Scheme 14c) is strongly supported by biochemical and structural data.

The DNA binds with the G-segment (gate segment) to the gyrase at the interface of the N-terminus of the GyrA dimer and the TOPRIM domain of GyrB. Upon binding of two molecules adenosin triphosphate (ATP) the GyrB subunits dimerize, which closes the N-gate and traps the so-called T-segment (transport segment) of DNA. Then the DNA gyrase transiently cleaves the G-segment via a double-strand breakage of the DNA by the formation of covalent DNA-phosphotyrosyl bonds. The hydrolysis of one ATP molecule leads to the rotation of GyrB, opening of the DNA gate, and the T-segment is transported through the cleaved DNA site. The subsequent religation of the G-segment introduces two negative supercoils into the

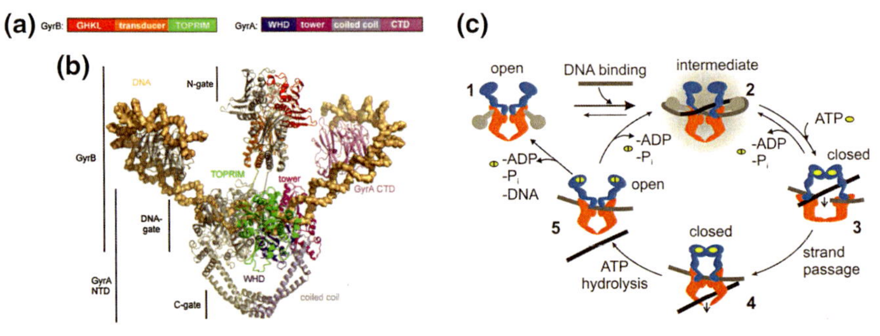

Scheme 14 Structural architecture of the DNA gyrase and schematic mechanism. Adapted from Gubaev and Klostermeier (2014) and Costenaro et al. (2007)

DNA. Finally, the release of the T-segment and the hydrolysis of the second ATP molecule sets the gyrase into its starting position (Costenaro et al. 2007; Collin et al. 2011).

In general, inhibitors of the DNA gyrase can be divided into two groups: Those blocking the ATP binding site and those interfering with the DNA binding or the DNA strand passage within the gyrase. The latter are called catalytic site inhibitors or gyrase poisons. Due to the close structural similarity of DNA gyrase and topoisomerase IV, inhibitors possessing these mechanisms of action usually show activity against both targets. This dual-targeting phenomenon often results in low mutation frequencies for drug resistance and has probably significantly contributed to the success of antibacterial drugs targeting topoisomerases.

The first ATP-site inhibitor discovered was novobiocin **32** (Scheme 15), a bacterial metabolite produced by *Streptomyces niveus*, introducing the class of aminocoumarins as gyrase-inhibiting compounds.

The compounds bind to the GyrB subunit near the ATP binding site with very high affinity and thereby block the functionality of the gyrase. Although the aminocoumarins show potent antibacterial activity against Gram-positive bacteria, their therapeutic application is limited due to poor activity against most Gram-negative pathogens and in particular due to poor water solubility and in vivo toxicities (Maxwell 1993; Maxwell and Lawson 2003). Therefore, novel ATP-site inhibitors possessing more drug-like scaffolds have been investigated in recent years, with some of them already demonstrating their efficacy in vivo (Scheme 15) (Ehmann and Lahiri 2014).

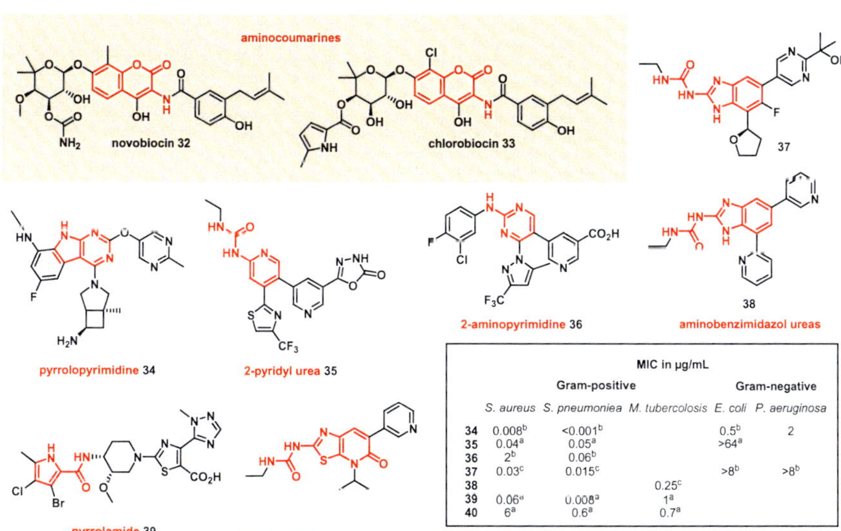

	MIC in µg/mL				
	Gram-positive			Gram-negative	
	S. aureus	*S. pneumoniea*	*M. tubercolosis*	*E. coli*	*P. aeruginosa*
34	0.008[b]	<0.001[b]		0.5[b]	2
35	0.04[a]	0.05[a]		>64[a]	
36	2[b]	0.06[b]			
37	0.03[c]	0.015[c]		>8[b]	>8[b]
38			0.25[c]		
39	0.06[a]	0.008[a]	1[a]		
40	6[a]	0.6[a]	0.7[a]		

[a] MIC from 1 strain. [b] MIC_{90} from < 30 strains. [c] MIC_{90} from >30 strains

Scheme 15 Structures of the aminocoumarins novobiocin and chlorobiocin as well as advanced ATP-site inhibitors with novel scaffolds (name-giving structural motifs are highlighted in red) and their antibacterial activities, adapted from Ehmann and Lahiri (2014)

Tari et al. presented the pyrrolopyrimidine **34** showing broad-spectrum antibacterial activity with impressive MIC_{90} values of 0.008 μg/mL against *S. aureus* (Tari et al. 2013). Compound **34** displays also good activity against Gram-negative pathogens representing a major advancement for ATP-site inhibitors. Furthermore, the 2-pyridinyl urea **35** (Basarab et al. 2013), the 2-aminopyrimidine **36** (Uria-Nickelsen et al. 2013), aminobenzimidazols **37** (Finn 2013) and **38** (Chopra et al. 2012), the pyrrolamide **39** (Shahul et al. 2014) and the thiazolopyridone **40** (Kale et al. 2013, 2014) were reported as efficient ATP-site inhibitors against Gram-positive bacteria, and, for some of them (**38**, **39** and **40**) also moderate activities against *M. tuberculosis* were revealed.

The first catalytic site inhibitor of the DNA gyrase was nalidixic acid **41** (Scheme 16), introducing the antibacterial class of quinolones. The first generation of quinolones had relatively weak antimicrobial activity, but with the introduction of the fluoroquinolones such as ciprofloxacin **42** or levofloxacin **43** very potent antimicrobials were available, displaying broad-spectrum activity against Gram-positive and Gram-negative pathogens (Collin et al. 2011). Such catalytic site inhibitors stabilize the DNA cleavage complex of the gyrase or topoisomerase IV and thereby poison the enzyme. This has been shown to be a very effective mode of inhibition of type II topoisomerases, since very low concentration of the inhibiting compounds bound to their target can lead to sufficient protein-stabilized DNA breaks, initiating a cascade which cumulates in cell death (Anderson and Osheroff 2001; Kohanski et al. 2010). Nevertheless, significant occurrence of resistance toward fluoroquinolones has been reported (Jacoby 2005), mostly associated with mutations in the quinolone resistance-determining regions in GyrA and GyrB (or ParC and ParE, the equivalents in topoisomerase IV) (Yoshida et al. 1990, 1991; Jacoby 2005).

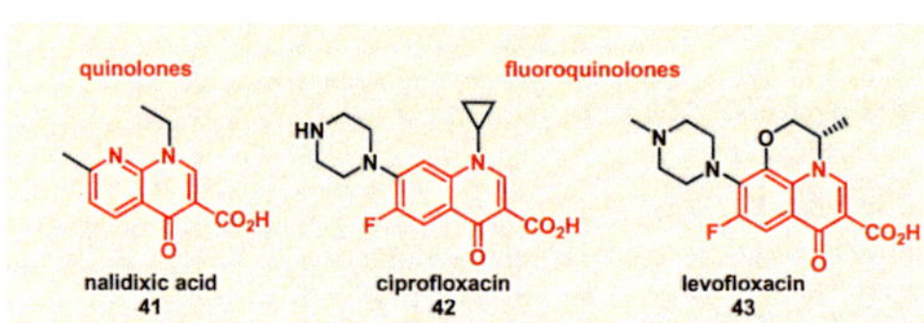

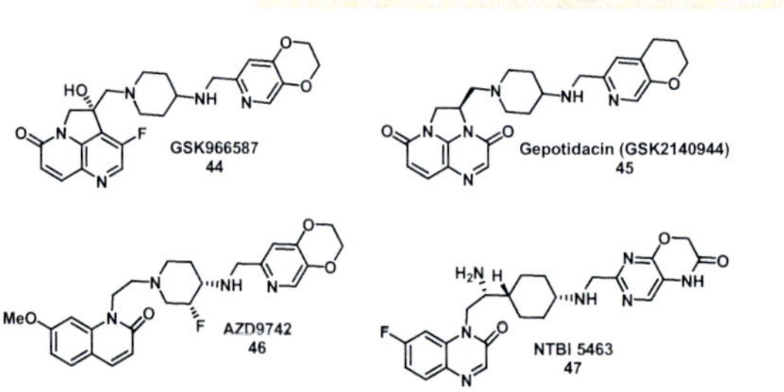

	MIC in μg/mL			
	Gram-positive		Gram-negative	
	S. aureus	*S. pneumoniea*	*E. coli*	*P. aeruginosa*
44	0.13[b]	0.13[b]		
45	0.5[c]	0.25[c]	4[c]	
46	0.06[a]	0.13[a]	4[a]	>8[a]
47	1[a]		2[b]	2[b]

[a] MIC from 1 strain. [b] MIC_{90} from < 30 strains.
[c] MIC_{90} from >30 strains

Scheme 16 Structures of nalidixic acid and the fluoroquinolones ciprofloxacin and moxifloxacin as well as advanced catalytic site inhibitors with novel scaffolds and their antibacterial activities (adapted from *Ehmann* et al.) (Ehmann and Lahiri 2014)

In the recent years, intensive research led to the discovery of a group of several non-fluoroquinolone inhibitors of the gyrase that bind near the catalytic site, but are mechanistically and microbiologically distinct from the fluoroquinolones (Ehmann and Lahiri 2014).

These compounds are simply called novel bacterial topoisomerase inhibitors (NBTIs). Most NTBIs possess a 4-aminopiperidine moiety bridging two aromatic motifs. Investigations on the mechanism of action have been shown that NBTIs also stabilize the DNA–protein complex, but in contrast to the fluoroquinolones they bind to the gyrase in the presence of intact, unbroken DNA strands (Bax et al. 2010).

The first frontrunner advanced to clinical trials was GSK966587 **44** developed by researchers at GlaxoSmithKline (Miles et al. 2013). The compound bears a tricyclic 3-fluoro-4,5-dihydro-7H-pyrrolo[3,2,1-de][1,5]naphthyridin-7-one core and showed potent activity against Gram-positive bacteria (Ehmann and Lahiri 2014). A follower of the same series, gepotidacin **45** (Mayer and Janin 2014), is in several clinical phase II trials investigating the treatment of respiratory tract infections caused by *S. pneumoniae*, acute bacterial skin and skin structure infections caused by *S. aureus* and uncomplicated urogenital gonorrhea (Paris 2015). Additionally, researchers at AstraZeneca reported further NTBIs showing potent activity against Gram-positive bacteria such as AZD9742 **46** (Reck et al. 2012) and NTBI 5463 **47** (Dougherty et al. 2014). The latter presents a 4-aminocyclohexyl bridge between the two aromatic regions with promising activity against Gram-negative pathogens. Noteworthy, all of these compounds show activity against fluoroquinolone-resistant bacterial strains.

Recently, several catalytic site inhibitors of the DNA gyrase with novel unique scaffolds different from the early NBTIs (Scheme 16) have been discovered. Researchers at AstraZeneca reported compound ETX0914 **48** (Scheme 17) with an unusual spiropyrimidinetrione scaffold that showed potent broad-spectrum activity against Gram-positive as well as Gram-negative bacteria (Jacobsson et al. 2014; Alm et al. 2015; Kern et al. 2015; Su et al. 2016). The compound was active against fluoroquinolone-resistant strains, suggesting that the exact mode of inhibition is also different from the fluoroquinolones (Basarab et al. 2015). Genetic analysis of

	Pathogen	MIC µg/mL 48
Gram-positive	S. aureus	0.10
	S. pneumoniae	0.19
	S. pyogenes	0.19
Gram-negative	E. coli	3.0
	H. influenzae	0.10
	N. gonorrhoeae	0.19

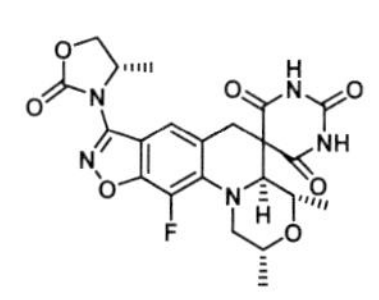

spiropyrimidinetrione ETX0914
48

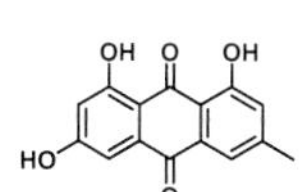

emodin
50

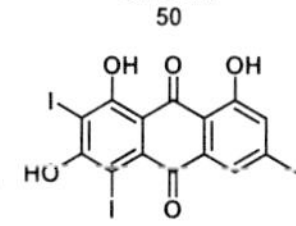

2,4-diiodoemodin
49

Gram-positive pathogen	MIC µg/mL 49
S. aureus	0.002
MRSA	0.004
MRSE	0.004
E. faecalis	0.016
VRE	0.008
B. cereus	0.004
B. laterosporus	0.004
S. pneumoniae	0.061
Streptococcus pneumoniae	0.032

Scheme 17 Structures and antibacterial activities of spiropyrimidinetrione ETX0914 and 2,4-diiodoemodin

resistant mutants showed that mutations in the conserved GyrB TOPRIM domain are correlated to decreased susceptibilities to ETX0914 (Alm et al. 2015). Currently, the compound is in clinical phase II trials for the treatment of uncomplicated gonorrhea (Paris 2015). Another compound, which has recently been shown to have extremely potent antibacterial activities (up to MIC = 0.002 μg/mL) against Gram-positive pathogens including several multi-drug-resistant and also fluoroquinolone-resistant strains, is 2,4-diiodoemodin **49** (Chen et al. 2014), a semi-synthetic derivative of the secondary metabolite emodin **50**, which naturally occurs in plants and fungi. Interestingly, **49** has been shown to be an inhibitor of DNA gyrase and bacterial topoisomerase I, which is a unique dual-targeting mode of this compound. Additionally, unlike emodin, which caused cytotoxic effects through strong inhibition of the human topoisomerase IIα, 2,4-diiodoemodin showed only very little inhibitory effect against this enzyme. Nevertheless, a further understanding of this new class of antibacterial haloemodins, for example with respect to target promiscuity and the associated side effects, is needed to advance this structural template.

A novel structural class of gyrase inhibitors of proteobacterial origin, consisting of oligomeric pseudopeptides formed by chains of coupled *para*-aminobenzoic acids, has recently been discovered by two research groups from two different Gram-negative producer strains.

Albicidin **51** (Scheme 18) has been isolated from *Xanthomonas albilineans* (Cociancich et al. 2015), a xylem-invading plant pathogen that causes leaf scald

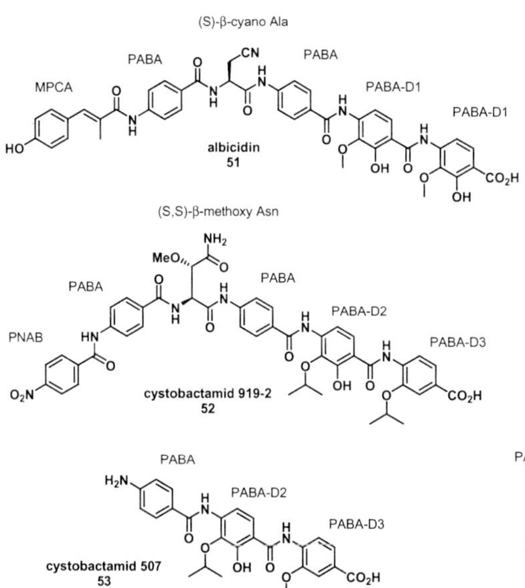

	pathogens	51	52	53	54
		MIC μg/mL			
Gram-positive	*E. faecalis*		0.1	>128	3.7-7.4
	S. aureus	16	0.1	64	32.5
	S. pneumoniae		0.1	16	14.7
	M. luteus	1		64	
Gram-negative	*A. baumannii*		7.4		>59
	E. coli	0.031	0.9	>128	14.7-29.4
	S. enteriditis	0.5			
	P. aeruginosa	1			

PNBA = *para*-nitrobenzoic acid
PABA = *para*-aminobenzoic acid
PABA-D = *para*-aminobenzoic acids derivative
MPCA = 2-methyl-*para*-coumaric acid

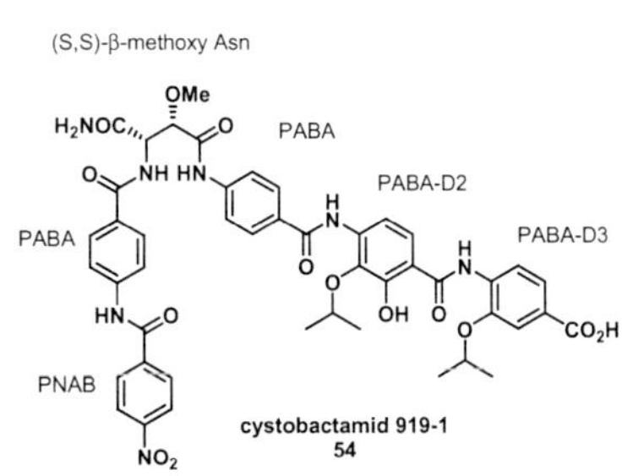

Scheme 18 Structures and antibacterial activities of albicidin and the cystobactamids, novel unusual PABA-oligomeric pseudopeptides

disease in sugarcane (Royer et al. 2004). The compound is a phytotoxin that blocks the DNA gyrase in sugarcane chloroplasts (Birch and Patil 1987a, b). Additionally, albicidin has been demonstrated to be a nanomolar inhibitor of bacterial DNA gyrase with a broad-spectrum antibacterial activity against Gram-positive and Gram-negative pathogens (Hashimi et al. 2007; Cociancich et al. 2015; Kretz et al. 2015). Albicidin stabilizes the DNA gyrase cleavage complex with relaxed or supercoiled DNA in the presence of ATP via binding to the GyrA subunit, leading to a similar DNA fragmentation pattern as ciprofloxacin. Nevertheless, only low levels of cross resistance to albicidin in some fluoroquinolone- and couvermycin-resistant strains have been observed (Hashimi et al. 2007). The existence of a potent gyrase inhibitor produced by *X. albilineans* has been known for decades, but since the bacterium produces only very small amounts of the compound, the compound's structure could be elucidated only recently, when *Süssmuth* and coworkers expressed heterologously in *X. axonopodis vp. vesicatoria* and established an extended purification protocol (Cociancich et al. 2015).

The structure of albicidin consists of the non-proteinogenic α-amino acid (*S*)-β-cyanoalanine, the two aromatic δ-amino acids *para*-aminobenzoic acid (PABA) and 2-hydroxy-3-methoxy-*para*-aminobenzoic acid (PABA-D1) as well as 2-methyl-*para*-coumaric acid (MPCA) that represents the N-terminus of this oligomeric pseudopeptide. Biosynthetically albicidin is a product of the polyketide synthase-nonribosomal peptide synthase (PKS-NRPS) machinery (Huang et al. 2001; Vivien et al. 2007; Piel 2010). Albicidin has been successfully targeted by a total synthesis by *Süssmuth* and coworkers, thus confirming the unusual structure (Kretz et al. 2015).

In parallel, the structure of the cystobactamid 919-2 **52** (Scheme 18) was published by *Müller* and coworkers (Baumann et al. 2014; Herrmann et al. 2016).

The compound was isolated together with the truncated homolog cystobactamid 507 **53** (Moreno et al. 2015) and the isomer cystobactamid 919-1 **54** from cultivations of myxobacterium *Cystobacter* sp. Cbv34. Cystobactamid 919-2 shares the general PABA-derived oligomeric, pseudopeptidic structure of albicidin, but the central amino acid is a non-proteinogenic (*S,S*)-β-methoxy asparagine, and it possesses a different oxidation pattern on the right-hand site of the molecule. Additionally, there is a *para*-nitrobenzoic acid (PNBA) unit at the N-terminus. In accordance with the structural similarity of cystobactamids and albicidin, also cystobactamid 919-2 displayed very potent broad-spectrum antibacterial activity against Gram-positive and Gram-negative pathogens due to inhibition of the DNA gyrase at the GyrA subunit with low levels of cross resistance to fluoroquinolone-resistant strains (Baumann et al. 2014). Both compounds were regarded as "natural quinolones" representing the first natural products inhibiting the GyrA subunit of DNA gyrase (Baumann et al. 2014; Johnston and Magarvey 2015). Currently, chemical optimization studies of albicidin and cystobactamid 919-2 to improve their properties as antibacterial drugs are ongoing (personal communication, M. Brönstrup).

6 Bacterial DnaN Polymerase Sliding Clamp—A Highly Conserved, Multi-enzyme Interacting Core Architecture of the Bacterial Replisome

As mentioned earlier, multi-enzyme complexes or structures synthesized by multi-enzymatic pathways are less prone to bacterial resistance development, since their malfunctioning is hard to bypass.

In this context the bacterial replisome machinery, which consists of at least twelve interacting enzymes that are highly conserved in bacteria, is a good structure to target (Robinson et al. 2012). Next to the DNA gyrase, which also belongs to the bacterial replisome and has already been discussed, the bacterial DnaN polymerase sliding clamp has recently attracted attention as an antibacterial target (Georgescu et al. 2008b; Kjelstrup et al. 2013; Wolff et al. 2014; Yin et al. 2014a, b; Holzgrabe 2015; Kling et al. 2015; Yin et al. 2015).

The bacterial DnaN or β-clamp is a ring-shaped homodimer, with each monomer composed of three globular domains, which functions as a crucial subunit of the DNA polymerase III holoenzyme (Scheme 19). The general protein–DNA interactions between polymerases and DNA templates are weak (Mizrahi et al. 1985). Therefore, the association of the polymerase to the DNA strand is one of the rate-limiting steps of the replication. The assembly of the two β-subunits of the bacterial β-clamp around the DNA, mediated by ATP hydrolysis, leads to strong and specific protein–protein interactions between the β-clamp and the polymerase III (Pol III), thereby preventing polymerase dissociation and dramatically increasing the polymerase activity (Stukenberg et al. 1991; Argiriadi et al. 2006; Georgescu et al. 2008a; Donnell et al. 2013; Cho et al. 2014). Therefore, the

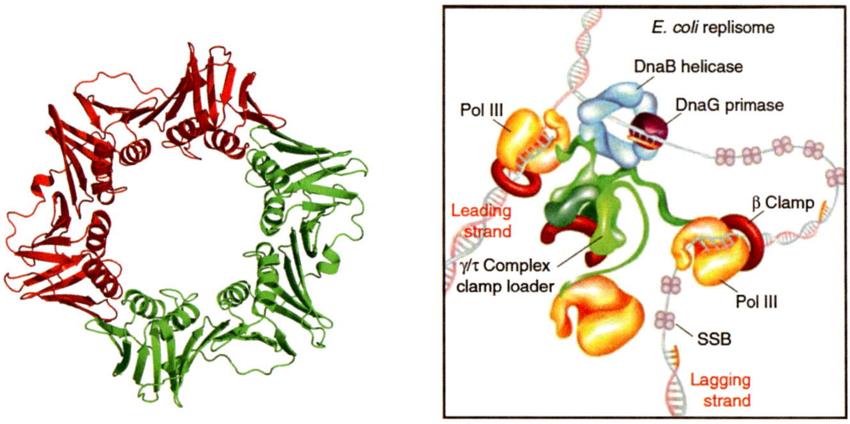

Scheme 19 Structure of the DnaN/β-clamp in *E. Coli* (*left*) and architecture of the bacterial replisome (*right*). Adapted from Oakley et al. (2003) and Donnell et al. (2013); SSB stands for single-strand binding protein

β-clamp serves as a processivity-promoting factor in DNA replication (Stukenberg et al. 1991).

Furthermore, as the β-clamp shows direct protein–protein interactions with a multitude of other enzymes involved in the replication process such as the α-polymerase, the ε-proofreading subunits (Ozawa et al. 2013), or the clamp loader γ/τ complex (Stukenberg et al. 1991), it can be seen as the linchpin of the replisome (Robinson et al. 2012). The important functionality within the replication, its moderate copy number per cell (330–600 homodimers) (Leu et al. 2000), its highly conserved structure across the bacterial species (Burnouf et al. 2004; Argiriadi et al. 2006; Georgescu et al. 2008a; Gui et al. 2011; Wolff et al. 2014; Niiranen et al. 2015), and its significant structural divergence to the mammalian counterpart (proliferating cell nuclear antigen, PCNA) (Bloom 2009; Robinson et al. 2012; Donnell et al. 2013) make the β-clamp a particularly attractive antibacterial target.

In 2001, *Jennings* and coworkers discovered that the general protein–protein interaction between the β-clamp and several β-clamp-interacting proteins is based on a consensus sequence of amino acids in a linear pentapeptidic-binding motif, given as QL[S/D]LF and QLxLx[L/F][S/D] (Scheme 20) (Dalrymple et al. 2001).

Jennings and coworkers were able to identify the first synthetic nonapeptides containing these linear binding motifs that inhibited either the binding of the α-subunit or of the δ-subunit of *E. coli* Pol III to the corresponding *E. coli* β-clamp, displaying IC_{50} values of 1.33–14.6 μM or 8.8–12.9 μM, respectively.

It was demonstrated in further SAR studies in 2004 (Wijffels et al. 2004) and 2011 (Wijffels et al. 2011) that lipophilic N-terminal extensions of the pentapeptidic linear binding motif as well as substitutions with more lipophilic phenylalanine derivatives led to tremendous increases of the inhibitory effect of the binding of the α-subunit of *E. coli* Pol III to the corresponding *E. coli* β-clamp (IC_{50} up to 20 nM). These linear motifs (LM) bind to the LM-binding pocket, which is highly conserved across Gram-negative and Gram-positive bacteria (Argiriadi et al. 2006) and consists of two subsites (I and II) at the interface of the two domains on each

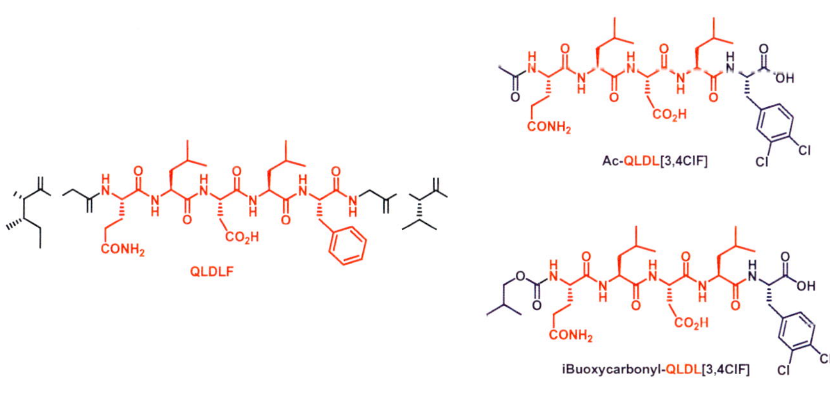

Scheme 20 Early and advanced peptidic inhibitors of α-$β_2$ binding

Scheme 21 Small molecule inhibitors of the β-clamp

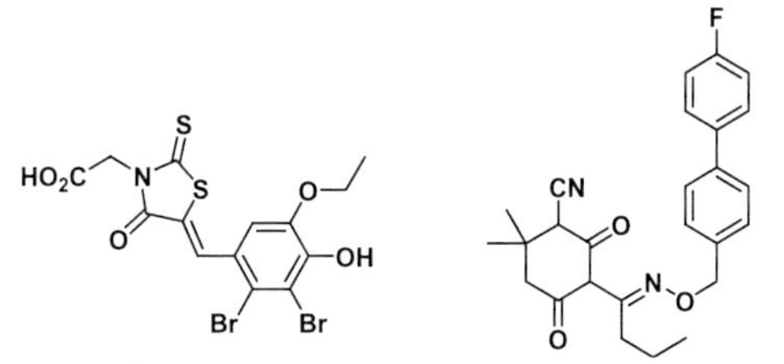

RU-7 **58**
K_i(*E. coli*) = 10 μM

59
IC_{50}(α-$β_2$-binding) = 40 μM

monomer of the β-clamp homodimer (Dalrymple et al. 2001; Bunting et al. 2003; Burnouf et al. 2004; Georgescu et al. 2008a, b).

In 2008, *O'Donnell* and coworkers published the first non-peptidic small molecule inhibitor of the β-clamp, RU-7 **58** (Scheme 21) (Georgescu et al. 2008b). The compound was identified by screening a library of 30000 polar small molecules for their inhibitory effect on α-$β_2$-binding by a fluorescence anisotropy titration assay. RU-7 inhibited the *E. coli* relication system with a K_i of 10 μM by binding to the subunit I of the LM-binding pocket, but showed no inhibitory effect on the PCNA-dependent eukaryotic system in the yeast *S. cereviase*.

An other small molecule, the oxim ether **59**, was identified by an in silico screen of 32000 compounds followed by an α-$β_2$-binding assay (Wijffels et al. 2011). The compound, which also bound to the subunit I of the LM-binding pocket, displayed an IC_{50} value for inhibition of α-$β_2$-binding of 40 μM.

In 2014, *Oakley* and coworkers reported that the weak antibacterial effects of non-steroidal anti-inflammatory drugs (NSAIDs) such as (*R*)-vedaprofen **60**, bromfenac **61**, or (*S*)-carprofen **62** (Scheme 22) are due to their binding to the subunit I of the LM-binding pocket of the β-clamp (Yin et al. 2014a).

Recently, a fragment-based screening of 352 compounds utilizing *X*-ray crystallography led to the discovery of tetrahydrocarbazol **63** (Scheme 22) as a β-clamp binding small molecule, which is structurally closely related to carprofen and was

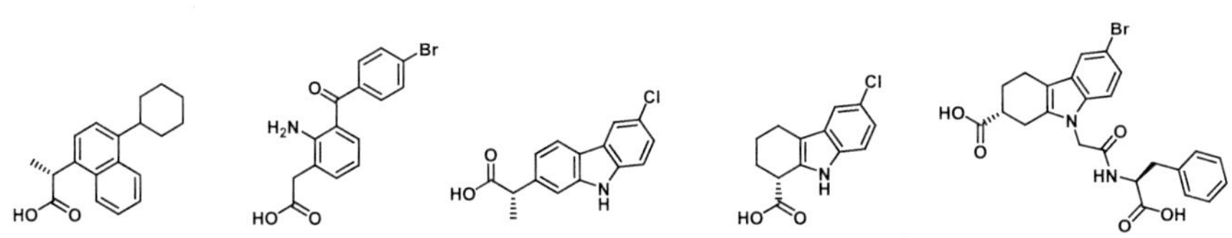

(R)-vedaprofen **60**; bromfenac **61**; (S)-carprofen **62**; tetrahydrocarbazol (THC) **63**; N-acetamido-THC **64**

E. coli β-clamp inhibition IC_{50}= 115 μM (**63**)

E. coli β-clamp inhibition IC_{50}= 27 μM (**64**)

NSAID	*E. coli* β-clamp inhibition IC_{50} [μM]	*S. aureus* MIC [μM]
vedaprofen	222	156
bromfenac	328	2500
carprofen	481	313

Scheme 22 Structures of NSAIDs, their antibacterial activities, and tetrahydrocarbazoles as β-clamp binding lead compounds

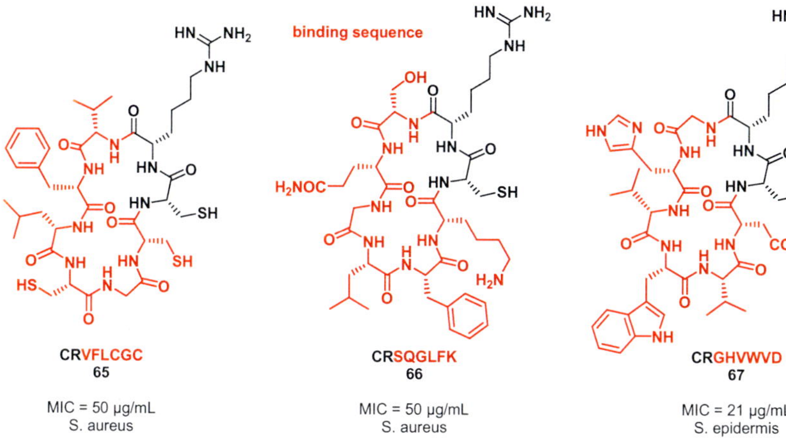

Scheme 23 Cyclic octapeptides inhibiting the DnaN–DnaN protein interaction

also shown to bind into the subunit I of the β-clamp (Yin et al. 2014b). Further, SAR studies led to the identification of N-acetamido-THC **64** with an 10-fold decreased IC_{50} for the inhibition of the β-clamp (Yin et al. 2015).

Whereas all compounds discussed so far were binding to the LM-binding pocket of the β-clamp, *Lobner-Olesen* and coworkers recently reported cyclic octapeptides (**65–67**, Scheme 23) inhibiting the DnaN–DnaN monomer interaction of the homodimer (Kjelstrup et al. 2013).

The compounds were identified using the SICLOPPS technology (Scott et al. 1999) for the intracellular in vivo production of cyclic 21-mer peptides. A six-amino-acid-long randomized sequence was used to build a library of over 900000 combinations, which were subsequently screened for their ability to inhibit DnaN–DnaN interactions in the *S. aureus* replisome. The initial six-amino-acid-long hit sequences were incorporated in an octapeptide and synthesized utilizing the Impact Twin System (New England Biolabs) and evaluated for their antibacterial activity. Interestingly, while cyclic octapeptides displayed MIC values of 50 µg/mL and 21 µg/mL, respectively, the corresponding linear peptides were totally inactive.

An issue with all of the synthetic DnaN binders described above is that their antibacterial potency is mostly weak. This again illustrates that the ability to penetrate into the bacterial cell is a crucial hurdle for new antibiotic templates.

This is fundamentally different for griselimycin (GM) **68** (Scheme 24), a long known natural product from bacterial ferments that attracted renewed interest as an anti-tuberculosis drug (Holzgrabe 2015; Kling et al. 2015). Griselimycin is a cyclic decadepsipeptide containing two unusual L-(4*R*)-4-methylproline and one unsubstituted L-proline in the linear sequence, which is cyclized with its C-terminal glycine onto the hydroxyl group of an internal threonine side chain. Additionally, the compound possesses several posttranslational N-methylations, and the

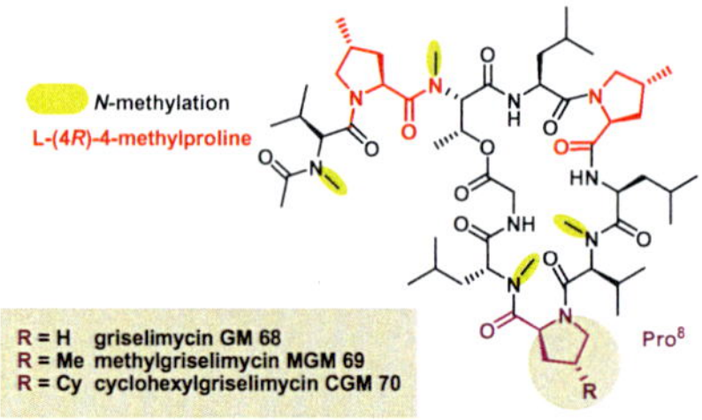

	GM	MGM	CGM
MIC (μg/mL) for *M. tuberculosis* in liquid culture	1	0.6	0.06
MIC (μg/mL) for *M. tuberculosis* in liquid culture containing 25% human serum	1.2	0.9	0.3
MIC (μg/mL) for *M. tuberculosis* in macrophages	6.2	2.1	0.2
Unbound fraction (%) in human plasma	nd	7.9	0.3
Metabolic stability in human liver microsomes (% remaining compound after 20 min)	62	100	86
Metabolic degradation/clearance for human liver microsomes (μL/min per mg)	187.5	3	6.75
C_{max} (ng/mL) in plasma	3900	2820	2620
C_{max} (ng/mL) in lung tissue	2580	8550	8850
AUC (ng·hour/mL) in plasma	5100	6600	23000
AUC (ng·hour/mL) in lung tissue	5900	12000	70000
Clearance (liters/hour per kg)	2.4	2.1	1.1
V_{ss} (liters/kg)	1.2	1.6	5.5
$t_{1/2}$ (hours) in plasma	2.0	2.4	4.3
$t_{1/2}$ (hours) in lung tissue	3.9	4.3	4.3
Oral bioavailability (F)(%)	48	47	89

C_{max}, maximum concentration; **AUC**, area under the concentration curve; **V_{ss}**, volume of distribution; **$t_{½}$**, half-life; **nd**, not determined. Pharmacokinetic parameters (C_{max},AUC,and $t_{½}$) were determined after a single oral administration of 30mg/kg of the test compound in mice. For oral bioavailability, a single oral dosewas compared to a single intravenous dose of 3 mg/kg. F =[AUCoral]/[AUCiv], the ratio of exposure of an equivalent dose after nonintravenous (in this case, oral) and intravenous administration as ameasure of bioavailability

Scheme 24 Griselimycin, novel analogs, and their pharmacokinetic properties and anti-TB activities

N-terminus is acetylated. Griselimycin was isolated from two strains of *Streptomyces* in the 1960s (Terlain and Thomas 1969, 1971a, b), and was found to exhibit antibacterial activity specifically against *Corynebacteria* including *M. tuberculosis*, though the exact mode of action and target were unknown. The pharmaceutical company Rhône-Poulenc started with early investigations for the development of griselimycin as an anti-tuberculosis drug. The first studies in humans were promising, despite poor pharmacokinetics of the compound (Bénazet et al. 1966; Noufflard-Guy-Loé and Berteaux 1965). But first derivatization programs to find analogs with improved pharmacokinetics (Bouchaudon 1964; Jolles 1971) were soon terminated in the 1970s after rifampin became available for tuberculosis treatment. Since novel tuberculosis drugs are desperately needed and GM was reported to be highly active against multi-drug-resistant *M. tuberculosis*, researchers from Sanofi-Aventis, *Müller,* and coworkers re-investigated the use of griselimycin as anti-tuberculosis drug (Kling et al. 2015).

First, a metabolic stability profiling of natural, less-abundant analogs of GM led to the identification of Pro^8 as main site of metabolic degradation responsible for the poor pharmacokinetic parameters of griselimycin. This result was supported by the finding that the natural derivative methylgriselimycin (MGM) **69** (Scheme 24) showed a remarkably increased stability toward degradation compared to griselimycin. Since MGM is produced in only small amounts, a solid-supported total synthesis for griselimycins was established to provide access to several GM analogs bearing Pro^8 modifications, supposed to improve potency and metabolic stability of the lead structure. From biological evaluation of these analogs cyclohexylgriselimycine (CGM) **70** (Scheme 24) was identified, displaying MIC values of 0.06 μg/mL and 0.2 μg/mL in the drug susceptible *M. tuberculosis* strain H37Rv and within macrophage-like RAW264.7 cells, respectively. CGM revealed a time-dependent in vitro activity, and although the unbound fraction in plasma was very low, the MIC dropped only fivefold in the presence of human plasma. CGM showed high-level activity against a panel of *M. tuberculosis* strains covering a broad geographical and evolutionary diversity, including strains mono-resistant to first- and second-line anti-tuberculosis drugs. The overall ADMET properties of CGM also improved; in particular CGM displayed a high-level oral bioavailability of 89 %, a moderate clearance (1.1 L/h/kg), and an expanded half-life and drug exposure, enabling a once daily administration.

In vivo activity studies in a mouse model of tuberculosis revealed a minimum-effective dose (MED) of 50 mg/kg, and mice treated with 600 mg/kg daily dose were proven culture-negative after 4 weeks of treatment. Giving the promising result of CGM administered in monotherapy, a combination therapy with first-line anti-tuberculosis drugs such as rifampin (RIF), pyrazinamide (PZA), and isoniazid (INH) was investigated. In an in vivo mouse model of TB, CGM was administered at MED of 100 mg/kg alone and in combination with other TB drugs in a model of chronic TB. CGM alone was demonstrated to be as active as INH, the most bactericidal first-line anti-tuberculosis drug, and showed significantly improved activity in combination with RIF compared to the standard combination therapy (INH/RIF/PZA).

The analysis of the genes of a GM- and MGM-producing *Streptomyces* strain, which is naturally resistant to GM, led to the identification of *griR*, a homolog of the dnaN gene (51 % of identity). Introduction of the *griR* gene to *Streptomyces coelicolor*, a strain susceptible to GM, allowed the strain to survive in the presence of GM, suggesting that overexpression of GriR mediates GM resistance and that the griselimycins interact with DnaN polymerase sliding clamp (β-clamp) of the replisome. Further, genomic-based investigations on the resistance mechanisms in vitro and in vivo in *M. smegmatis* and *M. tuberculosis* revealed that resistance in these species is achieved by overexpression of DnaN. However, this process occurred with very low rate and was accompanied by considerable loss of fitness, expressed in a negative correlation of growth to increasing level of resistance.

Finally, the interaction between GM, MGM, and CGM with DnaN from *M. smegmatis*, *M. tuberculosis*, and *E. coli* as well as human sliding clamp (PCNA) was analyzed and characterized by surface plasmon resonance (SPR) and by a crystal structure analysis. SPR analysis demonstrated a high affinity binding of GM, MGM, and CGM to mycobacterial sliding clamps in the picomolar range, whereas significantly lower binding to the sliding clamp of *E. coli* and no binding between PCNA and griselimycins was detected. The analysis of co-crystals of GM and CGM with DnaN from *M. smegmatis* and *M. tuberculosis* revealed that the griselimycins bind to the LM-binding pocket, filling both subsite I and II. The linear N-terminal segment and the adjacent half of the macrocycle of GM and CGM superimpose very well with the natural linear peptidic-binding motif of the LM-binding pocket.

Since there is no preexisting resistance of griselimycins due to the different modes of action compared to the approved anti-tuberculosis drugs, and since resistance occurs at an extremely low frequency associated with fitness loss of the pathogens, the griselimycins are attractive candidates for the treatment of drug-sensitive and multi-drug-resistant tuberculosis. Clinical trials with CGM have not been reported yet, though.

7 Menaquinone as a Small Molecule Antibacterial Target

Isoprenoid quinones such as phylloquinone (vitamin K_1) **71**, ubiquinone (coenzyme Q_{10}) **72,** and menaquinone **73** (Scheme 25) are present in almost all living organisms as plasma membrane-anchored electron carriers or antioxidants (Kawamukai 2002; Aguilaniu et al. 2005; Kurosu and Begari 2010; Nowicka and Kruk 2010; Cervellati and Greco 2016). These compounds are involved in several essential, live-sustaining metabolic processes such as the generation of adenosine triphosphate (ATP) in the respiratory chain. The general structure of these compounds consists of a polar, quinone-containing head and an unpolar, hydrophobic side chain, which allows the molecules to anchor in the phospholipid bilayer of cell membranes.

While phylloquinone and menaquinone show a methylnaphtoquinone core, the polar head group of ubiquinone consists of a dimethoxy-benzoquinone. The length and saturation of the side chain can differ in the particular species. The quinone moiety in these compounds is crucial for their biological activity as electron carriers in cellular redox processes, as it is reduced by a two-step reversible reduction process via the intermediate semiquinone to the corresponding hydroquinone. While ubiquinone is utilized as coenzyme for the ATP generation in the mammalian respiratory chain, in most Gram-positive and anaerobic Gram-negative bacteria menaquinone is the major isoprenoid quinone, responsible for the electron transfer in the bacterial respiratory chain and a multitude of other metabolic redox processes (Kurosu and Begari 2010; Fujimoto et al. 2012).

The biosynthesis of menaquinone in *E. coli* is catalyzed by seven menaquinone-specific enzymes, MenA-G, which are encoded in two gene clusters (Young 1975;

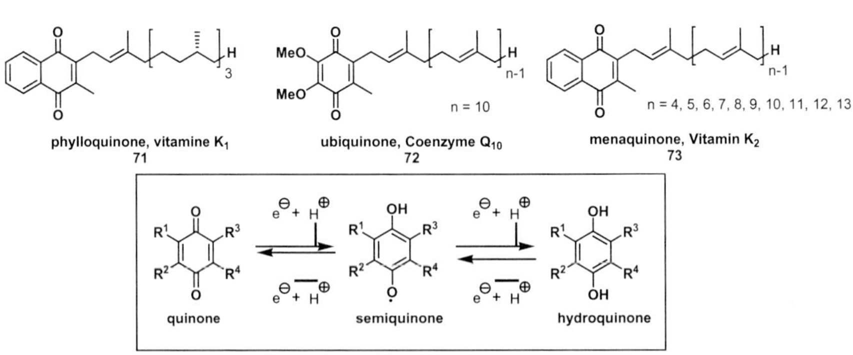

Scheme 25 Structures of menaquinone, phylloquinone, and ubiquinone

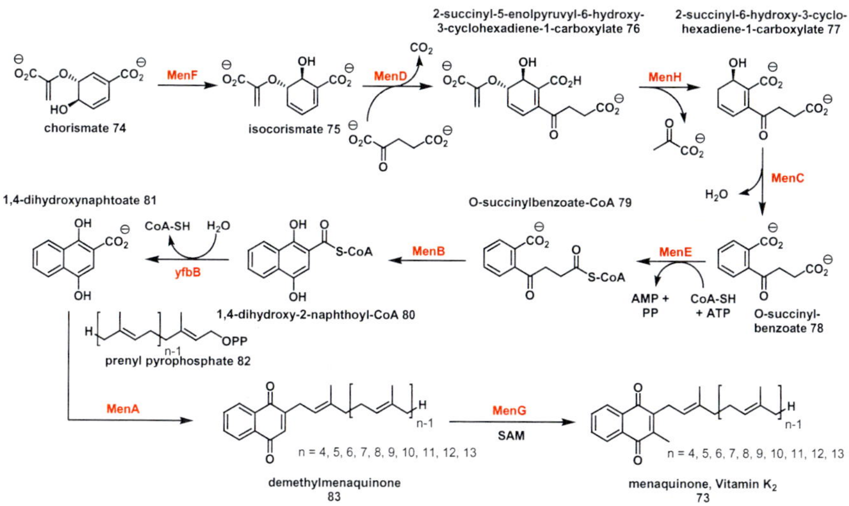

Scheme 26 Biosynthesis of menaquinone

Schoepp-Cothenet et al. 2009). It starts from chorismate **74** (Scheme 26), an intermediate of the shikimate pathway. First, MenF isomerizes chorismate **74** to isochorismate **75**. MenD, a thiamin diphosphate-dependent enzyme, then mediates the *Stetter*-type 1,4-conjugate addition of α-ketoglutarate to isochorismate, leading to the formation of **76**. The elimination of the pyruvate moiety is then catalyzed by MenH, yielding 2-succinyl-6-hydroxy-3-cyclohexadiene-1-carboxylate **77**, and subsequently MenC forces the elimination of water providing *O*-succinylbenzoate **78**. The *O*-succinyl-CoA ligase MenE attaches **78** to coenzyme A and a *Diekmann*-type cyclization catalyzed by MenB and subsequent thioester hydrolysis by yfbB leads to formation of 1,4-dihydroxy-2-naphthoate **81**. Finally, prenylation by MenA and methylation mediated by MenG gives menaquinone **73** (Kurosu and Begari 2010). The most common menaquinones contain 7, 8, or 9 isoprene units; however, menaquinones with 4, 5, 6, 10, 11, 12, or 13 isoprene units have been isolated from bacteria.

Since it has been demonstrated that menaquinone is crucial for bacterial growth (Dhiman et al. 2009; Fujimoto et al. 2012) and colony development (Pelchovich et al. 2013), it is not a surprise that menaquinone biosynthesis has already been investigated as a target for the development of novel antibiotics (Kurosu and Begari 2010).

In the last years, a few inhibitors of individual enzymes of menaquinone biosynthesis have been reported (Kurosu and Begari 2010). For example, allylaminomethanone A **84** and phenylethylaminomethanone A **85** (Scheme 27) are inhibitors of MenA, which mimic the structure of the product of MenA catalysis, demethylmenaquinone, and display MIC values of 1.5 and 12.5 μg/mL, respectively, against *M. tuberculosis* (Kurosu et al. 2007; Debnath et al. 2012).

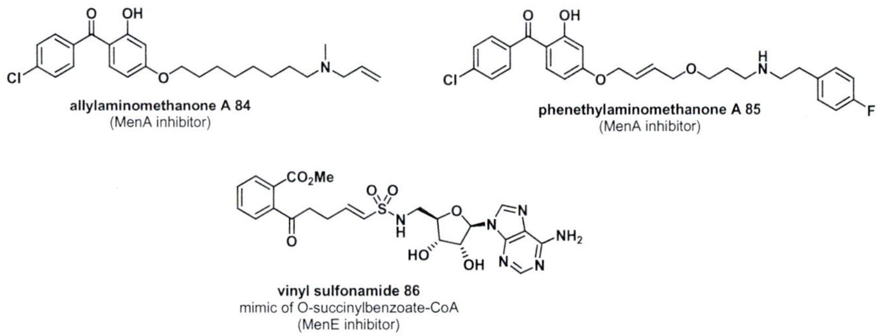

Scheme 27 Structures of representative inhibitors of menaquinone biosynthesis

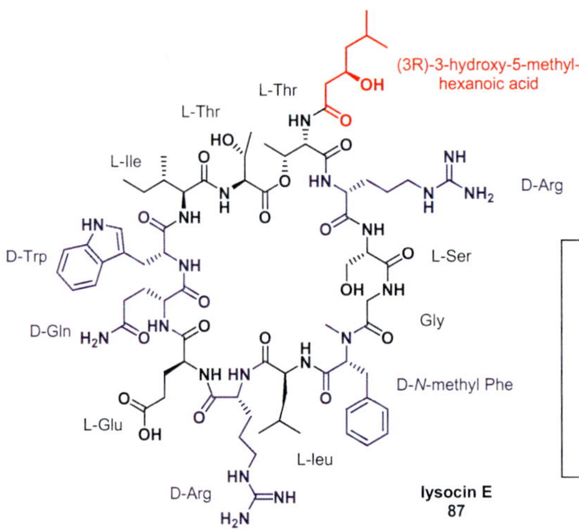

Gram-positive pathogens	MIC μ/mL 87
MRSA	4
S. simulans	4
S. haemolyticus	1
S. pseudointermedius	4
B. subtilis	4
B. cereus	2
L. monocytogenes	1

Scheme 28 Lysocin E and its antibacterial activity

Furthermore, the vinylsulfonamide **86**, which was designed as mimic of O-succinylbenzoate-CoA ester, was reported to efficiently inhibit MenE at low nanomolar concentrations (Lu et al. 2012).

Recently, the novel natural cyclodepsipeptide lysocin E **87** (Scheme 28) was reported to target the redox metabolism in bacteria by directly binding to menaquinone (Hamamoto et al. 2014). Since the biosynthesis of menaquinone is a multi-enzymatic pathway as described above and the role of menaquinone is crucial for viability of bacteria, menaquinone seems to be a much smarter choice for an antibacterial target than the individual enzymes of the biosynthesis, even more when considering that an alternative biosynthetic pathway for menaquinones starting from futalosine exists in several Gram-positive bacteria (Hiratsuka et al. 2008).

The cyclic structure of **87** displays a head-to-tail lactonization of the hydroxyl function of the N-terminal threonine to the carboxylic acid of the C-terminal threonine. Furthermore, the compound contains five D-amino acids, (3R)-3-hydroxy-5-methyl-hexanoic acid capping the N-terminal amino group and a N-methylation at the present phenylalanine.

Lysocin E was discovered by an extensive screening campaign in invertebrates, where culture supernatants of soil bacteria were tested in a silkworm infection model. The compound has shown good antibacterial activity against several Gram-positive pathogens including MRSA.

Lysocin E induced potassium leakage through the bacterial membrane, which led to a rapid loss of membrane potential and bacteriolysis. The hemolytic activity against mammalian cells was very low, and further investigations demonstrated that the MIC value of **87** against *S. aureus* increased in a dose-dependent manner upon addition of menaquinone. The direct binding of menaquinone in a 1:1 complex was further indicated by micro-calorimetric measurements, detecting an exothermic response upon addition of lysocin to menaquinone. Interestingly, no binding of the compound to the structurally similar ubiquinone was observed, suggesting that the naphthoquinone core in menaquinone is crucial for the interaction.

Furthermore, the first in vivo studies in *S. aureus* infected mice showed a promising ED_{50} value of 0.5 mg/kg body weight, even smaller than that of vancomycin. This is remarkable in view of the moderate–good MIC value in vitro. Even a dose of 400 mg/kg body weight administered by intraperitoneal injection was tolerated, and no increase of biochemical markers for tissue damage of liver and kidneys was observed. Lysocin E has already been successfully addressed in a solid-supported total synthesis by *Inoue* and coworkers, and the first structure–activity relationship studies have revealed that the *N*-methylation at the phenylalanine in lysocin E is crucial for its activity (Murai et al. 2015).

In summary, lysocin E demonstrates that the depletion of essential small molecule mediators of energy metabolism is a viable, but yet underexplored antibiotics strategy. It also highlights that in vivo phenotypic assays, a highly successful discovery strategy of pharmaceutical research before the advent of in vitro assay systems, are still a powerful way to disclose urgently needed, novel antibiotic lead compounds.

8 Conclusions

Antibiotic lead finding suffers from insufficient volume and scale since the pharmaceutical industry has considerably reduced its efforts in this indication, inter alia due to an unsatisfying outcome of a target-centric, high throughput screening-based antibiotic research strategy (Payne et al. 2007; Tommasi et al. 2015), and due to uncertain economic revenues. But it does not suffer from a shortage of validated biological mechanisms, or from an exhaustion of chemical matter, as demonstrated by the showcases of this review. Microbial natural products continue to play a dominant role as sources of novel antibiotics, and their discovery is facilitated by multiple advances in natural product research (Harvey et al. 2015): Classical strategies like the screening of extracts in relevant phenotypic assays (e.g., lysocin), or a re-investigation of neglected compounds of the past with modern chemical and genetic methods (as with griselimycin), are complemented by the cultivation of underexplored sources (e.g., teixobactin, cystobactamid), the heterologous expression of (almost) silent gene clusters (e.g., albicidin) or genome mining techniques (Doroghazi et al. 2014). Apart from being promising leads and drug candidates, natural products also teach important lessons and question existing 'rules' of drug discovery. Griselimycin, for example, demonstrates that peptides of elevated (>1000 Da) molecular weight can not only disrupt protein–protein interactions at picomolar concentrations, but at the same time exhibit high oral bioavailability, and high stability in plasma. Teixobactin, on the other hand, confirmed the rule that a dual molecular mode of action is advantageous with respect to suppression of resistance. It also underlines the common principle that natural antibiotics preferably target hotspots in complex biosynthetic machineries, as their malfunctioning is hard to circumvent by bacterial mutations.

New structural templates also arise from synthetic libraries that are screened against innovative targets, as exemplified by the inhibitors of WTA biosynthesis, the DNA sliding clamp, the DNA gyrase, or the β-lactamases. While many synthetic hits from the past suffered from insufficient translocation or permeation across bacterial membranes, the examples above demonstrate how an early incorporation of whole cell assays in the optimization cascade (or directly in the primary screening assay) assured the successful optimization of synthetic structures to antibiotics.

Our selection of case studies has been subjective and non-comprehensive, as it did not discuss the numerous additional innovative approaches to fight bacterial infections that include antibiotics (e.g., translation inhibitors, antimicrobial peptides, or actively transported conjugates), non-antibiotic pathogenicity-blocking small molecules [e.g., secretion system inhibitors, quorum sensing inhibitors, immune stimulants (Rasko and Sperandio 2010)], biomolecules [e.g., lysins, toxin binders (Morrison 2015)], or even larger entities [e.g., bacteriophages, probiotics] (Czaplewski et al. 2016). However, the few examples provide sufficient arguments that from a scientific point of view, the discovery and development of novel antibiotics is highly rewarding.

Acknowledgments We thank Dr. Dr. Werner Tegge, Giambattista Testolin, Isabell Schneider and Dr. Verena Fetz for fruitful discussions and for proofreading of the manuscript.

References

Aguilaniu H, Durieux J, Dillin A (2005) Metabolism, ubiquinone synthesis, and longevity. Genes Dev 19:2399–2406. doi:10.1101/gad.1366505

Aldred KJ, Kerns RJ, Osheroff N (2014) Mechanism of quinolone action and resistance. Biochemistry 53:1565–1574. doi:10.1021/bi5000564

Alm RA, Lahiri SD, Kutschke A et al (2015) Characterization of the novel DNA gyrase inhibitor AZD0914: low resistance potential and lack of cross-resistance in Neisseria gonorrhoeae. Antimicrob Agents Chemother 59:1478–1486. doi:10.1128/AAC.04456-14

Amaral L, Martins A, Spengler G, Molnar J (2014) Efflux pumps of Gram-negative bacteria: what they do, how they do it, with what and how to deal with them. Front Pharmacol 4:Article168. doi:10.3389/fphar.2013.00168

Anderson VE, Osheroff N (2001) Type II topoisomerases as targets for quinolone antibacterials: turning Dr. Jekyll into Mr. Hyde. Curr Pharm Des 7:337–353. doi:10.2174/1381612013398013

Anderson RJ, Groundwater PW, Todd A, Worsley AJ (2012) Antibacterial agents: chemistry, mode of action, mechanisms of resistance and clinical applications. Wiley, New York

Argiriadi MA, Goedken ER, Bruck I, et al (2006) Crystal structure of a DNA polymerase sliding clamp from a Gram-positive bacterium. BMC Struct Biol 6:2. doi:10.1186/1472-6807-6-2

Askoura M, Mottawea W, Abujamel T, Taher I (2011) Efflux pump inhibitors (EPIs) as new antimicrobial agents against *Pseudomonas aeruginosa*. Libyan J Med 6:1–8. doi:10.3402/ljm.v6i0.5870

Atilano ML, Pereira PM, Yates J et al (2010) Teichoic acids are temporal and spatial regulators of peptidoglycan cross-linking in *Staphylococcus aureus*. Proc Natl Acad Sci USA 107:18991–18996. doi:10.1073/pnas.1004304107

Baltrus DA (2013) Exploring the costs of horizontal gene transfer. Trends Ecol Evol 28:489–495. doi:10.1016/j.tree.2013.04.002

Baltz RH (2008) Renaissance in antibacterial discovery from actinomycetes. Curr Opin Pharmacol 8:557–563. doi:10.1016/j.coph.2008.04.008

Basarab GS, Manchester JI, Bist S et al (2013) Fragment-to-hit-to-lead discovery of a novel pyridylurea scaffold of ATP competitive dual targeting type II topoisomerase inhibiting antibacterial agents. J Med Chem 56:8712–8735. doi:10.1021/jm401208b

Basarab GS, Kern GH, McNulty J et al (2015) Responding to the challenge of untreatable gonorrhea: ETX0914, a first-in-class agent with a distinct mechanism-of-action against bacterial Type II topoisomerases. Sci Rep 5:11827. doi:10.1038/srep11827

Bates AD, Maxwell A (2007) Energy coupling type II topoisomerase why do they hydrolyze ATP? Biochemistry 46:7929–7941. doi:10.1021/bi700789g

Bauer A, Brönstrup M (2014) Industrial natural product chemistry for drug discovery and development. Nat Prod Rep 31:35–60. doi:10.1039/c3np70058e

Bauer R, Dicks LMT (2005) Mode of action of lipid II-targeting lantibiotics. Int J Food Microbiol 101:201–216. doi:10.1016/j.ijfoodmicro.2004.11.007

Baumann S, Herrmann J, Raju R et al (2014) Cystobactamids: myxobacterial topoisomerase inhibitors exhibiting potent antibacterial activity. Angew Chem Int Ed 53:14605–14609. doi:10.1002/anie.201409964

Bax BD, Chan PF, Eggleston DS et al (2010) Type IIA topoisomerase inhibition by a new class of antibacterial agents. Nature 466:935–940. doi:10.1038/nature09197

Bénazet et al (1966) Antibiotics—advances in research, production and clinical use. In: Herold M, Gabriel Z (eds) Proceedings of the congress on antibiotics held in Prague, 15–19 June, 1964. Butterworths, London, pp 262–264

Bengtsson B, Greko C (2014) Antibiotic resistance–consequences for animal health, welfare, and food production. Ups J Med Sci 119:96–102. doi:10.3109/03009734.2014.901445

Bérdy J (2012) Thoughts and facts about antibiotics: where we are now and where we are heading. J Antibiot 65:441. doi:10.1038/ja.2012.54

Berleman J, Auer M (2013) The role of bacterial outer membrane vesicles for intra- and interspecies delivery. Environ Microbiol 15:347–354. doi:10.1111/1462-2920.12048

Bierbaum G, Sahl HG (1985) Induction of autolysis of staphylococci by the basic peptide antibiotics Pep 5 and nisin and their influence on the activity of autolytic enzymes. Arch Microbiol 141:249–254. doi:10.1007/BF00408067

Birch RG, Patil SS (1987a) Correlation between albicidin production and chlorosis induction by Xanthomonas albilineans, the sugarcane leaf scald pathogen. Physiol Mol Plant Pathol 30:199–206. doi:10.1016/0885-5765(87)90033-6

Birch RG, Patil SS (1987b) Evidence that an albicidin-like phytotoxin induces chlorosis in sugarcane leaf scald disease by blocking plastid DNA replication. Physiol Mol Plant Pathol 30:207–214. doi:10.1016/0885-5765(87)90034-8

Biswas S, Brunel J-M, Dubus J-C et al (2012) Colistin: an update on the antibiotic of the 21st century. Expert Rev Anti Infect Ther 10:917–934. doi:10.1586/eri.12.78

Blair JMA, Richmond GE, Piddock LJV (2014) Multidrug efflux pumps in Gram-negative bacteria and their role in antibiotic resistance. Future Microbiol 9:1165–1177. doi:10.2217/fmb.14.66

Bloom LB (2009) Loaing clamps for DNA replication and repair. DNA Repair 8:570–578. doi:10.1016/j.dnarep.2008.12.014.Loading

Blunt JW, Copp BR, Munro MHG et al (2011) Marine natural products. Nat Prod Rep 28:196–268. doi:10.1039/c005001f

Boneca IG, Chiosis G (2003) Vancomycin resistance: occurrence, mechanisms and strategies to combat it. Expert Opin Ther Targets 7:311–328. doi:10.1517/14728222.7.3.311

Bouchaudon J (1964) Nouveau cyclopeptide, sa préparation et les médicaments qui le contiennent

Boucher HW, Talbot GH, Bradley JS et al (2009) Bad bugs, no drugs: no ESKAPE! An update from the Infectious Diseases Society of America. Clin Infect Dis 48:1–12. doi:10.1086/595011

Boucher HW, Talbot GH, Benjamin DK et al (2013) 10 x '20 Progress–development of new drugs active against gram-negative bacilli: an update from the Infectious diseases society of America. Clin Infect Dis 56:1685–1694. doi:10.1093/cid/cit152

Bozdogan B, Appelbaum PC (2004) Oxazolidinones: activity, mode of action, and mechanism of resistance. Int J Antimicrob Agents 23:113–119. doi:10.1016/j.ijantimicag.2003.11.003

Breu F, Guggenbichler S, Wollmann J (2001) Human as the world's greatest evolutionary force. Science 293:1786–1790

Breukink E, de Kruijff B (2006) Lipid II as a target for antibiotics. Nat Rev Drug Discov 5:321–332. doi:10.1038/nrd2004

Brötz H, Josten M, Wiedemann I et al (1998) Role of lipid-bound peptidoglycan precursors in the formation of pores by nisin, epidermin and other lantibiotics. Mol Microbiol 30:317–327. doi:10.1046/j.1365-2958.1998.01065.x

Brötz-Oesterhelt H, Brunner NA (2008) How many modes of action should an antibiotic have? Curr Opin Pharmacol 8:564–573. doi:10.1016/j.coph.2008.06.008

Brown S, Xia G, Luhachack LG et al (2012) Methicillin resistance in *Staphylococcus aureus* requires glycosylated wall teichoic acids. Proc Natl Acad Sci USA 109:18909–18914. doi:10.1073/pnas.1209126109

Brown S, Santa Maria JP, Walker S (2013) Wall teichoic acids of gram-positive bacteria. Annu Rev Microbiol 67:313–336. doi:10.1146/annurev-micro-092412-155620

Brown DG, Lister T, May-Dracka TL (2014) New natural products as new leads for antibacterial drug discovery. Bioorg Med Chem Lett 24:413–418. doi:10.1016/j.bmcl.2013.12.059

Brown L, Wolf JM, Prados-Rosales R, Casadevall A (2015) Through the wall: extracellular vesicles in Gram-positive bacteria, mycobacteria and fungi. Nat Rev Microbiol 13:620–630. doi:10.1038/nrmicro3480

Bunting KA, Roe SM, Pearl LH (2003) Structural basis for recruitment of translesion DNA polymerase Pol IV/DinB to the β-clamp. EMBO J 22:5883–5892. doi:10.1093/emboj/cdg568

Burnouf DY, Olieric V, Wagner J et al (2004) Structural and biochemical analysis of sliding clamp/ligand interactions suggest a competition between replicative and translesion DNA polymerases. J Mol Biol 335:1187–1197. doi:10.1016/j.jmb.2003.11.049

Bush K (2010) Bench-to-bedside review: the role of beta-lactamases in antibiotic-resistant Gram-negative infections. Crit Care 14:224. doi:10.1186/cc8892

Bush K (2015) Investigational agents for the treatment of Gram-negative bacterial infections: a reality check. ACS Infect Dis 1:509–511. doi:10.1021/acsinfecdis.5b00100

Butler EK, Davis RM, Bari V et al (2013a) Structure-function analysis of MurJ reveals a solvent-exposed cavity containing residues essential for peptidoglycan biogenesis in *Escherichia coli*. J Bacteriol 195:4639–4649. doi:10.1128/JB.00731-13

Butler MS, Blaskovich MA, Cooper MA (2013b) Antibiotics in the clinical pipeline in 2013. J Antibiot 66:571–591. doi:10.1038/ja.2013.86

Butler MS, Robertson AAB, Cooper MA (2014) Natural product and natural product derived drugs in clinical trials. Nat Prod Rep 31:1612–1661. doi:10.1039/c4np00064a

Campbell J, Singh AK, Swoboda JG et al (2012) An antibiotic that inhibits a late step in wall teichoic acid biosynthesis induces the cell wall stress stimulon in *Staphylococcus aureus*. Antimicrob Agents Chemother 56:1810–1820. doi:10.1128/AAC.05938-11

Carattoli A (2013) Plasmids and the spread of resistance. Int J Med Microbiol 303:298–304. doi:10.1016/j.ijmm.2013.02.001

Casacuberta E, Gonzalez J (2013) The impact of transposable elements in environmental adaptation. Mol Ecol 22:1503–1517. doi:10.1111/mec.12170

Cervellati R, Greco E (2016) In vitro antioxidant activity of ubiquinone and ubiquinol, compared to vitamin E. Helv Chim Acta 99:41–45. doi:10.1002/hlca.201500124

Champoux JJ (2001) DNA topoisomerases: structure, function, and mechanism. Annu Rev Biochem 70:369–413. doi:10.1161/01.RES.0000255691.76142.4a

Chan WC, Bycroft BW, Lian L-Y, Roberts GCK (1989) Isolation and characterisation of two degradation products derived from the peptide antibiotic nisin. FEBS Lett 252:29–36. doi:10.1016/0014-5793(89)80884-1

Chen Q, Duan F, Li X et al (2014) Haloemodin as novel antibacterial agent inhibiting DNA gyrase and bacterial topoisomerase i. J Med Chem 57:3707–3714. doi:10.1021/jm401685f

Cho WK, Jergic S, Kim D et al (2014) Loading dynamics of a sliding DNA clamp. Angew Chem Int Ed 53:6768–6771. doi:10.1002/anie.201403063

Chopra I, Roberts M (2001) Tetracycline antibiotics: mode of action, applications, molecular biology, and epidemiology of bacterial resistance tetracycline antibiotics: mode of action, applications, molecular biology, and epidemiology of bacterial resistance. Microbiol Mol Biol Rev 65:232–260. doi:10.1128/MMBR.65.2.232

Chopra S, Matsuyama K, Tran T et al (2012) Evaluation of gyrase B as a drug target in *Mycobacterium tuberculosis*. J Antimicrob Chemother 67:415–421. doi:10.1093/jac/dkr449

Chung BC, Zhao J, Gillespie RA et al (2013) Crystal structure of MraY, an essential membrane enzyme for bacterial cell wall synthesis. Science 341:1012–1016. doi:10.1126/science.1236501

Clardy J, Fischbach MA, Walsh CT (2006) New antibiotics from bacterial natural products. Nat Biotechnol 24:1541–1550. doi:10.1038/nbt1266

Cleveland J, Montville TJ, Nes IF, Chikindas ML (2001) Bacteriocins: Safe, natural antimicrobials for food preservation. Int J Food Microbiol 71:1–20. doi:10.1016/S0168-1605(01)00560-8

Coates AR, Halls G, Hu Y (2011) Novel classes of antibiotics or more of the same? Br J Pharmacol 163:184–194. doi:10.1111/j.1476-5381.2011.01250.x

Cociancich S, Pesic A, Petras D et al (2015) The gyrase inhibitor albicidin consists of p-aminobenzoic acids and cyanoalanine. Nat Chem Biol 11:195–197. doi:10.1038/nchembio.1734

Cohen GN (2011) Microbial biochemistry, 2nd edn. Springer, Heidelberg

Cole ST, Riccardi G (2011) New tuberculosis drugs on the horizon. Curr Opin Microbiol 14:570–576. doi:10.1016/j.mib.2011.07.022

Collin F, Karkare S, Maxwell A (2011) Exploiting bacterial DNA gyrase as a drug target: Current state and perspectives. Appl Microbiol Biotechnol 92:479–497. doi:10.1007/s00253-011-3557-z

Connor EE (1998) Sulfonamide antibiotics. Prim Care Update Ob Gyns 5:32–35

Costenaro L, Grossmann JG, Ebel C, Maxwell A (2007) Modular structure of the full-length DNA gyrase B subunit revealed by small-angle X-ray scattering. Structure 15:329–339. doi:10.1016/j.str.2007.01.013

Cragg GM, Newman DJ (2013) Natural products: a continuing source of novel drug leads. Biochim Biophys Acta-Gen Subj 1830:3670–3695. doi:10.1016/j.bbagen.2013.02.008

Czaplewski L, Bax R, Clokie M et al (2016) Alternatives to antibiotics-a pipeline portfolio review. Lancet Infect Dis 16:239–251. doi:10.1016/S1473-3099(15)00466-1

D'Costa VM, King CE, Kalan L et al (2011) Antibiotic resistance is ancient. Nature 477:457–461. doi:10.1038/nature10388

D'Elia MA, Millar KE, Beveridge TJ, Brown ED (2006a) Wall teichoic acid polymers are dispensable for cell viability in Bacillus subtilis. J Bacteriol 188:8313–8316. doi:10.1128/JB.01336-06

D'Elia MA, Pereira MP, Chung YS et al (2006b) Lesions in teichoic acid biosynthesis in *Staphylococcus aureus* lead to a lethal gain of function in the otherwise dispensable pathway. J Bacteriol 188:4183–4189. doi:10.1128/JB.00197-06

D'Elia MA, Henderson JA, Beveridge TJ et al (2009a) The N-acetylmannosamine transferase catalyzes the first committed step of teichoic acid assembly in Bacillus subtilis and *Staphylococcus aureus*. J Bacteriol 191:4030–4034. doi:10.1128/JB.00611-08

D'Elia MA, Pereira MP, Brown ED (2009b) Are essential genes really essential? Trends Microbiol 17:433–438. doi:10.1016/j.tim.2009.08.005

Dalrymple BP, Kongsuwan K, Wijffels G et al (2001) A universal protein-protein interaction motif in the eubacterial DNA replication and repair systems. Proc Natl Acad Sci U S A 98:11627–11632. doi:10.1073/pnas.191384398

Davies J (2006) Where have all the antibiotics gone? Can J Infect Dis Med Microbiol 17:287–290

Davis BD (1987) Mechanism of bactericidal action of aminoglycosides. Microbiol Rev 51:341–350 doi:0146-0749/87/030341-10$02.00/0

de Kruijff B, van Dam V, Breukink E (2008) Lipid II: a central component in bacterial cell wall synthesis and a target for antibiotics. Prostaglandins Leukot Essent Fat Acids 79:117–121. doi:10.1016/j.plefa.2008.09.020

Debnath J, Siricilla S, Wan B et al (2012) Discovery of selective menaquinone biosynthesis inhibitors against *Mycobacterium tuberculosis*. J Med Chem 55:3739–3755. doi:10.1021/jm201608g

Dhiman RK, Mahapatra S, Slayden RA et al (2009) Menaquinone synthesis is critical for maintaining mycobacterial viability during exponential growth and recovery from non-replicating persistence. Mol Microbiol 72:85–97. doi:10.1111/j.1365-2958.2009.06625.x

Diez J, Martinez JP, Mestres J et al (2012) Myxobacteria: natural pharmaceutical factories. Microb Cell Fact 11:52–54. doi:10.1186/1475-2859-11-52

Donnell MO, Langston L, Stillman B et al (2013) Principles and concepts of DNA replication in bacteria, archea, and eukarya. Cold Spring Harb Perspect Biol 5:a010180. doi:10.1101/cshperspect.a010108

Doroghazi JR, Albright JC, Goering AW et al (2014) A roadmap for natural product discovery based on large-scale genomics and metabolomics. Nat Chem Biol 10:963–968. doi:10.1038/nchembio.1659

Dougherty TJ, Nayar A, Newman JV et al (2014) NBTI 5463 is a novel bacterial type II topoisomerase inhibitor with Gram-negative antibacterial activity and in vivo efficacy. Antimicrob Agents Chemother. doi:10.1128/aac.02778-13

Dubée V, Chau F, Arthur M et al (2011) The in vitro contribution of autolysins to bacterial killing elicited by amoxicillin increases with inoculum size in Enterococcus faecalis. Antimicrob Agents Chemother 55:910–912. doi:10.1128/AAC.01230-10

Edwards DI (1993) Nitroimidazole drugs–action and resistance mechanisms. I. Mechanisms of action. J Antimicrob Chemother 31:9–20. doi:10.1093/jac/31.1.9

Ehmann DE, Lahiri SD (2014) Novel compounds targeting bacterial DNA topoisomerase/DNA gyrase. Curr Opin Pharmacol 18:76–83. doi:10.1016/j.coph.2014.09.007

Eichberg MJ (2015) Public funding of clinical-stage antibiotic development in the United States and European Union. Heal Secur 13:156–165. doi:10.1089/hs.2014.0081

Elshahawi SI, Shaaban KA, Kharel MK, Thorson JS (2015) A comprehensive review of glycosylated bacterial natural products. Chem Soc Rev 44:7591–7697. doi:10.1039/C4CS00426D

Epand RM, Walker C, Epand RF, Magarvey NA (2016) Molecular mechanisms of membrane targeting antibiotics. BBA Biomembranes 1858:980–987. doi:10.1016/j.bbamem.2015.10.018
European Centre for Disease Prevention and Control (2012) Annual report of the European antimicrobial resistance surveilance network (EARS-Net). ECDC, Stockholm
Farha MA, Leung A, Sewell EW et al (2013) Inhibition of WTA synthesis blocks the cooperative action of pbps and sensitizes MRSA to β-lactams. ACS Chem Biol 8:226–233. doi:10.1021/cb300413m
Finn J (2013) Evaluation of WO2012177707 and WO2012097269: Vertex's phosphate prodrugs of gyrase and topoisomerase antibacterial agents. Expert Opin Ther Pat 23:1233–1237. doi:10.1517/13543776.2013.820707
Fischbach MA, Walsh CT (2009) Antibiotics for emerging pathogens. Science 325:1089–1093. doi:10.1126/science.1176667
Floss HG, Yu T-W (2005) Rifamycin-mode of action, resistance, and biosynthesis. Chem Rev 105:621–632. doi:10.1021/cr030112j
Forsberg KJ, Reyes A, Wang B et al (2012) The shared antibiotic resistome of soil bacteria and human pathogens. Science 337:1107–1111. doi:10.1126/science.1220761
Frankel RB, Kalmijn AJ, Amann R et al (2006) Sampling the antibiotic resistome. Science 311:374–378. doi:10.1126/science.1120800
Fujimoto N, Kosaka T, Yam M (2012) Menaquinone as well as ubiquinone as a crucial component in the *Escherichia coli* respiratory chain. In: Ekinci D (ed) Chemical biology, pp 187–208
Georgescu RE, Kim SS, Yurieva O et al (2008a) Structure of a sliding clamp on DNA. Cell 132:43–54. doi:10.1016/j.cell.2007.11.045
Georgescu RE, Yurieva O, Kim S-S et al (2008b) Structure of a small-molecule inhibitor of a DNA polymerase sliding clamp. Proc Natl Acad Sci USA 105:11116–11121. doi:10.1073/pnas.0804754105
Gerth K, Pradella S, Perlova O et al (2003) Myxobacteria: Proficient producers of novel natural products with various biological activities—past and future biotechnological aspects with the focus on the genus Sorangium. J Biotechnol 106:233–253. doi:10.1016/j.jbiotec.2003.07.015
Gibbons S (2004) Anti-staphylococcal plant natural products. Nat Prod Rep 21:263–277. doi:10.1039/b212695h
Giguère S (2013) Lincosamides, pleuromutilins, and streptogramins. Antimicrob Ther Vet Med 199–210. doi:10.1016/0007-1935(89)90107-3
Gilbert GL (2015) Knowing when to stop antibiotic therapy. Med J Aust 202:121–123. doi:10.5694/mja14.01201
Gillings MR (2013) Evolutionary consequences of antibiotic use for the resistome, mobilome, and microbial pangenome. Front Microbiol 4:1–10. doi:10.3389/fmicb.2013.00004
Godtfredsen WO, Jahnsen S, Lorck H et al (1962) Fusidic acid: a new antibiotic. Nature 193:987
Górska A, Sloderbach A, Marszałł MP (2014) Siderophore-drug complexes: Potential medicinal applications of the "Trojan horse" strategy. Trends Pharmacol Sci 1–8. doi:10.1016/j.tips.2014.06.007
Gubaev A, Klostermeier D (2014) Reprint of "the mechanism of negative DNA supercoiling: a cascade of DNA-induced conformational changes prepares gyrase for strand passage". DNA Repair (Amst) 20:130–141. doi:10.1016/j.dnarep.2014.06.006
Gui W-J, Lin S-Q, Chen Y-Y et al (2011) Crystal structure of DNA polymerase III β sliding clamp from Mycobacterium tuberculosis. Biochem Biophys Res Commun 405:272–277. doi:10.1016/j.bbrc.2011.01.027
Hamamoto H, Urai M, Ishii K et al (2014) Lysocin E is a new antibiotic that targets menaquinone in the bacterial membrane. Nat Chem Biol 11:127–133. doi:10.1038/nchembio.1710
Hammes WP, Neuhaus FC (1974) On the mechanism of action of vancomycin: inhibition of peptidoglycan synthesis in Gaffkya homari. Antimicrob Agents Chemother 6:722–728. doi:10.1128/AAC.6.6.722
Harrison E, Brockhurst MA (2012) Plasmid-mediated horizontal gene transfer is a coevolutionary process. Trends Microbiol 20:262–267. doi:10.1016/j.tim.2012.04.003

Harvey AL, Edrada-Ebel R, Quinn RJ (2015) The re-emergence of natural products for drug discovery in the genomics era. Nat Rev Drug Discov 14:111–129. doi:10.1038/nrd4510

Hashimi SM, Wall MK, Smith AB et al (2007) The phytotoxin albicidin is a novel inhibitor of DNA gyrase. Antimicrob Agents Chemother 51:181–187. doi:10.1128/AAC.00918-06

Hawkey PM, Livermore DM (2012) Carbapenem antibiotics for serious infections. BMJ 344:1–7. doi:10.1136/bmj.e3236

Hawser S, Lociuro S, Islam K (2006) Dihydrofolate reductase inhibitors as antibacterial agents. Biochem Pharmacol 71:941–948. doi:10.1016/j.bcp.2005.10.052

Hecker SJ, Reddy KR, Totrov M et al (2015) Discovery of a cyclic boronic acid β-lactamase inhibitor (RPX7009) with utility vs class A serine carbapenemases. J Med Chem 58:3682–3692. doi:10.1021/acs.jmedchem.5b00127

Heide L (2014) New aminocoumarin antibiotics as gyrase inhibitors. Int J Med Microbiol 304:31–36. doi:10.1016/j.ijmm.2013.08.013

Henry RJ (1943) The mode of action of sulfonamides. Bacteriol Rev 7:175–262. doi:10.1128/AEM.02766-10

Herrmann J, Lukežič T, Kling A, et al (2016) Strategies for the discovery and development of new antibiotics from natural products: three case studies. Curr Top Microbiol Immunol

Hesterkamp T (2016) Antibiotics clinical development and pipeline. Curr Top Microbiol Immunol. doi:10.1007/82_2015_451

Hiratsuka T, Furihata K, Ishikawa J et al (2008) An alternative menaquinone biosynthetic pathway operating in microorganisms. Science 321:1670–1673. doi:10.1126/science.1160446

Hirsch EB, Ledesma KR, Chang KT et al (2012) In vitro activity of MK-7655, a novel beta-lactamase inhibitor, in combination with imipenem against carbapenem-resistant Gram-negative bacteria. Antimicrob Agents Chemother 56:3753–3757. doi:10.1128/AAC.05927-11

Holzgrabe U (2015) New griselimycins for treatment of tuberculosis. Chem Biol 22:981–982. doi:10.1016/j.chembiol.2015.08.002

Hooper DC, Wolfson JS, McHugh GL et al (1982) Effects of novobiocin, coumermycin-a1, clorobiocin, and their analogs on *Escherichia-coli* Dna gyrase and bacterial-growth. Antimicrob Agents Chemother 22:662–671

Hsu S-TD, Breukink E, Tischenko E et al (2004) The nisin-lipid II complex reveals a pyrophosphate cage that provides a blueprint for novel antibiotics. Nat Struct Mol Biol 11:963–967. doi:10.1038/nsmb830

Huang G, Zhang L, Birch RG (2001) A multifunctional polyketide-peptide synthetase essential for albicidin biosynthesis in Xanthomonas albilineans. Microbiology 147:631–642

Jacobsson S, Golparian D, Alm RA et al (2014) High in vitro activity of the novel spiropyrimidinetrione AZD0914, a DNA gyrase inhibitor, against multidrug-resistant Neisseria gonorrhoeae isolates suggests a new effective option for oral treatment of gonorrhea. Antimicrob Agents Chemother 58:5585–5588. doi:10.1128/AAC.03090-14

Jacoby GA (2005) Mechanisms of bacterial resistance to quinolones. 41:S120–S126

Jad YE, Acosta GA, Naicker T et al (2015) Synthesis and biological evaluation of a teixobactin analogue. Org Lett 17:6182–6185. doi:10.1021/acs.orglett.5b03176

Ji C, Juárez-Hernández RE, Miller MJ (2012) Exploiting bacterial iron acquisition: siderophore conjugates. Future Med Chem 4:297–313. doi:10.4155/fmc.11.191

Jia Z, O'Mara ML, Zuegg J et al (2013) Vancomycin: ligand recognition, dimerization and super-complex formation. FEBS J 280:1294–1307. doi:10.1111/febs.12121

Johnston CW, Magarvey NA (2015) Natural products: untwisting the antibiotic'ome. Nat Chem Biol 11:177–178. doi:10.1038/nchembio.1757

Johnston CW, Skinnider MA, Dejong CA et al (2016) Assembly and clustering of natural antibiotics guides target identification. Nat Chem Biol 12:233–239. doi:10.1038/nchembio.2018

Johnstone TC, Nolan EM (2015) Beyond iron: non-classical biological functions of bacterial siderophores. Dalton Trans 44:6320–6339. doi:10.1039/c4dt03559c

Jolles G (1971) New cyclopeptides, GB Patent 1,252,553

Kahne D, Leimkuhler C, Lu W, Walsh C (2005) Glycopeptide and lipoglycopeptide antibiotics. Chem Rev 105:425–448. doi:10.1021/cr030103a

Kåhrström CT (2015) Antimicrobials: a new drug for resistant bugs. Nat Rev Microbiol 13:126–127. doi:10.1038/nrmicro3429

Kale MG, Raichurkar A, Shahul Hameed P et al (2013) Thiazolopyridine ureas as novel antitubercular agents acting through inhibition of DNA gyrase B. J Med Chem 56:8834–8848. doi:10.1021/jm401268f

Kale RR, Kale MG, Waterson D et al (2014) Thiazolopyridone ureas as DNA gyrase B inhibitors: optimization of antitubercular activity and efficacy. Bioorg Med Chem Lett 24:870–879. doi:10.1016/j.bmcl.2013.12.080

Kalman D, Barriere S (1990) Review of the pharmacology, pharmacokinetics, and clinical use of cephalosporins. Tex Heart Inst J 17:203–215

Kannan K, Mankin AS (2011) Macrolide antibiotics in the ribosome exit tunnel: species-specific binding and action. Ann N Y Acad Sci 1241:33–47. doi:10.1111/j.1749-6632.2011.06315.x

Kawamukai M (2002) Biosynthesis, Bioproduction and Novel Roles of Ubiquinone. J Biosci Bioeng 94:511–517

Keller S, Schadt HS, Ortel I, Süssmuth RD (2007) Action of atrop-abyssomicin C as an inhibitor of 4-amino-4-deoxychorismate synthase PabB. Angew Chem Int Ed 46:8284–8286. doi:10.1002/anie.200701836

Kern G, Palmer T, Ehmann DE, et al (2015) Inhibition of Neisseria gonorrhoeae type II Topoisomerases by the Novel Spiropyrimidinetrione AZD0914. J Biol Chem M115.663534. doi:10.1074/jbc.M115.663534

Kirst HA (2013) Developing new antibacterials through natural product research. Expert Opin Drug Discov 8:479–493. doi:10.1517/17460441.2013.779666

Kjelstrup S, Hansen PMP, Thomsen LE et al (2013) Cyclic Peptide Inhibitors of the β-Sliding Clamp in *Staphylococcus aureus*. PLoS ONE. doi:10.1371/journal.pone.0072273

Kling A, Lukat P, Almeida DV et al (2015) Targeting DnaN for tuberculosis therapy using novel griselimycins. Science 348:1106–1112. doi:10.1126/science.aaa4690

Koehn FE, Carter GT (2005) The evolving role of natural products in drug discovery. Nat Rev Drug Discov 4:206–220. doi:10.1038/nrd1657

Kohanski MA, Dwyer DJ, Collins JJ (2010) How antibiotics kill bacteria: from targets to networks. Nat Rev Microbiol 8:423–435. doi:10.1038/nrmicro2333

Kotra LP, Haddad J, Mobashery S (2000) Aminoglycosides: perspectives on mechanisms of action and resistance and strategies to counter resistance. Antimicrob Agents Chemother 44:3249–3256. doi:10.1128/AAC.44.12.3249-3256.2000.Updated

Kramer NE, Smid EJ, Kok J et al (2004) Resistance of Gram-positive bacteria to nisin is not determined by Lipid II levels. FEMS Microbiol Lett 239:157–161. doi:10.1016/j.femsle.2004.08.033

Kretz J, Kerwat D, Schubert V et al (2015) Total synthesis of albicidin: A lead structure from xanthomonas albilineans for potent antibacterial gyrase inhibitors. Angew Chem Int Ed 54:1969–1973. doi:10.1002/anie.201409584

Kuipers OP, Rollema HS, de Vos WM, Siezen RJ (1993) Biosynthesis and secretion of a precursor of nisin Z by Lactococcus lactis, directed by the leader peptide of the homologous lantibiotic subtilin from Bacillus subtilis. FEBS Lett 330:23–27. doi:10.1016/0014-5793(93)80911-D

Kurosu M, Begari E (2010) Vitamin K2 in electron transport system: are enzymes involved in vitamin K2 biosynthesis promising drug targets? Molecules 15:1531–1553. doi:10.3390/molecules15031531

Kurosu M, Narayanasamy P, Biswas K et al (2007) Discovery of 1,4-didydroxy-2-naphthoate prenyltransferase inhibitors: new drug leads for multidrug-resistant gram-positive pathogens. J Med Chem 50:3973–3975. doi:10.1021/jm070638m

Lapierre P, Gogarten JP (2009) Estimating the size of the bacterial pan-genome. Trends Genet 25:107–110. doi:10.1016/j.tig.2008.12.004

Lapuebla A, Abdallah M, Olafisoye O et al (2015a) Activity of imipenem with relebactam against gram-negative pathogens from New York City: Table 1. Antimicrob Agents Chemother 59:5029–5031. doi:10.1128/AAC.00830-15

Lapuebla A, Abdallah M, Olafisoye O, et al (2015b) Activity of meropenem combined with RPX7009, a novel β-lactamase inhibitor, against Gram-negative clinical isolates in New York City. Antimicrob Agents Chemother 59:AAC.00843–15. doi:10.1128/AAC.00843-15

Lazarevic V, Karamata D (1995) The tagGH operon of Bacillus subtilis 168 encodes a two-component ABC transporter involved in the metabolism of two wall teichoic acids. Mol Microbiol 16:345–355. doi:10.1111/j.1365-2958.1995.tb02306.x

Lee K, Campbell J, Swoboda JG et al (2010) Development of improved inhibitors of wall teichoic acid biosynthesis with potent activity against *Staphylococcus aureus*. Bioorg Med Chem Lett 20:1767–1770. doi:10.1016/j.bmcl.2010.01.036

Lee SH, Wang H, Labroli M, et al (2016a) TarO-specific inhibitors of wall teichoic acid biosynthesis restore b -lactam efficacy against methicillin-resistant staphylococci. Sci Transl Med 8:329ra32. doi:10.1126/scitranslmed.aad7364

Lee W, Schaefer K, Qiao Y et al (2016b) The mechanism of action of lysobactin. J Am Chem Soc 138:100–103. doi:10.1021/jacs.5b11807

Leski TA, Tomasz A (2005) Role of penicillin-binding protein 2 (PBP2) in the antibiotic susceptibility and cell wall cross-linking of *Staphylococcus aureus*. J Bacteriol 2:1815–1824. doi:10.1128/JB.187.5.1815

Leu FP, Hingorani MM, Turner J, O'Donnell M (2000) The δ subunit of DNA polymerase III holoenzyme serves as a sliding clamp unloader in *Escherichia coli*. J Biol Chem 275:34609–34618. doi:10.1074/jbc.M005495200

Lewis K (2013) Platforms for antibiotic discovery. Nat Rev Drug Discov 12:371–387. doi:10.1038/nrd3975

Lim D, Strynadka NCJ (2002) Structural basis for the beta lactam resistance of PBP2a from methicillin-resistant *Staphylococcus aureus*. Nat Struct Biol 9:870–876. doi:10.1038/nsb858

Lin AH, Murray RW, Vidmar TJ, Marotti KR (1997) The oxazolidinone eperezolid binds to the 50S ribosomal subunit and competes with binding of chloramphenicol and lincomycin. Antimicrob Agents Chemother 41:2127–2131

Ling LL, Schneider T, Peoples AJ et al (2015) A new antibiotic kills pathogens without detectable resistance. Nature 517:455–459. doi:10.1038/nature14098

Lipsky BA, Baker CA (1999) Fluoroquinolone toxicity profiles: a review focusing on newer agents. Clin Infect Dis 28:352–364. doi:10.1086/515104

Littmann J, Buyx A, Cars O (2015) Antibiotic resistance: An ethical challenge. Int J Antimicrob Agents 46:359–361. doi:10.1016/j.ijantimicag.2015.06.010

Liu LF, Liu CC, Alberts BM (1980) Type II DNA topoisomerases: enzymes that can unknot a topologically knotted DNA molecule via a reversible double-strand break. Cell 19:697–707. doi:10.1016/S0092-8674(80)80046-8

Livermore DM, Warner M, Mushtaq S (2013) Activity of MK-7655 combined with imipenem against enterobacteriaceae and Pseudomonas aeruginosa. J Antimicrob Chemother 68:2286–2290. doi:10.1093/jac/dkt178

Livermore DM, Mushtaq S, Warner M, Woodford N (2015) Activity of OP0595/β-lactam combinations against Gram-negative bacteria with extended-spectrum, AmpC and carbapenem-hydrolysing β-lactamases. J Antimicrob Chemother 70:3032–3041. doi:10.1093/jac/dkv239

Lowther GE (1979) The united states food and drug administration and the practice of optometry. J Am Optom Assoc 50:579–582

Lu X, Zhou R, Sharma I et al (2012) Stable analogues of OSB-AMP: potent Inhibitors of MenE, the o-succinylbenzoate-CoA synthetase from bacterial menaquinone biosynthesis. ChemBioChem 13:129–136. doi:10.1002/cbic.201100585

Maki H, Yamaguchi T, Murakami K (1994) Cloning and characterization of a gene affecting the methicillin resistance level and the autolysis rate in *Staphylococcus aureus*. J Bacteriol 176:4993–5000

Maren TH (1976) Relatons between structure and biological activity of sulfonamides. Annu Rev Pharmacol Toxicol 16:309–327. doi:10.1146/annurev.pa.16.040176.001521

Martin NI, Breukink E (2007) Expanding role of lipid II as a target for lantibiotics. Future Microbiol 2:513–525. doi:10.2217/17460913.2.5.513

Mast Y, Wohlleben W (2014) International Journal of Medical Microbiology Streptogramins – Two are better than one ! 304:44–50

Maxwell A (1993) The interaction between coumarin drugs and DNA gyrase. Mol Microbiol 9:681–686. doi:10.1111/j.1365-2958.1993.tb01728.x

Maxwell A, Lawson DM (2003) The ATP-binding site of type II topoisomerases as a target for antibacterial drugs. Curr Top Med Chem 3:283–303. doi:10.2174/1568026033452500

Mayer C, Janin YL (2014) Non-quinolone inhibitors of bacterial type IIA topoisomerases: a feat of bioisosterism. Chem Rev 114:2313–2342

Mccomas CC, Mccomas CC, Crowley BM et al (2003) Partitioning the loss in vancomycin binding affinity for D-Ala-D-Lac into lost H-bond and repulsive lone pair contributions. J Am Chem Soc 125:9314–9315

Medini D, Donati C, Tettelin H et al (2005) The microbial pan-genome. Curr Opin Genet Dev 15:589–594. doi:10.1016/j.gde.2005.09.006

Meeske AJ, Sham L-T, Kimsey H et al (2015) MurJ and a novel lipid II flippase are required for cell wall biogenesis in Bacillus subtilis. Proc Natl Acad Sci U S A 112:6437–6442. doi:10.1073/pnas.1504967112

Meredith TC, Swoboda JG, Walker S (2008) Late-stage polyribitol phosphate wall teichoic acid biosynthesis in *Staphylococcus aureus*. J Bacteriol 190:3046–3056. doi:10.1128/JB.01880-07

Michalopoulos AS, Livaditis IG, Gougoutas V (2011) The revival of fosfomycin. Int J Infect Dis 15:e732–e739. doi:10.1016/j.ijid.2011.07.007

Miles TJ, Hennessy AJ, Bax B et al (2013) Novel hydroxyl tricyclics (e.g., GSK966587) as potent inhibitors of bacterial type IIA topoisomerases. Bioorg Med Chem Lett 23:5437–5441. doi:10.1016/j.bmcl.2013.07.013

Miller EL (2002) The penicillins: a review and update. J Midwifery Women's Heal 47:426–434. doi:10.1016/S1526-9523(02)00330-6

Mislin GL, Schalk IJ (2014) Siderophore-dependent iron uptake systems as gates for antibiotic Trojan horse strategies against *Pseudomonas aeruginosa*. Metallomics 6:408–420. doi:10.1039/c3mt00359k

Mital A (2009) Synthetic nitroimidazoles: biological activities and mutagenicity relationships. Sci Pharm 77:497–520. doi:10.3797/scipharm.0907-14

Mizrahi V, Henrie RN, Marlier JF et al (1985) Rate-limiting steps in the DNA polymerase I reaction pathway. Biochemistry 24:4010–4018. doi:10.1021/bi00336a031

Mohammadi T, Karczmarek A, Crouvoisier M et al (2007) The essential peptidoglycan glycosyltransferase MurG forms a complex with proteins involved in lateral envelope growth as well as with proteins involved in cell division in *Escherichia coli*. Mol Microbiol 65:1106–1121. doi:10.1111/j.1365-2958.2007.05851.x

Mohammadi T, van Dam V, Sijbrandi R et al (2011) Identification of FtsW as a transporter of lipid-linked cell wall precursors across the membrane. EMBO J 30:1425–1432. doi:10.1038/emboj.2011.61

Moreno M, Elgaher WAM, Herrmann J et al (2015) Synthesis and biological evaluation of cystobactamid 507: a bacterial topoisomerase inhibitor from Cystobacter sp. Synlett 26:1175–1178. doi:10.1055/s-0034-1380509

Morrison C (2015) Antibacterial antibodies gain traction. Nat Rev Drug Discov 14:737–738. doi:10.1038/nrd4770

Mullane KM, Gorbach S (2011) Fidaxomicin: first-in-class macrocyclic antibiotic. Expert Rev Anti Infect Ther 9:767–777. doi:10.1586/eri.11.53

Münch D, Roemer T, Lee SH et al (2012) Identification and in vitro analysis of the GatD/MurT enzyme-complex catalyzing lipid II amidation in *Staphylococcus aureus*. PLoS Pathog 8:1–11. doi:10.1371/journal.ppat.1002509

Münch D, Engels I, Müller A et al (2015) Structural variations of the cell wall precursor lipid II and their influence on binding and activity of the lipoglycopeptide antibiotic oritavancin. Antimicrob Agents Chemother 59:772–781. doi:10.1128/AAC.02663-14

Murai M, Kaji T, Kuranaga T et al (2015) Total synthesis and biological evaluation of the antibiotic lysocin E and its enantiomeric, epimeric, and N-demethylated analogues. Angew Chem Int Ed 54:1556–1560. doi:10.1002/anie.201410270

Nesme J, Simonet P (2015) The soil resistome: a critical review on antibiotic resistance origins, ecology and dissemination potential in telluric bacteria. Environ Microbiol 17:913–930. doi:10.1111/1462-2920.12631

Newman DJ, Cragg GM (2012) Natural products as sources of new drugs over the 30 years from 1981 to 2010. J Nat Prod 75:311–335. doi:10.1021/np200906s

Nichols D, Cahoon N, Trakhtenberg EM et al (2010) Use of ichip for high-throughput in situ cultivation of "uncultivable" microbial species. Appl Environ Microbiol 76:2445–2450. doi:10.1128/AEM.01754-09

Nicolau DP (2008) Carbapenems: a potent class of antibiotics. Expert Opin Pharmacother 9:23–37. doi:10.1517/14656566.9.1.23

Niiranen L, Lian K, Johnson KA, Moe E (2015) Crystal structure of the DNA polymerase III β subunit (β-clamp) from the extremophile Deinococcus radiodurans. BMC Struct Biol. doi:10.1186/s12900-015-0032-6

Nikaido H (2003) Molecular basis of bacterial outer membrane permeability revisited. Microbiol Mol Biol Rev 67:593–656. doi:10.1128/MMBR.67.4.593

Nöllmann M, Crisona NJ, Arimondo PB (2007) Thirty years of *Escherichia coli* DNA gyrase: From in vivo function to single-molecule mechanism. Biochimie 89:490–499. doi:10.1016/j.biochi.2007.02.012

Noufflard-Guy-Loé H, Berteaux S (1965) Rev Tuberc Pneumol 29:301–326

Novak R (2011) Are pleuromutilin antibiotics finally fit for human use? Ann N Y Acad Sci 1241:71–81. doi:10.1111/j.1749-6632.2011.06219.x

Nowicka B, Kruk J (2010) Occurrence, biosynthesis and function of isoprenoid quinones. Biochim Biophys Acta Bioenerg 1797:1587–1605. doi:10.1016/j.bbabio.2010.06.007

O'Connell KMG, Hodgkinson JT, Sore HF et al (2013) Combating multidrug-resistant bacteria: Current strategies for the discovery of novel antibacterials. Angew Chem Int Ed 52:10706–10733. doi:10.1002/anie.201209979

O'Neill J (2015) Securing new drugs for future generations : the pipeline of antibiotics. Wellcome Trust, London

Oakley AJ, Prosselkov P, Wijffels G et al (2003) Flexibility revealed by the 1.85 Å crystal structure of the $β$ sliding-clamp subunit of *Escherichia coli* DNA polymerase III. Acta Crystallogr Sect D 59:1192–1199. doi:10.1107/S0907444903009958

Oliphant CM, Green GM (2002) Quinolones: a comprehensive review. Am Fam Physician 65:455–464

Opperman TJ, Nguyen ST (2015) Recent advances toward a molecular mechanism of efflux pump inhibition. Front Microbiol 6:1–16. doi:10.3389/fmicb.2015.00421

Ozawa K, Horan NP, Robinson A et al (2013) Proofreading exonuclease on a tether: the complex between the *E. coli* DNA polymerase III subunits α, ε, θ and β reveals a highly flexible arrangement of the proofreading domain. Nucleic Acids Res 41:5354–5367. doi:10.1093/nar/gkt162

Page MGP (2013) Siderophore conjugates. Ann N Y Acad Sci 1277:115–126. doi:10.1111/nyas.12024

Papp-Wallace KM, Endimiani A, Taracila MA, Bonomo RA (2011) Carbapenems: past, present, and future. Antimicrob Agents Chemother 55:4943–4960. doi:10.1128/AAC.00296-11

Paris L (2015) Report of the Antibiotics resistance project: antibiotics currently in clinical development. http://www.pewtrusts.org/en/multimedia/data-visualizations/2014/antibiotics-currently-in-clinical-development

Parmar A, Iyer A, Vincent CS et al (2016) Efficient total syntheses and biological activities of two teixobactin analogues. Chem Commun 52:6060–6063. doi:10.1039/C5CC10249A

Payne DJ, Gwynn MN, Holms DJ, Pompliano DL (2007) Drugs for bad bugs: confronting the challanges of antibacterial discovery. Nat Rev Drug Discov 6:29–40

Pelchovich G, Omer-Bendori S, Gophna U (2013) Menaquinone and iron are essential for complex colony development in *Bacillus subtilis*. PLoS ONE 8:1–14. doi:10.1371/journal.pone.0079488

Pendleton JN, Gorman SP, Gilmore BF (2013) Clinical relevance of the ESKAPE pathogens. Expert Rev Anti Infect Ther 11:297–308. doi:10.1586/eri.13.12

Pereira MP, Schertzer JW, D'Elia MA et al (2008) The wall teichoic acid polymerase TagF efficiently synthesizes poly(glycerol phosphate) on the TagB product lipid III. ChemBioChem 9:1385–1390. doi:10.1002/cbic.200800026

Perez-Cruz C, Delgado L, Lopez-Iglesias C, Mercade E (2015) Outer-inner membrane vesicles naturally secreted by gram-negative pathogenic bacteria. PLoS ONE 10:1–18. doi:10.1371/journal.pone.0116896

Perry JA, Westman EL, Wright GD (2014) The antibiotic resistome: what' s new? Curr Opin Microbiol 21:45–50. doi:10.1016/j.mib.2014.09.002

Peterson LR (2009) Bad bugs, no drugs: no ESCAPE revisited. Clin Infect Dis 49:992–993. doi:10.1086/605540

Piddock LJV (2012) The crisis of no new antibiotics-what is the way forward? Lancet Infect Dis 12:249–253. doi:10.1016/S1473-3099(11)70316-4

Piddock LJV (2015) Teixobactin, the first of a new class of antibiotics discovered by ichip technology? J Antimicrob Chemother 70:2679–2680. doi:10.1093/jac/dkv175

Piel J (2010) Biosynthesis of polyketides by trans-AT polyketide synthases. Nat Prod Rep 27:996–1047. doi:10.1039/b816430b

Pinho MG, Errington J (2005) Recruitment of penicillin-binding protein PBP2 to the division site of *Staphylococcus aureus* is dependent on its transpeptidation substrates. Mol Microbiol 55:799–807. doi:10.1111/j.1365-2958.2004.04420.x

Pinho MG, de Lencastre H, Tomasz A (2001) An acquired and a native penicillin-binding protein cooperate in building the cell wall of drug-resistant staphylococci. Proc Natl Acad Sci USA 98:10886–10891. doi:10.1073/pnas.191260798

Policy IP (2010) The 10 × '20 initiative: pursuing a global commitment to develop 10 new antibacterial drugs by 2020. Clin Infect Dis 50:1081–1083. doi:10.1086/652237

Polz MF, Alm EJ, Hanage WP (2013) Horizontal gene transfer and the evolution of bacterial and archaeal population structure. Trends Genet 29:170–175. doi:10.1016/j.tig.2012.12.006

Powers JH (2003) Development of drugs for antimicrobial-resistant pathogens. Curr Opin Infect Dis 16:547–551. doi:10.1097/01.qco.0000104294.87920.b4

Projan SJ (2003) Why is big Pharma getting out of antibacterial drug discovery? Curr Opin Microbiol 6:427–430. doi:10.1016/j.mib.2003.08.003

Pucci MJ, Bush K (2013) Investigational antimicrobial agents of 2013. Clin Microbiol Rev 26:792–821. doi:10.1128/CMR.00033-13

Rasko DA, Sperandio V (2010) Anti-virulence strategies to combat bacteria-mediated disease. Nat Rev Drug Discov 9:117–128. doi:10.1038/nrd3013

Rawat D, Nair D (2010) Extended-spectrum beta-lactamases in gram negative bacteria. J Glob Infect Dis 2:263–274. doi:10.4103/0974-777X.68531

Reck F, Alm RA, Brassil P et al (2012) Novel N-linked aminopiperidine inhibitors of bacterial topoisomerase type II with reduced p K a: antibacterial agents with an improved safety profile. J Med Chem 55:6916–6933. doi:10.1021/jm300690s

Redgrave LS, Sutton SB, Webber MA, Piddock LJV (2014) Fluoroquinolone resistance: mechanisms, impact on bacteria, and role in evolutionary success. Trends Microbiol 22:438–445. doi:10.1016/j.tim.2014.04.007

Rex JH (2014) ND4BB: addressing the antimicrobial resistance crisis. Nat Rev Microbiol 12:231–232. doi:10.1038/nrmicro3245

Robinson A, Causer RJ, Dixon NE (2012) Architecture and conservation of the bacterial DNA replication machinery, an underexploited drug target. Curr Drug Targets 13:352–372. doi:10.2174/138945012799424598

Roca J, Wang JC (1992) The capture of a DNA double helix by an ATP-dependent protein clamp: a key step in DNA transport by type II DNA topoisomerases. Cell 71:833–840. doi:10.1016/0092-8674(92)90558-T

Roca J, Wang JC (1994) DNA transport by a type II DNA topoisomerase: evidence in favor of a two-gate mechanism. Cell 77:609–616. doi:10.1016/0092-8674(94)90222-4

Roca J, Berger JM, Harrison SC, Wang JC (1996) DNA transport by a type II topoisomerase: direct evidence for a two-gate mechanism. Proc Natl Acad Sci U S A 93:4057–4062. doi:10.1073/pnas.93.9.4057

Rodríguez-Rojas A, Rodríguez-Beltrán J, Couce A, Blázquez J (2013) Antibiotics and antibiotic resistance: A bitter fight against evolution. Int J Med Microbiol 303:293–297. doi:10.1016/j.ijmm.2013.02.004

Royer M, Costet L, Vivien E et al (2004) Albicidin pathotoxin produced by Xanthomonas albilineans is encoded by three large PKS and NRPS genes present in a gene cluster also containing several putative modifying, regulatory, and resistance genes. Mol Plant Microbe Interact 17:414–427

Saha R, Saha N, Donofrio RS, Bestervelt LL (2013) Microbial siderophores: a mini review. J Basic Microbiol 53:303–317. doi:10.1002/jobm.201100552

Sanchez S, Demain AL (eds) (2015) Antibiotics: current innovations and future trends. Caister Academic Press, Norfolk

Sauvage E, Kerff F, Terrak M et al (2008) The penicillin-binding proteins: Structure and role in peptidoglycan biosynthesis. FEMS Microbiol Rev 32:234–258. doi:10.1111/j.1574-6976.2008.00105.x

Savoia D (2012) Plant-derived antimicrobial compounds: alternatives to antibiotics. Future Microbiol 7:979–990. doi:10.2217/fmb.12.68

Schäberle TF, Lohr F, Schmitz A, König GM (2014) Antibiotics from myxobacteria. Nat Prod Rep 31:953. doi:10.1039/c4np00011k

Schaefer B (2014) Natural Products in the Chemical Industry. Springer, Berlin & Heidelberg

Schoeffler AJ, Berger JM (2008) DNA topoisomerases: harnessing and constraining energy to govern chromosome topology. Q Rev Biophys 41:41–101. doi:10.1017/S003358350800468X

Schoepp-Cothenet B, Lieutaud C, Baymann F et al (2009) Menaquinone as pool quinone in a purple bacterium. Proc Natl Acad Sci U S A 106:8549–8554. doi:10.1073/pnas.0813173106

Schueffler A, Anke T (2014) Fungal natural products in research and development. Nat Prod Rep 1425–1448. doi:10.1039/C4NP00060A

Schwartz B, Markwalder JA, Wang Y (2001) Lipid II: total synthesis of the bacterial cell wall precursor and utilization as a substrate for glycosyltransfer and transpeptidation by penicillin binding protein (PBP) 1b of *Eschericia coli*. J Am Chem Soc 123:11638–11643. doi:10.1021/ja0166848

Schwechheimer C, Kuehn MJ (2015) Outer-membrane vesicles from Gram-negative bacteria: biogenesis and functions. Nat Rev Microbiol 13:605–619. doi:10.1038/nrmicro3525

Scott LJ (2013) Fidaxomicin: a review of its use in patients with clostridium difficile infection. Drugs 73:1733–1747. doi:10.1007/s40265-013-0134-z

Scott CP, Abel-Santos E, Wall M et al (1999) Production of cyclic peptides and proteins in vivo. Proc Natl Acad Sci U S A 96:13638–13643. doi:10.1073/pnas.96.24.13638

Sewell EWC, Brown ED (2014) Taking aim at wall teichoic acid synthesis: new biology and new leads for antibiotics. J Antibiot 67:43–51. doi:10.1038/ja.2013.100

Shahul HP, Solapure S, Mukherjee K et al (2014) Optimization of pyrrolamides as mycobacterial gyrb atpase inhibitors: structure-activity relationship and in vivo efficacy in a mouse model of tuberculosis. Antimicrob Agents Chemother 58:61–70. doi:10.1128/AAC.01751-13

Sham L-T, Butler EK, Lebar MD et al (2014) MurJ is the flippase of lipid-linked precursors for peptidoglycan biogenesis. Science 345:220–222

Sharma M, Chauhan PM (2012) Dihydrofolate reductase as a therapeutic target for infectious diseases: opportunities and challenges. Future Med Chem 4:1335–1365. doi:10.4155/fmc.12.68

Shiva F (2015) Antibiotic prescription and bacterial resistance. Arch Pediatr Infect Dis 3:e21540. doi:10.5812/pedinfect.21540
Silver LL (2007) Multi-targeting by monotherapeutic antibacterials. Nat Rev Drug Discov 6:41–55. doi:10.1038/nrd2202
Singh GS (2004) β-Lactams in the new millennium. Part-I : monobactams and carbapenems. Med Chem (Los Angeles) 4:62–92
Smillie C, Garcillan-Barcia MP, Francia MV et al (2010) Mobility of plasmids. Microbiol Mol Biol Rev 74:434–452. doi:10.1128/MMBR.00020-10
Spellberg B, Powers JH, Brass EP et al (2004) Trends in antimicrobial drug development: implications for the future 90502:1279–1286
Stadler M, Hoffmeister D (2015) Fungal natural products-the mushroom perspective. Front Microbiol 6:1–4. doi:10.3389/fmicb.2015.00127
Storm DR, Strominger JL (1974) Binding of bacitracin to cells and protoplasts of *Micrococcus lysodeikticus*. J Biol Chem 249:1823–1827
Stukenberg PT, Studwell-Vaughan PS, O'Donnell M, O'Donnell M (1991) Mechanism of the sliding beta-clamp of DNA polymerase III holoenzyme. J Biol Chem 266:11328–11334. doi:10.1074/jbc.M111.338145
Su X-H, Wang B-X, Le W-J et al (2016) Multidrug-resistant neisseria gonorrhoeae Isolates from Nanjing, China, are sensitive to killing by a novel DNA gyrase inhibitor, ETX0914 (AZD0914). Antimicrob Agents Chemother 60:621–623. doi:10.1128/AAC.01211-15
Swoboda JG, Meredith TC, Cambell J et al (2010) Discovery of a small molecule that blocks wall teichoic acid biosynthesis in *Stphylococcus aureus*. ACS Chem Biol 5:839–849
Swoboda JG, Mylonakis E, Wilkinson BJ, Walker S (2011) Synthetic lethal compound combinations reveal a fundamental connection between wall teichoic acid and peptidoglycan biosynthesis in *Staphylococcus aureus*. ACS Chem Biol 6:106–116
Syvanen M (2012) Evolutionary implications of horizontal gene transfer. Annu Rev Genet 46:341–358. doi:10.1146/annurev-genet-110711-155529
Tari LW, Li X, Trzoss M et al (2013) Tricyclic GyrB/ParE (TriBE) inhibitors: a new class of broad-spectrum dual-targeting antibacterial agents. PLoS ONE 8:e84409. doi:10.1371/journal.pone.0084409
Taubes G (2008) The bacterial fight back. Science 321:356–361
Tegos GP, Haynes M, Strouse JJ et al (2011) Microbial efflux pump inhibition: tactics and strategies. Curr Pharm Des 17:1291–1302. doi:10.1038/nature11130.Reduced
Terlain B, Thomas JP (1969) C R Hebd Seances Acad Sci Ser C 269:1546–1549
Terlain B, Thomas JP (1971a) Bull Chim Soc Fr 6:2349–2356
Terlain B, Thomas JP (1971b) Bull Chim Soc Fr 6:2357–2362
Tettelin H, Riley D, Cattuto C, Medini D (2008) Comparative genomics: the bacterial pan-genome. Curr Opin Microbiol 11:472–477. doi:10.1016/j.mib.2008.09.006
't Hart P, Oppedijk SF, Breukink E, Martin NI (2016) New insights into nisin's antibacterial mechanism revealed by binding studies with synthetic lipid II analogues. Biochemistry 55:232–237. doi:10.1021/acs.biochem.5b01173
Thaker M, Spanogiannopoulos P, Wright GD (2010) The tetracycline resistome. Cell Mol Life Sci 67:419–431. doi:10.1007/s00018-009-0172-6
Tommasi R, Brown DG, Walkup GK et al (2015) ESKAPEing the labyrinth of antibacterial discovery. Nat Rev Drug Discov 14:529–542. doi:10.1038/nrd4572
Torres C (2010) Up against the wall. Nat Med 16:628–631. doi:10.1038/nm0610-628
Toussaint KA, Gallagher JC (2015) β-Lactam/β-lactamase inhibitor combinations: from then to now. Ann Pharmacother 49:86–98. doi:10.1177/1060028014556652
Upadhyay A, Upadhyaya I, Kollanoor-Johny A, Venkitanarayanan K (2014) Combating pathogenic microorganisms using plant-derived antimicrobials: a minireview of the mechanistic basis. Biomed Res Int Article ID 761741. doi:10.1155/2014/761741
Uria-Nickelsen M, Neckermann G, Sriram S et al (2013) Novel topoisomerase inhibitors: microbiological characterisation and in vivo efficacy of pyrimidines. Int J Antimicrob Agents 41:363–371. doi:10.1016/j.ijantimicag.2012.12.001

Venter H, Mowla R, Ohene-Agyei T, Ma S (2015) RND-type drug efflux pumps from Gram-negative bacteria: molecular mechanism and inhibition. Front Microbiol 6:377. doi:10.3389/fmicb.2015.00377

Ventola CL (2015) The antibiotic resistance crisis: part 1: causes and threats. P & T 40:277–283

Vivien E, Pitorre D, Cociancich S et al (2007) Heterologous production of albicidin: a promising approach to overproducing and characterizing this potent inhibitor of DNA gyrase. Antimicrob Agents Chemother 51:1549–1552. doi:10.1128/AAC.01450-06

Wang JC (2002) Cellular roles of DNA topoisomerases: a molecular perspective. Nat Rev Mol Cell Biol 3:430–440. doi:10.1038/nrm831

Wang H, Gill CJ, Lee SH et al (2013) Discovery of wall teichoic acid inhibitors as potential anti-MRSA beta-lactam combination agents. Chem Biol 20:272–284. doi:10.1016/j.chembiol.2012.11.013

Wang W, Qiu Z, Tan H, Cao L (2014) Siderophore production by actinobacteria. Biometals 27:623–631. doi:10.1007/s10534-014-9739-2

Watanakunakorn C (1984) Mode of action and in-vitro activity of vancomycin. J Antimicrob Chemother 14:7–18. doi:10.1093/jac/14.suppl_D.7

Wattal C, Goel N (2011) Tackling antibiotic resistance. Nat Rev Microbiol 9:894–896. doi:10.1038/nrmicro2693

Whitman WB, Coleman DC, Wiebe WJ (1998) Prokaryotes: the unseen majority. Proc Natl Acad Sci USA 95:6578–6583. doi:10.1073/pnas.95.12.6578

Wiedemann I, Breukink E, Van Kraaij C et al (2001) Specific binding of nisin to the peptidoglycan precursor lipid II combines pore formation and inhibition of cell wall biosynthesis for potent antibiotic activity. J Biol Chem 276:1772–1779. doi:10.1074/jbc.M006770200

Wijffels G, Dalrymple BP, Prosselkov P et al (2004) Inhibition of protein interactions with the beta 2 sliding clamp of *Escherichia coli* DNA polymerase III by peptides from beta 2-binding proteins. Biochemistry 43:5661–5671. doi:10.1021/bi036229j

Wijffels G, Johnson WM, Oakley AJ et al (2011) Binding inhibitors of the bacterial sliding clamp by design. J Med Chem 54:4831–4838. doi:10.1021/jm2004333

Williams DH, Bardsley B (1999) The vancomycin group of antibiotics and the fight against resistant bacteria. Angew Chem Int Ed 38:1172–1193. doi:10.1002/(SICI)1521-3773(19990503)38:9<1172:AID-ANIE1172>3.0.CO;2-C

Wolff P, Amal I, Oliéric V et al (2014) Differential modes of peptide binding onto replicative sliding clamps from various bacterial origins. J Med Chem 57:7565–7576. doi:10.1021/jm500467a

Wright GD (2007) The antibiotic resistome: the nexus of chemical and genetic diversity. NatRevMicrobiol 5:175–186. doi:10.1038/nrmicro1614

Wright GD (2010) The antibiotic resistome. Exp Opin Drug Disc 5:779–788. doi:10.1517/17460441.2010.497535

Wright GD (2012) The origins of antibiotic resistance. In: Coates ARM (ed) Handbook of experimental pharmacology, pp 45–65

Wright G (2015a) An irresistible newcomer. Nature 517:442–443. doi:10.1038/nature14193

Wright GD (2015b) Solving the antibiotic crisis. ACS Infect Dis 1:80–84. doi:10.1021/id500052s

Wright GD, Poinar H (2012) Antibiotic resistance is ancient: implications for drug discovery. Trends Microbiol 20:157–159. doi:10.1016/j.tim.2012.01.002

Wright PM, Seiple IB, Myers AG (2014) The evolving role of chemical synthesis in antibacterial drug discovery. Angew Chem Int Ed 53:8840–8869. doi:10.1002/anie.201310843

Wyke AW, Ward JB, Hayes MV, Curtis NAC (1981) A role in vivo for penicillin-binding protein-4 of *Staphylococcus aureus*. Eur J Biochem 119:389–393. doi:10.1111/j.1432-1033.1981.tb05620.x

Xu Z-Q, Flavin MT, Flavin J (2014a) Combating multidrug-resistant Gram-negative bacterial infections. Expert Opin Investig Drugs 23:163–182. doi:10.1517/13543784.2014.848853

Xu Z-Q, Flavin MT, Flavin J (2014b) Combating multidrug-resistant Gram-negative bacterial infections. Expert Opin Investig Drugs 23:163–182. doi:10.1517/13543784.2014.848853

Yahav D, Farbman L, Leibovici L, Paul M (2011) Colistin: new lessons on an old antibiotic: EBSCOhost. Clin Microbiol Infect 18:18–29

Yin Z, Wang Y, Whittell LR et al (2014a) DNA replication is the target for the antibacterial effects of nonsteroidal anti-inflammatory drugs. Chem Biol 21:481–487. doi:10.1016/j.chembiol.2014.02.009

Yin Z, Whittell LR, Wang Y et al (2014b) Discovery of lead compounds targeting the bacterial sliding clamp using a fragment-based approach. J Med Chem 57:2799–2806. doi:10.1021/jm500122r

Yin Z, Whittell LR, Wang Y et al (2015) Bacterial sliding clamp inhibitors that mimic the sequential binding mechanism of endogenous linear motifs. J Med Chem 58:4693–4702. doi:10.1021/acs.jmedchem.5b00232

Yoshida H, Bogaki M, Nakamura M, Nakamura S (1990) Quinolone resistance-determining region in the DNA gyrase gyrA gene of *Escherichia coli*. Antimicrob Agents Chemother 34:1271–1272. doi:10.1128/aac.34.6.1271

Yoshida H, Bogaki M, Nakamura M et al (1991) Quinolone resistance-determining region in the DNA gyrase gyrB gene of *Escherichia coli*. Antimicrob Agents Chemother 35:1647–1650. doi:10.1128/AAC.35.8.1647

Young IG (1975) Biosynthesis of bacterial menaquinones. Menaquinone mutants of *Escherichia coli*. Biochemistry 14:399–406. doi:10.1021/bi00673a029

Yunis AA (1988) Chloramphenicol: relation of structure to activity and toxicity. Annu Rev Pharmacol Toxicol 28:83–100 doi:10.1146/annurev.pa.28.040188.000503

Zaffiri L, Gardner J, Toledo-Pereyra LH (2012) History of antibiotics. From salvarsan to cephalosporins. J Invest Surg 25:67–77. doi:10.3109/08941939.2012.664099

Zapun A, Philippe J, Abrahams KA et al (2013) In vitro reconstitution of peptidoglycan assembly from the gram-positive pathogen streptococcus pneumoniae. ACS Chem Biol 8:2688–2696. doi:10.1021/cb400575t

Zetts R (2014) Report of the Antibiotics resistance project: antibiotics currently in clinical development. http://www.pewtrusts.org/en/multimedia/data-visualizations/2014/antibiotics-currently-in-clinical-development

Zgurskaya HI, López CA, Gnanakaran S (2015) Permeability barrier of Gram-negative cell envelopes and approaches to bypass it. ACS Infect Dis 1:512–522. doi:10.1021/acsinfecdis.5b00097

Zhanel GG, Walkty AJ, Karlowsky JA (2015) Fidaxomicin: a novel agent for the treatment of clostridium difficile infection. Can J Infect Dis Med Microbiol 26:305–313

Zhang G, Meredith TC, Kahne D (2013) On the essentiality of lipopolysaccharide to Gram-negative bacteria. Curr Opin Microbiol 16:779–785. doi:10.1016/j.mib.2013.09.007

Zhao M, Goedecke T, Gunn J, et al (2013) Protostane and fusidane triterpenes: a mini review. Molecules 18:4054–4080. doi:10.3390/molecules18044054

Synthesis of Antibiotics

Markus Kalesse, Andreas Böhm, Andi Kipper and Vanessa Wandelt

Abstract The synthesis of β-lactams, tetracyclines, and erythromycins as three of the major families of antibiotics will be described herein. We will describe why these antibiotics were the ultimate synthetic targets in the past and how modern synthetic organic chemistry has evolved to address these challenges with new, improved strategies and methods. An additional aspect we would like to highlight here is the fact that these first syntheses had to be particularly creative as most of the modern synthetic methods were not available at that time, or were developed in the course of these syntheses.

Contents

M. Kalesse (✉) · A. Böhm · A. Kipper · V. Wandelt
Institute for Organic Chemistry and Centre of Biomolecular Drug Research, Leibniz Universität Hannover, Schneiderberg 1B, 30167 Hannover, Germany
e-mail: Markus.Kalesse@oci.uni-hannover.de

M. Kalesse
Helmholtz Centre for Infection Research, Inhoffenstraße 7, 38124 Brunswick, Germany

Current Topics in Microbiology and Immunology (2016) 398: 419–446
DOI 10.1007/82_2016_502

Published Online: 22 September 2016

1 Introduction

The history of antibiotics goes hand-in-hand with the synthesis of natural and unnatural substances. The first synthetic antibiotic, arsphenamine (Fig. 1) was prepared in 1907 and used for the treatment of syphilis (Alt 1909). However, arsenic-based drugs had limited use and were toxic to the human hosts as well as to the bacteria. The compounds that truly revolutionized mankind's fight against pathological bacteria were penicillins, discovered in 1928 (Fleming 1929). It took some time though, for the world to realize their importance and meanwhile another class of antibiotics, sulfonamides, was identified and marketed (Domagk 1935). The Second World War brought penicillin and antibiotic research into focus, and during the following two decades a number of different classes of antibiotics had been discovered: Aminoglycosides (Schatz et al. 1944), tetracyclines (Duggar 1948), macrolides (McGuire et al. 1952), glycopeptides (McCormick et al. 1956), quinolones (Lesher et al. 1962) etc.

A comprehensive overview on the synthesis of antibiotics is not in the scope of this chapter and there are a number of reviews written on this subject matter (Lukacs and Ohno 1990; Lukacs 1993; Krohn et al. 1993; Bruggink 2001; Nicolaou et al. 2009). Here, we have chosen to cover some of the more prominent antibiotics and describe the challenges accompanying their synthesis as well as recent advances in that field.

2 Synthesis of β-Lactam Antibiotics

The β-lactams (Fig. 2) constitute the biggest and most important class of antibiotics. Their chemistry and biology has been thoroughly reviewed on many occasions (Morin and Gorman 1982; Page 1992; Georg 1993). They all have the

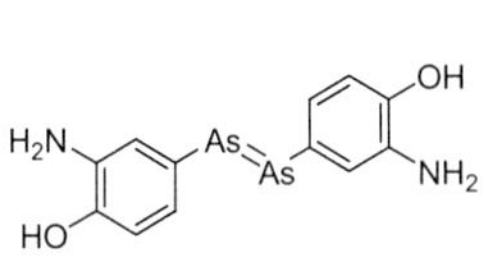

arsphenamine (1)

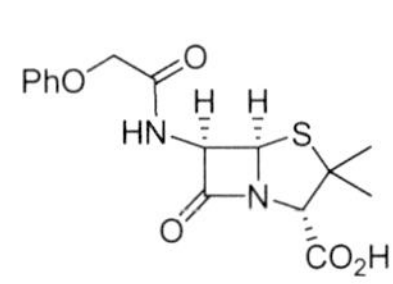

penicillin V (2)

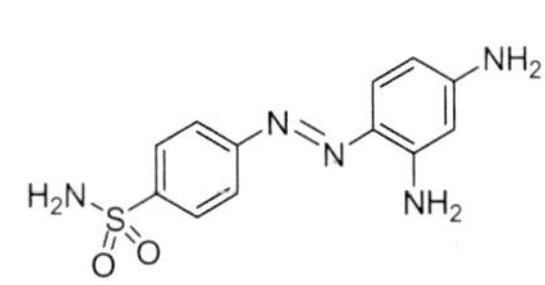

Prontosil (3)

Fig. 1 First antibiotics

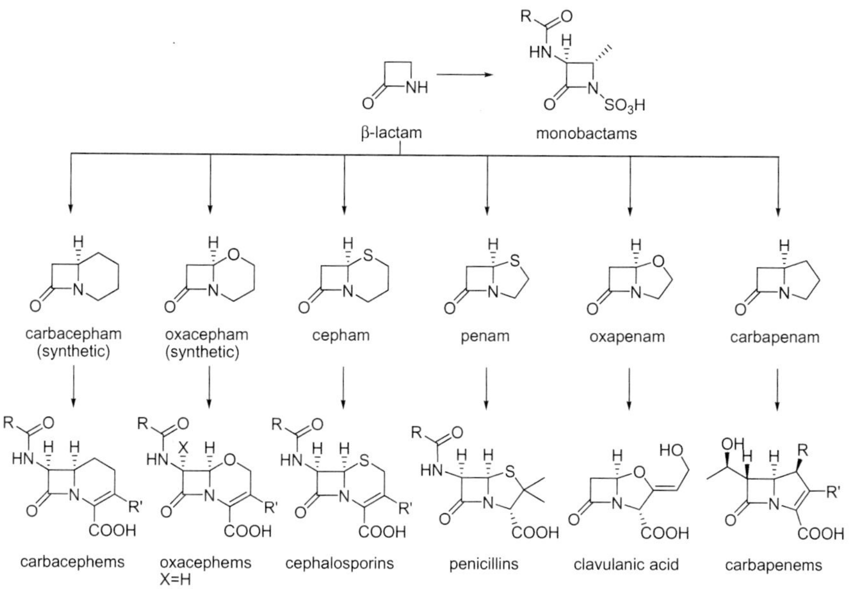

Fig. 2 β-lactam family

strained four-membered β-lactam ring in common which serves as the pharmacophoric group. The individual members of this family differ in their ring size, and substitution pattern as depicted (Fig. 2).

2.1 *Synthesis of Penicillin V (Sheehan 1957)*

Even though the penicillins were first discovered in the late 1920s (Fleming 1929), their structure remained a topic of controversy for more than 15 years. Finally, in 1945, the X-ray crystal structure of penicillin G was obtained by Dorothy Crowfoot Hodgkin and the existence of the characteristic β-lactam ring was unambiguously proven (Clarke 1949). However, the first total synthesis was not achieved until 1957, when Sheehan et al. completed a decade of work on this topic with the synthesis of the potassium salt of penicillin V (**2**) (Sheehan and Henery-Logan 1957, 1959, 1962; Sheehan 1982).

The most challenging feature of penicillin is its four-membered β-lactam ring fused to a thiazolidine ring. The fused ring system disrupts the conjugation of nitrogen's free electron-pair to the carbonyl group. That, together with high ring-strain makes the amide bond labile to both acidic and basic conditions. Penicillin rapidly opens at the β-lactam ring and degrades, e.g., into pseudopenicillin (**6**) (under acidic conditions), which can further undergo rearrangement to stable inactive compounds (Scheme 1) (Chain 1948; Deshpande et al. 2004). The

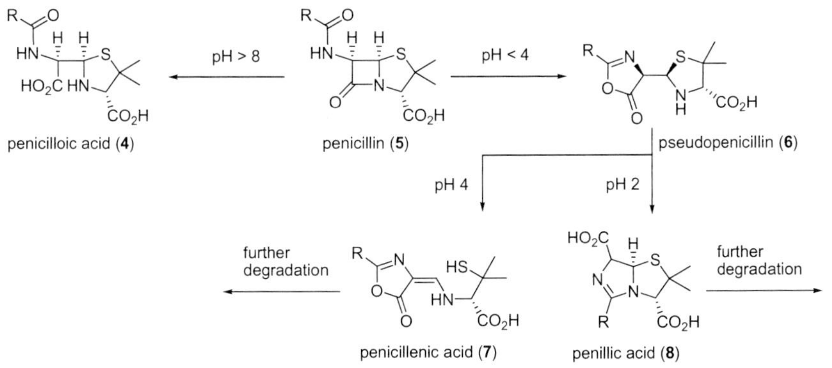

Scheme 1 Decomposition of the β-lactam ring

aforementioned difficulty required devising completely new mild amide coupling conditions to effect the synthesis of penicillin.

Sheehan's synthesis of penicillin V (**2**) utilized penicillamine **12** and aldehyde **11** as key intermediates (Scheme 2). Penicillamine was synthesized in eight steps from racemic valine and obtained in enantiopure form through crystallization as a brucine salt (Clarke et al. 1949). Aldehyde **11** was prepared in three steps from glycine.

Sheehan's total synthesis of penicillin started with glycine derived aldehyde **13**, where the carboxylic acid moiety was protected as a *tert*-butyl ester and the primary amine was masked as a phthalimide. Thioaminal formation under basic conditions (Sheehan and Johnson 1954) with penicillamine (**12**) was followed by removal of phthalimide protecting group with hydrazine and recrystallization of the hydrochloride salt to give compound **15** in optically pure form (Scheme 3). Next, the acyl side chain was installed to C-6 nitrogen followed by removal of *tert*-butyl protecting group from the acid and crystallization of ammonium salt of acid **17** from aqueous acetone. The crucial lactamization was achieved by employing the mild coupling reagent *N,N'*-dicyclohexylcarbodiimide (DCC) developed by Sheehan a few years earlier for coupling of amino acids under pH neutral conditions (Sheehan and Hess 1955). With that, the first total synthesis of the potassium salt of penicillin V (**2**) was achieved.

Penicillins for pharmaceutical use are nowadays, as they were back in the 1950s, produced by fermentation and semi-synthesis. Nevertheless, the first total synthesis of penicillin V (**2**) serves as a benchmark in β-lactam chemistry and as a starting

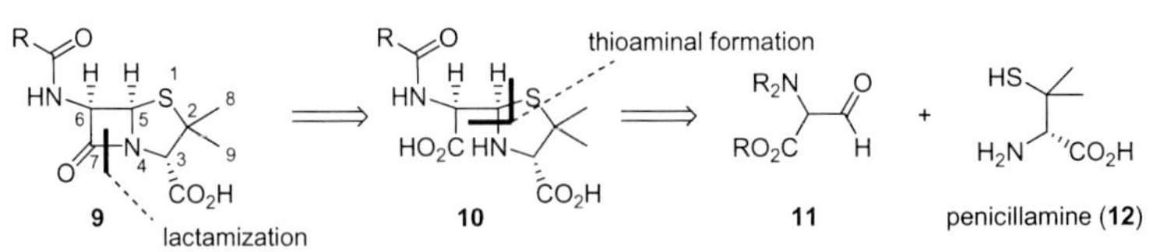

Scheme 2 Sheehan's retrosynthesis of penicillin V (**2**)

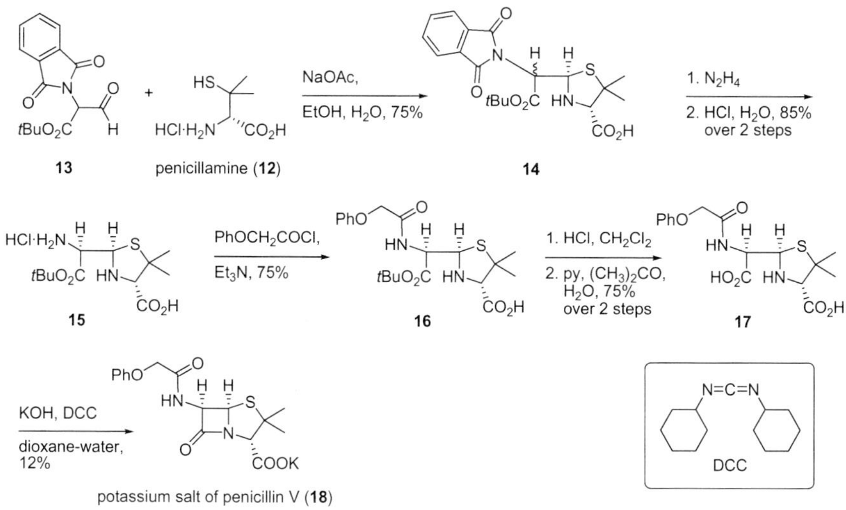

Scheme 3 Sheehan's synthesis of potassium salt of penicillin V (**18**)

point for the synthesis of other β-lactam derivatives. Sheehan utilized for the first time the *tert*-butoxy group as an acid labile protecting group for carboxylic acids and introduced the idea of using phthalimide as a protecting group for primary amines. Nevertheless, the most influential aspect of this synthesis would be the employment of the DCC reagent for coupling of carboxylic acids with amines. This reagent has enabled the solid-phase peptide coupling chemistry and is still in use nowadays—60 years after the initial introduction.

2.2 *Synthesis of Cephalosporin C (Woodward 1966)*

Cephalosporin C was first discovered in 1955 from a species of *Acremonium* (Newton and Abraham 1955), and its structure was determined in 1961 (Loder et al. 1961; Hodgkin and Maslen 1961). It belongs to a class of β-lactams called cephalosporins bearing a close resemblance to the penicillins but differing in the size of the ring connected to the β-lactam. It attracted close interest because of its antibacterial activity, which, though moderate, persisted through different classes of bacteria that had become resistant to penicillins (Chauvette et al. 1962).

The challenges in the cephalosporin C synthesis were somewhat different from the challenges of the synthesis of penicillin. Whereas the latter was usually prepared by a late stage β-lactam formation from parent β-amino acid (e.g., **10**), which was known to be a decomposition product of naturally occurring penicillin; the

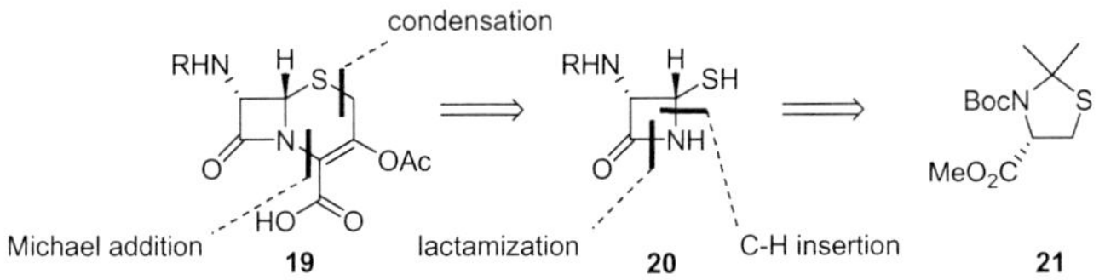

Scheme 4 Woodward's retrosynthesis of cephalosporin C (**32**)

corresponding β-amino acid for the former was not a known compound. Therefore, a different retrosynthetic approach was chosen by Woodward et al., in which they decided to form the β-lactam ring early in the synthesis and then generate the dihydrothiazine ring system at a later stage (Scheme 4) (Woodward et al. 1966; Woodward 1966). The β-lactam system was prepared from naturally occurring *S*-cysteine by stereoselective amination of its β-position.

Woodward's synthesis of cephalosporin C (**32**) (Scheme 5) started from *S*-cysteine (**22**), which was protected at different functionalities by first treating it with acetone, then *tert*-butanol and phosgene in the presence of pyridine, and finally the acid moiety was converted into methyl ester **23** by treatment with diazomethane. The key step of this synthetic strategy followed, which was the stereoselective functionalization of β-position of the parent amino acid. This was achieved by treatment with dimethyl azodicarboxylate (**24**) (DMAD). Supposedly, suphur adds to DMAD, which at the same time deprotonates the carbon next to the sulphur, and rearrangement of the sulfonium ylid leads to the formation of C–N bond. The reaction proceeded in a stereospecific manner, though unfortunately leading to the undesired stereoisomer. Therefore, an inversion of configuration at that centre was required, which was achieved by sequence of oxidation, mesylation, and azidation. The azide was reduced to the amine **27** in the presence of aluminum amalgam and a Lewis acid-induced lactamization gave the β-lactam, which was directly coupled with the Michael acceptor **28** to give **29**. Treating the latter with TFA removed the Boc-protecting group, which induced opening of the thiazolidine ring and liberated the sulphur for the attack of the carbonyl moiety of the side chain. Dehydration gave the cephem core **30**. Compound **31** was obtained after two DCC mediated couplings—the first functionalizing the amine nitrogen and the second protecting the free carboxylic acid moiety in the side chain. Reduction of the carbonyl group was followed by acetylation and double bond isomerization in anhydrous pyridine. Finally, the trichloroethyl (Troc) protecting groups were removed under reductive conditions with Zn in acetic acid giving cephalosporin C (**32**).

With this, the first total synthesis of cephalosporin derivatives from simple starting materials was achieved. The synthesis featured a novel stereoselective method for C–H functionalization of the β-position of the cysteine and one of the first uses of the Troc group as a protecting group for carboxylic acids.

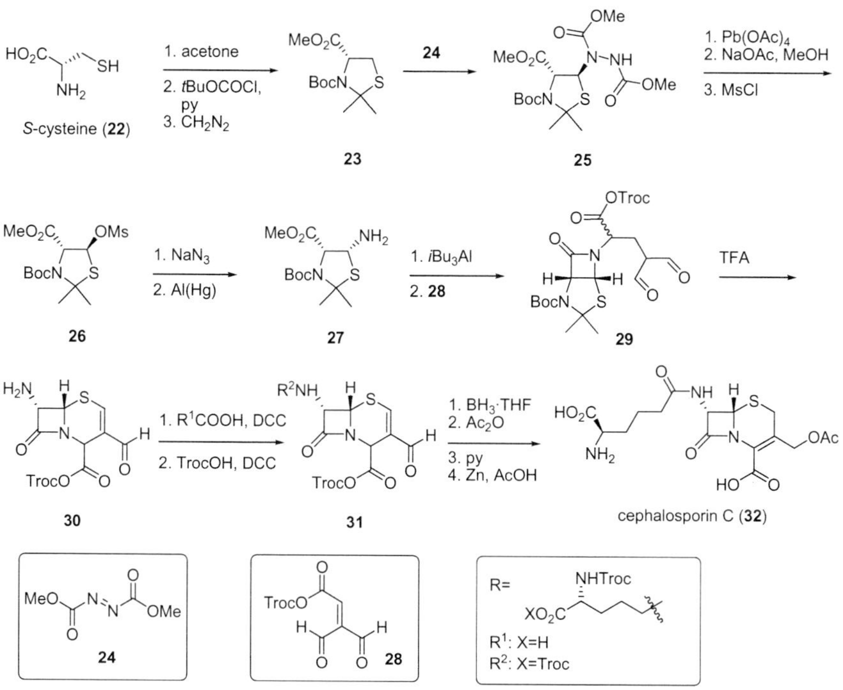

Scheme 5 Woodward's synthesis of cephalosporin C (32)

2.3 The Merck Synthesis of Thienamycin (Mellio 1980, 1986)

The first member belonging to the class of carbapenems was thienamycin (**33**), which was discovered by scientists at Merck in late 1970s (Albers-Schoenberg et al. 1978). It attracted immediate interest by displaying both—broad-spectrum antibacterial and β-lactamase inhibitory activity. As opposed to penicillins and cephalosporins, biosynthetic production of carbapenems was inefficient and therefore synthetic approach to these structures gained a great importance. A variety of different syntheses have been carried out (Hanessian et al. 1990; Ishibashi et al. 1995; Jacobi et al. 1996; Cabri and Di Fabio 2000).

Carbapenems differ from aforementioned penicillins and cephalosporins in several distinct aspects. First of all, the ring connected to the β-lactam does not contain any additional heteroatoms. Secondly, varying appendage in the α-position to the amide group of the β-lactam ring, which plays great roll in the activity of penicillins and cephalosporins, does not significantly affect the activity of carbapenems. And third, the hydrogens attached to the β-lactam ring are *anti* to each other, whereas in the other two classes they have to be *syn* to display desired antibacterial activity.

Researchers from Merck were the first to achieve the total synthesis of thienamycin (**33**) (Johnston et al. 1978) as well as to make it amenable for industrial production (Melillo et al. 1980, 1986). Their optimized retrosynthetic approach (Scheme 6) relied on carbene chemistry for the creation of the pyrroline ring. Asymmetric induction was achieved by chiral auxiliary controlled stereoselective hydrogenation.

The Merck synthesis (Scheme 7) started from cheap and readily available dimethyl acetonediacarboxylate (**36**). Reaction with (*R*)-(+)-α-methylbenzylamine followed by treatment of the formed enamine with ketene gas gave the ketone **38**. Introduction of the chiral amine allowed a stereoselective reduction to take place upon hydrogenation with a platinum catalyst. Treatment with strong acid lead to lactonization and compound **39** could be obtained as optically pure crystals. The lactone opening in aqueous HCl was followed by hydrogenolysis, which exposed the β-amino and acid

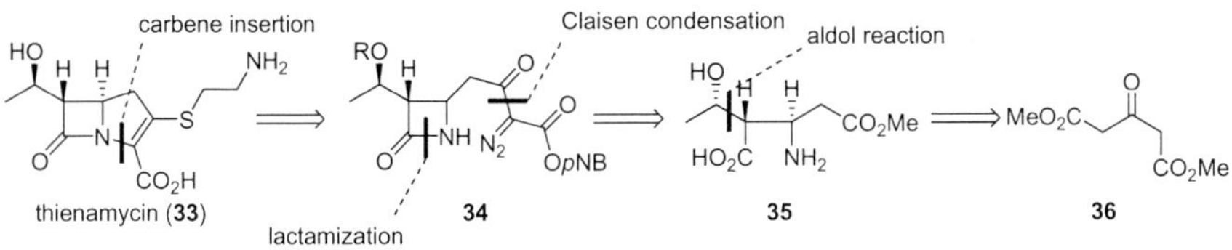

Scheme 6 Merck's retrosynthesis of thienamycin (**33**)

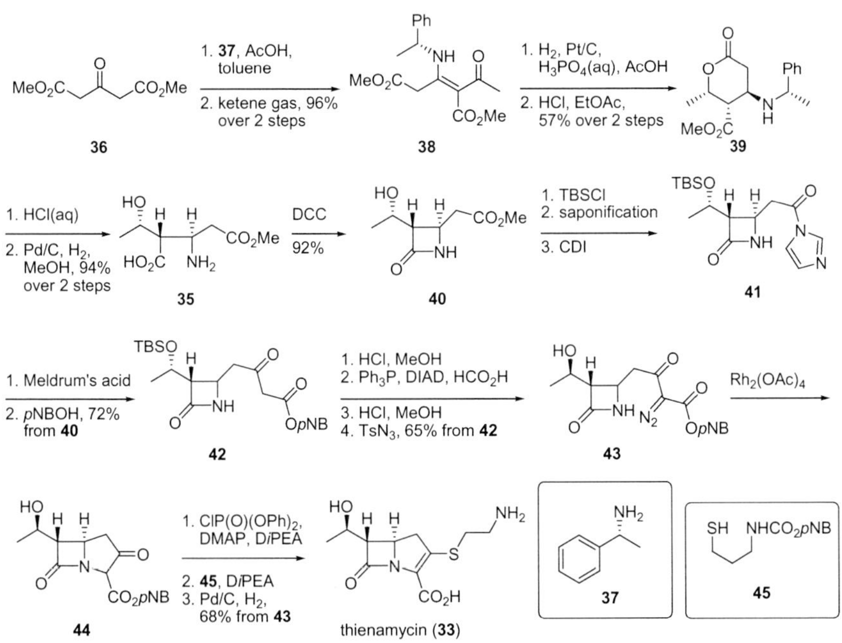

Scheme 7 Merck's synthesis of thienamycin (**33**)

functionalities in **35**. A DCC mediated coupling gave the β-lactam core and a few functional group manipulations provided the activated amide **41**. Treatment with Meldrum's acid followed by opening the acyl-intermediate with *para*-nitrobenzyl alcohol, gave the β-keto ester **42**. The stereoselective hydrogenation of compound **38** had provided a single diastereomer but unfortunately the stereochemistry of the hydroxy group in the appendage adjacent to the carbonyl group of β-lactam was incorrect. Therefore, inversion of that stereocentre had to be undertaken by first deprotecting the compound **42**, and then performing the inversion under Mitsunobu conditions. Regitz diazotransfer yielded the diazo compound **43**. The pyrroline ring was closed through a rhodium catalyzed carbene insertion into the N–H bond giving compound **44**. Ketone **44** was converted to enol phosphate, then treated with N-protected cysteamine and after hydrogenation thienamycin (**33**) was obtained.

3 Synthesis of Tetracycline Antibiotics

Tetracyclines are antibiotics produced by strains of *Streptomyces*. The first compound of that class, chlortetracycline (aureomycin) (**46**), was isolated in 1948 by Duggard but the structure was first elucidated for another tetracycline—terramycin (**46**)—in 1950 by researchers from *Chas. Pfizer & Co.* together with Robert Burns Woodward (Muxfeldt 1962b). They found, that the structure of tetracyclines is based on polyoxygenated hydronaphthacene backbone, which can contain different other functionalities around the periphery (Fig. 3) (Korst et al. 1968). The unique structure of the tetracyclines in combination with their remarkable antibiotic activity attracted interest of several research groups. The synthetic challenges of preparing these compounds though were considerable—even Woodward stated in 1952 that it would hardly ever be possible to synthesize a molecule of such complexity (Der Spiegel 1968).

3.1 *Racemic Total Synthesis of Terramycin (Muxfeldt 1968)*

Muxfeldt et al. were the first to report the total synthesis of racemic terramycin in 1968. Retrosynthetically, they divided terramycin into three basic building blocks

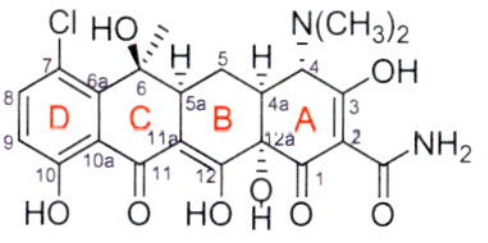

chlorotetracycline (**46**)

Fig. 3 Structure of chlorotetracycline (**46**)

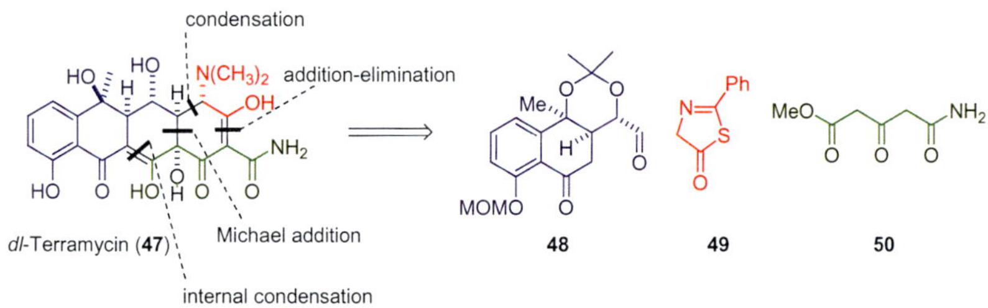

Scheme 8 Muxfeldt's retrosynthetic analysis of terramycin (**47**)

using four retrosynthetic cuts (Scheme 8). These three building blocks were the aldehyde **48**, thiazolone **49,** and methyl-3-oxoglutarate **50** (Muxfeldt et al. 1968).

The preparation of thiazolone **49** began with thiobenzoylglycine, which was transformed to the corresponding hydrobromide and then converted to thiazolone **49** by treatment with sodium acetate (Muxfeldt et al. 1967). The synthetic benefits derived from this building block were threefold. First, it exhibited high acidity due to its cyclic unsaturated structure, leading to mild reaction conditions in the condensation step. Second, it served as an internal protecting group for the amino group; and third, the thioester did act as an activated carboxylate in the final condensation step. Methyl-3-oxoglutarate **50**, was readily available from dimethyl 3-oxoglutarate in two steps (Muxfeldt et al. 1968). The synthesis of aldehyde **48**, on the other hand, was a more complex endeavour. Its synthesis started with the cycloaddition reaction between 1-acetoxybutadiene (**52**) and juglone acetate (**51**), which provided **53** as a 3:1 mixture of regioisomers (Scheme 9). A series of protecting group manipulations, generation of the tertiary alcohol and an ozonolysis, followed by an aldol condensation, provided aldehyde **54** in seven steps. At this stage, both the D- and C-ring of terramycin (**47**) were already in place (Muxfeldt 1962a, b; Muxfeldt et al. 1979) but to obtain compound **48**, two carbon atoms had to be removed from the third ring of **54**. This task was achieved by another ozonolysis, followed by a basic hydrolysis, which induced a retro-aldol transformation. However, epimerization at C-5 under the aforementioned reaction conditions required a subsequent equilibration, which was performed via the formation of the corresponding enamine using piperidine. Then, the phenolic alcohol was protected with MOMCl at C-10 and enamine functionality was removed in the presence of deactivated silica providing **48**. These transformations set the stage for the condensation with thiazolone **49**, which could be achieved in the presence of basic lead acetate, building up thiazolone **55**. Finally, the A and B rings were formed through the Michael addition of methyl 3-oxoglutarate (**50**), followed by an intramolecular addition-elimination reaction giving intermediate **56** which then condensed to the tetracyclic compound **57**. This impressive transformation established the carbon skeleton. The remaining transformations which were required to complete the synthesis, were deprotections, oxidation of the carbon C-12a and alkylation of the amine nitrogen (Muxfeldt et al. 1968). The racemic synthesis of

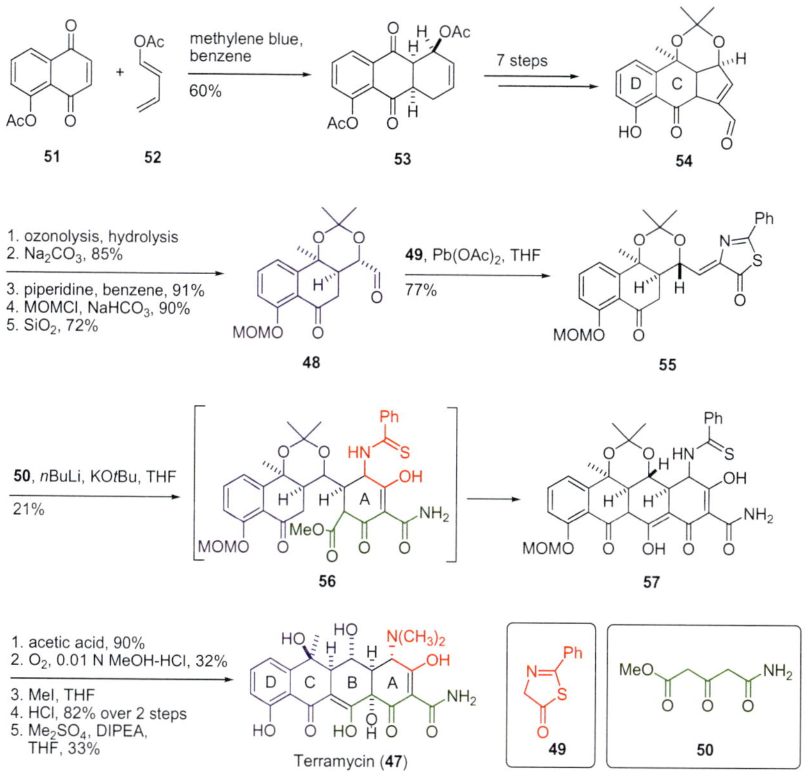

Scheme 9 Muxfeldt's racemic total synthesis of terramycin (**47**)

terramycin (**47**) was completed in 22 steps with an overall yield of 0.06 %. Regardless of the low overall yield, this synthesis is considered one of the milestones in organic synthesis as it showed for the first time that molecules of even that complexity can be generated by laboratory synthesis.

3.2 Racemic Total Synthesis of 6-Demethyl-6-Deoxytetracycline (Woodward 1968)

The racemic total synthesis of 6-demethyl-6-deoxytetracycline (**58**) by Woodward, which was also developed in 1968, followed a slightly different strategy than the one developed by Muxfeldt. Both syntheses started from the D-ring, but in contrast to Muxfeldt's approach (B ring was joined last), Woodward built the rest of the rings successively from C to A (Korst et al. 1968; Muxfeldt et al. 1968). Woodward's retrosynthetic approach featured four major disconnections (Scheme 10) and the A-ring, which contains three of the four stereogenic centres, was constructed late in

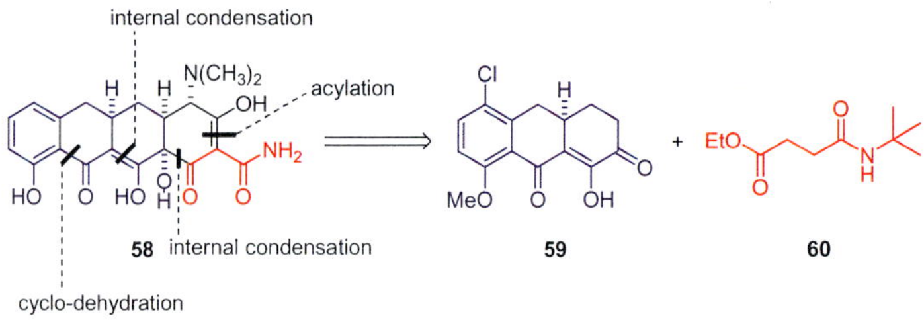

Scheme 10 Woodward's retrosynthetic analysis of 6–demethyl–6–deoxytetracycline (**58**)

the synthesis. Therefore, **58** can be reduced to tricyclic triketone **59** and ethyl *N-tert*-butylmalonate (**60**) (Korst et al. 1968).

Methyl *m*-methoxybenzoate (**61**) was used as the starting material for the tricyclic triketone **59**. It was first transformed in three steps into keto triester **64** (Scheme 11), then hydrolyzed and concomitantly decarboxylated to give the

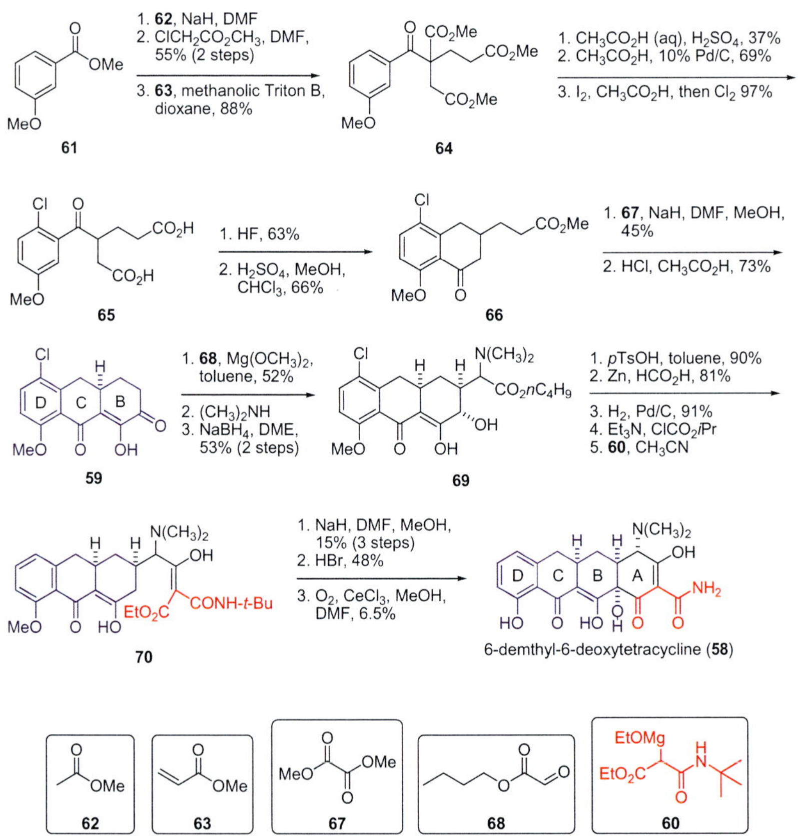

Scheme 11 Woodward's racemic total synthesis of 6-demethyl-6-deoxytetracycline (**58**)

corresponding diacid which in turn was re-esterified and the keto carbonyl group removed by hydrogenolysis. Then, chlorination was performed at C-7, (here and subsequently the chlortetracycline numbering is used, Fig. 3) giving **65**, to block cyclization in the *para*-position to the methoxy group. The cyclodehydration established the C-ring and esterification of the acid generated tetralone derivate **66**. Next, the B-ring was installed by an intermolecular condensation with dimethyl oxalate (**67**) giving the tricyclic ester, which could be decarboxylated to the key intermediate tricyclic triketone **59**. This key intermediate was reacted with aldehyde **68** to give aldol condensation product. This aldol product is an α,β-unsaturated 1,4-dicarbonyl compound, which could be attacked by nucleophiles in Michael manner on two different carbon atoms. In fact, due to steric reasons reaction with dimethylamine leads to only one regioisomeric product and at the same time also establishes *syn* relationship between hydrogens of the B-ring. Direct reduction of the intermediate carbonyl compound leads to **69**. Next, the hydroxy and chloro groups were removed: the former by lactonization-reduction sequence and the latter by catalytic hydrogenation. Thus, the liberated acid moiety was activated by forming the mixed anhydride, which was then acylated with the ethoxymagnesium derivate of ethyl *N-tert*-butyl-malonamate (**60**) giving the A-ring precursor **70**. The synthesis was completed by building up the A-ring by treatment of **70** with sodium hydride obtaining the tetracycline derivate, which was then deprotected with hydrobromic acid and oxygenated in the presence of cerium(III) chloride. This concluded the synthesis of 6-demethyl-6-deoxytetracycline (**58**) (Korst et al. 1968) in 25 steps with an overall yield of 0.02 %.

3.3 Benzyloxyisoxazole Systems in Construction of Tetracyclines

In the aforementioned syntheses of tetracyclines, a Claisen cyclisation was used to form a bond between C-1 and C-12a. This transformation could interfere with the sensitive functionalities at the A-ring (Korst et al. 1968). To circumvent this problem, Stork et al. used isoxazoles that served as internal protecting groups for the sensitive β-keto amide moiety (Scheme 12) (Stork and Hagedorn 1978).

The benefit of this strategy was that the tricyclic dienolone **75** was easily accessible, as demonstrated in the Muxfeldt's synthesis of terramycin (**47**)

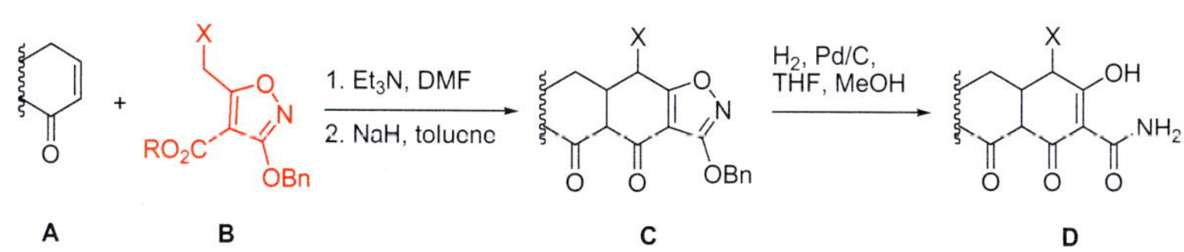

Scheme 12 General method to protect the β-keto amide moiety as 3-benzoxy isoxazoles

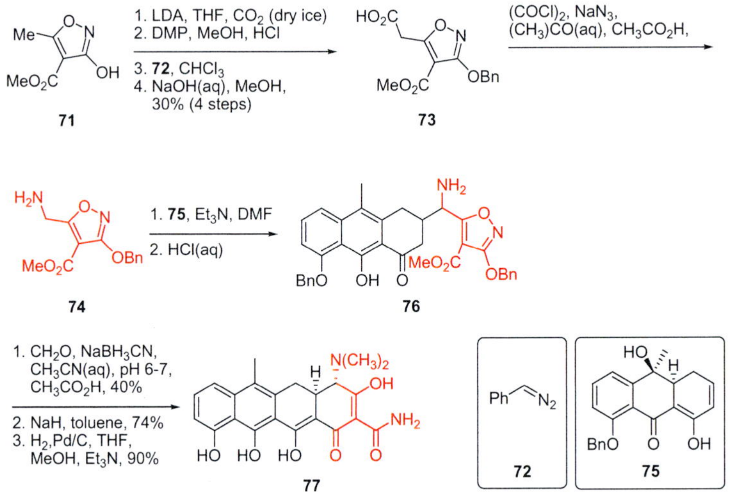

Scheme 13 Introduction of the benzyloxyisoxazole systems

(Muxfeldt 1962a), and the corresponding isoxazole derivate **74** could be obtained in five steps from 3-hydroxy-5-methylisoxazole-4-carboxylate (**71**) (Scheme 13). A Michael addition between **74** and **75** followed by dehydration gave **76** as a mixture of epimers at the amino group. The tertiary amine was obtained under reductive amination conditions and the A ring was closed by Claisen cyclization. Liberation of the β-keto amide functionality was achieved smoothly by hydrogenolysis giving the 12-deoxyanhydrotetracycline (**77**), which could then be transformed into tetracyclines (Stork and Hagedorn 1978).

3.4 Tetracycline Synthesis (Myers 2005)

During the next four decades, a number of novel tetracycline syntheses were developed. The common characteristic of those syntheses was the linear construction of the cyclic system from the D-ring to the A-ring. This was, however, not ideal for the synthesis of new active antibiotics, since substituents on the D-ring affected the antibiotic properties. Therefore, in 2005 Myers et al. reported a convergent three-step approach to tetracyclines, starting from building blocks **79** and **80** (Scheme 14). In this case, the C-ring is built up by a stereocontrolled Michael-Dieckmann cyclization, which forms two new C–C bonds and two stereogenic centres (Charest et al. 2005; Wright et al. 2014).

Myers' synthesis started from benzoic acid (**81**), which was dihydroxylated using a mutant strain of *Alcaligenes eutrophus*; subsequent epoxidation established the highly functionalized acid **82**. Esterification with trimethylsilyldiazomethane (Scheme 15) followed by double silylation and isomerization of this epoxide

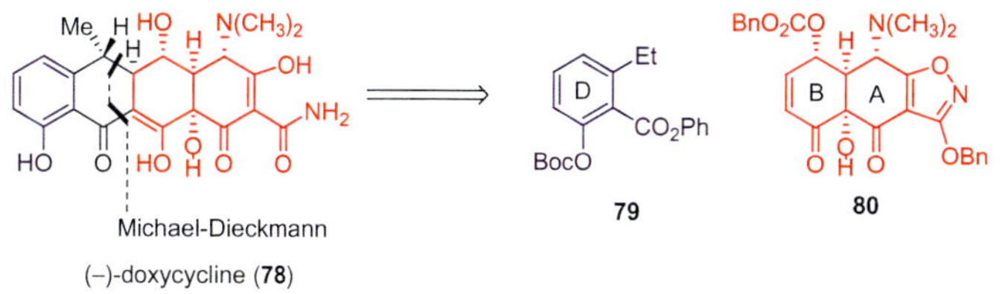

Scheme 14 Myers' retrosynthetic analysis of (−)–doxycycline (**78**)

afforded the regioisomeric epoxy ester **83** (Charest et al. 2005), which was then attacked by organolithium reagent **84** (prepared in four steps from glyoxylic acid) giving desired ketone **85**. It is worth mentioning that here, similarly to Stork's synthesis, the carboxamide function was introduced as an internally protected 3-benzyloxyisoxazole (Stork and Hagedorn 1978). Ketone **85** provided the stage for the key transformation of this synthesis: the cyclization establishing the tricyclic AB-intermediate **88**. The cyclization is initiated by an S_N2' epoxide opening by the tertiary amine, which is followed by formation of ylide **87** and a [2,3]-sigmatropic Sommelet-Hauser-type rearrangement (Pine 2004). Next, the secondary hydroxy group of **88** was replaced by a thiophenyl group with retention of configuration. The allylic alcohol **91** was formed by a diastereoselective oxidation of the sulfide followed by a Mislow-Evans rearrangement and four further steps led to enone **80**. The D-ring building block **79** could be prepared in five steps from anisic acid (Myers et al. 2001). Michael-Dieckmann cyclization between **79** and **80** formed two carbon–carbon bonds and two stereogenic centres, therefore giving full tetracyclic core. It is worth mentioning, that remarkably high diastereoselectivities was observed in this transformation, even though up to this date no mechanistic rationale has been provided. Finally, the removal of protecting groups provided (−)–doxycycline (**78**) in 8.3 % yield and 18 steps (Charest et al. 2005; Wright et al. 2014).

The Michael-Dieckmann reaction sequence for the synthesis of tetracyclines has proven to be robust and effective with a variety of different D-ring substitution patterns. This reaction can be performed on kilogram scale in >90 % yield and only deprotection steps are required to obtain the final product. That makes it a highly valuable method for the synthesis of new tetracyclines with different substitutions patterns which were not accessible before (Wright et al. 2014; Sun et al. 2008). Following this approach, more than 3000 new antibiotic candidates have been synthesized and tested. Currently, some of them are in different stages of clinical trials. The leading compound thereof is eravacycline (**96**) a broad-spectrum antibiotic, which is currently in phase III clinical trials (Wright et al. 2014). It can be synthesized in only four steps from **93** and **94** by the strategy of Myers et al. (Scheme 16) (Ronn et al. 2013).

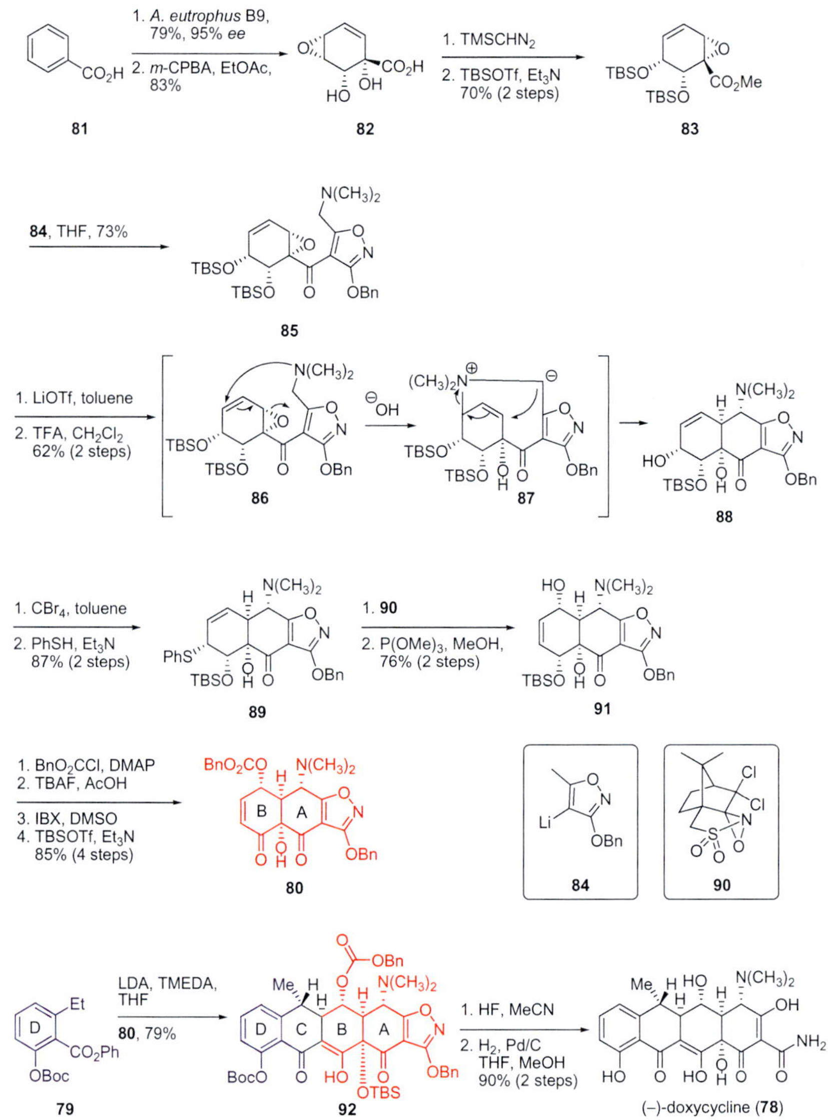

Scheme 15 Myers' total synthesis of (−)–doxycycline (**78**)

4 Synthesis of Macrolide Antibiotics

Another very important class of antibiotic agents is the macrolides. The structure of these polyketidic natural products exhibits a lactone core (usually 14- to 16-membered) to which desoxy sugars may be attached. The biological activity within this class is extremely diverse and includes antibacterial, antifungal as well as

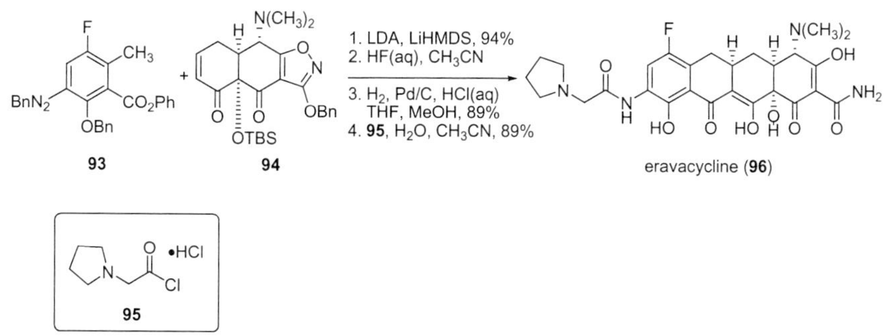

Scheme 16 Synthesis of eravacycline (**96**)

immunosuppressive properties. One of the most prominent and important representatives of macrolide antibiotics is erythromycin A, which itself as well as a variety of its derivatives are used as pharmaceutical drugs.

4.1 Synthesis of Erythromycin A (Woodward 1981)

Erythromycin A (**97**) is the most important representative of the macrolide family of antibiotics. It was isolated over 60 years ago from *Saccharopolyspora erythraea* (formerly *Strepotmyces erythraeus*) and consists of a 14-membred macrocycle to which L-cladinose and D-desosamine are attached (Scheme 17). It is in clinical use as a drug to cure infections caused by Gram-negative as well as Gram-positive bacteria infecting the skin or the upper respiratory tract. When resistance to β-lactams occurs or if the patient is allergic to penicillins, erythromycin stays the medication of choice. The importance of erythromycin A as a drug is underlined by the World Health Organisation putting it on its List of Essential Medicines.

Even though erythromycin A is not produced by chemical synthesis, but by cultivation of erythromycin producing bacterial strains, the first total synthesis of this complex natural product accomplished by Woodward et al. in 1981a, b, c remains a milestone in organic chemistry (Woodward et al. 1981a, b, c).

The synthesis started with the connection and cyclization of building blocks **101** and **102** (Scheme 18), which were already literature known by that time (Gais 1977; Bennett and Hock 1927; Sudo et al. 1967). The so-gained α,β-unsaturated ketone **103** was stereoselectively reduced, MOM-protected, and dihydroxylated to give **104** after acetal protection. The bicyclic product was the key intermediate of the synthesis. Removal of the methoxymethyl protecting group and subsequent oxidation using "activated" DMSO lead to ketone **99**. On the other hand, hydrogenation of **104** with Raney–Ni simultaneously removed the thioethers and the benzyl protecting group. Selenoxide elimination of the alcohol followed by ozonolysis provided the aldehyde **98**.

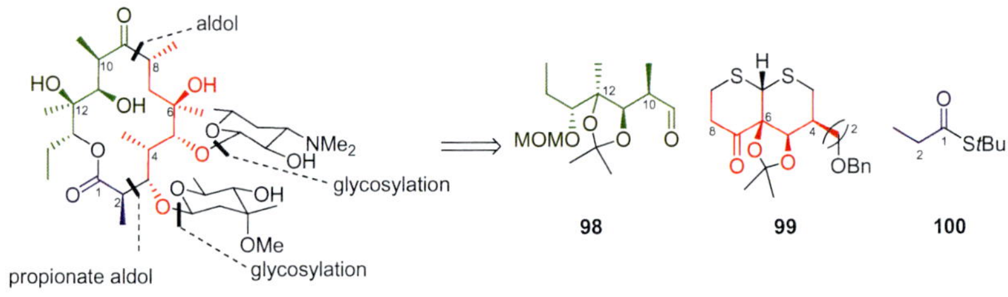

Scheme 17 Structure of erythromycin A (**97**) and retrosynthetic analysis by Woodward

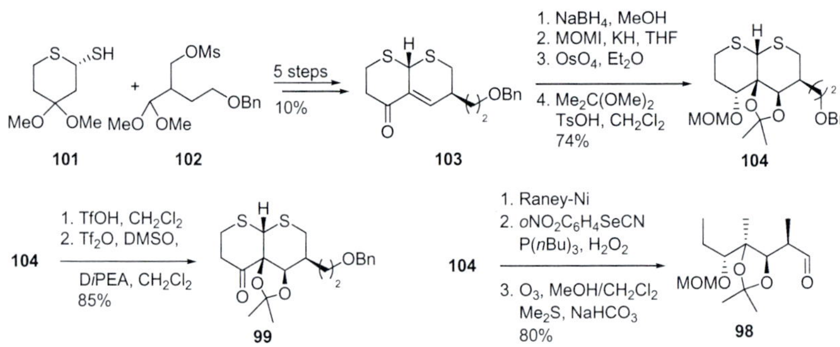

Scheme 18 Synthesis of the key intermediates in Woodward's erythromycin A synthesis

Ketone **99** and the aldehyde **98** were connected in the next step via an aldol reaction (Scheme 19) followed by oxidation of the resulting alcohol to diketone **105**. The stereocentre at C-8 (all atom numbers as in the erythromycin A) between the two keto groups of **105** was introduced by a substrate controlled sodium borohydride reduction of the enol acetate, which was generated by treatment of **105** with a base and subsequent quench with acetic anhydride. The hydroxy group at C-7 was then converted to the corresponding thioether **106** in two steps. Reduction of the remaining ketone functionality at C-9 with lithium aluminum hydride was followed by protection of that alcohol as an acetate. Hydrogenation with Raney-Ni resulted in (I) deprotection of the benzyl protecting group, (II) defunctionalization at C-7 and (III) liberation of the C-4, C-6 and C-8 methyl groups by cleavage of the cyclic thioethers. A sequence of selenoxide elimination and ozonolysis finished the synthesis of aldehyde **107**.

Introduction of the two missing carbon atoms of erythromycin's skeleton was accomplished by an aldol reaction between aldehyde **107** and the deprotonated propionyl thiolate (Scheme 20). This aldol reaction favoured the undesired *anti*-configured product. This could be corrected by a kinetically controlled inversion upon deprotonation/protonation at −100 °C giving the desired *syn*-product. Transformation of the C-9 acetate to the mesylate was carried out with interim selective protection of the C-3 alcohol as its phenyloxymethyl carbonate to afford

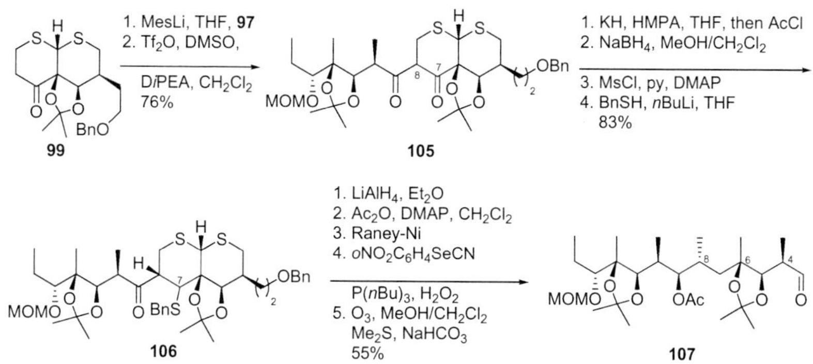

Scheme 19 Connection of the two key building blocks

the linear erythromycin skeleton **108** with all stereocentres in place. Next, the mesylate was converted to the azide with inversion of configuration at that centre (vide infra for the importance of the configuration there). The azide was reduced to the amine and directly protected as a carbamate, before the acetonides were cleaved and hexaol **110** was obtained. Upon treatment with trimethylamine, the cyclic carbamate protecting the functional groups at C-9 and C-11 respectively was introduced. The hydroxy groups at C-3 and C-5 were protected as the acetal of mesitylaldehyde, and the methyl ester transferred in two steps to its 2-pyridyl thio

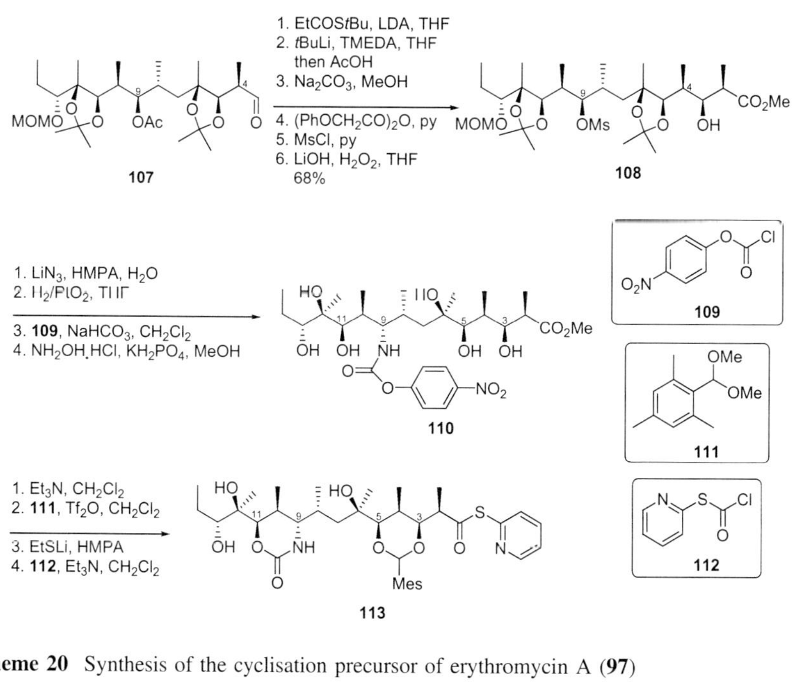

Scheme 20 Synthesis of the cyclisation precursor of erythromycin A (**97**)

analogue **113**. The necessity of a cyclic protecting group linking C-3 and C-5 was shown by a large screening of substrates for their ability to undergo cyclisation under the conditions developed by Corey (Corey and Nicolaou 1974). This screening also showed that the (*S*)-configuration at C-9 is crucial.

Cyclisation of **113** (Scheme 21) proceeded smoothly to yield **114** in 70 %. The carbamate was then *N*-acylated with phenyl benzoylchloride and subsequently cleaved under basic conditions. Deprotection of C-3 and C-5 gave **116**, the precursor for the remaining two glycosylations. It was anticipated that the C-3 OH would undergo glycosylation more rapid than C-5 OH. As the opposite was observed, C-5 was first glycosylated with D-desosamine **117** by a modified Koenigs-Knorr procedure affording the expected β-configuration (Hanessian and Banoub 1977). As methanolysis after the reaction removed the C-4″ carbonate it needed to be reattached prior to glycosylation with L-cladinoside **118**. The use of acetonitrile as solvent for this second glycosylation resulted in double inversion of configuration and gave the desired diglycosylated **119**.

The end game started with the removal of the protecting groups at the C-9 nitrogen and at the C-4″ oxygen with Na–Hg. Transformation to the amine

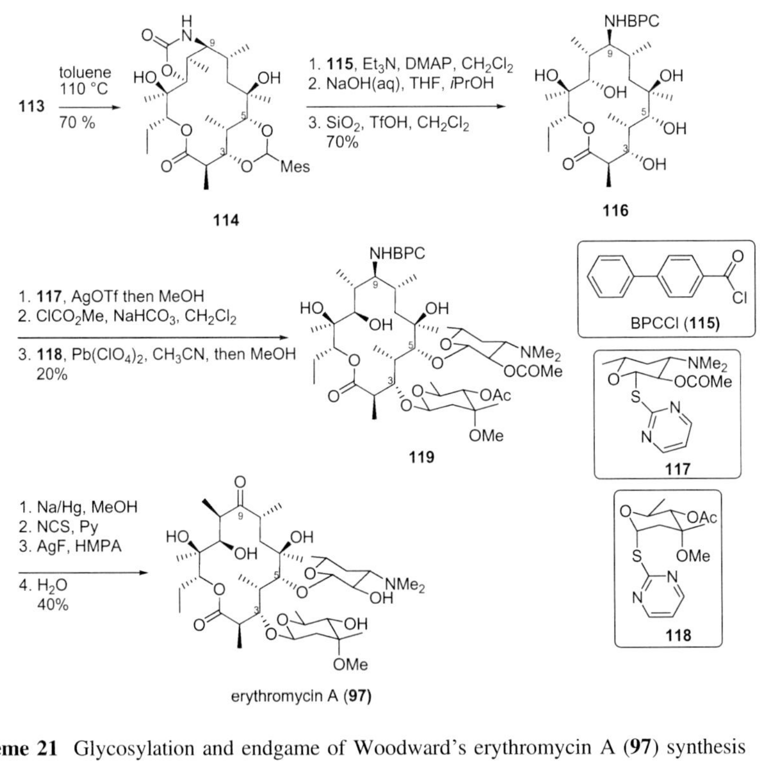

Scheme 21 Glycosylation and endgame of Woodward's erythromycin A (**97**) synthesis

hydrochloride (with *N*-Chlorosuccinimide) was followed by dehydrochlorination with silver fluoride to give erythromycinimine. Treatment of this imine with water liberated the keto group and finished the total synthesis of erythromycin A (**97**).

4.2 Recent Strategies for the Erythromycin Core

After the synthesis of erythromycin by Woodward a variety of different approaches were undertaken to synthesize other members of the erythromycin family as well as their aglycons (erythronolides). A very elegant and efficient approach, especially regarding the ring closure, was presented in 2009 by White (Stang and White 2009). The cyclisation precursor **120** was prepared in 18 steps from propionylated pseudoephedrine using standard transformations for the construction of polyketidic structures. The key step (Scheme 22) was the one-pot allylic oxidation of **120** at C-13 and macrolactonization catalyzed by a Pd(II)/bis(sulfoxide) catalyst. This transformation is highly atom economic as well as stereoselective, producing the desired stereoisomer at C-13 with a diastereomeric ratio of >40:1. From **120**, the synthesis of 6-desoxyerythronolide B was finished in three further steps.

A novel approach for the synthesis of polyketidic structures by nitrile oxide cycloaddition to double bonds of allylic alcohols was presented by Carreira et al. in 2001 and applied in the synthesis of erythronolide A in 2005 (Bode et al. 2001; Muri et al. 2005). Generation of the nitrile oxides was accomplished by treatment of an oxime with *tert*-butyl hypochlorite at low temperatures. Isoxazolines are the products of the cycloadditions and their N–O bond can be cleaved reductively to yield β-hydroxyketones. As the reaction proceeds via a transition state involving coordination of Mg(II) to the allylic alcohol, the newly generated hydroxy group is on the same face as the OH of the allylic alcohol. The configuration of the double bond defines the stereochemistry in α-position of the carbonyl group. Thus, using different double bond configurations as well as different configurations of the allylic alcohol, all possible combinations of stereochemical arrangements at the new stereocenters are accessible, making this a very versatile method.

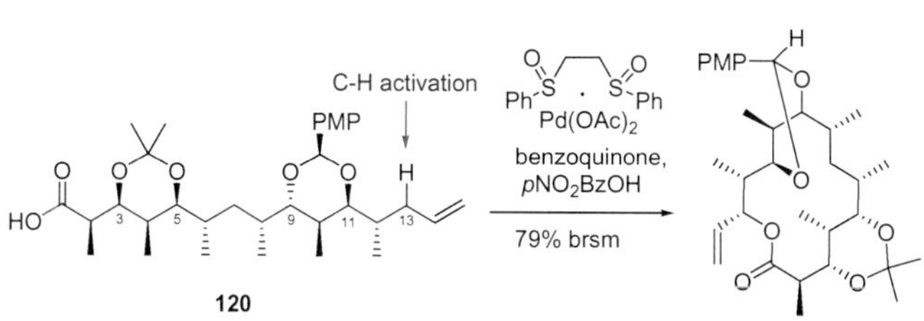

120

121

Scheme 22 Catalytic allylic oxidation/macrolactamization as key step in White's synthesis of 6-desoxyerythronolide B

Carreira's synthesis of erythronolide A (Scheme 23) started with the cycloaddition of readily available oxime **122** to allylic alcohol **123**. As expected, the stereogenic centres of the product were in an all-*syn* relationship (*d.r.* >98:2). The carbon chain was then extended by a sequence of oxidation and Grignard addition, and the resulting tertiary alcohol was protected as its triethylsilyl ether. The isoxazoline was subsequently cleaved using Raney–Ni to give the β-hydroxyketone **126**. Reduction of the ketone with zinc borohydride resulted in the *syn*-configuration of the diol, which was subsequently protected as the acetal of benzaldehyde. Deprotection and oxidation of the primary alcohol to the aldehyde, followed by treatment with hydroxylamine hydrochloride gave oxime **127**, which served as the precursor for another nitrile oxide cycloaddition to give **128**. Ten more steps finished the synthesis of erythronolide A (**129**).

4.3 Semisynthetic Antibiotics Derived from Erythromycin

For erythromycins the chemical synthesis was never a competitive way in order to provide the world supply. There are mainly two reasons for that: On the one hand, erythromycin is a complex natural product containing ten stereogenic centres in its 14-membered lactone. On the other hand, cultivation of the erythromycin producing bacterial strains and extraction of the desired product has always been comparably

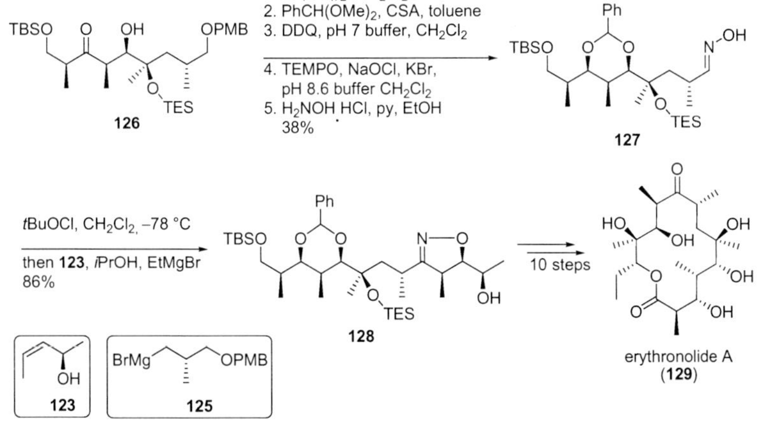

Scheme 23 Nitrile oxide cycloaddition in Carreira's synthesis of erythronolide A (**129**)

easy and inexpensive. This rapid access to such a privileged structure, in contrast, makes erythromycin an interesting starting point for semisynthetic approaches to obtain derivatives that are superior either in terms of metabolic stability or exhibit a broader range of antibacterial activity. The antibiotics azithromycin, clarithromycin, roxithromycin, dirithromycin as well as flurithromycin, the last not being on the market, are important examples of structures semisynthetically derived from erythromycin.

Clarithromycin (trade name Biaxin) for example is chemically one of the simplest derivatives of erythromycin as it differs only by methylation at the C-6 hydroxy group. It was first synthesized by Taisho Pharmaceuticals in 1980 (Watanabe et al. 1981), launched to the Japanese market in 1991 and FDA approved in 1995. Due to its improved bioavailability, compared to the rather acid labile erythromycin, it can be administered orally. It is considered a broad-spectrum antibiotic due to its activity against Gram-positive and Gram-negative bacteria (Hamilton 2014).

From erythromycin A (**97**), the synthesis starts with the introduction of two protecting groups (Scheme 24). The C-9 keto group is first protected as its oxime, followed by the transformation of the C-11 and C-12 hydroxy groups to an acetonide using 2-methoxypropene. Methylation is then carried out with iodomethane using potassium hydroxide as a base. Global deprotection with formic acid and sodium bisulfite gives then clarithromycin (**131**).

An agent that is similar in its pharmacological behaviour to clarithromycin is roxithromycin (**133**). The drug, which is sold under the trade name Rudil (or Rudile in Germany, Austria and Switzerland) is easily accessible from erythromycin in two steps (Krishna et al. 1998).

As the first step of the synthesis (Scheme 25), the C-9 ketone of erythromycin is converted to the oxime. The 2-methoxyethoxymethyl side chain is subsequently introduced via S_N2 displacement.

Erythromycylamine (**134**), an erythromycin derivative in which the C-9 keto group is replaced by an amine (Scheme 26), has very similar antibacterial activity to erythromycin itself, but can be administered orally. Dirithromycin (**136**) is the commercially available prodrug that is metabolized in the body to erythromycylamine (**134**).

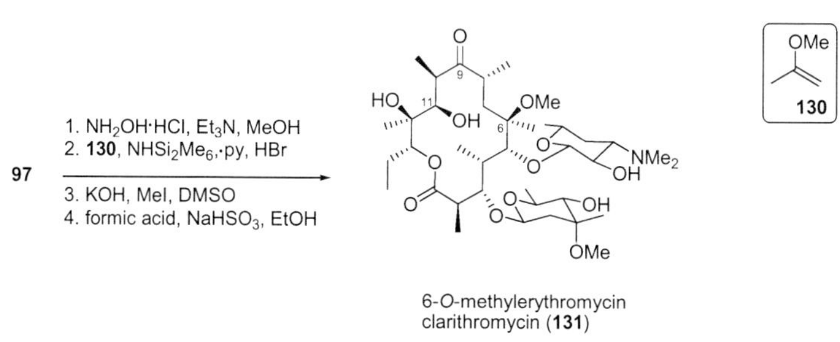

6-*O*-methylerythromycin
clarithromycin (**131**)

Scheme 24 Synthesis of clarithromycin (**131**) from erythromycin A (**97**)

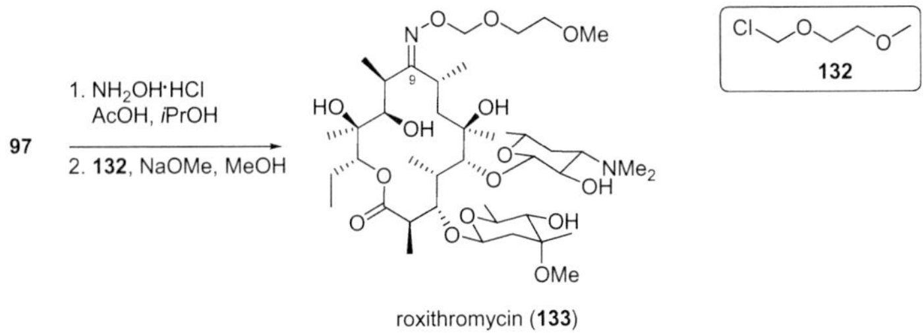

Scheme 25 Synthesis of roxithromycin (**133**) from erythromycin A (**97**)

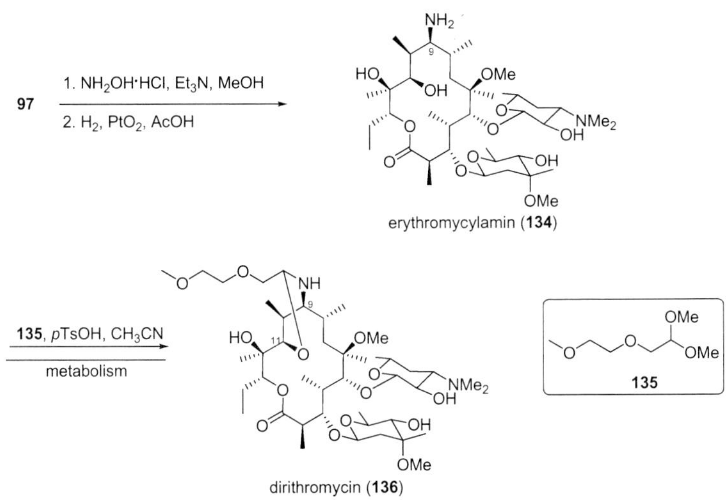

Scheme 26 Synthesis of dirithromycin (**136**) via erythromycylamine (**134**) from erythromycin A (**97**)

Dirithromycin (**136**) is synthesized from erythromycin (**97**) in three steps (Mcgill 1992; Massey et al. 1970). The sequence starts with the formation of the hydroxylimine from the C-9 keto group (Scheme 26). After reduction of the amine with hydrogen and platinum(IV) oxide as a catalyst, the hemiaminal ether of **136** is formed under acidic conditions from the corresponding dimethyl acetal **135**.

5 Summary

As the aforementioned antibiotics are still the pivotal pillars of anti-infectives therapy and resistant strains are evolving worldwide, the synthetic contributions described above provide the essential background for today's anti-infectives research. As described for the tetracyclines, new derivatives are still being synthesized today in order to fight emerging and resistant pathogens. The advent of new synthetic strategies might allow us to provide more rapid and efficient access to new and improved antibiotics. As one of the prerequisites of antibiotics research, synthetic chemistry will always play a key role in this area.

References

Albers-Schoenberg G, Arison BH, Hensens OD et al (1978) Structure and absolute configuration of thienamycin. J Am Chem Soc 100:6491–6499

Alt K (1909) Treatment experiments with arsenophenylglycine in paralytics. Muenchener Medizinische Wochenschrift 56:1457–1459

Bennett GM, Hock AL (1927) Benzyl δ-chlorobutyl ether: a new unsymmetrical derivative of tetramethylene glycol. J Chem Soc 472

Bode JW, Fraefel NM, Muri D, Carreira EM (2001) A general solution to the modular synthesis of polyketide building blocks by kanemasa hydroxy-directed nitrile oxide cycloadditions. Angew Chem Int Ed 40:2082–2085

Bruggink A (ed) (2001) Synthesis of β-lactam antibiotics. Chemistry, biocatalysis & process integration. Springer, Netherlands

Cabri W, Di Fabio R (2000) From bench to market. The evolution of chemical synthesis. Oxford University Press, Oxford

Chain E (1948): The chemistry of penicillin. Annu Rev Biochem:657–704

Charest MG, Lerner CD, Brubaker JD, Siege DR, Myers AG (2005) A convergent enantioselective route to structurally diverse 6-deoxytetracycline antibiotics. Science 308(5720):395–398

Chauvette RR, Flynn EH, Jackson BG, Lavagnino ER, Morin RB, Mueller RA, Pioch RP, Roeske RW, Ryan CW, Spencer JL, Van Heyninge E (1962) Chemistry of cephalosporin antibiotics. II. Preparation of a new class of antibiotics and the relation of structure to activity. J Am Chem Soc 84:3401–3402

Clarke HT (ed) (1949) The chemistry of penicillin. Princeton University Press, New Jersey

Clarke HT, Johnson JR, Robinson R (1949) The chemistry of penicillin. Report on a collaborative investigation by american and british chemists under the joint sponsorship of the office of scientific research and development, Washington, D.C., and the Medical Research Council. Princeton University Press, Princeton

Corey EJ, Nicolaou KC (1974) Efficient and mild lactonization method for the synthesis of macrolides. J Am Chem Soc 96:5614–5616

Der Spiegel (1968) Diabolische kette. 38:190 191

Deshpande AD, Baheti KG, Chatterjee NR (2004) Degradation of β-lactam antibiotics. Curr Sci India 87(12):1684–1695

Domagk G (1935) A contribution to the chemotherapy of bacterial infections. Deut Med Wochenschr 61:250–253

Duggar BM (1948) Aureomycin. A production of the continuing search for new antibiotics. Ann NY Acad Sci 51(2):177–181

Fleming A (1929) On the antibacterial action of cultures of a *penicillum*, with special reference to their use in the isolation of *B. influenzae*. Brit J Exp Path 10:226–236

Gais HJ (1977) Cyclic dithiohemiacetals—synthesis and properties. Angew Chem Int Ed 16:196–197

Georg GI (1993) The organic chemistry of β-lactams. VCH Publishers, New York

Hamilton R (ed) (2014) Tarascon pocket pharmacopoeia 2015. Jones & Bartlett Learning, Sudbury

Hanessian S, Banoub J (1977) Chemistry of the glycosidic linkage. An efficient synthesis of 1,2-trans-di-saccharides. Carbohydr Res 53:C13

Hanessian S, Desilets D, Bennani YL (1990) A novel ring-closure strategy for the carbapenems. The total synthesis of (+)-thienamycin. J Org Chem 55:3098–3103

Hodgkin DC, Maslen EN (1961) The X-ray analysis of the structure of cephalosporin C. Biochem J 79(2):393–402

Ishibashi H, Kameoka C, Iriyama H, Kodama K, Sato T, Ikeda M (1995) Sulfur-directed regioselective radical cyclization leading to β-lactams. Formal synthesis of (±)-ps-5 and (+)-thienamycin. J Org Chem 60:1276–1284

Jacobi PA, Murphree S, Rupprecht F, Zheng WJ (1996) Formal total syntheses of the β-lactam antibiotics thienamycin and ps-5. J Org Chem 61:2413–2427

Johnston DBR, Schmitt SM, Bouffard FA, Christensen BG (1978) Total synthesis of (±)-thienamycin. J Am Chem Soc 100:313–315

Korst JJ, Johnston JD, Woodward RB (1968) The total synthesis of dl-6-demethyl-6-deoxytetracycline. J Am Chem Soc 90:439–457

Krishna MM, Babu MS, Vyas KD, Kulkarni AK (1998) Process for preparing roxithromycin and derivatives thereof. Eur Pat EP1004591 A1

Krohn K, Kirst H, Maag H (eds) (1993) Antibiotics and antiviral compounds. Chemical synthesis and modification. VCH, Weinheim, New York

Lesher GY, Gruett MD, Froelich EJ, Brundage RP, Bailey JH (1962) 1,8-Naphthyridine derivatives. A new class of chemotherapeutic agents. J Med Chem 5:1063–1065

Loder B, Newton GGF, Abraham EP (1961) Sutdies on vitamin A esterase. 4. The hydrolysis and synthesis of vitamin A esters by rat intestinal mucosae. Biochem J 79:408–416

Lukacs G (ed) (1993) Recent progress in the chemical synthesis of antibiotics and related microbial products. Springer, Berlin Heidelberg

Lukacs G, Ohno M (1990) Recent progress in the chemical synthesis of antibiotics. Springer, Berlin Heidelberg

Massey EH, Kitchell B, Martin LD, Gerzon K, Murphy HW (1970) Erythromycylamine. Tetrahedron Lett 11:157–160

McCormick MH, McGuire JM, Pittenger GE, Pittenger RC, Stark WM (1956) Vancomycin, a new antibiotic. I. Chemical and biologic properties. Antibiot Ann 1955–1956:606–611

Mcgill JM (1992) One step process for production of dirithromycin. CA2079430 A1, 29 Sep 1992

McGuire JM, Bunch RL, Anderson RC, Boaz HE, Flynn EH, Powell HM, Smith JW (1952) Ilotycin, a new antibiotic. Antibiot Chemother 2:281–283

Melillo DG, Shinkai I, Liu T, Ryan K, Sletzinger M (1980) A practical synthesis of (±)-thienamycin. Tetrahedron Lett 21:2783–2786

Melillo DG, Cvetovich RJ, Ryan KM, Sletzinger M (1986) An enantioselective approach to (+)-thienamycin from dimethyl 1,3-acetonedicarboxylate and (+)-α-methylbenzylamine. J Org Chem 51:1498–1504

Morin RB, Gorman M (1982) Chemistry and biology of β-lactam antibiotics. Academic Press, New York London

Muri D, Lohse-Fraefel N, Carreira EM (2005) Total synthesis of erythronolide A by MgII-mediated cycloadditions of nitrile oxides. Angew Chem Int Ed 117:4104–4106

Muxfeldt H (1962a) Synthese eines terramycin-bausteins. Angew Chem 74:825–828

Muxfeldt H (1962b) Synthesen in der tetracyclin-reihe. Angew Chem 74:443–478

Muxfeldt H, Behling J, Grethe G, Rogalski W (1967) Tetracyclines. VI. Some new aspects in the chemistry of oxazolones and their sulfur analogs. J Am Chem Soc 89:4991–4996

Muxfeldt H, Hardtmann G, Kathawala F, Vedejs E, Mooberry JB (1968) Tetracyclines. VII. Total synthesis of dl-terramycin. J Am Chem Soc 90:6534–6536

Muxfeldt H, Haas G, Hardtmann G, Kathawala F, Mooberry JB, Vedejs E (1979) Tetracyclines. 9. Total synthesis of dl-terramycin. J Am Chem Soc 101:689–701

Myers AG, Siegel DR, Buzard DJ, Charest MG (2001) Synthesis of a broad array of highly functionalized, enantiomerically pure cyclohexanecarboxylic acid derivatives by microbial dihydroxylation of benzoic acid and subsequent oxidative and rearrangement reactions. Org Lett 3:2923–2926

Newton GGF, Abraham EP (1955) Cephalosporin C, a new antibiotic containing sulphur and D-α-aminoadipic acid. Nature 175:548

Nicolaou KC, Chen JS, Edmonds DJ, Estrada AA (2009) Recent advances in the chemistry and biology of naturally occurring antibiotics. Angew Chem Int Ed 48:660–719

Page MI (1992) The chemistry of β-lactams. Springer, Netherlands

Pine SH (2004): The base-promoted rearrangements of quaternary ammonium salts. Org React. (Wiley, Hoboken, NJ, USA)

Ronn M, Zhu Z, Hogan PC et al (2013) Process R&D of eravacycline. The first fully synthetic fluorocycline in clinical development. Org Process Res Dev 17(5):838–845

Schatz A, Bugle E, Waksman SA (1944) Streptomycin, a substance exhibiting antibiotic activity against gram-positive and gram-negative bacteria. Exp Biol Med 55(1):66–69

Sheehan JC (1982) The enchanted ring. The untold story of penicillin. MIT Press, Cambridge

Sheehan JC, Henery-Logan KR (1957) The total synthesis of penicillin V. J Am Chem Soc 79:1262–1263

Sheehan JC, Henery-Logan KR (1959) The total synthesis of penicillin V. J Am Chem Soc 81:3089–3094

Sheehan JC, Henery-Logan KR (1962) The total and partial general syntheses of the penicillins. J Am Chem Soc 84:2983–2990

Sheehan JC, Hess GP (1955) A new method of forming peptide bonds. J Am Chem Soc 77:1067–1068

Sheehan JC, Johnson DA (1954) The synthesis of substituted penicillins and simpler structural analogs. VIII. Phthalimidomalonaldehydic esters. Synthesis and condensation with penicillamine. J Am Chem Soc 76:158–160

Stang EM, White MC (2009) Total synthesis and study of 6-deoxyerythronolide B by late-stage C–H oxidation. Nat Chem 1(7):547–551

Stork G, Hagedorn AA (1978) 3-Benzyloxyisoxazole system in construction of tetracyclines. J Am Chem Soc 100:3609–3611

Sudo R, Kaneda A, Itoh N (1967) Synthesis of carolic acid. J Org Chem 32:1844–1846

Sun C, Wang Q, Brubaker J, Wright PM, Lerner CD, Noson K, Charest M, Siegel DR, Wang YM, Myers AG (2008) A robust platform for the synthesis of new tetracycline antibiotics. J Am Chem Soc 130:17913–17927

Watanabe Y, Morimoto S, Omura S (1981) Novel erythromycin compounds. Applied for by Taisho Pharmaceutical Co. EP0041355 A1, 4 May 1980

Woodward RB (1966) Recent advances in the chemistry of natural products. Science 153 (3735):487–493

Woodward RB, Heusler K, Gosteli J, Naegeli P, Oppolzer W, Ramage R, Ranganat S, Vorbrugg H (1966) The total synthesis of cephalosporin C. J Am Chem Soc 88:852–853

Woodward RB et al (1981a) Asymmetric total synthesis of erythromycin. 2. Synthesis of an erythronolide A lactone system. J Am Chem Soc 103:3213–3215

Woodward RB et al (1981b) Asymmetric total synthesis of erythromcin. 1. Synthesis of an erythronolide a secoacid derivative via asymmetric induction. J Am Chem Soc 103:3210–3213

Woodward RB et al (1981c) Asymmetric total synthesis of erythromycin. 3. Total synthesis of erythromycin. J Am Chem Soc 103:3215–3217

Wright PM, Seiple IB, Myers AG (2014) The evolving role of chemical synthesis in antibacterial drug discovery. Angew Chem Int Ed 53:8840–8869

Antibiotics Clinical Development and Pipeline

Thomas Hesterkamp

Abstract There is a constant need for resupply with resistance-breaking antibiotics. Governmental programs and updated regulatory guidance have incentivized mainly small- and medium-sized biopharmaceutical companies to develop novel antibiotics up to market licensure, while major pharma players, with exceptions, have abandoned the space for a perceived lack of a return on their investment. The portfolio of approved drugs has improved over recent years for gram-positive infections, including infections caused by methicillin-resistant *Staphylococcus aureus*. On the other hand, unmet medical need has surfaced in indications dominated by gram-negative pathogens including complicated intra-abdominal and bloodstream infections as well as hospital-acquired and ventilator-associated pneumonia. Few if any treatment options are left for extended-spectrum beta-lactamase- and carbapenemase-producing Enterobacteriaceae, e.g., *Klebsiella pneumoniae*, and the multi-drug-resistant non-fermenting gram-negative bacteria *Pseudomonas aeruginosa* and *Acinetobacter baumannii*. The present paper summarizes and reviews the clinical pipeline of novel antibiotics by clinical indication and identifies the unmet medical need in the space.

Contents

T. Hesterkamp (✉)
Helmholtz-Zentrum für Infektionsforschung and Deutsches Zentrum für Infektionsforschung, Inhoffenstraße 7, 38124 Braunschweig, Germany
e-mail: thomas.hesterkamp@helmholtz-hzi.de

Current Topics in Microbiology and Immunology (2016) 398: 447–474
DOI 10.1007/82_2015_451

Published Online: 09 July 2015

1 Introduction

There is a widely held concern over the lack of innovation and productivity in the research and development of novel antibiotics. While this concern is not new, some fundamental aspects are indeed calling for action. First, hospital-acquired infections of vulnerable patient groups are frequent and causing a significant economic burden (Bassetti and Righi 2013). *Staphylococcus aureus* remains a major culprit in this setting, but other pathogens from the gram-negative spectrum including Enterobacteriaceae and non-fermenters like *Pseudomonas aeruginosa* and *Acinetobacter baumannii* are increasingly causing outbreaks, too. Second, due to the intense use of antibiotics in particular in intensive care units (ICUs; for abbreviations and glossary, see Table 1), rates of resistance formation by pathogens are high. Some of the (first-line) drug-resistant pathogens are more difficult and resource-intensive to eradicate, thus adding to the disease and cost burden (Calfee 2012). Methicillin-resistant *Staphylococcus aureus* (MRSA) has become a synonym for hospital-acquired infections in major parts of the world and the so-called antibiotic crisis (Spellberg 2011). A series of new drug approvals in the 2010–2014 period in particular for MRSA have addressed this problem in part. Third, MRSA is making its way into the community, at least in the USA, thereby threatening broader parts of the general population (Nimmo 2012). Fourth, there is a growing concern over the emergence and global spread of extended-spectrum beta-lactamase (ESBL) and carbapenem-resistant Enterobacteriaceae (CRE). While infrequent in Germany as of today, resistance may indeed spread rapidly through mobile genetic elements shared horizontally between Enterobacteriaceae (for recent review, see Hawkey 2015). Unlike MRSA, the research and development of resistance-breaking novel drug candidates for ESBL/CRE will take longer and incur higher cost. In addition, the current clinical pipeline is considered as critically thin in this regard. There remains a very heavy reliance on β-lactam antibiotics (including cephalosporins, carbapenems, and mono-bactams) for treatment of serious gram-negative infections, and this is not expected to change in the foreseeable future. The response to the above must indeed be to reinvigorate drug discovery and development, starting with the discovery of novel antibacterial chemotypes from natural and synthetic sources. At the same time, increasing hygiene standards and introducing antibiotic stewardship programs in hospitals and nursing homes should be considered at least equally effective toward reducing disease and cost burden.

Table 1 Abbreviations and glossary in alphabetical order

AE/SAE	Adverse event/serious adverse event
BARDA	Biomedical Advanced Research and Development Authority
BLI	β-Lactamase inhibitor
CDAD	*Clostridium difficile*-associated diarrhea
CRE	Carbapenem-resistant Enterobacteriaceae
DTRA	Defense Threat Reduction Agency
EFPIA	European Federation of Pharmaceutical Industries and Associations
EMA	European Medicines Agency
ESBL	Extended-spectrum β-lactamase
ESKAPE	*Enterococcus faecium*, *Staphylococcus aureus*, *Klebsiella pneumoniae*, *Acinetobacter baumannii*, *Pseudomonas aeruginosa*, *Enterobacter* spp.
FDA	US Food and Drug Agency
GMP, GLP	Good manufacturing practice, good laboratory practice
ICU	Intensive care unit
IDSA	Infectious Disease Society of America
IMP	Investigational medicinal product
IMI	Innovative Medicines Initiative
IND	Investigational new drug
LPAD	Limited Population Antibacterial Drug
MRSA	Methicillin-resistant *Staphylococcus aureus*
MDR	Multi-drug resistant
MIC	Minimum inhibitory concentration
NDA	New drug application
QIDP	Qualified infectious disease product
PK/PD	Pharmacokinetics/pharmacodynamics
RTI	Respiratory tract infection
XDR	Extremely drug resistant

The industry's commitment to research and development of novel antibiotics is inconsistent and rather low. Major players have left the arena approximately 10–15 years ago, and very few have returned. A current estimate is that less than 10 of the 50 leading pharmaceutical houses have active programs in the space.[1] Antibiotic development has become a business of specialized small- and medium-sized bio-pharmaceutical companies, many of which are based in the USA. To name a few, Achaogen, AiCuris, Basilea, Cempra, Cubist (now and hereinafter Merck & Co), Durata (now and hereinafter Actavis), Nabriva, Paratek, Rempex, Tetraphase, and The Medicines Company, among others, have built a reputation of sustaining developmental programs up to drug licensure and market introduction. While the

[1] http://www.pewtrusts.org/en; http://www.vfa.de/de/arzneimittel-forschung/woran-wir-forschen/neue-antibiotika-den-vorsprung-wahren.html.

overall pharmaceutical blockbuster concept has been questioned and to an extent abandoned by the major pharmaceutical houses (Kumar Kakkar and Dahiya 2014), the current financial rewards for antibiotic development are not attractive enough for them to reenter the space (McKellar and Fendrick 2014; Spellberg et al. 2012). Some authors are speaking of a market inefficiency and calling for governmental programs to incentivize the developers either through shared cost of development (the so-called push incentives) or through extended/enhanced remuneration through extended market exclusivity (the so-called pull incentives; Spellberg et al. 2012; Spellberg 2014). The public–private partnership of EU commission and six European Federation of Pharmaceutical Industries and Associations (EFPIA) firms, the Innovative Medicines Initiative (IMI), is tackling the issue through shared cost of development whereby European tax payers are cofunding selected preclinical and clinical programs. Points in case are the New Drugs For Bad Bugs initiative (ND4BB), Combatting Bacterial Resistance in Europe (COMBACTE), European Gram-Negative Antibacterial Engine (ENABLE), and Molecules Against Gram-Negative Infections (MAGNET) programs.[2] The USA has taken a different approach, summarized under the Generating Antibiotic Incentives Now (GAIN) act, enacted by US congress in 2012. Herein, selected clinical programs (drugs, formulations, indications) are granted a qualified infectious disease product (QIDP) status under which new drug applications are fast-tracked by the US Food and Drug Agency (FDA) for approval, and sponsors are granted an extra 5-year market exclusivity (Keener 2014; Spellberg 2014). The extra 5-year market exclusivity is based on the Hatch–Waxman market exclusivity to foster research and development in the USA, not to be seen as a simple increment to patent protection but rather to salvage products that are nearing patent expiry by the time of registration. In addition, the USA is quite heavily funding antibiotic development through the Biomedical Advanced Research and Development Authority (BARDA) grants and its Defense Threat Reduction Agency (DTRA) agency, where respective human pathogens are subject of the developmental program. Collectively, these 'push-and-pull' measures have advanced the clinical pipeline.

2 Recent Approvals and Research Agenda

Recent market additions (period ~ 2010 to 2014) are two cephalosporins (ceftaroline fosamil and ceftobiprole) by Forest Laboratories (now Actavis) and Basilea Pharmaceutica, respectively, for indications ABSSSI, CAP, and CAP/HAP, but excluding VAP (indications see Table 2). Ceftolozane–tazobactam, a cephalosporin–beta-lactamase inhibitor (BLI) combination (Merck & Co), received marketing authorization for cIAI and cUTI in late 2014 in the USA and is submitted for approval in the EU. The lipoglycopeptides telavancin (Theravance Biopharma),

[2] http://www.imi.europa.eu/.

Table 2 Clinical indications, abbreviations, and bacterial pathogen spectrum encountered

Abbreviation	Indication	Pathogen spectrum
CAP	Community-acquired pneumonia	*Streptococcus pneumoniae, Haemophilus influenza, Mycobacterium tuberculosis, Enterobacteriaceae, Legionella* spp., *Staphylococcus aureus, Chlamydia pneumoniae*
HAP/VAP	Hospital-acquired pneumonia, ventilator-associated pneumonia	*Pseudomonas aeruginosa, Staphylococcus aureus, Klebsiella* spp., *Escherichia coli, Enterobacter* spp., *Acinetobacter* spp., *Stenotrophomonas maltophilia, Enterococcus* spp., *Serratia* spp.
IAI	Intra-abdominal infections	*Escherichia coli, Klebsiella* spp., *Pseudomonas aeruginosa, Proteus mirabilis, Enterobacter* spp., *Bacteroides fragilis* and other spp., *Clostridium* spp., *Prevotella* spp., *Peptostreptococcus* spp., *Fusobacterium* spp., *Eubacterium* spp., *Streptococcus* spp., *Enterococcus faecalis* and *faecium, Enterococcus* spp., *Staphylococcus aureus*
UTI	Urinary tract infections	*Escherichia coli, Enterococcus* spp., *Pseudomonas aeruginosa, Klebsiella* spp., *Enterobacter* spp., *Proteus* spp., *Acinetobacter* spp., *Citrobacter* spp., *Enterobacteriaceae*
ABSSSI	Acute bacterial skin and skin structure infections	*Streptococcus pyogenes, Staphylococcus aureus, Streptococcus* spp., *Enterococcus faecalis, Clostridium* spp., *Vibrio vulnificus, Aeromonas hydrophila, Peptostreptococcus* spp., *Pasteurella* spp., *Pseudomonas aeruginosa*
u/c	Uncomplicated/complicated	

dalbavancin (Actavis), and oritavancin (The Medicines Company) have been registered for infections due to *S. aureus* including MRSA. In case of dalbavancin and oritavancin, the registration is for indication ABSSSI, and in case of telavancin, the drug is in addition to ABSSSI indicated for HAP/VAP through susceptible *S. aureus* including MRSA. A second-generation oxazolidinone, tedizolid by Merck & Co, has been registered for ABSSSI by MRSA, too. In addition, fidaxomicin has been registered for *Clostridium difficile* infection, and two drugs against *Mycobacterium tuberculosis*, delamanid (Otsuka) and bedaquiline (Janssen/TB Alliance), have received conditional registration to accelerate full clinical development. With this, the general output of the collective R&D effort of the industry in the antibiotic space has been significantly above average for the reporting period. Below, I shall review the status of the pipeline at clinical stages phase I through to phase III stratified by major indications.

In 2008 and 2009, Louis B Rice and the Infectious Disease Society of America (IDSA) released the concept of the ESKAPE (standing for the pathogens *Enterococcus faecium*, *S. aureus*, *Klebsiella pneumoniae*, *A. baumannii*, *P. aeruginosa*, *Enterobacter* species) organisms to focus the global R&D efforts for antibiotics to the key human multi-drug-resistant pathogens in the hospital setting (Rice 2008; Boucher et al. 2009). Since then this concept has been further refined to explore more treatment options for multi-drug-resistant (MDR) gram-negative bacteria including Enterobacteriaceae and non-fermenters (Boucher et al. 2013; Tacconelli et al. 2014; Bassetti et al. 2013; Spellberg and Shlaes 2014). Short-to-mid-term protection from degradation arguably is the most realistic prospect for developing new β-lactam antibiotics with resistance-breaking gram-negative activity (ceftolozane–tazobactam, ceftazidime–avibactam, aztreonam–avibactam, meropenem–RPX7009; for review, see Drawz et al. 2014). Going forward, however, the concern is that it will be increasingly difficult to protect β-lactams against the dozens of diverse β-lactamases that have emerged and spread (Bush 2013), thus leading to a further dried-up gram-negative pipeline. Additional exploitation of the tetracycline and aminoglycoside scaffolds is taking place with successes (omadacycline and eravacycline for the former group, plazomicin within the latter; for review, see Karaiskos and Giamarellou 2014). Fundamentally though, in order to reprime the pipeline, we have to go all the way back to screening either untapped resources of natural compounds or, using refined methodology, to rescreen the available ones. Entirely novel scaffolds may indeed stand a chance to only succumb to resistance later and attract extra funding, e.g., from the private side, thus giving leeway to the field. In addition, finding and developing narrow-spectrum drugs for selected MDR pathogens and mono-microbial infections should receive attention, too. It is interesting to note that not only classical pathogen-specific indications like *C. difficile* and *M. tuberculosis* infections are witnessing narrow-spectrum drug development, but also the opportunistic non-fermentative gram-negative pathogen *P. aeruginosa* (Srinivas et al. 2010). A successful development of, e.g., POL7080 (Polyphor/Roche; see Table 3) may indeed impact the industry, in that the development of rapid diagnostic techniques and drug development are intimately intertwined. This challenges the current empirical treatment paradigm, which tends to be broad spectrum in nature and infrequently de-escalated after pathogen identification and susceptibility testing (Montero et al. 2015). In addition, narrow-spectrum drugs, with their reduced potential to disrupt the general microflora, may also provide the impetus to develop further the clinical end points of, e.g., recurrent infections due to rapid recolonization after antibiosis and establish these more firmly as outcome measures for clinical trials of antibiotics. Third, again under the view of reducing unnecessary resistance formation, preventative treatments should be developed to reduce the burden of infection in the hospital setting. *S. aureus* may be decolonized by topical treatments of skin and mucosa, thus reducing the number of infections due to the patient's own colonizing strain. This is being explored broadly for intensive care unit patients (Huang et al. 2014). Decolonizing MDR gram negatives from the intestinal flora would require a harsher, systemic treatment with antibiotics, but

Table 3 Clinical-stage pipeline for antibacterial agents targeting RTIs

Drug candidate	Sponsor	Indication	Latest stage of development	Pathogen coverage	Source
Nemonoxacin	TaiGen Biotechnology	CAP	Phase III recruiting for i.v. formulation under US IND; phase III oral formulation complete	MRSA, VRE, *A. baumannii*	NCT02205112; NCT01529476
Solithromycin	Cempra Pharmaceuticals	CAP	Phase III complete (oral); phase III ongoing (i.v. to oral); phase I ongoing (pediatric formulation)	Gram positive	http://www.cempra.com/products/pipeline/; NCT01968733
Zabofloxacin	Dong-Wha Pharmaceutical	CAP	Phase II		
JNJ-Q2 (QIDP)	Furiex Pharmaceuticals	CAP	Phase II complete (i.v. to oral switch study)	Broad spectrum incl MRSA and MDR *S. pneumoniae*	http://www.furiex.com/pipeline/discoverydevelopment-pipeline/fluoroquinolone/
Omadacycline (QIDP)	Paratek Pharmaceuticals	CAP	Phase II complete; phase III planned under FDA SPA	Broad spectrum incl MDR	http://www.paratekpharm.com/content/pipeline/index.htm
Radezolid (QIDP)	Melinta Therapeutics	CAP	Phase II complete (oral formulation)	Gram positive + fastidious gram negatives	http://melinta.com/early_stage_pipeline.php
Lefamulin	Nabriva Therapeutics	CAP	Phase II complete	Broad spectrum	http://www.nabriva.com/programs/pipeline/
Eravacycline	Tetraphase Pharmaceuticals	CAP	Phase I (i.v.)	Broad spectrum incl MDR gram negatives	http://tphase.com/pipeline/overview
GSK-2140944 (topo II inhibitor)	GlaxoSmithKline	Pulmonary	Phase I pulmonary PK complete, i.v. route [NCT01934205]; phase II bacterial infections [www]		http://www.gsk.com/media/280387/product-pipeline-2014.pdf

(continued)

Table 3 (continued)

Drug candidate	Sponsor	Indication	Latest stage of development	Pathogen coverage	Source
WCK 771 (2349)	Wockhardt	Pulmonary	Phase I	Broad spectrum	
Plazomicin (QIDP)	Achaogen partly under BARDA contract	HAP/VAP due to CRE	Phase III i.v. in combination with meropenem or tigecycline; open-label superiority trial versus colistin combination	MDR gram negatives	http://www.achaogen.com/pipeline-overview/; NCT01970371
Tedizolid (Sivextro)	Merck & Co	HAP/VAP	Phase III in progress	Gram positive	http://www.cubist.com/research-and-development/pipeline
Ceftolozane–tazobactam (Zerbaxa)	Merck & Co	HAP/VAP	Phase III in progress	Broad spectrum incl *Pseudomonas aeruginosa*	http://www.cubist.com/research-and-development/pipeline
Ceftazidime–avibactam	AstraZeneca/Actavis	HAP/VAP	Phase III (REPROVE) in progress	Broad spectrum incl MDR gram negatives	NCT01808092
Meropenem-RPX7009 (Carbavance)	Rempex Pharmaceuticals (The Medicines Company)	HAP/VAP due to CRE	Phase III ongoing	CRE	NCT02168946
POL7080 (QIDP)	Polyphor/Roche	VAP, bronchiectasis, LRTI	Phase II	*Pseudomonas aeruginosa*	NCT02096328; NCT02096315
MEDI4893	MedImmune	HAP/VAP prevention	Phase II	*Staphylococcus aureus*	NCT02296320
MEDI3902	MedImmune	HAP/VAP prevention	Phase I	*Pseudomonas aeruginosa*	NCT02255760

should be considered, too (Hawkey 2015). Last not least, vaccine development should continue to be explored as well. Unfortunately, with some exceptions, this has remained difficult for the major human pathogens in widespread circulation.

3 Preclinical and Clinical Development of Antibiotics

The preclinical and clinical developmental program for antibiotics does not fundamentally differ from programs performed for other groups of therapeutics. Preclinical development ends on the submission of a clinical trial application for phase I testing and the opening of an investigational new drug (IND) file at the national competent authorities or European Medicines Agency (EMA). This involves extensive documentation on the manufacturing under good manufacturing practice (GMP) and pharmaceutical quality of the investigational medicinal product (IMP) material to be used in the clinical development as well as safety pharmacology and toxicology data obtained typically from a rodent and non-rodent animal species in compliance with conditions of good laboratory practice (GLP). Information on pharmacological efficacy in preclinical models is included with a view to develop a benefit proposition over standard-of-care treatments and an initial dose estimate for human testing. Regulators further request new drug candidates to be tested in vitro on several hundred contemporary clinical isolates with minimum inhibitory concentration (MIC) distributions being reported by bacterial species.[3] Where the risk/benefit assessment is plausible from a regulatory perspective and the study protocol deemed acceptable, permission is granted to study the drug candidate in a well-defined group of healthy volunteers as part of phase I work. Key end points of phase I trials are the safety, tolerability, and pharmacokinetics (PK) of the treatment in the chosen dose range and dosing route. Systemic drug exposure levels in human need to be backed by exposure levels previously studied in preclinical animal models, and proper safety margins need to be maintained. Safety and tolerability information is expressed in the form of adverse event and serious adverse event (AE/SAE) grading (as per protocol) and reporting to the authorities. A decision is made whether the observed side effect profile of the tested dose offers benefits over risk for testing the drug candidate in a small, select group of patients in phase II studies. This may entail dose finding and drug combination studies. While programs frequently strive to meet the proof-of-concept efficacy data in phase II trials as early as possible, multiple additional phase I studies may be performed in a staggered mode to test different preparations, formulations, dosing schedules, drug combinations, food effects, and vulnerable groups for safety, tolerability, and PK.

Modeling relationships of PK and pharmacodynamics (PD) is of paramount importance for preclinical and clinical antibiotic development. With the significant progress in the field of PK/PD modeling in rodents and the recognition that the

[3]EMA Committee for Medicinal Products for Human Use 2011, CPMP/EWP/558/95 rev 2.

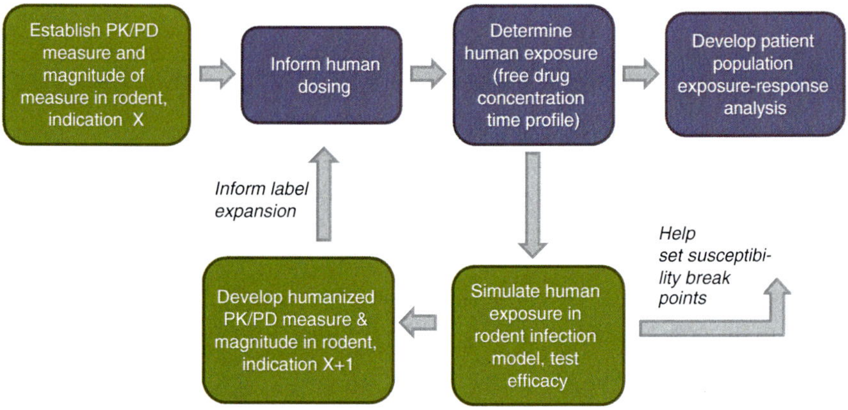

Fig. 1 Using rodent PK/PD measures to predict human efficacy

resulting PK/PD indices are broadly predictive of clinical efficacy (Andes and Craig 2002; Ambrose et al. 2007), authorities have developed policy to accept such preclinical information instead of clinical dose finding studies, provided that the relationship is straightforward to be interpreted, i.e., causal. The typical PK/PD indices correlating with efficacy are $AUC_{0\text{-}24}$/MIC, C_{max} > MIC, and T > MIC.[4] Interestingly, PK/PD modeling in rodents is not only used in a forward paradigm, whereby preclinical studies precede and inform clinical testing, but in addition and as depicted in Fig. 1, clinically achieved exposure levels of drugs are used to predict, in the form of human-simulated doses in rodents, the efficacy of treatments against specific MDR pathogens with the view to develop break points and expand label for the drugs under development (Crandon and Nicolau 2013; MacVane et al. 2014).

The standard paradigm for running pivotal, drug registration-relevant phase III trials is to test the drug candidate versus active control in blinded, controlled study designs for statistical non-inferiority of the investigative treatment. A single well-powered phase III trial may suffice to support registration, but typically two phase III studies with up to 1000 patients each are conducted in parallel in different parts of the world. The majority of phase III trials are for a disease or body-site infection like IAI, UTI, or CAP (see Table 2). These are disease conditions caused by a spectrum of human pathogens, frequently covering gram-positive, gram-negative, atypical, and anaerobic bacterial species in the form of a mono- or polymicrobial infections.[5] The antibacterial spectrum of the investigative drug should, obviously, support the testing in the disease including coverage of atypical

[4] $AUC_{0\text{-}24}$/MIC, the ratio of total free drug exposure integrated over an 24-h time interval and minimal inhibitory concentration in vitro; C_{max}/MIC, the ratio of maximum free plasma concentration and minimal inhibitory concentration in vitro; T > MIC, time of free plasma concentration above minimal inhibitory concentration in vitro.

[5] European Centre for Disease Prevention and Control 2013, Annual epidemiological report on 2011 surveillance data and 2012 epidemic intelligence data.

pathogens, but in addition, the cases enrolled into phase III trials should be representative of the bacterial pathogen spectrum encountered in the later clinical use and broader population. However, phase III trials have been conducted and drug labels obtained on individual pathogen infections, irrespective of body-site infection (see, for instance, Table 8 for current phase III study design for plazomicin; clinicaltrials.gov identifier NCT01970371). This is a valid strategy for prevalent pathogens with high morbidity or mortality, for example MRSA, vancomycin-resistant enterococci (VRE) and CRE in the nosocomial setting.

4 Updated Regulatory Guidance

When it became apparent to the broader public that the resupply with novel drug candidates and the rate of registration of new drugs was dwindling, that pharmaceutical companies were leaving the marketplace and that bacterial resistance was emerging fast in certain countries, regulatory authorities responded. Prompted and challenged by interventions by IDSA (Infectious Disease Society of America 2012) and others, requesting a more realistic regulatory path for novel antibiotics, FDA and EMA released guidance documents[6] providing flexibility to sponsors to achieve drug registration more realistically and economically but with an initially narrower drug labeling and hence reduced market potential (Ambrose et al. 2012; Nambiar et al. 2014 and regulatory references therein; Spellberg 2014; Bax and Green 2015). Industry representatives in return interpreted the guidance in the form of a new 'regulatory framework' postulate that I shall briefly outline herein. Accordingly, the regulatory framework comprises four tiers, A–D, in decreasing order of comprehensiveness of data supplied in support of a drug registration (Rex et al. 2013; Tomayko et al. 2014). The choice of tier should be driven by considerations including the extent of the unmet medical need, the quality and quantity of preclinical characterization of the drug candidate, and the feasibility of finding patients to be enrolled in clinical trials and to complete a clinical trial within reasonable time frame. In any case, the developmental path for a product candidate would be developed in a close dialogue between the responsible competent authority and the program's sponsor.

Tier A is the above disease- or indication-oriented conduct of two phase III trials in support of a broad drug label. This puts the highest demand on the drug candidate in terms of bacterial spectrum covered and statistical non-inferiority to active comparators. This may also be viewed as the most expensive regulatory path but the one with the highest market potential. On the other extreme, drugs against rare lethal pathogens for which human testing is not appropriate (example offered is anthrax or plague) should be developed under Tier D which follows the 'animal rule'.

---

[6]EMA Committee for Human Medicinal Products 2013, EMA/CHMP/351889/2013; http://www.fda.gov/ucm/groups/fdagov-public/@fdagov-drugs gen/documents/document/ucm359184.pdf.

The drug may obtain a conditional registration based on phase I safety and PK data in conjunction with efficacy data obtained from an animal model. Tier C is a novel developmental path whereby instead of two phase III trials, two or three smaller, prospective, randomized, open-label studies are performed, comparing the new treatment versus standard of care against a specific human pathogen or a specific drug-resistant human pathogen within and outside traditional indications or body sites. Underpowered for statistical non-inferiority testing, the development hinges on case quality (unambiguous cases of serious infection where spontaneous improvement is unlikely) or, possibly, historical controls. Additional observational data from patients receiving inappropriate treatment due to MDR may complement the picture. As such, Tier C is tailored for narrow-spectrum, innovative niche treatments typically sponsored by smaller private or public enterprises striving to gain market access as soon as possible in order to financially sustain the further clinical development. Tier C is also very much in line with the former IDSA postulate of a Limited Population Antibacterial Drug (LPAD) course of development (IDSA 2012). Tier B is a blend of Tiers A and C in that both one statistically powered phase III trial and two or three smaller, open-label prospective studies are combined in a data package with a view to accelerate the development and reaching the marketplace at lower cost. Patients enrolled in clinical studies following Tier B or C are expected to donate blood samples for analysis of systemic drug levels and strengthening the PK/PD model developed on the basis of preclinical and phase I data. Due to the less-than-standard safety database by the time of the new drug application (NDA), drugs registered under Tiers B and C would be second- or third-line treatments to be used only if other alternatives are known or suspected to be less suitable.

The overall clinical-stage pipeline of antibiotics has been elegantly reviewed by key opinion leaders in the field. Pucci and Bush have organized their review around compound classes and mechanism of action of drug candidates to precisely delineate the pros and cons of compounds, in the context of their peers within class, with a view to prospectively position these in the marketplace (Pucci and Bush 2013). The review by Page and Bush is based on a similar concept (Page and Bush 2014). Herein, in contrast, developmental candidates are reviewed by indication for which they are developed with a view to identify areas of unmet medical need. In consequence, individual compounds may appear more than once but with a precise developmental status for the respective medical indication.

5 New Drug Candidates for Key Indications: Respiratory Tract Infections (RTIs)

RTIs are a broad and diverse spectrum of diseases that may be caused by bacterial or viral pathogens. A key discriminator for bacterial pneumonia is whether the disease was contracted in the community (CAP) or nosocomial setting (HAP). A particular form of HAP is pneumonia acquired in the context of use of invasive ventilators in the intensive care unit (VAP). Due to the significant morbidity and mortality associated with HAP/VAP, the significant economical burden of the disease, and the frequently polymicrobial nature, HAP/VAP has become a high-priority, high value indication for the biopharmaceutical industry. It is estimated that in Germany alone, 40,000 cases of HAP are diagnosed per year of which approximately 15,000 affect patients in the Intensive Care Unit (ICU) or require treatment in the ICU. The overall mortality has been estimated with 16 % for ICU patients with HAP/VAP.[7] At the same time, the pipeline of novel antibiotics for the HAP/VAP indication is considered to be critically thin (Boucher et al. 2013).

The pathogen most frequently encountered in the community setting is *Streptococcus pneumoniae*, followed, significantly less frequently by *Haemophilus influenzae*, *M. tuberculosis*, and Enterobacteriaceae, occasionally or sporadically by *Legionella* spp., *S. aureus* or *Chlamydia pneumoniae*. HAP/VAP patients show a different spectrum of pathogens including *S. aureus*, *P. aeruginosa*, and Enterobacteriaceae including *Enterobacter* spp. (see also Table 2). Both indications suffer from approximately one-quarter of infections remaining without unambiguous pathogen identification.

According to German treatment guideline, the first-line treatment for CAP in the ambulatory setting is an aminopenicillin. Second-line treatments may be selected from the groups of macrolide or tetracycline antibiotics, while fluoroquinolones are not recommended.[8] Patients with risk factors may be treated by a combination of a β-lactam with a BLI. Treatment is for 5–7 days. CAP in hospitalized patients may be treated with a combination of a β-lactam and a macrolide antibiotic. Alternatively, here a fluoroquinolone may be used instead. For CAP patients in ICUs, the same combination and fallback treatment may be used unless there is an indication or diagnosis of *P. aeruginosa* infection, which, according to the guideline, would prompt the coadministration of a *Pseudomonas* active β-lactam with a *Pseudomonas* active fluoroquinolone. Treatment courses of 8–10 days, with *Pseudomonas* up to 15 days, are typical (ibid.). The current German treatment guideline for HAP[7] recommends treatment onset to be as rapid as possible, even before taking biosamples for diagnosis. Where there is no suspicion for infection by MDR pathogens, a third-generation cephalosporin, a β-lactam–BLI combination, ertapenem, or a fluoroquinolone may be used. For HAP cases suspected to be caused by MDR pathogens, piperacillin–tazobactam is recommended as the first-line treatment. Alternatively, *Pseudomonas* active carbapenem or cephalosporin antibiotics may be used in combination with an aminoglycoside or *Pseudomonas* active fluoroquinolone. Where there is a further suspicion of a

---

[7]Dalhoff K, Abele-Horn M, Andreas S, Bauer T, von Baum H, Deja M, Ewig S, Gastmeier P, Gatermann S, Gerlach H, Grabein B, Höffken G, Kern WV, Kramme E, Lange C, Lorenz J, Mayer K, Nachtigall I, Pletz M, Rohde G, Rosseau S, Schaaf B, Schaumann R, Schreiter D, Schütte H, Seifert H, Sitter H, Spies C, Welte T (2012). http://www.awmf.org/leitlinien/detail/ll/020-013.html.

[8]Höffken G, Lorenz J, Kern WV, Welte T, Bauer T, Dalhoff K, Ewig S, Grabein B, Halle E, Kolditz M, Marre R, Sitter H (2005) http://www.awmf.org/leitlinien/detail/ll/082-001.html.

contribution by MRSA, an MRSA active agent may be coadministered. The overall treatment course is 8 days with de-escalation from broad-spectrum combination treatment after 3 days, if indicated.

Infections in the hospital setting of comorbid pneumonia patients by MDR gram-negative bacilli may require resorting to drugs or drug combinations previously discontinued due to inconsistent efficacy or side effects including tigecycline, polymyxin, and fosfomycin, thus further pinpointing to the problem of lack of innovation. Table 3 provides a summary of treatments in late-stage clinical development for either one of the pneumonia indications. As pointed out before, HAP/VAP is key indication to the pharmaceutical industry with major pharma houses backing the indication in early- and late-stage trials. Other than POL7080, which is a narrow-spectrum drug candidate with a novel mechanism of action, two monoclonal antibodies are being trialed for the prevention of HAP/VAP by *P. aeruginosa* (MEDI3902, phase I) or *S. aureus* (MEDI4893, phase II). The prevention of bacterial infection product positioning and trial principle is innovative, with a clinical end point of incidence of *S. aureus* infection within 31 days of onset of mechanical ventilation. Where standard interventional phase II trials enroll 60–80 patients, MedImmune expects to enroll 462 patients in the study to cope with the unfavorable statistics of the trial design.

6 New Drug Candidates for Key Indications: Intra-Abdominal Infections (IAIs)

There is no dedicated German treatment guideline for IAIs but coverage of the topic as part of the sepsis guideline.[9] Accordingly, the first-line treatment may be selected from the group of gram-negative active, in particular *Pseudomonas* active β-lactams and β-lactam-BLI combinations, in particular piperacillin-tazobactam, 3rd or 4th generation cephalosporins like ceftazidime or cefepime or carbapenems like imipenem or meropenem. Suspicion of a gram-positive contribution, e.g., by *S. aureus* or MRSA, may lead to treatment by a combination of the above with linezolid or daptomycin. Ceftazidime should be combined with a gram-positive agent. Among the investigational agents (Table 4), ceftazidime–avibactam (CAZ-AVI) is in advanced registration (it received licensure in the USA in February of 2015 for cIAI and cUTI). CAZ-AVI was combined with metronidazole versus standard-of-care meropenem for coverage of the anaerobic pathogen spectrum. Merck & Co's BLI MK7655 is being trialed in phase II in combination with imipenem and cilastatin. The tetracycline-based eravacycline is being trialed in phase III (status ongoing but

[9] Reinhart K, Brunkhorst FM, Bone HG, Bardutzky J, Dempfle CE, Mannheim, Forst H, Gastmeier P, Gerlach H, Gründling M, John S, Kern W, Kreymann G, Krüger G, Kujath P, Marggraf G, Martin J, Mayer K, Meier-Hellmann A, Oppert M, Putensen C, Quintel M, Ragaller M, Rossaint R, Seifert H, Spies C, Stüber F, Weiler N, Weimann A, Werdan K, Welte T (2010) http://www.awmf.org/leitlinien/detail/ll/079-001.html.

Table 4 Clinical-stage pipeline for antibacterial agents targeting IAIs

Drug candidate	Sponsor	Indication	Latest stage of development	Pathogen coverage	Source
Ceftazidime–avibactam	AstraZeneca/Actavis	cIAI	Phase III complete (combination with metronidazole vs. meropenem)	Broad spectrum incl MDR gram negatives	NCT01726023; NCT01499290; NCT01500239
Eravacycline	Tetraphase Pharmaceuticals	cIAI	Phase III ongoing, not recruiting; versus ertapenem	Broad spectrum incl MDR gram negatives	http://tphase.com/pipeline/overview; NCT01844856
MK7655	Merck & Co	cIAI	Phase II ongoing (combination with imipenem/cilastin vs. imipenem/cilastatin alone)	Gram negatives	NCT01506271

not recruiting) versus ertapenem. Further regarding CAZ-AVI, this investigational agent will have completed a total of seven phase III trials (indications cIAI, cUTI, HAP/VAP) by the time of its registration with an impressive total of more than 4600 patients enrolled in phase III studies.

7 New Drug Candidates for Key Indications: Urinary Tract Infections (UTIs)

UTIs are another indication with bias for gram-negative pathogens. Uncomplicated and complicated UTIs may involve upper urinary tract infections, e.g., acute pyelonephritis. Ambulatory UTIs are usually considered uncomplicated (uUTI), while complicated UTIs (cUTI) are those of hospitalized patients with, e.g., catheters and exposure to MDR pathogens. Three-quarters of uncomplicated UTIs are caused by *Escherichia coli*, followed by *Staphylococcus saprophyticus*, *K. pneumoniae*, and *Proteus mirabilis*. *Enterococcus* spp. may also be encountered (see also Table 2). The German treatment guideline recommends use of fosfomycin or nitrofurantoin as first-line treatments of uncomplicated lower urinary tract infections.[10] Fluoroquinolones (ciprofloxacin, levofloxacin) may be used as a first-line option for uncomplicated pyelonephritis. More severe cases may be treated with cephalosporins third generation, alternatively β-lactam–BLI combinations, carbapenems, or aminoglycosides. In the absence of a German treatment guideline for cUTIs, the Dutch guideline recommends i.v. therapy with amoxicillin with or without clavulanic acid, alternatively a second- or third-generation cephalosporin. Both first-line drug recommendations are to be combined with gentamycin (Koningstein et al. 2014). With a few exceptions, sponsors are trialing their investigational drugs in parallel in cIAI and cUTI patients, given the overlapping bacterial pathogen spectrum. With CAZ-AVI, eravacycline and Carbavance (a meropenem–RPX7009 combination), three product candidates have advanced to phase III trials (Table 5). Finafloxacin, a novel fluoroquinolone with FDA approval for otitis externa infection, did complete phase II trials in cUTI and pyelonephritis and may progress further. S-649266, a novel siderophore containing cephalosporin for parenteral treatment by Shionogi, is recruiting for phase II studies versus comparator imipenem–cilastatin, and further phase III studies are scheduled for 2015.[11]

---

[10]Naber KG, Vahlensieck W, Wagenlehner FME, Hummers-Pradier E, Schmiemann G, Hoyme U, Watermann D, Kaase M, Kniehl E, Fünfstück R, Sester U (2010) http://www.awmf.org/leitlinien/detail/ll/043-044.html.

[11]http://www.shionogi.com/newsroom/article.html#122491.

Table 5 Clinical-stage pipeline for antibacterial agents targeting UTIs

Drug candidate	Sponsor	Indication	Latest stage of development	Pathogen coverage	Source
Ceftazidime–avibactam	AstraZeneca/Actavis	cUTI	Phase III ongoing (vs. doripenem)	Broad spectrum incl MDR gram negatives	NCT01595438; NCT01599806
Eravacycline	Tetraphase Pharmaceuticals	cUTI	Phase III (i.v. and oral)	Broad spectrum incl MDR gram negatives	http://tphase.com/pipeline/overview
Carbavance (meropenem–RPX7009) QIDP	Rempex Pharmaceuticals (The Medicines Company)	cUTI and pyelonephritis	Phase III recruiting	Gram-negative pathogens incl CRE	NCT02166476; NCT02168946
Finafloxacin	MerLion Pharmaceuticals	cUTI and pyelonephritis	Phase II complete		NCT01928433; http://www.merlionpharma.com/?q=node/228
Plazomicin	Achaogen	cUTI and pyelonephritis	Phase II complete	MDR gram negatives	NCT01096849
Ceftaroline–avibactam	AstraZeneca (Cerexa/Forest Labs)	cUTI	Phase II complete		NCT01281462
MK7655	Merck & Co	cUTI	Phase II ongoing (in combination with imipenem/cilastatin vs. imipenem/cilastatin alone)		NCT01505634
S649266	Shionogi (GSK)	cUTI	Phase II ongoing (vs. imipenem/cilastin)	Gram-negative pathogens	NCT02321800

8 New Drug Candidates for Key Indications: Acute Bacterial Skin and Skin Structure Infections (ABSSSIs)

Acute bacterial skin and skin structure infections, short ABSSSI, are very frequent both in the ambulatory setting and in the hospital setting. Due to the availability of patients to be enrolled in clinical trials, rather straightforward diagnosis and the rarely life-threatening condition, ABSSSI trials and product positioning are widely pursued for investigational agents with restriction or bias for gram-positive bacteria, in particular *S. aureus* ('entry indication'). Unfortunately, also broad-spectrum agents are undergoing development in this indication (e.g., fluoroquinolones), and overuse or misuse of such agents for non-life-threatening infections may inappropriately foster resistance development without added benefit. The current standard of care according to German treatment guideline[12] for systemic therapy of uncomplicated ABSSSI are cephalosporins of the first generation (cephalexin, cefazolin), alternatively clindamycin or macrolide antibiotics. Deeper but uncomplicated infections may also be treated with flucloxacillin or oxacillin. Topical treatment of superficial infections may be accomplished with fusidic acid (not registered in Germany), mupirocin, retapamulin, bacitracin, or others. Severe or complicated ABSSSIs are recommended to be treated with an acylaminopenicillin–BLI combination (e.g., amoxicillin–clavulanic acid), a carbapenem with or without clindamycin or where the infection is suspected to be caused by MRSA, cotrimoxazol, linezolid, vancomycin, daptomycin, or few others. Burn wounds or diabetic foot ulcers may also be infected with *P. aeruginosa*, thus requiring a combination treatment with a Pseudomonas active third-generation cephalosporin or carbapenem. Table 6 lists the investigational drugs for treatment of ABSSSI. Accordingly, the majority of products are pursued by small- or medium-sized biopharmaceutical companies in phase II. Spellberg argues that a further population of the clinical pipeline with ABSSSI-directed treatments, including drug candidates that may in principle cater to serious broad-spectrum infections, may neither meet unmet medical need nor meet significant market shares (Spellberg 2014).

9 New Drug Candidates for Other Infections and Single-Pathogen Indications

While the majority of investigational agents are developed within above traditional indications, few agents are developed outside. These are summarized in Table 7. Finafloxacin received licensure for the treatment of otitis externa by *P. aeruginosa* or *S. aureus* ('swimmer's ears') in the USA in February 2015. Cempra

[12]Schöfer H, Bruns R, Effendy I, Hartmann M, Jappe U, Plettenberg A, Reimann H, Seifert H, Shah P, Sunderkötter C, Weberschock T, Wichelhaus TA, Nast A (2011) http://www.awmf.org/leitlinien/detail/ll/013-038.html.

Table 6 Clinical-stage pipeline for antibacterial agents targeting ABSSSIs

Drug candidate	Sponsor	Indication	Latest stage of development	Pathogen coverage	Source
Delafloxacin	Melinta Therapeutics	ABSSSI	Phase III recruiting	Broad spectrum	NCT01984684
Ozenoxacin	Ferrer Internacional SA	Impetigo	Phase III recruiting (topical formulation)	Gram positive	NCT02090764
Omadacycline (QIDP)	Paratek Pharmaceuticals	ABSSSI	Phase III opened, not yet recruiting	Broad spectrum incl MDR	NCT02378480
Pexiganan acetate	Dipexium Pharmaceuticals	Diabetic foot infection	Phase III		NCT01594762, NCT01590758
JNJ-Q2 (QIDP)	Furiex Pharmaceuticals	ABSSSI	Phase II complete	Broad spectrum incl MRSA and MDR *S. pneumoniae*	NCT01128530
Radezolid (QIDP)	Melinta Therapeutics	ABSSSI	Phase II complete	Gram positive + fastidious gram negatives	NCT00646958
Lefamulin (BC-3781) (QIDP)	Nabriva Therapeutics	ABSSSI	Phase II complete	Broad spectrum	NCT01119105
AFN1252 (QIDP)	Affinium Pharmaceuticals/Debiopharm	ABSSSI	Phase II complete	*Staphylococcus* spp.	NCT01519492
Fusidic acid (CEM-102)	Cempra Pharmaceuticals	ABSSSI	Phase II complete	Gram positives	NCT00948142
TD1792	Theravance Biopharma	ABSSSI	Phase II complete	Gram positives	NCT00442832
Nemonoxacin	TaiGen Biotechnology	Diabetic foot infections	Phase II (oral formulation)	MRSA, VRE, *A. baumannii*	NCT00685698
GSK-2140944 (topo II inhibitor)	GlaxoSmithKline	ABSSSI	Phase II recruiting	Gram positive	NCT02045797
Brilacidin	Cellceutix Corporation	Serious skin infections	Phase II ongoing, not recruiting		NCT02052388

(continued)

Table 6 (continued)

Drug candidate	Sponsor	Indication	Latest stage of development	Pathogen coverage	Source
MRX-1	MicuRx Pharmaceuticals	ABSSSI	Phase II opened, not yet recruiting		NCT02269319
DEBIO-1452 (1450)	Debiopharm	ABSSSI	Phase II	*Staphylococcus* spp.	
CG-400549	CrystalGenomics		Phase II		
BC-7013	Nabriva Therapeutics	ABSSSI, eye and wound infections	Phase I complete	Broad spectrum	http://www.nabriva.com/programs/pipeline/

Table 7 Clinical-stage pipeline for antibacterial agents targeting other infections

Drug candidate	Sponsor	Indication	Latest stage of development	Pathogen coverage	Source
Finafloxacin	MerLion Pharmaceuticals	Otitis externa	Phase III complete, FDA approval	*Pseudomonas aeruginosa* and *Staphylococcus aureus*	
Ceftazidime–avibactam	AstraZeneca/Actavis	Other infections	Phase III complete	Ceftazidime-resistant gram-negative pathogens	NCT01644643
Carbavance (meropenem-RPX7009) QIDP	The Medicines Company (Rempex Pharmaceuticals)	Bacteremia	Phase III recruiting	Gram-negative pathogens incl CRE	NCT02168946
Fusidic acid (CEM-102)	Cempra Pharmaceuticals	Refractory bone and joint infections	Phase II terminated (oral formulation in combination with rifampicin)	Gram positives	NCT01756924
CG-400549	CrystalGenomics	Osteomyelitis	Phase II	*Staphylococcus aureus*	

Pharmaceuticals took fusidic acid into development for refractory bone and joint infections. A phase II trial (NCT01756924), trialing fusidic acid in combination with rifampicin, was terminated before completion, and alternative study designs are being considered. Rempex's/The Medicines Company's Carbavance, mentioned above, is specifically trialed in phase III in a variety of indications, including bacteremia specifically caused by CRE. The trial design compares Carbavance with best available therapy. For the same problem organisms, CRE, but using a different phase III superiority trial design with a primary end point of all-cause mortality at day 28, Achaogen is trialing plazomicin versus colistin, both in combination with either meropenem or tigecycline. Patients with CRE bacteremia or nosocomial pneumonia are being recruited into this trial. Pursuing the same or closely related unmet medical need of MDR gram-negative hospital infections, sponsors are recruiting into phase I trials for aztreonam–avibactam, BAL30072, and OP0595. BAL30072 and OP0595 are innovative agents that should be monitored closely throughout the clinical development. It will be interesting to see how updated regulatory policy will influence the clinical developmental path and individual trial design for these high-profile drug candidates. A group of drug candidates in Table 8 are developed for the single-pathogen indication uncomplicated urogenital gonorrhea. This group comprises the agents solithromycin, AZD0914, and GSK-2140944 and is tailored for coverage of MDR *Neisseria gonorrhoeae* which had been flagged as a serious, growing, and global issue (Unemo and Shafer 2014). Healthcare- and antibiotic use-associated diarrhea is frequently caused by *C. difficile* infection and intoxication, *Clostridium difficile*-associated diarrhea (CDAD). Treatments perturbing the intestinal microbiota, thus fostering *C. difficile* overgrowth, are cephalosporins of the third generation, proton pump inhibitors, histamine receptor antagonists, and others. The first-line treatments for *C. difficile* infection are oral vancomycin (which is non-absorbable and does not leave the intestinal compartment) and metronidazole with cure rates in the 60–80 % range but frequent recurrence (Bagdasarian et al. 2015). The recently registered *C. difficile* drug fidaxomicin (Merck & Co) will find application in recurrent *C. difficile* infection. There are at least 6 additional drug candidates in various stages of the clinical development (including surotomycin, cadazolid, LFF571, NVB302), including two open phase III trials. Both phase III products have secondary outcome measures of sustained cure, in case of surotomycin for 50 days past dosing, thus pointing to where the future value proposition of the investigational agents may be. Table 8 further lists drugs under development for pulmonary tuberculosis by MDR (delamanid, PA-824) or drug-susceptible *M. tuberculosis* (posizolid, sutezolid, SQ-109). While the pipeline has improved with the conditional registrations of delamanid and bedaquiline for MDR mycobacteria, and the therefore expanded options of combination trials with second-line drugs, the disease burden and resistance situation warrant further investment and clinical trial work, e.g., through private–public development partnerships like the TB Alliance.

Table 8 Clinical-stage pipeline for antibacterial agents targeting single-pathogen indications

Drug candidate	Sponsor	Indication	Latest stage of development	Pathogen coverage	Source
Plazomicin	Achaogen	Infections due to CRE including bloodstream infections	Phase III recruiting (open-label superiority study)	Carbapenem-resistant Enterobacteriaceae	NCT01970371
Carbavance (meropenem-RPX7009) QIDP	The Medicines Company (Rempex Pharmaceuticals)	Serious infections	Phase III recruiting	Gram-negative pathogens incl carbapenem-resistant Enterobacteriaceae	NCT02168946
BAL30072	Basilea Pharmaceutica	Hospital infections due to MDR gram negatives	Phase I ongoing	MDR *Acinetobacter baumannii* and *Pseudomonas aeruginosa*	http://www.basilea.com/Portfolio/BAL30072/
Aztreonam–avibactam	AstraZeneca/IMI ND4BB	Hospital infections due to MDR gram negatives	Phase I ongoing	Metallo-beta-lactamase producing Enterobacteriaceae	NCT01689207
OP0595/RG6080	Meiji/Fedora/Roche	Hospital infections due to MDR gram negatives	Phase I complete	Enterobacteriaceae	NCT02134834
ACHN-975 (LpxC inhibitor)	Achaogen	Tbd	Phase I complete	Gram negative	NCT01597947

(continued)

Table 8 (continued)

Drug candidate	Sponsor	Indication	Latest stage of development	Pathogen coverage	Source
Finafloxacin	MerLion Pharmaceuticals	H. pylori infections	Phase II complete (combination treatment with amoxicillin or esomeprazole)	*Helicobacter pylori*	NCT00723502
Solithromycin	Cempra Pharmaceuticals	Uncomplicated urogenital gonorrhea	Phase III recruiting	*Neisseria gonorrhoeae*	NCT02210325
AZD0914	NIAID/AstraZeneca	Gonorrhea	Phase II recruiting	*Neisseria gonorrhoeae*	NCT02257918
GSK-2140944 (topo II inhibitor)	GlaxoSmithKline	Uncomplicated urogenital gonorrhea	Phase II opened, not yet recruiting	*Neisseria gonorrhoeae*	NCT02294682
Surotomycin	Merck & Co	CDAD	Phase III recruiting	*Clostridium difficile*	NCT01598311
Actoxumab (MK3415)	Merck & Co	CDAD	Phase III	*Clostridium difficile*	NCT01513239, NCT01241552
Bezlotoxumab (MK3415)	Merck & Co	CDAD	Phase III	*Clostridium difficile*	NCT01513239, NCT01241552
Cadazolid	Actelion	CDAD	Phase III recruiting	*Clostridium difficile*	NCT01987895; NCT01983683
LFF571	Novartis	CDAD	Phase II complete	*Clostridium difficile*	NCT01232595
SMT19969	Summit	CDAD	Phase II	*Clostridium difficile*	
NVB302	Novacta Biosystems Ltd	CDAD	Phase I complete	*Clostridium difficile*	http://www.novactabio.com/news.php
Delamanid (OPC-67683)	Otsuka Pharmaceutical Co	Pulmonary tuberculosis	Registered and phase III	MDR *Mycobacterium tuberculosis*	http://www.otsuka.com/en/rd/pharmaceuticals/pipeline/pdf.php?financial=344; NCT01424670

(continued)

Table 8 (continued)

Drug candidate	Sponsor	Indication	Latest stage of development	Pathogen coverage	Source
Bedaquiline (TMC207)	Global Alliance for TB Drug Development; Janssen	Pulmonary tuberculosis	Registered and phase III (combination treatment with PA-824 and linezolid)	MDR *Mycobacterium tuberculosis*	NCT02333799
Pretomanid (PA-824)	Global Alliance for TB Drug Development; Johns Hopkins University	Pulmonary tuberculosis	Phase III ongoing for combination treatments; phase II for stand alone	MDR *Mycobacterium tuberculosis*	NCT02333799; NCT02342886; NCT02256696
Posizolid (AZD5847)	NIAID/AstraZeneca	Smear-positive pulmonary tuberculosis	Phase II complete	*Mycobacterium tuberculosis*	NCT01516203
Sutezolid (PNU-100480)	Sequella Inc	Pulmonary tuberculosis	Phase II complete	*Mycobacterium tuberculosis*	NCT01225640
SQ-109	M Hoelscher/ Sequella Inc	Pulmonary tuberculosis	Phase II complete, phase III in preparation	*Mycobacterium tuberculosis*	NCT01218217
Carbavance (meropenem-RPX7009) QIDP	The Medicines Company/BARDA	Melioidosis and glanders	N/A	*Burkholderia* spp.	http://www.hhs.gov/news/press/2014pres/02/20140205b.html

10 Conclusions and Outlook

I herein reviewed the antibiotic clinical pipeline by antibacterial indication. While the number of clinical trials may impress, many of these trials are for the same compound, being explored in multiple indications, or within established antibiotic drug classes, e.g., fluoroquinolones. The pipeline shows a bias toward community-acquired infections like CAP and ABSSSI with a predominant gram-positive spectrum. As noted by others (Spellberg 2014), not all these compounds will compete in the marketplace with current first-line or generic treatments, and not all will address the unmet medical need. There are significantly fewer options for the serious hospital-acquired infections by gram-negative bacteria as encountered in cIAI, cUTI, and HAP/VAP indications. Here, the focus remains on resistance-breaking β-lactam/BLI combinations in a head-to-head race with the rapidly emerging and globally spreading β-lactamases and carbapenemases. Three novel β-lactam/BLI class antibiotics are in the early clinical development: S-649266 (a siderophore containing cephalosporin), BAL30072 (a siderophore containing mono-sulfactam), and OP0595, a novel BLI. AIC-499, another novel β-lactam antibiotic, is completing preclinical studies. These compounds were carefully selected to address the MDR gram-negative pathogen space (e.g., Higgins et al. 2012). Significant innovations may also come from the further analoging of the aminoglycoside and tetracycline scaffolds. With POL7080, Polyphor/Roche is positioning a drug candidate for treatment of *P. aeruginosa* infections in HAP/VAP and bronchiectasis. This is an interesting step toward the development of a narrow-spectrum drug for a key pathogen associated with significant morbidity and mortality that will also prompt the further development of real-time diagnostic capabilities. Preventative or adjunctive therapy is being explored for *P. aeruginosa* (MEDI3902, with funding from the newly launched IMI-MAGNET program) and *S. aureus* (MEDI4893) with monoclonal antibodies and an elaborate clinical trial design. There is a short list of a few decolonizing agents other than antibodies for *S. aureus* that I have not reviewed herein but that may help reduce the burden of hospital-acquired bacteremias.

There is meanwhile a broad consensus in the field that the threat by MDR or XDR gram-negative Enterobacteriaceae and non-fermenters is the biggest these days, forcing physicians to resort to drugs previously abandoned for either poor efficacy or side effects. This pressing unmet medical need unfortunately meets the space with the highest technical challenges in the discovery and development of novel antibiotics. Pull, and in particular push, incentives from governmental side are needed to foster and fund product development partnerships between academia, small- and medium-sized enterprises, and pharmaceutical players. Among the pull incentives, new business models for the stockpiling of novel resistance-breaking antibiotics are needed that uncouple revenues from clinical use. More flexible and realistic regulatory guidelines for the development of antibiotics targeting MDR/XDR bacteria may exert additional pull forces and facilitate entry into clinical trials.

Stating the obvious, the discovery and development of novel, differentiated antibiotics, no matter if targeting gram-positive or gram-negative bacteria, requires a constant level of commitment from all sectors.

References

Ambrose PG, Bhavnani SM, Rubino CM, Louie A, Gumbo T, Forrest A, Drusano GL (2007) Pharmacokinetics-pharmacodynamics of antimicrobial therapy: it's not just for mice anymore. Clin Infect Dis 44:79–86

Ambrose PG, Bhavnani SM, Dudley MN, Ellis-Grosse E, Rubino CM, Drusano GL (2012) New EMA guideline for antimicrobial development. Lancet Infect Dis 412:265–266

Andes D, Craig WA (2002) Animal models of pharmacokinetics and pharmacodynamics: a critical review. Int J Antimicrob Agents 19:261–268

Bagdasarian N, Rao K, Malani PN (2015) Diagnosis and treatment of *Clostridium difficile* in adults: a systematic review. JAMA 313:398–408

Bassetti M, Righi E (2013) Multidrug-resistant bacteria: what is the threat? Hematol Am Soc Hematol Educ Prog 2013:428–432

Bassetti M, Merelli M, Temperoni C, Astilean A (2013) New antibiotics for bad bugs: where are we? Ann Clin Microbiol Antimicrob 12:22

Bax R, Green S (2015) Antibiotics: the changing regulatory and pharmaceutical industry paradigm. J Antimicrob Chemother 70:1281–1284

Boucher HW, Talbot GH, Bradley JS, Edwards JE, Gilbert D, Rice LB, Scheld M, Spellberg B, Bartlett J (2009) Bad bugs, no drugs: no ESKAPE! an update from the Infectious Diseases Society of America. Clin Infect Dis 48:1–12

Boucher HW, Talbot GH, Benjamin DK Jr, Bradley J, Guidos RJ, Jones RN, Murray BE, Bonomo RA, Gilbert D (2013) Infectious Diseases Society of America. 10 × '20 progress development of new drugs active against gram-negative bacilli: an update from the Infectious Diseases Society of America. Clin Infect Dis 56:1685–1694

Bush K (2013) Proliferation and significance of clinically relevant β-lactamases. Ann N Y Acad Sci 1277:84–90

Calfee DP (2012) Crisis in hospital-acquired, healthcare-associated infections. Annu Rev Med 3:359–371

Crandon JL, Nicolau DP (2013) Human simulated studies of aztreonam and aztreonam-avibactam to evaluate activity against challenging gram-negative organisms, including metallo-β-lactamase producers. Antimicrob Agents Chemother 57:3299–3306

Drawz SM, Papp-Wallace KM, Bonomo RA (2014) New β-lactamase inhibitors: a therapeutic renaissance in an MDR world. Antimicrob Agents Chemother 58:1835–1846

Garnacho-Montero J, Escoresca-Ortega A, Fernández-Delgado E (2015) Antibiotic de-escalation in the ICU: how is it best done? Curr Opin Infect Dis 28:193–198

Hawkey PM (2015) Multidrug-resistant gram-negative bacteria: a product of globalization. J Hosp Infect 89:241–247

Higgins PG, Stefanik D, Page MG, Hackel M, Seifert H (2012) In vitro activity of the siderophore monosulfactam BAL30072 against meropenem-non-susceptible *Acinetobacter baumannii*. J Antimicrob Chemother 67:1167–1169

Huang SS, Septimus E, Avery TR, Lee GM, Hickok J, Weinstein RA, Moody J, Hayden MK, Perlin JB, Platt R, Ray GT (2014) Cost savings of universal decolonization to prevent intensive care unit infection: implications of the REDUCE MRSA trial. Infect Control Hosp Epidemiol 35(Suppl 3):S23–S31

Infectious Diseases Society of America (2012) White paper: recommendations on the conduct of superiority and organism-specific clinical trials of antibacterial agents for the treatment of infections caused by drug-resistant bacterial pathogens. Clin Infect Dis 55:1031–1046

Karaiskos I, Giamarellou H (2014) Multidrug-resistant and extensively drug-resistant gram-negative pathogens: current and emerging therapeutic approaches. Expert Opin Pharmacother 15:1351–1370

Keener AB (2014) First QIDP drug approved, but designation may fail urgent needs. Nat Med 20:690–691

Koningstein M, van der Bij AK, de Kraker ME, Monen JC, Muilwijk J, de Greeff SC, Geerlings SE, van Hall MA (2014) ISIS-AR Study Group. Recommendations for the empirical treatment of complicated urinary tract infections using surveillance data on antimicrobial resistance in the Netherlands. PLoS ONE 9:e86634

Kumar Kakkar A, Dahiya N (2014) The evolving drug development landscape: from blockbusters to niche busters in the orphan drug space. Drug Dev Res 75:231–234

MacVane SH, Crandon JL, Nichols WW, Nicolau DP (2014) In vivo efficacy of humanized exposures of Ceftazidime-Avibactam in comparison with Ceftazidime against contemporary Enterobacteriaceae isolates. Antimicrob Agents Chemother 58:6913–6919

McKellar MR, Fendrick AM (2014) Innovation of novel antibiotics: an economic perspective. Clin Infect Dis 59(Suppl 3):S104–S107

Nambiar S, Laessig K, Toerner J, Farley J, Cox E (2014) Antibacterial drug development: challenges, recent developments, and future considerations. Clin Pharmacol Ther 96:147–149

Nimmo GR (2012) USA300 abroad: global spread of a virulent strain of community-associated methicillin-resistant *Staphylococcus aureus*. Clin Microbiol Infect 18:725–734

Page MG, Bush K (2014) Discovery and development of new antibacterial agents targeting Gram-negative bacteria in the era of pandrug resistance: is the future promising? Curr Opin Pharmacol 18:91–97

Pucci MJ, Bush K (2013) Investigational antimicrobial agents of 2013. Clin Microbiol Rev 26:792–821

Rex JH, Eisenstein BI, Alder J, Goldberger M, Meyer R, Dane A, Friedland I, Knirsch C, Sanhai WR, Tomayko J, Lancaster C, Jackson J (2013) A comprehensive regulatory framework to address the unmet need for new antibacterial treatments. Lancet Infect Dis 13:269–275

Rice LB (2008) Federal funding for the study of antimicrobial resistance in nosocomial pathogens: no ESKAPE. J Infect Dis 197:1079–1081

Spellberg B (2011) The antibiotic crisis: can we reverse 65 years of failed stewardship? Arch Intern Med 171:1080–1081

Spellberg B (2014) The future of antibiotics. Crit Care 18:228

Spellberg B, Shlaes D (2014) Prioritized current unmet needs for antibacterial therapies. Clin Pharmacol Ther 96:151–153

Spellberg B, Sharma P, Rex JH (2012) The critical impact of time discounting on economic incentives to overcome the antibiotic market failure. Nat Rev Drug Discov 11:168

Srinivas N, Jetter P, Ueberbacher BJ, Werneburg M, Zerbe K, Steinmann J, Van der Meijden B, Bernardini F, Lederer A, Dias RL, Misson PE, Henze H, Zumbrunn J, Gombert FO, Obrecht D, Hunziker P, Schauer S, Ziegler U, Käch A, Eberl L, Riedel K, DeMarco SJ, Robinson JA (2010) Peptidomimetic antibiotics target outer-membrane biogenesis in *Pseudomonas aeruginosa*. Science 327:1010–1013

Tacconelli E, Cataldo MA, Dancer SJ, De Angelis G, Falcone M, Frank U, Kahlmeter G, Pan A, Petrosillo N, Rodríguez-Baño J, Singh N, Venditti M, Yokoe DS, Cookson B (2014) European Society of Clinical Microbiology. ESCMID guidelines for the management of the infection control measures to reduce transmission of multidrug-resistant Gram-negative bacteria in hospitalized patients. Clin Microbiol Infect 20, Suppl 1:1–55

Tomayko JF, Rex JH, Tenero DM, Goldberger M, Eisenstein BI (2014) The challenge of antimicrobial resistance: new regulatory tools to support product development. Clin Pharmacol Ther 96:166–168

Unemo M, Shafer WM (2014) Antimicrobial resistance in Neisseria gonorrhoeae in the 21st century: past, evolution, and future. Clin Microbiol Rev 27:587–613

Anti-infectives in Drug Delivery—Overcoming the Gram-Negative Bacterial Cell Envelope

Florian Graef, Sarah Gordon and Claus-Michael Lehr

Abstract Infectious diseases are becoming a major menace to the state of health worldwide, with difficulties in effective treatment especially of nosocomial infections caused by Gram-negative bacteria being increasingly reported. Inadequate permeation of anti-infectives into or across the Gram-negative bacterial cell envelope, due to its intrinsic barrier function as well as barrier enhancement mediated by resistance mechanisms, can be identified as one of the major reasons for insufficient therapeutic effects. Several in vitro, in silico, and in cellulo models are currently employed to increase the knowledge of anti-infective transport processes into or across the bacterial cell envelope; however, all such models exhibit drawbacks or have limitations with respect to the information they are able to provide. Thus, new approaches which allow for more comprehensive characterization of anti-infective permeation processes (and as such, would be usable as screening methods in early drug discovery and development) are desperately needed. Furthermore, delivery methods or technologies capable of enhancing anti-infective permeation into or across the bacterial cell envelope are required. In this respect, particle-based carrier systems have already been shown to provide the opportunity to overcome compound-related difficulties and allow for targeted delivery. In addition, formulations combining efflux pump inhibitors or antimicrobial peptides with anti-infectives show promise in the restoration of antibiotic activity in resistant bacterial strains. Despite considerable progress in this field however, the design of carriers to specifically enhance transport across the bacterial

---

F. Graef · S. Gordon · C.-M. Lehr
Department of Drug Delivery, Helmholtz Institute for Pharmaceutical Research Saarland (HIPS), Helmholtz Center for Infection Research (HZI), Saarbrücken, Germany
e-mail: Florian.Graef@helmholtz-hzi.de

S. Gordon
e-mail: Sarah.Gordon@helmholtz-hzi.de

C.-M. Lehr (✉)
Department of Pharmacy, Saarland University, Saarbrücken, Germany
e-mail: Claus-Michael.Lehr@helmholtz-hzi.de

Current Topics in Microbiology and Immunology (2016) 398: 475–496
DOI 10.1007/82_2016_491

Published Online: 24 February 2016

envelope or to target difficult-to-treat (e.g., intracellular) infections remains an urgently needed area of improvement. What follows is a summary and evaluation of the state of the art of both bacterial permeation models and advanced anti-infective formulation strategies, together with an outlook for future directions in these fields.

Contents

1 Introduction

The effective treatment of infectious diseases by means of anti-infective drug therapies is currently associated with a significant and increasing level of difficulty. The incidence of nosocomial infections caused in particular by pathogenic bacteria is an indicator of this problem. In Germany alone, 400,000–600,000 hospital-acquired, bacterial infections occur per year; 7500–15,000 of such cases are in fact lethal (Wissenschaften and Deutsche Akademie der Naturforscher 2013). These statistics are mainly due to the increasing incidence of bacterial resistance to drug therapy, leading to a lack of sufficiently active anti-infective treatment options. Gram-negative bacteria are particularly problematic in this respect: As an example, carbapenem-resistant Enterobacteriaceae (CRE, for abbreviations see Table 1) are capable of evading the action of almost all currently available antibiotics. This dire trend leads to the occurrence of nearly untreatable infections, with only two "last resort" antibiotics available (tigecycline and colistin)—neither of which are effective in every patient (McKenna 2013). We are therefore faced with a major global challenge with respect to the successful treatment of Gram-negative bacterial infections (Wellington et al. 2013).

While resistance to anti-infective drug therapies is without doubt the primary threat to effective infectious disease treatment, the evolution of resistance is compounded by a number of additional factors. Firstly, the successful delivery of anti-infectives to their site of action constitutes a challenging and complicated task,

Table 1 Abbreviations, in alphabetical order

AMPs	Antimicrobial peptides
CRE	Carbapenem-resistant Enterobacteriaceae
DUV	Deep ultraviolet
EPIs	Efflux pump inhibitors
IM	Inner membrane
LB	Langmuir–Blodgett
LS	Langmuir–Schaefer
LPS	Lipopolysaccharide
MD	Molecular dynamics
MIC	Minimal inhibitory concentration
MRSA	Methicillin-resistant *Staphylococcus aureus*
OD	Optical density
OM	Outer membrane
OMPs	Outer membrane proteins
PaßN	Phenylalanine arginyl ß-naphthylamine
PC	Phosphatidylcholine
PE	Phosphatidylethanolamine
PG	Phosphatidylglycerol
PL	Phospholipid
PS	Periplasmic space
QSARs	Quantitative structure–activity relationships
QSI	Quorum-sensing inhibitor
RND	Resistance–nodulation–cell division
SLBs	Supported lipid bilayers
SLNs	Solid lipid nanoparticles

even in the case of a wild-type bacterium. This is due to the fact that the bacterial cell envelope, especially that of Gram-negative bacteria, works intrinsically as a complex and significant biologic barrier to the effective delivery of anti-infective compounds and formulations (see Sect. 2.1, Nelson et al. 2009). The occurrence of several resistance mechanisms such as upregulation of efflux pump expression, downregulation or alteration of the expression of transport and channel-forming proteins (e.g., porins), and the production of enzymes (e.g., ß-lactamase) within this envelope structure therefore acts to compound an already existing problem for anti-infectives which must penetrate into or entirely through the bacterial envelope in order to reach their site of action (Dever and Dermody 1991). As a further factor for consideration, from the so-called golden age of antibiotic discovery—lasting from the 1950s to the 1960s (Fischbach and Walsh 2009)—until the introduction of the oxazolidinones in 2000, no new anti-infective class was able to successfully reach the market. This low flow within the antibiotic development pipeline continues today, meaning that the diminishing pool of effective therapies is not being replenished by newly emerging treatment options.

The above-described factors contributing toward the problematic nature of effective infection treatment can collectively be regarded as symptoms of a bacterial bioavailability problem. Such a bioavailability issue draws attention to two significant necessities in the area of anti-infective research.

The first is the desperate need for new models and strategies to better investigate and characterize the trafficking of anti-infectives into or across the bacterial cell envelope, in order to increase the collective knowledge of the envelope as a barrier which needs to be overcome. As a second need, novel anti-infective candidates with new modes of action are required, as are new delivery strategies which enable effective penetration into or across the Gram-negative bacterial cell envelope. The current document will attempt to address aspects of both research needs, outlining the state of the art in each area as well as potential or actual future research directions. Specific emphasis will continue to be given to Gram-negative bacteria as particularly problematic pathogens.

2 The Gram-Negative Bacterial Cell Envelope as a Bioavailability Barrier to Anti-infectives

As already mentioned, the intrinsic structure of the Gram-negative bacterial cell envelope presents a significant barrier to the successful delivery of anti-infectives. Therefore, a brief overview of the major structural components of the cell envelope, including details of envelope modifications responsible for the occurrence and evolution of resistance, is first given here.

2.1 The Intrinsic Bacterial Barrier

The Gram-negative bacterial cell envelope can be divided into three major parts, each of which constitutes a significant obstacle to anti-infective penetration (Fig. 1). Starting from the bacterial cytoplasm and proceeding outward, the inner membrane (IM) represents the first layer of the envelope barrier. It consists of a phospholipid (PL) bilayer mainly composed of phosphatidylethanolamine (PE), phosphatidylglycerol (PG), and cardiolipin, with incorporated transmembrane proteins and lipoproteins. The periplasmic space (PS) with the peptidoglycan cell wall constitutes the second layer. Peptidoglycan, a polymer of repeating disaccharides, is responsible for the maintenance of cell shape and structure. The surrounding area is a highly viscous, aqueous compartment, densely packed with proteins (Mullineaux et al. 2006). Furthermore, defense mechanisms including enzymes (e.g., ß-lactamase) are also located within this space. The outer membrane (OM) forms the third envelope substructure. The membrane itself is asymmetric in nature, being subdivided into a PL (mainly PE) containing inner leaflet, and an outer leaflet mainly consisting of

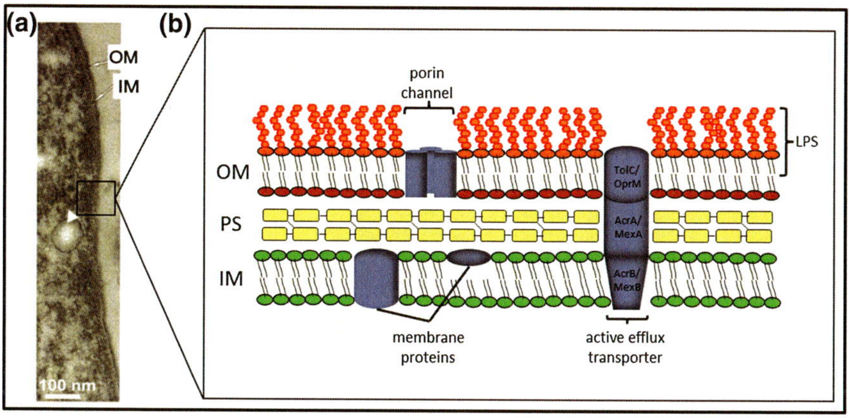

Fig. 1 Transmission electron microscopy image of the cell envelope of *Legionella dumoffii* (**a**, adapted from Palusinska-Szysz et al. 2012) and schematic overview of the Gram-negative bacterial cell envelope, highlighting the most important structural components; **b** the inner membrane (*IM*) incorporating transmembrane and lipoproteins, the periplasmic space (*PS*) housing the peptidoglycan cell wall, and the asymmetric outer membrane (*OM*) with its lipopolysaccharide (*LPS*) outer leaflet and porins. The general structure of an efflux transporter is also shown

lipopolysaccharide (LPS). LPS in turn is composed of the so-called lipid A, a phosphorylated glucosamine with six to seven acyl chains which anchors LPS to the inner leaflet of the OM; a core oligosaccharide; and the outermost portion of the molecule, the O-antigen. LPS acyl chains are mainly saturated, which confers a gel-like structure on the molecule. Association of gel-like LPS molecules within the outer leaflet is additionally strengthened by the presence of divalent cations being present in the surrounding medium, which neutralize the negative charge of LPS phosphate groups. This further contributes to the formation of a viscous structure which limits the permeation of hydrophobic compounds, including many anti-infectives and detergents. The OM also incorporates outer membrane proteins (OMPs) such as the porins (e.g., OmpF), which span the entire OM. Porins allow for and control the passive diffusion of hydrophilic compounds, for example ß-lactam antibiotics, with a size limit for such permeation of approximately 700 dalton (Silhavy et al. 2010). The combination of LPS and porins is therefore responsible for the strong permeability-limiting properties of the OM, which acts to restrict the permeation of hydrophobic as well as hydrophilic compounds.

In addition, efflux transporters, most commonly belonging to the resistance–nodulation–cell division (RND) superfamily, feature prominently within the cell envelope. Substructures of these transporters are present in each of the three major envelope subsections (e.g., AcrB-AcrA-TolC and MexB-MexA-OprM where subunits are located in the IM-PS-OM), meaning that the pump as a whole spans the entire envelope structure. Such efflux pumps are responsible for the active excretion of compounds (e.g., anti-infectives) in an energy-dependent manner (Kumar and Schweizer 2005).

2.2 The Role of the Envelope in Mediating Resistance Mechanisms

In principle, we can differentiate between three major, antimicrobial resistance strategies of bacteria: (i) degradation of anti-infective compounds by bacterial enzymes, (ii) protection of anti-infective targets by, e.g., structural or expression modification, and, of most relevance to the current document, (iii) alteration of the cell envelope barrier function (Davin-Regli et al. 2008), which will here be further described. Modifications to barrier properties result in an increased efflux in combination with a reduced uptake of anti-infectives, leading to inadequate intracellular anti-infective levels. The increased efflux of anti-infectives occurs due to an overexpression of efflux pumps (as mentioned above), which have a broad range of action and, as such, are able to mediate resistance to a variety of anti-infective classes (Tenover 2006). Resistance in the context of reduced uptake arises due to bacterial modification of OMP copy numbers or conformation, and/or alterations in LPS structure. The expression of OMPs, in particular porins, can be downregulated within the OM structure, or can alternatively be completely abrogated (Nikaido and Rosenberg 1981). The latter case is, for example, known from *Escherichia coli* isolates, which are resistant against cefoxitin due to the absence of the major OmpF porin channel (Tenover 2006). Furthermore, bacteria can modify the structure of their porins as a strategy to prevent anti-infective entry. Such structural modification can, for example, consist of a narrowing of the porin channel, which decreases the permeation of larger, hydrophilic compounds (De et al. 2001). The structure of LPS molecules can additionally be altered, in order to facilitate an increase in the barrier properties of the OM. The most effective mechanism by which LPS alteration leads to increased barrier function is via a reduction of negative net charge, leading to a reduced permeation of cationic anti-infectives (Kumar and Schweizer 2005).

2.3 Implications for Anti-infective Drug Delivery

Clearly, the unique structure of the Gram-negative bacterial cell envelope, together with the ability of bacteria to alter the structure and resulting functional activity of various envelope components, creates a considerable hurdle to the cellular permeation of anti-infectives. The development and application of models in order to facilitate an increased understanding of envelope permeation processes as well as the investigation of new anti-infective delivery approaches are therefore introduced and discussed below, as two research strategies required in order to address the issue of inadequate anti-infective permeation.

3 Strategies to Combat Intrinsic Difficulties/Bacterial Resistance Mechanisms Related to Anti-infective Transport

3.1 Models for Characterization of Drug Transport Across the Bacterial Cell Envelope

As detailed above, the Gram-negative bacterial cell envelope works as an effective biologic barrier to the successful delivery of anti-infectives to their target site. The fundamental existing barrier properties of the envelope are also able to be further increased through the upregulation of resistance mechanisms. Therefore, in addition to well-established and commonly used efficacy testing approaches, it is of considerable interest to obtain a greater and more detailed level of knowledge regarding rate, extent, and mechanisms of the processes by which anti-infectives permeate (actively or passively) across the envelope. Models which mimic the cell envelope and so enable provision of such information can thus help to facilitate the rational design of anti-infective agents, capable of overcoming intrinsic delivery difficulties/bacterial resistance mechanisms. Such models could additionally contribute useful information to early anti-infective drug discovery processes. The currently existing and employed models of the envelope structure, used in order to provide permeation and transport information, will be described in the following section. The needs which are unmet by these existing models will also be mentioned.

3.1.1 Electrophysiological Studies

Electrophysiological studies are applied to obtain information about the transport of anti-infectives through single porins. The principle of electrophysiology is based on the reconstitution of such channel-forming proteins—mostly OmpF, as the main porin responsible for the passive OM permeation of many anti-infectives such as the ß-lactams and quinolones—into planar lipid bilayers (Fig. 2a). Such bilayers mostly consist of phosphatidylcholine (PC) and are made, for example, by bursting porin-containing proteoliposomes across an aperture within a solid support (Kreir et al. 2008). An external voltage is then applied across the aperture-spanning membrane, which causes an ion flux through the inserted porin channel. The strength of the resulting current allows for the provision of information regarding the channel structure and its functional properties in a variety of experimental settings (e.g., ranges of salt concentration, pH). The technique is additionally able to be automated (Mach et al. 2008a) and can be further optimized, for example, by applying the porin-containing supported lipid bilayer system into glass nanopipettes (Gornall et al. 2011). In addition to providing information on porin structure and function, anti-infective passage kinetics through the bilayer-reconstituted porins can be studied by the use of high-resolution ion-current fluctuation analysis (Pages et al. 2008).

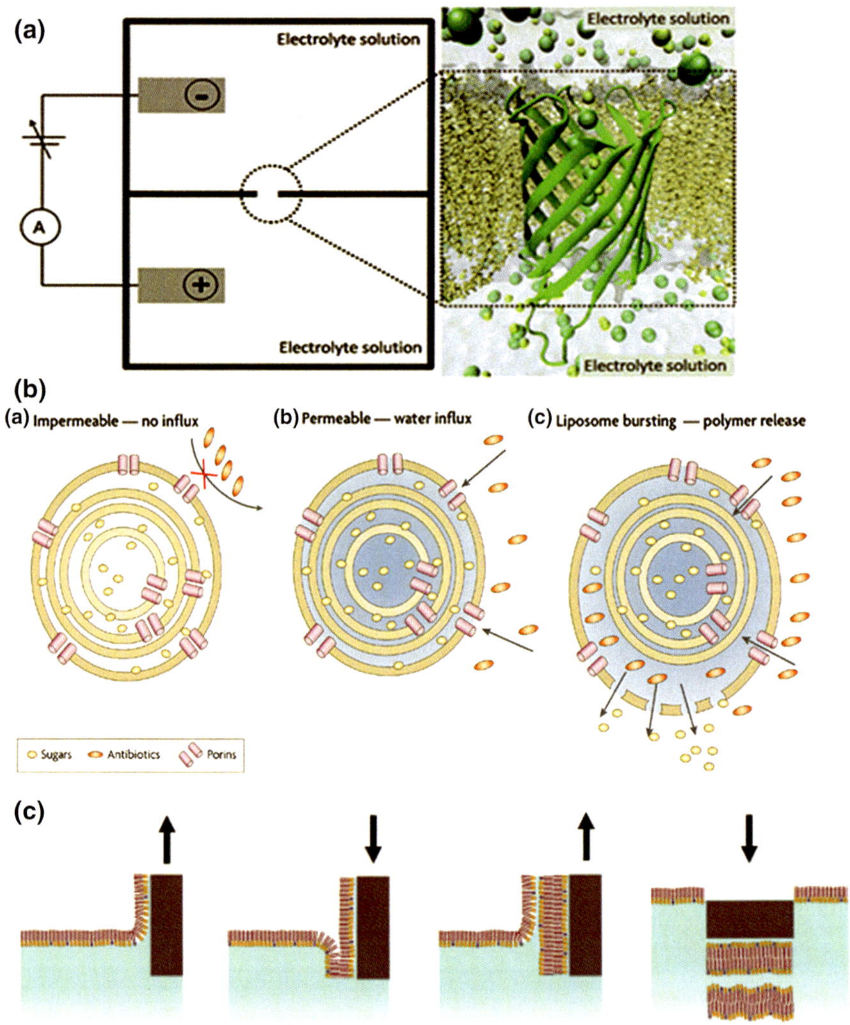

Fig. 2 Schematic overview of in vitro approaches to produce bacterial membrane models, for the characterization of drug transport. **a** The experimental setup for electrophysiological studies to investigate anti-infective transport through porins, incorporated in planar lipid bilayers (adapted from Modi et al. 2012 with the permission from The Royal Society of Chemistry); **b** the principle of a liposome swelling assay, employed to assess permeabilization processes mediated by anti-infective compounds (reprinted from Pages et al. 2008 with the permission from Macmillan Publishers Ltd: (Nature Reviews Microbiology) Pages et al. 2008, copyright 2008); **c** the Langmuir-based preparation procedure for preparing floating lipid bilayers (from Fragneto et al. 2012, with the kind permission from Springer Science+Business Media) is additionally depicted

In general, the permeation of anti-infective compounds through porins is detected by a decrease in current due to an occlusion of the porin channel by the permeating compound. Electrophysiological studies therefore facilitate the determination of the direct translocation of charged molecules through porin channels. They additionally allow for the evaluation of the interaction of anti-infectives with the constriction zone of porins (the narrowest part of the porin channel, which mediates the sizewise exclusion of molecule permeation across the OM) in particular. The relative affinity of different anti-infective compounds for specific porins can also be elucidated using electrophysiological studies—for example, enrofloxacin has been shown to have the strongest recorded affinity for OmpF. Combining the information obtained from electrophysiological studies with molecular dynamic (MD) simulations enables the identification of the specific anti-infective pathway across porin channels (Danelon et al. 2006; Mach et al. 2008b; Nestorovich et al. 2002). This further contributes to understanding the occurrence of porin structure-related resistance mechanisms. Again taking the case of enrofloxacin, a relatively simple modification to the OmpF channel (a single-point mutation in the constriction zone) has been shown by such a combination electrophysiology–MD approach to lead to a drastic decrease in anti-infective translocation (Mahendran et al. 2010).

3.1.2 Liposome-Based Assays

Assays utilizing liposomes as artificial membrane models are also employed to investigate permeabilization effects of anti-infectives, as well as the extent of anti-infective permeation into such model membranes. The existing liposome-based approaches can be basically differentiated into two major categories. On the one hand, the so-called liposome swelling assays or leakage studies must be mentioned. These approaches are based on uni- or multilamellar vesicles made of a single PL species (e.g., PC, PE) with or without incorporated porins (Nikaido and Rosenberg 1983), PL mixtures, or PL-LPS [rough (without O-antigen) or smooth LPS (with O-antigen)] mixtures, utilized in an attempt to mimic the OM components. Polymers or fluorescent dyes are further incorporated into the central aqueous compartment or within the bilayers of such liposomes. This provides for an indirect detection method for permeation of the analyzed anti-infective, by means of tracking changes in optical density (OD), or via fluorescence analysis. As an example of such a setup, the anti-infective of interest is mixed with polymer-containing liposome dispersions (which often also have inserted porins) under isosmotic conditions. If the anti-infective is not able to penetrate into the vesicles, the measured OD will remain unaltered. If, however, the anti-infective compound is able to permeate into the liposomes, a swelling of the vesicles occurs due to an influx of water, caused by the presence of an osmotic gradient as mediated by the permeating anti-infective compound. Anti-infective permeation and liposome swelling can ultimately result in bursting of the liposomes, leading to a release of the incorporated polymer, which is then detectable as a decrease in OD (Fig. 2b). Liposome swelling or leakage assays also facilitate the study of direct membrane disrupting effects of proteins (e.g., lamB or surfactant protein A) and antimicrobial peptides (AMPs, e.g., aurein 1.2) on artificial membrane systems (Fernandez et al. 2012, 2013; Kuzmenko et al. 2006; Luckey and Nikaido 1980). Of considerable current interest with respect to such assays is the use of liposomes which imitate more closely Gram-negative bacterial membrane compositions. In this respect, liposomes made of LPS or PL-LPS mixtures as a more realistic OM mimic are also often used for swelling or leakage studies, or to investigate PL-LPS interactions within the model membrane. Such studies could help to improve the understanding of the OM organization or modulation of the OM during the development of resistance mechanisms (D'errico et al. 2009; Kubiak et al. 2011).

On the other hand, liposomes can be used to study the accumulation and uptake of anti-infective compounds in liposomal membranes by the direct analysis of anti-infectives themselves. In this respect, anti-infectives of interest, which are either autofluorescent or fluorescently labeled, are incubated with liposomes. The relation between anti-infective structural characteristics/modifications and interaction with the artificial liposomal membrane can then be studied, as can anti-infective uptake into such model membranes. This is achieved by determining the accumulation of anti-infectives within the lipid membrane or their uptake into the vesicles, analyzed via nucleic magnetic resonance spectroscopy or fluorescence microscopy (Ries et al. 2015; Rodrigues et al. 2003).

3.1.3 Langmuir Trough-Based Approaches

Mono- and/or bilayers prepared from PL and/or LPS in various combinations are also employed as models in anti-infective research. Such approaches facilitate the investigation of the organization and interactions within artificial membrane systems and, with respect to permeation, study of the influence of antimicrobial proteins or peptides in particular on membrane integrity. In general, these experimental setups were originally developed to study the interactions within mammalian-mimicking membrane structures, or interaction of external entities with such structures. Imitation of the double bilayer nature of the Gram-negative bacterial envelope, as well as the OM, with particular emphasis being placed on its structural components and asymmetric nature, is, however, of considerable interest in the current application of such approaches.

Preparation of lipid monolayers on the surface of water or buffer is achieved by using a Langmuir film balance or trough, whereas lipid bilayers are mostly prepared as supported lipid bilayers (SLBs) on silicon—less often mica or gold—surfaces as solid supports. SLBs in turn can be prepared from lipid vesicles which are fused onto the solid support, via Langmuir–Blodgett (LB) or a combination of LB and Langmuir–Schaefer (LS) deposition techniques (Peetla et al. 2009). An advancement of SLBs resulting in the production of more bacteriomimetic models is represented by the additional incorporation of floating lipid bilayers. Production of such a model involves combining three vertical LB depositions with one horizontal LS deposition (Fig. 2c), resulting in the formation of a lipid bilayer which floats at a distance of 2–3 nm from the supported lipid bilayer (Charitat et al. 1999; Fragneto et al. 2012). Langmuir-derived lipid bilayers are often prepared from PC or PC-PG mixtures (in order to mimic the negative charge of bacterial membranes) and used to study the membrane insertion potential of AMPs, as well as their disordering and transmembrane pore-forming abilities. Results show a higher affinity and disruptive effect on models composed of negatively charged PLs (Fernandez et al. 2012, 2013). Langmuir-produced monolayers made of PG with or without incorporated OmpF have also been used to investigate the interaction of antibacterial proteins such as colicins with the OM, as well as their pathway across the OM (Clifton et al. 2012). As mentioned above, many such Langmuir-based models additionally take the impact of LPS as the major OM structural component into account. In one instance, stable Langmuir monolayers were prepared at the air–liquid interface using rough strain LPS. Such a model provided valuable information about LPS structure at the air–liquid interface and therefore constitutes a further step to a more accurate model of the OM (Le Brun et al. 2013). In a further approach, a realistic mimic of the OM structure was prepared by combining a PC bilayer deposited via LB on a solid support (representing the inner OM leaflet) with an overlying rough strain LPS bilayer (outer OM leaflet), introduced via LS deposition. This model successfully mimics the asymmetric nature of the OM and was first employed to describe the molecular mechanisms of the well-known OM destabilization effect occurring as a result of the removal of divalent cations from surrounding media (Clifton et al. 2015).

3.1.4 In Silico Methods

Besides the so-far described in vitro models, in silico approaches are also utilized to investigate the impact and interaction of various lipid species within simulated bilayers (which in turn may have a bearing on anti-infective permeability). They are furthermore applied to inform the development of membrane models which more closely and accurately reflect the structure and components of the IM and OM. In silico approaches may also be employed to determine the affinity and/or translocation of anti-infectives with or across porin channels, as alluded to previously (see Sect. 3.1.1) They may additionally be used to screen compounds for anti-infective activity based on quantitative structure–activity relationships (QSARs), defined via topological descriptors [numerical values correlating chemical properties with biologic activity (Mayers 2009)] and physicochemical parameters. In one example, MD simulations have been used to mimic the IM—consisting of PE and PG in a 3:1 ratio, closely reflecting the in cellulo composition—in order to evaluate the intra-bilayer PL interactions. Conducted simulations showed that interactions between these specific PLs are mainly based on H-bond formation and chiefly occur between PE and PG, less often between only PE molecules and almost never between PG molecules alone. As a consequence, the presence of PG within the membrane leads to a decrease in PE headgroup protrusion and a reduced motion along the artificial membrane; this results in an enhanced membrane stability, leading to a strengthening of the IM permeation barrier (Murzyn et al. 2005; Zhao et al. 2008). In silico studies which even more accurately reflect the PL composition of the IM have also been conducted. As a first step, bilayers consisting of CL alone were simulated, to determine its biophysical role within membranes via evaluation of its charge-dependent lipid packing (Lemmin et al. 2013). Further, IM models which additionally include heterogeneous lipids, exhibiting different acyl chain lengths and cyclopropane rings, can be considered as yet further improvements toward an accurate IM mimic (Pandit and Klauda 2012).

The OM has also been simulated in various in silico studies, starting with models consisting of LPS alone and followed by simulations using a combination of a PL inner leaflet and LPS outer leaflet to more accurately reflect the asymmetric OM structure. These models have largely been used to study properties such as interactions between LPS molecules in the OM, the stabilization effect of divalent cations on the membrane structure (and resulting barrier properties), the effect of electroporation on the barrier function of protein-free, asymmetric membrane structures, and the impact of OM enzymes as well as proteins on membrane integrity (Lins and Straatsma 2001; Piggot et al. 2011; Wu et al. 2014). Simulations of the OM as well as the IM have additionally been employed to study the interaction of AMPs with such artificial systems, highlighting the way in which AMPs are able to pass through and disrupt the bacterial envelope—firstly due to a self-promoted uptake across the OM and subsequently as a result of disruption of the IM via the formation of micelle-like aggregates (Berglund et al. 2015). MD simulations have further been used to determine the molecular and rate-limiting interactions occurring during anti-infective permeation through porins on an atomic scale. Such studies allow for a better understanding of the translocation pathway and estimated permeation time of anti-infective compounds, as well as the way in which modifications in the porin channel constriction zone can affect and reduce anti-infective permeation (Hajjar et al. 2010; Mach et al. 2008b; Mahendran et al. 2010; Singh et al. 2012). In silico screening has furthermore been employed to define QSARs of anti-infectives by evaluating the impact of physicochemical properties such as lipophilicity and molecular weight on anti-infective activity (Cronin et al. 2002; O'Shea and Moser 2008). The definition of topological descriptors together with the performance of linear discriminant analysis further enables the attainment of discriminant functions, which allow for the differentiation between active and non-active anti-infectives. Such functions can subsequently be applied to screen compound libraries for new lead structures showing promising anti-infective activity (Murcia-Soler et al. 2003, 2004).

3.1.5 In Cellulo Approaches

In cellulo approaches which give information about permeability processes by facilitating the determination of intrabacterial accumulated anti-infectives are of enormous interest, as such approaches of course constitute the most accurate representation of the envelope structure. Within the scope of these approaches, a large number of bacteria are usually incubated with the anti-infective compound of interest. This is followed by washing to remove remaining extracellular and/or adherent compound, lysis of the bacterial cells, and subsequent quantification of the amount of intracellular drug. LC-MS/MS methods are generally employed in order to quantify what often proves to be a very low level of anti-infective compound. Such quantification methods are also frequently applied to examine the permeation of various different anti-infectives tested on distinct bacterial strains (Cai et al. 2009; Davis et al. 2014). As such an approach is possibly error-prone due to the potential for inadequate removal of extracellular/adherent anti-infective, as well as the population-based rather than single-cell nature of the quantification process, approaches with direct single-cell resolution based on deep ultraviolet (DUV) fluorescence or the combination of a DUV fluorescence microscope with a synchrotron beamline have been employed. These approaches allow for quantifying fluorescent or fluorescently labeled compounds and for example have been used to compare anti-infective uptake in wild-type and mutant/resistant bacterial strains (Kascakova et al. 2012; Pages et al. 2013). It must be mentioned here, however, that such an approach is still limited to an "inside/outside" distinction of anti-infective location, and determination of anti-infective permeation with any higher degree of spatial resolution remains extremely difficult.

3.1.6 Shortcomings of Existing Models and Future Directions

The current modeling approaches discussed in this section help to get a better understanding of permeation processes across various substructures of the Gram-negative bacterial cell envelope. However, drawbacks and unmet needs can be mentioned for each of the above categories of models available to date. As a general comment, the in vitro and in silico modeling approaches described here mostly focus on producing or simulating structures which approximate either the IM or OM, and not the cell envelope as a whole—or in the small number of cases where the overall envelope structure is approximated, the resulting model is often tailored to the examination of intramembrane interactions or causes of membrane disruption. In addition, many such models consist of a phospholipid composition which deviates from that found in cellulo, and while it has been mentioned that attempts are made in some cases to represent the asymmetric structure of the OM in models of this membrane component, many models still neglect to feature this important aspect. Furthermore, due to considerable difficulties associated with scale and resolution, the vast majority of models to date allow for a qualitative prediction of anti-infective permeation and transport, rather than for quantification of such processes. In cellulo approaches where multiple planktonic cells rather than single cells are used have proven very useful in order to provide detailed and, in some cases, quantitative insights into permeation processes; however, as mentioned, such methods generally rely on an average permeation within a bacterial population to draw conclusions regarding single-cell permeation. Furthermore, current in cellulo approaches do not allow for the evaluation of the specific extent of anti-infective permeation into the envelope structure. Hence, models which mimic the overall envelope in terms of their PL composition and structure, which are designed to explicitly investigate and quantify transport and permeation processes, and which are able to discriminate between active and passive permeation of anti-infective compounds and delivery systems in both a spatially and kinetically resolved manner, are desperately needed.

3.2 Improving Bacterial Bioavailability Using Advanced Delivery Strategies

In addition to employment and development of bacterial permeation models, a direct research focus is also placed on anti-infective therapies themselves in an attempt to overcome the cell envelope structure, achieve an increase in intrabacterial drug concentrations, and, in doing so, improve bioavailability in bacteria. In this respect, the search for new anti-infective drug candidates as well as the investigation of alternative approaches to antibiotic therapy continues, as presented and discussed in detail elsewhere. Additional strategies, such as the reformulation of currently available anti-infectives with permeation-enhancing excipients or the application of advanced carrier systems, also represent promising research directions. Such strategies are particularly valuable in instances where bacterial bioavailability issues cannot be directly resolved by the introduction of new molecules, or through the modification of existing anti-infective structures using medicinal chemistry approaches. As such, a number of currently investigated advanced formulation strategies are presented below.

3.2.1 Efflux Pump Inhibitors

As mentioned in Sect. 2.1, efflux in wild-type as well as drug-resistant Gram-negative bacteria is mainly mediated by the RND superfamily of efflux transporters. The use of formulations incorporating efflux pump inhibitors (EPIs) which are able to interact with such pumps, decreasing anti-infective efflux and subsequently leading to higher intracellular drug levels, therefore represents a useful strategy to restore anti-infective potency. The inhibition of pumps as mediated by EPIs can be described as occurring by two major modes of action. One can be classified as biological, in which EPIs act to decrease the expression of the pumps themselves by inhibiting transcription or translation via antisense oligonucleotides. A pharmacological mechanism represents the second mode of action, in which EPIs operate through direct interaction with the pump affinity site, acting, for example, to collapse the efflux energy or to competitively or non-competitively inhibit the efflux process (Van Bambeke et al. 2010). EPIs can be further differentiated into inhibitors with a narrow spectrum of activity, being used as diagnostic tools to detect active efflux, or inhibitors with a broad spectrum of action, which could be potentially useful in clinical settings. The further ability of EPIs to restore the activity of simultaneously applied anti-infectives [being visible, for example, in a decrease of minimum inhibitory concentration (MIC)] makes them an even more promising approach as a means to increase anti-infective bacterial bioavailability. Examples of known EPIs include analogues or lead structures of tetracyclines or fluoroquinolones, arylpiperidine, and phenothiazine derivatives as well as peptidomimetics (Pages and Amaral 2009). Peptidomimetics with phenylalanine arginyl ß-naphthylamine (PAßN) as lead compound represent the first efflux inhibiting group which showed an effective blocking of fluoroquinolone efflux in a RND overexpressing strain of *Pseudomonas aeruginosa* (Renau et al. 2002). Currently, EPIs are used primarily as in vitro screening tools; their potential use in the clinic is still under investigation due to the existence of several challenging factors. The primary obstacle to the use of EPIs in a clinical setting is that of toxicity—most of the known EPIs to date need to be used in high concentrations, which may lead to possible toxic effects. Their use in combination with anti-infectives also demands the absence of interactions between the EPI and the anti-infective compound, as well as comparability in their pharmacokinetic profiles.

3.2.2 Antimicrobial Peptides

The use of the previously mentioned AMPs, either alone or especially in synergistic combinations with conventional anti-infectives, represents a further strategy to overcome anti-infective bioavailability problems by enhancing their transport across the bacterial cell envelope. AMPs can be further used as stimuli for the innate immune system, or as endotoxin-neutralizing agents (Gordon et al. 2005). AMPs in themselves are generally small cationic peptides which can be derived from humans, bacteria, or even viruses (Yount and Yeaman 2004). Their mode of action as bioavailability-potentiating agents is primarily based on the initiation of bacterial membrane perturbation, an effect mainly mediated by electrostatic interactions between the positively charged peptide and the negatively charged LPS of the OM. Such interactions lead to a destabilization of the OM by displacing present divalent cations, which facilitates the penetration of AMPs and any other associated compounds through the OM structure. Following this self-promoted uptake through the OM, the further association of AMPs with the outer leaflet of the IM followed by the formation of micelle-like aggregates finally leads to a rupture of the bacterial envelope. This allows either for bacterial killing, or for an even further enhanced uptake of the simultaneously administered anti-infective. A nondestructive action of AMPs, facilitated by binding to DNA or RNA, is also further described (Hancock 1997; Hancock and Chapple 1999). Several studies report the potentiating effect of AMP–anti-infective combinations, resulting in an increased anti-infective activity even in hard to treat bacterial strains and biofilm-forming species. In this respect, the synergistic effect of AMPs together with a wide range of anti-infectives with different modes of action could be demonstrated by the effective treatment of *Clostridium difficile* (Nuding et al. 2014). Furthermore, combinations of AMPs with anti-infectives have shown to result in an increased activity against the biofilm formation of methicillin-resistant *Staphylococcus aureus* (MRSA) and have demonstrated a successful inhibition of *Pseudomonas fluorescens* (Mataraci and Dosler 2012; Naghmouchi et al. 2012). Hence, AMPs represent a promising approach to improve anti-infective bioavailability in Gram-negative as well as Gram-positive bacteria, as well as in particularly problematic bacterial infections involving biofilm formation. Several clinical trials, especially for topical application of AMPs to human subjects, are ongoing, but are associated with several challenges. In addition to the potential for toxic effects which could, for example, result from the non-specific membrane disruption, the fast degradation and short half-life of AMPs constitute the main obstacles to generalized use (Park et al. 2011). The incorporation of AMPs into particulate carrier systems could potentially help to reduce or overcome these difficulties—such approaches are further discussed below.

3.2.3 Nanoparticulate Drug Carriers ("Nanopharmaceuticals")

Anti-infectives as free drugs may show low water solubility, unfavorable pharmacokinetics, side effects, or stability problems (Xiong et al. 2014)—all factors which intrinsically create problems for penetration into and effective action within bacteria. The incorporation of anti-infectives into carrier systems, such as liposomes, polymeric nanoparticles, solid lipid nanoparticles (SLNs), or dendrimers, may help to reduce the impact of such characteristics and as such presents several advantages compared to the use of free anti-infectives. In light of their typical size range, these carriers are nowadays also regarded as nanomaterials or nanoparticles, and with respect to their specific application also referred to as nanomedicines or nanopharmaceuticals.

The incorporation of anti-infectives into nanoparticulate carrier systems may allow for a high drug loading in some cases, facilitating an increase in effective drug solubility; a masking of undesirable drug effects; a tailoring of anti-infective pharmacokinetics; or a directly increased permeability. Modifications, for example, to the particle surface may allow for further improvements, such as a targeted delivery. One of the first examples of a nanoparticulate anti-infective formulation which was granted access to the market is a liposomal formulation of amphotericin B—this formulation remains widely used in clinical settings due to the exhibition of many of the above-mentioned advantages (Walsh et al. 1998). Polymeric nanoparticles are also extensively investigated as carriers for anti-infective drugs in several laboratories around the globe. The protective function of particulate carriers and the possibility for coloading is also a considerable advantage with respect to the delivery of readily degraded compounds like AMPs. The possibility to incorporate more than one anti-infective compound into particulate carriers, or to combine anti-infective loaded carriers with particles of known antimicrobial substances like gold or silver, constitutes a further advantage to the use of such delivery systems (Huh and Kwon 2011). Significant progress in the development of nanotechnology-based approaches specifically to treat bacterial infections has been made in recent years, leading to the existence of several sophisticated carrier systems. For example, Trojan horse systems made of nanoparticles tagged with folic acid have been shown to mediate an increased activity of the incorporated anti-infective vancomycin in resistant *Staphylococcus aureus* (Chakraborty et al. 2012). The linkage of penicillin G to surface-functionalized silica nanoparticles has also shown a restored anti-infective activity even in formerly resistant MRSA (Wang et al. 2014).

Infection-activated delivery systems are another promising approach, being, for example, composed of chitosan-modified gold nanoparticles which are attached to liposomes or polymeric triple-layered nanogels. Substances like toxins or enzymes which are present in the local environment of a bacterial infection work as a trigger for the release of carrier-incorporated anti-infective, allowing for the reduction of potential side effects resulting from the systemic anti-infective administration as well as the achievement of high local drug concentrations at the site of infection (Pornpattananangkul et al. 2011; Xiong et al. 2012). Anionic liposomes have also been successfully used to incorporate and deliver plasmid DNA and antisense oligonucleotides into inner bacterial compartments in order to inhibit the gene expression in resistant strains (Fillion et al. 2001; Meng et al. 2009). Recently, an SLN-based formulation was successfully used to incorporate and deliver high amounts of a novel quorum-sensing inhibitor (QSI), which acts as anti-virulence factors by interfering with bacterial cell–cell communication via action on intracellular targets (Miller and Bassler 2001). SLNs with incorporated QSI showed a prolonged release, mucus-penetrating ability, and an effective delivery to the pulmonary region, as well as an enhanced anti-virulence activity against *Pseudomonas aeruginosa* as compared to the compound alone (Nafee et al. 2014). As these examples illustrate, innovative delivery strategies (along with the search for and optimization of novel anti-infective targets and compounds) offer the potential for overcoming bacterial absorption problems.

3.2.4 Evaluation of Current Status and Future Directions

The combination of EPIs and AMPs with conventional or even new anti-infectives may result in a reduction of undesirable intrinsic anti-infective properties as well as an increased bacterial permeation, leading to higher intracellular drug levels and so an enhanced bacterial bioavailability. Furthermore, carrier systems are able to provide a means of circumventing compound-related difficulties, such as unfavorable pharmacokinetics, and to achieve high intracellular drug levels. In this manner, such advanced formulation strategies may act to increase the bioavailability of anti-infectives and for this reason continue to be employed and developed. The treatment of intracellular infections as well as the specific development of permeability-enhancing carriers constitutes an important direction of future applications.

4 Conclusion and Outlook

This paper has aimed to give an overview of current difficulties in the treatment of infectious diseases, in particular those caused by Gram-negative bacteria. In this respect, the significant bioavailability problems of anti-infective compounds—defined as an inadequate delivery to their (mainly intrabacterial) sites of action—largely stem from the complex nature of the cell envelope and its formidable barrier function. This barrier function may be even further enhanced by the evolution of resistance mechanisms. Numerous models—in vitro, in silico as well as in cellulo in nature—may be used in order to increase the understanding of permeation processes into or across the envelope, as well as to enable the evaluation of how the cell envelope in its entirety or as its individual substructures acts as a permeation-limiting factor. However, a paucity of quantitative approaches which accurately mimic the overall envelope structure has to be mentioned, meaning that obtained information may lack comprehensiveness. Therefore, new permeation models which more accurately represent the various structural components of the Gram-negative bacterial cell envelope, and which are further able to provide quantitative, kinetically and spatially resolved permeation data are desperately needed. Such models would also ideally allow for discrimination between active and passive transport processes and would be applicable as high-throughput screening methods in early drug discovery. With respect to anti-infective compounds themselves, the combination of EPIs or AMPs with conventional anti-infectives presents a promising strategy in overcoming bacterial bioavailability problems, enabling the restoration of anti-infective activity even in resistant strains. Particulate delivery systems may similarly facilitate an increase in anti-infective bioavailability, by acting to overcome drawbacks related to the free drug itself; such carrier systems may additionally facilitate a targeted delivery of anti-infectives. Anti-infective formulations which are designed to particularly increase the permeation or transport of anti-infectives into or across the bacterial cell envelope, or to treat particularly problematic bacteria (such as those which reside within mammalian cells) are still urgently needed however and would constitute a further significant improvement in anti-infective therapy.

References

Berglund NA, Piggot TJ, Jefferies D, Sessions RB, Bond PJ, Khalid S (2015) Interaction of the antimicrobial peptide polymyxin B1 with both membranes of *E. coli*: a molecular dynamics study. PLoS Comput Biol 11:1004180

Cai H, Rose K, Liang LH, Dunham S, Stover C (2009) Development of a liquid chromatography/mass spectrometry-based drug accumulation assay in *Pseudomonas aeruginosa*. Anal Biochem 385:321–325

Chakraborty SP, Sahu SK, Pramanik P, Roy S (2012) In vitro antimicrobial activity of nanoconjugated vancomycin against drug resistant *Staphylococcus aureus*. Int J Pharm 436:659–676

Charitat T, Bellet-Amalric E, Fragneto G, Graner F (1999) Adsorbed and free lipid bilayers at the solid-liquid interface. Eur Phys J B 8:583–593

Clifton LA, Johnson CL, Solovyova AS, Callow P, Weiss KL, Ridley H, Le Brun AP, Kinane CJ, Webster JR, Holt SA, Lakey JH (2012) Low resolution structure and dynamics of a colicin-receptor complex determined by neutron scattering. J Biol Chem 287:337–346

Clifton LA, Skoda MW, Le Brun AP, Ciesielski F, Kuzmenko I, Holt SA, Lakey JH (2015) Effect of divalent cation removal on the structure of gram-negative bacterial outer membrane models. Langmuir 31:404–412

Cronin MT, Aptula AO, Dearden JC, Duffy JC, Netzeva TI, Patel H, Rowe PH, Schultz TW, Worth AP, Voutzoulidis K, Schuurmann G (2002) Structure-based classification of antibacterial activity. J Chem Inf Comput Sci 42:869–878

Danelon C, Nestorovich EM, Winterhalter M, Ceccarelli M, Bezrukov SM (2006) Interaction of zwitterionic penicillins with the OmpF channel facilitates their translocation. Biophys J 90:1617–1627

Davin-Regli A, Bolla JM, James CE, Lavigne JP, Chevalier J, Garnotel E, Molitor A, Pages JM (2008) Membrane permeability and regulation of drug "influx and efflux" in enterobacterial pathogens. Curr Drug Targets 9:750–759

Davis TD, Gerry CJ, Tan DS (2014) General platform for systematic quantitative evaluation of small-molecule permeability in bacteria. ACS Chem Biol 9:2535–2544

De E, Basle A, Jaquinod M, Saint N, Mallea M, Molle G, Pages JM (2001) A new mechanism of antibiotic resistance in Enterobacteriaceae induced by a structural modification of the major porin. Mol Microbiol 41:189–198

Der Wissenschaften Akademie, Hamburg Deutsche Akademie, Leopoldina Der Naturforscher (eds) (2013) Antibiotika-Forschung: Probleme und Perspektiven. De Gruyter, Berlin

D'Errico G, Silipo A, Mangiapia G, Molinaro A, Paduano L, Lanzetta R (2009) Mesoscopic and microstructural characterization of liposomes formed by the lipooligosaccharide from *Salmonella minnesota* strain 595 (Re mutant). Phys Chem Chem Phys 11:2314–2322

Dever LA, Dermody TS (1991) Mechanisms of bacterial resistance to antibiotics. Arch Intern Med 5:886–895

Fernandez DI, Le Brun AP, Whitwell TC, Sani MA, James M, Separovic F (2012) The antimicrobial peptide aurein 1.2 disrupts model membranes via the carpet mechanism. Phys Chem Chem Phys 14:15739–15751

Fernandez DI, Le Brun AP, Lee TH, Bansal P, Aguilar MI, James M, Separovic F (2013) Structural effects of the antimicrobial peptide maculatin 1.1 on supported lipid bilayers. Eur Biophys J 42:47–59

Fillion P, Desjardins A, Sayasith K, Lagace J (2001) Encapsulation of DNA in negatively charged liposomes and inhibition of bacterial gene expression with fluid liposome-encapsulated antisense oligonucleotides. Biochim Biophys Acta 1515:44–54

Fischbach MA, Walsh CT (2009) Antibiotics for emerging pathogens. Science 325:1089–1093

Fragneto G, Charitat T, Daillant J (2012) Floating lipid bilayers: models for physics and biology. Eur Biophys J 41:863–874

Gordon YJ, Romanowski EG, McDermott AM (2005) A review of antimicrobial peptides and their therapeutic potential as anti-infective drugs. Curr Eye Res 30:505–515

Gornall JL, Mahendran KR, Pambos OJ, Steinbock LJ, Otto O, Chimerel C, Winterhalter M, Keyser UF (2011) Simple reconstitution of protein pores in nano lipid bilayers. Nano Lett 11:3334–3340

Hajjar E, Mahendran KR, Kumar A, Bessonov A, Petrescu M, Weingart H, Ruggerone P, Winterhalter M, Ceccarelli M (2010) Bridging timescales and length scales: from macroscopic flux to the molecular mechanism of antibiotic diffusion through porins. Biophys J 98:569–575

Hancock RE (1997) Peptide antibiotics. Lancet 349:418–422

Hancock RE, Chapple DS (1999) Peptide antibiotics. Antimicrob Agents Chemother 43: 1317–1323

Huh AJ, Kwon YJ (2011) Nanoantibiotics: a new paradigm for treating infectious diseases using nanomaterials in the antibiotics resistant era. J Control Release 156:128–145

Kascakova S, Maigre L, Chevalier J, Refregiers M, Pages JM (2012) Antibiotic transport in resistant bacteria: synchrotron UV fluorescence microscopy to determine antibiotic accumulation with single cell resolution. PLoS ONE 7:e38624

Kreir M, Farre C, Beckler M, George M, Fertig N (2008) Rapid screening of membrane protein activity: electrophysiological analysis of OmpF reconstituted in proteoliposomes. Lab Chip 8:587–595

Kubiak J, Brewer J, Hansen S, Bagatolli LA (2011) Lipid lateral organization on giant unilamellar vesicles containing lipopolysaccharides. Biophys J 100:978–986

Kumar A, Schweizer HP (2005) Bacterial resistance to antibiotics: active efflux and reduced uptake. Adv Drug Deliv Rev 57:1486–1513

Kuzmenko AI, Wu H, McCormack FX (2006) Pulmonary collectins selectively permeabilize model bacterial membranes containing rough lipopolysaccharide. Biochemistry 45:2679–2685

Le Brun AP, Clifton LA, Halbert CE, Lin B, Meron M, Holden PJ, Lakey JH, Holt SA (2013) Structural characterization of a model gram-negative bacterial surface using lipopolysaccharides from rough strains of *Escherichia coli*. Biomacromolecules 14:2014–2022

Lemmin T, Bovigny C, Lançon D, Dal Peraro M (2013) Cardiolipin models for molecular simulations of bacterial and mitochondrial membranes. J Chem Theory Comput 9:670–678

Lins RD, Straatsma TP (2001) Computer simulation of the rough lipopolysaccharide membrane of *Pseudomonas aeruginosa*. Biophys J 81:1037–1046

Luckey M, Nikaido H (1980) Specificity of diffusion channels produced by lambda phage receptor protein of *Escherichia coli*. Proc Natl Acad Sci U S A 77:167–171

Mach T, Chimerel C, Fritz J, Fertig N, Winterhalter M, Futterer C (2008a) Miniaturized planar lipid bilayer: increased stability, low electric noise and fast fluid perfusion. Anal Bioanal Chem 390:841–846

Mach T, Neves P, Spiga E, Weingart H, Winterhalter M, Ruggerone P, Ceccarelli M, Gameiro P (2008b) Facilitated permeation of antibiotics across membrane channels-interaction of the quinolone moxifloxacin with the OmpF channel. J Am Chem Soc 130:13301–13309

Mahendran KR, Hajjar E, Mach T, Lovelle M, Kumar A, Sousa I, Spiga E, Weingart H, Gameiro P, Winterhalter M, Ceccarelli M (2010) Molecular basis of enrofloxacin translocation through OmpF, an outer membrane channel of *Escherichia coli*-when binding does not imply translocation. J Phys Chem B 114:5170–5179

Mataraci E, Dosler S (2012) In vitro activities of antibiotics and antimicrobial cationic peptides alone and in combination against methicillin-resistant *Staphylococcus aureus* biofilms. Antimicrob Agents Chemother 56:6366–6371

Mayers D (ed) (2009) Antimicrobial drug resistance: mechanisms of drug resistance. Springer, Berlin

McKenna M (2013) The last resort. Nature 499:394–396

Meng J, Wang H, Hou Z, Chen T, Fu J, Ma X, He G, Xue X, Jia M, Luo X (2009) Novel anion liposome-encapsulated antisense oligonucleotide restores susceptibility of methicillin-resistant *Staphylococcus aureus* and rescues mice from lethal sepsis by targeting mecA. Antimicrob Agents Chemother 53:2871–2878

Miller MB, Bassler BL (2001) Quorum sensing in bacteria. Annu Rev Microbiol 55:165–199

Modi N, Winterhalter M, Kleinekathofer U (2012) Computational modeling of ion transport through nanopores. Nanoscale 4:6166–6180

Mullineaux CW, Nenninger A, Ray N, Robinson C (2006) Diffusion of green fluorescent protein in three cell environments in *Escherichia coli*. J Bacteriol 188:3442–3448

Murcia-Soler M, Perez-Gimenez F, Garcia-March FJ, Salabert-Salvador MT, Diaz-Villanueva W, Medina-Casamayor P (2003) Discrimination and selection of new potential antibacterial compounds using simple topological descriptors. J Mol Graph Model 21:375–390

Murcia-Soler M, Perez-Gimenez F, Garcia-March FJ, Salabert-Salvador MT, Diaz-Villanueva W, Castro-Bleda MJ, Villanueva-Pareja A (2004) Artificial neural networks and linear discriminant analysis: a valuable combination in the selection of new antibacterial compounds. J Chem Inf Comput Sci 44:1031–1041

Murzyn K, Rog T, Pasenkiewicz-Gierula M (2005) Phosphatidylethanolamine-phosphatidylglycerol bilayer as a model of the inner bacterial membrane. Biophys J 88:1091–1103

Nafee N, Husari A, Maurer CK, Lu C, De Rossi C, Steinbach A, Hartmann RW, Lehr CM, Schneider M (2014) Antibiotic-free nanotherapeutics: ultra-small, mucus-penetrating solid lipid nanoparticles enhance the pulmonary delivery and anti-virulence efficacy of novel quorum sensing inhibitors. J Control Release 192:131–140

Naghmouchi K, Le Lay C, Baah J, Drider D (2012) Antibiotic and antimicrobial peptide combinations: synergistic inhibition of *Pseudomonas fluorescens* and antibiotic-resistant variants. Res Microbiol 163:101–108

Nelson ML, Grier MC, Barbaro SE, Ismail MY (2009) Polyfunctional antibiotics affecting bacterial membrane dynamics. Anti-Infect Agents Med Chem 8:3–16

Nestorovich EM, Danelon C, Winterhalter M, Bezrukov SM (2002) Designed to penetrate: time-resolved interaction of single antibiotic molecules with bacterial pores. Proc Natl Acad Sci U S A 99:9789–9794

Nikaido H, Rosenberg EY (1981) Effect on solute size on diffusion rates through the transmembrane pores of the outer membrane of *Escherichia coli*. J Gen Physiol 77:121–135

Nikaido H, Rosenberg EY (1983) Porin channels in Escherichia coli: studies with liposomes reconstituted from purified proteins. J Bacteriol 153:241–252

Nuding S, Frasch T, Schaller M, Stange EF, Zabel LT (2014) Synergistic effects of antimicrobial peptides and antibiotics against *Clostridium difficile*. Antimicrob Agents Chemother 58:5719–5725

O'Shea R, Moser HE (2008) Physicochemical properties of antibacterial compounds: implications for drug discovery. J Med Chem 51:2871–2878

Pages JM, Amaral L (2009) Mechanisms of drug efflux and strategies to combat them: challenging the efflux pump of Gram-negative bacteria. Biochim Biophys Acta 1794:826–833

Pages JM, James CE, Winterhalter M (2008) The porin and the permeating antibiotic: a selective diffusion barrier in Gram-negative bacteria. Nat Rev Microbiol 6:893–903

Pages JM, Kascakova S, Maigre L, Allam A, Alimi M, Chevalier J, Galardon E, Refregiers M, Artaud I (2013) New peptide-based antimicrobials for tackling drug resistance in bacteria: single-cell fluorescence imaging. ACS Med Chem Lett 4:556–559

Palusinska-Szysz M, Zdybicka-Barabas A, Pawlikowska-Pawlega B, Mak P, Cytrynska M (2012) Anti-*Legionella dumoffii* activity of Galleria mellonella defensin and apolipophorin III. Int J Mol Sci 13:17048–17064

Pandit KR, Klauda JB (2012) Membrane models of *E. coli* containing cyclic moieties in the aliphatic lipid chain. Biochim Biophys Acta 1818:1205–1210

Park SC, Park Y, Hahm KS (2011) The role of antimicrobial peptides in preventing multidrug-resistant bacterial infections and biofilm formation. Int J Mol Sci 12:5971–5992

Peetla C, Stine A, Labhasetwar V (2009) Biophysical interactions with model lipid membranes: applications in drug discovery and drug delivery. Mol Pharma 6:1264–1276

Piggot TJ, Holdbrook DA, Khalid S (2011) Electroporation of the *E. coli* and *S. aureus* membranes: molecular dynamics simulations of complex bacterial membranes. J Phys Chem B 115:13381–13388

Pornpattananangkul D, Zhang L, Olson S, Aryal S, Obonyo M, Vecchio K, Huang CM, Zhang L (2011) Bacterial toxin-triggered drug release from gold nanoparticle-stabilized liposomes for the treatment of bacterial infection. J Am Chem Soc 133:4132–4139

Renau TE, Leger R, Yen R, She MW, Flamme EM, Sangalang J, Gannon CL, Chamberland S, Lomovskaya O, Lee VJ (2002) Peptidomimetics of efflux pump inhibitors potentiate the activity of levofloxacin in *Pseudomonas aeruginosa*. Bioorg Med Chem Lett 12:763–766

Ries O, Carnarius C, Steinem C, Ducho C (2015) Membrane-interacting properties of the functionalised fatty acid moiety of muraymycin antibiotics. Med Chem Comm 6:879–886

Rodrigues C, Gameiro P, Prieto M, De Castro B (2003) Interaction of rifampicin and isoniazid with large unilamellar liposomes: spectroscopic location studies. Biochim Biophys Acta 1620:151–159

Silhavy TJ, Kahne D, Walker S (2010) The bacterial cell envelope. Cold Spring Harb Perspect Biol 2:a000414

Singh PR, Ceccarelli M, Lovelle M, Winterhalter M, Mahendran KR (2012) Antibiotic permeation across the OmpF channel: modulation of the affinity site in the presence of magnesium. J Phys Chem B 116:4433–4438

Tenover FC (2006) Mechanisms of antimicrobial resistance in bacteria. Am J Med 119:S3–10

Van Bambeke F, Pages LM, Lee VJ (2010) Inhibitors of bacterial efflux pumps as adjuvants in antibacterial therapy and diagnostoc tools for detection of resistance by efflux. In: Atta-ur-Rahman Choudary MI (ed) Frontiers in anti-infective drug discovery. Bentham, Sharjah

Walsh TJ, Yeldandi V, Mcevoy M, Gonzalez C, Chanock S, Freifeld A, Seibel NI, Whitcomb PO, Jarosinski P, Boswell G, Bekersky I, Alak A, Buell D, Barret J, Wilson W (1998) Safety, tolerance, and pharmacokinetics of a small unilamellar liposomal formulation of amphotericin B (AmBisome) in neutropenic patients. Antimicrob Agents Chemother 42:2391–2398

Wang L, Chen YP, Miller KP, Cash BM, Jones S, Glenn S, Benicewicz BC, Decho AW (2014) Functionalised nanoparticles complexed with antibiotic efficiently kill MRSA and other bacteria. Chem Commun (Camb) 50:12030–12033

Wellington EM, Boxall AB, Cross P, Feil EJ, Gaze WH, Hawkey PM, Johnson-Rollings AS, Jones DL, Lee NM, Otten W, Thomas CM, Williams AP (2013) The role of the natural environment in the emergence of antibiotic resistance in gram-negative bacteria. Lancet Infect Dis 13:155–165

Wu EL, Fleming PJ, Yeom MS, Widmalm G, Klauda JB, Fleming KG, Im W (2014) *E. coli* outer membrane and interactions with OmpLA. Biophys J 106:2493–2502

Xiong MH, Bao Y, Yang XZ, Wang YC, Sun B, Wang J (2012) Lipase-sensitive polymeric triple-layered nanogel for "on-demand" drug delivery. J Am Chem Soc 134:4355–4362

Xiong MH, Bao Y, Yang XZ, Zhu YH, Wang J (2014) Delivery of antibiotics with polymeric particles. Adv Drug Deliv Rev 78:63–76

Yount NY, Yeaman MR (2004) Multidimensional signatures in antimicrobial peptides. Proc Natl Acad Sci U S A 101:7363–7368

Zhao W, Rog T, Gurtovenko AA, Vattulainen I, Karttunen M (2008) Role of phosphatidylglycerols in the stability of bacterial membranes. Biochimie 90:930–938

Printed in the United States
By Bookmasters